FUNDAMENTALS OF ENGINEERING ELECTROMAGNETICS

SUNIL BHOOSHAN

Professor and Head
Department of Electronics and Communication Engineering
Jaypee University of Information Technology
Waknaghat

UNIVERSITY PRESS

OXFORD
UNIVERSITY PRESS

Oxford University Press is a department of the University of Oxford.
It furthers the University's objective of excellence in research, scholarship,
and education by publishing worldwide. Oxford is a registered trade mark of
Oxford University Press in the UK and in certain other countries.

Published in India by
Oxford University Press
22 Workspace, 2nd Floor, 1/22 Asaf Ali Road, New Delhi 110002, India

ISBN-13: 978-0-19-807794-7
ISBN-10: 0-19-807794-7

Typeset in Times New Roman
by BeSpoke Integrated Solutions, Puducherry
Printed at Manipal Technologies Limited, Manipal

To

my family

and

the living memory of my parents

Foreword

It gives me great pleasure to pen down a foreword for Prof. Sunil Bhooshan's book, *Fundamentals of Engineering Electromagnetics*. Our long association at the Jaypee University of Information Technology, Waknaghat, since 2002, has made me a firm believer that as far as the clarity of fundamentals of the subject of electromagnetics is concerned, Prof Bhooshan is in a league by himself. Further, his teaching qualities make him one of the favourite teachers of the University.

I firmly believe that through his book he is going to help a large number of students to learn the fundamentals of one of the most difficult subjects of electronics engineering (as perceived by most students).

I wish him best of luck and look forward to see students understanding the subject better.

Yaj Medury
Chief Operating Officer (Education)
Jaypee University of Information Technology

Preface

Learning without thought is labour lost;
thought without learning is perilous.
—Confucius

Electromagnetics is a branch of physics which primarily deals with the analysis and applications of electric and magnetic fields. MRI scanners use a strong magnetic field and radio frequency fields to produce a detailed image of the inside of a body. Radars use very high frequency radio waves to detect and locate objects. Without antennas and electromagnetic waves, there would be no radio, television, or mobile phones.

The bending of rays of light in dielectrics, the laws of reflection, and the existence of metal mirrors were known since long. The use of lodestone in compasses was well known to the Chinese and convex lenses had been in use in Carthage (present day Tunisia). Though investigation of electromagnetic phenomena started many centuries ago, most of the laws of electromagnetic theory were discovered in the eighteenth and nineteenth century. Some prominent investigators of that time include Count Alessandro Giuseppe Volta (1745–1827), Andre Marie Ampere (1775–1836), Michael Faraday (1791–1867), James Clerk Maxwell (1831–1879), and Heinrich Hertz (1857–1894).

This period was followed by intense research in the engineering applications of the theory in the twentieth century. During World War II, electromagnetic theory received a sudden burst of attention with applications in radar and communications. After World War II, most of the prominent universities and corporations pursued research in this area and published a lot of data on electromagnetics.

Applications of Electromagnetics

Electromagnetics today is being rediscovered and applied in an interdisciplinary sense to many areas of engineering such as wireless and wire-line transmission and communication, computer interconnects, optical fibre links and components, antennas, plasmas, wave propagation in the ionosphere, and lasers.

Electromagnetics finds applications in military defence, high-speed electronics, ultrahigh-speed photonic integrated circuits, micro-cavity laser design, and light switching and imaging of the human body. Furthermore, it is applied in the study of microwave circuits consisting of microstrip lines, couplers, ferrite circulators, microwave filters antennas, transistors, and many other components.

Note to the Student

Electromagnetics is a subject where the importance of complicated equations and their formulation is very important, and the general tendency of students is to try and memorise as many of them as possible without really understanding. We are surrounded by electromagnetic waves and fields that is why it is imperative to study their characteristics. Almost every theorem of electromagnetics has a real-world application. In a way, the subject puts into perspective how things function the way they do. Once the readers see the equations in this light, all the complex equations will turn friendly.

Before the study of this subject is commenced, one needs to find answers to the following questions:

1. What will I study in this subject?
2. Where will I apply it?
3. Of what use is it?

Once these questions are answered, one will know the goals to be attained. It has been found that this subject is difficult to grasp due to the underlying mathematical complexity. To overcome this, the following must be kept in mind.

1. Due attention must be paid to the mathematical concepts which have been explained in Part I of the book.
2. Whenever any new equation or concept is studied, it should be immediately applied to some 'thought situation'. Any simple problem may be conjured and the equation or concept should then be applied to it.
3. One must solve the problems at the end of each chapter for truly grasping the essential concepts.

About the Book

This book is designed for a course in electromagnetics at the undergraduate level for students of electrical engineering and electronics and communication engineering. The book may also be used as a ready reference on electromagnetic theory by students of varied disciplines and practising engineers.

The motive of this book is to arouse the interest of the reader. Therefore, the subject is presented as a whole, starting from the mathematical principles involved, to the theory in general. The detailed treatment of the mathematics required, followed by the concepts are dealt with in detail. In order to explain the application of mathematics to engineering situations, both theory and solved problems are included in this book.

Each chapter includes a variety of pedagogical features which will make learning and revising an enjoyable experience.

Features of the Book

- The book comprehensively covers all the important topics taught in this course across various universities.
- Chapter objectives are clearly defined at the beginning of each chapter giving readers a peep into the contents of the chapter.
- The chapters contain extended information on various topics which will help generate interest in learning the subject.
- A list of formulae and a point-wise summary can be found at the end of each chapter.
- The book presents over 250 solved examples with detailed mathematical steps.
- It also contains more than 250 illustrations, many of which attempt a three-dimensional view of the situation depicted. For antenna patterns, computer

generated three-dimensional views of the patterns are presented. Furthermore, to aid understanding streamline flows are shown for the fields.

- Each chapter contains exercises at the end, comprising over 300 practice problems,150 objective type questions, and over 200 short-answer questions and review questions.
- In order to provide practice immediately after learning, practice problems (with answers) have also been included within the text.
- The book contains a chapter each on antennas and radio wave propagation.

Online Resources

The Online Resource Centre of the book contains MATLAB exercises, colour illustrations from the book, and some JAVA applets. The content is closely linked to the main text. An ORC icon in the left margin of the text indicates digital support for the topic.

Systems requirements

- MATLAB
- Web Browser
- JAVA (JRE)

Content and Coverage

The book is divided into four parts: Introduction, Electrostatics, Magnetostatics, and Time-varying Fields, Radiation, and Propagation.

Part I contains three chapters on basic mathematics. Chapter 1 explains scalars and vectors. Chapter 2 discusses coordinate systems and fields. Chapter 3 describes vector analysis and vector calculus.

Part II, electrostatics, consists of four chapters. In this part, Chapter 4 treats the various topics such as Coulomb's law, the electric field, and Gauss's law. Chapter 5 deals with electrostatic energy, potential, and concept of potential energy. Chapter 6 explains how electrostatic fields interact with material media. It explains current density and its role in the continuity equation. In this chapter, we also learn how to calculate capacitance. Chapter 7 derives two important partial differential equations, namely the Laplace's and Poisson's equations which are applied to various situations.

Part III consists of two chapters on the steady magnetic fields. Chapter 8 deals with the Biot-Savart law, Ampere's law, the scalar magnetic potential, the vector potential, and the magnetic flux density. Chapter 9 explains the Lorentz force, torque on a current loop, inductance, magnetisation, and magnetic circuits.

Part IV treats time-varying fields, radiation, and propagation. There are six chapters in this part. Chapter 10 is on Faraday's law and the wave equation. Chapter 11 explains about plane wave propagation and its interaction with matter such as dielectrics and metals. Chapter 12 discusses transmission lines and waveguides. Chapter 13 introduces the concept of radiation from currents and also explains how these concepts apply to the half-wave dipole. Chapter 14 is an introduction to antennas such as linear arrays and aperture antennas. Chapter 15 is about radio wave propagation on the surface of the earth, the troposphere, and the ionosphere.

The book contains appendices at the end comprising information on coordinate systems, important mathematical formulae, a quick recap of some key equations, and four sample question papers.

Acknowledgements

I must acknowledgeVinay Kumar, Rohit Sharma, and Shipra Sharma for giving me valuable suggestions when I was developing this book as also for their good wishes.

Salman Tallury, Agastya Bhooshan, Gagan Gupta, and Ravi Pandey require special mention as they made major contributions in terms of drawing of figures, and the designing and solving of problems at the end of each chapter. Parimal Tiwari who helped in preparing the MATLAB exercises and Vinay Kumar who helped with the ORC content, sample question papers, and MATLAB exercises also deserve a very special mention.

I must also acknowledge members of various teams who wrote the software for Linux without which this book would not have been possible.

I also acknowledge with thanks the help and support I received from the editorial team of Oxford University Press India.

I welcome readers to give me feedback on this book through e-mail at svb.emtbook@gmail.com. Please feel free to point out errors and any other lapses.

SUNIL BHOOSHAN

Features of the Book

Illustrations

Well-labelled illustrations such as these will help → students easily understand concepts explained in the chapter.

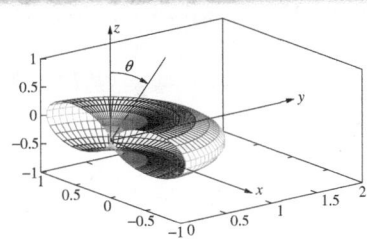

Fig. 13.6 Polar plot of the normalised power radiated by an infinitesimal current element

Points to Remember

At the end of every chapter key formulae → and concepts have been recapitulated to help students in revision.

POINTS TO REMEMBER

- There are various types of charges such as the idealised point charge, Q; the line charge, ρ_l; surface charge, ρ_s; and volume charge, ρ_v.
- The SI unit of charge is Coulomb which is also 1 A-s.

SELF ASSESSMENT

Objective Type Questions

1. Ground wave propagation has the following factors:
   (a) a direct path between Tx and Rx
   (b) a reflection from the sky
   (c) a surface wave
   (d) a reflected path between Tx and Rx

Short-Answer Questions

1. Find the radius of curvature of rays in the troposphere where the height is 5 km and the partial pressure of water vapour is 40 millibars. The temperature is 0°C.

Review Questions

1. Discuss the Rayleigh roughness criterion.
2. What is the concept of complex permittivity?

Answers

Objective Type Questions

1. (c)   2. (c)   3. (a)   4. (b)   5. (b)   6. (a)   7. (c)
8. (b) and (c)   9. (d)   10. (a) and (d)

Short-Answer Questions

1. No
2. No

Self-Assessment

A Self-Assessment section can be found at the end of every chapter comprising objective type questions, short-answer questions, review questions, numerical problems, and answers to selected questions as well.

Brief Contents

Detailed Contents

Part II Electrostatics

Part III Magnetostatics

Part IV Time Varying Fields, Radiation, and Propagation

Frequently Used Reference Material

Fundamental Constants

Symbol	Name	Value
c	Velocity of light	2.9979×10^9 (m s^{-1})
e	Charge of an electron	1.6022×10^{-19} (C)
g	Acceleration due to gravity	9.8067 (m s^{-2})
G	Newton's gravitational constant	6.6726×10^{-11} (N m^2 kg^{-2})
ε_0	Permittivity of vacuum	8.8542×10^{-12} (F m^{-1})
Z_0	Characteristic impedance of free space	376.99 (Ω)
h	Planck's constant	6.6261×10^{-34} (J s)
k	Boltzmann's constant	1.3807×10^{-23} (J $^\circ$K^{-1})
m_e	Electron mass	9.1094×10^{-31} (kg)
m_n	Neutron mass	1.6749×10^{-27} (kg)
m_p	Proton mass	1.6726×10^{-27} (kg)
μ_0	Permeability of free space	1.2566×10^{-6} (H m^{-1})
N_A	Avogadro's constant	6.022×10^{23} (mole^{-1})
r_B	Bohr radius	0.53×10^{-10} (m)

Units

Base quantity	Name	Symbol
Length	metre	m
Mass	kilogramme	kg
Time	second	s
Electric current	ampere	A
Thermodynamic temperature	kelvin	$^\circ$K
Amount of substance	mole	mole
Luminous intensity	candela	cd

Greek Alphabets

α, A, alpha	β, B, beta	γ, Γ, gamma	δ, Δ, delta	ε, ϵ, E epsilon
ζ, Z, zeta	η, H, eta	θ, υ, Θ, theta	ι, I, iota	κ, K, kappa
λ, Λ, lambda	μ, M, mu	ν, N, nu	ξ, Ξ, xi ('zi')	π, ω, Π, pi
ρ, R, rho	σ, $\varsigma\Sigma$, sigma	τ, T, tau	υ, Υ, upsilon	ϕ, Φ, phi
χ, X, chi	ψ, Ψ, psi ('si')	ω, Ω, omega	ϱ, varrho	

SI Prefixes

Symbol	Name	10^n	Symbol	Name	10^n
a	atto	10^{-18}	h	hecto	10^2
f	femto	10^{-15}	k	kilo	10^3
p	pico	10^{-12}	M	mega	10^6
n	nano	10^{-9}	G	giga	10^9
μ	micro	10^{-6}	T	tera	10^{12}
m	milli	10^{-3}	P	peta	10^{15}
d	deca	10^1	E	exa	10^{18}

Dielectric Constants of Materials

Material	ε_r	Loss tangent	Material	ε_r	Loss tangent
Air	1.00054		Paraffin	2–3	
Alumina	9.6–10	0.0002	Polystyrene	2.55	0.0001–0.0003
Bakelite	3.7		Sea water	80	4–5
Fused quartz	3.8	0.00006–0.0002	Silicon	11.7–12.9	
Fused silica (glass)	3.8		Soil	2.55–2.59	0.0017–0.0062
			Teflon	2.1	0.00028
Gallium Arsenide (GaAs)	13.1	0.0016	Vacuum (free space)	1.00000	
Germanium	16		Vinyl	2.8–4.5	
Glass	4–10		Water	80–88	0.04–0.15
Ice (pure distilled water)	3.2–4.15	0.0009–0.12	Wood	1.2–2.1	0.03–0.04
Paper	3.0				

Relative Permeabilities of Materials

Material	μ_r	Material	μ_r
Electrical steel	$\cong 4000$	Nickel	$\cong 100$
Mu-metal	$\cong 20{,}000$	Platinum	$\cong 1.0$
Permalloy	$\cong 8000$	Aluminium	$\cong 1.0$
Ferrite (NiZn)	6–650	Vacuum	1.0
Ferrite (MnZn)	>650	Copper	$\cong 1.0$
Steel	$\cong 700$	Water	$\cong 1.0$

Conductivities of Materials

Material	Conductivity $\sigma(\mho/m)$	Material	Conductivity $\sigma(\mho/m)$
Silver	6.30×10^7	Nickel	1.43×10^7
Copper	5.8×10^7	Iron	1.0×10^7
Gold	4.52×10^7	Tin	9.17×10^6
Alluminum	3.5×10^7	Nichrome	9.09×10^5
Silicon	2.52×10^{-6}	Germanium	1.45×10^{-2}
Chromium	7.74×10^4	Zinc	1.66×10^5
Mercury	1.04×10^4	Lead	4.81×10^4

List of Symbols

A	ground-wave attenuation factor *526*	$d\mathbf{S}$	a differential surface vector element
$\hat{A}$	unit vector, see Eq (1.8) *10*	dV	a differential volume element
$[A]$	a matrix	$\mathbf{D}$	electric flux density, see Eq
$[A]^t$	transpose of a matrix		(3.95) *83*
$\|\mathbf{A}\|$	magnitude of a vector, see Eq (1.9) *10*	E	array factor of a linear array, see Eq (14.12) *492*
$\mathbf{A}$	vector potential, see Eq (8.69) *266*	E	energy. The total energy = K.E. + P.E., see Eq (5.65) *160*
$\mathbf{A}$	a typical vector, see Eq (1.7) *9*	$\mathbf{E}$	electric field, see Eq (3.95)
$\mathbf{A\cdot B}$	dot-product between two vectors, see Eq (1.16) *14*	ε	*83* induced emf generated in a closed loop immersed in a
$\mathbf{A\times B}$	cross-product between two vectors		time varying magnetic field
$\mathbf{A(r)}$	vector field, see Eq (2.1) *32*	e	charge on an electron (C), see Eq (4.0) *94*
$\tilde{\mathbf{A}}(\mathbf{R}, t)$	a sinusoidal vector field, see Eq (10.61) *332*	$E_n(r, \theta, \phi)$	normalised electric field pattern of an antenna, see Eq
$\mathbf{a}$	acceleration (m/s^2), see Eq (1.35) *22*		(13.65) *472*
$\mathbf{a}_x$	unit vector in the x direction, see Eq (1.4) *9*	$\mathbf{F}$	force (N), see Eq (1.36) *22*
$a, b,...$	typical scalars, see Eq (1.1) *4*	$\mathbf{F(r)}$	a vector field, see Eq (3.9) *60*
$\tilde{a}\,(t)$	a sine or cosine function, see Eq (10.56) *330*	$\mathcal{F}$	magnetomotive force
A_e	effective area of an antenna, see Eq (13.85) *478*	f	frequency
		f_{cr}	critical frequency, see Eq (15.109) *544*
$\mathbf{a}_\rho$	unit vector along the ρ direction in cylindrical coordinates, see Eq (2.9) *41*	$f_n(\theta, \phi)$	normalised function in spherical coordinates which describes the electric or
$\mathbf{B}$	magnetic flux density, see Eq (3.95) *83*		magnetic field, see Eq (13.62) *471*
b	phase constant, see Eq (15.19) *526*	G	distributed conductance per meter in a transmission line,
C	distributed capacitance per meter in a transmission line, see Eq (12.1) *386*	g	see Eq (12.1) *386* acceleration due to gravity (m/s^2), see Eq (2.1) *31*
c	velocity of light	$\mathbf{H}$	magnetic field, see Eq (3.95)
D	directivity, dissipation factor, see Eq (13.70) *473*		*83*
d	a distance or separation	$I(z)$	total current flowing in a conductor (with the higher
$d\mathbf{l}$	a differential linear vector element		potential) of a transmission line, see Eq (12.42) *395*

I, i current. (A)

I_+, I_- amplitudes of the forward and backward current waves, see Eq (12.31) *393*

$\mathbf{J}$ current density, see Eq (3.95) *83*

j $\sqrt{-1}$, see Eq (1.38) *23*

$J_n(x)$ Bessel function of the first kind of order n, see Eq (12.160) *437*

K kinetic energy

k free-space proagation constant ($= 2\pi/\lambda$), see Eq (11.0) *342*, a general constant

k_x, k_y propagation constants in the x and y directions in a rectangular waveguide, see Eq (12.123) *430*

L distributed inducance per meter in a transmission line, see Eq (12.0) *386*

$\mathcal{L}$ a line over which an integration is performed

L inductance (H), see Eq (1.38) *23*

l, L lengths along a line, see Eq (4.1) *95*

M mass (kg), *31*

m mass (kg), see Eq (1.36) *22*

m_e mass of an electron, see Eq (5.71) *161*

m_p mass of a proton, see Eq (5.71) *161*

N_A Avagadro's number, see Eq (4.10) *100*

N electron density in the ionosphere, see Eq (15.103) *544*

N refractivity, see Eq (15.52) *533*

n refractive index, see Eq (15.53) *533*

n carrier charge density

$\mathbb{P}$ magnitude of the Poynting vector, see Eq (13.85) *479*

$\mathbb{P}_{target}$ power density at the target in the case of a radar, see Eq (13.90) *480*

$P(r, \theta, \phi)$ magnitude of the poynting vector of an antenna as a function of spherical coordinates (r, θ, ϕ), see Eq (13.54) *467*

P_+, P_- power in the forward and reverse waves respectively in a transmission line, see Eq (12.39) *394*

P_{av} average power flow in a transmission line

P_{ave} average power density radiated by an antenna (W/m^2)

$P_n(\theta, \varphi)$ normalised power pattern of an antenna

$\mathbf{P}$ polarisation density

$\mathbf{P}$ the Poynting vector, see Eq (11.127) *372*

P potential energy

p numerical distance, see Eq (15.19) *526*

p pressure, see Eq (15.54) *533*

ppec   perfect electric conductor. A conductor in which $\sigma \rightarrow \infty$, see Eq (11.65) *362*

P_T total power radiated by an antenna, see Eq (13.50) *467*

P_t, P_r transmitted and received power by antennas, see Eq (13.84) *478*

Q, q charges (C), see Eq (4.0) *94*

R distributed resistance per meter in a transmission line, see Eq (12.1) *386*, Radius of the Earth, resistance. (Ω)

$\mathcal{R}$ reluctance

R_r Radiation resistance.

$\mathbf{r'}$ position vector where charges or currents lie

$\mathbf{r}$ position vector of a point in 3-space, see Eq (1.5) *9*, position vector, see Eq (2.4) *36*

r normalised resistance. The real part of Z_{in},

r, θ, ϕ spherical coordinate system, see Eq (2.30) *47*

$\mathcal{S}$ a surface over which an integration is performed

T absolute temperature, see Eq (13.71) *473*, one-period of time ($= 2\pi/\omega$), see Eq (11.25) *349*

$\hat{t}$ unit vector which is tangent to a curve, see Eq (3.3) *58*

t a parameter, see Eq (2.5) *36*, time, see Eq (2.5) *36*

u the real part of the reflection coefficient, Γ

$\mathsf{U}(\theta, \varphi)$	radiation intensity	∇^2	the Laplacian
$V(z)$	total voltage across the two conductors in a transmission line, see Eq (12.42) *395*	∇	the nabla operator, see Eq (3.40) *70*
V_+, V_-	amplitudes of the forward and backward voltage waves. $\Gamma_L = V_-/V_+$, see Eq (12.30) *393*	$\nabla S, \widehat{\nabla S}, \nabla\mathbf{S}$	concerning differential surface elements, see Eq (3.27) *64*
ΔV	differential volume element, see Eq (3.35) *67*	α, β, γ	direction cosines, see Eq (2.4) *34*
$\mathbf{v}$	velocity (m/s), see Eq (1.34) *22*	β	the propagation constant of a transmission line or wave. It may or may not be equal to k
$\mathcal{V}$	a volume over which an integration is performed	χ_e	electric susceptibility
V	voltage or potential unit. (V)	δ	skin depth
v	volume when the symbol V is being used for the potential in the same equation, see Eq (6.119) *197*, the imaginary part of the reflection coeffcient, Γ	Δx	a small increment in x, see Eq (3.0) *58*
		δ_{ij}	$\delta_{ij} = 1$ for $i = j$ otherwise it is zero, see Eq (A.2) *552*
W	work done to move charges, see Eq (6.117) *197*	ε'	real part of the complex permittivity
W_e	energy stored in the electric field, see Eq (6.123) *199*	ε''	imaginary part of the complex permittivity
w_e	energy density of the electric field, see Eq (6.123) *199*	ε_C	complex permittivity
W_m	energy stored in the magnetic field, see Eq (9.71) *296*	ε	permittivity, see Eq (3.96) *83*
w_m	energy density of the magnetic field	ε_r	relative dielectric constant
		$\Gamma(z)$	the reflection coefficient anywhere along the line for the z-coordinate, z, see Eq (12.36) *394*
x	normalised conductance. The imaginary part of $\mathbb{Z}_{in}$		
x, y, z	Cartesian coordinates, see Eq (1.4) *9*	γ	complex propagation constant of transmission lines, see Eq (12.18) *389*
Y	admittance ($\mho$), see Eq (1.39) *23*	Γ_L	reflection coefficient at the load impedance of a transmission line, see Eq (12.44) *396*
Y	shunt distributed admittance in a transmission line ($\mho/\psi$), see Eq (12.12) *388*	λ	free space wavelength
$\mathbb{Z}_{in}, \mathbb{Z}$	normalised input impedance of transmission lines	λ_g	wavelength in the guide or transmission line, see Eq (12.30) *393*
Z	impedance (Ω), see Eq (1.38) *23*, series distributed impedance in a transmission line ($\Omega/$m), see Eq (12.12) *388*	λ	total flux linkages in an inductor, see Eq (9.55) *292*
		μ	permeability, see Eq (3.97) *83*
		ω_p	plasma frequency, see Eq (15.83) *540*
Z, Z_0	characteristic impedance of a medium or free space respectively	Ω_A	beam area, see Eq (13.72) *474*
		ω	radian frequency (rad/s), see Eq (1.38) *23*, rotational speed in rad/sec, see Eq (1.31) *20*
Z_L	load impedance of a transmission line, see Eq (12.43) *396*		
Z_s	surface impedance	ω_{cmn}	cutoff (radian) frequency of a waveguide, see Eq (12.148) *434*
α	real part of the complex propagation constant, the attenuation constant of transmission lines (Neppers/m), see Eq (12.18) *389*	$\Phi(\mathbf{r})$	a scalar field, see Eq (3.9) *60*
		Ψ	flux

xxiv | List of Symbols

Ψ_m magnetic flux, see Eq (9.54) *292*

$\mathbf{a}\rho$ unit vector along the ρ direction in cylindrical coordinates

ρ, ϕ, z cylindrical coordinates, see Eq (2.8) *40*

ρ_m mobile volume charge density

ρ_v charge density, see Eq (3.95) *83*

ρ_l linear charge density (C/m), see Eq (4.1) *95*

ρ_s surface charge density $(C/m)^2$, see Eq (4.2) *95*

Σ a system of charges, see Eq (4.20) *106*

σ conductivity of a material

ζ $\zeta = \sqrt{k^2 - \beta^2}$, see Eq (12.167) *438*

$\nabla\Phi, \nabla\cdot\mathbf{F}, \nabla \times \mathbf{F}, \nabla^2\Phi$ gradient, divergence, curl and laplacian, see Eq (12.161) *437*

$\int_{\mathcal{L}} ...dl$ line integral over a line $\mathcal{L}$, see Eq (4.59) *122*

$\oint_{\mathcal{L}} ...dl$ line integral over a loop $\mathcal{L}$, see Eq (4.59) *122*

$\iint_S ...dS$ surface integral over a surface S, see Eq (4.59) *122*

$\oint_S ...dS$ surface integral over a closed surface S, see Eq (4.59) *122*

$\iiint_V ...dV$ volume integral over a volume V, see Eq (4.59) *122*

$\Phi = \iint_S \mathbf{F}\cdot d\mathbf{S}$ flux through a surface, see Eq (3.34) *67*

$\iint_{ABCD} \Phi\ (\mathbf{r})dS$ surface integral see *65*

$\int_A^B \mathbf{F}(\mathbf{r})\cdot d\mathbf{r}$ line integral over a vector field, see Eq (3.10) *60*

$\int_A^B \Phi(\mathbf{r})dR$ line integral over a scalar field

∂_x $\partial/\partial x$

$\begin{pmatrix} a \\ b \end{pmatrix}, \begin{bmatrix} a \\ b \end{bmatrix}$ column vector

(a b), [a b]   row vector

$\begin{vmatrix} a & b \\ c & d \end{vmatrix}, \begin{Vmatrix} a & b \\ c & d \end{Vmatrix}$ determinant

$\begin{pmatrix} a & b \\ c & d \end{pmatrix}, \begin{bmatrix} a & b \\ c & d \end{bmatrix}$ matrix

PART I

Introduction

1

Scalars and Vectors

Advice is seldom welcome;
and those who need it the most always want it the least.
— Earl of Chesterfield

CHAPTER OBJECTIVES

To enable the students to understand the following:

- Scalars with special emphasis on units
- Basic concept of a vector and the unit vector
- Addition and subtraction of two vectors
- Scalar and vector products of two vectors
- Units and dimensions of engineering quantities

1.1 | Introduction

Fundamental to the study of electromagnetism is the study and use of vector and scalar fields. One must therefore acquire a good knowledge of vector analysis.

As has often been quoted—mathematics is the language of science. When engineers need to apply the results of any subject, they first need to describe it in general terms through equations and then apply those equations to concrete situations. These statements apply to all engineering sciences, and in this particular case, to electromagnetics.

1.2 | Scalars

A scalar is a real number (and sometimes a complex one) which describes a physical quantity. For example, 10.2 kg describes the mass of something which can be sand, steel, or liquid gas; 32.4 million metric tonnes may describe the mass of rice produced by some country; and 22.4 l of gas at STP describes the volume occupied by one mole of a gas at STP. Here 10.2 kg, 32.4 million metric tonnes, and 22.4 l are

Table 1.1 The basic SI units

Base quantity	Name	Symbol
Length	metre	m
Mass	kilogram	kg
Time	second	s
Electric current	ampere	A
Thermodynamic temperature	kelvin	°K
Amount of substance	mole	mole
Luminous intensity	candela	cd

all scalars. The units used in this book are the SI units and Table 1.1 lists the basic SI units.

1.2.1 Rules for Manipulation of Scalars

Scalars are manipulated according to the well known rules:

Rule 1 Scalars $a, b, \ldots$ having same units may be added together to yield a third scalar. For example,

$$c = a + b \tag{1.1}$$

It is essential to note some points

- If a is 20 grams of sand and b equal to 30 grams of sand then their addition gives us c equal to 50 grams of sand. The word *grams* implies that both the scalars possess the *same units*. So before adding two scalar numbers *make sure that they both have the same units*.
- 20 grams of sand cannot be added to 22.4 l of gas, because the two scalars are of different types.
- 20 grams of some material cannot be added to 0.5 pounds of the same material (even though both are of same type) because both these numbers belong to different *systems of units*.

Hence when adding scalars $a, b, c, \ldots$ attention must be paid to the fact that all the scalars must be of the *same type*: that is all masses or all volumes etc. and, furthermore, all the scalars must be quantified in the same system of units, for example (g) or (kg) or (l), etc. Barrow (1966) is a good reference for a quick glance at the SI system of units.[1]

Rule 2 Scalars can be multiplied by other scalars or real (complex) numbers. *But here, even though the two scalars may not necessarily be of the same type (i.e. having the same units and dimensions) they must belong to the same system of units*, i.e., the British system, or the cgs system or the SI system. For example

$$c = ab \tag{1.2}$$

Here

- a can be equal to 22.4 (dimensionless) and $b = 1$ l. So $c = 22.4$ l.
- a can be 10 m/s and $b = 10$ s making c equal to 100 m.

An important point to be noted is that *not only the two scalars get multiplied, but also the two units do*. In the first example c has the dimensions of litres since: (*no units*) × (litres) gives litres. In the second case (metres/sec) × (sec) is metres.

---

[1] The reader may refer to the URL— http://physics.nist.gov/cuu/Units/units.html.

Rule 3 Scalars may be subtracted and for it the guidelines given in rule 1 must be followed. Besides, on getting negative numbers one must be clear about the physical interpretation of a negative number. For example

$$c = a - b \qquad (1.3)$$

If $a = 10$ l and $b = 20$ l then $c = -10$ l does not have physical meaning.

Thus when any answer with a negative sign occurs in the equations, the answer is admissible only if it has *physical meaning*.

Rule 4 Scalars may be divided irrespective of the units of the two quantities involved in the division but the units of the result is equal to the division 'of the units' of the two numbers. For example

$$c = a/b \qquad (1.4)$$

and if a is 10 m and b is 2 s then c is 5 m/s.

1.2.2 Keeping Track of Calculations

Mistakes may crop up while doing scalar manipulations and calculations.

When doing a calculation it is important to keep in mind the order of magnitude of various terms. It is better to do a calculation with only two or three decimal places of accuracy to get the correct result rather than doing a calculation with eight or nine decimal places and obtaining an answer which is completely off the mark. Sometimes while using a calculator one may punch in wrong digits and get absurd results. It is important to carry out calculations with utmost care and check the results obtained at every step as the following example shows.

Example 1.1 A substance A of 10 lbs having volume 1.233 l is added to 9.221 kg of a substance B having volume 2555 cm^3. If on adding the two substances there is a contraction of volume by 12%, what is the density of the material at the end?

Solution To start with do a preliminary computation with only one decimal point (or only whole numbers) to get an idea of what the answer should be.

Step 1. First, we need to quantify everything to a set of common units. Hence

$$10 \text{ lbs} = \frac{10 \text{ lbs}}{2.2 \text{ lbs/kg}} = 4.546 \text{ kg}$$

Is this answer reasonable? Yes, because 10 rounded off is 10, and 2.2 rounded off is 2, hence $10/2 \approx 5$ which is close to 4.546.

Step 2. Next we add the two masses

$$\text{Total mass} = 4.546 + 9.221 = 14.767 \text{ kg}$$

Is the answer reasonable? Because 4.546 round off to 5 + 9.221 rounded off to 9 is 14, but 14.767 rounded off is 15. Something has gone wrong. Doing the calculation again

$$\text{Total mass} = 4.546 + 9.221 = 13.767 \text{ kg}$$

Step 3. Calculate the total volume in litres. All values must be converted to litres

$$1.233 + 0.2555 = 1.4885 \text{ l}$$

Is this okay? 2555 cm^3 is 2.555 l, not 0.2555 l. So re-calculating

$$1.233 + 2.555 = 3.788 \text{ l}$$

$1 + 3 = 4$. Going to first decimal places, $0.2 + 0.5$ is 0.7. The answer is reasonable.

Step 4. Calculate the 12% contraction.

$$10\% \text{ of } 3.788 \text{ l is } 0.3788 \text{ l}$$

The answer should be around 0.3788 l

$$12 \% \text{ of } 3.788 = 0.4546$$

0.1 of 3.7 is 0.37; 0.02 is around 0.07. So 0.37+0.07 = 0.44, which is close to the answer given above. So the answer is reasonable. Now calculate the volume after contraction.

$$3.788 - 0.4546 = 3.3335 \text{ l}$$

3.8 − 0.5 is 3.3. The answer seems correct.

Step 5. Calculate the density

$$\text{Density} = \frac{13.767 \text{ kg}}{3.3335 \text{ l}} = 4.13 \text{ kg/l}$$

14/3 = 4.66. Possibility of an error? Recalculate. We get the same result. Therefore this is the correct answer.

Example 1.2 If the circumference of a circle is 1600 m, calculate the diameter.

Solution What is the maximum and minimum value of the diameter? Compute these values, then proceed to do an accurate calculation. *It is important to get some idea of the answer before doing an accurate calculation.* So make an estimate of the maximum and minimum values of the diameter and then proceed to make accurate calculation.

Step 1. Write the formula to be used.
If d is the diameter of a circle and c the circumference then

$$\pi d = c$$

Step 2. Get an idea of the value of the diameter. To estimate the diameter, since $3 \leq \pi \leq 4$ the diameter lies between

$$\frac{1600}{4} (= 400 \text{ m}) \leq d \leq \frac{1600}{3} (= 533.3 \text{ m})$$

We can say that the value of the diameter will be close to 533 m since $\pi \approx 3$.

Step 3. Doing an accurate calculation

$$d = \frac{1600}{\pi} = 509.3 \text{ m}$$

Practice Problem 1.1
A car races around a circular wedge consisting of a circular arc connected by two radii to the centre of the circle. The angle subtended by the arc is 40°. The length of the arc is 10 miles, and the car takes one hour to complete one circuit. What is the speed of the car in km/hr? Follow the procedure given in the previous examples.

[Ans: 61.95 km/hr]

Archemedes of Syracuse
Archimedes (287–212 BC), a student of Euclid, was one of the greatest Greek mathematicians of his time. He did a great deal of work in the fields of mathematics and engineering. One of his contributions is to provide the approximation to the value of π. He concluded that $3\frac{10}{71} < \pi < 3\frac{1}{7}$.

1.2.3 Order of Magnitude of Calculations

Example 1.3 Multiply two complex numbers: $a = 3 + j2$ and $b = 15 + j3$ using their polar forms.

Solution

Step 1. Estimate the magnitudes of a and b and then calculate their exact values:

$$|a| = \sqrt{9 + 4} = \sqrt{13} \approx 4$$

$$|a| = \sqrt{9 + 4} = \sqrt{13} = 3.606 \quad \text{(exact value)}$$

$$|b| = \sqrt{225 + 9} \approx \sqrt{225} = 15$$

$$|b| = \sqrt{225 + 9} = 15.297 \quad \text{(exact value)}$$

Step 2. Estimate their phase angles and then calculate the exact values:

$$\angle a = \tan^{-1}\left\{\frac{2}{3}\right\} \approx 45°$$

$$\angle a = \tan^{-1}\left\{\frac{2}{3}\right\} = 33.7° \quad \text{(exact value)}$$

$$\angle b = \tan^{-1}\left\{\frac{3}{15}\right\} \approx 0°$$

$$\angle b = \tan^{-1}\left\{\frac{3}{15}\right\} = 11.31° \quad \text{(exact value)}$$

Step 3. Estimate and then calculate the product.

$$ab \approx 60\angle45°$$

$$ab = 3.606 \times 15.297\angle33.7° + 11.3° = 55.154\angle45° \quad (\text{exact value})$$

Practice Problem 1.2
If the major and minor axes of an ellipse are: $a = 20$ m and $b = 10$ m, estimate its area.

[Ans: $150 < A < 200$; $A = 157.07$ m^2]

1.2.4 Approximations

The following example illustrates the usefulness of approximations.

Example 1.4 How will you estimate the result of the sum $S = \sum_{n=1}^{\infty} \frac{1}{n^2}$?

Solution Let us think of a quick way to make an estimation of this infinite sum. Think of what is similar to summation. It is integration which gives summation of differential quantities. Seeing the analog between $1/n^2$ and $1/x^2$ it seems possible to estimate the sum of the series.

Step 1. First compute the first three numbers. These are: 1, 0.25, 0.11, etc.

Step 2. Plot these points and draw a graph.

Figure 1.1 shows a graph of the function $1/x^2$.

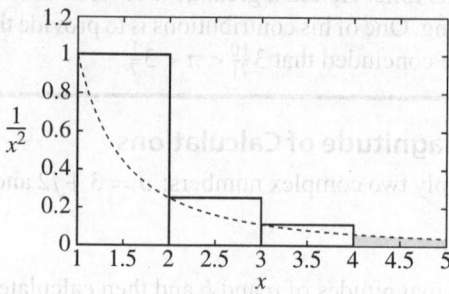

Fig. 1.1

Draw three rectangles as shown in Fig. 1.1. The three rectangles have areas 1, $1/2^2 = 0.25$, and $1/3^2 = 0.111$. Note that the sum of the area of the rectangles is greater than the area under the curve. But after the third term the area of the rectangles and that of the area under the curve tend to become comparable.

Step 3. The sum of the series can therefore be approximated by

$$S \cong 1 + (0.25) + (0.111) + \text{Shaded area}$$

The shaded area is given by using integration as:

$$\int_4^\infty \frac{dx}{x^2} = 0.25$$

Hence the sum can be approximated as

$$S \cong 1 + 0.25 + 0.111 + 0.25 \cong 1.611$$

Therefore $\quad \sum_{n=1}^\infty \frac{1}{n^2} \cong 1.611$

The actual answer is $\pi^2/6 = 1.645$. The approximation is only 2% lower than the correct answer.

1.3 | Vectors

One of the fundamental mathematical entities used in the study of electromagnetic theory, are vectors (Spiegel 1974). Vectors are defined to be entities which possess both magnitude (like scalars) *and* direction. The depiction of these entities is typically in the form of *directed line segments,* as shown in Fig. 1.2.

Paralellogram Law

The parallelogram law for the addition of vectors is probably part of the work of Aristotle (384–322 BC), a Greek philosopher. This law was formulated in the book *Mechanics* of Heron of Alexandria around first century AD and is mentioned as a corollary in the famous *Principia Mathematica* by Isaac Newton. Though Newton dealt almost wholly with forces and velocities (which are vectors), in the *Principia*, he never formally proposed the concept of a vector. The systematic study of vectors was carried out in the 19th and early 20th centuries.

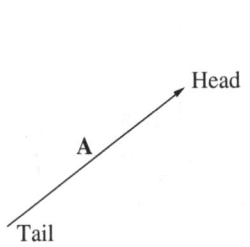

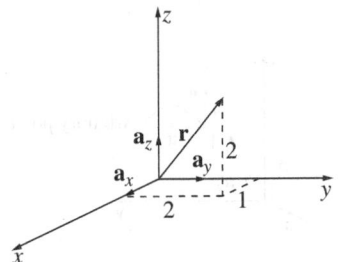

Fig. 1.2 A vector   **Fig. 1.3** Rectangular coordinate system

The line segment is shaped like an arrow, labelled **A** with a tail and a head. The tail–head combination describes a direction while the length of the arrow gives the magnitude of the vector. As mentioned before, vectors are depicted *as directed line segments*. Vectors are in fact abstract entities whose *representation* takes the form of directed line segments.

Take the concept of 'force'. As we all know, a force cannot be seen, nor heard, nor smelt, but only its effects can be felt. Just like you cannot see the number '1', similarly vectors cannot be seen. But their mathematical manipulation leads to real results which can be corroborated by experiments.

Next we consider the rudiments of the rectangular coordinate system in three-dimensional space with which, it is hoped, that the reader is familiar.

The rectangular coordinate system is shown in Fig. 1.3. A point in 3D-space consists of three numbers (x, y, z) which correspond to the three distances cut off by perpendiculars from the point in question to the three axes: x, y, and z. The position vector of the point is the vector joining the origin to the point.

The unit vector in rectangular coordinates in the x-direction is $\mathbf{a}_x$. Similarly, we can specify unit vectors $\mathbf{a}_y$ and $\mathbf{a}_z$ in the y- and z-directions, respectively.

Referring to Fig. 1.3, the position vector of the point $(1, 2, 2)$ is notationally the vector

$$\mathbf{r}(1, 2, 2) \equiv (\mathbf{a}_x + 2\mathbf{a}_y + 2\mathbf{a}_z) \tag{1.5}$$

The magnitude of $\mathbf{r} \equiv r$ is given by the well-known distance formula:

$$r = \sqrt{x^2 + y^2 + z^2}$$

giving $\qquad\qquad r = \sqrt{1^2 + 2^2 + 2^2} = 3 \tag{1.6}$

which is the distance of the point $(1, 2, 2)$ from the origin.

It is also important to point out the nature of the unit vectors along the *rectangular* coordinate axes: (i) The unit vectors have fixed directions no matter which point in 3-space is chosen. (ii) The unit vectors are orthonormal (a set of orthonormal vectors have the property that they are perpendicular to each other), and (iii) each of the vectors of the orthonormal set has unit magnitude. This is shown in Fig. 1.4.

Any vector **A** specified in rectangular coordinates, has the representation:

$$\mathbf{A} = A_x \mathbf{a}_x + A_y \mathbf{a}_y + A_z \mathbf{a}_z \tag{1.7}$$

where A_x, A_y, and A_z are real or complex numbers. With this notation, the force $\mathbf{F} = \mathbf{a}_x + 2\mathbf{a}_y + 2\mathbf{a}_z$ (N), is shown in Fig. 1.5. Thus we can see that $A_x \equiv F_x = 1\,\text{N}$, $A_y \equiv F_y = 2\,\text{N}$, and $A_z \equiv F_z = 2\,\text{N}$. That is, the unit vector $\mathbf{a}_x$ is multiplied by 1, $\mathbf{a}_y$ is multiplied by 2 and $\mathbf{a}_z$ is multiplied by 2. Or in other words the component

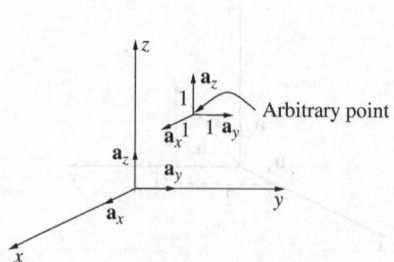
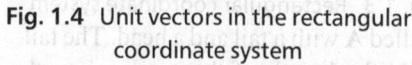

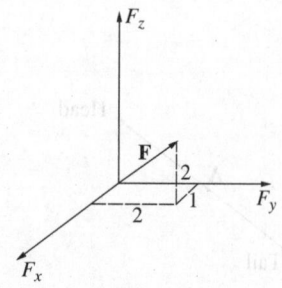

Fig. 1.4 Unit vectors in the rectangular coordinate system

Fig. 1.5 Force in rectangular coordinates

of the force in the $\mathbf{a}_x$ direction is 1 N, that in the $\mathbf{a}_y$ direction is 2 N, and that in a_z direction is 2 N.

1.3.1 Unit Vector in General Direction

The unit vector in any direction is a vector of unit magnitude.

It is clear from the above discussion that *scalar multiplication of vectors is the natural outcome of the definition of unit vectors*. This is so because unit vectors have to be multiplied by real numbers to give us vectors which are directed in any general direction. It may be noted that unit vectors are *dimensionless*, but the scalars have dimensions. Another way to corroborate this statement is by observing that if a vector **A** is given, then the unit vector in the direction of **A** is

$$\hat{\mathbf{A}} = \mathbf{A}/|\mathbf{A}| \tag{1.8}$$

where the 'hat' notation is used to denote a general unit vector, and $|\mathbf{A}| \equiv A$ is the *magnitude or length* of **A**

$$|\mathbf{A}| = \sqrt{A_x^2 + A_y^2 + A_z^2} \tag{1.9}$$

In this text, A may often be written instead of $|\mathbf{A}|$, when there is no ambiguity. For example for the force which we just considered, the magnitude of the force is

$$F = \sqrt{1^2 + 2^2 + 2^2} = 3$$

Therefore the unit vector in the direction of the force F is

$$\hat{\mathbf{F}} = \mathbf{F}/F = (1\mathbf{a}_x + 2\mathbf{a}_y + 2\mathbf{a}_z)/3$$

 See MATLAB programs `UnitVectA.m` and `MagA.m` in Chapter 1

Practice Problem 1.3
A vector $\mathbf{A} = 5\mathbf{a}_x + 5\mathbf{a}_y + 5\mathbf{a}_z$ is given. Find A and $\hat{\mathbf{A}}$.

[Ans: $A = 5\sqrt{3}$ and $\hat{\mathbf{A}} = (\mathbf{a}_x + \mathbf{a}_y + \mathbf{a}_z)/\sqrt{3}$]

1.3.2 Vector Addition and Subtraction

Vectors can be manipulated according to well-defined rules. Examples of operations between scalars are $+, -, \times,$ and $\div$. The simplest operation is the scalar multiplication of a vector. Thus if **A** is a vector, then $5\mathbf{A}$ is another vector, whose length (or more accurately, magnitude) is five times A, but its direction is unchanged. In rectangular coordinates, if

$$\mathbf{A} \equiv [A_x, A_y, A_z] \text{ then } 5\mathbf{A} \equiv [5A_x, 5A_y, 5A_z] \tag{1.10}$$

A particularly important binary operation between two or more vectors is vector addition. Vectors can be 'added'. Here the concept of '+' (vector addition) is different from addition of scalars. To form $\mathbf{A} + \mathbf{B}$, the vector $\mathbf{B}$ is translated parallel to itself and the tail of $\mathbf{B}$ is attached to the head of $\mathbf{A}$ as shown in Fig. 1.6. The vector equation of addition reads:

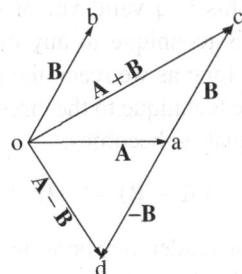

$$C = A + B \qquad (1.11)$$

We can perform the operation the other way, attaching the tail of $\mathbf{A}$ to the head of $\mathbf{B}$ and get the same result.[2] This last statement says that vector addition is commutative:

Fig. 1.6 Vector addition and subtraction

$$C = A + B = B + A \qquad (1.12)$$

that is, whether $\mathbf{A}$ is added to $\mathbf{B}$ or $\mathbf{B}$ is added to $\mathbf{A}$, it gives the same result. Thus if $\mathbf{A} = 2\mathbf{a}_x + 3\mathbf{a}_y + 4\mathbf{a}_z$ and $\mathbf{B} = 5\mathbf{a}_x + 6\mathbf{a}_y + 7\mathbf{a}_z$ then

$$\mathbf{A} + \mathbf{B} = \left(2\mathbf{a}_x + 3\mathbf{a}_y + 4\mathbf{a}_z\right) + \left(5\mathbf{a}_x + 6\mathbf{a}_y + 7\mathbf{a}_z\right) = 7\mathbf{a}_x + 9\mathbf{a}_y + 11\mathbf{a}_z$$

and $\quad \mathbf{B} + \mathbf{A} = \left(5\mathbf{a}_x + 6\mathbf{a}_y + 7\mathbf{a}_z\right) + \left(2\mathbf{a}_x + 3\mathbf{a}_y + 4\mathbf{a}_z\right) = 7\mathbf{a}_x + 9\mathbf{a}_y + 11\mathbf{a}_z$

Vector addition is associative as well. For three vectors $\mathbf{A}$, $\mathbf{B}$, and $\mathbf{C}$:

$$(A + B) + C = B + (A + C) \qquad (1.13)$$

Equation (1.13) indicates that the order of addition is unimportant because it leads to the same result, that is, if we add $\mathbf{A}$ to $\mathbf{B}$ first and add the result to $\mathbf{C}$, we get the same result as when we add $\mathbf{A}$ to $\mathbf{C}$ and then add the result to $\mathbf{B}$. This property is analogous to associativity in scalars. To get an idea of the correct nature of this result, choose any three arbitrary vectors and apply the previous equation.

Similarly one can subtract vectors. $-\mathbf{B}$ is $\mathbf{B}$ *reversed in direction* (and this is just another vector) and so when $-\mathbf{B}$ is added to $\mathbf{A}$:

$$C = A + (-B) = A - B \qquad (1.14)$$

This is also shown in Fig. 1.6. It must be remembered however that vectors which are added or subtracted must of the *same type*, just as with scalars.

1.3.2.1 Simple Technique

Whenever we look at a polygon of vectors as in addition or subtraction we can easily compute the final result by using a certain technique which is explained below. In Fig. 1.6 we can start from any point, say 'a', and go to any other neighbouring point connected by a vector, say point 'd'. Going from 'a' to 'd' the vector involved is $-\mathbf{B}$ and since we are going in the direction of the arrow we write this vector as it is that is $-\mathbf{B}$.

From point 'd' we go to point 'o' going against the direction of the arrow, so we change the sign of the vector and add it to the previous vector:

$$(-B) - (A - B)$$

Note the change in sign of the second term. From 'o' let us go back to 'a', hence we add $\mathbf{A}$ to the previous sum. Since we have come back to the starting point we make the sum of the terms equal to zero:

$$(-B) - (A - B) + A = 0$$

---

[2] Left as an exercise for students.

This is a valid vector equation. We can apply this technique to any other polygon of vectors, as long as we reach the starting point. Applying the technique to the circuit: o-d-a-c-o[3] the vector equation becomes:

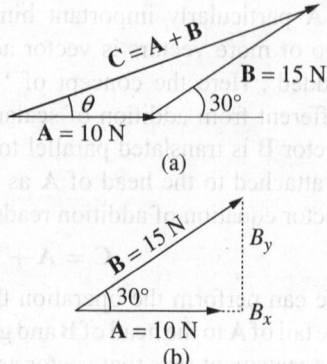

$$(\mathbf{A} - \mathbf{B}) - (-\mathbf{B}) + (\mathbf{B}) - (\mathbf{A} + \mathbf{B}) = 0$$

The reader may be wondering how this technique helps formulation of vector equations? Well, let us apply this technique to the circuit 'o-a-c' and let us call the vector 'o-c' '**C**'. Then we find that

$$\mathbf{C} = \mathbf{A} + \mathbf{B}$$

Fig. 1.7 Addition of two vectors

1.3.2.2 Calculations with Vector Addition

Example 1.5   Let us add two vectors, **A** of 10 N to **B** of 15 N at an angle of 30° to **A** as shown in Fig. 1.7(a) and let the result be denoted as **C**. Find the magnitude and direction of **C**.

Solution   Method (a)

Step 1.   First draw a sketch of the two vectors in the notebook [see Fig. 1.7(a)].

Step 2.   From Fig. 1.7 we can see that vectors **A**, **B**, and **C** form a triangle. We calculate the length of the third side based on the law of cosines. The third side gives us the magnitude of **C**:

$$|\mathbf{C}| = \sqrt{|\mathbf{A}|^2 + |\mathbf{B}|^2 - 2\,|\mathbf{A}|\,|\mathbf{B}|\cos(180° - 30°)}$$

$$= \sqrt{10^2 + 15^2 - 2 \times 10 \times 15\cos 150°}$$

$$|\mathbf{C}| = 24.18 \text{ N}$$

Step 3.   If we apply the law of sines to the triangle again, the angle which **C** makes with the horizontal is 18.01°

$$\frac{\sin\theta}{|\mathbf{B}|} = \frac{\sin 150°}{|\mathbf{C}|}$$

$$\angle C(= \sin\theta) = 18.01°$$

The angle is measured with reference to **A**.

Method (b)

Step 1.   Draw a sketch as shown in Fig. 1.7(b)

Step 2.   Decompose the vector **B** into two perpendicular vectors, the vector parallel to the horizontal is given by

$$\mathbf{B}_x = 15 \times \cos(30°)\,\mathbf{a}_x \text{ N} = 12.93\mathbf{a}_x \text{ N}$$

The vector in the vertical direction is given by

$$\mathbf{B}_y = 15 \times \sin(30°)\,\mathbf{a}_y \text{ N} = 7.5\mathbf{a}_y \text{ N}$$

---

[3] o-d→(**A** − **B**); d-a→ −(−**B**); a-c→ **B**; c-o → −(**A** + **B**).

Step 3. We add $\mathbf{B}_x$ to $\mathbf{A}$ giving $\mathbf{C}_x$ and $\mathbf{C}_y = \mathbf{B}_y$

$$C_x = 22.93\mathbf{a}_x \text{ N}$$
$$C_y = B_y = 7.5\mathbf{a}_y \text{ N}$$
$$C = |\mathbf{C}| = \sqrt{22.93^2 + 7.5^2} = 24.18 \text{ N}$$

which gives us the same answer as before. The angle that $\mathbf{C}$ makes with the horizontal is given by

$$\tan^{-1}(C_y/C_x) = 18.01°$$

Example 1.6 Add two vectors $\mathbf{A}$ and $\mathbf{B}$ such that the angle from $\mathbf{A}$ to $\mathbf{B}$ is 150°, $A = 10$ N and $B = 20$ N (see Fig. 1.8).

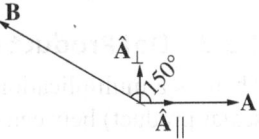

Fig. 1.8

Solution This example shows us how to split a vector into two perpendicular components, one along the direction of a second vector and the other component perpendicular to it.

Step 1. Referring to a sketch (Fig. 1.8), $\mathbf{B}$ is split into two parts one along $\mathbf{A}$:

$$\mathbf{B}_{\| \text{ to } \mathbf{A}} = B\cos\theta \ \hat{\mathbf{A}}$$
$$= 20 \times \cos(150°)\hat{\mathbf{A}}$$
$$= -17.32 \ \hat{\mathbf{A}} \text{ N}$$

and the other perpendicular to $\mathbf{A}$

$$\mathbf{B}_{\perp \text{ to } \mathbf{A}} = B\sin\theta \ \hat{\mathbf{A}}_{\perp}$$
$$= 10 \ \hat{\mathbf{A}}_{\perp}\text{N}$$
$$= \mathbf{B} - \mathbf{B}_{\|\text{to } \mathbf{A}}$$

where $\hat{\mathbf{A}}$ is the unit vector in the direction of $\mathbf{A}$ and $\hat{\mathbf{A}}_{\perp}$ is the unit vector perpendicular to $\mathbf{A}$ but lying in the plane enclosed by $\mathbf{B}$ and $\mathbf{A}$.

Step 2. Since $\qquad\qquad \mathbf{A} + \mathbf{B} = \mathbf{C}$

Adding $\mathbf{B}_\|$ to $\mathbf{A}$, we get

$$\mathbf{C}_\| = (10 - 17.32) \ \hat{\mathbf{A}} = -7.32 \ \hat{\mathbf{A}}$$
$$\mathbf{C}_\perp = 10 \ \hat{\mathbf{A}}_\perp$$

Now $\qquad\qquad \mathbf{C} = \mathbf{C}_\| + \mathbf{C}_\perp$
$$\mathbf{C} = -7.32 \ \hat{\mathbf{A}} + 10 \ \hat{\mathbf{A}}_\perp\text{N}$$

Example 1.7 Find the difference of the two vectors: $\mathbf{A}$ and $\mathbf{B}$ given that $\mathbf{A} = 1\mathbf{a}_x + 3\mathbf{a}_y + 5\mathbf{a}_z$ N and $\mathbf{B} = 5\mathbf{a}_y$ N in rectangular coordinates.

Solution
$$\mathbf{C} = \mathbf{A} - \mathbf{B} = (1\mathbf{a}_x + 3\mathbf{a}_y + 5\mathbf{a}_z) - (5\mathbf{a}_y)$$
$$= (1 - 0) \ \mathbf{a}_x + (3 - 5) \ \mathbf{a}_y + (5 - 0) \ \mathbf{a}_z$$
$$= \mathbf{a}_x - 2\mathbf{a}_y + 5\mathbf{a}_z$$

Practice Problem 1.4

Subtract **B** from **A**. The angle from **A** to **B** is 150°, $A = 10$ N and $B = 20$ N.

[Ans: Magnitude 29.09; $\angle -22.32°$ with **A**]

Practice Problem 1.5

Find vector **B**, which when added to $\mathbf{A} = 30\mathbf{a}_x$ gives us $\mathbf{C} = 15\mathbf{a}_x + 15\mathbf{a}_y$ N.

[Ans: $-15\mathbf{a}_x + 15\mathbf{a}_v$ N]

Practice Problem 1.6

If $\mathbf{C} = \mathbf{A} + \mathbf{B}$, and **A** has a magnitude of 10, **C** also has a magnitude of 10 and is at an angle of 90° to **A**, find **B**.

[Ans: $B = 10\sqrt{2}$; $\angle \mathbf{B} = 135°$]

1.3.3 Dot Product or Scalar Product

There is a multiplication operation between the vectors. This is 'dot' product (or scalar product) between two vectors which results in a scalar.

The dot or scalar product is given by

$$\mathbf{A} \cdot \mathbf{B} = |\mathbf{A}||\mathbf{B}| \cos \theta \qquad (1.15)$$

where · is the operator representative of the scalar product and θ is the angle between **A** and **B** in accordance with Fig. 1.9. Notice that though two *vectors* are involved in the product, the result is a *scalar*. From the definition it is clear that the dot product is commutative:

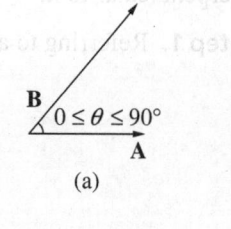

(a)

$$\mathbf{A} \cdot \mathbf{B} = |\mathbf{A}||\mathbf{B}| \cos (\theta) = \mathbf{B} \cdot \mathbf{A} \qquad (1.16)$$

When we examine Fig. 1.9 we can see that the product is positive for values of $0° \leq |\theta| \leq 90°$ and it is negative for $90° \leq |\theta| \leq 180°$. When θ is zero, that is, when the vectors are in the same direction the magnitudes of the two vectors get multiplied. On the other hand, when the two vectors are opposite in direction then though the magnitudes of the two vectors

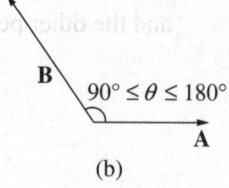

(b)

Fig. 1.9 Dot product between two vectors

get multiplied but the sign is negative. When the two vectors are perpendicular to each other ($\theta = 90°$) then the scalar product is zero. From the definition of the scalar product (substituting **A** for **B**):

$$\mathbf{A} \cdot \mathbf{A} = |\mathbf{A}||\mathbf{A}| \cos 0° = |\mathbf{A}|^2$$

or

$$|\mathbf{A}| = \sqrt{\mathbf{A} \cdot \mathbf{A}} \qquad (1.17)$$

1.3.3.1 Work and Scalar Product

The scalar product is very useful in calculating the work done by a force. The following example illustrates it.

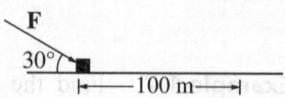

Fig. 1.10 Work done and dot product

Example 1.8

(a) Find the work done when a constant force **F** of magnitude 100 N is applied on a mass as shown (Fig. 1.10), moving it over a distance of 100 m. The force is applied at an angle of 30° with the horizontal.

(b) Calculate the work done to move the mass by 100 m again, but here the magnitude of the force varies with displacement according to the law: $|\mathbf{F}(x)| = 10x$ N, the direction remaining the same as before.

Solution Part (a)

Step 1. We know that the work done by a force is a dot product:

$$W = \mathbf{F} \cdot \mathbf{d}$$

where $\mathbf{F}$ is the force and $\mathbf{d}$ is the displacement vector.

Step 2. So to work out this part we take the dot product of $\mathbf{F}$ with the displacement $\mathbf{d} = 100\mathbf{a}_x$ m. Note that $F = 100$ N.

$$W = \text{work done} = \mathbf{F} \cdot \mathbf{d}$$
$$= \mathbf{F} \cdot (100\mathbf{a}_x)$$
$$= F \cdot 100 \cos (30°) \text{ joules}$$
$$= 8.660 \text{ kJ}$$

Part (b)

Step 1. A infinitesimal amount of work dW is equal to $\mathbf{F} \cdot d\mathbf{l}$ where $\mathbf{F}$ is the force and $d\mathbf{l}$ is the infinitesimal displacement. So

$$\text{Total work done} \quad W = \int \mathbf{F} \cdot d\mathbf{l}$$

Step 2.
$$W = \text{work done} = \int_{x=0}^{x=100} \mathbf{F}(x) \cdot dx$$
$$= \int_{x=0}^{x=100} (10x) \times \cos(30°) \, dx$$
$$= |(5x^2)\Big|_{x=0}^{x=100} \times 0.866$$
$$= 43.3 \text{ kJ}$$

Practice Problem 1.7
(a) A force of $10y^2(\mathbf{a}_x + \mathbf{a}_y)$ N is applied to a mass to move it by a distance of 15 m in the $\mathbf{a}_y$ direction starting from $(0,0)$. Find the work done. (b) If the resistance from the surroundings is such that the mass does not move at all, what is the work done?

[Ans: (a) 11.25 kJ (b) 0]

1.3.3.2 Scalar Products of Orthogonal Unit Vectors

Example 1.9 Find the scalar products between all pairs of unit vectors belonging to the rectangular coordinate system, $\mathbf{a}_x$, $\mathbf{a}_y$, and $\mathbf{a}_z$.

Solution

Step 1. We first write all the possible dot products. These dot products are $\mathbf{a}_x \cdot \mathbf{a}_x$, $\mathbf{a}_x \cdot \mathbf{a}_y$, $\mathbf{a}_x \cdot \mathbf{a}_z$, $\mathbf{a}_y \cdot \mathbf{a}_x$, $\mathbf{a}_y \cdot \mathbf{a}_y$, $\mathbf{a}_y \cdot \mathbf{a}_z$, $\mathbf{a}_z \cdot \mathbf{a}_x$, $\mathbf{a}_z \cdot \mathbf{a}_y$, and $\mathbf{a}_z \cdot \mathbf{a}_z$.

Step 2. We know that

(i) All the unit vectors are perpendicular to each other, and
(ii) Each unit vector is of magnitude equal to 1.

Step 3. From the definition of the scalar product (Eq. (1.15)) it is clear that for two vectors if $\theta = \pm 90°$, $\cos \theta = 0$, the dot product is zero. Hence, we can conclude that

$$\mathbf{a}_x \cdot \mathbf{a}_y = \mathbf{a}_x \cdot \mathbf{a}_z = 0$$
$$\mathbf{a}_y \cdot \mathbf{a}_x = \mathbf{a}_y \cdot \mathbf{a}_z = 0$$
$$\mathbf{a}_z \cdot \mathbf{a}_x = \mathbf{a}_z \cdot \mathbf{a}_y = 0$$

We equate all these products to zero in our list.

Step 4. Since the magnitude of each of the the unit vectors ($\mathbf{a}_x, \ldots$) is 1 and taking the dot product of each unit vector with itself, we get

$$\mathbf{a}_x \cdot \mathbf{a}_x = \mathbf{a}_y \cdot \mathbf{a}_y = \mathbf{a}_z \cdot \mathbf{a}_z = 1 \times 1 \text{ since } \cos 0° = 1$$

Using the results of the last example we can find the dot product between two vectors when they are specified in rectangular coordinates. If $\mathbf{A} = (A_x, A_y, A_z)$ and $\mathbf{B} = (B_x, B_y, B_z)$ then

$$\mathbf{A} \cdot \mathbf{B} = (A_x \mathbf{a}_x + A_y \mathbf{a}_y + A_z \mathbf{a}_z) \cdot (B_x \mathbf{a}_x + B_y \mathbf{a}_y + B_z \mathbf{a}_z) \qquad (1.18)$$

Since all dot products of the unit vectors with *other unit vectors* are zero and dot products of the unit vectors with *themselves* are 1 then

$$(A_x \mathbf{a}_x + A_y \mathbf{a}_y + A_z \mathbf{a}_z) \cdot (B_x \mathbf{a}_x + B_y \mathbf{a}_y + B_z \mathbf{a}_z) = A_x \mathbf{a}_x \cdot (B_x \mathbf{a}_x + B_y \mathbf{a}_y + B_z \mathbf{a}_z)$$
$$+ A_y \mathbf{a}_y \cdot (B_x \mathbf{a}_x + B_y \mathbf{a}_y + B_z \mathbf{a}_z)$$
$$+ A_z \mathbf{a}_z \cdot (B_x \mathbf{a}_x + B_y \mathbf{a}_y + B_z \mathbf{a}_z)$$

Considering the right-hand side term by term

$$A_x \mathbf{a}_x \cdot (B_x \mathbf{a}_x + B_y \mathbf{a}_y + B_z \mathbf{a}_z) = A_x \mathbf{a}_x \cdot B_x \mathbf{a}_x + A_x \mathbf{a}_x \cdot B_y \mathbf{a}_y + A_x \mathbf{a}_x \cdot B_z \mathbf{a}_z = A_x B_x$$
$$A_y \mathbf{a}_y \cdot (B_x \mathbf{a}_x + B_y \mathbf{a}_y + B_z \mathbf{a}_z) = A_y \mathbf{a}_y \cdot B_x \mathbf{a}_x + A_y \mathbf{a}_y \cdot B_y \mathbf{a}_y + A_y \mathbf{a}_y \cdot B_z \mathbf{a}_z = A_y B_y$$
$$A_z \mathbf{a}_z \cdot (B_x \mathbf{a}_x + B_y \mathbf{a}_y + B_z \mathbf{a}_z) = A_z \mathbf{a}_z \cdot B_x \mathbf{a}_x + A_z \mathbf{a}_z \cdot B_y \mathbf{a}_y + A_z \mathbf{a}_z \cdot B_z \mathbf{a}_z = A_z B_z$$

Therefore $\qquad\qquad \mathbf{A} \cdot \mathbf{B} = A_x B_x + A_y B_y + A_z B_z \qquad\qquad (1.19)$

In Eqn (1.18) if we substitute $\mathbf{a}_x$ for $\mathbf{B}$ then

$$A_x = \mathbf{A} \cdot \mathbf{a}_x \qquad (1.20)$$

Similarly $\qquad\qquad A_y = \mathbf{A} \cdot \mathbf{a}_y \qquad\qquad (1.21)$

$$A_z = \mathbf{A} \cdot \mathbf{a}_z \qquad (1.22)$$

1.3.4 Cross Product or Vector Product

Another binary operation used by mathematicians and which is useful in the study of natural phenomena is the 'cross product' or 'vector product'. In this case the vector product of two vectors is a third vector. Thus (see Fig. 1.11) in symbolic notation:

$$\mathbf{C} = \mathbf{A} \times \mathbf{B} \qquad (1.23)$$

where the magnitude of $\mathbf{C}$ is given by

$$|\mathbf{A} \times \mathbf{B}| = |\mathbf{A}||\mathbf{B}| \sin \theta \qquad (1.24)$$

The direction of $\mathbf{C}$ is given by the well-known 'right-hand thumb rule'. The right-hand thumb rule states that the direction of the vector product is perpendicular to

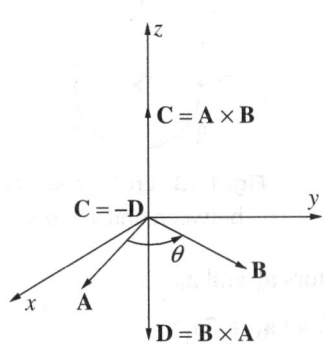

Fig. 1.11 Vectors resulting from the cross product

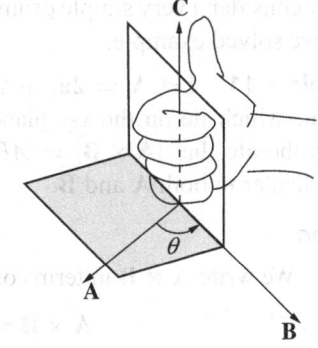

Fig. 1.12 Right-hand thumb rule and the cross product

both $\mathbf{A}$ and $\mathbf{B}$ and is given by the direction of the thumb when the right hand is held in a position as it were holding an imaginary stick with the thumb along the direction of the stick. The hand is held in such a way that the fingers are curled from $\mathbf{A}$ to $\mathbf{B}$. This can be seen from Fig. 1.12. It is important to note that the cross product is anti-commutative.

$$\mathbf{C} = \mathbf{A} \times \mathbf{B} = -\mathbf{B} \times \mathbf{A} \tag{1.25}$$

1.3.4.1 Cross Products of Orthogonal Unit Vectors

Example 1.10 Find the cross products between all unit vectors in rectangular coordinates: $\mathbf{a}_x$, $\mathbf{a}_y$, and $\mathbf{a}_z$.

Solution

Step 1. As in the previous example, we need to find the various cross products:
$\mathbf{a}_x \times \mathbf{a}_x$, $\mathbf{a}_x \times \mathbf{a}_y$, $\mathbf{a}_x \times \mathbf{a}_z$, $\mathbf{a}_y \times \mathbf{a}_x$, $\mathbf{a}_y \times \mathbf{a}_y$, $\mathbf{a}_y \times \mathbf{a}_z$,
$\mathbf{a}_z \times \mathbf{a}_x$, $\mathbf{a}_z \times \mathbf{a}_y$ and $\mathbf{a}_z \times \mathbf{a}_z$.

Step 2. We know that the cross product is proportional to $\sin\theta$, where θ is the angle between the two vectors.

Hence $\quad \mathbf{a}_x \times \mathbf{a}_x = \mathbf{a}_y \times \mathbf{a}_y = \mathbf{a}_z \times \mathbf{a}_z = 1 \times 1 \sin 0° = 0 \qquad (1.26)$

Step 3. Further with reference to Fig. 1.12, $\mathbf{a}_x \times \mathbf{a}_y$ is in the direction of $\mathbf{a}_z$.

Step 4. Since the magnitudes of both $\mathbf{a}_x$ and $\mathbf{a}_y$ are both 1, therefore:

$$\left|\mathbf{a}_x \times \mathbf{a}_y\right| = \left|\mathbf{a}_x\right| \left|\mathbf{a}_y\right| \sin\theta = 1 \times 1 \sin 90° = 1$$

Step 5. Since $\mathbf{a}_x \times \mathbf{a}_y$ has a magnitude 1 and is in the direction of $\mathbf{a}_z$, we realise that the product is itself the unit vector $\mathbf{a}_z$.

$$\mathbf{a}_x \times \mathbf{a}_y = \mathbf{a}_z \tag{1.27}$$

Step 6. Since the cross product is anti-commutative

$$\mathbf{a}_y \times \mathbf{a}_x = -\mathbf{a}_z \tag{1.28}$$

Similarly by this type of reasoning we can easily obtain the other products.

Memory aid: We can also memorise these relations by using Fig. 1.13. If we go along the direction of the arrow, $\mathbf{a}_x \times \mathbf{a}_y = \mathbf{a}_z$ or $\mathbf{a}_z \times \mathbf{a}_x = \mathbf{a}_y$. But if we go *against* the direction of the arrow, a negative sign is required: $\mathbf{a}_y \times \mathbf{a}_x = -\mathbf{a}_z$ or $\mathbf{a}_x \times \mathbf{a}_z = -\mathbf{a}_y$.

We now consider a very simple example based on the above solved example.

Example 1.11 Let $\mathbf{A} = 2\mathbf{a}_x + 4\mathbf{a}_y$ and $\mathbf{B} = \mathbf{a}_x + 7\mathbf{a}_y$ which lie on the x-y plane. Find $\mathbf{A} \times \mathbf{B}$. Corroborate that $|\mathbf{A} \times \mathbf{B}| = AB\sin\theta$ and is perpendicular to both $\mathbf{A}$ and $\mathbf{B}$.

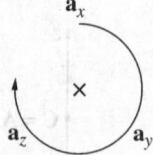

Fig. 1.13 Cross products between unit vectors

Solution

Step 1. We write $\mathbf{A} \times \mathbf{B}$ in terms of the unit vectors $\mathbf{a}_x$ and $\mathbf{a}_y$.

$$\mathbf{A} \times \mathbf{B} = (2\mathbf{a}_x + 4\mathbf{a}_y) \times (\mathbf{a}_x + 7\mathbf{a}_y)$$

Step 2. We next multiply term by term

$$\mathbf{A} \times \mathbf{B} = (2\mathbf{a}_x \times \mathbf{a}_x) + (2\mathbf{a}_x \times 7\mathbf{a}_y) + (4\mathbf{a}_y \times \mathbf{a}_x) + (4\mathbf{a}_y \times 7\mathbf{a}_y)$$

Step 3. We use the results just obtained: $\mathbf{a}_x \times \mathbf{a}_x = 0$; $\mathbf{a}_x \times \mathbf{a}_y = \mathbf{a}_z$, etc.

$$\mathbf{A} \times \mathbf{B} = 0 + 14\mathbf{a}_z + (-4\mathbf{a}_z) + 0 = 10\mathbf{a}_z$$

It is clear from this example that $\mathbf{A} \times \mathbf{B}$ is perpendicular to both $\mathbf{A}$ and $\mathbf{B}$ and has a magnitude of 10.

Step 4. To calculate the value of θ, the angle between $\mathbf{A}$ and $\mathbf{B}$, we proceed as follows:

$$\mathbf{A} \cdot \mathbf{B} = (2\mathbf{a}_x + 4\mathbf{a}_y) \cdot (\mathbf{a}_x + 7\mathbf{a}_y) = 30 = AB\cos\theta$$

with $\quad |A| = \sqrt{2^2 + 4^2} = \sqrt{20}$

$$|B| = \sqrt{1^2 + 7^2} = \sqrt{50}$$

Step 5. We calculate θ using the dot product formula

$$\mathbf{A} \cdot \mathbf{B} = A_x \cdot B_x + A_y \cdot B_y + A_z \cdot B_z = 30$$

and $\mathbf{A} \cdot \mathbf{B} = |A| \, |B| \cos\theta$

$$\theta = \angle AB = \cos^{-1}[(\mathbf{A} \cdot \mathbf{B})/AB] = \cos^{-1}(30/\sqrt{1000}) = 18.41°$$

Step 6. We now use the cross product formula. $\sin(18.41°) = 0.3159$ with A and B calculated earlier

$$|\mathbf{A} \times \mathbf{B}| = AB\sin\theta = \sqrt{50 \times 20} \times 0.3159$$
$$= 31.62 \times 0.3159 = 10$$

1.3.4.2 Cross Product in Rectangular Coordinates

Based on the previous example we are in a position to calculate the cross or vector product in rectangular coordinates.

Let $\mathbf{A} = A_x\mathbf{a}_x + A_y\mathbf{a}_y + A_z\mathbf{a}_z$ and $\mathbf{B} = B_x\mathbf{a}_x + B_y\mathbf{a}_y + B_z\mathbf{a}_z$ then

$$\mathbf{A} \times \mathbf{B} = (A_x\mathbf{a}_x + A_y\mathbf{a}_y + A_z\mathbf{a}_z) \times (B_x\mathbf{a}_x + B_y\mathbf{a}_y + B_z\mathbf{a}_z)$$
$$= A_x\mathbf{a}_x \times (B_x\mathbf{a}_x + B_y\mathbf{a}_y + B_z\mathbf{a}_z) + A_y\mathbf{a}_y \times (B_x\mathbf{a}_x + B_y\mathbf{a}_y + B_z\mathbf{a}_z)$$
$$+ A_z\mathbf{a}_z \times (B_x\mathbf{a}_x + B_y\mathbf{a}_y + B_z\mathbf{a}_z) \tag{1.29}$$

The first term of Eqn (1.29) read

$$A_x\mathbf{a}_x \times (B_x\mathbf{a}_x + B_y\mathbf{a}_y + B_z\mathbf{a}_z)$$

The first product, $A_x B_x \mathbf{a}_x \times \mathbf{a}_x$ is zero. The second product, $A_x \mathbf{a}_x \times B_y \mathbf{a}_y$ is equal to $A_x B_y \mathbf{a}_z$ and the third product, $A_x \mathbf{a}_x \times B_z \mathbf{a}_z$ equals $-A_x B_z \mathbf{a}_y$. Hence the first term is

$$A_x \mathbf{a}_x \times (B_x \mathbf{a}_x + B_y \mathbf{a}_y + B_z \mathbf{a}_z) = A_x B_y \mathbf{a}_z - A_x B_z \mathbf{a}_y$$

Similarly, the second and third terms respectively are

$$A_y \mathbf{a}_y \times (B_x \mathbf{a}_x + B_y \mathbf{a}_y + B_z \mathbf{a}_z) = -A_y B_x \mathbf{a}_z + A_y B_z \mathbf{a}_x$$
$$A_z \mathbf{a}_z \times (B_x \mathbf{a}_x + B_y \mathbf{a}_y + B_z \mathbf{a}_z) = A_z B_x \mathbf{a}_y - A_z B_y \mathbf{a}_x$$

Finally collecting terms

$$\mathbf{A} \times \mathbf{B} = (A_y B_z - B_y A_z)\mathbf{a}_x + (A_z B_x - B_z A_x)\mathbf{a}_y + (A_x B_y - B_x A_y)\mathbf{a}_z \quad (1.30)$$

The cross product may be written as

$$\mathbf{A} \times \mathbf{B} = \mathbf{a}_x \begin{vmatrix} A_y & A_z \\ B_y & B_z \end{vmatrix} + \mathbf{a}_y \begin{vmatrix} A_z & A_x \\ B_z & B_x \end{vmatrix} + \mathbf{a}_z \begin{vmatrix} A_x & A_y \\ B_x & B_y \end{vmatrix}$$

simplifying this becomes

$$\mathbf{A} \times \mathbf{B} = \begin{vmatrix} \mathbf{a}_x & \mathbf{a}_y & \mathbf{a}_z \\ A_x & A_y & A_z \\ B_x & B_y & B_z \end{vmatrix} \quad (1.31)$$

1.3.4.3 Memorizing Cross-product Calculations

To memorise these equations we examine Fig. 1.14 which shows a triangle with its vertices marked 'x', 'y', and 'z' in an anti-clockwise order. This figure is the same one used previously. If $\mathbf{C} = \mathbf{A} \times \mathbf{B}$ and we want to calculate C_x then we go to the point marked 'x' and write C_x then we go towards 'y' along the arrow and write A_y. We have up to now:

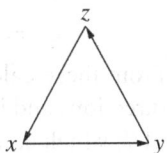

Fig. 1.14 Method to memorise calculation cross products

$$C_x = A_y$$

We continue on to the point marked z and multiply A_y by B_z. We have so far:

$$C_x = A_y B_z$$

Since we are at the point marked z we write A_z.

$$C_x = A_y B_z \quad A_z$$

From here we come back to point marked y (moving against the direction of the arrow) we multiply A_z by B_y.

$$C_x = A_y B_z \quad A_z B_y$$

Since we have moved *against* the direction of the arrow, a negative sign must be placed in front of this product.

$$C_x = A_y B_z - A_z B_y$$

We can continue to write the other components in the same manner.

Example 1.12 Find the vector product $(\mathbf{A} \times \mathbf{B})$ of two vectors in rectangular coordinates given by $\mathbf{A} = (4, 3, 0)$ and $\mathbf{B} = (3, 4, 0)$. Confirm the formula that $|\mathbf{A} \times \mathbf{B}|$ is given by $AB \sin \theta$.

Solution

Step 1. First we can see that both vectors have no z-component. The magnitudes of A and B are therefore:

$$|A| = \sqrt{4^2 + 3^2} = 5$$

$$|B| = \sqrt{3^2 + 4^2} = 5$$

Step 2. The dot product between A and B is used to calculate the angle between the two vectors:

$$\mathbf{A} \cdot \mathbf{B} = 4 \times 3 + 3 \times 4 = 24 \equiv AB \cos \theta$$

where θ is the angle between the two vectors. Hence

$$\cos \theta = 24/(AB) = 24/25$$

$$\theta = \cos^{-1}(24/25) = 16.26°$$

Step 3. We now use $|\mathbf{A} \times \mathbf{B}| = AB \sin\theta$:

$$AB \sin(\theta) = 5 \times 5 \times 0.28 = 7$$

The cross product has this magnitude and its direction is given by the right-hand thumb rule and therefore perpendicular to both (Fig. 1.12).

Using Eqn (1.30), we get

$$C_x = (\mathbf{A} \times \mathbf{B})_x = A_y B_z - A_z B_y = 3 \times 0 - 0 \times 4 = 0$$
$$C_y = (\mathbf{A} \times \mathbf{B})_y = A_z B_x - A_x B_z = 0 \times 3 - 4 \times 0 = 0$$
$$C_z = (\mathbf{A} \times \mathbf{B})_z = A_x B_y - A_y B_x = 4 \times 4 - 3 \times 3 = 7$$

From these calculations it is clear that C has no components in x and y directions and is thus perpendicular to both A and B and has a magnitude of 7, which confirms our result.

Practice Problem 1.8
The vertices of a regular tetrahedron are $O = (0,0,0)$, $A = (0,2,0)$, $B = (\sqrt{3}, 1, 0)$, and $C = (1/\sqrt{3}, 1, 2\sqrt{2}/3)$. Find the unit vector perpendicular to face ABC.

[Ans: $0.41\mathbf{a}_x - 0.71\mathbf{a}_y + 0.29\mathbf{a}_z$]

Example 1.13 A spinning object O is spinning at the rate of $\omega = 60$ rpm about the vertical axis as shown in Fig. 1.15. Find the velocity of any particle at any point $\mathbf{r}(x,y,z)$.

Solution

Step 1. Characterising the rotation as a vector, the angular velocity ω may be written as:

$$\omega = 60\mathbf{a}_z \ (\text{rev/minute}) = 2\pi \, \mathbf{a}_z \ (\text{rad/sec})$$

Step 2. The position vector of any point is given by $\mathbf{r} = x\mathbf{a}_x + y\mathbf{a}_y + z\mathbf{a}_z$ (m)

Step 3. In scalar notation $v = \omega\rho$, but we are dealing with vectors so the correct formula is $\mathbf{v} = \omega \times \mathbf{r}$ as $\mathbf{v}$ is perpendicular to both ω as well as $\mathbf{r}$

$$\mathbf{v} = \omega \times \mathbf{r} = (2\pi\mathbf{a}_z) \times (x\mathbf{a}_x + y\mathbf{a}_y + z\mathbf{a}_z)$$
$$= 2\pi x(\mathbf{a}_z \times \mathbf{a}_x) + 2\pi y(\mathbf{a}_z \times \mathbf{a}_y) + 2\pi (\mathbf{a}_z \times \mathbf{a}_z)$$
$$= 2\pi(x\mathbf{a}_y - y\mathbf{a}_x) \ \text{m/s}$$

Verify whether your answer is correct! Notice that the velocity of the particle is independent of the z-coordinate. The equation implies that all particles on a straight line parallel to the z-axis possess the same velocity, which from our intuition we can justify.

Take the magnitude of both sides of the equation.

$$|\mathbf{v}| = 2\pi\sqrt{x^2 + y^2} = 2\pi\rho\,(=\omega\rho)\text{ m/s}$$

ρ being the perpendicular distance of the point from the z-axis. This is also intuitively correct.

Consider a point on the x-axis and find out the direction of the velocity. If

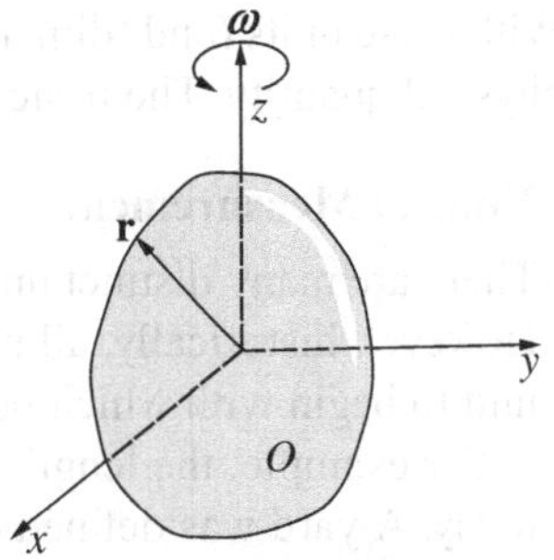

$$\mathbf{r} = x\mathbf{a}_x$$

then

$$\mathbf{v} = 2\pi x\mathbf{a}_y$$

Note the direction which is correct.

Fig. 1.15 A spinning object

Cross product is not associative: We must be sure to remember that the cross product is *not* associative. For example

$$(\mathbf{a}_x \times \mathbf{a}_x) \times \mathbf{a}_z = \mathbf{0} \times \mathbf{a}_z = 0$$

and

$$\mathbf{a}_x \times (\mathbf{a}_x \times \mathbf{a}_z) = \mathbf{a}_x \times (-\mathbf{a}_y) = -\mathbf{a}_z$$

So clearly

$$(\mathbf{a}_x \times \mathbf{a}_x) \times \mathbf{a}_z \neq \mathbf{a}_x \times (\mathbf{a}_x \times \mathbf{a}_z)$$

In general,

$$(\mathbf{A} \times \mathbf{B}) \times \mathbf{C} \neq \mathbf{A} \times (\mathbf{B} \times \mathbf{C}) \tag{1.32}$$

1.3.4.4 Scalar Triple Product

$\mathbf{A} \cdot \mathbf{B} \times \mathbf{C}$ is known as scalar triple product.

Note that here the cross product has to be taken first, and then only the dot product. Since

$$\mathbf{B} \times \mathbf{C} = \begin{vmatrix} \mathbf{a}_x & \mathbf{a}_y & \mathbf{a}_z \\ B_x & B_y & B_z \\ C_x & C_y & C_z \end{vmatrix}$$

or

$$\mathbf{B} \times \mathbf{C} = \mathbf{a}_x(B_yC_z - B_zC_y) + \cdots$$

therefore

$$\mathbf{A} \cdot \mathbf{B} \times \mathbf{C} = A_x(B_yC_z - B_zC_y) + \cdots$$

comparing the two previous equations, we can see that $\mathbf{a}_x, \mathbf{a}_y$, and $\mathbf{a}_z$ has been replaced by A_x, A_y, and A_z. Therefore,

$$\mathbf{A} \cdot \mathbf{B} \times \mathbf{C} = \begin{vmatrix} A_x & A_y & A_z \\ B_x & B_y & B_z \\ C_x & C_y & C_z \end{vmatrix} \tag{1.33}$$

1.4 | Units and Dimensions

Units and dimensions play an important role in electromagnetic theory. Throughout this book, the fundamental units used will that of the SI units (Le Système International d'Unités). These fundamental units are: metres (m), for the dimension of length [L], kilogram (kg) for the 'dimension' of mass [M], seconds (s), for the dimension of time [T] and Coulombs (C), for the dimension of charge [Q].

With these units (and 'dimensions') we can derive the units and dimensions of any physical quantity. The basic units of the SI scheme are given in Table 1.1.

Units of Measurement

There are many distinct units of measurement, such as metres, feet, and yards. However, historically, all measurement required a standard, a rough and ready unit to begin with which people could use in day to day transactions.

For example, the length of a human foot lead ultimately to the foot we have today. A yard was defined differently at different times but was eventually defined as the distance from the tip of the nose to the tip of the middle finger of the outstretched arm of king Henry I of England. Similarly, for the Romans, one thousand paces was a 'millia' (Latin for a thousand) from where we get our mile.

1.4.1 Derivation of Other Units from Basic Units

Let us find the unit and dimension of velocity.

$$\mathbf{v} = \text{velocity} = d\mathbf{r}/dt \tag{1.34}$$

where $\mathbf{r}$ is the position vector of a particle in metres and t is time in seconds. Since we are dividing $d\mathbf{r}$ which is in metres by dt which is in seconds, hence the units of velocity is

$$\frac{\text{meters}}{\text{second}} \equiv (\text{meters})\,(\text{second})^{-1}$$

and the dimension of velocity is $[L][T]^{-1}$. Observe that the *unit* and *dimension* of velocity is obtained *through a formula*. Similarly, the unit of acceleration is derived from the formula:

$$\mathbf{a} = \text{acceleration} = \frac{d\mathbf{v}}{dt} = \frac{d}{dt}\left(\frac{d\mathbf{r}}{dt}\right) \tag{1.35}$$

Which is d/dt (units of s^{-1}) $\times$ $d\mathbf{r}/dt$ (having the units of ms^{-1}), therefore the unit of acceleration is therefore $s^{-1} \times ms^{-1} \equiv ms^{-2}$, and the dimension is $[L][T]^{-2}$. Similarly, from the formula

$$\mathbf{F} = m\mathbf{a} \tag{1.36}$$

the unit of force is $(kg)(m)(s)^{-2}$, and the dimension of force is $[M][L][T]^{-2}$.[4] Due to the complexity of the unit of force, it is given a new name: newton (N).

Students often get confused about units of various physical quantities and this leads to their confusion about the subject of electromagnetics in general. Here the units used for the various physical quantities are in terms of Coulomb (C), amperes ($A \equiv C/s$), volts (V), ohms (Ω), farads (F) and henrys (H). It is important to note that the short form of a unit is written in capital letters when the unit is named after a person; for example (C) or (N), but the unit itself is written in small letters: for example newton. In this book, when the physical quantity is discussed, then we shall also discuss its units. In the meanwhile, we will consider ampere and volt to be the more fundamental of these five quantities, then from Ohm's law

$$V = IR \tag{1.37}$$

The unit of ohms Ω is volt/ampere.

---

[4]It can be taken as an exercise.

Let us look at the impedance equation

$$Z = j\omega L \tag{1.38}$$

where Z is the impedance in Ω, ω is the radian frequency in rad/sec (which is essentially s^{-1}, since radians is dimensionless) and L is the inductance in henry. Then unit of henry (H) is thus $\Omega/\sec^{-1} = \Omega \sec$, since radians have no dimensions.

Finally we can derive the unit of farad from the admittance equation:

$$Y = j\omega C \tag{1.39}$$

where Y is the admittance in $\mho$ and C is the capacitance in farad (F). The unit of farad is then $\mho/\sec^{-1} = \mho \sec$, and so on.

Example 1.14 Why is the radian dimensionless?

Solution By definition, the angle subtended by an arc of length l at its centre is θ radians. The complete circumference, c, subtends an angle of 2π radians and the length of the circumference is $c = 2\pi r$. Or, the angle subtended is c/r. Therefore $\theta = l/r$ which is the ratio of two lengths. Therefore, the units of θ are dimensionless.

POINTS TO REMEMBER

- When adding or subtracting two numbers both numbers should have the same units. a kg $\pm$ b kg is fine; *but not a kg $\pm$ b lbs.*
- Pay attention to the order of magnitude of both terms, and most importantly, the order of magnitude of the result. $1 \times 10^4 \pm 2 \times 10^{-4}$ is $\approx 1 \times 10^4$.
- When multiplying or dividing two numbers, they must belong to the same system of units. a (m) $\div$ b (s) $\times c$ (hours) is incorrect.
- The position vector $\mathbf{r}$ of a point (x, y, z) in Cartesian coordinates is

$$\mathbf{r}(x, y, z) \equiv x\mathbf{a}_x + y\mathbf{a}_y + z\mathbf{a}_z$$

- Any vector $\mathbf{A}$ is given by $\mathbf{A} = A_x\mathbf{a}_x + A_y\mathbf{a}_y + A_z\mathbf{a}_z$

  Its magnitude is given by $|\mathbf{A}| = \sqrt{A_x^2 + A_y^2 + A_z^2}$

  and its direction by the unit vector $\hat{\mathbf{A}} = \mathbf{A}/|\mathbf{A}|$

- When a vector $\mathbf{A}$ is multiplied by a scalar (say by 5) then

$$5\mathbf{A} \equiv 5A_x\mathbf{a}_x + 5A_y\mathbf{a}_y + 5A_z\mathbf{a}_z$$

- When two vectors are added

$$\mathbf{A} + \mathbf{B} = \left(A_x + B_x\right)\mathbf{a}_x + \left(A_y + B_y\right)\mathbf{a}_y + \left(A_z + B_z\right)\mathbf{a}_z$$

- Vector addition is commutative, $\mathbf{A} + \mathbf{B} = \mathbf{B} + \mathbf{A}$
- Vector addition is associative, $\left(\mathbf{A} + \mathbf{B}\right) + \mathbf{C} = \mathbf{A} + \left(\mathbf{B} + \mathbf{C}\right)$
- The vector $-\mathbf{B}$ is the vector $\mathbf{B}$ in reversed direction.
- Vector subtraction, $\mathbf{A} - \mathbf{B} = \left(A_x - B_x\right)\mathbf{a}_x + \left(A_y - B_y\right)\mathbf{a}_y + \left(A_z - B_z\right)\mathbf{a}_z$
- The dot or scalar product is given by $\mathbf{A} \cdot \mathbf{B} = |\mathbf{A}||\mathbf{B}|\cos\theta$
  where $\cdot$ is the operator representative of the scalar product and θ is the angle between $\mathbf{A}$ and $\mathbf{B}$. Also, $\mathbf{A} \cdot \mathbf{B} = A_xB_x + A_yB_y + A_zB_z$
- The dot product is commutative, $\mathbf{A} \cdot \mathbf{B} = \mathbf{B} \cdot \mathbf{A}$
- Work done by a force in causing displacement of a body

$$W = \text{work done} = \mathbf{F} \cdot \mathbf{d}$$

W is the work done, $\mathbf{F}$ is the force, $\mathbf{d}$ is the displacement. Or W expressed as an integral is $W = \int \mathbf{F} \cdot d\mathbf{l}$.

- $\mathbf{a}_x \cdot \mathbf{a}_x = \mathbf{a}_y \cdot \mathbf{a}_y = \mathbf{a}_z \cdot \mathbf{a}_z = 1$. All other products are zero
- For $i, j = x, y, z$ then $\mathbf{a}_i \cdot \mathbf{a}_j = 1$ if $i = j$. Otherwise it is zero.
- The vector product is a vector perpendicular to both vectors in the direction given by the right-hand thumb rule

$$|\mathbf{A} \times \mathbf{B}| = |\mathbf{A}||\mathbf{B}|\sin\theta \qquad \text{or} \qquad \mathbf{A} \times \mathbf{B} = \begin{vmatrix} \mathbf{a}_x & \mathbf{a}_y & \mathbf{a}_z \\ A_x & A_y & A_z \\ B_x & B_y & B_z \end{vmatrix}$$

where $|\ldots|$ represents a determinant.

- The cross product is anti-commutative, $\mathbf{A} \times \mathbf{B} = -\mathbf{B} \times \mathbf{A}$
- The cross product does not follow associative law, $(\mathbf{A} \times \mathbf{B}) \times \mathbf{C} \neq \mathbf{A} \times (\mathbf{B} \times \mathbf{C})$
- Scalar triple product, $\mathbf{A} \cdot \mathbf{B} \times \mathbf{C} = \begin{vmatrix} A_x & A_y & A_z \\ B_x & B_y & B_z \\ C_x & C_y & C_z \end{vmatrix}$

SELF ASSESSMENT

Objective Type Questions

Note: In the following questions one or more choices may be correct. Sometimes choices may be separated with an 'or' and sometimes with an 'and'.

1. A unit vector $\hat{\mathbf{A}}$ is equal to
   (a) A/A   (b) A/A   (c) A-A   (d) None of these
2. A boat is travelling across a N-S river, perpendicular to the bank, with a velocity 10 m/s E. The current of water is moving at 10 m/s N. If the width of the river is 100 m, where does the boat reach on the other bank?
   (a) Directly opposite, E   (b) 50 mN of the opposite point
   (c) 100 mS of the opposite point   (d) 100 mN from the opposite point
3. Which of the following vectors are perpendicular to each other:
   (a) $\mathbf{a}_x, \mathbf{a}_y$ (b) $(\mathbf{a}_x + \mathbf{a}_y), (\mathbf{a}_x - \mathbf{a}_y)$ (c) $(\mathbf{A} + \mathbf{B}), (\mathbf{A} - \mathbf{B})$
   (a) a and b   (b) a and c   (c) c only   (d) all
4. A top is spinning at 2 rev/s, what is the velocity of a point on the axis of rotation?
   (a) 0 m/s   (b) 1 m/s   (c) 2 m/s   (d) 3 m/s
5. If two vectors $\mathbf{A}$ and $\mathbf{B}$ are such that $\mathbf{A} \cdot \mathbf{B} = 0$. Neither $\mathbf{A}$ nor $\mathbf{B}$ is $\mathbf{0}$. What is the angle between the vectors?
   (a) $0°$ (b) $45°$ (c) $90°$ (d) $270°$
6. If $|\mathbf{E}_1| = 1$ and $|\mathbf{E}_2| = 4$ and $\mathbf{E}_1 \cdot \mathbf{E}_2 = 6$ then
   (a) The angle is $0°$ (b) Impossible answer
   (c) The angle is either $45°$ or $-45°$ (d) The maximum value of $\mathbf{E}_1 \cdot \mathbf{E}_2 = 3$
7. If $|\mathbf{E}_1| = 1$ and $|\mathbf{E}_2| = 4$ and $\mathbf{E}_1 \cdot \mathbf{E}_2 = 4$ then
   (a) The angle is $0°$ (b) Impossible answer
   (c) The angle is either $45°$ or $-45°$ (d) The maximum value of $\mathbf{E}_1 \cdot \mathbf{E}_2 = 3$
8. If $|\mathbf{E}_1| = 1$ and $|\mathbf{E}_2| = 4$ and $\mathbf{E}_1 \cdot \mathbf{E}_2 = 2.828$ then
   (a) The angle is $0°$ (b) Impossible answer
   (c) The angle is either $45°$ or $-45°$ (d) The maximum value of $\mathbf{E}_1 \cdot \mathbf{E}_2 = 3$
9. If $\mathbf{E} \cdot \mathbf{B}_1 = \mathbf{E} \cdot \mathbf{B}_2$ then
   (a) $\mathbf{E} = 0$ (b) $\mathbf{E} \cdot (\mathbf{B}_1 - \mathbf{B}_2) = 0$ (c) $\mathbf{B}_1 = \mathbf{B}_2$ (d) All of the above

10. A force $\mathbf{F} = 3\mathbf{a}_x + 4\mathbf{a}_y + 5\mathbf{a}_z$ (N) moves a particle from $P = (1, 1, 1)$ to $Q = (4, 6, 2)$. What is the work done?

   (a) 35 J   (b) 34 J   (c) 33 J   (d) None of the above

11. If **A** and **B** are two non co-linear vectors with an angle θ from **A** to **B** and if **C** is defined by $\mathbf{C} = \dfrac{\mathbf{A} \times \mathbf{B}}{AB \sin\theta}$

   (a) **C** is perpendicular to **A** and **B**   (b) **C** is a unit vector
   (c) $\mathbf{A} \times \mathbf{B}$ is not a vector   (d) all of the above

12. $\mathbf{A} = 2\mathbf{a}_x + \mathbf{a}_y + 2\mathbf{a}_z$ and $\mathbf{r} = x\mathbf{a}_x + y\mathbf{a}_y + z\mathbf{a}_z$ then $\mathbf{r} \cdot \mathbf{A} = 5$ represents

   (a) a curve   (b) a region of 3-space   (c) a plane   (d) a sphere

Short-Answer Questions

1. The addition of a scalar of magnitude 5 to 2 g of iron is 7. Is this correct? Justify your answer.
2. The product of a scalar of magnitude 5 and 2 g of iron is 10 g. Is this correct? Justify your answer.
3. The difference of 5 g of iron and 2 g of iron is −3 g. Is this correct? Justify your answer.
4. The dot product of a vector **A** of unit 'm' with a vector **B** of unit 's' gives a result **C** (m s). Is this correct? Justify your answer.
5. The cross product of a vector **A** of unit 'm' with a vector **B** of unit 's' gives a result **C** (m s). Is this correct? Justify your answer.
6. In the case where a planet is moving around the Sun, show how and why the planet speeds up and slows down.
   Hint: See Ex. 1.8. The force is directed towards the Sun. When $\mathbf{F} \cdot \Delta\mathbf{l}$ is positive the planet gains potential energy and loses kinetic energy, etc.

Review Questions

1. Name two common applications of electromagnetic theory and write a short note (about 10 lines) on each application.
2. Why is the study of scalars operations imperative in engineering electromagnetics? In particular, why must we pay attention to small details (like units and order of magnitude) in problem solving?
3. Define a vector and a unit vector.
4. Why do we need to know about vector operations? Write down the rule for
   (a) Scalar multiplication of a vector. Does scalar multiplication of a vector change the direction of the vector?
   (b) Addition of two vectors.
   (c) Subtraction of two vectors.
5. Why is the dot product important in vector operations? (Discuss with respect to work done) Why is it called scalar product? Discuss the sign of the dot product with respect to angle between the two vectors. Why is the dot product commutative based on the definition?
6. How do you arrive at the cross product between two vectors? Explain the right-hand thumb rule.
7. When we take the cross product of vectors given in terms of $\mathbf{a}_x$, $\mathbf{a}_y$, and $\mathbf{a}_z$, explain how we obtain the cross product using the determinant notation. Using the same notation explain why the cross product is anti-commutative.

8. State the law of triangle of velocities.

   Hint: See Section 1.3.2

9. When considering the position vector of two points P_1 and P_2. What part of the vector OP_1 is in the direction of OP_2?

   Hint: Consider the dot product and the unit vector of $\overrightarrow{OP_2}$!

10. Using the right-hand thumb rule show why the directions of $\mathbf{A} \times \mathbf{B}$ and $\mathbf{B} \times \mathbf{A}$ are opposite to each other. Why is $\mathbf{A} \times \mathbf{A} = 0$ even though $\mathbf{A} \neq \mathbf{0}$.

11. Explain how there are two types of units: fundamental and derived.

Numerical Problems

1. An electron moves with a velocity of 2×10^7 m/s, find its kinetic energy. (a) First write the formula, (b) then write the units of the KE, (c) estimate the order of magnitude of the result and the answer, (d) then use the calculator to calculate the answer. Follow this procedure every time you do a calculation. This procedure seems very involved, but in the end it will save you a lot of trouble.

2. How many metres are there in a mile? Follow some procedure to estimate and check your answer.

3. Calculate the following two integrations (a) $\int_0^{\pi} \sin(x)\, dx$ and (b) $\int_0^{(3/2)\pi} \cos(x)\, dx$. How will you check whether your results are correct?

4. If a ball is dropped from a certain height from rest and takes 1 s to hit the floor and then return to its initial position, estimate the velocity with which it hits the floor. Also estimate the height from which it is dropped. Think of various ways to do this estimation. Finally do an accurate calculation.

5. Is the sum $\sum_{n=1}^{\infty} \frac{1}{n}$ convergent or divergent?

   Hint: See Ex. 1.4

6. What is the result of the integral $\int_{-\infty}^{\infty} xe^{-x^2}\, dx$?

   Hint: What is the type of function which is being integrated? This is an example of how some time spent initially on thinking of the solution will save you a lot of trouble. In simple integrations of this type a graph of the integrand is very helpful. This method of approach in the long run will be very beneficial.

7. Vector $\mathbf{A}$ has a magnitude of 1 m , $\mathbf{B}$ has a magnitude of 2 m at an angle of $30°$ to $\mathbf{A}$. Find the magnitude and angle of the resultants $\mathbf{A} + \mathbf{B}$ and $\mathbf{A} - \mathbf{B}$.

8. Vector $\mathbf{A}$ has a magnitude of 1 m, $\mathbf{B}$ has a magnitude of 2 m. The angle between $\mathbf{A}$ and $\mathbf{B}$ is not specified. Find the minimum and maximum value of $|\mathbf{A} + \mathbf{B}|$ and $|\mathbf{A} - \mathbf{B}|$ for all possible angles.

9. $\mathbf{A}$ is vector of length 1 m. $\mathbf{B}$ is $\perp$ to $\mathbf{A}$. (a) What should be the length of $\mathbf{B}$ to make $|\mathbf{A} + \mathbf{B}| = 2$ m? (b) $|\mathbf{A} - \mathbf{B}| = 2$ m?

10. Vector $\mathbf{A}$ has a magnitude of 1 m, $\mathbf{B}$ has a magnitude of 2 m. The angle between $\mathbf{A}$ and $\mathbf{B}$ is not specified. Find the angle between $\mathbf{A}$ and $\mathbf{B}$ to make $|\mathbf{A} + \mathbf{B}| = 1.5$ m.

11. Show that $-\mathbf{B}$ is $\mathbf{B}$ reversed in direction, and then show that $\mathbf{B} + (-\mathbf{B}) = 0$.

12. Find the position vectors of the two points $P_1 = (1, -1, 2)$, $P_2 = (2, 3, 2)$, and the vectors $P_1 P_2$ and $P_2 P_1$.

13. Show that for two vectors $\mathbf{A}$ and $\mathbf{B}$ using the diagrammatic representation of vectors that $\mathbf{A} + \mathbf{B} = \mathbf{B} + \mathbf{A}$.

14. Find the position vector of the mid point of $P_1 P_2$ where P_1 and P_2 are two points given in Problem 12. Let this point be P_m.

15. Find the angles between the three vectors OP_1, OP_2, and OP_m where O is the origin and where P_1 and P_2 are two points given in Problem 12.

16. For three vectors **A**, **B**, and **C** using the diagrammatic representation of vectors show that $(\mathbf{A} + \mathbf{B}) + \mathbf{C} = \mathbf{B} + (\mathbf{A} + \mathbf{C})$.

17. Find the unit vector which is perpendicular to the plane OP_1P_2 where P_1 and P_2 are two points given in Problem 12.

18. Show that the work done by a force **F** operating over the displacement **d** must be $\mathbf{F} \cdot \mathbf{d}$.

19. Show that in the case of a planet having a circular orbit around the Sun, no work is done at any time.

20. Show that in the case of a planet having an elliptical orbit around the Sun, no work is done over one cycle of the orbit.

21. Show that $|\mathbf{A} \times \mathbf{B}|$ is equal to the parallelogram whose two sides are **A** and **B**.
    Hint: Calculate the area of a parallelogram.

22. Given $\mathbf{A} = 5\mathbf{a}_x + 3\mathbf{a}_y$, $\mathbf{B} = 5\mathbf{a}_x - 3\mathbf{a}_y$, and $\mathbf{C} = 6\mathbf{a}_z$. Compute $\mathbf{A} \cdot \mathbf{B} \times \mathbf{C}$ and $\mathbf{A} \times (\mathbf{B} \times \mathbf{C})$. Show that $\mathbf{A} \times (\mathbf{B} \times \mathbf{C}) = (\mathbf{A} \cdot \mathbf{C})\mathbf{B} - (\mathbf{A} \cdot \mathbf{B})\mathbf{C}$ using the values given.

23. Show that $\mathbf{A} \cdot \mathbf{B} \times \mathbf{C}$ is equal to the volume of the parallelopiped whose sides are **A**, **B**, and **C**.

24. Explain why if three vectors **A**, **B**, and **C** lie on a plane then the determinant

$$\begin{vmatrix} A_x & A_y & A_z \\ B_x & B_y & B_z \\ C_x & C_y & C_z \end{vmatrix} = 0$$

Hint: Consider the product: $\mathbf{A} \cdot (\mathbf{B} \times \mathbf{C})$.

25. Show that the unit of force (N) is $(\text{kg})(\text{m})(\text{s})^{-2}$.
    Hint: What are the units of mass and acceleration?

26. Find the units of the angular velocity, torque, energy, and power in SI units.

27. Derive the units of refractive index of a dielectric. Start with the definition of refractive index.

28. Investigate the units of heat and temperature.

29. What is the unit of pressure in SI unit?

Answers

Objective Type Questions

1. (a)   2. (d)   3. (a) and (b)   4. (a)   5. (c) or (d)   6. (b)
7. (a)   8. (c)   9. (d)   10. (b)   11. (a) and (b)   12. (c)

Short-Answer Questions

1. $5 + 2$ g is not equal to 7 since the dimensions of the two added quantities are not the same. 5 is dimensionless and 2 is in grams.
2. 5×2 g $= 10$ g is correct. The interpretation is 'Five 2 g pieces of iron ...' etc.
3. What is the meaning of -3 g of iron? Physically, there is no such thing. Therefore -3 g is incorrect.
4. No the answer is incorrect, $\mathbf{A} \cdot \mathbf{B}$ is always a scalar, and **C** is a vector.
5. Yes, **C** (m-s) may be the correct answer if $\mathbf{C} = \mathbf{A} \times \mathbf{B}$.

Numerical Problems

1. (a) $1/2\left(m_e v^2\right)$ (b) Joules   (c) 18×10^{-17} (J)   (d) 1.8218×10^{-16} (J)
2. 1609.3 m to a mile
3. (a) 2   (b) -1
4. 9.8 m/s; 4.9 m

5. Divergent
6. 0
7. $|\mathbf{A} + \mathbf{B}| = 2.909$; $\angle (\mathbf{A} + \mathbf{B}) = 20.1°$ with $\mathbf{A}$; $|\mathbf{A} - \mathbf{B}| = 1.239$;
   $\angle (\mathbf{A} - \mathbf{B}) = 233.8°$ with $\mathbf{A}$
8. $|\mathbf{A} + \mathbf{B}|_{max} = 3.0$; $\angle (\mathbf{A} + \mathbf{B})_{max} = 0°$; $|\mathbf{A} + \mathbf{B}|_{min} = 1$;
   $\angle (\mathbf{A} + \mathbf{B})_{min} = 180°$; $|\mathbf{A} - \mathbf{B}|_{max} = 3.0$; $\angle (\mathbf{A} - \mathbf{B})_{max} = 180°$;
   $|\mathbf{A} - \mathbf{B}|_{min} = 1$; $\angle (\mathbf{A} - \mathbf{B})_{min} = 0°$
9. (a) $\sqrt{3}$ (b) $\sqrt{3}$
10. $46.56°$ with $\mathbf{A}$
12. $\mathbf{P}_1 = \mathbf{a}_x - \mathbf{a}_y + 2\mathbf{a}_z$; $\mathbf{P}_2 = 2\mathbf{a}_x + 3\mathbf{a}_y + 2\mathbf{a}_z$; vector $P_1P_2 = \mathbf{P}_2 - \mathbf{P}_1$; vector $P_2P_1 = \mathbf{P}_1 - \mathbf{P}_2$
14. $\mathbf{P}_m = (1/2)(\mathbf{P}_1 + \mathbf{P}_2) = 1.5\mathbf{a}_x + \mathbf{a}_y + 2\mathbf{a}_z$
15. Angle between OP_1 and $OP_2 = 72.71°$. Angle between OP_1 and $OP_m = 46.9°$. Angle between OP_2 and $OP_m = 25.7°$
17. $\pm (\mathbf{P}_1 \times \mathbf{P}_2)/|\mathbf{P}_1 \times \mathbf{P}_2| = \pm(1/\sqrt{93})[-8, 2, 5]$
22. $\mathbf{A} \cdot \mathbf{B} \times \mathbf{C} = -180$; $\mathbf{A} \times (\mathbf{B} \times \mathbf{C}) = -96\mathbf{a}_z$
26. Angular velocity = rad/sec; Torque = N-m; Energy = N-m; Power = N-m/sec
27. Dimensionless
28. Heat = Energy, Temperature = dimensionless, $...°$ K
29. N/m^2

2

Coordinate Systems and Fields

I hear, and I forget.
I see, and I remember.
I do, and I understand.
— Chinese Proverb

CHAPTER OBJECTIVES

To enable the students to understand the following:

- Scalar and vector fields
- Right- and left-handed coordinate systems
- Rectangular coordinate system with emphasis on
  - unit vectors of a rectangular coordinate system, $\mathbf{a}_x, \mathbf{a}_y,$ and $\mathbf{a}_z$
  - distance between two points
  - equation of a straight line
  - equation of a plane
- Cylindrical coordinate system with emphasis on
  - unit vectors of a cylindrical coordinate system, $\mathbf{a}_\rho, \mathbf{a}_\phi,$ and $\mathbf{a}_z$

  - conversion between rectangular coordinates and cylindrical coordinates
  - equations of lines and surfaces in cylindrical coordinates
- Spherical coordinate system, with emphasis on
  - unit vectors of a spherical coordinate system, $\mathbf{a}_r, \mathbf{a}_\theta,$ and $\mathbf{a}_\phi$
  - conversion between rectangular coordinates and spherical coordinates
  - conversion between cylindrical coordinates and spherical coordinates

2.1 | Introduction

The study of coordinate systems is important because electromagnetic vectors (electric or magnetic) are functions of coordinates. The most useful coordinate systems from the viewpoint of electromagnetic theory are the rectangular, the cylindrical, and the spherical coordinate systems.

The need for the use of more than one coordinate system arises from the fact that electromagnetic phenomena are easier calculated or understood in a system that is appropriate to the application. Frequently, it is necessary to transform from one coordinate system to another.

The rectangular coordinate system in three dimensional space, is based upon three mutually perpendicular 'lines' or axes. The labelling of these axes is done in two ways: a right-handed and left-handed orthogonal sets.

2.2 | Scalar and Vector Fields

In the electromagnetic theory, various vectors and scalars—the electric field, magnetic field and the electric potential—are defined over some region of space. In this example, the first two are vectors, while the third is a scalar.

A vector or scalar *field* is a vector or a scalar which is a *function of space, i.e, coordinates and in some cases time also*. To talk about a vector or scalar field therefore one has to first define an origin, a coordinate system and then the function or functions which define the vector for a vector field, and scalar in case of a scalar field. The electric and magnetic fields are the vector fields and the electric potential is a scalar field.

2.2.1 Scalar Fields

Consider the example of atmospheric pressure. We know that

1. Pressure above the surface of the earth is a function of the distance from the surface. The pressure higher up is lower than the pressure closer to the ground.
2. As we move transversely at any level, the pressure does not change.

To write an equation describing the pressure, we must define a coordinate system whose origin is at some point on the ground. Obviously, the pressure must be a function of the height, h. This is shown in Fig. 2.1. Therefore, when we write

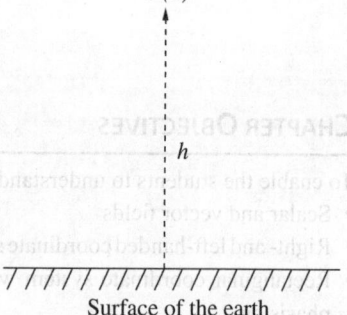

Fig. 2.1 Coordinates for the pressure scalar field

the equation for the pressure, both conditions (1) and (2) must be satisfied. Since only the height is involved, only one coordinate is needed and we can safely say that the pressure is only a function of the height, $P \equiv P(h)$ where h is the height from the origin placed at some point on the surface of the earth (generally at mean sea level). The equation for the pressure is

$$P(h) = P_0 e^{-\alpha h}$$

where h is the height above the surface of the earth (within about a km), α is some constant which has to be experimentally determined, or theoretically calculated ($\alpha \approx 1.19278 \times 10^{-4}$) and P_0 is the pressure on the surface of the earth, that is, $P = P_0$ when the height, $h = 0$.

First let us see whether both conditions (1) and (2) given above are satisfied. Condition (1) is satisfied since the pressure does indeed reduce when we go up, but the exact form of the equation is to be corroborated experimentally. Condition (2) is satisfied since when we move transversely, the pressure remains the same. $P(h)$ is a *scalar field* because pressure is a scalar quantity which is a function of coordinates and in this case function of a single coordinate, h.

Example 2.1 At what height h is the pressure 9/10 of the pressure at the surface of the earth?

Solution

Step 1. Write out the formula:

$$P(h) = P_0 e^{-1.193 \times 10^{-4} h} = 0.9 P_0$$

Step 2. Calculate h:

$$h = \ln(0.9)/(-1.193 \times 10^{-4}) = 883 \text{ m}$$

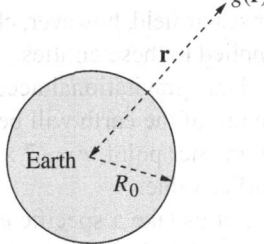

Fig. 2.2 $g(r)$ for earth

In another example, the origin of the coordinate system is located at the centre of the earth (see Fig. 2.2). r is the distance of the observation point from the origin, where the scalar field is being evaluated. A little thought tells us that the coordinate system to be used is essentially the spherical coordinate system. Using this coordinate system, the acceleration due to gravity g for the earth may be defined as:

$$g(r) = GM_E/r^2, \quad R_0 \leq r$$

where $g(r)$ is the acceleration due to gravity of the earth at a distant point from the earth's centre, in m/s^2; r is the distance of a distant point from the earth's centre in meters; R_0 is the radius of the earth, in meters; G is the gravitational constant, in SI units; and M_E is the mass of the earth, in kg. Note that though the acceleration due to gravity is a vector, here it is *modelled* as a scalar.

From these two examples we can see that a scalar field is essentially a scalar which is some physical quantity whose value changes from point to point in space. Or in the language of mathematics, it is a physical quantity whose value is a function of coordinates. In general, we can characterise a scalar field by the equation

$$\Phi \equiv \Phi(\mathbf{r}) \tag{2.1}$$

where Φ is obviously a scalar and $\mathbf{r}$ is the position vector of the point where the scalar exists. Thus in our two examples, in the first case we are interested in the pressure of the atmosphere somewhere above the clouds; and in the second case we are interested in the acceleration due to gravity at points much above the surface of the earth.

For rectangular coordinates, a scalar field may be written as:

$$\Phi \equiv \Phi(x, y, z)$$

and in the case of cylindrical and spherical coordinates the scalar field are described by the equations:

$$\Phi \equiv \Phi(\rho, \phi, z)$$
$$\Phi \equiv \Phi(r, \theta, \phi)$$

where the sets (ρ, ϕ, z) and (r, θ, ϕ) are the cylindrical and spherical coordinates, respectively.

2.2.2 Vector Fields

We define vector fields as below:

A vector which is a function of coordinates is a vector field. Generally, the co-ordinate system will be three-dimensional but sometimes it could be one- or two-dimensional in nature. The difference between just a vector and a vector field is that a vector field maybe differentiated or integrated with respect to the coordinates, but *a simple vector may not be differentiated or integrated (with respect to the coordinates) because it is a constant, and does not change from point to point.* A vector or scalar field, however, changes from point to point and therefore calculus may be applied to these entities.

Thus gravitational acceleration field g at a distance of $r = 7 \times 10^6$ m from the centre of the earth will be given by the formula given above. When g is evaluated at another point, $r = (7 \times 10^6 + 1)$ m, its value will be slightly different from the earlier value.

Let us take a specific example of a vector field. In rectangular coordinates, the vector field $\mathbf{E}$ at a point in space may be specified by the equation:

$$\mathbf{E} = \frac{K(x\mathbf{a}_x + y\mathbf{a}_y + z\mathbf{a}_z)}{\sqrt[3/2]{x^2 + y^2 + z^2}}$$

where K is a constant, and x, y, and z are the coordinates of the point in question. This equation may be split into three equations:

$$E_x = \frac{Kx}{\sqrt[3/2]{(x^2 + y^2 + z^2)}}$$

$$E_y = \frac{Ky}{\sqrt[3/2]{(x^2 + y^2 + z^2)}}$$

$$E_z = \frac{Kz}{\sqrt[3/2]{(x^2 + y^2 + z^2)}}$$

Let us proceed to examine this vector field. At the coordinate point (1,1,1). $\mathbf{E}$ is given by

$$\mathbf{E} = \left. \frac{K(x\mathbf{a}_x + y\mathbf{a}_y + z\mathbf{a}_z)}{\sqrt[3/2]{x^2 + y^2 + z^2}} \right|_{(x,y,z)=(1,1,1)}$$

$$= \frac{K}{\sqrt{27}}(\mathbf{a}_x + \mathbf{a}_y + \mathbf{a}_z) = 0.1925(\mathbf{a}_x + \mathbf{a}_y + \mathbf{a}_z)K$$

And at the point (2,2,2), the $\mathbf{E}$-field is

$$\mathbf{E} = 0.04811(\mathbf{a}_x + \mathbf{a}_y + \mathbf{a}_z)K$$

So it is clear that at different points, the field has different values and as we move away from the origin, the $\mathbf{E}$ field reduces drastically.

Generalising these results, we can write the $\mathbf{E}$ vector field in general terms as

$$\mathbf{E} \equiv \mathbf{E}(\mathbf{r})$$

where $\mathbf{r}$ is the position vector of any general point. As we did for scalars, we can write the representations of $\mathbf{E}(\mathbf{r})$ in different coordinate systems as:

$$\mathbf{E} \equiv \mathbf{E}(x, y, z)$$

for rectangular coordinates system.

$$\mathbf{E} \equiv \mathbf{E}(\rho, \phi, z)$$

for cylindrical coordinates system. And

$$\mathbf{E} \equiv \mathbf{E}(r, \theta, \phi)$$

for spherical coordinates system.

Vector Fields

The mathematical concept of vector fields was originated in the Nineteenth century physics. It was proposed by Michael Faraday in his conceptualisation of magnetism in terms of 'lines of force'. Faraday also underscored the fact that the field concept is very useful in explaining the physical phenomena. Presently, concept of vector fields is used in practically all branches of engineering and physics.

2.3 | Rectangular Coordinate System

As already stated, coordinate systems play crucial role in the application of electromagnetic theory to various engineering situations. A suitable coordinate system is chosen in which the equations reduce to a particularly simple ones. Though the rectangular coordinate system is most often used, the cylindrical and spherical coordinate systems also have proved to be useful in various applications. We first take a closer look at the rectangular coordinate system and then we shall proceed to other coordinate systems.

There are two versions of coordinate systems known as left-handed and right-handed coordinate systems (Fig. 2.3). We can write our equations in either coordinate system as both are equally valid. In the right-handed rectangular coordinate system [Fig. 2.3(a)], if we go from $\mathbf{a}_x$ to $\mathbf{a}_y$ then we get $\mathbf{a}_z$, following the *right-hand thumb rule*, while in the left-handed rectangular system when we go

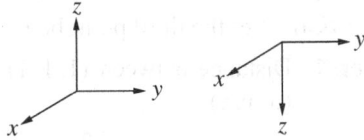

(a) Right-handed   (b) Left-handed
coordinate system   coordinate system

Fig. 2.3 Right- and left-handed coordinate systems

from $\mathbf{a}_x$ to $\mathbf{a}_y$ then we get $\mathbf{a}_z$, but following the *left-hand thumb rule*. In this book, however, we shall be using right-handed coordinate systems only.

2.3.1 Distance Between Two Points

We will be required to do various mathematical manipulations in the rectangular (or Cartesian) system. For example, if two points are given, then how do we find the distance between the points? Similarly if two lines are given, then how do we find the angles between the two lines?

To answer these questions we use vector algebra with great advantage. Referring to Fig. 2.4 with $(x_0, y_0, z_0) = (1, 2, 2)$ and $(x_1, y_1, z_1) = (2, 3, 2)$ the difference vector is given by

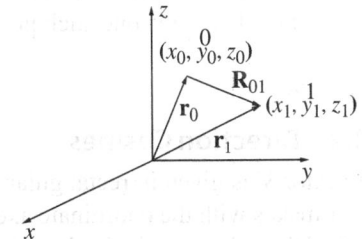

Fig. 2.4 Difference vector for two points in the rectangular coordinates

$$\mathbf{R}\{\text{point 0 to point1}\} = \mathbf{r}\{\text{point 1}\} - \mathbf{r}\{\text{point 0}\}$$

or
$$\mathbf{R}_{01} = \mathbf{r}(x_1, y_1, z_1) - \mathbf{r}(x_0, y_0, z_0)$$

$$= \mathbf{r}_1 - \mathbf{r}_0 \qquad (2.2)$$

For two points $(x_0, y_0, z_0) = (1, 2, 2)$ and $(x_1, y_1, z_1) = (2, 3, 2)$

$$\mathbf{R}_{01} = \mathbf{r}_1 - \mathbf{r}_0 = (2, 3, 2) - (1, 2, 2) = (1, 1, 0)$$

The distance between point 1 and point 0 is

$$\mathbf{R}_{01} = |\mathbf{r}_1 - \mathbf{r}_0| = \sqrt{2}$$

The length of the two position vectors are

$$r_0 = \sqrt{1^2 + 2^2 + 2^2} = 3$$

and

$$r_1 = \sqrt{1^2 + 2^2 + 2^2} = \sqrt{17}$$

If we take the dot product of $\mathbf{R}_{01}$ with $\mathbf{r}_1$, we get

$$\mathbf{R}_{01} \cdot \mathbf{r}_1 = (1, 1, 0) \cdot (2, 3, 2) = 5$$

then we can find the angle between the two vectors,

$$\sqrt{2} \times \sqrt{17} \times \cos\theta = 5$$

or

$$\cos\theta = 5/\sqrt{34} = 0.8549$$

or

$$\theta = 30.9°$$

Similarly, we can find the angle with the other position vector.

Example 2.2   Find a third point equidistant from $(1, 1, 1)$ and $(2, 2, 2)$.

Solution   Let the third point be (x, y, z).

Step 1. Distance between $(1, 1, 1)$ and $(x, y, z) =$ distance between $(2, 2, 2)$ and (x, y, z)

$$(1 - x)^2 + (1 - y)^2 + (1 - z)^2 = (2 - x)^2 + (2 - y)^2 + (2 - z)^2 \quad (2.3)$$

There is one equation and three unknowns, so there are many such points. After cancelling terms on both sides, we get

$$1 - 2x + 1 - 2y + 1 - 2z = 4 - 4x + 4 - 4y + 4 - 4z$$

or

$$2x + 2y + 2z = 9$$

This equation describes the plane on which all points are equidistant from $(1,1,1)$ and $(2,2,2)$. In this equation, putting $x = 1$, $y = 1$ then $z = 5/2$. So $(1, 1, \frac{5}{2})$ is one such point.

 See `MagDirVect.m` in Chapter 2

2.3.2   Direction Cosines

If a vector $\mathbf{V}$ is given in rectangular coordinates and we wish to find the angle the vector makes with the coordinate axes, then we can do so in the following way. We first find the unit vector in the direction of $\mathbf{V}$ ($\hat{\mathbf{V}} = \mathbf{V}/V$) and then take its dot product with the coordinate unit vectors $\mathbf{a}_x$, $\mathbf{a}_y$, and $\mathbf{a}_z$. Then the three direction cosines are these three dot products. Therefore if the unit vector $\hat{\mathbf{V}}$ makes angles of θ, ϕ, and ψ with the x, y, and z axes respectively then the direction cosines (denoted by α, β, γ) of $\mathbf{V}$ are given by

$$\alpha = \hat{\mathbf{V}} \cdot \mathbf{a}_x = \cos\theta, \ \beta = \hat{\mathbf{V}} \cdot \mathbf{a}_y = \cos\phi, \ \gamma = \hat{\mathbf{V}} \cdot \mathbf{a}_z = \cos\psi \quad (2.4)$$

Example 2.3   Show that $\cos^2\theta + \cos^2\phi + \cos^2\psi = 1$.

Solution   We know that $\hat{\mathbf{V}}$ is a unit vector so

$$\hat{\mathbf{V}}_x = \hat{\mathbf{V}} \cdot \mathbf{a}_x = \cos\theta$$

$$\hat{V}_y = \hat{V} \cdot \mathbf{a}_y = \cos\phi$$

$$\hat{V}_z = \hat{V} \cdot \mathbf{a}_z = \cos\psi$$

and $|\hat{V}| = 1$. So

$$\hat{V}_x^2 + \hat{V}_y^2 + \hat{V}_z^2 = 1$$

so

$$\cos^2\phi + \cos^2\theta + \cos^2\psi = 1$$

Example 2.4　Find the direction cosines and angles that the position vector $\mathbf{r} = (2, 3, 4)$ makes with the coordinate axes.

Solution

Step 1. The unit vector $\hat{\mathbf{r}}$ is

$$\hat{\mathbf{r}} = \frac{(2\mathbf{a}_x + 3\mathbf{a}_y + 4\mathbf{a}_z)}{\sqrt{2^2 + 3^2 + 4^2}} = \frac{(2\mathbf{a}_x + 3\mathbf{a}_y + 4\mathbf{a}_z)}{5.3852}$$

$$= 0.37139\mathbf{a}_x + 0.55709\mathbf{a}_y + 0.74278\mathbf{a}_z$$

Step 2. The direction cosines are the cosines of the angles which the unit vector makes with the three axes. They are therefore

$$\alpha = 0.37139, \ \beta = 0.55709, \text{ and } \gamma = 0.74278$$

Step 3. The corresponding angles are $\cos^{-1}\alpha$, $\cos^{-1}\beta$, and $\cos^{-1}\gamma$: which are $68.2°$, $56.2°$, and $42°$.

 See `DirCosines.m` in Chapter 2

Practice Problem 2.1

Find the direction cosines and angles of the point $(-5, 0, 5)$.

[Ans: Direction cosines: $-1/\sqrt{2}, \ 0, \ 1/\sqrt{2}$; Angles: $3\pi/4, \ \pi/2, \ \pi/4$]

2.3.3　Vector Equation of a Straight Line

Very often one requires the vector equation of a straight line passing through a vector, or through a unit vector which is attached to some point in the 3-D space. As we know, any straight line can have only one degree of freedom.[1] Generally, the simplest equation of a straight line is the parametric equation and the simplest example of a parametric equation is that of one of the axes. The parametric equation of the x-axis is

$$\mathbf{r} = \mathbf{a}_x t$$

where t is the parameter, and $-\infty \le t \le \infty$. Examining the previous equation in some more detail we find that $\mathbf{r} = -\mathbf{a}_x$ for $t = -1$; $\mathbf{r} = 0$ for $t = 0$; and $\mathbf{r} = \mathbf{a}_x$ for $t = 1$. This defines the x-axis.

　　To obtain the parametric equation of a straight line in a more systematic manner, we examine Fig. 2.5. The figure shows that if a vector $\mathbf{r}_0$ is given, attached to a

---

[1]The number of degrees of freedom corresponds to the number of independent variables in any equation or mathematical description. For example in two dimensions the general equation of a straight line: $y = mx + c$, where m and c are constants has only one degree of freedom. If x is specified, y is uniquely determined; and, on the other hand if y is specified then x is uniquely determined.

point (x_0, y_0, z_0) in 3-D space, then

$$tR + r_0 \quad (-\infty < t < \infty)$$

is another point which lies on the straight line shown. t is a parameter whose positive values lead to points towards the head of the vector $\mathbf{R}$ while negative values of t lead to points on the straight line towards the tail of the vector. The position vector of a point on the straight line is

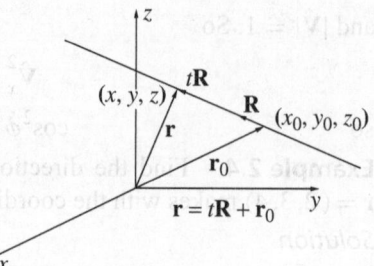

Fig. 2.5 A straight line in rectangular coordinates

$$\mathbf{r}(x, y, z) = t\mathbf{R} + \mathbf{r_0} \quad (-\infty < t < \infty) \quad (2.5)$$

Note: *This vector equation consists of three equations: the first having x on the left-hand side, and two more having y and z on the left-hand side*

$$\left. \begin{array}{l} x = tR_x + x_0 \\ y = tR_y + y_0 \\ z = tR_z + z_0 \end{array} \right\} \quad (-\infty < t < \infty)$$

Example 2.5 Find the parametric equation of the line which passes through $(0, 0, 0)$ and $(1, 1, 1)$.

Solution

Step 1. In this example $(x_0, y_0, z_0) = (0, 0, 0)$ and $(x_1, y_1, z_1) = (1, 1, 1)$
Step 2. Then $\mathbf{R} = \mathbf{r_1} - \mathbf{r_0} = (1, 1, 1)$
Step 3. Using the theory which has been developed above

$$\mathbf{r} = \mathbf{r_0} + t\mathbf{R}$$
$$x = x_0 + t(x_1 - x_0) = t$$
$$y = y_0 + t(y_1 - y_0) = t$$
$$z = z_0 + t(z_1 - z_0) = t$$

Example 2.6 The two points $(1,1,1)$ cm and $(1,2,3)$ cm are connected by a straight line and extended on both sides. (a) Find a point on the straight line which is at a distance of 5 cm from $(1,1,1)$ towards $(1,2,3)$ and also away from $(1,2,3)$ (in the other direction). (b) Find the co-ordinates of the points of intersection of this line with the x-, y- and z-planes.

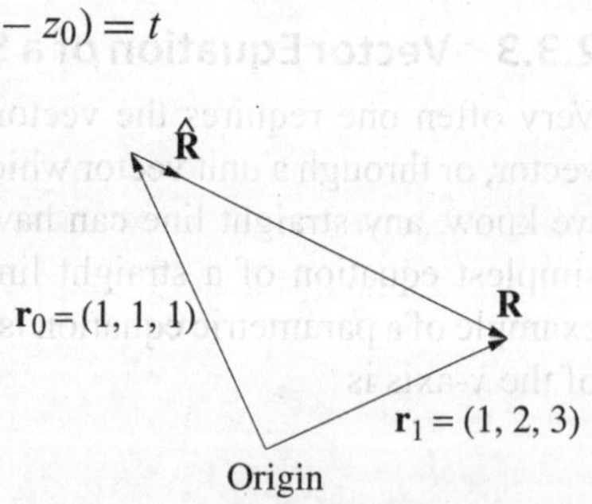

Fig. 2.6 Vector equation of a straight line

Solution

Part (a)

Step 1. First of all calculate $\mathbf{R}$ (see Fig. 2.6). It is given by

$$\mathbf{R} = \mathbf{r_1} - \mathbf{r_0}$$
$$= (1, 2, 3) - (1, 1, 1) = (0, 1, 2) \text{ cm}$$

Step 2. Next we write the vector equation for the straight line joining the two points:

$$\mathbf{r}(x, y, z) = \tau \times (0, 1, 2) + (1, 1, 1) \quad (-\infty < \tau < \infty) \quad (2.6)$$

Step 3. Notice that for $\tau = 0$ we are at the point $(1,1,1)$ and for $\tau = 1$ we land up at $(1,2,3)$. This equation is not helpful for finding the distance between the points.

Step 4. To get the suitable equation we need to calculate the unit vector $\hat{\mathbf{R}}$ in the direction of $\mathbf{R}$. The unit vector is calculated by

$$\hat{\mathbf{R}} = \mathbf{R}/R = (0, 1, 2)/\sqrt{5} = (0, 1/\sqrt{5}, 2/\sqrt{5})$$

Step 5. We can now write a *new* equation involving the parameters t and $\mathbf{a}_r$:

$$\mathbf{r}(x, y, z) = t \times (0, 1/\sqrt{5}, 2\sqrt{5}) + (1, 1, 1) \qquad (-\infty < t < \infty)$$

We can clearly see that the parameter t is the *distance parameter*. When t is 0, we are at $(1,1,1)$, but when *t=1 then we are at one unit vector away (along the line) from (1,1,1) or 1 cm away.*

Step 6. Now we can get the correct answer for this part: Put $t = 5\,\text{cm}$ in the previous equation.

$$\begin{aligned}
\mathbf{r}(x_a, y_a, z_a) &= 5 \times (0, 1/\sqrt{5}, 2\sqrt{5}) + (1, 1, 1) \\
&= [0, (5/\sqrt{5}), (2 \times 5/\sqrt{5})] + (1, 1, 1) \\
&= (0, \sqrt{5}, 2\sqrt{5}) + (1, 1, 1) \\
&= [1, (1 + \sqrt{5}), (1 + 2\sqrt{5})]\,\text{cm}
\end{aligned}$$

When we put $t = -5\,\text{cm}$ we go in the *other direction* and get the coordinates of the second point:

$$\begin{aligned}
\mathbf{r}(x_b, y_b, z_b) &= -5 \times (0, 1/\sqrt{5}, 2\sqrt{5}) + (1, 1, 1) \\
&= [0, -5/\sqrt{5}, (-2 \times 5/\sqrt{5})] + (1, 1, 1) \\
&= (0, -\sqrt{5}, -2\sqrt{5}) + (1, 1, 1) \\
&= [1, (1 - \sqrt{5}), (1 - 2\sqrt{5})]\,\text{cm}
\end{aligned}$$

Part (b)

Step 1. In this part we have to find where the line meets the three planes. Let us consider the *x-y* plane. The *x-y* plane is described by the equation $z = 0$ (*Note* that when $z = 0$, x and y can be anything). Here we have to put $z = 0$ on the left-hand side of the equation:

$$\mathbf{r}(x, y, z) = t \times (0, 1/\sqrt{5}, 2\sqrt{5}) + (1, 1, 1) \qquad (-\infty < t < \infty)$$

Then z equation becomes $0 = t \times 2/\sqrt{5} + 1$ or $t = -\sqrt{5}/2$

Step 2. Using this value of t and utilising the other two equations:

$$x_{x\text{-}y\ \text{plane}} = 1\,\text{cm} \quad \text{and} \quad y_{x\text{-}y\ \text{plane}} = 1/2\,\text{cm}$$

Hence point of intersection of the line with *x-y* plane is $(1, \frac{1}{2}, 0)$.

Step 3. Let us proceed and calculate where this line meets the *x-z* plane described by $y = 0$ and substituted into the second equation:

$$0 = t \times 1/\sqrt{5} + 1 \quad \text{or} \quad t = -\sqrt{5}$$

With which we calculate:

$$x_{x\text{-}y\ \text{plane}} = 1\,\text{cm}, \ z_{x\text{-}z\ \text{plane}} = -1\,\text{cm}$$

Here point of intersection with *x-z* plane is $(1, 0, -1)$.

Step 4. In the same manner we proceed to calculate the meeting point of the line with the third plane—the y-z plane described by $x = 0$. The appropriate equation is $0 = t \times 0 + 1$ or $0 = 1$.

Here after careful thought we conclude that the *line does not meet the y-z plane!*

Practice Problem 2.2

Find the parametric equation of the line which passes through the points $(-1, -1, -1)$ and $(1,1,1)$.

[Ans: $\mathbf{r} = (-1, -1, -1) + t(2, 2, 2)$]

2.3.4 Equation of a Plane

Very often during the application of the electromagnetic equations, the equation of a plane is desired. How can we obtain the equation of one of the coordinate planes? Take the y-z plane. All over the y-z plane the value of x is zero. Hence, the equation of the y-z plane is $x = 0$.

Referring to Fig. 2.7, suppose $\mathbf{R}_1$ and $\mathbf{R}_2$ are two vectors given in 3-D space, then we would like to calculate the equation of a plane which encloses both $\mathbf{R}_1$ and $\mathbf{R}_2$ and also passes through the point $\mathbf{r}_0(x_0, y_0, z_0)$. Let the co-ordinates of any point lying on the plane be $\mathbf{r}(x, y, z)$. To proceed further we first calculate two vectors: (shown in Fig. 2.7)

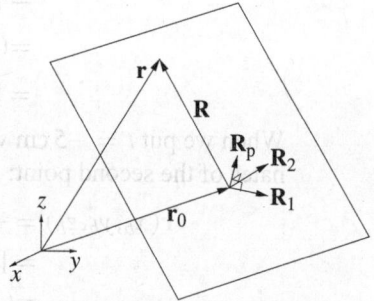

$$\mathbf{R}_p = \mathbf{R}_1 \times \mathbf{R}_2 \text{ and } \mathbf{R} = \mathbf{r} - \mathbf{r}_0$$

We can clearly see that $\mathbf{R}_p$ is perpendicular to the plane, and $\mathbf{R}$ *must lie on the plane*. To make $\mathbf{R}$ lie on the plane, we put the restriction that it should be perpendicular to $\mathbf{R}_p$:

Fig. 2.7 Equation of a plane

$$\mathbf{R} \cdot \mathbf{R}_p = 0 \quad \text{or} \quad (\mathbf{r} - \mathbf{r}_0) \cdot \mathbf{R}_p = 0$$

i.e.
$$(\mathbf{r} - \mathbf{r}_0) \cdot (\mathbf{R}_1 \times \mathbf{R}_2) = 0 \tag{2.7}$$

This equation is a single equation in three variables: x, y, and z and therefore has two degrees of freedom. The equation applies particularly in two cases: (i) When two vectors $\mathbf{R}_1$, $\mathbf{R}_2$, and $\mathbf{r}_0$ are given (a plane parallel to two given vectors $\mathbf{R}_1$ and $\mathbf{R}_2$ and passing through a point given by $\mathbf{r}_0$) and (ii) When $\mathbf{R}_p$ and $\mathbf{r}_0$ are given (a plane passing through a point and perpendicular to a given vector.)

Example 2.7 Find the equation of a plane passing through $(1, 0, 0)$, $(0, 1, 0)$, and $(0, 0, 1)$ in the rectangular coordinate system. Also find the unit vector perpendicular to this plane (Fig. 2.8).

Solution

Step 1. Two vectors which lie on this plane are

$$\mathbf{R}_1 = (0, 1, 0) - (1, 0, 0) = (-1, 1, 0)$$

and
$$\mathbf{R}_2 = (0, 0, 1) - (1, 0, 0) = (-1, 0, 1)$$

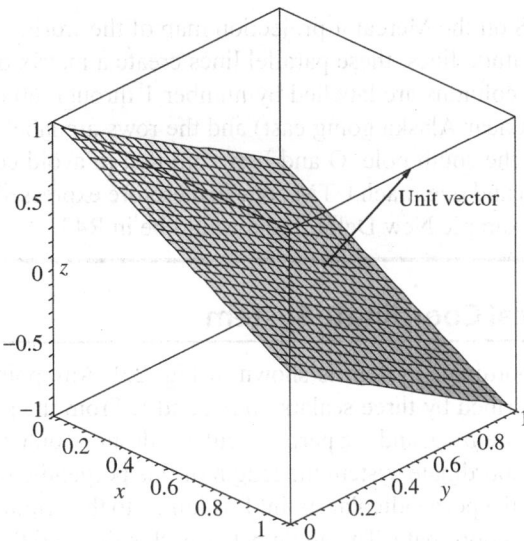

Fig. 2.8 3-D plot of the plane of the example

Step 2. If we take the cross product $\mathbf{R}_1 \times \mathbf{R}_2$, we get a vector perpendicular to the plane:

$$\mathbf{R}_p = \mathbf{R}_1 \times \mathbf{R}_2 = (-1,1,0) \times (-1,0,1)$$
$$= -\mathbf{a}_x \times (-1,0,1) + \mathbf{a}_y \times (-1,0,1)$$

On further simplification, we get $\mathbf{R}_p = \mathbf{a}_y + \mathbf{a}_z + \mathbf{a}_x \equiv (1,1,1)$

Step 3. By inspection, the unit vector is $\hat{\mathbf{R}}_p = (1/\sqrt{3})(\mathbf{a}_x + \mathbf{a}_y + \mathbf{a}_z)$

Step 4. Using Eq. (2.7), we get

$$[(x,y,z) - (1,0,0)] \cdot (1,1,1) = (x-1) + y + z = 0$$

Hence the desired equation is $x + y + z = 1$

Example 2.8 Find the plane perpendicular to the unit vector $\hat{\mathbf{n}} = (\mathbf{a}_x + \mathbf{a}_y + \mathbf{a}_z)/\sqrt{3}$ and passing through the point (1,1,1).

Solution From the theory which we have just discussed, the required plane is

$$[\mathbf{r} - (1,1,1)] \cdot \hat{\mathbf{n}} = 0 \quad \text{or} \quad [(x+y+z) - 3]/\sqrt{3} = 0$$

which gives the equation of the desired plane as $x + y + z = 3$

Practice Problem 2.3
Find the plane passing through $(0,0,0)$, $(1,0,0)$, and $(0,1,0)$.

[Ans: $z = 0$]

 See `UnitVectNormPlne.m` in Chapter 2

Global Positioning System

A practical application of the rectangular coordinate system as applied to the Earth Global Positioning System (GPS) is the Universal Transverse Mercator (UTM) locator system. With UTM, the earth is divided by parallel lines which are E-W and N-S on the Mercator projection map of the world. Similar to the latitude and longitude lines, these parallel lines create a matrix of 60 columns by 20 rows. The columns are labelled by number 1 through 60 (starting from the tip of Siberia near Alaska going east) and the rows are labelled C through X, starting from the south pole. O and I, are omitted to avoid confusion with the numbers zero and one. Each UTM zone is therefore expressed by a number and a letter, for example New Delhi, India would be in R43.

2.4 | Cylindrical Coordinate System

The cylindrical coordinate system is shown in Fig. 2.9. Any point in cylindrical coordinates is described by three scalars: ρ, ϕ, and z. From the point in question whose coordinates are ρ, ϕ, and z, a perpendicular is dropped onto the x-y plane. As in the rectangular coordinate system the length of this perpendicular is the z coordinate. The foot of the perpendicular (point f) is joined to the origin (O). The length of this line is the ρ coordinate. The angle between this line and the x-axis is the ϕ coordinate. The z coordinate is the same as in the case of the rectangular coordinate system. By observing the figure which shows the cylindrical system superimposed on the rectangular coordinate system, we get the following set of equations showing relationship between the coordinates of the two systems,

$$x = \rho \cos\phi, \ y = \rho \sin\phi, \ z = z \tag{2.8}$$

To obtain the inverse transformation, we do the following manipulations: ρ is obtained by squaring and adding the first two equations, while ϕ is obtained by dividing the second equation by the first. The inverse transformation equations are

$$\rho = \sqrt{x^2 + y^2}, \ \phi = \tan^{-1}(y/x), \ z = z \tag{2.9}$$

Example 2.9 Find the cylindrical coordinates of the point $(x = 1, y = 1, z = 1)$ given in rectangular coordinates.

Solution

Step 1. The ρ coordinate is given by

$$\rho = \sqrt{x^2 + y^2} = \sqrt{2} = 1.4142$$

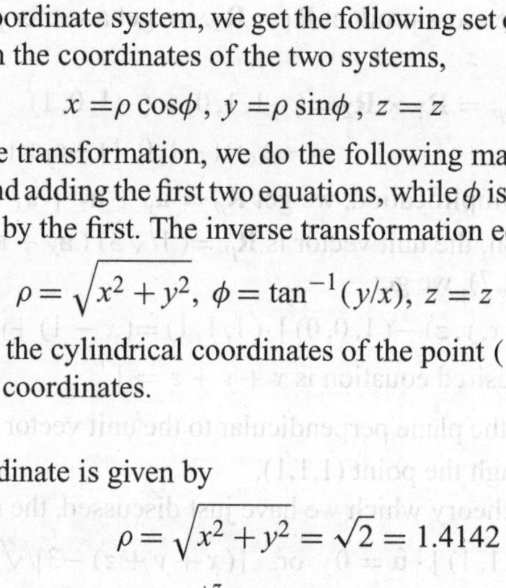

Fig. 2.9 Cylindrical coordinate system

$$\phi = \tan^{-1}\left(\frac{y}{x}\right) = \frac{\pi}{4} = 0.78540$$

and
$$z = 1$$

Hence the coordinates in cylindrical coordinates are

$$(\rho, \phi, z) = (1.4142, 0.78540, 1)$$

Example 2.10 Find the rectangular coordinates of the point $(\rho = 1,\ \phi = 1,\ z = 1)$.

Solution

Step 1. The cylindrical coordinates of the point in question are

$$(\rho, \phi, z) = (1, 1, 1)$$

Step 2. The three rectangular coordinates are

$$x = \rho \cos\phi = 1 \times 0.54030$$
$$y = \rho \sin\phi = 1 \times 0.84147$$
$$z = 1$$

Practice Problem 2.4

Find the cylindrical coordinates of the points $(x = 1, y = 1, z = 1)$, $(-1, 1, 1)$, $(1, -1, 1)$, and $(-1, -1, 1)$

[Ans: In all cases, $z = 1$ and $\rho = \sqrt{2}, \phi = \pi/4$; $\rho = \sqrt{2}, \phi = 3\pi/4$, $\rho = \sqrt{2}, \phi = 5\pi/4$, $\rho = \sqrt{2},\ \phi = -\pi/4$]

Examining Fig. 2.9, we see that along the straight line d-e only ρ changes. On this line the z and ϕ are constants. The unit vector $\mathbf{a}_\rho$ is parallel to this line. The circle 'a' 'b' 'c' is a circular curve on which only the ϕ coordinate changes while z and ρ remain constant. The unit vector $\mathbf{a}_\phi$ is tangent to the circle at the point. Similarly, on the straight line f–g only the z coordinate changes and the unit vector $\mathbf{a}_z$ lies parallel to it.

Note: The unit vectors in the cylindrical coordinate system are dependent on the position of the point in space and are orthogonal to each other.

Referring again to Fig. 2.9 we find that $\mathbf{a}_\rho$ is parallel to the x-y plane and also parallel to the line joining the origin to the perpendicular from (ρ, ϕ, z) to the x-y plane. It makes an angle ϕ with the x-axis and an angle $(\pi/2 - \phi)$ with the y-axis.

Since $\mathbf{a}_\rho$ is parallel to the $x - y$ plane, it is perpendicular to $\mathbf{a}_z$. Further, unit vector $\mathbf{a}_\phi$ is perpendicular to $\mathbf{a}_\rho$ in the *increasing direction* of ϕ and again parallel to the x-y plane. By observing the figure, we can see that $\mathbf{a}_\phi = \mathbf{a}_z \times \mathbf{a}_\rho$. Therefore, $\mathbf{a}_\rho$ and $\mathbf{a}_\phi$ are perpendicular to $\mathbf{a}_z$. Thus we see that the $(\mathbf{a}_\rho, \mathbf{a}_\phi, \mathbf{a}_z)$ form a right-handed, *orthonormal* set of vectors.

$$\mathbf{a}_\rho \times \mathbf{a}_\phi = \mathbf{a}_z,\ \mathbf{a}_\phi \times \mathbf{a}_z = \mathbf{a}_\rho,\ \mathbf{a}_z \times \mathbf{a}_\rho = \mathbf{a}_\phi \qquad (2.10)$$

For recapitulation, the relations given above are put down in the form of Table 2.1 and Fig. 2.10. Looking at the figure we can see that $\mathbf{a}_\rho \times \mathbf{a}_\phi$ moving along the arrow gives us $\mathbf{a}_z$— and so on.

$\mathbf{a}_\rho$, $\mathbf{a}_\phi$, $\mathbf{a}_x$, and $\mathbf{a}_y$ lie on the same plane, a plane parallel to the x-y plane.

Table 2.1 Cross products of the cylindrical coordinate unit vectors

×	$\mathbf{a}_\rho$	$\mathbf{a}_\phi$	$\mathbf{a}_z$
$\mathbf{a}_\rho$	0	$\mathbf{a}_z$	$-\mathbf{a}_\phi$
$\mathbf{a}_\phi$	$-\mathbf{a}_z$	0	$\mathbf{a}_\rho$
$\mathbf{a}_z$	$\mathbf{a}_\phi$	$-\mathbf{a}_\rho$	0

The three unit vectors for the cylindrical coordinates are related to the unit vectors for rectangular coordinates. Let us start with $\mathbf{a}_\rho$ (refer to Fig. 2.11).

$\mathbf{a}_\rho$ makes an angle of ϕ with $\mathbf{a}_x$ and ($\pi/2-\phi$) with $\mathbf{a}_y$, therefore

$$\mathbf{a}_\rho = \cos\phi\, \mathbf{a}_x + \sin\phi\, \mathbf{a}_y$$

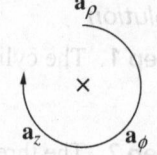

Fig. 2.10 $\mathbf{a}_\rho$, $\mathbf{a}_\phi$, and $\mathbf{a}_z$

Also $\mathbf{a}_\phi = \mathbf{a}_z \times \mathbf{a}_\rho$, therefore

$$\mathbf{a}_\phi = \cos\phi\, \mathbf{a}_y - \sin\phi\, \mathbf{a}_x$$

These relations are given below

$$\mathbf{a}_\rho = \cos\phi\, \mathbf{a}_x + \sin\phi\, \mathbf{a}_y \tag{2.11}$$
$$\mathbf{a}_\phi = \cos\phi\, \mathbf{a}_y - \sin\phi\, \mathbf{a}_x \tag{2.12}$$
$$\mathbf{a}_z = \mathbf{a}_z \tag{2.13}$$

And they may be written in the form of a matrix, $[\mathbf{T}]_{rc}$ as

$$\begin{pmatrix} \mathbf{a}_\rho \\ \mathbf{a}_\phi \\ \mathbf{a}_z \end{pmatrix} = \begin{pmatrix} \cos\phi & \sin\phi & 0 \\ -\sin\phi & \cos\phi & 0 \\ 0 & 0 & 1 \end{pmatrix} \begin{pmatrix} \mathbf{a}_x \\ \mathbf{a}_y \\ \mathbf{a}_z \end{pmatrix} \tag{2.14}$$

$$[\mathbf{T}]_{rc}$$

The transformation equations in the reverse direction (with the transformation matrix $\mathbf{T}_{rc}^{-1}$ between rectangular–cylindrical coordinates) in matrix form are

$$\begin{pmatrix} \mathbf{a}_x \\ \mathbf{a}_y \\ \mathbf{a}_z \end{pmatrix} = \begin{pmatrix} \cos\phi & -\sin\phi & 0 \\ \sin\phi & \cos\phi & 0 \\ 0 & 0 & 1 \end{pmatrix} \begin{pmatrix} \mathbf{a}_\rho \\ \mathbf{a}_\phi \\ \mathbf{a}_z \end{pmatrix} \tag{2.15}$$

$$[\mathbf{T}]_{rc}^{-1}$$

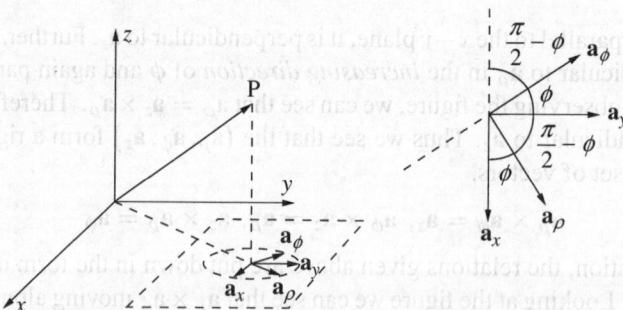

Fig. 2.11 ($\mathbf{a}_x$, $\mathbf{a}_y$) and ($\mathbf{a}_\rho$, $\mathbf{a}_\phi$) of rectangular and cylindrical coordinates, respectively

With the inverse related to $[\mathbf{T}]_{rc}$ by $[\mathbf{T}]_{rc}^{-1} = [\mathbf{T}]_{rc}^{t}$

The superscript t implies transpose. Figure 2.11 shows these relations in 3-D space.

The position vector $\mathbf{r}$ is given by

$$\mathbf{r} = x\mathbf{a}_x + y\mathbf{a}_y + z\mathbf{a}_z = \rho \cos\phi \, \mathbf{a}_x + \rho \sin\phi \, \mathbf{a}_y + z\mathbf{a}_z \tag{2.16}$$

or $$\mathbf{r} = \rho \, \mathbf{a}_\rho + z\mathbf{a}_z \tag{2.17}$$

If we take the dot product of $\mathbf{a}_\rho$ with $\mathbf{a}_x$ and $\mathbf{a}_y$ using Eqn (2.11), we get

$$\mathbf{a}_\rho \cdot \mathbf{a}_x = \cos\phi$$
$$\mathbf{a}_\rho \cdot \mathbf{a}_y = \sin\phi \tag{2.18}$$

Similarly dot product of $\mathbf{a}_\phi$ with $\mathbf{a}_x$ and $\mathbf{a}_y$ using Eqn (2.12) gives us

$$\mathbf{a}_\phi \cdot \mathbf{a}_x = -\sin\phi$$
$$\mathbf{a}_\phi \cdot \mathbf{a}_y = \cos\phi \tag{2.19}$$

These two equations are extremely useful because if some vector $\mathbf{A}$ is given both in rectangular as well as cylindrical coordinates then

$$\mathbf{A} \equiv A_\rho \mathbf{a}_\rho + A_\phi \mathbf{a}_\phi + A_z \mathbf{a}_z = A_x \mathbf{a}_x + A_y \mathbf{a}_y + A_z \mathbf{a}_z$$

Then $$A_\rho = \mathbf{a}_\rho \cdot (A_\rho \mathbf{a}_\rho + A_\phi \mathbf{a}_\phi + A_z \mathbf{a}_z) = \mathbf{a}_\rho \cdot (A_x \mathbf{a}_x + A_y \mathbf{a}_y + A_z \mathbf{a}_z)$$

$$= A_x(\mathbf{a}_\rho \cdot \mathbf{a}_x) + A_y(\mathbf{a}_\rho \cdot \mathbf{a}_y) + A_z(\mathbf{a}_\rho \cdot \mathbf{a}_z)$$

$$= A_x \cos\phi + A_y \sin\phi \tag{2.20}$$

Using this technique, we can now write the general matrix equation for conversion as:

$$\begin{pmatrix} A_\rho \\ A_\phi \\ A_z \end{pmatrix} = \begin{pmatrix} \cos\phi & \sin\phi & 0 \\ -\sin\phi & \cos\phi & 0 \\ 0 & 0 & 1 \end{pmatrix} \begin{pmatrix} A_x \\ A_y \\ A_z \end{pmatrix} \tag{2.21}$$

and $$\begin{pmatrix} A_x \\ A_y \\ A_z \end{pmatrix} = \begin{pmatrix} \mathbf{a}_x \cdot \mathbf{a}_\rho & \mathbf{a}_x \cdot \mathbf{a}_\phi & \mathbf{a}_x \cdot \mathbf{a}_z \\ \mathbf{a}_y \cdot \mathbf{a}_\rho & \mathbf{a}_y \cdot \mathbf{a}_\phi & \mathbf{a}_y \cdot \mathbf{a}_z \\ \mathbf{a}_z \cdot \mathbf{a}_\rho & \mathbf{a}_z \cdot \mathbf{a}_\phi & \mathbf{a}_z \cdot \mathbf{a}_z \end{pmatrix} \begin{pmatrix} A_\rho \\ A_\phi \\ A_z \end{pmatrix} \tag{2.22}$$

or $$\begin{pmatrix} A_x \\ A_y \\ A_z \end{pmatrix} = \begin{pmatrix} \cos\phi & -\sin\phi & 0 \\ \sin\phi & \cos\phi & 0 \\ 0 & 0 & 1 \end{pmatrix} \begin{pmatrix} A_\rho \\ A_\phi \\ A_z \end{pmatrix} \tag{2.23}$$

The individual terms of the matrix have been evaluated from Eqns (2.18) and (2.19) where the dot products are given. These dot products can be put down in tabular form (Table 2.2).

Table 2.2 Dot products between $\mathbf{a}_x$, $\mathbf{a}_y$, $\mathbf{a}_z$ and $\mathbf{a}_\rho$, $\mathbf{a}_\phi$, $\mathbf{a}_z$

·	$\mathbf{a}_x$	$\mathbf{a}_y$	$\mathbf{a}_z$
$\mathbf{a}_\rho$	$\cos(\phi)$	$\sin(\phi)$	0
$\mathbf{a}_\phi$	$-\sin(\phi)$	$\cos(\phi)$	0
$\mathbf{a}_z$	0	0	1

Example 2.11   Compute $\mathbf{a}_x$, $\mathbf{a}_y$, and $\mathbf{a}_z$ in terms of $\mathbf{a}_\rho$, $\mathbf{a}_\phi$, and $\mathbf{a}_z$.

Solution

Step 1. $\mathbf{a}_x = (1,0,0)$ in rectangular coordinates. In Eqn (2.21), we insert $(1,0,0)$ in place of (A_x, A_y, A_z) and get

$$\begin{pmatrix} A_\rho \\ A_\phi \\ A_z \end{pmatrix} = \begin{pmatrix} \cos\phi & \sin\phi & 0 \\ -\sin\phi & \cos\phi & 0 \\ 0 & 0 & 1 \end{pmatrix} \begin{pmatrix} 1 \\ 0 \\ 0 \end{pmatrix} = \begin{pmatrix} \cos\phi \\ -\sin\phi \\ 0 \end{pmatrix}$$

Hence $\mathbf{a}_x = \underbrace{\cos\phi}_{A_\rho}\mathbf{a}_\rho - \underbrace{\sin\phi}_{A_\phi}\mathbf{a}_\phi + \underbrace{0}_{A_z}\mathbf{a}_z$

Step 2. We notice that the unit vector $\mathbf{a}_x$ is a function of cylindrical coordinates. Thus it depends on the coordinates of the point where the unit vector is evaluated. For example, if we take the point $(1,0,z)$ in cylindrical coordinates then $\mathbf{a}_x = \mathbf{a}_\rho$. On the other hand at the point $(1, \pi/2, z)$, $\mathbf{a}_x = -\mathbf{a}_\phi$.

Step 3. In a similar manner, $\mathbf{a}_y = \sin\phi\,\mathbf{a}_\rho + \cos\phi\,\mathbf{a}_\phi$ and $\mathbf{a}_z$ is the same in both coordinate systems.

Example 2.12   The vector field $\mathbf{A}(x,y,z)$ at the point $(1,1,1)$ in rectangular coordinates is $(1,1,1)$. Find the value of the vector field at the same point in cylindrical coordinates.

Solution

Step 1. The point $(1,1,1)$ (rectangular coordinates) translates to $(\rho,\phi,z) = (1.4142, 0.78540, 1)$ (cylindrical coordinates, see Example 2.9).

Step 2. The point $(1,1,1)$ in rectangular coordinates, using the matrix to convert to cylindrical coordinates is

$$\begin{pmatrix} A_\rho \\ A_\phi \\ A_z \end{pmatrix} = \begin{pmatrix} \cos\phi & \sin\phi & 0 \\ -\sin\phi & \cos\phi & 0 \\ 0 & 0 & 1 \end{pmatrix} \begin{pmatrix} 1 \\ 1 \\ 1 \end{pmatrix}$$

Step 3. Since $\phi = 0.7854^r$, $\cos(0.7854) = \sin(0.7854) = 0.7071$.

So $\qquad\qquad A_\rho = 1.4142,\; A_\phi = 0,\; A_z = 1$

are the components of the vector field at the point.

Example 2.13   The vector field $\mathbf{A}(\rho, \phi, z)$ (cylindrical coordinates) in the region $0 \le \rho \le 1$ is given by $\mathbf{A} = \rho\,\mathbf{a}_\phi$. Find the value of the vector field in the same region in rectangular coordinates.

Solution

Step 1. The components of the vector field $\mathbf{A}$ are $A_\rho = 0, A_\phi = \rho, A_z = 0$. We need to find A_x, A_y, A_z. So using the conversion matrix

$$\begin{pmatrix} A_x \\ A_y \\ A_z \end{pmatrix} = \begin{pmatrix} \cos\phi & -\sin\phi & 0 \\ \sin\phi & \cos\phi & 0 \\ 0 & 0 & 1 \end{pmatrix} \begin{pmatrix} A_\rho \\ A_\phi \\ A_z \end{pmatrix}$$

$$= \begin{pmatrix} \cos\phi & -\sin\phi & 0 \\ \sin\phi & \cos\phi & 0 \\ 0 & 0 & 1 \end{pmatrix} \begin{pmatrix} 0 \\ \rho \\ 0 \end{pmatrix} = \begin{pmatrix} -\rho\sin\phi \\ \rho\cos\phi \\ 0 \end{pmatrix}$$

Step 2. We use $\rho = \sqrt{x^2 + y^2}$; $\cos\phi = x/\rho$, and $\sin\phi = y/\rho$ then we have

$$\mathbf{A} = \begin{pmatrix} A_x \\ A_y \\ A_z \end{pmatrix} = \begin{pmatrix} -\sqrt{x^2+y^2} \, \dfrac{y}{\sqrt{x^2+y^2}} \\ \sqrt{x^2+y^2} \, \dfrac{x}{\sqrt{x^2+y^2}} \\ 0 \end{pmatrix} = \begin{pmatrix} -y \\ x \\ 0 \end{pmatrix}$$

or $\qquad \mathbf{A} = -y\mathbf{a}_x + x\mathbf{a}_y.$

Example 2.14 Transform the following vector field to Cartesian coordinates.

$$\mathbf{A}(\rho,\phi,z) = \mathbf{a}_\rho \rho z \sin\phi \qquad \text{(DTU-NSIT, May 2009)}$$

Solution As a first step, $\quad \mathbf{a}_\rho = \mathbf{a}_x \cos\phi + \mathbf{a}_y \sin\phi$

so $\qquad \mathbf{A}(\rho,\phi,z) = \left(\mathbf{a}_x \cos\phi + \mathbf{a}_y \sin\phi \right) \rho z \sin\phi$

$$= \mathbf{a}_x \cos\phi \sin\phi \, \rho z + \mathbf{a}_y \sin^2\phi \, \rho z$$

Now putting $x = \rho \cos\phi$ and $y = \rho \sin\phi$, we get

$$\mathbf{A}(x,y,z) = \mathbf{a}_x x \sin\phi \, z + \mathbf{a}_y \sin\phi \, yz = \mathbf{a}_x \frac{xyz}{\sqrt{x^2+y^2}} + \mathbf{a}_y \frac{y^2 z}{\sqrt{x^2+y^2}}$$

 See `DistCylCoord.m` in Chapter 2

2.4.1 Equations of Surfaces and Lines in Cylindrical Coordinates

The simplest surface in cylindrical coordinates corresponds to $\rho = constant$. A little reflection will tell us that this is the equation of the curved surface of cylinder. An example of such a surface is shown in Fig. 2.12. The figure shows three surfaces defined by the equations $\rho = constant$, $z = constant$, and $\phi = constant$. In this figure, $z = constant$ are planes parallel to the x-y plane. When we put $\phi = \phi_0 = constant$ then it describes a half-plane passing through the z-axis and at an angle of ϕ_0.

Example 2.15 Find the parametric equation of the cylinder $\rho = 2$ in rectangular coordinates.

Solution

Step 1. The equation set for transformation from ρ, ϕ, and z to x, y, and z are given by Eqn set (2.8):

$$x = \rho \cos\phi, \ y = \rho \sin\phi, \ z = z$$

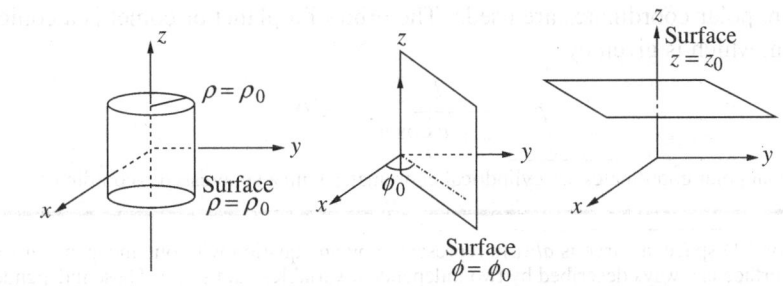

Fig. 2.12 Surfaces in cylindrical coordinates

These are in fact the parametric equations of x, y, and z for any value of (or in terms of) ρ, ϕ, and z.

Step 2. In these equations, we substitute $\rho = 2$ and we get the parametric equations of the cylinder in rectangular coordinates as a function of ϕ and z as parameters. These are

$$x = 2\cos\phi, \ y = 2\sin\phi, \ z = z$$

Step 3. Since these equations have two degrees of freedom in 3-D space, they constitute a surface.[2]

Let us now take a look at the mathematical description of lines in cylindrical coordinates. Generally, the lines which can easily be described are the lines which lie on the surface of a cylinder. As in the case of rectangular coordinates, lines in cylindrical coordinates possess only one degree of freedom. Thus, equations of the type

$$\rho \equiv \rho(t), \ \phi \equiv \phi(t), \ z \equiv z(t) \tag{2.24}$$

represent curves. Here t is the parameter in question. A simple example is that of a helix (spiral, screw). A helix is a curve in space which lies on a cylinder and advances spirally (see Thomas & Finney, 1996). A little reflection tells us that while the x- and y-values lie on a circle, the z-coordinate advances in such a way that the whole curve lies on a cylinder. When the curve advances in the ϕ direction by 2π, then the z-coordinate advances in the z-direction by L. The parametric equation for a right-handed helix, where the distance between turns is 2π and the radius of the helix is 0.5, is given by the equations (in cylindrical coordinates):

$$\rho = 0.5, \ \phi = t, \ z = t \tag{2.25}$$

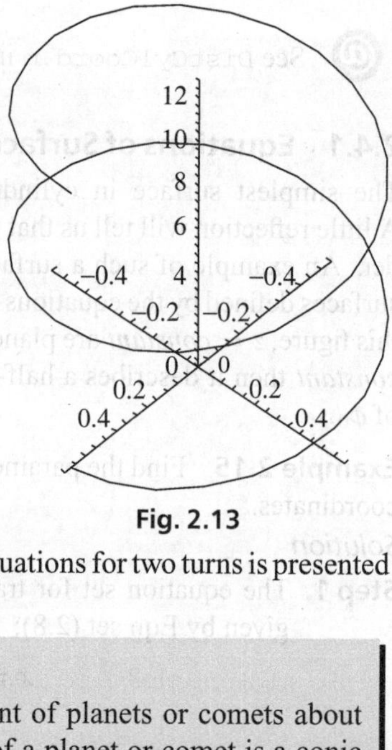

Fig. 2.13

A plot of the helix described by these parametric equations for two turns is presented in Fig. 2.13.

Polar Coordinates

In the mathematical description of the movement of planets or comets about the sun, polar coordinates are used.[a] The orbit of a planet or comet is a conic section, which is given by

$$\rho = \frac{l}{1 + e\cos\phi}; \ z = 0$$

[a]Note that polar coordinates are cylindrical coordinates with a suppressed z-coordinate.

---

[2]In 3-D or 2-D space, a curve is *always* represented by an equation with one independent variable. A surface is always described by two independent variables, and so on. These independent variables are called parameters. Thus equations in one parameter represents a line (in 2-D or 3-D space); in 3-D space, equations in two parameters represent a surface.

where e is the eccentricity and l is the semi-latus rectum (the chord through a focus parallel to the conic section directrix is called the latus rectum,). If $e > 1$, this equation defines a hyperbola; if $e = 1$, it is a parabola; and if $e < 1$, the curve is an ellipse. When $e = 0$ we obtain a circle of radius l.

2.5 | Spherical Coordinate System

The spherical coordinates are represented by three parameters: (r, θ, ϕ) which are shown in Fig. 2.14.

To describe any point, the r parameter is the distance from the point in question to the origin. This is the length Op. Therefore,

$$r \equiv |\mathbf{r}| = \sqrt{x^2 + y^2 + z^2} \tag{2.26}$$

where x, y, and z are the rectangular coordinates of the point. The θ coordinate is the angle between the two lines: the z-axis and the line Op (i.e., the line joining the origin to the point 'p'). From observing the figure, it is clear that

$$z = r\cos\theta \tag{2.27}$$

or
$$\theta = \cos^{-1}(z/r) = \cos^{-1}\left(\frac{z}{\sqrt{x^2 + y^2 + z^2}}\right) \tag{2.28}$$

similarly it is clear that the ϕ coordinate is angle between the x-axis and the line joining the origin to the foot of the perpendicular (the perpendicular is p–f or g–f) from the point p to the x-y plane. Therefore,

$$\phi = \cos^{-1}(x/\rho) = \cos^{-1}\left(\frac{x}{\sqrt{x^2 + y^2}}\right) \tag{2.29}$$

where $\rho = \sqrt{x^2 + y^2}$. These relations give the values of r, θ, and ϕ in terms of the of x, y, and z coordinates. The inverse transformation is given by

$$x = r\sin\theta\,\cos\phi\,,\ y = r\sin\theta\,\sin\phi\,,\ z = r\cos\theta \tag{2.30}$$

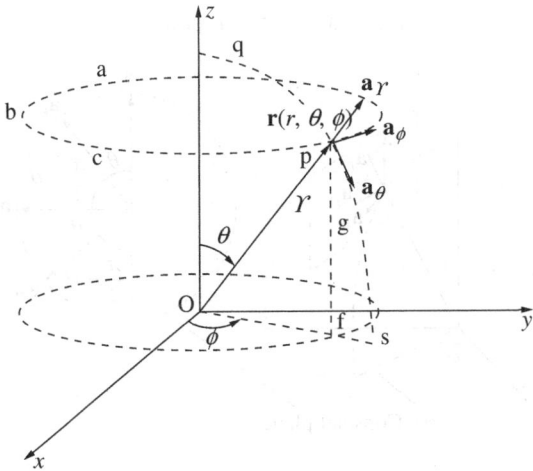

Fig. 2.14 Spherical coordinate system

The unit vectors in the spherical coordinate system are depicted in Fig. 2.14. $\mathbf{a}_r$ is the outward extension of the straight line 'Op'. On the line 'Op', θ and ϕ are both constant while r varies. On the circle 'abc' r and θ are both constants, while ϕ is variable. Tangent to this circle is the unit vector $\mathbf{a}_\phi$. The third unit vector $\mathbf{a}_\theta$ is tangent to the circle 'qps' on which r and ϕ are both constants and θ is variable. The circle 'abc' is like the line of latitude on the surface of the earth, while the circle 'qps' is like a line of longitude.

These three unit vectors are perpendicular to each other with

$$\mathbf{a}_r \times \mathbf{a}_\theta = \mathbf{a}_\phi, \ \mathbf{a}_\theta \times \mathbf{a}_\phi = \mathbf{a}_r, \ \mathbf{a}_\phi \times \mathbf{a}_r = \mathbf{a}_\theta \qquad (2.31)$$

To compute the unit vectors of the spherical coordinates in terms of the unit vectors of the rectangular and cylindrical coordinate systems, let us take a look at Fig. 2.15.

The unit vectors $\mathbf{a}_\rho$, $\mathbf{a}_r$, $\mathbf{a}_\theta$, and $\mathbf{a}_z$ lie on the same plane as shown in the figure.

The unit vector $\mathbf{a}_r$ is connected to $\mathbf{a}_z$ and $\mathbf{a}_\rho$ by

$$\mathbf{a}_r = \mathbf{a}_z \cos\theta + \mathbf{a}_\rho \sin\theta \qquad (2.32)$$

Since

$$\mathbf{a}_\rho = \mathbf{a}_x \cos\phi + \mathbf{a}_y \sin\phi \qquad (2.33)$$

Therefore,

$$\mathbf{a}_r = \mathbf{a}_z \cos\theta + \underbrace{(\mathbf{a}_x \cos\phi + \mathbf{a}_y \sin\phi)}_{\mathbf{a}_\rho} \sin\theta \qquad \text{(rectangular to spherical)} \quad (2.34)$$

The important point to remember is that the spherical and cylindrical coordinates share the same ϕ and $\mathbf{a}_\phi$ coordinate and unit vector. Thus observing the plane $\phi =$ constant the unit vectors $\mathbf{a}_z$, $\mathbf{a}_\rho$ (which are perpendicular to each other) and $\mathbf{a}_r$, $\mathbf{a}_\theta$ (which are also perpendicular to each other) lie on this plane as diagrammed. Since $\mathbf{a}_\theta = \mathbf{a}_\phi \times \mathbf{a}_r$ and using the results of the cross products of cylindrical coordinates:

$$\mathbf{a}_\theta = \mathbf{a}_\phi \times \mathbf{a}_r = \mathbf{a}_\phi \times (\mathbf{a}_z \cos\theta + \mathbf{a}_\rho \sin\theta)$$

$$= \underbrace{\mathbf{a}_\phi \times \mathbf{a}_z}_{\mathbf{a}_\rho} \cos\theta + \underbrace{\mathbf{a}_\phi \times \mathbf{a}_\rho}_{-\mathbf{a}_z} \sin\theta$$

$$=(\mathbf{a}_x \cos\phi + \mathbf{a}_y \sin\phi) \cos\theta - \mathbf{a}_z \sin\theta \qquad (2.35)$$

The third unit vector $\mathbf{a}_\phi$ is as for cylindrical coordinates,

$$\mathbf{a}_\phi = \mathbf{a}_y \cos\phi - \mathbf{a}_x \sin\phi \qquad (2.36)$$

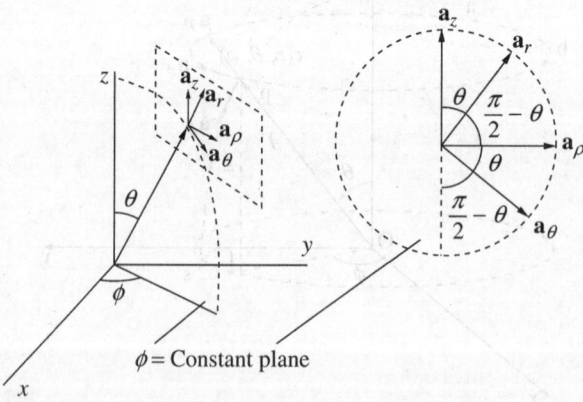

Fig. 2.15 $(\mathbf{a}_\rho, \mathbf{a}_z)$ and $(\mathbf{a}_r, \mathbf{a}_\theta)$ are the cylindrical and spherical coordinate systems

Therefore, the unit vectors of the spherical coordinate system in terms of the unit vectors of the rectangular coordinates system are given by

$$\mathbf{a}_r = \sin\theta \cos\phi \, \mathbf{a}_x + \sin\theta \sin\phi \, \mathbf{a}_y + \cos\theta \, \mathbf{a}_z$$
$$\mathbf{a}_\theta = \cos\theta \cos\phi \, \mathbf{a}_x + \cos\theta \sin\phi \, \mathbf{a}_y - \sin\theta \, \mathbf{a}_z$$
$$\mathbf{a}_\phi = -\sin\phi \, \mathbf{a}_x + \cos\phi \, \mathbf{a}_y \tag{2.37}$$

These results can be put in the form of a matrix equation (with the transformation matrix $\mathbf{T}_{rs}$):

$$\begin{pmatrix} \mathbf{a}_r \\ \mathbf{a}_\theta \\ \mathbf{a}_\phi \end{pmatrix} = \underbrace{\begin{pmatrix} \sin\theta \cos\phi & \sin\theta \sin\phi & \cos\theta \\ \cos\theta \cos\phi & \cos\theta \sin\phi & -\sin\theta \\ -\sin\phi & \cos\phi & 0 \end{pmatrix}}_{\mathbf{T}_{rs}} \begin{pmatrix} \mathbf{a}_x \\ \mathbf{a}_y \\ \mathbf{a}_z \end{pmatrix} \tag{2.38}$$

We can also find the inverse transformation (with $\mathbf{T}_{rs}^{-1}$) as

$$\begin{pmatrix} \mathbf{a}_x \\ \mathbf{a}_y \\ \mathbf{a}_z \end{pmatrix} = \underbrace{\begin{pmatrix} \sin\theta \cos\phi & \cos\theta \cos\phi & -\sin\phi \\ \sin\theta \sin\phi & \cos\theta \sin\phi & \cos\phi \\ \cos\theta & -\sin\theta & 0 \end{pmatrix}}_{\mathbf{T}_{rs}^{-1}} \begin{pmatrix} \mathbf{a}_r \\ \mathbf{a}_\theta \\ \mathbf{a}_\phi \end{pmatrix} \tag{2.39}$$

Note that
$$\mathbf{T}_{rs}^{-1} = \mathbf{T}_{rs}^t \tag{2.40}$$

The three unit vectors of the spherical coordinate system are orthonormal, and a table of their dot products with $\mathbf{a}_x$ etc., are shown in Table 2.3.

We can compute the components of a vector in either of the two coordinate systems

$$\begin{pmatrix} A_r \\ A_\theta \\ A_\phi \end{pmatrix} = \begin{pmatrix} \mathbf{a}_r \cdot \mathbf{a}_x & \mathbf{a}_r \cdot \mathbf{a}_y & \mathbf{a}_r \cdot \mathbf{a}_z \\ \mathbf{a}_\theta \cdot \mathbf{a}_x & \mathbf{a}_\theta \cdot \mathbf{a}_y & \mathbf{a}_\theta \cdot \mathbf{a}_z \\ \mathbf{a}_\phi \cdot \mathbf{a}_x & \mathbf{a}_\phi \cdot \mathbf{a}_y & \mathbf{a}_\phi \cdot \mathbf{a}_z \end{pmatrix} \begin{pmatrix} A_x \\ A_y \\ A_z \end{pmatrix} \tag{2.41}$$

$$= \begin{pmatrix} \sin\theta \cos\phi & \sin\theta \sin\phi & \cos\theta \\ \cos\theta \cos\phi & \cos\theta \sin\phi & -\sin\theta \\ -\sin\phi & \cos\phi & 0 \end{pmatrix} \begin{pmatrix} A_x \\ A_y \\ A_z \end{pmatrix}$$

$$\begin{pmatrix} A_x \\ A_y \\ A_z \end{pmatrix} = \begin{pmatrix} \mathbf{a}_x \cdot \mathbf{a}_\rho & \mathbf{a}_x \cdot \mathbf{a}_\theta & \mathbf{a}_x \cdot \mathbf{a}_\phi \\ \mathbf{a}_y \cdot \mathbf{a}_\rho & \mathbf{a}_y \cdot \mathbf{a}_\theta & \mathbf{a}_y \cdot \mathbf{a}_\phi \\ \mathbf{a}_z \cdot \mathbf{a}_\rho & \mathbf{a}_z \cdot \mathbf{a}_\theta & \mathbf{a}_z \cdot \mathbf{a}_\phi \end{pmatrix} \begin{pmatrix} A_r \\ A_\theta \\ A_\phi \end{pmatrix} \tag{2.42}$$

$$= \begin{pmatrix} \sin\theta \cos\phi & \cos\theta \cos\phi & -\sin\phi \\ \sin\theta \sin\phi & \cos\theta \sin\phi & \cos\phi \\ \cos\theta & -\sin\theta & 0 \end{pmatrix} \begin{pmatrix} A_r \\ A_\theta \\ A_\phi \end{pmatrix}$$

Table 2.3 Dot products of spherical coordinate unit vectors

·	$\mathbf{a}_r$	$\mathbf{a}_\theta$	$\mathbf{a}_\phi$
$\mathbf{a}_x$	$\sin\theta \cos\phi$	$\cos\theta \cos\phi$	$-\sin\phi$
$\mathbf{a}_y$	$\sin\theta \sin\phi$	$\cos\theta \sin\phi$	$\cos\phi$
$\mathbf{a}_z$	$\cos\theta$	$-\sin\theta$	0

Example 2.16 Compute $\mathbf{a}_x$, $\mathbf{a}_y$, and $\mathbf{a}_z$ in terms of $\mathbf{a}_r$, $\mathbf{a}_\theta$, and $\mathbf{a}_\phi$.

Solution

Step 1. $\mathbf{a}_x$ is $(1,0,0)$ in rectangular coordinates. Therefore in Eqn (2.41), we insert $(1,0,0)$ instead of (A_x, A_y, A_z) and get

$$\begin{pmatrix} A_r \\ A_\theta \\ A_\phi \end{pmatrix} = \begin{pmatrix} \mathbf{a}_r \cdot \mathbf{a}_x & \mathbf{a}_r \cdot \mathbf{a}_y & \mathbf{a}_r \cdot \mathbf{a}_z \\ \mathbf{a}_\theta \cdot \mathbf{a}_x & \mathbf{a}_\theta \cdot \mathbf{a}_y & \mathbf{a}_\theta \cdot \mathbf{a}_z \\ \mathbf{a}_\phi \cdot \mathbf{a}_x & \mathbf{a}_\phi \cdot \mathbf{a}_y & \mathbf{a}_\phi \cdot \mathbf{a}_z \end{pmatrix} \begin{pmatrix} 1 \\ 0 \\ 0 \end{pmatrix}$$

which translates to

$$\begin{pmatrix} A_r \\ A_\theta \\ A_\phi \end{pmatrix} = \begin{pmatrix} \sin\theta\,\cos\phi & \sin\theta\,\sin\phi & \cos\theta \\ \cos\theta\,\cos\phi & \cos\theta\,\sin\phi & -\sin\theta \\ -\sin\phi & \cos\phi & 0 \end{pmatrix} \begin{pmatrix} 1 \\ 0 \\ 0 \end{pmatrix} = \begin{pmatrix} \sin\theta\,\cos\phi \\ \cos\theta\,\cos\phi \\ -\sin\phi \end{pmatrix}$$

Step 2. Hence $\quad \mathbf{a}_x = \underbrace{\sin\theta\,\cos\phi\,\mathbf{a}_r}_{A_r} + \underbrace{\cos\theta\,\cos\phi\,\mathbf{a}_\theta}_{A_\theta} - \underbrace{\sin\phi\,\mathbf{a}_\phi}_{A_\phi}$

Step 3. In the same manner, we get

$$\mathbf{a}_y = \sin\theta\,\sin\phi\,\mathbf{a}_r + \cos\theta\,\sin\phi\,\mathbf{a}_\theta + \cos\phi\,\mathbf{a}_\phi$$

$$\mathbf{a}_z = \cos\theta\,\mathbf{a}_r - \sin\theta\,\mathbf{a}_\theta$$

@ See `DistSphCoord.m` in Chapter 2

The spherical coordinate system is especially suited for mathematical descriptions of spheres, cones or half planes passing through the origin. Thus $r = 3$ is a sphere whose radius is three; $\theta = \pi/4$ is a cone whose cone angle is $\pi/2$; and $\phi = \pi/6$ is a half plane passing through the origin at an angle of $30°$ to x–z plane. These surfaces are shown in Fig. 2.16. As discussed earlier, any equation with two degrees of freedom is a surface. Hence we can rewrite the equation of a sphere as

$$r = 3, \quad \phi = \text{anything}, \quad \theta = \text{anything} \tag{2.43}$$

which has only two degrees of freedom (as only two variables ϕ and θ may be specified independently while r is fixed and equal to 3). Just as in the case of the cylindrical coordinate system, we can write parametric equations of a space curve which are functions of a single variable in the spherical coordinate system:

$$r \equiv r(t); \quad \theta \equiv \theta(t); \quad \text{and} \quad \phi \equiv \phi(t) \tag{2.44}$$

where t is the parameter and $r(t)$, $\theta(t)$, and $\phi(t)$ are functions of t.

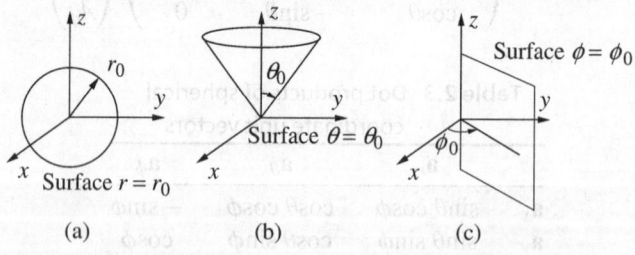

Fig. 2.16 Coordinate surfaces in spherical coordinates

POINTS TO REMEMBER

- We express our equations in terms of scalar and vector fields as electromagnetism is expressed in terms of vectors and scalars which change from point to point in space.
- Scalar and vector fields are scalars and vectors respectively which are functions of coordinates and time.
- Scalar and vector fields may be differentiated and integrated with respect to coordinates and time.
- The type of coordinate system is chosen which aids in making the equations simple in the particular coordinate system.
- $\mathbf{R}_{01}$, the position vector from point 0 at $\mathbf{r}_0$ to point 1 at $\mathbf{r}_1$ is given by

$$\mathbf{R}_{01} = \mathbf{r}_1 - \mathbf{r}_0$$

- The direction cosines of a vector $\mathbf{V}$ are given by

$$\alpha = \hat{\mathbf{V}} \cdot \mathbf{a}_x = \cos\theta$$
$$\beta = \hat{\mathbf{V}} \cdot \mathbf{a}_y = \cos\phi$$
$$\gamma = \hat{\mathbf{V}} \cdot \mathbf{a}_x = \cos\psi$$

- The parametric equation of a straight line where $\mathbf{R}$ is vector parallel to the line, $\mathbf{r}_0$ is the position vector of a specific point on the line, and $\mathbf{r}$ is the position vector of any point on the line is

$$\mathbf{r}(x, y, z) = t\mathbf{R} + \mathbf{r}_0 \quad (-\infty < t < \infty)$$

where t is a parameter.
- The equation of a plane in rectangular coordinates is given by

$$(\mathbf{r} - \mathbf{r}_0) \cdot (\mathbf{R}_1 \times \mathbf{R}_2) = 0$$

where $\mathbf{r}$ is the position vector of a point on the plane, $\mathbf{r}_0$ is the position vector of a specific point on the plane, and $\mathbf{R}_1$ and $\mathbf{R}_2$ are two vectors lying on the plane.
- If ρ, ϕ, and z are the cylindrical coordinates of a point then the rectangular coordinates are given by

$$x = \rho\cos\phi, \quad y = \rho\sin\phi, \quad z = z$$

- If x, y, and z are the rectangular coordinates of a point then the cylindrical coordinates are given by

$$\rho = \sqrt{x^2 + y^2}, \quad \phi = \tan^{-1}(y/x), \quad z = z$$

- The cross-product of the unit vectors in cylindrical coordinates are given by

$$\mathbf{a}_\rho \times \mathbf{a}_\phi = \mathbf{a}_z, \quad \mathbf{a}_\phi \times \mathbf{a}_z = \mathbf{a}_\rho, \quad \mathbf{a}_z \times \mathbf{a}_\rho = \mathbf{a}_\phi$$

- The position vector of a point in cylindrical coordinates is

$$\mathbf{r} = \rho\,\mathbf{a}_\rho + z\mathbf{a}_z$$

- When the same vector is expressed in cylindrical and rectangular coordinates then

$$\begin{pmatrix} A_\rho \\ A_\phi \\ A_z \end{pmatrix} = \begin{pmatrix} \cos\phi & \sin\phi & 0 \\ -\sin\phi & \cos\phi & 0 \\ 0 & 0 & 1 \end{pmatrix} \begin{pmatrix} A_x \\ A_y \\ A_z \end{pmatrix}$$

and
$$\begin{pmatrix} A_x \\ A_y \\ A_z \end{pmatrix} = \begin{pmatrix} \cos\phi & -\sin\phi & 0 \\ \sin\phi & \cos\phi & 0 \\ 0 & 0 & 1 \end{pmatrix} \begin{pmatrix} A_\rho \\ A_\phi \\ A_z \end{pmatrix}$$

- Spherical coordinates use three coordinates: r, θ, and ϕ. In terms of rectangular coordinates these are

$$r = \sqrt{x^2 + y^2 + z^2}$$

$$\theta = \arccos\left(\frac{z}{\sqrt{x^2 + y^2 + z^2}}\right)$$

$$\phi = \arccos\left(\frac{x}{\sqrt{x^2 + y^2}}\right)$$

- If the coordinates are expressed in spherical coordinates then rectangular coordinates are given by the equations

$$x = r\sin\theta\cos\phi, \quad y = r\sin\theta\sin\phi, \quad z = r\cos\theta$$

- The cross-product of the unit vectors in spherical coordinates are

$$\mathbf{a}_r \times \mathbf{a}_\theta = \mathbf{a}_\phi, \quad \mathbf{a}_\theta \times \mathbf{a}_\phi = \mathbf{a}_r, \quad \mathbf{a}_\phi \times \mathbf{a}_r = \mathbf{a}_\theta$$

- If a vector is given in rectangular coordinates, then in spherical coordinates its components are

$$\begin{pmatrix} A_r \\ A_\theta \\ A_\phi \end{pmatrix} = \begin{pmatrix} \sin\theta\cos\phi & \sin\theta\sin\phi & \cos\theta \\ \cos\theta\cos\phi & \cos\theta\sin\phi & -\sin\theta \\ -\sin\phi & \cos\phi & 0 \end{pmatrix} \begin{pmatrix} A_x \\ A_y \\ A_z \end{pmatrix}$$

- If a vector is given in spherical coordinates then in rectangular coordinates its components are

$$\begin{pmatrix} A_x \\ A_y \\ A_z \end{pmatrix} = \begin{pmatrix} \sin\theta\cos\phi & \cos\theta\cos\phi & -\sin\phi \\ \sin\theta\sin\phi & \cos\theta\sin\phi & \cos\phi \\ \cos\theta & -\sin\theta & 0 \end{pmatrix} \begin{pmatrix} A_r \\ A_\theta \\ A_\phi \end{pmatrix}$$

- Cylindrical coordinates are used where the geometry is either a cylinder, $\rho = \rho_0$, or a semi-infinite half-plane, $\phi = \phi_0$ or a plane $z = z_0$ or a combination of these three.
- Spherical coordinates are used where the geometry is either a sphere, $r = r_0$, or a cone, $\theta = \theta_0$, or a semi-infinite half-plane, $\phi = \phi_0$ or a combination of these three.

SELF ASSESSMENT

Objective Type Questions

Note: In the following questions one or more choices may be correct. Sometimes choices may be separated with an 'or' and sometimes with an 'and'.

1. $\mathbf{a}_y$ in spherical coordinates at the point $P(r = 4, \theta = 0.2\pi, \phi = 0.8\pi)$ is
   (a) $0.4\mathbf{a}_r + 0.5\mathbf{a}_\theta - 0.81\mathbf{a}_\phi$ (b) $0.35\mathbf{a}_r + 0.48\mathbf{a}_\theta - 0.81\mathbf{a}_\phi$
   (c) $-0.48\mathbf{a}_r + 0.48\mathbf{a}_\theta - 0.81\mathbf{a}_\phi$ (d) $0.48\mathbf{a}_r + 0.35\mathbf{a}_\theta + 0.81\mathbf{a}_\phi$

2. The surface $\rho = 2$, $\rho = 4$, $\phi = 45°$, $\phi = 135°$, $z = 3$ and $z = 4$ form a closed surface. The enclosing area is
   (a) 34.29   (b) 32.27   (c) 20.7   (d) 16.4

3. The vector $\mathbf{B} = (10/r) \, \mathbf{a}_r + r \cos\theta \, \mathbf{a}_\theta + \mathbf{a}_\phi$ in rectangular coordinates at the point $(-3, 4, 0)$ is
   (a) $\mathbf{a}_x - 2\mathbf{a}_y$ (b) $-2\mathbf{a}_x + \mathbf{a}_y$ (c) $1.36\mathbf{a}_x - 2.72\mathbf{a}_y$ (d) $-2.72\mathbf{a}_x + 1.36\mathbf{a}_y$
4. A vector field in Cartesian form is given as
   $$\mathbf{D} = x\mathbf{a}_x + y\mathbf{a}_y$$
   find the cylindrical form.
   (a) $\mathbf{a}_\rho/\rho$ (b) $\rho \, \mathbf{a}_\rho$ (c) $\phi \, \mathbf{a}_\phi$ (d) $-\rho \, \mathbf{a}_\phi$
5. A vector field in Cartesian form is given as
   $$\mathbf{D} = y\mathbf{a}_x - x\mathbf{a}_y$$
   find the cylindrical form.
   (a) $\mathbf{a}_\rho/\rho$ (b) $\rho \, \mathbf{a}_\rho$ (c) $\phi \, \mathbf{a}_\phi$ (d) $-\rho \, \mathbf{a}_\phi$
6. A vector field in Cartesian form is given as
   $$\mathbf{D} = x\mathbf{a}_x - y\mathbf{a}_y$$
   find the cylindrical form.
   (a) $\mathbf{a}_\rho/\rho + \rho \, \mathbf{a}_\rho$ (b) $\cos(2\phi) \, \rho \, \mathbf{a}_\rho - \sin(2\phi) \, \mathbf{a}_\phi$
   (c) $\cos\phi \, \mathbf{a}_\rho + \sin\phi \, \mathbf{a}_\phi$ (d) None of these
7. A vector field in Cartesian form is given as
   $$\mathbf{D} = y\mathbf{a}_x + x\mathbf{a}_y$$
   find the cylindrical form.
   (a) $\cos(2\phi) \, \rho \, \mathbf{a}_\rho + \sin(2\phi) \, \mathbf{a}_\phi$ (b) $\cos(2\phi) \, \rho \, \mathbf{a}_\rho - \sin(2\phi) \, \mathbf{a}_\phi$
   (c) $\cos\phi \, \mathbf{a}_\rho + \sin\phi \, \mathbf{a}_\phi$ (d) None of these
8. A vector field is given by $\mathbf{D} = k/r^2 \mathbf{a}_r$ in spherical coordinates. Find the D_z component of the vector field in rectangular coordinates.
   (a) $kz/ \left(x^2 + y^2 + z^2\right)$ (b) $kxy/ \left(x^2 + y^2 + z^2\right)$
   (c) $kz/ \left(x^2 + y^2 + z^2\right)^{3/2}$ (d) None of these
9. A vector field is given by $\mathbf{D} = k/r^2 \mathbf{a}_r$ in spherical coordinates. Find the D_z component of the vector field in cylindrical coordinates.
   (a) $kz/ \left(\rho^2 + z^2\right)$ (b) $k\rho^2 \sin\phi \cos\phi / \left(\rho^2 + z^2\right)$
   (c) $kz/ \left(\rho^2 + z^2\right)^{3/2}$ (d) None of these
10. The unit of $\mathbf{a}_r$ in spherical coordinates is
    (a) m   (b) no dimension   (c) ft   (d) None of these

Short-Answer Questions

1. When do we use a particular coordinate system? To describe a plane, which coordinate system will we use? Give example.
2. Give one example each of (a) a scalar field and (b) a vector field.
   What is the advantage of using fields in electromagnetic theory?
3. If we define a vector u by connecting the points $(x = 0, y = 0, z = k)$ and $(\rho, \phi, 0)$, find the unit vector $\hat{u}$.
4. Transform $3\mathbf{a}_x - 3\mathbf{a}_y - 2\mathbf{a}_z$ into spherical coordinates at the point $(x = -3, y = 1, z = 3)$.
5. The equation for $\mathbf{A}$ written in spherical coordinates is
   $$\mathbf{A} = \frac{e^{-jkr}}{r} \mathbf{a}_r$$
   Rewrite it in rectangular coordinates.

6. Express $\mathbf{A} = k/r^2\mathbf{a}_r$ in cylindrical coordinates.
7. The vector field $\mathbf{A} = k\mathbf{a}_\phi/\rho$ given in cylindrical coordinates is to be expressed in spherical coordinates.

Review Questions

1. Define (a) a scalar field and (b) a vector field.
   Why do we need scalar and vector fields to study electromagnetic theory?
2. Why is the pressure above the earth's surface a scalar field? Is the dielectric constant of air above the earth's surface a scalar field? Research this question.
3. Are scalar and vector fields differentiable with respect to coordinates? With respect to time?
4. What is the difference between a right-handed and left-handed coordinate system? Are equations written in the two coordinate systems exactly same? Explain your answer with an example.
5. What is the difference between the equation of a curve and that of a surface?
6. Explain with examples where you would use Cartesian, cylindrical, and spherical co-ordinate system.
7. Are the unit vectors constant vectors in the spherical coordinate system? Explain with an example.
8. What are the possible ways of defining the straight line, if we want to find the equation of a straight line in 3-D space?
9. Why do we choose the cylindrical coordinate system for defining a helix?
10. Which are those surfaces which are special in terms of coordinate = constant in spherical coordinates?

Numerical Problems

1. The scalar field $V(z) = -100/z$ for $z > 1000$ km. Find V at 1000 and 1010 km.
2. Find dV/dz of Prob. 1 at 1000 km.
3. A vector field is given by $\mathbf{A} = 3xy\mathbf{a}_x + 3yz\mathbf{a}_y + 3zx\mathbf{a}_z$. Compute

$$\frac{\partial A_x}{\partial x} + \frac{\partial A_y}{\partial y} + \frac{\partial A_z}{\partial z}$$

   what is the result?
4. For the vector field $\mathbf{A}$ of Prob. 3, Compute

$$\begin{vmatrix} \mathbf{a}_x & \mathbf{a}_y & \mathbf{a}_z \\ \partial/\partial x & \partial/\partial y & \partial/\partial z \\ A_x & A_y & A_z \end{vmatrix}$$

5. A scalar field is given by $\Phi = 3xyz$
   Compute $\mathbf{A} = \dfrac{\partial \Phi}{\partial x}\mathbf{a}_x + \dfrac{\partial \Phi}{\partial y}\mathbf{a}_y + \dfrac{\partial \Phi}{\partial z}\mathbf{a}_z$ at the point (1,1,1).
6. In rectangular coordinates, find the equation of a straight line which passes through the point (1,1,1) and is parallel to the x-axis.
7. In rectangular coordinates, find the equation of a plane which passes through the point (1,1,1) and is parallel to the x–y plane.
8. Find the equation of the straight line joining the origin to (1,1,1).
9. Find the equation of a plane passing through (1,2,3) and perpendicular to the vector $2\mathbf{a}_x + 3\mathbf{a}_y + 4\mathbf{a}_z$.
10. Find the equation of a line passing through the points (1,2,3) and (2,3,4).

11. Find the equation of a circle in the x-y plane with radius 5.
12. What does $(x, y, z) = (5 \cos t, 5 \sin t, 4)$ with $0 \leq t \leq 2\pi$ represent? Justify.
13. Find the equation of an infinite cylinder of radius 2 with its axis coincident with the z-axis in cylindrical coordinates.
14. (a) A field vector is given by the expression $\mathbf{A} = (x/R) \mathbf{a}_x + (y/R) \mathbf{a}_y + (z/R) \mathbf{a}_z$ where $R = \sqrt{x^2 + y^2 + z^2}$. Transform the vector into the cylindrical system of coordinates.
    (b) Give the cylindrical coordinates of the point whose Cartesian coordinates are given by $x = 3; y = 4; z = 5$ units and show the same on a sketch. [DTU-NSIT, June 2010]
15. Find the equation of a circle of radius 2 with its centre coincident with the origin and lying on the x-y plane in cylindrical coordinates
16. Find the equation of the half-plane $x = 0$, $y \geq 0$ and $-\infty < z < \infty$ in cylindrical coordinates.
17. For the rectangular coordinate system find the point $(x = 1, y = 2, z = 3)$ in cylindrical coordinates.
18. The points $(0,0,0)$, $(1,0,0)$, and $(0,1,0)$ form a triangle. Find the unit vector normal to the surface of the triangle.
19. Find the unit vector to the surface $x^2 + y^2 + z^2 = 25$ at the point $(0,0,5)$.
20. Find the unit vector to the surface $x^2 + y^2 + 2z = 5$ at the point $(0,0,5/2)$ and $(1,1,3/2)$
21. Consider the vector field $\mathbf{A} = \sin\phi\, \mathbf{a}_\rho + \cos z \mathbf{a}_\phi + \rho\, \mathbf{a}_z$.
    Find the normal and tangential component at the point $(\rho = 2, \phi = 1^r, z = 1)$ on the cylinder $\rho = 2$.

Answers

Objective Type Questions

1. (b)   2. (b)   3. (b)   4. (b)   5. (d)   6. (b)   7. (a)   8 (c)   9. (c)   10. (b)

Short-Answer Questions

3. $\hat{u} = (\rho \cos\phi\, \mathbf{a}_x + \rho \sin\phi\, \mathbf{a}_y - k\mathbf{a}_z)/\sqrt{\rho^2 + k^2}$ is the unit vector in the rectangular coordinates. On further conversion $\hat{u} = (\rho\, \mathbf{a}_\rho - k\mathbf{a}_z)/\sqrt{\rho^2 + k^2}$
4. We observe the point $(-3, 1, 3)$. At this point $\theta = 0.8117^r$ and $\phi = 2.8198^r$. We now use Eqn (2.41) to convert from rectangular to spherical coordinates. After conversion, the vector is $-4.129\mathbf{a}_r + 1.451\mathbf{a}_\theta + 1.897\mathbf{a}_\phi$.
5. $r = \sqrt{x^2 + y^2 + z^2}$. $\mathbf{a}_r = \cos\theta\, \mathbf{a}_z + \sin\theta \cos\phi\, \mathbf{a}_x + \sin\theta \sin\phi\, \mathbf{a}_y$. $\cos\theta = z/r$; $\sin\theta = \sqrt{x^2 + y^2}/r$; $\cos\phi = x/\sqrt{x^2 + y^2}$; $\sin\phi = y/\sqrt{x^2 + y^2}$.
   Using these results

$$\mathbf{A} = \frac{e^{-jk\sqrt{x^2+y^2+z^2}}}{(x^2 + y^2 + z^2)}(x\mathbf{a}_x + y\mathbf{a}_y + z\mathbf{a}_z)$$

   or $\mathbf{a}_r = \mathbf{r}/r$.
6. $r^2 = \rho^2 + z^2$; $\mathbf{a}_r = (\rho\, \mathbf{a}_\rho + z\mathbf{a}_z)/\sqrt{\rho^2 + z^2}$ giving

$$\mathbf{A} = \frac{k(\rho\, \mathbf{a}_\rho + z\mathbf{a}_z)}{(\rho^2 + z^2)^{3/2}}$$

7. $\rho = \sqrt{r^2 - z^2}$; the unit vector $\mathbf{a}_\phi$ remains the same. So $\mathbf{A} = \dfrac{k\mathbf{a}_\phi}{\sqrt{r^2 - z^2}}$

Numerical Problems

1. -10^{-4} and -0.9901×10^{-4}
2. 10^{-10}

3. $3(x + y + z)$; a scalar field

4. $-3y\mathbf{a}_x - 3z\mathbf{a}_y - 3x\mathbf{a}_z$; a vector field

5. $\mathbf{A} = 3(\mathbf{a}_x + \mathbf{a}_y + \mathbf{a}_z)$

6. $(x, y, z) = (1 + t, 1, 1)$. Other solutions are also possible

7. $z - 1 = 0$

8. $(x, y, z) = (1 + t, 1 + t, 1 + t)$. Other solutions are also possible

9. $2x + 3y + 4z - 20 = 0$

10. $(x, y, z) = (1 + t, 2 + t, 3 + t)$

11. $(x, y, z) = (5\cos t, 5\sin t, 0)$ with $0 \leq t \leq 2\pi$

12. The equation of a circle with radius 5 on the plane $z = 4$

13. $\rho = 2$

15. $\rho = 2$, $z = 0$, $\phi = t$ $\quad 0 \leq t \leq 2\pi$

16. $\phi = \pi/2$

17. $\rho = \sqrt{5}$, $\phi = 63.4°$, $z = 3$

18. $\pm\mathbf{a}_z$

19. $\mathbf{a}_z$

20. $\mathbf{a}_z$ and $(\mathbf{a}_x + \mathbf{a}_y + \mathbf{a}_z)/\sqrt{3}$

21. Normal component $0.8415\mathbf{a}_\rho$; Tangential component $0.5403\mathbf{a}_\phi + 2\mathbf{a}_z$

3

Vector Calculus

Do not worry about your difficulties in Mathematics.
I can assure you mine are still greater.
— Albert Einstein

CHAPTER OBJECTIVES

To enable the students to understand the following:

- Concepts of line integral, surface integral, and volume integral starting from the fundamentals
- Concepts of grad, curl, and div and their physical interpretation
- Divergence theorem and Stokes's theorem
- Different Maxwell's equations and their applications
- Units of various electromagnetic quantities

3.1 | Introduction

The analysis of vector and scalar fields is facilitated through the use of vector calculus. Scalar and vector fields are the words of the mathematical language which describe the majority of electromagnetic phenomena. Vector analysis rests on a few important but simple concepts, which once understood lead to proficiency in the subject.

 See `ConversionofCoordinates.m` in Chapter 3

3.2 | Differential Element of a Line

To understand the vector calculus, a clear concept of differential elements must be acquired. The differential elements generally occur in all the coordinate systems which we have considered previously. We need to know the different types of differential elements which occur in the description of EM fields and how to manipulate them. To start with, we will concentrate on linear elements. Figure 3.1 shows a curve in space on which two closely placed points are shown. The coordinates of point A are (x, y, z) in rectangular coordinates and a point B which is very close to point A, has the coordinates $(x+\Delta x, y+\Delta y, z+\Delta z)$. The position vector $\mathbf{r}$ of point A is $x\mathbf{a}_x + y\mathbf{a}_y + z\mathbf{a}_z$. In particular, $\mathbf{r}(A) \equiv \mathbf{r}(x, y, x)$ in rectangular coordinates, $\mathbf{r}(\rho, \phi, z)$ in cylindrical coordinates, and so on. Similarly, the position vector of the point B is $\mathbf{r}(B) \equiv \mathbf{r}(x+\Delta x, y+\Delta y, z+\Delta z)$ in rectangular coordinates, and $\mathbf{r}(\rho+\Delta \rho, \phi+\Delta \phi, z+\Delta z)$ in cylindrical coordinates, etc. Point B is so close to point A that the linear element AB may be considered to be a straight line *even though both points may lie on a curve*. The differential elements of the *coordinates* in the rectangular coordinates is the set $(\Delta x, \Delta y, \Delta z)$ where Δx may be read as 'a very small increment of x'. Note that $\lim(\Delta x \to 0) = dx$, $\lim(\Delta y \to 0) = dy$ and $\lim(\Delta z \to 0) = dz$.

The differential element of a line (the vector $\overrightarrow{AB}$) is the vector

$$\Delta \mathbf{l} \equiv \Delta \mathbf{r} = \Delta x\mathbf{a}_x + \Delta y\mathbf{a}_y + \Delta z\mathbf{a}_z \quad (3.1)$$

The equation given above says that the vector $\Delta \mathbf{l}$ is the sum of three vectors: a small vector $\Delta \mathbf{x}$ in the $\mathbf{a}_x$ direction, a small vector $\Delta \mathbf{y}$ in the $\mathbf{a}_y$ direction, and a small vector $\Delta \mathbf{z}$ in the $\mathbf{a}_z$ direction, where $\Delta \mathbf{x} = \Delta x\mathbf{a}_x$, $\Delta \mathbf{y} = \Delta y\mathbf{a}_y$, and $\Delta \mathbf{z} = \Delta z\mathbf{a}_z$. The length of this vector is

$$\Delta l \equiv |\Delta \mathbf{r}| = \sqrt{\Delta x^2 + \Delta y^2 + \Delta z^2} \quad (3.2)$$

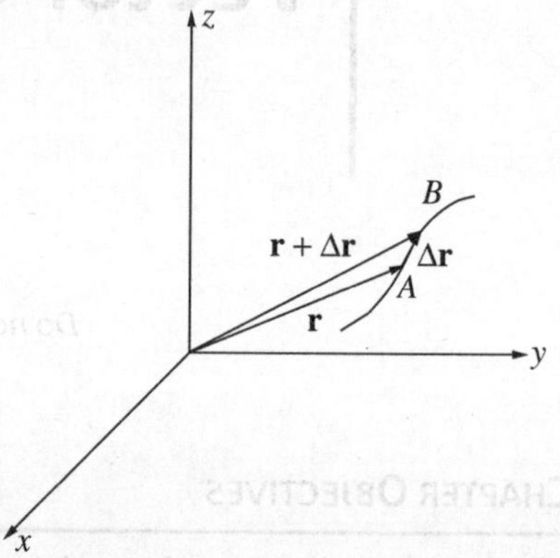

Fig. 3.1 Differential element of a line

It is obvious that the unit vector along the line is

$$\hat{\mathbf{t}} = \frac{\Delta \mathbf{l}}{\Delta l} = \frac{\Delta x\mathbf{a}_x + \Delta y\mathbf{a}_y + \Delta z\mathbf{a}_z}{\sqrt{\Delta x^2 + \Delta y^2 + \Delta z^2}} \quad (3.3)$$

On a little reflection it should be clear that this unit vector is along the direction of the tangent to the curve on which the differential element lies. In the same way, for the cylindrical coordinates shown in Fig. 3.2, the differential elements of the *coordinates* are $(\Delta \rho, \rho\Delta \phi, \Delta z)$. Referring to Fig. 3.2, the differential element Δz is the same as that in rectangular coordinates; the element $\Delta \rho$ is a small linear element in the direction of the unit vector $\mathbf{a}_\rho$ and which is a short extension of the line d-e in the figure; while $\rho\Delta \phi$ is a small linear element tangent to the circle

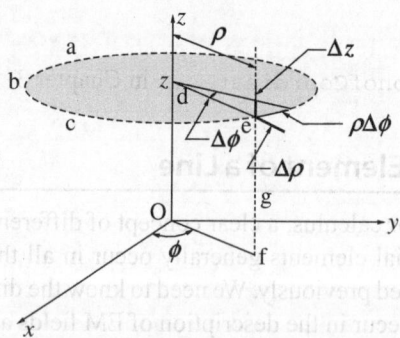

Fig. 3.2 Differential linear elements in cylindrical coordinates

a-b-c and which is in the direction of $\mathbf{a}_\phi$. The differential element, the vector **AB** of Fig. 3.1, is the vector

$$\Delta \mathbf{l} \equiv \Delta \mathbf{r} = \Delta \rho\, \mathbf{a}_\rho + \rho \Delta \phi\, \mathbf{a}_\phi + \Delta z \mathbf{a}_z \tag{3.4}$$

From the equation given above, we can see that the vector $\Delta \mathbf{l}$ is the sum of three small vectors: $\Delta \rho\, \mathbf{a}_\rho$, $\rho \Delta \phi\, \mathbf{a}_\phi$, and $\Delta z \mathbf{a}_z$. By varying the values of $\Delta \rho$, $\rho \Delta \phi$, and Δz we can reach any neighbouring point. The length of this vector is

$$|\Delta \mathbf{r}| \equiv \Delta l = \sqrt{\Delta \rho^2 + \rho^2 \Delta \phi^2 + \Delta z^2} \tag{3.5}$$

and therefore the unit tangent vector is

$$\hat{\mathbf{t}} = \frac{\Delta \rho\, \mathbf{a}_\rho + \rho \Delta \phi\, \mathbf{a}_\phi + \Delta z \mathbf{a}_z}{\sqrt{\Delta \rho^2 + \rho^2 \Delta \phi^2 + \Delta z^2}} \tag{3.6}$$

In spherical coordinates, shown in Fig. 3.3, the differential coordinates are (Δr, $\Delta\theta$, $\Delta\phi$). The spherical coordinate system is exactly the same system which is used to describe the geography of the earth. q-p-s in this figure is a line of longitude passing through the point with position vector $\mathbf{r}$. On this circle r and ϕ are constant. a-b-c is a line of latitude passing through our point and on it r and θ are constant. The element Δr is a linear element which is in the direction of the unit vector $\mathbf{a}_r$ and an extension of the r coordinate keeping θ and ϕ constant. $r\Delta\theta$ is a small element of the line of longitude q-p-s. Finally, the differential linear element $r \sin\theta \Delta\phi$ is an element of the line of latitude a-b-c.

The general differential linear element, the vector **AB** of Fig. 3.1 is

$$\Delta \mathbf{l} \equiv \Delta \mathbf{r} = \Delta r \mathbf{a}_r + r\Delta\theta\, \mathbf{a}_\theta + r \sin(\theta)\, \Delta\phi\, \mathbf{a}_\phi \tag{3.7}$$

Its length is

$$\Delta l \equiv |\Delta \mathbf{r}| = \sqrt{\Delta r^2 + r^2\Delta\theta^2 + r^2 \sin^2(\theta)\, \Delta\phi^2} \tag{3.8}$$

and the unit tangent vector to the curve at this point is

$$\hat{\mathbf{t}} = \frac{\Delta r \mathbf{a}_r + r\Delta\theta\, \mathbf{a}_\theta + r \sin(\theta)\, \Delta\phi\, \mathbf{a}_\phi}{\sqrt{\Delta r^2 + r^2\Delta\theta^2 + r^2 \sin^2(\theta)\, \Delta\phi^2}} \tag{3.9}$$

 See `LengthOfLine.m` in Chapter 3

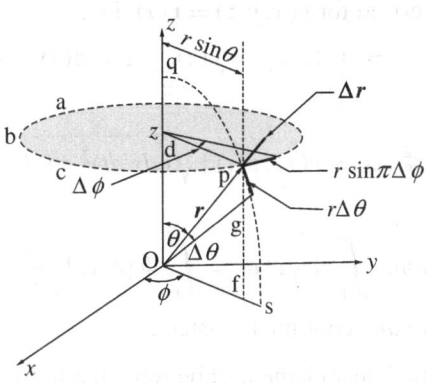

Fig. 3.3 Differential linear elements in spherical coordinates

3.2.1 Line Integral

Given a scalar or vector field, $\Phi(\mathbf{r})$ or $\mathbf{F}(\mathbf{r})$ respectively, we can define two types of line integrals:

$$\int_A^B \Phi(\mathbf{r})\, dl \quad \text{the scalar line integral}$$

and $\displaystyle\int_A^B \mathbf{F}(\mathbf{r}) \cdot d\mathbf{l}$ the vector line integral

$$(3.10)$$

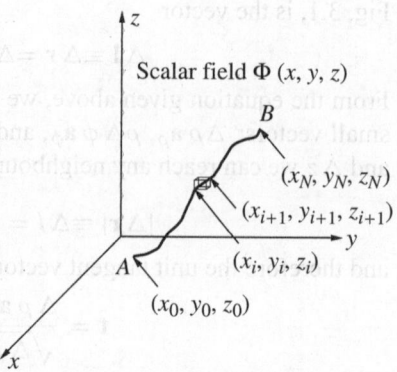

Fig. 3.4 The scalar line integral

To understand the scalar line integral let us look at Fig. 3.4.

In reference to this figure, we wish to compute the integral

$$\int_A^B \Phi(\mathbf{r})\, dl$$

over the curve $A \to B$. To do this, we divide the curve into $N+1$ number of points starting from $\mathbf{r}_0 = (x_0, y_0, z_0)$ to $\mathbf{r}_N = (x_N, y_N, z_N)$. Two typical points on this curve are shown, $\mathbf{r}_i \equiv (x_i, y_i, z_i)$ and $\mathbf{r}_{i+1} \equiv (x_{i+1}, y_{i+1}, z_{i+1})$. The integral is computed according to

$$\lim_{N \to \infty} \left\{ \sum_{i=0}^{i=N-1} \Phi(x_i, y_i, z_i)\, \Delta l_i \right\} \qquad (3.11)$$

where
$$\Delta l_i = \left|(x_{i+1} - x_i)\,\mathbf{a}_x + (y_{i+1} - y_i)\,\mathbf{a}_y + (z_{i+1} - z_i)\,\mathbf{a}_z\right| \qquad (3.12)$$

$$= \sqrt{(x_{i+1} - x_i)^2 + (y_{i+1} - y_i)^2 + (z_{i+1} - z_i)^2} \qquad (3.13)$$

In the limit as $N \to \infty$, the sum (Eqn (3.11)) becomes an integral:

$$\lim_{N \to \infty} \left\{ \sum_{i=0}^{i=N-1} \Phi(x_i, y_i, z_i)\, \Delta l_i \right\} = \int_A^B \Phi(\mathbf{r})\, dl$$

$$= \int_A^B \Phi(x, y, z)\, \sqrt{dx^2 + dy^2 + dz^2} \qquad (3.14)$$

If the curve is parameterised for $\mathbf{r}(x, y, z) = \mathbf{r}(t)$, i.e.,

$$x = x(t), \quad y = y(t), \quad z = z(t) \qquad (3.15)$$

Then

$$\int_A^B \Phi(\mathbf{r})\, dr = \int_A^B \Phi[x(t), y(t), z(t)]\, \sqrt{(dx/dt)^2 + (dy/dt)^2 + (dz/dt)^2}\, dt$$

$$(3.16)$$

or in short-hand notation: $\displaystyle\int_A^B \Phi(\mathbf{r})\, dl = \int_A^B \Phi[\mathbf{r}(t)] \left|\frac{d\mathbf{r}}{dt}\right| dt \qquad (3.17)$

This equation applies to *any* coordinate system.

Example 3.1 Find the linear element $d\mathbf{l}$ between the following points:

1. $x = 1, y = 1, z = 1$ and $x = 1, y = 1, z = 1.001$

2. $\rho = 1, \phi = 1, z = 1$ and $\rho = 1, \phi = 1.01, z = 1.001$
3. $r = 1, \theta = 1, \phi = 1$ and $r = 1, \theta = 1.01, \phi = 1.001$

Solution

1. **Step 1.** From the question, it is clear that we are dealing with rectangular coordinates. By inspection, $\Delta x = 0$; $\Delta y = 0$; $\Delta z = 0.001$

   Step 2. Hence $\Delta \mathbf{l} = 0.001 \mathbf{a}_z$

2. **Step 1.** Since we are dealing with cylindrical coordinates

$$\Delta \rho = 0; \quad \Delta \phi = 0.01; \quad \Delta z = 0.001$$

   Step 2. The differential linear elements are (since $\rho = 1$)

$$\Delta \rho = 0; \quad \rho \Delta \phi = 0.01; \quad \Delta z = 0.001$$

   Step 3. Therefore $\Delta \mathbf{l} = 0.01 \mathbf{a}_\phi + 0.001 \mathbf{a}_z$

3. **Step 1.** In spherical coordinates,

$$\Delta r = 0; \quad \Delta \theta = 0.01; \quad \Delta \phi = 0.001$$

   Step 2. Since $r = 1$, $\theta = 1$, and $\phi = 1$

$$\Delta r = 0; \quad r \Delta \theta = 0.01; \quad r \sin\theta \Delta\phi = 8.415 \times 10^{-4}$$

   Step 3. Therefore $\Delta \mathbf{l} = 0.01 \mathbf{a}_\theta + 8.415 \times 10^{-4} \mathbf{a}_\phi$

Example 3.2 Find the length of the straight line from $A(0,0,0)$ to $B(1,1,1)$.

Solution

Step 1. First we need to find the parametric equation of the straight line. By the methods we have discussed in Section 2.3.3, in Chapter 2, this will be

$$x = t, \; y = t, \; z = t$$

As a check, we notice that at $t = 0$ the coordinates are $(0,0,0)$ and at $t = 1$ the coordinates become $(1,1,1)$.

Step 2. The length of the line is given by $\int_A^B dl$ where $A(0,0,0)$ is the point when $t = 0$, and the point $B(1,1,1)$ is reached when $t = 1$. Using Eqn (3.17) we can identify by inspection that

$$\Phi(\mathbf{r}) = 1, \; \mathbf{r}(t) = t\mathbf{a}_x + t\mathbf{a}_y + t\mathbf{a}_z$$

Then $\dfrac{d\mathbf{r}}{dt} = \mathbf{a}_x + \mathbf{a}_y + \mathbf{a}_z$ and $\left| \dfrac{d\mathbf{r}}{dt} \right| = \sqrt{3}$

Step 3. We substitute these values in the line integral. Therefore,

$$\text{Required length} = \int_0^1 \sqrt{3} dt = \sqrt{3}$$

Example 3.3 Find the length of one turn of the helix

$$\rho = 1, \quad \phi = 2\pi t, \quad z = 2t$$

Solution

Step 1. Just as in the previous example, one turn implies that t goes from $t = 0$ to $t = 1$. Therefore, ϕ goes from $\phi = 0$ to $\phi = 2\pi$. Then using Eqn (3.17), we find that

$$\Phi(\mathbf{r}) = 1$$

or

$$\frac{d\mathbf{r}(t)}{dt} = \frac{d\rho}{dt}\mathbf{a}_\rho + \rho\frac{d\phi}{dt}\mathbf{a}_\phi + \frac{dz}{dt}\mathbf{a}_z = 0\mathbf{a}_\rho + (1)(2\pi)\mathbf{a}_\phi + 2\mathbf{a}_z$$

Hence $\left|\dfrac{d\mathbf{r}(t)}{dt}\right| = \left|0\mathbf{a}_\rho + (1)(2\pi)\mathbf{a}_\phi + 2\mathbf{a}_z\right| = \sqrt{4\pi^2 + 4}$

Step 2. Therefore $\displaystyle\int_A^B dl = \int_A^B \left|\frac{d\mathbf{r}}{dt}\right| dt = \int_{t=0}^{t=1} \sqrt{4\pi^2 + 4}\ dt = \sqrt{4\pi^2 + 4}$

We now proceed to the vector variety of the line integral. By scrutinising Fig. 3.5, the vector line integral $\int_A^B \mathbf{F}(\mathbf{r}) \cdot d\mathbf{r}$ can be written as the limit of a sum:

$$\lim_{N\to\infty} \left\{ \sum_{i=0}^{i=N-1} \mathbf{F}(x_i, y_i, z_i) \cdot \Delta\mathbf{l}_i \right\} \tag{3.18}$$

where $\quad \Delta\mathbf{l}_i = (x_{i+1} - x_i)\mathbf{a}_x + (y_{i+1} - y_i)\mathbf{a}_y + (z_{i+1} - z_i)\mathbf{a}_z \tag{3.19}$

or

$$\lim_{N\to\infty} \left\{ \sum_{i=0}^{i=N-1} \mathbf{F}(x_i, y_i, z_i) \cdot \Delta\mathbf{l}_i \right\}$$

$$= \int_A^B \mathbf{F}(\mathbf{r}) \cdot d\mathbf{l} = \int_A^B (F_x\mathbf{a}_x + F_y\mathbf{a}_y + F_z\mathbf{a}_z) \cdot (dx\mathbf{a}_x + dy\mathbf{a}_y) \tag{3.20}$$

or using the parametric form:

$$\int_A^B \mathbf{F}(\mathbf{r}) \cdot d\mathbf{r} = \int_A^B \left(F_x\frac{dx}{dt} + F_y\frac{dy}{dt} + F_z\frac{dz}{dt} \right) dt \tag{3.21}$$

which can be written as

$$\int_A^B \mathbf{F}(\mathbf{r}) \cdot d\mathbf{l} = \int_A^B \mathbf{F} \cdot \frac{d\mathbf{r}}{dt} dt \tag{3.22}$$

An example of the vector line integral is to find the 'work' done by a force in displacing a particle in a field. Thus for instance,

$$W = \int_A^B \mathbf{F}(\mathbf{r}) \cdot d\mathbf{l} \tag{3.23}$$

is the work done by a variable force $\mathbf{F}$ exerted on a particle. Note that the expression is the integral generalisation of the formula

$$W = \overrightarrow{\text{Force}} \cdot \overrightarrow{\text{Distance}}$$

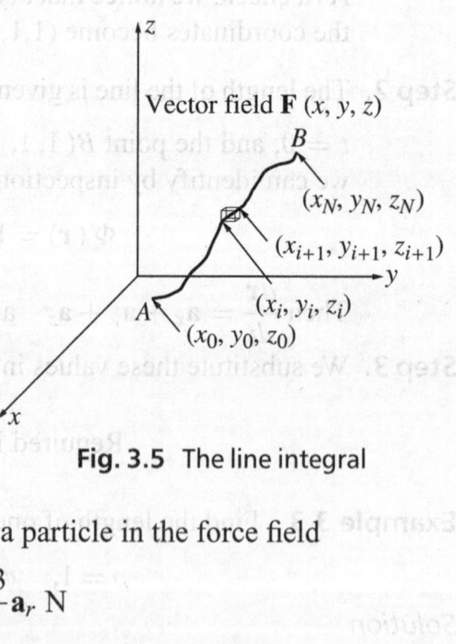

Vector field $\mathbf{F}(x, y, z)$

B

(x_N, y_N, z_N)

$(x_{i+1}, y_{i+1}, z_{i+1})$

A

(x_i, y_i, z_i)

(x_0, y_0, z_0)

Fig. 3.5 The line integral

Example 3.4 Find the work done to move a particle in the force field

$$\mathbf{F}(\mathbf{r}) = -\frac{103}{r^2}\mathbf{a}_r\ \text{N}$$

along the straight line from $\mathbf{r} = (10, 0, 0)$ m in spherical coordinates to $\mathbf{r} = (100, 0, 0)$ m.

Solution

Step 1. The straight line can be characterised by the parametric set of equations

$$r = 10t, \; \theta = 0, \; \phi = 0 \tag{3.24}$$

where t varies from $t = 1$ to $t = 10$.

Step 2. By examining Eqn (3.22) we can identify (in the spherical coordinate system) that

$$\mathbf{F}(\mathbf{r}) = -\frac{103}{r^2}\mathbf{a}_r = -\frac{103}{100t^2}\mathbf{a}_r, \quad \text{for } 1 \le t \le 10 \tag{3.25}$$

$$\frac{d\mathbf{r}}{dt} = \frac{dr}{dt}\mathbf{a}_r + r\frac{d\theta}{dt}\mathbf{a}_\theta + r\sin\theta\,\frac{d\phi}{dt}\mathbf{a}_\phi = 10\mathbf{a}_r$$

Step 3. Hence $\displaystyle\int_A^B \mathbf{F}(\mathbf{r}) \cdot d\mathbf{r} = \int_A^B \mathbf{F}(\mathbf{r}) \cdot \frac{d\mathbf{r}}{dt}dt = -\int_{t=1}^{t=10}\left(\frac{1030}{100t^2}\right)dt$

The integral can now be evaluated

$$-\int_{t=1}^{t=10}\left(\frac{1030}{100t^2}\right)dt = \left[\frac{1030}{100t}\Big|_{t=1}^{t=10}\right] = \frac{1030}{1000} - \frac{1030}{100}$$

$$= 1.03 - 10.30 = -9.27 \text{ J}$$

Example 3.5 Given the vector field

$$\mathbf{G} = y\mathbf{a}_x - 2.5x\mathbf{a}_y + 3z\mathbf{a}_z$$

find the line integral of $\mathbf{G}$ along the straight line from $(1,1,1)$ to $(0,0,0)$.

Solution

Step 1. The parametric equations of the straight line are

$$x = 1 - t; \; y = 1 - t; \; z = 1 - t, \quad 0 \le t \le 1$$

Step 2. The line integral is

$$L = \int \mathbf{G} \cdot d\mathbf{l}$$

$$= \int_{(1,1,1)}^{(0,0,0)} \mathbf{G} \cdot (\mathbf{a}_x dx + \mathbf{a}_y dy + \mathbf{a}_z dz) = \int_{(1,1,1)}^{(0,0,0)} (y dx - 2.5x dy + 3z dz)$$

$$= \int_{t=0}^{1} [(1-t)\,d(1-t) - 2.5(1-t)\,d(1-t) + 3(1-t)\,d(1-t)]$$

$$= \int_{t=0}^{1} [(1.5 - 1.5t)\,d(-t)] = \left[\frac{1.5t^2}{2} - 1.5t\right]_{t=0}^{1} = -0.75$$

Integration in Ancient Times

Integration could have possibly been used to calculate the volume of frustum of square pyramid as exhibited in the Moscow Mathematical Papyrus of ancient Egypt perhaps written around 1850 BC. Archimedes used a method similar to integration to find the area of a circle by inscribing the circle with a polygon of a greater and greater number of sides. Eudoxus (about 370 BC), systematised such a method applying it to volumes. Similar methods were developed in ancient times in China and India. The Indian mathematician Aryabhata used a similar method to find the volume of a cube.

3.3 | Differential Element of a Surface

Before we take a look at surface integrals we must understand the concept of a small element of area. Looking at Fig. 3.6, *ABCD* is some curved surface on which lies a small element of area *abcd*. The element of area of magnitude ΔS is so small that though it lies on the curved surface, it can be considered locally as plane. This is like the case of a plane on the surface of the earth: though the earth is spherical, the plane appears flat. Since the surface *abcd* is plane-like, it has a unit vector $\hat{S}$ associated with it which is perpendicular to the elemental surface area. The surface element thus has vectorial properties. The magnitude of ΔS is

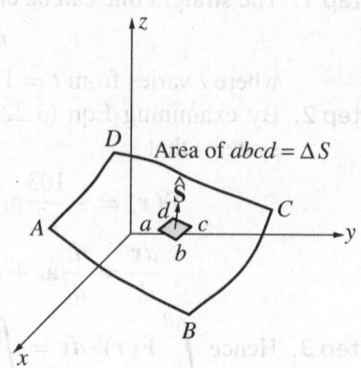

Fig. 3.6 Differential surface element

$$|\Delta \mathbf{S}| = \Delta S \tag{3.26}$$

while the unit vector associated with it is perpendicular to the surface at the point **r**.

$$\Delta \hat{\mathbf{S}} = \hat{\mathbf{S}}(\mathbf{r}) \tag{3.27}$$

Examples of such surfaces are given in Figs 3.7 and 3.8.

In rectangular coordinates one such basic surface element would be $\Delta S = \Delta x \Delta y$ and the unit vector associated with this element would be $\Delta \hat{\mathbf{S}} = \hat{\mathbf{S}} = \mathbf{a}_z$. Such an element would lie on a plane parallel to the *x-y* plane. Another example of a surface element lying on a plane parallel to the *y-z* plane would be $\Delta S = \Delta y \Delta z$ and its direction would be $\hat{\mathbf{S}} = \mathbf{a}_x$. These elements are shown in Fig. 3.7.

Similarly, in cylindrical coordinates the basic surface elements (ΔS) are multiplications of Δz, $\Delta \rho$, and $\rho \Delta \phi$. Figure 3.8 shows these surface elements.

Finally, in spherical coordinates the elemental lengths are Δr, $r \Delta \theta$, and $r \sin\theta \Delta \phi$. ΔS consists of multiplication of these elements. For example in Fig. 3.9, a surface element has been shown lying on the surface of a sphere. The linear elements are $r \sin\theta \Delta \phi$ and $r \Delta \theta$ which are perpendicular to each other, and hence $\Delta S = (r_0 \sin\theta \Delta \phi)(r_0 \Delta \theta)$ and $\hat{\mathbf{S}} = \mathbf{a}_r$. In another surface element, shown in Fig. 3.10

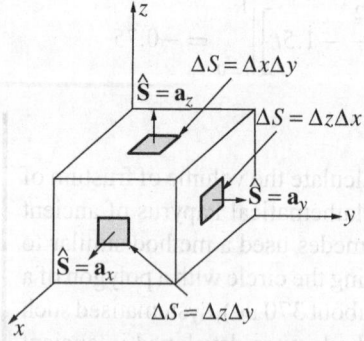

Fig. 3.7 Differential surface elements in rectangular coordinates

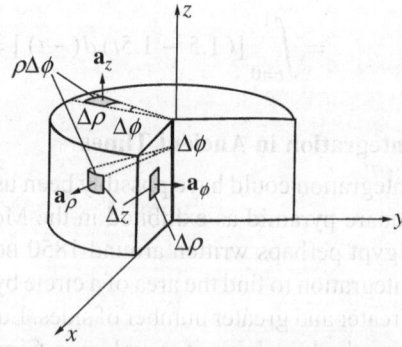

Fig. 3.8 Differential surface elements in cylindrical coordinates

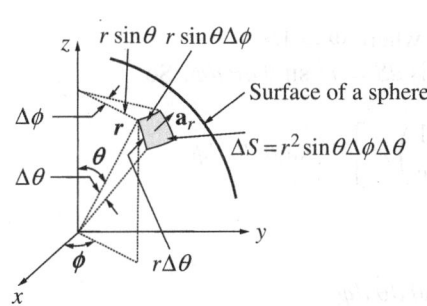

Fig. 3.9 Differential surface element in spherical coordinates of the surface of a sphere

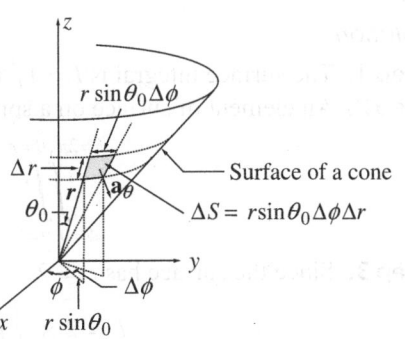

Fig. 3.10 Surface element lying on the surface of a cone

the two linear elements, perpendicular to each other, are $r \sin\theta_0 \, \Delta\phi$ and Δr, $\Delta S = (r \sin\theta_0 \, \Delta\phi)(\Delta r)$ and $\hat{\mathbf{S}} = \mathbf{a}_\theta$.

3.3.1 Surface Integral

The concept of a surface integral may now be introduced and in this section we will be concerned with only simple surface integrals. Referring to Fig. 3.11, we calculate the scalar surface integral

$$\iint_{ABCD} \Phi(\mathbf{r}) \, dS$$

where $ABCD$ is the surface S over which the integral is evaluated and $\Phi(\mathbf{r})$ is a scalar field.

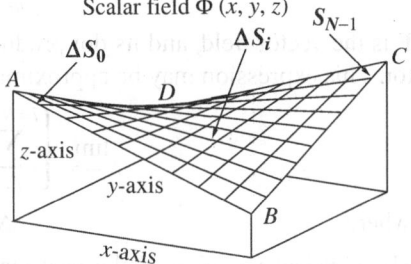

Fig. 3.11 Calculation of the scalar surface integral

The surface is first divided into a number of surface elements N starting with ΔS_0 to ΔS_{N-1}. A typical surface element ΔS_i is also shown. The integral is computed according to the sum

$$\lim_{N \to \infty} \left\{ \sum_{i=0}^{i=N-1} \Phi(\mathbf{r}_i) \, \Delta S_i \right\} \tag{3.28}$$

where $\Phi(\mathbf{r}_i)$ is the scalar field evaluated at the centre of the surface element, ΔS_i. (ΔS_i is a typical surface element, as discussed in the previous section.)

In the limit as $N \to \infty$ the sum becomes an integral:

$$\iint_{ABCD} \Phi(\mathbf{r}) \, dS \tag{3.29}$$

This expression applies to any coordinate system.

Example 3.6 For $\Phi(\mathbf{r}) = 1/r$ (spherical coordinates), find the surface integral of Φ over the whole surface of a sphere with radius equal to 2.

Solution

Step 1. The surface integral is $I = \iint \Phi \, dS$ where $\Phi = 1/r$.

Step 2. An element of surface on a sphere is $dS = r^2 \sin\theta \, d\theta \, d\phi$. So

$$I = \int\int\limits_{\phi=0,\theta=0}^{\phi=2\pi,\theta=\pi} \left[\left(\frac{1}{r}\right) r^2 \right]_{r=2} \sin\theta \, d\theta \, d\phi$$

Step 3. Since the surface has $r = 2$

$$I = 2 \times \iint \sin\theta \, d\theta \, d\phi$$

$$= 2 \left(\phi \,|_{\phi=0}^{\phi=2\pi} \right) \times \left(-\cos\theta \,|_{\theta=0}^{\theta=\pi} \right)$$

$$= -2 \times (2\pi) \times (-1 - 1) = 8\pi$$

The vector surface integral may be calculated using the same principles as:

$$\iint_{ABCD} \mathbf{F} \cdot d\mathbf{S} \tag{3.30}$$

$\mathbf{F}$ is the vector field, and its dot product is taken with the differential surface vector. This expression may be approximated by (see Fig. 3.12):

$$\lim_{N \to \infty} \left\{ \sum_{i=0}^{i=N-1} \mathbf{F}(\mathbf{r}_i) \cdot \Delta \mathbf{S}_i \right\} \tag{3.31}$$

where

$$\Delta \mathbf{S}_i = \Delta S_i \hat{\mathbf{S}}_i \tag{3.32}$$

The physical meaning of the vectorial surface integral may be understood from observation of Fig. 3.13 showing the differential flux as the total 'flow' of the field through the elemental surface. The figure shows that the field $\mathbf{F}$ is at an angle θ to the normal $\hat{\mathbf{S}}$. The field can therefore be split into two parts: one along $\hat{\mathbf{S}}$ (perpendicular to the surface) which is $F_\perp \equiv F\cos\theta$ and the other perpendicular to $\hat{\mathbf{S}}$ which is equal to $F_\parallel \equiv F\sin\theta$ (parallel to the surface). The parallel part does not pass through the surface at all, while the perpendicular part passes completely through the surface. Therefore the differential flux, $\Delta\Phi$ is

$$\Delta\Phi = F\cos(\theta)\,\Delta S = \mathbf{F} \cdot \Delta \mathbf{S} \tag{3.33}$$

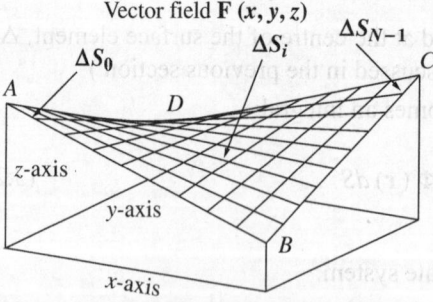

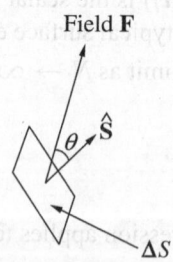

Fig. 3.12 Calculation of the vector surface integral

Fig. 3.13 Differential flux

The total flux through some surface S is given by an integration of the above differential

$$\Phi = \iint_S \mathbf{F} \cdot d\mathbf{S} \tag{3.34}$$

Example 3.7 Find the flux, Φ, through the surface of a sphere with unit radius for (i) $\mathbf{F} = (1/r)\, \mathbf{a}_\phi$, (ii) $\mathbf{F} = (1/r)\, \mathbf{a}_\theta$, and (iii) $\mathbf{F} = (1/r)\, \mathbf{a}_r$ in spherical coordinates.

Solution

Step 1. The flux is given by

$$\Phi = \iint_S \mathbf{F} \cdot d\mathbf{S}$$

$d\mathbf{S}$ in all the three cases is

$$d\mathbf{S} = \mathbf{a}_r\, r^2 \sin\theta\, d\theta\, d\phi \Big|_{r=1} = \mathbf{a}_r \sin\theta\, d\theta\, d\phi$$

Step 2. For Case (i)

$$\mathbf{F} \cdot d\mathbf{S} = \left[(1/r)\, \mathbf{a}_\phi\right] \cdot (\mathbf{a}_r r^2 \sin\theta\, d\theta\, d\phi) = 0$$

So the flux is zero.

For Case (ii) also

$$\mathbf{F} \cdot d\mathbf{S} = \left[(1/r)\, \mathbf{a}_\theta\right] \cdot (\mathbf{a}_r r^2 \sin\theta\, d\theta\, d\phi) = 0$$

But in the third case

$$\mathbf{F} \cdot d\mathbf{S} = \left[(1/r)\, \mathbf{a}_r\right] \cdot (\mathbf{a}_r r^2 \sin\theta\, d\theta\, d\phi)$$
$$= r \sin\theta\, d\theta\, d\phi\, |_{r=1}$$
$$= \sin\theta\, d\theta\, d\phi$$

The flux is therefore

$$\Phi = \int_{\phi=0,\theta=0}^{\phi=2\pi,\theta=\pi} \sin\theta\, d\theta\, d\phi = 4\pi$$

3.3.2 Volume Integral

Before we discuss the volume integral we must understand the meaning of an element of volume. In rectangular coordinates, a volume element is

$$\Delta V = \Delta x \Delta y \Delta z \tag{3.35}$$

and consists of the three linear elements multiplied together. Using the same concept the element of volume in the cylindrical coordinate system is

$$\Delta V = (\Delta \rho)\,(\rho\, \Delta \phi)\,(\Delta z) \tag{3.36}$$

Such a volume element is shown in Fig. 3.14. Similarly, in the spherical coordinate system the volume element is

$$\Delta V = (\Delta r)\,(r\Delta \theta)\,(r \sin\theta \Delta \phi) \tag{3.37}$$

This is shown in Fig. 3.15.

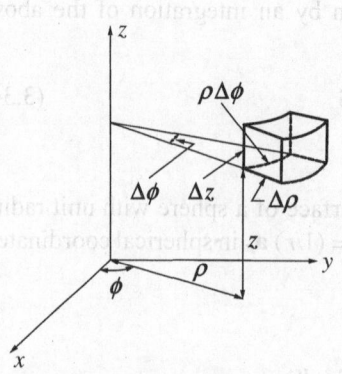

Fig. 3.14 Differential volume element in cylindrical coordinates

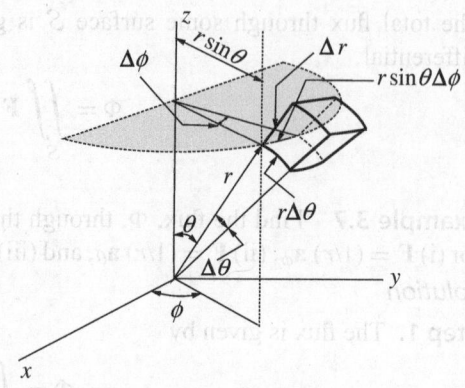

Fig. 3.15 Differential volume element in spherical coordinates

The volume integral usually integrates a scalar field $\Phi(\mathbf{r})$. Using the earlier method of converting a line and a surface into smaller elements dl and ds respectively, volume, $\mathcal{V}$, is also converted into smaller (N) regions having elemental volumes ΔV_i ($i = 0, \ldots, N-1$) and then summing over i, the volume integral is

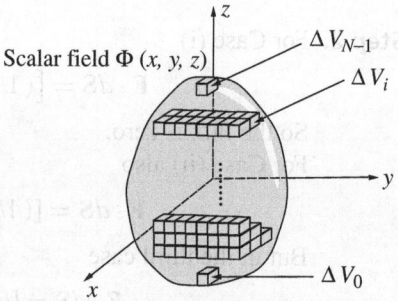

Scalar field $\Phi(x, y, z)$

Fig. 3.16 Calculation of the volume integral

$$\iiint_{\mathcal{V}} \Phi(\mathbf{r})\, dV = \lim_{N \to \infty} \left\{ \sum_{i=0}^{i=N-1} \Phi(\mathbf{r}_i)\, \Delta V_i \right\}$$

(3.38)

This is shown in Fig. 3.16. Though the elemental volumes are shown to be small cubes, they can be of arbitrary shapes, as long as they fit next to each other and cover the whole volume.

Example 3.8 Find the volume integral I of a scalar field $V = 1/r$ over a sphere with unit radius in spherical coordinates.

Solution

Step 1. The volume integral I is given by

$$I = \iiint_{\mathcal{V}} (1/r)\, dV$$

where $\mathcal{V}$ is the region occupied by a sphere of unit radius.

Step 2. A volume element in spherical coordinates is $dV = r^2 \sin\theta\, d\theta\, d\phi\, dr$. So

$$I = \iiint_{\mathcal{V}} (1/r)\, (r^2 \sin\theta\, d\theta\, d\phi\, dr) = \iiint_{\mathcal{V}} r \sin\theta\, d\theta\, d\phi\, dr$$

$$= \left(\frac{r^2}{2}\right)\Big|_{r=0}^{r=1} \times (-\cos\theta)\,|_{\theta=0}^{\theta=\pi} \times (\phi)\,|_{\phi=0}^{\phi=2\pi} = 2\pi$$

3.4 | Differential Calculus Concepts

Once the notion of a field has been understood, we can proceed to apply calculus concepts to fields. First let us take a problem in one-dimension in a very simple way. Let $f(x)$ be a function, which we evaluate at two points $f(x_0)$ and $f(x_1)$ with $x_1 > x_0$ but x_1 almost equal to x_0. Because $x_1 \approx x_0$ therefore $f(x_0) \approx f(x_1)$. For example let the function $f(x)$ be x^3 with $x_0 = 3$ and $x_1 = 3.01$ then

$$f(x_0) = f(3) = 3^3 = 27$$

and $$f(x_1) = f(3.01) = 3.01^3 = 27.271$$

If we set $\delta x = x_1 - x_0$, then

$$\delta x = x_1 - x_0 = 3.01 - 3 = 0.01$$

Then by the same token we can call $\delta [f(x)] = f(x_1) - f(x_0)$.

Thus, $\delta [f(x)] = f(x_1) - f(x_0) = 27.271 - 27.0 = 0.271$ (3.39)

By convention, $\delta [f(x)]/\delta x$ is an approximation to the derivative of $f(x)$ at $x = 3$. Here by *numerical computation*,

$$\delta [f(x)]/\delta x = 0.271/0.01 = 27.1$$

The exact answer, as we know is 27. We can do the same computation making x_1 to be still closer to x_0 and we will get a result which is still closer to the exact result. For instance,

$$x_0 = 3; \; x_1 = 3.0001; \; \delta x = 0.0001$$

$$f(x_0) = 27; \; f(x_1) = 27.0027001; \; \delta [f(x)] = 0.0027001$$

$$\delta [f(x)]/\delta x = 0.0027001/0.0001 = 27.001 \approx 27.0$$

Let us proceed to understand the meaning of the *partial derivative*. Look at the scalar field

$$f(x,y,z) = xe^{-y} \sin(z)$$

which is a function of three variables: x, y, and z. We can evaluate the function $f(x,y,z)$ at (x_0, y_0, z_0) and (x_1, y_0, z_0). The two points are assumed to be very 'close' (i.e. $x_1 > x_0$ but $x_1 \approx x_0$) to each other as shown in Fig. 3.17. If we call

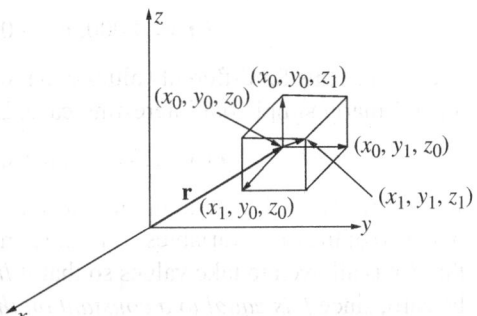

Fig. 3.17 Two neighbouring points

$$\delta_x [f(x,y,z)] = f(x_1,y_0,z_0) - f(x_0,y_0,z_0)$$

and $$\delta x = x_1 - x_0$$

then $$\frac{\delta_x [f(x,y,z)]}{\delta x} = \frac{f(x_1,y_0,z_0) - f(x_0,y_0,z_0)}{x_1 - x_0}$$

In the limit as $x_1 \to x_0$ (which means that x_1 becomes closer and closer to x_0 but x_1 *is never equal to x_0*) we can see that the definition leads to the partial derivative:

$$\frac{\partial f(x,y,z)}{\partial x}\bigg|_{x_0,y_0,z_0} = \frac{f(x_1,y_0,z_0) - f(x_0,y_0,z_0)}{x_1 - x_0}\bigg|_{x_1 \to x_0}$$

In the same manner we can define other partial derivatives:

$$\frac{\partial f(x,y,z)}{\partial y}\bigg|_{x_0,y_0,z_0} = \frac{f(x_0,y_1,z_0)-f(x_0,y_0,z_0)}{y_1-y_0}\bigg|_{y_1\to y_0}$$

and

$$\frac{\partial f(x,y,z)}{\partial z}\bigg|_{x_0,y_0,z_0} = \frac{f(x_0,y_0,z_1)-f(x_0,y_0,z_0)}{z_1-z_0}\bigg|_{z_1\to z_0}$$

We can put partial derivatives to use in the definition of 'nabla': ∇. Nabla is a vector operator whose definition is

$$\nabla \equiv \frac{\partial}{\partial x}\mathbf{a}_x + \frac{\partial}{\partial y}\mathbf{a}_y + \frac{\partial}{\partial z}\mathbf{a}_z \tag{3.40}$$

For the function $f(x,y,z)$

$$\delta f(x,y,z)\,|_{x_0,y_0,z_0}$$

$$= \frac{\partial f}{\partial x}\delta x + \frac{\partial f}{\partial y}\delta y + \frac{\partial f}{\partial z}\delta z\bigg|_{x_0,y_0,z_0}$$

$$= \left(\frac{\partial}{\partial x}\mathbf{a}_x + \frac{\partial}{\partial y}\mathbf{a}_y + \frac{\partial}{\partial z}\mathbf{a}_z\right)f(x,y,z)\cdot(\delta x\mathbf{a}_x + \delta y\mathbf{a}_y + \delta z\mathbf{a}_z)\bigg|_{x_0,y_0,z_0}$$

$$= \nabla f(x,y,z)\cdot\delta\mathbf{r}|_{x_0,y_0,z_0} \tag{3.41}$$

The vector $\delta\mathbf{r}$ of course is purely arbitrary. For example, one particular value of $\delta\mathbf{r}$ may be

$$\delta\mathbf{r} = 0.005\mathbf{a}_x + 0.00007\mathbf{a}_y + 0\mathbf{a}_z$$

Or another value may be

$$\delta\mathbf{r} = 0.0005\mathbf{a}_x + 0\mathbf{a}_y + 0.00009\mathbf{a}_z$$

and so on. Thus for different values of $\delta\mathbf{r}$ we get different values of δf. Let us take a particularly simple but interesting case. Let us look at the surface,

$$f(x,y,z) = f(x_0,y_0,z_0) \equiv (\text{constant})$$

where (x_0,y_0,z_0) is a point on the surface. Note that this defines a surface because only *two* of the three variables, x,y, and z may be independently specified. Suppose that $\delta\mathbf{r}$ is allowed to take values so that it *lies only on this surface*. Then δf, must be zero, since *f is equal to a constant on the surface*. More clearly:

$$\delta f(x,y,z)\,|_{x_0,y_0,z_0} = \nabla f(x,y,z)\cdot\delta\mathbf{r}|_{x_0,y_0,z_0} = 0$$

when $\delta\mathbf{r}$ lies on the surface $f(x,y,z) = f(x_0,y_0,z_0)$.

After a little reflection it is also clear that $\delta\mathbf{r}$ lying on $f(x,y,z) = f(x_0,y_0,z_0)$ really means that it lies on the tangent plane to the given surface at (x_0,y_0,z_0). And since

$$\nabla f(x,y,z)\cdot\delta\mathbf{r}|_{x_0,y_0,z_0} = 0$$

therefore these two vectors are perpendicular to each other. Hence $\nabla f(x,y,z)$ is perpendicular to the tangent plane to the surface at $\mathbf{r}_0 = (x_0,y_0,z_0)$. The mathematical entity ∇f is so important, that it is called the gradient of the scalar field $f(x,y,z)$. In cylindrical coordinates, it would be the gradient of $f(\rho,\phi,z)$, etc.

3.4.1 Del or Nabla Operator

An operator is a mathematical entity which 'operates' on a function or another entity. A very simple example of an operator is the differentiation operator

$$\frac{d}{dt} \tag{3.42}$$

Notice that d/dt in itself has no meaning. When it 'attacks' or operates on a function then only its meaning develops. Thus, when it operates on $x(t)$ then the operation has meaning

$$\frac{d}{dt}[x(t)] \tag{3.43}$$

which obviously is the derivative of $x(t)$. Another important point to be remembered is that an operator may operate on another operator to give us a completely new operator. For example,

$$\frac{d}{dt}\left[\frac{d}{dt}\right] = \frac{d^2}{dt^2} \tag{3.44}$$

which may now operate on a function to give us a double derivative. Another example is that of a matrix

$$\begin{pmatrix} 1 & 1 \\ 2 & 1 \end{pmatrix} \tag{3.45}$$

When this matrix operates on a vector it gives us another vector

$$\begin{pmatrix} 1 & 1 \\ 2 & 1 \end{pmatrix}\begin{pmatrix} 5 \\ 2 \end{pmatrix} = \begin{pmatrix} 7 \\ 12 \end{pmatrix} \tag{3.46}$$

Or it may operate on another matrix (an operator) to give us a new matrix (operator)

$$\begin{pmatrix} 1 & 1 \\ 2 & 1 \end{pmatrix}\begin{pmatrix} 3 & -2 \\ 2 & -4 \end{pmatrix} = \begin{pmatrix} 5 & -6 \\ 8 & -8 \end{pmatrix} \tag{3.47}$$

Let us go on to the case of the Del or Nabla operator which plays an extremely fundamental role in electromagnetic theory. The Del operator in rectangular coordinates takes the form

$$\nabla \equiv \mathbf{\nabla} = \frac{\partial}{\partial x}\mathbf{a}_x + \frac{\partial}{\partial y}\mathbf{a}_y + \frac{\partial}{\partial z}\mathbf{a}_z \tag{3.48}$$

Notice that the vector notation may be omitted without loss of understanding. ∇ may operate on mathematical entities in various ways:

1. ∇ (a vector) may operate on the scalar field Φ to give a vector field

$$\nabla[\Phi] \equiv \nabla\Phi \tag{3.49}$$

which is called the 'gradient (or grad) of Φ'. In rectangular coordinates, the gradient takes the form

$$
\begin{aligned}
\nabla\Phi &= \left(\frac{\partial}{\partial x}\mathbf{a}_x + \frac{\partial}{\partial y}\mathbf{a}_y + \frac{\partial}{\partial z}\mathbf{a}_z\right)\Phi\,(x,y,z) \\
&= \frac{\partial\Phi}{\partial x}\mathbf{a}_x + \frac{\partial\Phi}{\partial y}\mathbf{a}_y + \frac{\partial\Phi}{\partial z}\mathbf{a}_z
\end{aligned} \tag{3.50}
$$

2. The Del operator may operate on a vector field through the dot (·) product. In this case the result would be a scalar field. In rectangular coordinates the product becomes

$$\nabla \cdot \mathbf{F} = \left(\frac{\partial}{\partial x} \mathbf{a}_x + \frac{\partial}{\partial y} \mathbf{a}_y + \frac{\partial}{\partial z} \mathbf{a}_z \right) \cdot (F_x \mathbf{a}_x + F_y \mathbf{a}_y + F_z \mathbf{a}_z)$$

$$= \frac{\partial F_x}{\partial x} + \frac{\partial F_y}{\partial y} + \frac{\partial F_z}{\partial z} \qquad (3.51)$$

which is called the 'divergence' (or div) of the vector $\mathbf{F}$.

3. ∇ may operate on a vector through the vector or cross (×) product to give us another vector field. The product in rectangular coordinates is

$$\nabla \times \mathbf{F} = \begin{vmatrix} \mathbf{a}_x & \mathbf{a}_y & \mathbf{a}_z \\ \dfrac{\partial}{\partial x} & \dfrac{\partial}{\partial y} & \dfrac{\partial}{\partial z} \\ F_x & F_y & F_z \end{vmatrix}$$

$$= \mathbf{a}_x \left(\frac{\partial F_z}{\partial y} - \frac{\partial F_y}{\partial z} \right) + \mathbf{a}_y \left(\frac{\partial F_x}{\partial z} - \frac{\partial F_z}{\partial x} \right) + \mathbf{a}_z \left(\frac{\partial F_y}{\partial x} - \frac{\partial F_x}{\partial y} \right) \qquad (3.52)$$

which is called the 'circulation' (or curl) of the vector field $\mathbf{F}$.

4. The ∇ operator may operate on itself also in two ways, the first of which is

$$\nabla \cdot \nabla = \nabla^2 = \left(\frac{\partial}{\partial x} \mathbf{a}_x + \frac{\partial}{\partial y} \mathbf{a}_y + \frac{\partial}{\partial z} \mathbf{a}_z \right) \cdot \left(\frac{\partial}{\partial x} \mathbf{a}_x + \frac{\partial}{\partial y} \mathbf{a}_y + \frac{\partial}{\partial z} \mathbf{a}_z \right)$$

$$= \frac{\partial^2}{\partial x^2} + \frac{\partial^2}{\partial y^2} + \frac{\partial^2}{\partial z^2} \qquad (3.53)$$

which is a scalar operator and is called the 'Laplacian.' The Laplacian may operate on a scalar field to give us another scalar field

$$\nabla^2 \Phi = \frac{\partial^2 \Phi}{\partial x^2} + \frac{\partial^2 \Phi}{\partial y^2} + \frac{\partial^2 \Phi}{\partial z^2} \qquad (3.54)$$

The Laplacian may also operate on a vector

$$\nabla^2 \mathbf{A} = \nabla^2 A_x \mathbf{a}_x + \nabla^2 A_y \mathbf{a}_y + \nabla^2 A_z \mathbf{a}_z \qquad (3.55)$$

5. The ∇ operator may operate on itself through the cross product

$$\nabla \times \nabla \Phi = \mathbf{0} \qquad (3.56)$$

which is identically zero! Students may verify this for themselves.

3.4.2 Gradient

The *gradient (or grad)* is defined by the operation of the Del operator on a scalar field

$$\nabla \Phi (\mathbf{r}) \qquad (3.57)$$

The gradient in the rectangular coordinate system when applied to a scalar field which is a function of rectangular coordinates is

$$\nabla \Phi = \frac{\partial \Phi}{\partial x} \mathbf{a}_x + \frac{\partial \Phi}{\partial y} \mathbf{a}_y + \frac{\partial \Phi}{\partial z} \mathbf{a}_z \qquad (3.58)$$

In cylindrical coordinates, the form of $\nabla \Phi$ is

$$\nabla \Phi = \frac{\partial \Phi}{\partial \rho} \mathbf{a}_\rho + \frac{1}{\rho} \frac{\partial \Phi}{\partial \phi} \mathbf{a}_\phi + \frac{\partial \Phi}{\partial z} \mathbf{a}_z \qquad (3.59)$$

while in spherical coordinates $\nabla\Phi$ is

$$\nabla\Phi = \frac{\partial\Phi}{\partial r}\mathbf{a}_r + \frac{1}{r}\frac{\partial\Phi}{\partial\theta}\mathbf{a}_\theta + \frac{1}{r\sin\theta}\frac{\partial\Phi}{\partial\phi}\mathbf{a}_\phi \qquad (3.60)$$

During introduction of grad(Φ) in Section 3.4 [Eqn (3.40)], it was shown that *the gradient of a scalar field at a particular point* (x_0, y_0, z_0) *is perpendicular to the surface* $\Phi(x, y, z) = \Phi(x_0, y_0, z_0)$ ($\equiv$ constant). *This result holds good for all coordinate systems* (Fig. 3.18).

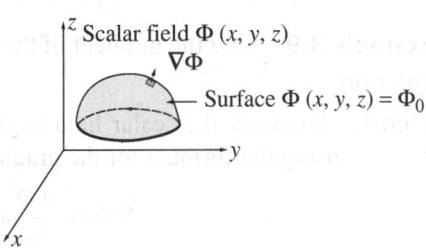

Fig. 3.18 $\nabla\Phi$ shown on the surface $\Phi(x, y, z) = \Phi(x_0, y_0, z_0)$

The second result which is very important is to do with the line integral of the gradient. Let

$$\nabla\Phi = \mathbf{E} \qquad (3.61)$$

be a vector field which we want to integrate between two points $A \equiv (x_i, y_i, z_i)$ and $B \equiv (x_f, y_f, z_f)$ over *any* curve $\mathcal{L}$. It does not matter what the shape of the curve is. Then the line integral

$$\int_{A\atop\mathcal{L}}^{B} [\nabla\Phi \cdot d\mathbf{l}] = \int_{A\atop\mathcal{L}}^{B} \left\{\left[\frac{\partial\Phi}{\partial x}\mathbf{a}_x + \frac{\partial\Phi}{\partial y}\mathbf{a}_y + \frac{\partial\Phi}{\partial z}\mathbf{a}_z\right] \cdot [dx\mathbf{a}_x + dy\mathbf{a}_y + dz\mathbf{a}_z]\right\}$$

$$= \int_{A\atop\mathcal{L}}^{B} \left[\frac{\partial\Phi}{\partial x}dx + \frac{\partial\Phi}{\partial y}dy + \frac{\partial\Phi}{\partial z}dz\right]$$

$$= \int_{A\atop\mathcal{L}}^{B} d\Phi = \Phi(x_f, y_f, z_f) - \Phi(x_i, y_i, z_i) \qquad (3.62)$$

In other words, *the line integral of the gradient of a scalar field is independent of the path of integration*. This is the most important result concerning the gradient.

$$\int_{A\atop\mathcal{L}}^{B} [\nabla\Phi \cdot d\mathbf{l}] = \Phi(B) - \Phi(A)$$

Some other results follow from this result!

1. From Eqn (3.61) we can say that if any field is the gradient of a scalar then its line integral from $A \equiv (x_i, y_i, z_i)$ to $B \equiv (x_f, y_f, z_f)$ is independent of the path of integration, i.e., if

$$\mathbf{E}(\equiv \nabla\Phi) \ then \int_{A\atop\mathcal{L}}^{B} [\mathbf{E} \cdot d\mathbf{l}] \ is \ the \ same \ for \ any \ \mathcal{L} \qquad (3.63)$$

2. When the gradient of a scalar is integrated over a closed curve, i.e., when $A \equiv (x_i, y_i, z_i)$ is equal to $B \equiv (x_f, y_f, z_f)$ (which means that $(x_i, y_i, z_i) = (x_f, y_f, z_f)$)

$$\oint_{\text{any } C} [\nabla\Phi \cdot d\mathbf{l}] = 0 \qquad (3.64)$$

This is true because from Eqn (3.62), $\Phi(x_f, y_f, z_f) = \Phi(x_i, y_i, z_i)$ and hence the result follows.

3. If the closed line integral of a vector field $\mathbf{E}$ is always identically zero then the vector is the gradient of some scalar field, i.e.

$$\text{If } \oint \mathbf{E} \cdot d\mathbf{l} = 0 \text{ then } \mathbf{E} = \nabla\Phi \tag{3.65}$$

Example 3.9 Find the gradient of the scalar field $\Phi = ze^{-\rho}\cos\phi$.

Solution

Step 1. Obviously the scalar field is given in cylindrical coordinates. Therefore using the formula for the gradient in cylindrical coordinates,

$$\nabla\Phi = \frac{\partial \Phi}{\partial \rho}\mathbf{a}_\rho + \frac{1}{\rho}\frac{\partial \Phi}{\partial \phi}\mathbf{a}_\phi + \frac{\partial \Phi}{\partial z}\mathbf{a}_z$$

therefore $\nabla\Phi = -ze^{-\rho}\cos\phi\,\mathbf{a}_\rho - \dfrac{z}{\rho}e^{-\rho}\sin\phi\,\mathbf{a}_\phi + e^{-\rho}\cos\phi\,\mathbf{a}_z$

Example 3.10 For scalar field $\Phi = xyz$, investigate the line integral of $\nabla\Phi$ along the straight line joining the points $(0,0,0)$ and $(1,1,1)$.

Solution

Step 1. Using the formula for the gradient in rectangular coordinates,

$$\nabla\Phi = \frac{\partial \Phi}{\partial x}\mathbf{a}_x + \frac{\partial \Phi}{\partial y}\mathbf{a}_y + \frac{\partial \Phi}{\partial z}\mathbf{a}_z$$

$$= yz\mathbf{a}_x + xz\mathbf{a}_y + xy\mathbf{a}_z$$

Step 2. Since the line integral is to be evaluated on the straight line joining $(0,0,0)$ and $(1,1,1)$

$$I = \int_{\mathcal{L}} \nabla\Phi \cdot d\mathbf{l} = \int_{\mathcal{L}} (yz\mathbf{a}_x + xz\mathbf{a}_y + xy\mathbf{a}_z) \cdot (dx\mathbf{a}_x + dy\mathbf{a}_y + dz\mathbf{a}_z)$$

Step 3. To perform this line integration, we need to use the parametric equations for the straight line

$$x(t) = t; \ y(t) = t; \ z(t) = t$$

giving
$$dx = dt; \ dy = dt; \ dz = dt$$

or
$$I = \int_{\mathcal{L}} (yz\mathbf{a}_x + xz\mathbf{a}_y + xy\mathbf{a}_z) \cdot (dx\mathbf{a}_x + dy\mathbf{a}_y + dz\mathbf{a}_z)$$

$$= \int_{\mathcal{L}} (yzdx + xzdy + xydz)$$

$$= \int_{t=0}^{t=1} (t^2 dt + t^2 dt + t^2 dt)$$

$$= t^3 \Big|_0^1 = 1$$

$$= \Phi(1,1,1) - \Phi(0,0,0)$$

Other results involving the gradient (where ϕ_1 and ϕ_2 are scalar fields and a and b are constants) are

$$\nabla(a\phi_1 + b\phi_2) = a\nabla\phi_1 + b\nabla\phi_2 \tag{3.66}$$

$$\nabla(\phi_1\,\phi_2) = \phi_2\,\nabla\phi_1 + \phi_1\,\nabla\phi_2 \tag{3.67}$$

$$\nabla(\phi_1/\phi_2) = \frac{\phi_2\,\nabla\phi_1 - \phi_1\,\nabla\phi_2}{(\phi_2)^2} \tag{3.68}$$

$$\nabla \times \nabla\phi = 0 \tag{3.69}$$

The last of these is an important result which we will re-examine shortly.

 See `GradOfScalarFldCylCoords.m` and `GradOfScalarFldCartCoords.m` in Chapter 3

3.4.3 Curl

The curl of a vector $\mathbf{A}$ $(\equiv \nabla \times \mathbf{A})$ plays a very important role in electromagnetic theory. In rectangular coordinates the curl of $\mathbf{A}$ is given by

$$\nabla \times \mathbf{A} = \begin{vmatrix} \mathbf{a}_x & \mathbf{a}_y & \mathbf{a}_z \\ \partial/\partial x & \partial/\partial y & \partial/\partial z \\ A_x & A_y & A_z \end{vmatrix}$$

$$= \mathbf{a}_x\left(\partial_y A_z - \partial_z A_y\right) + \mathbf{a}_y\left(\partial_z A_x - \partial_x A_z\right) + \mathbf{a}_z\left(\partial_x A_y - \partial_y A_x\right) \tag{3.70}$$

where $\partial_x \equiv \partial/\partial x$, etc. In other coordinate systems the curl is as given below.

$$\nabla \times \mathbf{A} = \left(\frac{1}{\rho}\frac{\partial A_z}{\partial\phi} - \frac{\partial A_\phi}{\partial z}\right)\mathbf{a}_\rho + \left(\frac{\partial A_\rho}{\partial z} - \frac{\partial A_z}{\partial\rho}\right)\mathbf{a}_\phi + \frac{1}{\rho}\left[\frac{\partial(\rho A_\phi)}{\partial\rho} - \frac{\partial A_\rho}{\partial y}\right]\mathbf{a}_z \tag{3.71}$$

in cylindrical coordinates, and

$$\nabla \times \mathbf{A} = \frac{1}{r\sin\theta}\left[\frac{\partial(\sin\theta\,A_\phi)}{\partial\theta} - \frac{\partial A_\theta}{\partial\phi}\right]\mathbf{a}_r$$

$$+ \frac{1}{r}\left[\frac{1}{\sin\theta}\frac{\partial A_r}{\partial\phi} - \frac{\partial(rA_\phi)}{\partial r}\right]\mathbf{a}_\theta + \frac{1}{r}\left[\frac{\partial(rA_\theta)}{\partial r} - \frac{\partial A_r}{\partial\theta}\right]\mathbf{a}_\phi \tag{3.72}$$

in spherical coordinates.

An important relation for the curl (see Spiegel, 1974 for more detail) is

$$\oint_L \mathbf{A}\cdot d\mathbf{l} = \iint_S (\nabla \times \mathbf{A})\cdot d\mathbf{S} \tag{3.73}$$

where (see Fig. 3.19(a)) L is a closed curve over which the line integral is calculated, and S is the surface enclosed by the curve. The result which we have given is due to Stokes, and therefore *is called Stokes's theorem*.

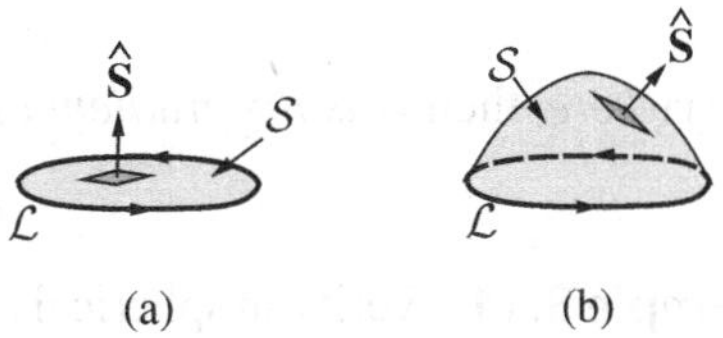

The line integral in Eqn (3.73) is evaluated in the counter-clockwise sense, while the vector associated with the surface is given by the right-hand thumb rule. The fingers curl in the direction of the line integral, while the vector associated with the enclosed surface is in the direction of the thumb.

Fig. 3.19 Properties of the curl. (a) When the surface enclosed is flat. (b) When the surface enclosed is bulging

We can apply the above equation to Fig. 3.19 (a). The figure shows a closed curve $\mathcal{L}$ enclosing a flat surface $\mathcal{S}$. The unit vector $\hat{\mathbf{S}}$ associated with the surface is given by the right-hand thumb rule. Then

$$\oint_{\mathcal{L}} \mathbf{A} \cdot d\mathbf{l} = \iint_{\mathcal{S}} (\nabla \times \mathbf{A}) \cdot d\mathbf{S}$$

We may also apply the above equation to the case where the surface concerned is bulging rather than flat as is the case in Fig. 3.19(b).

If $\mathbf{A}$ is the gradient of a scalar field Φ, i.e.,

$$\mathbf{A} = \nabla\Phi \quad \text{then} \quad \oint_{\mathcal{L}} \mathbf{A} \cdot d\mathbf{l} = 0 \tag{3.74}$$

and therefore
$$\nabla \times \mathbf{A} = \nabla \times (\nabla\Phi) = 0 \tag{3.75}$$

The proof of this is seen through Stokes's theorem. Consider a surface $\mathcal{S}$ bounded by a curve $\mathcal{L}$ as shown in Fig. 3.19. On this surface

$$\iint_{\mathcal{S}} \nabla \times \mathbf{A} \cdot d\mathbf{S} = \oint_{\mathcal{L}} \mathbf{A} \cdot d\mathbf{l}$$

$$= \oint_{\mathcal{L}} \nabla\Phi \cdot d\mathbf{l}$$

$$= 0$$

But this surface can be anywhere and of any dimension. Therefore the result should be independent of surface. Hence, $\nabla \times \mathbf{A}$ should be zero.

so that
$$\iint_{\mathcal{S}} \underbrace{\nabla \times \mathbf{A}} \cdot d\mathbf{S} = 0$$

or
$$\nabla \times \mathbf{A} = \nabla \times \nabla\Phi = 0 \tag{3.76}$$

Its inverse is also true. For every case if in some region $\mathcal{R}$,

$$\oint_{\mathcal{L}} \mathbf{A} \cdot d\mathbf{l} = 0 \quad \text{then } \mathbf{A} \text{ is the gradient of a scalar field} \tag{3.77}$$

in $\mathcal{R}$. Also if

$$\nabla \times \mathbf{A} = 0$$

everywhere, then **A** *is the gradient of a scalar field.*

$$\mathbf{A} = \nabla\Phi \tag{3.78}$$

Example 3.11 Verify in spherical coordinates that for any scalar field $\Phi(r, \theta, \phi)$, $\nabla \times \nabla\Phi = 0$.

Solution

Step 1. Using the formula for $\nabla\Phi$ in spherical coordinates

$$\nabla\Phi = \frac{\partial\Phi}{\partial r}\mathbf{a}_r + \frac{1}{r}\frac{\partial\Phi}{\partial\theta}\mathbf{a}_\theta + \frac{1}{r\sin\theta}\frac{\partial\Phi}{\partial\phi}\mathbf{a}_\phi$$

Step 2. $\nabla \times \mathbf{A}$ is

$$\nabla \times \mathbf{A} = \frac{1}{r\sin\theta} \left\{ \frac{\partial \left(\sin\theta \, A_\phi\right)}{\partial \theta} - \frac{\partial A_\theta}{\partial \phi} \right\} \mathbf{a}_r$$

$$+ \frac{1}{r} \left\{ \frac{1}{\sin\theta} \frac{\partial A_r}{\partial \phi} - \frac{\partial \left(r A_\phi\right)}{\partial r} \right\} \mathbf{a}_\theta + \frac{1}{r} \left\{ \frac{\partial \left(r A_\theta\right)}{\partial r} - \frac{\partial A_r}{\partial \theta} \right\} \mathbf{a}_\phi$$

$$(3.79)$$

Step 3. Substituting

$$A_r = \frac{\partial \Phi}{\partial r}; \quad A_\theta = \frac{1}{r} \frac{\partial \Phi}{\partial \theta}; \quad A_\phi = \frac{1}{r\sin\theta} \frac{\partial \Phi}{\partial \phi}$$

in the previous equation

$$\nabla \times \nabla\Phi = \frac{1}{r\sin\theta} \left\{ \frac{\partial \left(\sin\theta \, \frac{1}{r\sin\theta} \frac{\partial \Phi}{\partial \phi}\right)}{\partial \theta} - \frac{\partial \left(\frac{1}{r} \frac{\partial \Phi}{\partial \theta}\right)}{\partial \phi} \right\} \mathbf{a}_r$$

$$+ \frac{1}{r} \left\{ \frac{1}{\sin\theta} \frac{\partial \left(\frac{\partial \Phi}{\partial r}\right)}{\partial \phi} - \frac{\partial \left[r \left(\frac{1}{r\sin\theta} \frac{\partial \Phi}{\partial \phi}\right)\right]}{\partial r} \right\} \mathbf{a}_\theta$$

$$+ \frac{1}{r} \left\{ \frac{\partial \left[r \left(\frac{1}{r} \frac{\partial \Phi}{\partial \theta}\right)\right]}{\partial r} - \frac{\partial \left(\frac{\partial \Phi}{\partial r}\right)}{\partial \theta} \right\} \mathbf{a}_\phi$$

Step 4. Simplifying the above

$$\nabla \times \nabla\Phi = \frac{1}{r\sin\theta} \left\{ \frac{1}{r} \frac{\partial \left(\frac{\partial \Phi}{\partial \phi}\right)}{\partial \theta} - \frac{1}{r} \frac{\partial \left(\frac{\partial \Phi}{\partial \theta}\right)}{\partial \phi} \right\} \mathbf{a}_r$$

$$+ \frac{1}{r} \left\{ \frac{1}{\sin\theta} \frac{\partial^2 \Phi}{\partial \phi \partial r} - \frac{1}{\sin\theta} \frac{\partial \left[\left(\frac{\partial \Phi}{\partial \phi}\right)\right]}{\partial r} \right\} \mathbf{a}_\theta$$

$$+ \frac{1}{r} \left\{ \frac{\partial \left[\left(\frac{\partial \Phi}{\partial \theta}\right)\right]}{\partial r} - \frac{\partial \left(\frac{\partial \Phi}{\partial r}\right)}{\partial \theta} \right\} \mathbf{a}_\phi$$

which is zero.

Example 3.12 Apply Stokes's theorem to the vector field $\mathbf{A} = x^2 yz\mathbf{a}_x + y^2 zx\mathbf{a}_y + z^2 xy\mathbf{a}_z$ to the surface bounded by the four points $(0,0,1)$, $(1,0,1)$, $(1,1,1)$, and $(0,1,1)$ and verify the theorem.

Solution

Step 1. The formula for the curl in rectangular coordinates is

$$\nabla \times \mathbf{A} = \left(\frac{\partial A_z}{\partial y} - \frac{\partial A_y}{\partial z}\right) \mathbf{a}_x + \left(\frac{\partial A_x}{\partial z} - \frac{\partial A_z}{\partial x}\right) \mathbf{a}_y + \left(\frac{\partial A_y}{\partial x} - \frac{\partial A_x}{\partial y}\right) \mathbf{a}_z$$

where $A_x = x^2 yz; A_y = y^2 zx; A_z = z^2 xy$

Step 2. Substituting these into the previous formula:

$$\nabla \times \mathbf{A} = \left(\frac{\partial\,(z^2xy)}{\partial y} - \frac{\partial\,(y^2zx)}{\partial z} \right) \mathbf{a}_x + \left(\frac{\partial\,(x^2yz)}{\partial z} - \frac{\partial\,(z^2xy)}{\partial x} \right) \mathbf{a}_y$$

$$+ \left(\frac{\partial\,(y^2zx)}{\partial x} - \frac{\partial\,(x^2yz)}{\partial y} \right) \mathbf{a}_z$$

$$= \left(z^2x - y^2x \right) \mathbf{a}_x + \left(x^2y - z^2y \right) \mathbf{a}_y + \left(y^2z - x^2z \right) \mathbf{a}_z$$

Step 3. Stokes's theorem says

$$\iint_{S} \nabla \times \mathbf{A} \cdot d\mathbf{S} = \oint_{L} \mathbf{A} \cdot d\mathbf{l}$$

If we observe the region of integration we find that it is a plane area on the x-y plane, and when traversed in the counter-clockwise sense, an element of area is given by $d\mathbf{S} = dxdy\mathbf{a}_z$.

Step 4. So $\nabla \times \mathbf{A} \cdot d\mathbf{S} = (y^2z - x^2z)\,dxdy$ but the plane of integration is the $z = 1$ plane, so:

$$\nabla \times \mathbf{A} \cdot d\mathbf{S}|_{z=1} = (y^2z - x^2z)\,|_{z=1}dxdy = (y^2 - x^2)\,dxdy$$

Hence
$$\iint_{S} \nabla \times \mathbf{A} \cdot d\mathbf{S} = \int\!\!\!\int_{x=0,y=0}^{x=1,y=1} (y^2 - x^2)\,dxdy$$

$$= \int_{y=0}^{y=1} (y^2x - x^3/3) \Big|_{x=0}^{x=1} dy$$

$$= \int_{y=0}^{y=1} (y^2 - 1/3)\,dy$$

$$= 0$$

Step 5. We now compute the right side of the equation

$$I = \oint_{L} \mathbf{A} \cdot d\mathbf{l}$$

$$= \int_{(0,0,1)}^{(1,0,1)} A_x dx + \int_{(1,0,1)}^{(1,1,1)} A_y dy + \int_{(1,1,1)}^{(0,1,1)} A_x dx + \int_{(0,1,1)}^{(0,1,1)} A_y dy$$

and get

$$I = \int_{(0,0,1)}^{(1,0,1)} x^2y \Big|_{y=0} dx + \int_{(1,0,1)}^{(1,1,1)} y^2x \Big|_{x=1} dy$$

$$+ \int_{(1,1,1)}^{(0,1,1)} x^2y \Big|_{y=1} dx + \int_{(0,1,1)}^{(0,1,1)} y^2x \Big|_{x=0} dy$$

which simplifies to

$$I = \int_{y=0}^{y=1} y^2x\,dy + \int_{x=1}^{x=0} x^2 dx$$

$$= y^3/3 \Big|_{0}^{1} + x^3/3 \Big|_{1}^{0} = 0$$

We go another route and apply Stokes's theorem to the case when the line integral is very small (see Fig. 3.20), and the surface is bulging as in the figure. Then obviously,

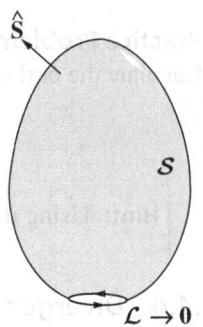

$$\oint_{\mathcal{L}\to 0} \mathbf{A} \cdot d\mathbf{l} = 0 \qquad (3.80)$$

This is true for any $\mathbf{A}$ since the length of the line $\mathcal{L}$ is zero. However by Stokes's theorem

$$\oint_{\mathcal{L}\to 0} \mathbf{A} \cdot d\mathbf{l} = \iint_{S \text{ for } \mathcal{L}\to 0} (\nabla \times \mathbf{A}) \cdot d\mathbf{S}$$

$$= \oiint_{S} (\nabla \times \mathbf{A}) \cdot d\mathbf{S} = 0 \qquad (3.81)$$

Fig. 3.20 Integration of the curl over a closed surface

But this holds for any $\mathbf{A}$, therefore the surface integral of the curl of $\mathbf{A}$ over any closed surface is identically zero!

$$\oiint_{S} (\nabla \times \mathbf{A}) \cdot d\mathbf{S} = 0 \qquad (3.82)$$

This result will be discussed in the next section.

Example 3.13   For the vector field $\mathbf{A} = r^2 \sin\theta \, \mathbf{a}_r + r^2 \sin\theta \cos\phi \, \mathbf{a}_\theta$, find the surface integral of $\nabla \times \mathbf{A}$ over a sphere of radius r_0.

Solution   The curl of $\mathbf{A}$ in the spherical coordinates is

$$\nabla \times \mathbf{A} = \frac{1}{r\sin\theta} \left\{ \frac{\partial (\sin\theta \, A_\phi)}{\partial \theta} - \frac{\partial A_\theta}{\partial \phi} \right\} \mathbf{a}_r + \frac{1}{r} \left\{ \frac{1}{\sin\theta} \frac{\partial A_r}{\partial \phi} - \frac{\partial (rA_\phi)}{\partial r} \right\} \mathbf{a}_\theta$$

$$+ \frac{1}{r} \left\{ \frac{\partial (rA_\theta)}{\partial r} - \frac{\partial A_r}{\partial \theta} \right\} \mathbf{a}_\phi$$

which becomes in our case

$$\mathbf{E} = \nabla \times \mathbf{A} = \frac{1}{r\sin\theta} \left\{ -\frac{\partial (r^2 \sin\theta \cos\phi)}{\partial \phi} \right\} \mathbf{a}_r + \frac{1}{r} \left\{ \frac{1}{\sin\theta} \frac{\partial (r^2 \sin\theta)}{\partial \phi} \right\} \mathbf{a}_\theta$$

$$+ \frac{1}{r} \left\{ \frac{\partial (r^3 \sin\theta \cos\phi)}{\partial r} - \frac{\partial (r^2 \sin\theta)}{\partial \theta} \right\} \mathbf{a}_\phi \qquad (3.83)$$

Since we will be integrating over a sphere, we need to consider only the $\mathbf{a}_r$ component, which simplifies to $E_r = r\sin\phi$.

Taking
$$\oiint_{S} \nabla \times \mathbf{A} \cdot d\mathbf{S} = \oiint_{\text{Sphere}} (r\sin\phi) \left(r^2 \sin\theta \, d\theta \, d\phi \right)$$

$$= \int_{\theta=0,\phi=0}^{\theta=\pi,\phi=2\pi} r_0^3 \sin\phi \sin\theta \, d\theta \, d\phi = 0$$

Practice Problem 3.1

Caculate the curl of the following vector:

$$\mathbf{A} = \frac{1}{r^2} \cos\theta \, \mathbf{a}_r \qquad \text{(DTU-NSIT, May 2009)}$$

$$\left[\text{\textbf{Hint:} Using the curl formula given in Section 3.3.3, } \nabla \times \mathbf{A} = \frac{\sin\left(\theta\right)}{r^3} \mathbf{a}_\phi \right]$$

3.4.4 Divergence

The divergence of vector field $\mathbf{A} = \nabla \cdot \mathbf{A}$ in rectangular coordinates is given by

$$\nabla \cdot \mathbf{A} = \frac{\partial A_x}{\partial x} + \frac{\partial A_y}{\partial y} + \frac{\partial A_z}{\partial z} \qquad (3.84)$$

In other coordinate systems the divergence of $\mathbf{A}$ is

$$\nabla \cdot \mathbf{A} = \frac{1}{\rho} \frac{\partial \left(\rho A_\rho\right)}{\partial \rho} + \frac{1}{\rho} \frac{\partial A_\phi}{\partial \phi} + \frac{\partial A_z}{\partial z} \qquad (3.85)$$

$$\nabla \cdot \mathbf{A} = \frac{1}{r^2} \frac{\partial \left(r^2 A_r\right)}{\partial r} + \frac{1}{r \sin\theta} \frac{\partial \left(\sin\theta \, A_\theta\right)}{\partial \theta} + \frac{1}{r \sin\theta} \frac{\partial A_\phi}{\partial \phi} \qquad (3.86)$$

in the cylindrical and spherical coordinate systems, respectively.

One of the most important relations for the divergence is the relation

$$\iiint_V \left(\nabla \cdot \mathbf{A}\right) dV = \oiint_S \mathbf{A} \cdot d\mathbf{S} \qquad (3.87)$$

and is *well known as the divergence theorem.* Here $\mathcal{V}$ is any volume and $\mathcal{S}$ is the surface enclosing the volume. The theorem is illustrated in Fig. 3.21. Referring to Eqn (3.87) and the figure, the integration here is over any volume $\mathcal{V}$ enclosed by the closed surface $\mathcal{S}$. The unit vector $\hat{\mathbf{S}}$ is the *outward* normal to the closed surface.

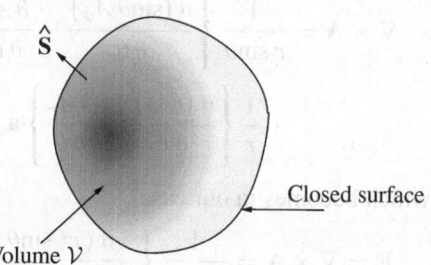

Closed surface

Volume $\mathcal{V}$

Fig. 3.21 Divergence theorem

Applying this theorem to the case when the vector field is the curl of *some* vector field, we find

$$\iiint_V \nabla \cdot \left(\nabla \times \mathbf{A}\right) dV = \oiint_S \left(\nabla \times \mathbf{A}\right) \cdot d\mathbf{S} = 0 \qquad (3.88)$$

The first equality is true due to the divergence theorem Eqn (3.87) and the second equality is true due to Eqn (3.82). Since this equation is valid over *any* volume, this must imply that

$$\nabla \cdot \left(\nabla \times \mathbf{A}\right) = 0 \qquad (3.89)$$

Conversely, if the divergence of a vector field is *always* zero, i.e.

$$\nabla \cdot \mathbf{A} = 0 \qquad (3.90)$$

then **A** must be the curl of another vector

i.e. $$\mathbf{A} = \nabla \times \mathbf{B} \tag{3.91}$$

since then $$\nabla \cdot (\nabla \times \mathbf{B}) = 0$$

Note: A vector field is fully specified only when both its divergence as well as its curl are specified.

Example 3.14 In cylindrical coordinates, show that $\nabla \cdot \nabla \times \mathbf{A} = 0$.

Solution

Step 1. In cylindrical coordinates,

$$\nabla \times \mathbf{A} = \left(\frac{1}{\rho} \frac{\partial A_z}{\partial \phi} - \frac{\partial A_\phi}{\partial z} \right) \mathbf{a}_\rho + \left(\frac{\partial A_\rho}{\partial z} - \frac{\partial A_z}{\partial \rho} \right) \mathbf{a}_\phi + \frac{1}{\rho} \left\{ \frac{\partial (\rho A_\phi)}{\partial \rho} - \frac{\partial A_\rho}{\partial \phi} \right\} \mathbf{a}_z$$

Step 2. $\nabla \cdot \mathbf{B}$ is given by

$$\nabla \cdot \mathbf{B} = \frac{1}{\rho} \frac{\partial (\rho B_\rho)}{\partial \rho} + \frac{1}{\rho} \frac{\partial B_\phi}{\partial \phi} + \frac{\partial B_z}{\partial z}$$

on comparing equations

$$B_\rho = \left(\frac{1}{\rho} \frac{\partial A_z}{\partial \phi} - \frac{\partial A_\phi}{\partial z} \right); \quad B_\phi = \left(\frac{\partial A_\rho}{\partial z} - \frac{\partial A_z}{\partial \rho} \right); \quad B_z = \frac{1}{\rho} \left\{ \frac{\partial (\rho A_\phi)}{\partial \rho} - \frac{\partial A_\rho}{\partial \phi} \right\}$$

Step 3. Substituting in the previous equation, we get

$$\nabla \times \mathbf{B} = \frac{1}{\rho} \frac{\partial \left[\rho \left(\frac{1}{\rho} (\partial A_z / \partial \phi) - (\partial A_\phi / \partial z) \right) \right]}{\partial \rho}$$

$$+ \frac{1}{\rho} \frac{\partial \left((\partial A_\rho / \partial z) - (\partial A_z / \partial \rho) \right)}{\partial \phi}$$

$$+ \frac{\partial \left[\frac{1}{\rho} \left\{ (\partial (\rho A_\phi) / \partial \rho) - (\partial A_\rho / \partial \phi) \right\} \right]}{\partial z}$$

$$= \frac{1}{\rho} \frac{\partial \left((\partial A_z / \partial \phi) - (\rho \, \partial A_\phi / \partial z) \right)}{\partial \rho} + \frac{1}{\rho} \underbrace{\left(\frac{\partial^2 A_\rho}{\partial z \partial \phi} - \frac{\partial^2 A_z}{\partial \rho \partial \phi} \right)}_{\text{cancels}}$$

$$+ \frac{\partial \left(\frac{1}{\rho} A_\phi \right)}{\partial z} + \frac{\partial^2 A_\phi}{\partial \rho \partial z} \underbrace{- \frac{1}{\rho} \frac{\partial^2 A_\rho}{\partial \phi \partial z}}_{\text{cancels}}$$

$$= \underbrace{\frac{1}{\rho} \frac{\partial^2 A_z}{\partial \phi \partial \rho}}_{1} \underbrace{- \frac{1}{\rho} \frac{\partial A_\phi}{\partial z}}_{2} \underbrace{- \frac{\partial^2 A_\phi}{\partial z \partial \rho}}_{3} \underbrace{- \frac{1}{\rho} \frac{\partial^2 A_z}{\partial \rho \partial \phi}}_{1} + \underbrace{\frac{1}{\rho} \frac{\partial A_\phi}{\partial z}}_{2} + \underbrace{\frac{\partial^2 A_\phi}{\partial \rho \partial z}}_{3}$$

All the terms cancel (1-1, 2-2, and 3-3 all cancel) and so we have proved our result.

Example 3.15 For a given vector in cylindrical coordinates,

$$\mathbf{A} = \rho^2 \cos^2\phi \, \mathbf{a}_\rho$$

and specified closed surface $\rho = 4$ and $0 < z < 1$, find both sides of the divergence theorem. Is the theorem satisfied? [DTU-NSIT, March 2008]

Solution Taking the divergence of $\mathbf{A}$

$$\nabla \cdot \mathbf{A} = \frac{1}{\rho} \frac{\partial (\rho A_\rho)}{\partial \rho} + \frac{1}{\rho} \frac{\partial A_\phi}{\partial \phi} + \frac{\partial A_z}{\partial z}$$

$$= 3\rho \cos^2(\phi)$$

integrating the divergence over the volume mentioned,

$$\iiint\limits_V \nabla \cdot \mathbf{A} dV = \int\limits_{\rho=0,\phi=0,z=0}^{\rho=4,\phi=2\pi,z=1} \!\!\!\!\!\!\!\!\!\! 3\rho \cos^2(\phi)(\rho \, d\rho \, d\phi \, dz)$$

$$= 3\pi \int\limits_{\rho=0}^{\rho=4} \rho^2 \, d\rho = 64\pi$$

Now integrating over the closed surface (only the sides of the cylinder are considered, since $A_z = 0$)

$$\oiint \mathbf{A} \cdot d\mathbf{S} = \left[\int\limits_{\phi=0,z=0}^{\phi=2\pi,z=1} \!\!\!\!\!\!\! \rho^2 \cos^2\phi (\rho \, d\phi \, dz) \right]_{\rho=4}$$

$$= 64 \int\limits_{\phi=0,z=0}^{\phi=2\pi,z=1} \!\!\!\!\!\!\! \cos^2\phi \, d\phi \, dz = 64\pi$$

therefore the theorem is satisfied.

3.5 | Maxwell's Equations

This section is a little advanced hence the student may read it as and when required.

Maxwell's equations, which are the fundamental equations on which the whole of electromagnetic theory is built, generally apply to four types of vector fields ($\mathbf{r}$ is the position vector of any point in space and t is time):

1. The *electric* field $\mathbf{E}(\mathbf{r}, t)$
2. The *magnetic* field $\mathbf{H}(\mathbf{r}, t)$
3. The *electric flux density* $\mathbf{D}(\mathbf{r}, t)$
4. The *magnetic flux density* $\mathbf{B}(\mathbf{r}, t)$

Why the first two are called 'fields' and the other two are called 'flux densities', will be explained later in the book. In the study of electrostatics, only the fields 1 and 3 will be considered ($\mathbf{E}(\mathbf{r}, t)$ and $\mathbf{D}(\mathbf{r}, t)$), and there these fields are *not* functions of time as $\mathbf{E}(\mathbf{r})$ and $\mathbf{D}(\mathbf{r})$. Similarly, in magnetostatics the fields 2 and 4 will be considered ($\mathbf{H}(\mathbf{r}, t)$ and $\mathbf{B}(\mathbf{r}, t)$), and there also these fields are time-independent functions as $\mathbf{H}(\mathbf{r})$ and $\mathbf{B}(\mathbf{r})$.

The general Maxwell's equations, which are both functions of position as well as time are given below, for future reference:

$$\nabla \cdot \mathbf{D} = \rho_v \tag{3.92}$$

$$\nabla \times \mathbf{E} = -\partial \, \mathbf{B}/\partial \, t \tag{3.93}$$

$$\nabla \cdot \mathbf{B} = 0 \tag{3.94}$$

and
$$\nabla \times \mathbf{H} = \partial \, \mathbf{D}/\partial \, t + \mathbf{J} \tag{3.95}$$

These equations are in SI units. Here ρ_v is the volume charge density in C/m^3 and $\mathbf{J}$ is the current density in A/m^2. In addition, there are two more equations:

$$\mathbf{D} = \varepsilon \, \mathbf{E} \tag{3.96}$$

which is a relation between $\mathbf{D}$ and $\mathbf{E}$ in a dielectric. Here ε is the permittivity of the dielectric medium, and its unit is F/m. Another equation applicable to magnetic material is

$$\mathbf{B} = \mu \, \mathbf{H} \tag{3.97}$$

where μ is the permeability of the magnetic material in H/m. The last equation (and which is very important) is the continuity equation:

$$\nabla \cdot \mathbf{J} = -\partial \, \rho_v /\partial \, t \tag{3.98}$$

3.6 | Units and Dimensions of EM Fields

Keeping the above discussion in mind, let us take a look at the units of the various terms of Maxwell's equations. The first of Maxwell's equations is

$$\nabla \cdot \mathbf{D} = \frac{\partial \, D_x}{\partial \, x} + \frac{\partial \, D_y}{\partial \, y} + \frac{\partial \, D_z}{\partial \, z} = \rho_v \tag{3.99}$$

The $\mathbf{D}$ field is called the electric flux density. ρ_v is the volume charge density and its unit is C/m^3. Hence we can say that $\mathbf{D}$ has the units of C/m^2.

Next, let us look at the equation:

$$\nabla \times \mathbf{H} = \partial \, \mathbf{D}/\partial \, t + \mathbf{J} \tag{3.100}$$

The unit of $\partial \, \mathbf{D}/\partial \, t$ is $Cm^{-2}s^{-1} = (Cs^{-1}) \, m^{-2} = A \, m^2$. Hence the unit of $\mathbf{J}$ is A/m^2 and of $\mathbf{H}$ is A/m.

Let us focus next on the relation between $\mathbf{D}$ and $\mathbf{E}$:

$$\mathbf{D} = \varepsilon \, \mathbf{E} \tag{3.101}$$

where the unit of ε is F/m. Hence the unit of $\mathbf{E}$ is $(C/m^2)/(F/m) = (C/F) \, m^{-1} = V/m$, since C/F is volt. Hence the unit of the $\mathbf{E}$-field is V/m. Re-looking at $\mathbf{D}$, it has the alternate units of (F/m) (V/m).

Finally, examination of the equation:

$$\nabla \times \mathbf{E} = -\partial \, \mathbf{B}/\partial \, t \tag{3.102}$$

straightaway gives us the unit of $\mathbf{B}$ as $(V \, s)/m^2 = (H/m)(A/m)$, where we have converted V to H by using $V = Ldi/dt$. The unit of $\mathbf{B}$ is also called tesla (T).[1]

The results of this discussion can be put down in the form of Table 3.1.

---

[1] 1 Tesla= 1 weber/m^2.

Table 3.1 Units of the various electromagnetic quantities

Symbol	Name	Unit
$\mathbf{D}$	Electric flux density	$C/m^2 = (F/m)\,(V/m)$
$\mathbf{E}$	Electric field	V/m
$\mathbf{H}$	Magnetic field	A/m
$\mathbf{B}$	Magnetic flux density	$(Vs)/m^2 = (H/m)\,(A/m) = Tesla$
ρ_v	Volume charge density	C/m^3
$\mathbf{J}$	Current density	A/m^2
ε	Permittivity	F/m
μ	Permeability	H/m

POINTS TO REMEMBER

- The differential element of a line in rectangular coordinates is

$$\Delta\mathbf{l} \equiv \Delta\mathbf{r} = \Delta x\mathbf{a}_x + \Delta y\mathbf{a}_y + \Delta z\mathbf{a}_z$$

- The differential element of a line in cylindrical coordinates is

$$\Delta\mathbf{l} \equiv \Delta\mathbf{r} = \Delta\rho\,\mathbf{a}_\rho + \rho\,\Delta\phi\,\mathbf{a}_\phi + \Delta z\mathbf{a}_z$$

- The differential element of a line in spherical coordinates is

$$\Delta\mathbf{l} \equiv \Delta\mathbf{r} = \Delta r\mathbf{a}_r + r\Delta\theta\,\mathbf{a}_\theta + r\sin(\theta)\,\Delta\phi\,\mathbf{a}_\phi$$

- The line integral of a scalar field is

$$\int_A^B \Phi(\mathbf{r})\,dl = \int_A^B \Phi[\mathbf{r}(t)]\left|\frac{d\mathbf{r}}{dt}\right|dt$$

- The line integral of a vector field is

$$\int_A^B \mathbf{F}(\mathbf{r})\cdot d\mathbf{l} = \int_A^B \mathbf{F}\cdot\frac{d\mathbf{r}}{dt}dt$$

- The flux out of a surface is

$$\Phi = \iint_S \mathbf{F}\cdot d\mathbf{S}$$

- The Nabla operator in rectangular coordinates is

$$\nabla = \frac{\partial}{\partial x}\mathbf{a}_x + \frac{\partial}{\partial y}\mathbf{a}_y + \frac{\partial}{\partial z}\mathbf{a}_z$$

- Grad ($\nabla\Phi$)

  - The gradient $\nabla\Phi$ of a scalar field $\Phi(\mathbf{r})$ at an arbitrary point $\mathbf{r}_0 = (x_0, y_0, z_0)$ is perpendicular to $\Phi(\mathbf{r}) = \Phi(\mathbf{r}_0)$.
  - The line integral of the gradient of a scalar field Φ,

$$\int_{\substack{A \\ \mathcal{L}}}^B [\nabla\Phi\cdot d\mathbf{l}]$$

  from point A to B is independent of the path of integration and will give the same result when integrated along any line $\mathcal{L}$ from A to B.

- When the gradient is integrated along any closed curve C, it always gives the same result which is zero.

$$\oint_{\text{any } C} [\nabla\Phi \cdot d\mathbf{l}] = 0$$

- The gradient $\nabla\Phi$ of a scalar field $\Phi(\mathbf{r})$ at an arbitrary point $\mathbf{r}_0 = (x_0, y_0, z_0)$ is perpendicular to $\Phi(\mathbf{r}) = \Phi(\mathbf{r}_0)$.
- The curl $\nabla \times \mathbf{A}$
  - The surface integral of the curl of a vector is connected to its line integral as

$$\oint_L \mathbf{A} \cdot d\mathbf{l} = \iint_S (\nabla \times \mathbf{A}) \cdot d\mathbf{S}$$

  where the line is a closed curve L which encloses the surface S. *This is Stokes's theorem.*
  - $\nabla \times \nabla\Phi = 0$. Here Φ is any scalar field.
  - The surface integral of the curl of *any* vector field over a *closed* surface S is *always* zero.

$$\oiint_S (\nabla \times \mathbf{A}) \cdot d\mathbf{S} = 0$$

- The surface integral of the curl of a vector is connected to its line integral as:

$$\oint_L \mathbf{A} \cdot d\mathbf{l} = \iint_S (\nabla \times \mathbf{A}) \cdot d\mathbf{S}$$

  where the line is a closed curve L which encloses the surface S. This is Stokes's theorem.
- $\nabla \times \nabla\Phi = 0$. Here Φ is any scalar field.
- $\oiint_S (\nabla \times \mathbf{A}) \cdot d\mathbf{S} = 0$. The surface integral of the curl of *any* vector field over a *closed* surface S is *always* zero.
- The divergence $\nabla \cdot \mathbf{A}$
  - The volume integral over the volume $\mathcal{V}$ of the divergence of a vector field $\nabla \cdot \mathbf{A}$ is always equal to the surface integral of that field over the *enclosing surface S*

$$\iiint_\mathcal{V} (\nabla \cdot \mathbf{A}) \, dV = \oiint_S \mathbf{A} \cdot d\mathbf{S}$$

  This *is the divergence theorem.*
  - The divergence of the curl of a vector is always zero. $\nabla \cdot (\nabla \times \mathbf{A}) = 0$
- To specify a vector field completely through equations, *both* its divergence and curl must be specified.

SELF ASSESSMENT

Objective Type Questions

1. The length of the vector $d\mathbf{l}$ along the curve $x = 2y = 4z^2$ at the point $(1, 1/2, 1/2)$ is
   (a) dx (b) $1.1dx$ (c) $1.263dx$ (d) $1.266dx$
2. The length of the curve $x = y = z$ from $(0,0,0)$ to $(1,1,1)$ is
   (a) 1   (b) $\sqrt{2}$ (c) $\sqrt{3}$ (d) 2
3. The length of the the the curve $x = 2y = 4z^2$ from $(0,0,0)$ to $(1,1/2,1/2)$ is
   (a) 2.661   (b) 3.661   (c) 1.268   (d) 2.2688

4. The line integral of the vector $A = 3a_x + 4a_y + 5a_z$ along the curve $x = y = z$ from $(0,0,0)$ to $(1,1,1)$ is
   (a) 12   (b) 11   (c) 10   (d) 9
5. The line integral of the vector $A = 3a_x + 4a_y + 5a_z$ along the curve $x = 2y = 4z^2$ from $(0,0,0)$ to $(1,1/2,1/2)$ is
   (a) 5.5   (b) 7.5   (c) 6.5   (d) 2
6. The magnitude of gradient of the scalar field $\Phi = ze^{-\rho} \cos\phi$ at the point $z = 1$, $\rho = 1$, $\phi = 1$ is
   (a) 2.2214   (b) 0.4181   (c) 0.5   (d) ∞
7. The surface integral of the vector $A = xya_x$ on the x-y plane for $-2 \le x \le 2$ and $-2 \le y \le 2$ is
   (a) 0   (b) 4   (c) -4 (d) 16
8. The surface integral of the vector $A = \rho \sin\phi\, a_z$ on the x-y plane for $0 \le \phi \le 2\pi$ and $0 \le \rho \le 2$ is
   (a) 2   (b) 0   (c) -4 (d) 16
9. The curl of the vector $A = \rho \sin\phi\, a_z$ is
   (a) 2   (b) $\cos\phi\, a_z - \sin\phi\, a_\rho$ (c) $\cos\phi\, a_\rho - \sin\phi\, a_\phi$ (d) $\cos\phi\, a_\rho + \sin\phi\, a_\phi$
10. The divergence of the vector $A = \rho \sin\phi\, a_z$ is
   (a) 0   (b) $\cos\phi\, a_z - \sin\phi\, a_\rho$ (c) ρ (d) $\sin\phi$

Short-Answer Questions

1. Apply the integral theorems to prove that

$$\nabla \cdot (\nabla \times A) = 0 \qquad \text{(Mumbai University, May 2009)}$$

2. Find the normal to the surface $z - x^2/4 = 8$ at the point $z = 9$, $x = 2$.
3. For the vector field $A = z \sin\phi\, a_\rho + z \cos\phi\, a_\phi + \rho \sin\phi\, a_z$, find the line integral of the vector field along the unit circle in $z = 1$ plane in the anti-clockwise direction.
4. Find the flux of $A = z \sin\phi\, a_\rho + z \cos\phi\, a_\phi + \rho \sin\phi\, a_z$ out of the closed cylinder bounded by the three surfaces $z = \pm 1$ and $\rho = 2$.
5. In the spherical coordinates a vector field is given by

$$A = \frac{K a_r}{r^2}$$

find the flux out of a closed sphere with radius R.
6. In rectangular coordinates show that $\nabla \cdot \nabla \times A = 0$ where A is any vector.

Review Questions

1. Define the differential linear elements in the rectangular, cylindrical, and spherical coordinate systems.
2. Explain the concept of (a) the gradient of a scalar field, ∇V, (b) divergence of a vector field, $\nabla \cdot A$.   (DTU-NSIT, March 2008)
3. Define the following in rectangular coordinates:
   (a) The Nabla operator
   (b) Gradient of a scalar field
   (c) The curl of a vector field
   (d) The divergence of a vector field
4. Give the physical significance of the gradient of a scalar field, and curl and divergence (of a vector field).

5. State the divergence and Stokes's theorems in rectangular coordinates.
6. What is the connection between the flux of a vector field and divergence?

Numerical Problems

1. Find the normal vectors to the surfaces $r = r_0$, $\theta = \theta_0$, and $\phi = \phi_0$ where r_0, θ_0, and ϕ_0 are constants.
2. The vector field $\mathbf{A} = y\mathbf{a}_x - x\mathbf{a}_y + z\mathbf{a}_z$ is given in rectangular coordinates. Find the surface where $|\mathbf{A}|$ is a constant.
3. Find the length of the curve $x = y^2 = z^2$ from $(1,1,1)$ to $(4,2,2)$.

   Hint: Consider the integral

   $$\int dl = \int_{(1,1,1)}^{(4,2,2)} \sqrt{dx^2 + dy^2 + dz^2}$$

   with $y = z = t$ and $x = t^2$.
4. In a particular application, the electric field in the spherical coordinate system is given by

   $$\mathbf{E} = K\frac{\mathbf{a}_r}{r^2}$$

   where K is a constant. Find $\nabla \cdot \mathbf{E}$ for this field.
5. In a particular application, the electric field in the spherical coordinate system is given by

   $$\mathbf{E} = K\frac{\mathbf{a}_r}{r^2}$$

   where K is a constant. Find $\nabla \times \mathbf{E}$ for this field.
6. A scalar field V in the spherical coordinate system is given by

   $$V = K\frac{1}{r}$$

   where K is a constant. Find $\partial V/\partial x$, $\partial V/\partial y$, and $\partial V/\partial z$.
7. The magnetic field $\mathbf{H}$ in cylindrical coordinates is given by

   $$\mathbf{H} = \frac{K}{\rho}\mathbf{a}_\phi$$

   where K is a constant. Compute $\nabla \times \mathbf{H}$.
8. The magnetic field $\mathbf{H}$ in cylindrical coordinates is given by

   $$\mathbf{H} = \frac{K}{\rho}\mathbf{a}_\phi$$

   where K is a constant. Compute $\nabla \cdot \mathbf{H}$.
9. What is the method of obtaining the equation of any plane in spherical coordinates? How would you go about doing it?
10. The equation of a cone in spherical coordinates is $\theta = \theta_0$. Find the equation in rectangular coordinates.
11. Find the volume of a sphere through integration using the spherical coordinate system.
12. Find the surface area of a cylinder of radius a and length l through integration.
13. A scalar field is given by

    $$V(\mathbf{r}) = e^{-(x^2+y^2+z^2)}$$

    find the volume integral of this field over all space.
14. The vector field in cylindrical coordinates is

    $$\mathbf{A} = \frac{K}{\rho}\mathbf{a}_\rho \quad \text{where } K \text{ is a constant}$$

It is to be integrated in the clockwise direction over the boundary of a unit circle lying on the x-y plane. Find the line integral of this vector field. By looking at the structure of **A**, can you guess the answer?

15. Find the gradient of $V = \dfrac{K}{r}$ where V is given in spherical coordinates and K is a constant. Find the line integral of this vector field over the boundary of a unit circle lying on the x-y plane.

16. Find $\nabla \times \nabla V$ of the function V of Problem 15.

17. Given the vector field $\mathbf{G} = y\mathbf{a}_x - 2.5x\mathbf{a}_y + 3z\mathbf{a}_z$. Find the line integral of **G** along the straight line from (1,1,1) to (2,3,4).
    Hint: Consider Example 5.

18. If $\mathbf{A} = 2xy\mathbf{a}_x + 3xy\mathbf{a}_y + 4xyz\mathbf{a}_z$, prove Stokes's theorem in the flat region enclosed by the straight lines (0,0,0) to (1,0,0) to (1,1,0) to (0,1,0) and back to (0,0,0).

19. Using any vector field $\mathbf{A}(x,y,z)$ show that $\nabla \cdot (\nabla \times \mathbf{A}) = 0$.

20. A vector field $\mathbf{A} = yz\mathbf{a}_x + xz\mathbf{a}_y + xy\mathbf{a}_z$. Find

    (a) The line integral of this vector field from $(1,1,1)$ to $(1,2,3)$
    (b) Find the surface integral of this field over the square region which is described by $\mathcal{R} = -1 \leq x \leq 1$ and $-1 \leq y \leq 1$.

21. For the vector field

$$\mathbf{D} = K\frac{x\mathbf{a}_x + y\mathbf{a}_y + z\mathbf{a}_z}{\left(x^2 + y^2 + z^2\right)^{3/2}}$$

find the divergence of **D**, K being a constant. Going over to spherical coordinates, find the surface integral of **D** over a sphere of any radius. Explain the results using

$$\iiint_V \nabla \cdot \mathbf{D}\, dV = \oiint_S \mathbf{D} \cdot d\mathbf{S}$$

22. If $\mathbf{A} = 2xy\mathbf{a}_x + 3xy\mathbf{a}_y + 4xyz\mathbf{a}_z$, prove the divergence theorem in the volume enclosed by the planes $z = 0\ z = 1;\ x = 0\ x = 1;$ and $y = 0\ y = 1$.

23. If the unit of the charge density ρ_v is C/m^3 then from the equation $\nabla \cdot \mathbf{D} = \rho_v$ show that the unit of **D** is C/m^2.

24. Using the unit of **E** as V/m, show that ε has the unit of F/m.

Answers

Objective Type Questions

1. (d)　2. (c)　3. (c)　4. (a)　5. (b)　6. (b)　7. (a)　8. (b)　9. (c)　10. (a)

Short-Answer Questions

2. The gradient gives the normal at any point. The gradient of the surface in general is

$$\mathbf{n} = \mathbf{a}_z - \frac{x}{2}\mathbf{a}_x$$

and the unit normal at the point in consideration is $\hat{\mathbf{n}} = (\mathbf{a}_z - \mathbf{a}_x)/\sqrt{2}$.

3. The line integral is

$$\int_c \mathbf{A} \cdot d\mathbf{l} = \int_c [z\cos\phi\, \rho]_{\rho=1,\, z=1}\, d\phi = \int_{\phi=0}^{2\pi} \cos\phi\, d\phi = 0$$

4. Outward flux from the $z = 1$ surface is

$$\int\limits_{\rho=0}^{\rho=2} \int\limits_{\phi=0}^{\phi=2\pi} (\rho \sin\phi)\, d\rho\, \rho\, d\phi = 2\pi \left[\rho^3/3\right]_0^2 = 16\pi/3$$

Outward flux from the $z = -1$ surface is

$$-\int\limits_{\rho=0}^{\rho=2} \int\limits_{\phi=0}^{\phi=2\pi} (\rho \sin\phi)\, d\rho\, \rho\, d\phi = -2\pi \left[\rho^3/3\right]_0^2 = -16\pi/3$$

Outward flux from the $\rho = 2$ surface is

$$\int\limits_{z=-1}^{z=1} \int\limits_{\phi=0}^{\phi=2\pi} [(z \sin\phi)\, \rho\, d\phi\, dz]_{\rho=2} = 0$$

So the total flux out of the closed surface is zero.

5. Outward flux from the $r = R$ surface is

$$\int\limits_{\theta=0}^{\theta=\pi} \int\limits_{\phi=0}^{\phi=2\pi} \left(\frac{K}{R^2}\right) R^2 \sin\theta\, d\theta\, d\phi = 2K\pi \left[-\cos\theta\right]_0^\pi = 4K\pi$$

6.
$$\nabla \times \mathbf{A} = \mathbf{a}_x \left(\partial_y A_z - \partial_z A_y\right) + \mathbf{a}_y \left(\partial_z A_x - \partial_x A_z\right) + \mathbf{a}_z \left(\partial_x A_y - \partial_y A_x\right)$$

and $\nabla \cdot \nabla \times \mathbf{A} = \partial_x \left(\partial_y A_z - \partial_z A_y\right) + \partial_y \left(\partial_z A_x - \partial_x A_z\right) + \partial_z \left(\partial_x A_y - \partial_y A_x\right)$

$$= \partial_{xy} A_z - \partial_{xz} A_y + \partial_{yz} A_x - \partial_{yx} A_z + \partial_{zx} A_y - \partial_{zy} A_x$$

$$= 0$$

The overbar, underbar, and underbracket terms cancel each other.

Numerical Problems

1. $\mathbf{a}_r$, $\mathbf{a}_\theta$, and $\mathbf{a}_\phi$
2. $x^2 + y^2 + z^2 =$ a constant
3. 3.326
4. Everywhere zero except at $r = 0$
5. Everywhere zero except at $r = 0$
6. Kx/r^3; Ky/r^3; Kz/r^3
7. Everywhere zero except at $\rho = 0$
8. Everywhere zero except at $\rho = 0$
9. Write the equation in rectangular coordinates, then substitute the values of x, y, z for spherical coordinates
10. $z = \cot\theta_0 \times \sqrt{x^2 + y^2}$
11. $(4/3)\, \pi r^3$
12. $2\pi a(a + l)$
13. $\pi\sqrt{\pi}$
14. 0
15. $-(K/r^2)\, \mathbf{a}_r$
16. 0
17. 17
20. (a) 5   (b) 0
21. $\nabla \cdot \mathbf{D} = 0$; $\int_S \mathbf{D} \cdot d\mathbf{S} = 0$

PART II

Electrostatics

PART II

Electrostatics

CHAPTER 4
The Electric Field and Gauss's Law

CHAPTER 5
Energy and Potential

CHAPTER 6
The Electric Field and Material Media

CHAPTER 7
Laplace's and Poisson's Equations

4

The Electric Field and Gauss's Law

> *If you wish to reach the highest,*
> *begin at the lowest*
> — Syrus: *Maxims*

CHAPTER OBJECTIVES

To enable the students to understand the following:

- Concept of charge and its types such as the idealised point charge and distributed charges like line charge, surface charge, and volume charge
- Coulomb's law and its application to the calculation of forces between point charges
- Concept of electric field

- Calculation of the electric field due to many charges, and its application to a dipole
- Calculation of the electric field for continuous charge distributions
- Concept of electric flux and the electric flux density vector
- Gauss's law

4.1 | Introduction

The knowledge of static electricity dates back to the earliest civilisations, but systematic study of this phenomenon started much later after the development of the scientific method. It was known to the ancient Greeks as early as 600 BC that when amber (the Greek word for amber is 'elektron') rubbed with wool, acquired the property of attracting light objects. In 1747, Benjamin Franklin (1706–1790) assumed that 'electric fire' (charge) is a common element existing in all bodies. If a body had more than its normal share of this fire it was positively charged, if less it was negatively charged. This was followed by experiments conducted by Joseph Priestley in 1766 who proposed that the force between electric charges follows an inverse square law. In 1777, Charles de Coulomb invented a torsion balance to measure the force between electrically charged objects.

Electrostatics is the study and analysis of the effects of stationary and almost stationary charges. These charges may be treated as point charges or distributed charges residing on the surfaces of conductors or dielectrics, or idealised as placed in some region of space. In most cases in the study of electrostatics we assume that the *effects due to the motion of charges are neglected* which in some cases may not be strictly true, but still gives us results which are corroborated by experiments.

When a material gets charged by rubbing or by other means, it has either a surplus or a deficit of electrons. A body with a surplus of electrons is said to be negatively charged and a body with a deficiency of electrons as positively charged. The amount or quantity of charge is expressed in Coulombs. A Coulomb is an enormous amount of charge, as will be clear through the examples given in the book, and in most electrostatic situations, levels of an extremely small fraction of a Coulomb give rise to significant effects. Electrostatic forces exist between charged bodies. Bodies with like charge experience repulsion between them, while oppositely-charged bodies experience attraction.

4.2 | Charge

From the study of atomic physics and chemistry, we know that the atom consists of electrons and protons which are negatively and positively charged respectively. Thus the smallest unit of charge is the charge on a proton or electron also called the elementary charge, is $e = 1.602 \times 10^{-19}$ C. We know that objects are made up of atoms which are neutral since protons and electrons are equal in numbers. When we rub an object with silk then due to transfer of charges between the objects a surplus or deficiency of charge develops, and the objects exhibit electrical properties, which is referred to as static electricity.

The SI unit of electric charge is Coulomb, which represents approximately 6.24×10^{18} elementary charges. One Coulomb is defined as the quantity of charge that passes through the cross-section of a conductor carrying 1 A of current in 1 s. The symbols Q and q are used to denote a quantity of electric charge.

The electrical charge is to the electric field as mass is to the gravitational field. Electrostatic fields are produced because of the existence of charge just like gravitational fields are produced due to the existence of mass. In electrostatic formulations charges have four types of idealisations, as shown in Fig. 4.1.

1. *Point charges* The idealisation here is that the whole charge Q is concentrated at a point. In this book such charges will be designated by Q or q.
2. *Line charges* The idealisation here is that the charge is distributed over a line or a curve. The unit of the charge distribution over a line is C/m and is referred to as 'linear charge density.' To calculate the line charge density, ρ_l, at a point we take a very small linear element Δl on the charged curve or line and find the

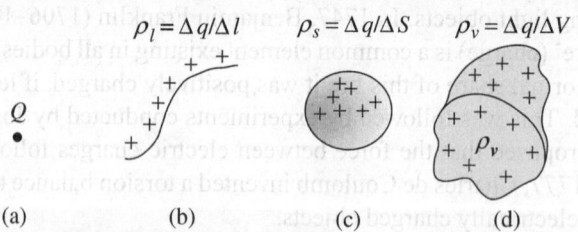

$$\rho_l = \Delta q/\Delta l \qquad \rho_s = \Delta q/\Delta S \qquad \rho_v = \Delta q/\Delta V$$

(a)   (b)   (c)   (d)

Fig. 4.1 Charge distributions: (a) point charge, (b) line charge, (c) surface charge, (d) volume charge

small charge, Δq Coulombs on it. It is obvious that the charge residing on the linear element Δl will be proportional to Δl itself. Thus, if Δl is comparatively large then the charge Δq residing on it, will also be large. And if Δl is small then Δq will be small. In essence, since the charge Δq will be proportional to Δl

$$\Delta q \approx \rho_l \, \Delta l$$

Or more precisely

$$\rho_l = \lim_{\Delta l \to 0} \left[\frac{\Delta q}{\Delta l} \right] \tag{4.1}$$

at that point. This can be done for all points on the curve to give a ρ_l which is a function of coordinates.

3. *Surface charge* Here the idealisation is that the charge is distributed over a surface. The unit of such a charge distribution is C/m². The mathematical definition of ρ_s at a point on the surface is obtained in a similar manner as earlier

$$\rho_s = \lim_{\Delta S \to 0} \left[\frac{\Delta q}{\Delta S} \right] \tag{4.2}$$

where Δq is the charge on a very small surface ΔS. This is done at all points on the surface to give a complete picture of the charge all over the surface.

4. *Volume charge* We take a small element of volume ΔV and find the total charge Δq contained in it. Then the volume charge density ρ_v in C/m³ at that point is

$$\rho_v = \lim_{\Delta V \to 0} \left[\frac{\Delta q}{\Delta V} \right] \tag{4.3}$$

4.2.1 Dirac Delta Function

The Dirac delta function is defined by two equations:

$$\delta(x) = \begin{cases} 0, & t \neq 0 \\ \infty, & t = 0 \end{cases} \tag{4.4}$$

$$\int_{-\infty}^{\infty} \delta(x)\,dx = 1 \tag{4.5}$$

The delta function can be written as the limit of some other function such as shown in Fig. 4.2. The pulse function is defined by

$$P(x) = \begin{cases} 1/e, & -e/2 < x < e/2 \\ 0, & \text{elsewhere} \end{cases} \tag{4.6}$$

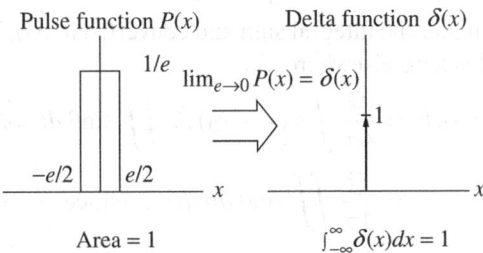

Fig. 4.2 Delta function as a limit of the pulse function

Notice that the area under the pulse is always equal to one, that is,

$$\int_{-\infty}^{\infty} P(x)\,dx = 1 \tag{4.7}$$

Now in the limit as $e \to 0$ the height of the pulse becomes infinity while the width of the pulse becomes zero. Hence

$$\lim_{e \to 0} P(x) = \delta(x) \tag{4.8}$$

Example 4.1   How would you express a point charge Q, mathematically, in rectangular coordinates using the Dirac delta function?

Solution   Having introduced the delta function let us look at the charge distribution

$$\rho_v = Q\delta(x)\,\delta(y)\,\delta(z)$$

which represents a point charge placed at the origin. The total charge of this distribution is

$$\iiint \rho_v\,dV = \iiint Q\delta(x)\,\delta(y)\,\delta(z)\,dxdydz$$

$$= Q\int \delta(x)\,dx \int \delta(y)\,dy \int \delta(z)\,dz$$

$$= Q \times 1 \times 1 \times 1$$

$$= Q$$

Having said this, let us look at the distribution

$$\rho_v = Q\delta(x - x')\,\delta(y - y')\,\delta(z - z')$$

where x', y', z' are constants. This represents a point charge of magnitude Q placed at (x', y', z').

Another application of the delta function is to specify the surface charge. For example, the surface charge

$$\rho_s = \frac{Q}{4\pi r_0^2}\delta(r - r_0)$$

is the uniform charge of magnitude Q distributed over a spherical surface with radius r_0. Let us integrate this charge

$$\iiint \frac{Q}{4\pi r_0^2}\delta(r - r_0)\,dV = \frac{Q}{4\pi r_0^2} \iiint \delta(r - r_0)\,r^2 dr\,\sin\theta\,d\theta\,d\phi$$

$$= \frac{Q}{4\pi r_0^2} \times r_0^2 \int \delta(r - r_0)\,dr \iint \sin\theta\,d\theta\,d\phi$$

We removed r from under the integral sign and converted it to r_0 because $\delta(r - r_0)$ exists only at r_0 and is zero elsewhere. So

$$\iiint \frac{Q}{4\pi r_0^2}\delta(r - r_0)\,dV = \frac{Q}{4\pi} \int \delta(r - r_0)\,dr \iint \sin\theta\,d\theta\,d\phi$$

$$= \frac{Q}{4\pi} \iint \sin\theta\,d\theta\,d\phi \quad (\text{since } \int \delta(r - r_0)\,dr = 1)$$

$$= Q \quad (\text{since } \iint \sin\theta\,d\theta\,d\phi = 4\pi)$$

Example 4.2 One million electrons are equally distributed over a linear region of 1 cm. Find the charge density.

Solution

Step 1. We are required to find the charge density, when 1 million (10^6) electrons (negative charges) are uniformly distributed over 1 cm (0.01 m).

Step 2. The charge of an electron is -1.602×10^{-19} C. So 10^6 electrons have a charge of

$$-1.602 \times 10^{-19} \times 10^6 = -1.602 \times 10^{-13} \text{ C}$$

Step 3. Since the charge is distributed over a region of 1 cm ($= 0.01$ m)

$$\text{Charge density } \rho_l = -1.602 \times 10^{-13} \div 0.01$$
$$= -1.602 \times 10^{-11} \text{ (C/m)}$$

Practice Problem 4.1

One million electrons are equally distributed over a region of 1 cm³. Find the charge density.

[Ans: $\rho_v = -1.602 \times 10^{-7}$ C/m³]

Example 4.3 The surface distribution of charges on a plane $z = 0$ is given by

$$\rho_s = 10^{-12} e^{-|x|} e^{-|y|}$$

Find the total charge in the region $\mathcal{R}$ bounded by $-5 \le x \le 5$ and $-3 \le y \le 3$.

Solution

Step 1. The total charge Q is given by

$$Q = \iint_{\mathcal{R}} \rho_s\,(x, y)\,dxdy$$

Step 2. Since $\mathcal{R}$ is the region bounded by $-5 \le x \le 5$ and $-3 \le y \le 3$

$$Q = \iint_{\mathcal{R}} \rho_s\,dxdy = 10^{-12} \iint_{\mathcal{R}} e^{-|x|} e^{-|y|}\,dxdy$$

Step 3. We have to take special care with each integral, namely:

$$\int_{-5}^{5} e^{-|x|}dx = \int_{-5}^{0} e^{x}dx + \int_{0}^{5} e^{-x}dx$$

$$= 2 \int_{0}^{5} e^{-x}dx \quad \text{(the two integrals are identical)}$$

Step 4. Therefore, we finally write

$$Q = 10^{-12} \left(\int_{-5}^{0} e^{x}dx + \int_{0}^{5} e^{-x}dx \right) \left(\int_{-3}^{0} e^{y}dy + \int_{0}^{3} e^{-y}dy \right)$$

$$= 10^{-12} \left[2\left(1 - e^{-5}\right) \right] \left[2\left(1 - e^{-3}\right) \right] \text{ C}$$

Practice Problem 4.2

The volume distribution of charges in a region $\mathcal{R}$ given by $-3 \leq x \leq 3$, $-3 \leq y \leq 3$, and $-3 \leq z \leq 3$ is given by

$$\rho_v = 10^{-14} \cos\left(\pi x/3\right) e^{-|x|} e^{-|y|} e^{-|z|}$$

Find the total charge in the region $\mathcal{R}$.

$$\left[\text{Ans: } 10^{-14}\left[2\left(1 - e^{-3}\right)\right]^2 \times 1.0014 \text{ C/m}^3\right]$$

Example 4.4 In a region in space, the charge density is given by

$$\rho_v = \begin{cases} e^{r_0} - e^r \ (\text{C/m}^3), & \text{when } 0 \leq r \leq r_0 \\ 0, & \text{when } r > r_0 \end{cases}$$

find the total charge enclosed in a volume of sphere of radius r.

Solution

Step 1. The total charge Q is given by

$$Q = \iiint_{\mathcal{R}} \rho_v \, dV$$

where $\mathcal{R}$ is a sphere of radius r.

Step 2. When $r \leq r_0$ there is a charge density (given by the statement of the problem) and when $r > r_0$ there is no charge density.

Step 3. Draw a rough sketch of the charge density. (Left for the students)

Step 4. Since we are considering spherical coordinates, we must write out the volume element in spherical coordinates, which is

$$dV = r^2 \sin\theta \, d\theta \, d\phi \, dr$$

as given in Section 4.1.5

Step 5. The amount of charge contained in such an element is $\rho_v \, r^2 \sin\theta \, d\theta \, d\phi \, dr$.

Step 6. Case (i) $r \leq r_0$ The total charge contained in a sphere of radius $r < r_0$ is

$$\iiint_{\text{Sphere, } r \leq r_0} \rho_v \, r^2 \sin\theta \, d\theta \, d\phi \, dr = \int_{\theta=0}^{\pi} \int_{\phi=0}^{2\pi} \int_{r=0}^{r} \left(e^{r_0} - e^r\right) r^2 \sin\theta \, d\theta \, d\phi \, dr$$

$$= \left(\int_{\theta=0}^{\pi} \sin\theta \, d\theta\right) \left(\int_{\phi=0}^{2\pi} d\phi\right) \left[\int_{r=0}^{r} \left(e^{r_0} - e^r\right) r^2 dr\right]$$

$$= (2)(2\pi) \left[\frac{r^3 e^{r_0}}{3} - \left(r^2 - 2r + 2\right) e^r + 2\right]\Bigg|_{r=0}^{r}$$

$$= 4\pi \left[\frac{r^3 e^{r_0}}{3} - \left(r^2 - 2r + 2\right) e^r + 2\right]$$

Case (ii) When $r > r_0$ the integral $\iiint \rho_v \, dv$ takes the value 0 as $\rho_v = 0$ for $r > r_0$. Therefore, the charge enclosed in a sphere of radius r_0 is given by putting $r = r_0$ in the expression obtained.

$$Q = 4\pi \left[\frac{r_0^3}{3} - \left(r_0^2 - 2r_0 + 2\right)\right] e^{r_0} + 8\pi$$

4.3 | Coulomb's Law and the Electric Field

It was proposed by Priestley, in 1767, that the law of electrical attraction was the same as that of gravitational attraction, namely that the strength of electrical attraction between oppositely charged bodies varies as the inverse square of the distance. It was much later, during the 1780's, that a French engineer, Charles Coulomb (1736–1806) investigated the quantitative relation of forces between charged objects. Using a torsion balance, created by Coulomb himself, he determined how an electric force varies as a function of the magnitude of the charges and the distance between them. Referring to Fig. 4.3, Coulomb found that

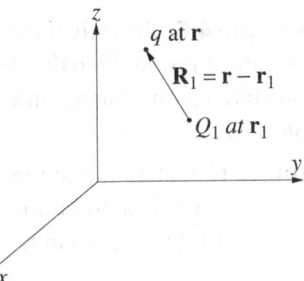

Fig. 4.3 Coulomb's law

1. The force exerted on a charge q by the charge Q was proportional to the product of the two charges and inversely proportional to the square of the distance between the two, i.e.,

$$\mathbf{F} \propto \frac{Qq}{R^2} \tag{4.9}$$

The above is known as Coulomb's law.

2. The force was repulsive when the two charges possessed the same polarity and attractive when the two charges were opposite in polarity. The direction of the force was along the line connecting the two bodies.

We can express these ideas, using vector notation, in an equation.

Figure 4.4 depicts two charges, Q_1 and q.

$\mathbf{r}_1$ is the position vector of Q_1.

$\mathbf{r}$ is the position vector of q, and

$\mathbf{R}_1 = \mathbf{r} - \mathbf{r}_1$

Then the force, $\mathbf{F}$, felt by charge q due to the presence of Q_1 is given by

$$\mathbf{F} = \frac{1}{4\pi\varepsilon}\left[\frac{Q_1 q}{R_1^2}\hat{\mathbf{R}}_1\right] = \frac{1}{4\pi\varepsilon}\left[\frac{Q_1 q\,(\mathbf{r}-\mathbf{r}_1)}{|\mathbf{r}-\mathbf{r}_1|^3}\right] \tag{4.10}$$

Fig. 4.4 Coulomb's law

where ε is called the permittivity of the medium in which the charges are placed. For vacuum, $\varepsilon = \varepsilon_0$ and $1/(4\pi\varepsilon_0) = 8.9898 \times 10^9$ with $\varepsilon_0 \approx (1/(36\pi)) \times 10^{-12}$ F/m. Note that air and vacuum have almost the same value of ε.

 See `CoulombForce.m` in Chapter 4

Electrostatic Spray Painting

In electrostatic spray painting or powder coating, the paint comprising tiny charged particles is sprayed. Similar charge on the particles causes them to repel each other while exiting the nozzle. The force of repulsion causes the particles to fan out and spread evenly on the surface. The spray is charged with one polarity while the surface is charged with the opposite polarity. Due to this, the paint is attracted to the surface and gives an even coat. The method also ensures that the paint adheres to the surface, reaches hard-to-reach areas and in general gives a far superior coating compared to ordinary spray painting.

Example 4.5 Find the force felt by a 1 nC test charge q due to a 1 nC charge Q at distances of 1 cm and 10 cm.

Solution The two charges *repel* each other with a force in accordance with Coulomb's law. Since no coordinate system is given we place one of the charges at the origin and the other at a distance of (i) 1 cm and (ii) 10 cm from it.

Case I:

$$F = \frac{q}{4\pi\varepsilon_0} \times \frac{Q}{R^2}$$

$$= \frac{1 \times 10^{-9}}{1.1127 \times 10^{-10}} \times \frac{1 \times 10^{-9}}{(0.01)^2}$$

$$= 8.987 \times 10^{-5} \text{ (N)}$$

Hence the force felt by either of the charges is 89.87 μN along the line joining the two charges.

Case II: At a distance of 10 cm the force will be 100 times weaker—due to the inverse square law. Hence the force on either of the two charges $= \frac{1}{100}(89.87)\,\mu N$ $= 0.899\,\mu N$ along the line joining the two charges.

Practice Problem 4.3
Find the force felt by a 1 pC test charge q due to a 1 μC charge Q at distances of (i) 1 cm and (ii) 10 cm.

[Ans: 89.87 μN at 1 cm and 898.7 nN at 10 cm]

Example 4.6 In a silver sphere weighing 1 g, 1 % of all the outermost single electrons are removed. Find the charge density inside the sphere. If the silver sphere was to act like a point charge, find the force on another charge of 1 nC placed 1 m from it.

Solution

Step 1. Calculate the number of atoms in 1 g of silver. We proceed as follows: silver has an atomic weight of 107.9; so 107.9 g of silver has 6.022×10^{23} $(= N_A)$ atoms. 1 g of silver has

$$N = \frac{6.022 \times 10^{23}}{107.9} = 5.581 \times 10^{21} \text{ atoms}$$

Step 2. 1 % of this number is $n = 5.581 \times 10^{19}$ atoms.

Step 3. Calculate the charge on the sphere. Due to removal of single electrons these atoms have an excess charge contributed by one proton $(-e = 1.602 \times 10^{-19}\text{C})$. Therefore, the excess charge is

$$Q = (5.581 \times 10^{19}) \times (1.602 \times 10^{-19}) = 8.542 \text{ C}$$

Step 4. Calculate the volume occupied by a gram of silver. The density of silver is 10.5×10^3 kg/m³. From this we compute that 1 g $(= 0.001$ kg) silver occupies a volume of

$$V = \frac{0.001}{10.5 \times 10^3} = 95.23 \times 10^{-9} \text{ m}^3$$

Step 5. Calculate volume charge density, ρ_v,

$$\rho_v = \frac{Q}{V} = \frac{8.852}{95.23 \times 10^{-9}} = 89.691 \text{ MC/m}^3$$

Step 6. Calculate the force. By Coulomb's law:

$$F = \frac{q}{4\pi\varepsilon_0} \times \frac{Q}{R^2} = \frac{1 \times 10^{-9}}{1.1127 \times 10^{-10}} \times \frac{8.542}{1^2} = 75.79 \text{ N}$$

which is quite a large force.

Example 4.7 Two metallic balls of 1 g each are suspended by 1 m long threads from a hook as shown in Fig. 4.5. Equal charges are placed on each of the two balls, and they are found to be separated from each other by 2 cm. Find the charge on each ball.

Solution

Step 1. Find the force on each ball. Using the diagram on the right of Fig. 4.5. If **T** is the tension in the string, mg the gravitational force on each ball, **F** the Coulomb force, then the two equations expressing the equilibrium of the forces on the ball are

$$T = mg \cos(\theta/2) + F \sin(\theta/2)$$
$$F \cos(\theta/2) = mg \sin(\theta/2)$$

Step 2. From the above equation

$$F = mg \tan(\theta/2)$$

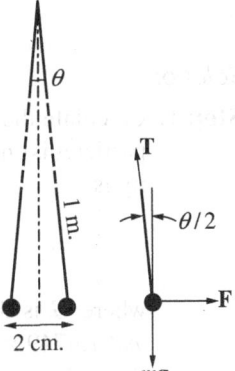

Fig. 4.5

Step 3. Calculate θ. From geometrical considerations and using $\sin(\theta/2) \approx \tan(\theta/2)$ when θ is small

$$F = mg \tan(\theta/2) \approx 0.001 \times 9.8 \times \frac{0.01}{1} = 9.8 \times 10^{-5} \text{ N}$$

Step 4. Calculate the charge Q. Using Coulomb's law

$$F = 9.8 \times 10^{-5} = \frac{Q^2}{4\pi\varepsilon_0 \times (0.02)^2}$$

giving $\qquad Q = 2.088 \text{ nC}$

This method may be used to accurately measure charges.

The Gold Leaf Electroscope

An application of measurement of charge based on Example 4.7 is the gold leaf electroscope by Exner (1910–1920). See Fig. 4.6. The basis on which the instrument works is the charging of two thin, extremely delicate gold leaves by an external charge introduced through a metal rod. Since similar charges cause the two leaves to repel each other, a scale is placed at the back of the assembly which directly reads off the charge residing on the leaves. Generally such an instrument would be calibrated through an experimental procedure rather than through analysis.

Example 4.8 In a hydrogen atom, compare the force of gravity between the electron and proton with the force of electrostatic attraction between them.

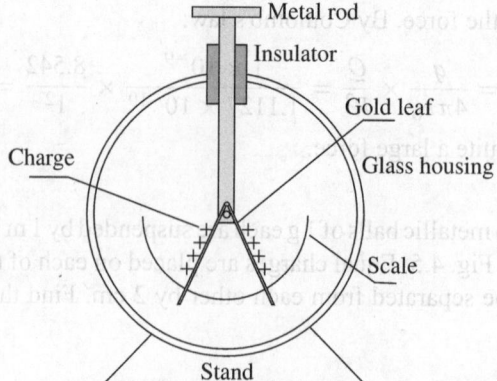

Fig. 4.6 Exner's gold-leaf electroscope

Solution

Step 1. Calculate the force of gravity between them. If the distance between the particles is one Bohr radius, $r_B = 0.53 \times 10^{-10}$ m then the force due to gravity is

$$F_G = G\frac{m_e m_p}{r_B^2}$$

where G is the gravitational constant and m_e (9.1094×10^{-31} kg) and m_p (1.6749×10^{-27} kg) are the masses of the electron and proton respectively. Using $G = 6.6726 \times 10^{-11}$ (mks units)

$$F_G = 3.61 \times 10^{-47} \text{ N}$$

This force is of attractive nature.

Step 2. Calculate the force of attraction between the charges. Using Coulomb's law

$$F_C = \frac{1}{4\pi\varepsilon_0}\frac{e^2}{r_B^2} = 8.212 \times 10^{-8} \text{ N}$$

Step 3. Compare the two types of forces

$$F_C/F_G = 2.27 \times 10^{39}$$

The electrostatic force of attraction is $\sim 10^{39}$ times the gravitational force.

4.3.1 Coulomb's Law and Superposition of Forces

When we examine Coulomb's law, we know that there is a force on one charge due to another charge. What happens when there are three charges: q, the test charge on which the force is felt, Q_1 and Q_2 as shown in Fig. 4.7. Now the force on q due to Q_1 is

$$F_1 = \frac{1}{4\pi\varepsilon_0}\left(\frac{Q_1 q}{R_1^3}R_1\right) = \frac{1}{4\pi\varepsilon_0}\left[\frac{Q_1 q(r-r_1)}{|r-r_1|^3}\right] \quad (4.11)$$

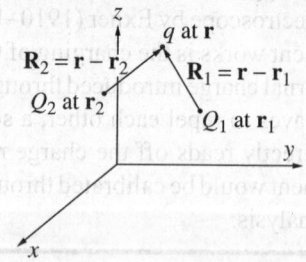

Fig. 4.7 Coulomb's law applicable to three charges

where $\mathbf{R}_1$ is the vector directed from Q_1 to q, $\mathbf{r}$ is the position vector of q, and $\mathbf{r}_1$ is the position vector of Q_1. Similarly

$$\mathbf{F}_2 = \frac{1}{4\pi\varepsilon_0}\left(\frac{Q_2 q}{R_2^3}\mathbf{R}_2\right) = \frac{1}{4\pi\varepsilon_0}\left[\frac{Q_2 q\,(\mathbf{r}-\mathbf{r}_2)}{|\mathbf{r}-\mathbf{r}_2|^3}\right] \qquad (4.12)$$

is the force on q due to Q_2. So the resultant force on the charge q is

$$\mathbf{F}_T = \mathbf{F}_1 + \mathbf{F}_2 = \frac{q}{4\pi\varepsilon_0}\left[\frac{Q_1\,(\mathbf{r}-\mathbf{r}_1)}{|\mathbf{r}-\mathbf{r}_1|^3} + \frac{Q_2\,(\mathbf{r}-\mathbf{r}_2)}{|\mathbf{r}-\mathbf{r}_2|^3}\right] \qquad (4.13)$$

Example 4.9 Find the force on a charge $q = 1\text{nC}$ located at the mid-point of two equal charges of $1\,\mu\text{C}$ located at $(1,1,1)$ and $(5,3,2)$ in rectangular coordinates.

Solution

Step 1. Find all the position vectors

$$\mathbf{r}_1(\text{location of } Q_1) = \mathbf{a}_x + \mathbf{a}_y + \mathbf{a}_z$$
$$\mathbf{r}_2(\text{location of } Q_2) = 5\mathbf{a}_x + 3\mathbf{a}_y + 2\mathbf{a}_z$$

$$\text{and } \mathbf{r}(\text{Location of the mid-point}) = \frac{(\mathbf{r}_1 + \mathbf{r}_2)}{2} = 3\mathbf{a}_x + 2\mathbf{a}_y + 1.5\mathbf{a}_z$$

Step 2. Draw a sketch and show all the vectors on the sketch.

So
$$\mathbf{R}_1 = \mathbf{r} - \mathbf{r}_1 = 2\mathbf{a}_x + \mathbf{a}_y + 0.5\mathbf{a}_z$$
$$\mathbf{R}_2 = \mathbf{r} - \mathbf{r}_2 = -2\mathbf{a}_x - \mathbf{a}_y - 0.5\mathbf{a}_z$$

where $\mathbf{R}_1, \mathbf{R}_2$ are the position vectors from Q_1, Q_2 to q.

Step 3. Since $\mathbf{R}_1 = -\mathbf{R}_2$ and $Q_1 = Q_2$, therefore, using Coulomb's law $\mathbf{F}_1 = -\mathbf{F}_2$ and $\mathbf{F}_T = F_1 + F_2 = 0$.

If we extend this concept to N charges, $Q_1, \ldots, Q_N$ located at $\mathbf{r}_1, \ldots, \mathbf{r}_N$ exerting Coulomb forces on q at $\mathbf{r}$ then

$$\mathbf{F}_T = \frac{q}{4\pi\varepsilon_0}\left[\frac{Q_1 q\,(\mathbf{r}-\mathbf{r}_1)}{|\mathbf{r}-\mathbf{r}_1|^3} + \frac{Q_1 q\,(\mathbf{r}-\mathbf{r}_2)}{|\mathbf{r}-\mathbf{r}_2|^3} + \cdots + \frac{Q_1 q\,(\mathbf{r}-\mathbf{r}_N)}{|\mathbf{r}-\mathbf{r}_N|^3}\right]$$
$$(4.14)$$

4.3.2 Electric Field

We know that the earth exerts a pull on the Moon, and the pull is mysterious, since, there is no visible physical connection between the two. Similarly, test charge q experiences a force when kept at a distance from Q_1. Rewriting Coulomb's law:

$$\mathbf{F}\{\text{Force on } q\} = q\left[\frac{1}{4\pi\varepsilon_0}\left(\frac{Q_1}{R_1^2}\hat{\mathbf{R}}_1\right)\right]$$
$$= q\mathbf{E}_1 \qquad (4.15)$$

The term in the square brackets, which we have called $\mathbf{E}_1$,

$$\mathbf{E}_1 = \frac{\mathbf{F}}{q} = \frac{1}{4\pi\varepsilon_0}\left(\frac{Q_1}{R_1^2}\hat{\mathbf{R}}_1\right) \qquad (4.16)$$

is independent of q. In the bracketed portion, there is no part played by q. However, $\hat{\mathbf{R}}_1$ being a unit vector $\mathbf{E}_1$ has the character of a vector.

Now as we move q around, it will always tend to feel a force 'created' by Q_1, no matter where it is moved. These arguments therefore make us tend to believe that the charge Q_1 produces an invisible 'force field' around itself, and any charges introduced into this 'field' are subjected to Coulomb's force. This *force field* is called the *electric field*.

Referring to Fig. 4.8, where no charge is placed at $\mathbf{r}$, the electric field at r due to Q_1 at r_1 is

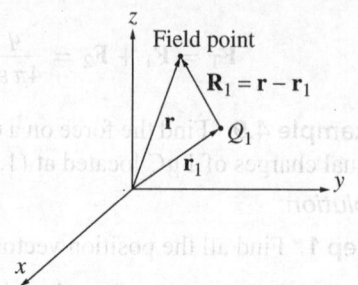

$$E = \frac{1}{4\pi\varepsilon_0} \times \frac{Q_1}{R_1^2}\hat{\mathbf{R}}_1 = \frac{1}{4\pi\varepsilon_0} \times \frac{Q_1\mathbf{R}_1}{R_1^3}$$

where $\mathbf{R}_1 = \mathbf{r} - \mathbf{r}_1$. The unit of $\mathbf{E}$ is

$$\frac{\text{Force}}{\text{Charge}} = \text{N/C in electrostatic units}$$

The electric field in electromagnetic units is V/m.

Fig. 4.8 Electric field at an arbitrary field point due to a point charge

Example 4.10 If two charges, one of 1 mC of mass 1 g and the other of 1 C of mass 1 kg, are placed in a uniform field of $\mathbf{E} = 1$ V/m, what will be forces and accelerations felt by them?
Solution Using $F = qE$, the 1 mC charge will experience a force of 1 mN and the 1 C charge, a force of 1 N. The accelerations of both will be the same. $1 \times (10^{-3}/10^{-3}) = 1$ m/s^2. $(\mathbf{a} = \mathbf{F}/m)$

Equation (4.16), also says that if the charge is *positive* then the electric field points radially *away* from the principal charge, in the direction $\mathbf{R}_1$ while if the charge is *negative* then the direction of the electric field points radially *towards* the principal charge, that is, in the direction $-\mathbf{R}_1$. The unit of the electric field leads to a relation between the force $\mathbf{F}$ which q experiences, and the external electric field $\mathbf{E}$

$$\mathbf{F} = q\mathbf{E} \tag{4.17}$$

This equation is valid even when the electric field is produced not only by a single point charge, but by *any external system of charges*. Note that a charge does not *feel a force due to the electric field produced by itself!* Figure 4.9 shows such a situation. There is an electric field, $\mathbf{E}(x, y, z)$, produced by *some* system of charges not shown in the figure. If charge q is immersed in this field, $\mathbf{E}$, it feels a force $\mathbf{F} = q\mathbf{E}(\mathbf{r})$.

 See `ForceEFld.m` in Chapter 4

Example 4.11 (a) Find the distance at which the E-field is 1.0 V/m due to a 1 C point charge. (b) If another 1.0 C point charge is placed at this point, find the force on it. See Fig. 4.10.
Solution
Step 1. (a) If we place the 1 C principal charge Q at the origin, the E-field due to it at a distance r from Q is given by

$$\mathbf{E}(\mathbf{r}) = \frac{1}{4\pi\varepsilon_0} \times \frac{Q}{r^2}\mathbf{a}_r$$

where $\mathbf{r}$ is the position vector of the field point.

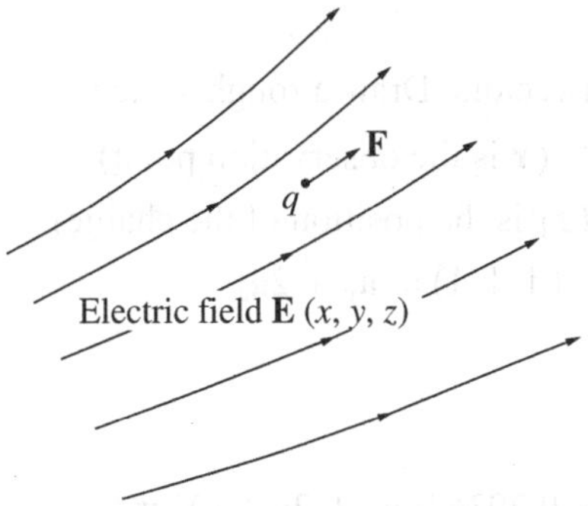

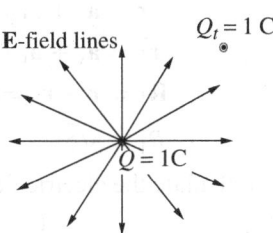

Fig. 4.9 Force felt on a charge q due to an external field $\mathbf{E}(x,y,z)$. $\mathbf{F}=q\mathbf{E}$

Fig. 4.10

Step 2. Find the distance R where $|\mathbf{E}|$ is 1.0 V/m.

$$|\mathbf{E}| = \left| \frac{1}{4\pi\,\varepsilon_0} \times \frac{Q}{R^2} \right| = \frac{1}{4\pi\,\varepsilon_0} \times \frac{Q}{R^2}$$

On putting the values of $E = 1$ V/m and $Q = 1$ C,

$$1 = \frac{1}{4\pi\,\varepsilon_0} \times \frac{1}{R^2}$$

giving $R^2 = 1/4\pi\,\varepsilon_0 = 8.988 \times 10^9$ (m), which gives $R = 94.8$ km.

From this result it is clear that 1 C is a very large charge indeed. A quick calculation shows that the E-field at a distance of 1 m from this charge is 8.988×10^9 V/m, which is a very intense field!

Step 3. (b) Calculate the force exerted on the 1 C test charge Q_t placed at a distance of 94.8 km.

$$F = Q_t E$$

Putting E as 1.0 V/m and Q as 1.0 C, $F = 1$ N.

___

Continuing with our discussion (refer to Fig. 4.4) the charge q also generates an electric field $\mathbf{E}_q$, of its own, and Q_1 feels a force due to q charge field. By the force equation:

$$\mathbf{F}_{\text{on } Q_1} = Q_1 \mathbf{E}_q \qquad (4.18)$$

where

$$\mathbf{E}_q = \frac{1}{4\pi\,\varepsilon_0} \times \frac{(-\hat{\mathbf{R}}_1)\,q}{R_1^2} \qquad (4.19)$$

Notice the negative sign on $\hat{\mathbf{R}}_1$. By examining the two equations, Eqns (4.16) and (4.19), we can see that though the $\mathbf{E}$ fields are different from each other but the forces on the two charges are equal and opposite to each other. The force that the principal charge, Q_1, exerts on the test charge, q, is equal and opposite to the force that test charge q exerts on the principle charge! This situation is the same as we find in the case of gravitational forces. The force that the Sun exerts on the Earth is equal and opposite to the force that the Earth exerts on the Sun.

Example 4.12 Find the electric field at $(1, 2, 3)$ (rectangular coordinates) due to a 1 nC point charge Q located at $(1, 1, 1)$.

Solution

Step 1. We first calculate all the position vectors. Draw a rough sketch.

$$\mathbf{r} = \mathbf{a}_x + 2\mathbf{a}_y + 3\mathbf{a}_z \qquad (\text{ } \mathbf{r} \text{ is the observation point})$$

$$\mathbf{r}_1 = \mathbf{a}_x + \mathbf{a}_y + \mathbf{a}_z \qquad (\text{ } \mathbf{r}_1 \text{ is the position of the charge})$$

$$\mathbf{R}_1 = \mathbf{r} - \mathbf{r}_1 = (1,2,3) - (1,1,1) = \mathbf{a}_y + 2\mathbf{a}_z$$

$$R_1 = \sqrt{5}$$

Step 2. Calculate the electric field.

$$\mathbf{E} = \frac{1}{4\pi\varepsilon_0} \times \frac{Q\mathbf{R}_1}{R_1^3} = 0.79364(\mathbf{a}_y + 2\mathbf{a}_z) \ (\text{V/m})$$

Measurement of the Electric Field

When we want to measure large electric fields from a moving vehicle, such as an aircraft, the 'field mill' is one of the best means for such a measurement. The field mill makes use of a rotating shutter or vane which exposes and then shields an electrostatic field-sampling probe cyclically.

This method produces a pulsating voltage on the probe that is proportional to the field strength. The signal is then averaged and amplified to give the relative field strength.

4.4 | Electric Field due to System of Point Charges

The electric field obeys the principle of superposition. This is so since the force is additive. Thus if more than one charge is present, then each charge produces its own electric field everywhere. The total field at any point in space is the *vector sum* of the electric fields produced by individual charges. To be specific, if single charge Q_1 produces an electric field $\mathbf{E}_1$ at the point $\mathbf{r}$ and the charge Q_2 produces an electric field $\mathbf{E}_2$ at the same point, then the total electric field at the point is the vector sum of the two fields

$$\mathbf{E}_{\text{Total}} = \mathbf{E}_1 + \mathbf{E}_2 \tag{4.20}$$

this is clear by examining Eqn (4.13).

This principle can be extended to the case of systems of charges. In other words, if one system of charges, Σ_1, produces an electric field $\mathbf{E}_1$ (no other fields being present) at the point $\mathbf{r}$ and another system of charges, Σ_2, produces an electric field $\mathbf{E}_2$ at the same point (no other fields being present), then the two systems of charges, Σ_1 and Σ_2, acting together, produce a field

$$\mathbf{E}_{\text{Total}} = \mathbf{E}_1 + \mathbf{E}_2 \tag{4.21}$$

at the same point $\mathbf{r}$. With the application of this principle we are able to compute the electric field due to any system of point charges distributed in space.

We now apply superposition to the case of the computation of the electric field due to the presence of only *two* point charges.

4.4.1 Electric Dipole

In this section we will examine the well-known problem of the electric dipole. The electric dipole consists of two charges of equal magnitude, one a positive charge Q, placed at $\mathbf{r}_1 = (0, d/2, 0)$ and the other a negative charge, $-Q$, placed at $\mathbf{r}_2 = (0, -d/2, 0)$. See Fig. 4.11. We have to find the electric field due to them. The study of the dipole has important consequences in the study of dielectrics.

Referring to the figure, we observe that there is symmetry about the y-axis. If we pass an arbitrary plane through the y-axis, the charge configuration on that plane will be identical to that in the x-y plane. Therefore, the fields produced will also be identical to that in the x-y plane. We conclude from this observation that the electric field will have a cylindrical symmetry about the y-axis.[1]

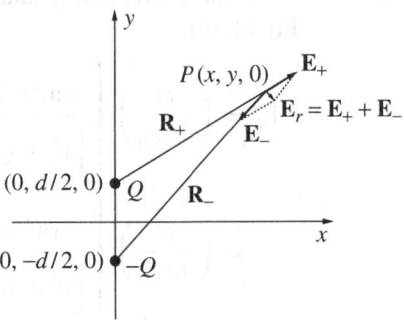

Fig. 4.11 A dipole

Hence the electric field is calculated only in the x-y plane ($z = 0$) where the observation point has the position vector $\mathbf{r} = (x, y, 0)$. We calculate the field due to each charge and then find the vector sum of the two at the field point. The electric field due to the positive charge Q is given by

$$\mathbf{E}_+ = \left(\frac{Q}{4\pi\varepsilon_0}\right)\frac{\hat{\mathbf{R}}_+}{R_+^2} \tag{4.22}$$

where
$$\mathbf{R}_+ = \mathbf{r} - \mathbf{r}_1 = [x, y, 0] - [0, d/2, 0]$$
$$= [x, (y - d/2), 0] \tag{4.23}$$

$$\hat{\mathbf{R}}_+ = \frac{[x, (y - d/2), 0]}{\sqrt{x^2 + (y - d/2)^2}} \tag{4.24}$$

and
$$R_+^2 = x^2 + (y - d/2)^2 \tag{4.25}$$

and the $\mathbf{E}$-field due to the negative Q charge is

$$\mathbf{E}_- = \left(\frac{-Q}{4\pi\varepsilon_0}\right)\frac{\hat{\mathbf{R}}_-}{R_-^2} \tag{4.26}$$

where
$$\mathbf{R}_- = \mathbf{r} - \mathbf{r}_2 = [x, y, 0] - [0, -d/2, 0]$$
$$= [x, (y + d/2), 0] \tag{4.27}$$

$$\hat{\mathbf{R}}_- = \frac{[x, (y + d/2), 0]}{\sqrt{x^2 + (y + d/2)^2}} \tag{4.28}$$

$$R_-^2 = x^2 + (y + d/2)^2 \tag{4.29}$$

The resultant electric field $\mathbf{E}_r$ is

$$\mathbf{E}_r = \mathbf{E}_+ + \mathbf{E}_- = \left(\frac{Q}{4\pi\varepsilon_0}\right)\left\{\frac{[x, (y - d/2), 0]}{\left[x^2 + (y - d/2)^2\right]^{3/2}} - \frac{[x, (y + d/2), 0]}{\left[x^2 + (y + d/2)^2\right]^{3/2}}\right\} \tag{4.30}$$

---

[1]Reader is advised to look for specific symmetries as that will lead to simplification of the problem.

Example 4.13 Find the resultant electric field on the x-axis and y-axis for a dipole.
Solution

Step 1. Calculate $\mathbf{E}_r$ on the x-axis:

On the x-axis both y and z are equal to zero. Putting $y = 0$, $z = 0$ in Eq. (4.30)

$$\mathbf{E}_r = \left(\frac{Q}{4\pi\varepsilon_0}\right) \left\{ \frac{x\mathbf{a}_x + (y - d/2)\,\mathbf{a}_y}{\left[x^2 + (y - d/2)^2\right]^{3/2}} - \frac{x\mathbf{a}_x + (y + d/2)\,\mathbf{a}_y}{\left[x^2 + (y + d/2)^2\right]^{3/2}} \right\}_{y=0}$$

$$= \left(\frac{Q}{4\pi\varepsilon_0}\right) \left\{ \frac{x\mathbf{a}_x - d/2\mathbf{a}_y}{\left[x^2 + (d/2)^2\right]^{3/2}} - \frac{x\mathbf{a}_x + d/2\mathbf{a}_y}{\left[x^2 + (d/2)^2\right]^{3/2}} \right\}$$

$$= -\left(\frac{Q}{4\pi\varepsilon_0}\right) \frac{d\mathbf{a}_y}{\left[x^2 + (d/2)^2\right]^{3/2}} \tag{4.31}$$

Step 2. Calculate $\mathbf{E}_r$ on y-axis

On the y-axis, $x = 0$, $z = 0$,

$$\mathbf{E}_r = \left(\frac{Q}{4\pi\varepsilon_0}\right) \left\{ \frac{x\mathbf{a}_x + (y - d/2)\,\mathbf{a}_y}{\left[x^2 + (y - d/2)^2\right]^{3/2}} - \frac{x\mathbf{a}_x + (y + d/2)\,\mathbf{a}_y}{\left[x^2 + (y + d/2)^2\right]^{3/2}} \right\}_{x=0}$$

$$= \left(\frac{Q}{4\pi\varepsilon_0}\right) \left\{ \frac{(y - d/2)\,\mathbf{a}_y}{(y - d/2)^3} - \frac{(y + d/2)\,\mathbf{a}_y}{(y + d/2)^3} \right\}$$

$$= -\left(\frac{Q}{4\pi\varepsilon_0}\right) \frac{d\mathbf{a}_y}{(y - d/2)^3}$$

It is difficult to visualise the electric field in all cases. If the charge distribution is simple, we can visualise the electric field. For example for the case of a lone positive charge at the origin, the electric field streams away in the $\mathbf{a}_r$ direction, that is, radially away from the charge. Since engineers need to get an idea of the field, numerical procedures exist which enable one to plot the direction of the field in any desired plane. These are called streamline or field plots which are defined by the equations

$$\frac{dx}{E_x} = \frac{dy}{E_y} = \frac{dz}{E_z} \tag{4.32}$$

For example, if we are interested in a field plot in the x-y plane, then we put $z = 0$ and plot only

$$\frac{dy}{dx} = \frac{E_y}{E_x} \tag{4.33}$$

Figure 4.12 shows the field plot of the electric field for a dipole with inter-charge distance $d = 1$ m. The $\mathbf{a}_x$ and $\mathbf{a}_y$ components of the electric field, namely E_x and E_y are computed on grid points which cover the whole region. The unit vector in the direction of the electric field is attached to the grid point as shown in the figure. In our case, the region of interest is a square of 2 m × 2 m in extent which is covered by a

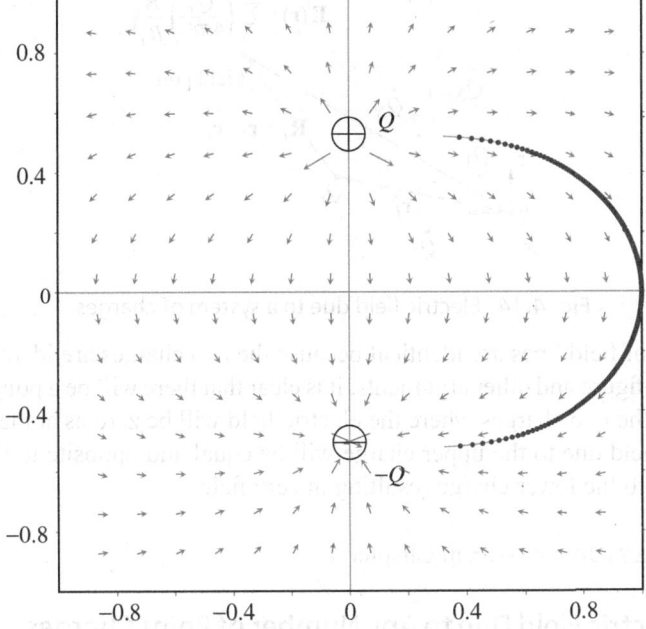

Fig. 4.12 Field plot for the dipole, $d = 1$ m

grid of 14×14. One streamline is drawn as an example. The streamline is calculated as follows. A small distance is moved in the direction of $\Delta y / \Delta x$ calculated from Eqn (4.33). Then the process is repeated at the new point till the streamline ends.

Let us see how the field lines look when two similar charges have equal values. With each of the two charges being a positive charge Q, the field plot is illustrated in Fig. 4.13. The following observations may be made from it.

1. The electric field streams out from each charge, but the lines do not cross each other.

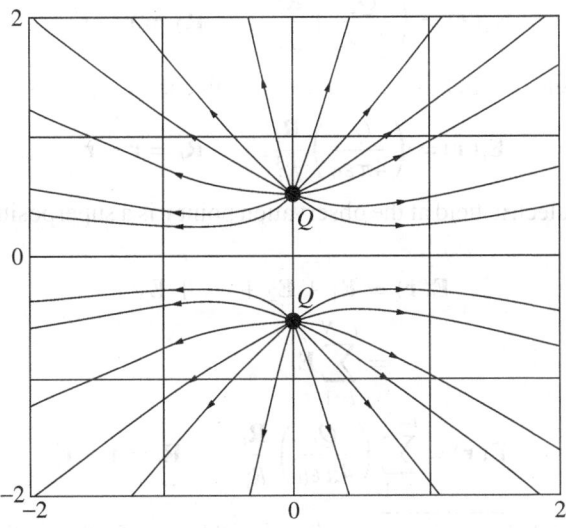

Fig. 4.13 Field plot for two equal charges of magnitude Q and $d = 1$ m

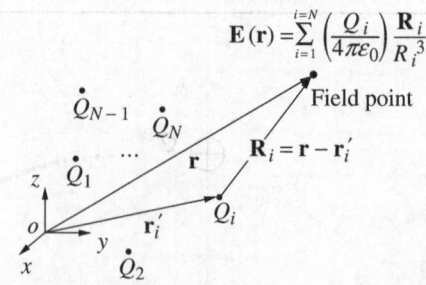

Fig. 4.14 Electric field due to a system of charges

2. Each set of field lines are identical because the two charges are identical.
3. From the figure and other arguments, it is clear that there will be a point, midway between the two charges where the electric field will be zero as at that point the electric field due to the upper charge will be equal and opposite to the electric field due to the lower charge resulting in zero field.

 See `EFldDipole.m` in Chapter 4

4.4.2 Electric Field Due to Any Number of Point Charges

We consider next, the total field due to a system of point charges at an arbitrary point $\mathbf{r}(=[x,y,z])$. Consider a set of point charges $Q_1, Q_2, \ldots, Q_N$ placed at positions $\mathbf{r}'_i = (x'_i, y'_i, z'_i), i = 1, \ldots, N$, as depicted in Fig. 4.14.[2]

The electric field due to Q_1 placed at $\mathbf{r}'_1$ and observed at the field point $\mathbf{r}$ is given by

$$E_1(\mathbf{r}) = \left(\frac{Q_1}{4\pi\varepsilon_0}\right)\frac{\mathbf{R}_1}{R_1^3}, \qquad \mathbf{R}_1 = \mathbf{r} - \mathbf{r}'_1$$

Similarly, the electric field due to the second charge, Q_2 located at $\mathbf{r}'_2$ and observed at the *same* field point $\mathbf{r}$ is given by

$$E_2(\mathbf{r}) = \left(\frac{Q_2}{4\pi\varepsilon_0}\right)\frac{\mathbf{R}_2}{R_2^3}, \qquad \mathbf{R}_2 = \mathbf{r} - \mathbf{r}'_2$$

In this way, the electric field E_i at $\mathbf{r}$, due to the ith charge Q_i whose location is $\mathbf{r}'_i$ is

$$E_i(\mathbf{r}) = \left(\frac{Q_i}{4\pi\varepsilon_0}\right)\frac{\mathbf{R}_i}{R_i^3}, \qquad \mathbf{R}_i = \mathbf{r} - \mathbf{r}'_i \qquad (4.34)$$

Hence the total electric field at the observation point $\mathbf{r}$ is a superposition of all these fields. Hence,

$$E(\mathbf{r}) = E_1 + E_2 + \cdots + E_N$$

$$= \sum_{i=1}^{i=N} E_i$$

or
$$E(\mathbf{r}) = \sum_{i=1}^{i=N} \left(\frac{Q_i}{4\pi\varepsilon_0}\right)\frac{\mathbf{R}_i}{R_i^3} \qquad \mathbf{R}_i = \mathbf{r} - \mathbf{r}'_i \qquad (4.35)$$

---

[2]Throughout the book the prime notation, $\mathbf{r}'_i$ or $\mathbf{r}'$, will be used for the position vectors of the charges which produce the field.

which, in rectangular coordinates becomes

$$E(\mathbf{r}) = \sum_{i=1}^{i=N} \left(\frac{Q_i}{4\pi\varepsilon_0}\right) \frac{\{(x-x_i')\,\mathbf{a}_x + (y-y_i')\,\mathbf{a}_y + (z-z_i')\,\mathbf{a}_z\}}{\left\{\sqrt{(x-x_i')^2 + (y-y_i')^2 + (z-z_i')^2}\right\}^3} \tag{4.36}$$

Example 4.14 Find the electric field at the centre of an equilateral triangle whose corners have equal charges Q.

Solution

Step 1. Draw a sketch of the arrangement. See Fig. 4.15. The charges are placed at the corner of an equilateral triangle. Set up the coordinate system with the origin at the centre of triangle.

Step 2. Let the three charges have position vectors $\mathbf{r}_1'$, $\mathbf{r}_2'$, and $\mathbf{r}_3'$. If $\mathbf{r}_1'$ makes an angle θ with the x-axis and its magnitude is A, then

$$\mathbf{r}_1' = A\cos\theta\,\mathbf{a}_x + A\sin\theta\,\mathbf{a}_y$$
$$\mathbf{r}_2' = A\cos(\theta + 2\pi/3)\,\mathbf{a}_x + A\sin(\theta + 2\pi/3)\,\mathbf{a}_y$$
$$\mathbf{r}_3' = A\cos(\theta + 4\pi/3)\,\mathbf{a}_x + A\sin(\theta + 4\pi/3)\,\mathbf{a}_y$$

Note that $\mathbf{r}_2'$ and $\mathbf{r}_3'$ are $\mathbf{r}_1'$ rotated by 120° and 240°, respectively.

Step 3. We calculate the electric field at the origin, $\mathbf{r} = 0$. The $\mathbf{R}_i$ $i = 1,\ldots,3$ of Eqn (4.35) are given by

$$\mathbf{R}_1 = -A\cos\theta\,\mathbf{a}_x - A\sin\theta\,\mathbf{a}_y$$
$$\mathbf{R}_2 = -A\cos(\theta + 2\pi/3)\,\mathbf{a}_x - A\sin(\theta + 2\pi/3)\,\mathbf{a}_y$$
$$\mathbf{R}_3 = -A\cos(\theta + 4\pi/3)\,\mathbf{a}_x - A\sin(\theta + 4\pi/3)\,\mathbf{a}_y$$

Therefore the electric field is

$$\mathbf{E} = \left(\frac{Q}{4\pi\varepsilon_0}\right)\frac{\mathbf{R}_1}{R_1^3} + \left(\frac{Q}{4\pi\varepsilon_0}\right)\frac{\mathbf{R}_2}{R_2^3} + \left(\frac{Q}{4\pi\varepsilon_0}\right)\frac{\mathbf{R}_3}{R_3^3}$$

$$= \left(\frac{Q}{4\pi\varepsilon_0 A^3}\right)\underbrace{(\mathbf{R}_1 + \mathbf{R}_2 + \mathbf{R}_3)}_{} = 0 \quad \text{as } \mathbf{R}_1 + \mathbf{R}_2 + \mathbf{R}_3 = 0$$

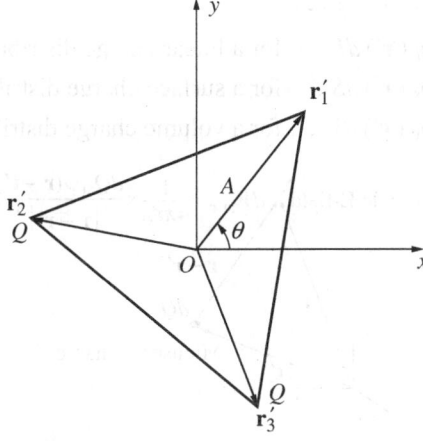

Fig. 4.15 Charges placed at the corners of an equilateral triangle

Step 4. To prove that $\mathbf{R}_1 + \mathbf{R}_2 + \mathbf{R}_3 = 0$, write out the expression for $\mathbf{R}_1 + \mathbf{R}_2 + \mathbf{R}_3$ and simplify using the following trigonometric relations:

$$\cos(\theta + 2\pi/3) = \cos\theta \cos(2\pi/3) - \sin\theta \sin(2\pi/3)$$

$$= -\frac{1}{2}\cos\theta - \frac{\sqrt{3}}{2}\sin\theta$$

$$\cos(\theta + 4\pi/3) = -\frac{1}{2}\cos\theta + \frac{\sqrt{3}}{2}\sin\theta$$

so, $$\cos\theta + \cos(\theta + 2\pi/3) + \cos(\theta + 4\pi/3) = 0$$

In the same way it can be shown:

$$\sin\theta + \sin(\theta + 2\pi/3) + \sin(\theta + 4\pi/3) = 0$$

Hence, $\mathbf{R}_1 + \mathbf{R}_2 + \mathbf{R}_3 = 0$.

4.5 | Electric Field due to Continuous Charge Distributions

We have obtained the electric field due to any arbitrary distribution of *point* charges, $Q_1 \ldots Q_N$ by applying the method of superposition. We can obtain, by using this result, the electric field due to any continuous charge distribution, such as any system of *volume, surface,* and *linear* charge distributions.

To proceed we have to first concentrate on the electric field produced at a point $\mathbf{r}$ by a minuscule point charge dQ placed at the position vector $\mathbf{r}'$ as shown in Fig. 4.16. Since the charge is minuscule, its field is also minuscule and given by

$$d\mathbf{E}_{\text{at } \mathbf{r}} = \frac{1}{4\pi\varepsilon_0} \times \frac{dQ_{\text{at } \mathbf{r}'}(\mathbf{r} - \mathbf{r}')}{|\mathbf{r} - \mathbf{r}'|^3}$$

If we compare this equation with the equation of a point charge, it is the same except that the charge and the field are both minuscule quantities. The field, as usual, is a radial field with the field lines emanating from the charge. The *charge dQ* will have different expressions for different kinds of charge distributions. If we want to compute the electric field due a linear charge distribution ρ_l, then $dQ = \rho_l\, dL'$; if we are considering on a surface with a surface charge distribution ρ_s, then, $dQ = \rho_s\, dS'$; and if we have a volume charge distribution, ρ_v, then $dQ = \rho_v\, dV'$. Putting down these facts in equation forms gives

$$dQ = \begin{cases} \rho_l\,(\mathbf{r}')\,dL', & \text{for a linear charge distribution} \\ \rho_s\,(\mathbf{r}')\,dS', & \text{for a surface charge distribution} \\ \rho_v\,(\mathbf{r}')\,dV', & \text{for a volume charge distribution} \end{cases} \qquad (4.37)$$

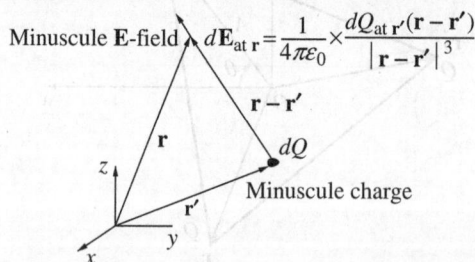

Fig. 4.16 Minuscule electric field, *dE* produced by a minuscule charge *dQ*

The resultant electric field due to whole charge distribution is obtained by integrating the expression for dE. Hence for the case of a line charge, with ρ_l being the linear charge density,

$$\mathbf{E}(\mathbf{r}) = \int_{\mathcal{L}'} \left(\frac{\rho_l \, dL'}{4\pi\varepsilon_0} \right) \frac{(\mathbf{r} - \mathbf{r}')}{|\mathbf{r} - \mathbf{r}'|^3} \tag{4.38}$$

$\rho_l \equiv \rho_l(\mathbf{r}')$ and $\mathcal{L}'$ is the linear region over which the integration is performed.

In the same manner we can derive the electric field at a distant point due to surface charge and volume charge distributions.

4.5.1 Infinite Line Charge

In this section we will apply the concepts and equations just discussed to a specific case, which is a infinite line charge with a linear charge density ρ_l (C/m) = constant. The line charge is placed along the z-axis of a coordinate system, as shown in Fig. 4.17. The infinite line charge is the most basic of all cylindrical structures. The other cylindrical structure like the charged infinite straight wire of finite thickness has a field which is similar to one produced by this structure. From this configuration we may also derive the electric field of other configurations—for example, for the two conductor lines.

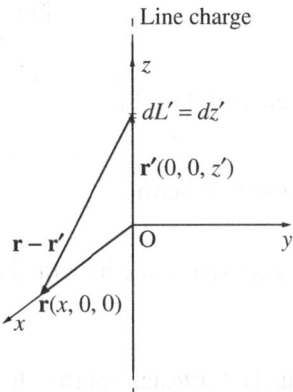

Fig. 4.17 Infinite line charge of density ρ_l C/m

To compute the electric field due to line charge at an arbitrary point, we have to apply Eqn (4.38) to the configuration shown in the figure.

In the most general case $\mathbf{r} = [x, y, z]$ is the field point and $\mathbf{r}' = [x', y', z']$ is the point where the charge producing the field lies. So $\mathbf{r} - \mathbf{r}' = [x - x', y - y', z - z']$. The three components of the electric field using the general Eqn (4.38) may be spelt out as

$$E_x = \int_{z'} \left(\frac{\rho_l \, dz'}{4\pi\varepsilon_0} \right) \frac{x - x'}{\left[(x - x')^2 + (y - y')^2 + (z - z')^2 \right]^{3/2}}$$

$$E_y = \int_{z'} \left(\frac{\rho_l \, dz'}{4\pi\varepsilon_0} \right) \frac{y - y'}{\left[(x - x')^2 + (y - y')^2 + (z - z')^2 \right]^{3/2}}$$

$$E_z = \int_{z'} \left(\frac{\rho_l \, dz'}{4\pi\varepsilon_0} \right) \frac{z - z'}{\left[(x - x')^2 + (y - y')^2 + (z - z')^2 \right]^{3/2}}$$

It may be seen that $\mathbf{r}'$ is constrained to lie on the z-axis since the line charge is along the z-axis. This translates to the equation $\mathbf{r}' = [x' = 0, y' = 0, z']$. The integration is performed over the z' coordinate ($\rho_l \, dL' \equiv \rho_l \, dz'$). Furthermore, from the symmetry of the problem it is clear that the electric field will be the same along any plane parallel to the x-y plane. Hence we can simply evaluate the field specifically, on the x-axis: $\mathbf{r} = [x, 0, z = 0]$ and obtain values of the field everywhere. Using these values: $\mathbf{r} - \mathbf{r}' = [x, 0, -z']$. The integration is performed in the interval $z' \equiv [-\infty, \infty]$,

$$\mathbf{E} = \int_{z'} \frac{\rho_l \, (\mathbf{r} - \mathbf{r}') \, dz'}{4\pi\varepsilon_0 \, |\mathbf{r} - \mathbf{r}'|^3} = \int_{-\infty}^{\infty} \frac{\rho_l \, dz'}{4\pi\varepsilon_0} \left[\frac{x}{(z'^2 + x^2)^{\frac{3}{2}}}, 0, -\frac{z'}{(z'^2 + x^2)^{\frac{3}{2}}} \right] \tag{4.39}$$

Looking at these terms one at a time, we examine the $\mathbf{a}_x$ part of the field. (Keep x constant!), the indefinite integral (see Appendix B, integral number 29)

$$\int \frac{dz'}{(x^2 + z'^2)^{3/2}} = \frac{z'}{x^2 \sqrt{z'^2 + x^2}} \tag{4.40}$$

Therefore

$$\int \frac{\rho_l \, x dz'}{\left(z'^2 + x^2\right)^{\frac{3}{2}}} = \rho_l \, x \int \frac{dz'}{\left(z'^2 + x^2\right)^{\frac{3}{2}}} = \frac{\rho_l \, x z'}{x^2 \sqrt{z'^2 + x^2}} \tag{4.41}$$

which the reader may verify by differentiation. Putting the integration limits at $z' = \pm\infty$, we get

$$\lim_{z' \to \pm\infty} \frac{\rho_l \, x z'}{x^2 \sqrt{z'^2 + x^2}} = \frac{\rho_l \, z'}{x \sqrt{z'^2}} = \pm\rho_l/x$$

and therefore E_x is

$$E_x = \frac{\rho_l}{2\pi \, \varepsilon_0 x} \tag{4.42}$$

The third term

$$- \int \frac{\rho_l \, z' dz'}{\left(z'^2 + x^2\right)^{\frac{3}{2}}} \tag{4.43}$$

is zero since the integrand is an odd function of z'. Thus, the total electric field is

$$\mathbf{E} = \frac{\rho_l}{2\pi \, \varepsilon_0 x} \mathbf{a}_x \tag{4.44}$$

in the Cartesian coordinate system. Due to the rotational symmetry of the problem, cylindrical coordinates will yield a simpler result

$$\mathbf{E} = \frac{\rho_l}{2\pi \, \varepsilon_0 \rho} \mathbf{a}_\rho \tag{4.45}$$

as due to symmetry, $\mathbf{a}_x$ will become $\mathbf{a}_\rho$; also x is the distance from the line charge to the observation point, and so x may be replaced by ρ. Note that the result is the same as the one obtained by application of Gauss's law.

To help the reader visualise the fields, the field plots for the infinite line charge in the x–y plane and also the plot for any plane passing through the z-axis are shown in Figs 4.18 and 4.19. The field is visualised as streaming away from the charged line cylindrically since there is one component, namely the ρ directed component.

Since

$$E_\rho \propto \frac{1}{\rho}$$

so as we approach the z-axis ($\rho \to 0$) the field becomes more and more intense and tends to infinity. As we increase the value of ρ, meaning moving away from the line charge, the field falls off towards zero.

4.5.1.1 Short Line Charge

Example 4.15 Compute the electric field at point P, on the x-y plane due to a short line charge of length $2a$ as shown in Fig. 4.20.

Solution

Step 1. The electric field is given by

$$\mathbf{E}(\mathbf{r}) = \int_{L'} \left(\frac{\rho_l \, dl'}{4\pi \varepsilon_0} \right) \frac{(\mathbf{r} - \mathbf{r}')}{|\mathbf{r} - \mathbf{r}'|^3} = \int_{z=-a}^{z=a} \left(\frac{\rho_l \, dz'}{4\pi \varepsilon_0} \right) \frac{(x\mathbf{a}_x - z'\mathbf{a}_z)}{\left(x^2 + z'^2\right)^{3/2}}$$

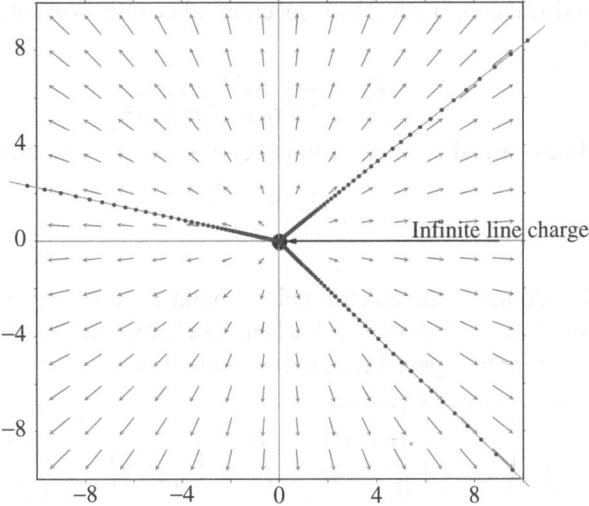

Fig. 4.18 Streamline plot of the infinite line charge in the *x*-*y* plane

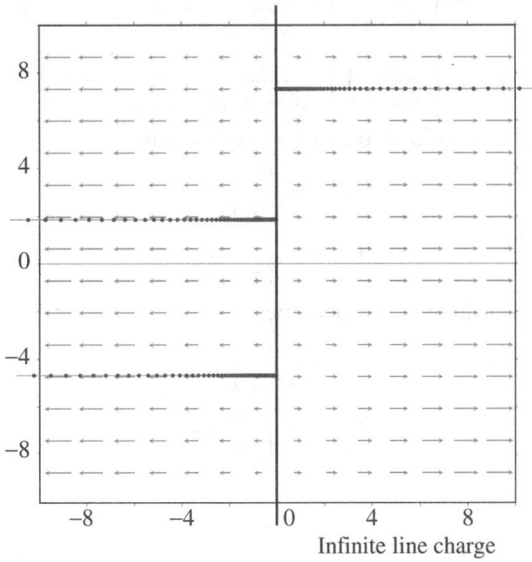

Infinite line charge

Fig. 4.19 Streamline plot for the infinite line charge in a plane passing through the *z*-axis

Step 2. The *x*-component of the electric field is

$$E_x = \frac{\rho_l}{4\pi\varepsilon_0} \int_{z=-a}^{z=a} \frac{x\,dz'}{\left(x^2 + z'^2\right)^{3/2}}$$

$$= \frac{\rho_l}{4\pi\varepsilon_0} \left. \frac{z'}{x\sqrt{z'^2 + x^2}} \right|_{z'=-a}^{z'=a} = \frac{\rho_l\,a}{2\pi\varepsilon_0\,x\sqrt{a^2 + x^2}}$$

using the tables of integrals given in Appendix B.

Step 3. Also from arguments given earlier

$$E_z = 0$$

Step 4. In cylindrical coordinates, by inspection, x becomes ρ and $\mathbf{a}_x$ becomes $\mathbf{a}_\rho$. Hence

$$E_\rho = \frac{\rho_l \, a}{2\pi \varepsilon_0 \, \rho \, \sqrt{a^2 + \rho^2}}$$

Step 5. If the line charged becomes infinitely long then $a \to \infty$ and

$$E_\rho = \frac{\rho_l}{2\pi \varepsilon_0 \, \rho}$$

Example 4.16 Compute the electric field at point P of a short line charge (as shown in Fig. 4.20) but of length $a + b$ when the charge exists from $-a \le z \le b$. Calculate the fields for the special case of $a = 0$ and $b = \infty$.

Solution The electric field is given by

$$\mathbf{E}(\mathbf{r}) = \int_{\mathcal{L}'} \left(\frac{\rho_l \, dl'}{4\pi \varepsilon_0}\right) \frac{(\mathbf{r} - \mathbf{r}')}{|\mathbf{r} - \mathbf{r}'|^3} = \int_{z=-a}^{z=b} \left(\frac{\rho_l \, dz'}{4\pi \varepsilon_0}\right) \frac{(x\mathbf{a}_x - z'\mathbf{a}_z)}{\left(x^2 + z'^2\right)^{3/2}}$$

and using the two integrals, we calculate the fields

$$E_x = \left(\frac{\rho_l}{4\pi \varepsilon_0 \, x}\right) \left(\frac{b}{\sqrt{b^2 + x^2}} - \frac{a}{\sqrt{a^2 + x^2}}\right)$$

and

$$E_z = \left(\frac{\rho_l}{4\pi \varepsilon_0}\right) \left(\frac{1}{\sqrt{b^2 + x^2}} - \frac{1}{\sqrt{a^2 + x^2}}\right)$$

Now for the special case of $a = 0$ and $b = \infty$, we have

$$E_x = \frac{\rho_l}{4\pi \varepsilon_0 \, x}$$

$$E_z = -\frac{\rho_l}{4\pi \varepsilon_0 \, x^2}$$

and in cylindrical coordinates this becomes

$$E_\rho = \frac{\rho_l}{4\pi \varepsilon_0 \, \rho}$$

$$E_z = -\frac{\rho_l}{4\pi \varepsilon_0 \, \rho^2}$$

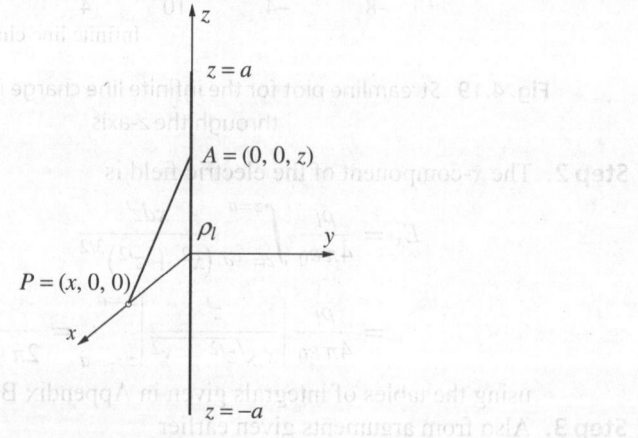

$z = a$

$A = (0, 0, z)$

ρ_l

y

$P = (x, 0, 0)$

x

$z = -a$

Fig. 4.20 A short line of charge

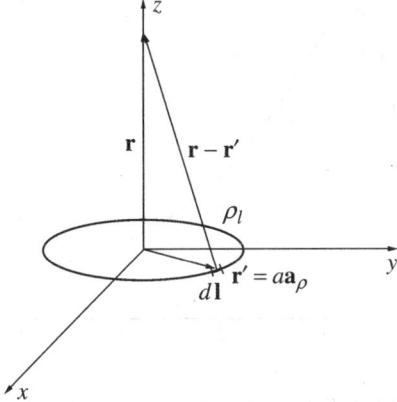

Fig. 4.21 A ring of charge

4.5.1.2 Ring of Charge

Example 4.17 Compute the electric field at a point on the axis of a ring of charge of radius a as shown in Fig. 4.21. What happens to the field when $a \ll z$?

Solution

Step 1. To calculate the electric field at $\mathbf{r} = (0, 0, z)$ due to a ring placed on the x-y plane, we use

$$\mathbf{E}(\mathbf{r}) = \int_{L'} \left(\frac{\rho_l \, dl'}{4\pi\varepsilon_0} \right) \frac{(\mathbf{r} - \mathbf{r}')}{|\mathbf{r} - \mathbf{r}'|^3} = \frac{\rho_l}{4\pi\varepsilon_0} \int \frac{a \, d\phi' \, (z\mathbf{a}_z - a\mathbf{a}_\rho)}{(a^2 + z^2)^{3/2}}$$

Step 2. Examine for symmetry: Due to symmetry the $\mathbf{a}_\rho$ component will be zero and therefore

$$\mathbf{E}(z) = \frac{\rho_l}{4\pi\varepsilon_0} \frac{2\pi \, az\mathbf{a}_z}{(a^2 + z^2)^{3/2}} = \frac{\rho_l \, az}{2\varepsilon_0 \, (a^2 + z^2)^{3/2}} \mathbf{a}_z$$

Step 3. When $a \ll z$ (but $a \neq 0$) the ring behaves like a point charge of magnitude $Q = 2\pi a\rho_l$. Then

$$\mathbf{E}(z) = \lim_{a \to 0} \frac{\rho_l \, az}{\varepsilon_0 \, (a^2 + z^2)^{3/2}} \mathbf{a}_z = \frac{Q}{4\pi\varepsilon_0 z^2} \mathbf{a}_z$$

which is the electric field at $(0, 0, z)$ of a charge located at the origin. A typical plot of the field is shown in Fig. 4.22.

Practice Problem 4.4

Do Ex. 4.17 using rectangular coordinates.

4.5.2 Infinite Sheet Charge

Let us consider the case of an infinite sheet of charge. We would like to compute the electric field produced by such a sheet. As shown in Fig. 4.23, the sheet coincides with the $z = 0$ plane and the charge on it is ρ_s C/m^2, which for the case under consideration, has a constant value. The sheet covers the entire x-y plane.

We expect the field to be the same in the x and y directions. As we move in either the x or y directions, but maintaining the same distance from it, the field should

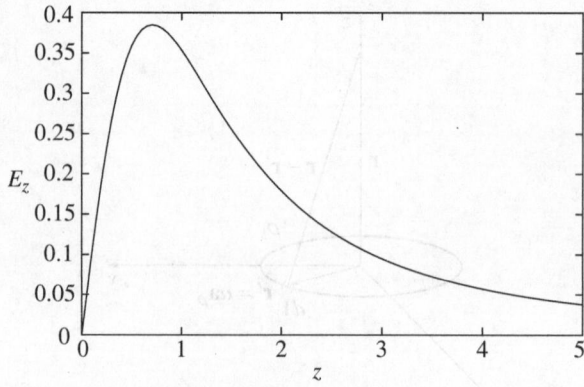

Fig. 4.22 E-field along the axis of a ring of charge

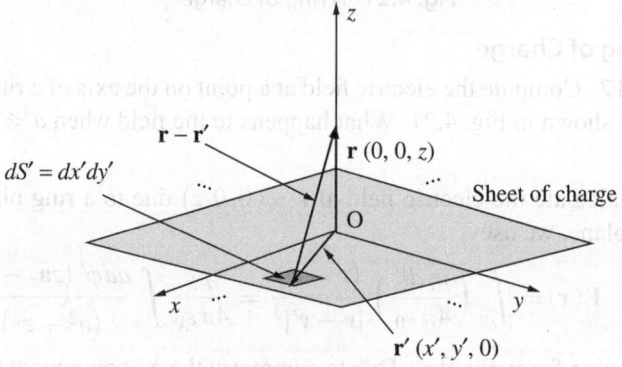

Fig. 4.23 An infinite sheet of charge

remain unchanged. Therefore the field will not be a function of either the x- or y-coordinates, but it will be a function of only the z-coordinate. To calculate the electric field, the equation

$$\mathbf{E}(\mathbf{r}) = \int_{S'} \left(\frac{\rho_s \, dS'}{4\pi\varepsilon_0} \right) \frac{(\mathbf{r} - \mathbf{r}')}{|\mathbf{r} - \mathbf{r}'|^3}$$

is to be applied in suitable manner. In this equation, $dS' = dx' dy'$ where dS' is a small element of area on the sheet of charge. The general form of the vectors $\mathbf{r}$ (the field point) and $\mathbf{r}'$ (position vector of the charges) are $\mathbf{r} = [x, y, z]$ and $\mathbf{r}' = [x', y', z']$. In our specific case $\mathbf{r}$ and $\mathbf{r}'$ reduce to $\mathbf{r} = [0, 0, z]$ and $\mathbf{r}' = [x', y', 0]$. Since the sheet occupies the $z = 0$ plane which is infinite in extent, we expect E_x and E_y to be zero. Using the two vectors $\mathbf{r}$ and $\mathbf{r}'$ for E_z

$$E_z = \iint_{x-y \text{ plane}} \left(\frac{\rho_s \, dx' dy'}{4\pi\varepsilon_0} \right) \frac{z}{\left[x'^2 + y'^2 + z^2 \right]^{3/2}}$$

This has to be integrated over the whole x-y plane and the variable z must be treated as a constant. We have to calculate the expression

$$\int_{x'=-\infty}^{\infty, \infty} \int_{y'=-\infty} \frac{z \, dx' dy'}{(z^2 + y'^2 + x'^2)^{\frac{3}{2}}}$$

Integrating with respect to x' (refer to the integral given in Eqn (4.40)):

$$\int \left(\frac{\rho_s \, dy'}{4\pi\varepsilon_0} \right) \int_{x'=-\infty}^{\infty} \frac{z \, dx'}{(z^2 + y'^2 + x'^2)^{\frac{3}{2}}}$$

$$= \int \left(\frac{\rho_s \, dy'}{4\pi\varepsilon_0} \right) \left[\frac{x' \, z}{(z^2 + y'^2) \sqrt{z^2 + y'^2 + x'^2}} \right]_{x'=-\infty}^{\infty} \qquad (4.46)$$

The value of this integral at its upper limit and lower limit $x' \to \pm\infty$ is $\pm z/(z^2 + y'^2)$. Substituting these limits, the integral becomes $2z/(z^2 + y'^2)$. We now integrate this expression with respect to y' between the limits $y = -y_0$ to $y = +y_0$ and then let $y_o \to \infty$. The integral

$$\left(\frac{\rho_s}{4\pi\varepsilon_0} \right) \int_{-y_0}^{+y_0} \frac{2z \, dy'}{z^2 + y'^2} = \left(\frac{\rho_s}{4\pi\varepsilon_0} \right) 4 \arctan \left(\frac{y_0}{z} \right) \qquad (4.47)$$

This result can be verified by referring to Appendix B. This expression becomes

$$\lim_{y_0 \to \infty} \left[4 \arctan \left(\frac{y_0}{z} \right) \right] = \begin{cases} 4 \times \pi/2, & \text{for } z > 0 \\ 4 \times -\pi/2, & \text{for } z < 0 \end{cases} \qquad (4.48)$$

Substituting the other constants

$$E_z = \begin{cases} \left(\dfrac{\rho_s}{4\pi\varepsilon_0} \right) 2\pi = \dfrac{\rho_s}{2\varepsilon_0}, & \text{when } z > 0 \\[2mm] \left(\dfrac{\rho_s}{4\pi\varepsilon_0} \right) (-2\pi) = -\dfrac{\rho_s}{2\varepsilon_0}, & \text{when } z < 0 \end{cases} \qquad (4.49)$$

The other components of the electric field are zero.

This is the same result which is obtained by application of Gauss's law. It is important to note that for all structures, the electric field points *away* from the positive charges. We shall return to these results later in this book.

Example 4.18 Two charged sheets, infinite in extent are placed at $z = \pm(d/2)$ with surface charge densities of ρ_s (at $d/2$) and $-\rho_s$ C/m^2 (at $-d/2$) respectively. Show using superposition that there is an electric field between the two sheets but none outside. Obtain the value of the electric field between the sheets and its direction. Remember this result for use in the case of a parallel plate capacitor, which will be considered later in the book.

Solution The electric field everywhere will be the superposition of the electric fields due to the two sheet charges. The upper sheet produces an **E** field given by

$$\mathbf{E}_u = \begin{cases} \rho_s/(2\varepsilon_0) \, \mathbf{a}_z, & z > d/2 \\ -\rho_s/(2\varepsilon_0) \, \mathbf{a}_z, & z < d/2 \end{cases}$$

Similarly, the lower sheet produces an **E** field given by

$$\mathbf{E}_l = \begin{cases} -\rho_s/(2\varepsilon_0) \, \mathbf{a}_z, & z > -d/2 \\ \rho_s/(2\varepsilon_0) \, \mathbf{a}_z, & z < -d/2 \end{cases}$$

With reference to Fig. 4.24, we see that there are three regions.
1. $z > d/2$. 2. $d/2 > z > -d/2$ and 3. $z < -d/2$.
Hence resultant electric field $\mathbf{E} = \mathbf{E}_u + \mathbf{E}_l$.

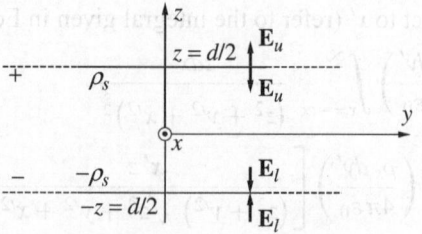

Fig. 4.24 Two sheet charges placed at $z = \pm d/2$

In Regions 1 and 3, the field is zero. In Region 2 the field is

$$\mathbf{E} = -\rho_s/\varepsilon_0\, \mathbf{a}_z, \quad d/2 > z > -d/2$$

4.5.2.1 Charged Disk

Example 4.19 Obtain the electric field along the axis of a charged disk with surface charge density of ρ_s C/m^2.

Solution

Step 1. From Fig. 4.25, $\mathbf{r}' = \rho'\, \mathbf{a}_\rho$ (with $0 \le \rho' \le a\ 0 \le \phi' \le 2\pi$) and $\mathbf{r} = z\mathbf{a}_z$.

Step 2. So $\mathbf{r} - \mathbf{r}' = z\mathbf{a}_z - \rho'\, \mathbf{a}_\rho$ and $|\mathbf{r} - \mathbf{r}'|^3 = \left(\rho'^2 + z^2\right)^{3/2}$.

Step 3. The electric field is given by (set $E_\rho = 0$)

$$E_z = \iint\limits_{\text{disk}} \left(\frac{\rho_s\, \rho'\, d\phi'\, d\rho'}{4\pi\varepsilon_0}\right) \frac{z}{\left[\rho'2 + z^2\right]^{3/2}}$$

$$= \int_{\rho'=0}^{a} \left(\frac{\rho_s\, \rho'\, d\rho'}{2\varepsilon_0}\right) \frac{z}{\left[\rho'^2 + z^2\right]^{3/2}} = \left(\frac{\rho_s}{2\varepsilon_0}\right)\left(1 - \frac{z}{\sqrt{z^2 + a^2}}\right)$$

4.5.2.2 Annular Ring of Charge

Example 4.20 Find the **E** field on the axis of a flat ring of charge with charge density ρ_s C/m^2 and with inner and outer radii equal to a and b respectively as shown in Fig. 4.26.

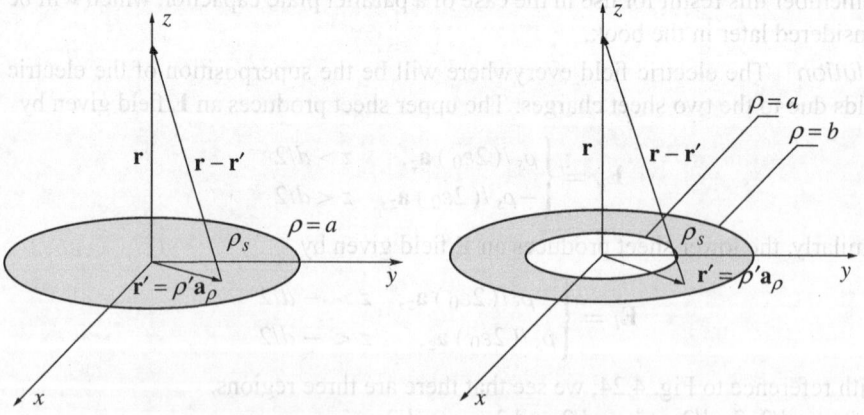

Fig. 4.25 Charged disk   **Fig. 4.26** Annular ring of charge

Solution

Step 1. From the figure, $\mathbf{r}' = \rho' \, \mathbf{a}_\rho$ (with $a \le \rho' \le b \, 0 \le \phi' \le 2\pi$) and $\mathbf{r} = z\mathbf{a}_z$

$$\mathbf{r} - \mathbf{r}' = z\mathbf{a}_z - \rho' \, \mathbf{a}_\rho$$

$$|\mathbf{r} - \mathbf{r}'|^3 = \left(\rho'^2 + z^2 \right)^{3/2}$$

Step 2. The electric field along the axis of the ring is given by

$$E_z = \underset{\text{ring}}{\iint} \left(\frac{\rho_s \, \rho' \, d\phi' \, d\rho'}{4\pi \varepsilon_0} \right) \frac{z}{\left[\rho'^2 + z^2 \right]^{3/2}}$$

$$= \int_{\rho'=a}^{b} \left(\frac{\rho_s \, \rho' \, d\rho'}{2\varepsilon_0} \right) \frac{z}{\left[\rho'^2 + z^2 \right]^{3/2}}$$

$$= \left(\frac{\rho_s \, z}{2\varepsilon_0} \right) \left(\frac{1}{\sqrt{z^2 + a^2}} - \frac{1}{\sqrt{z^2 + b^2}} \right)$$

4.6 | Electric Displacement Ψ and Flux Density D

Faraday conducted experiments with concentric spheres and found some astonishing results. A sphere with charge Q was placed within a larger sphere containing dielectric. The outer sphere was then earthed for a short while and the inner sphere was removed. It was found that the outer sphere contained the same amount of charge as the inner sphere but of opposite sign. These results held irrespective of the size of the spheres or whatever the dielectric. Faraday hypothesised that there was an 'electric flux' Ψ, which moved from the inner sphere to the outer being equal in magnitude to the charge and independent of the size of the spheres and the dielectric ε, that is,

$$\Psi = Q \; (\text{C}) \tag{4.50}$$

It was further hypothesised that there was an electric displacement from the inner to the outer sphere and on any sphere between the two (shown dotted in Fig. 4.27). The magnitude of this displacement density **D** is given by

$$D = \frac{\Psi}{4\pi r^2} = \frac{Q}{4\pi r^2} \; (\text{C/m}^2) \tag{4.51}$$

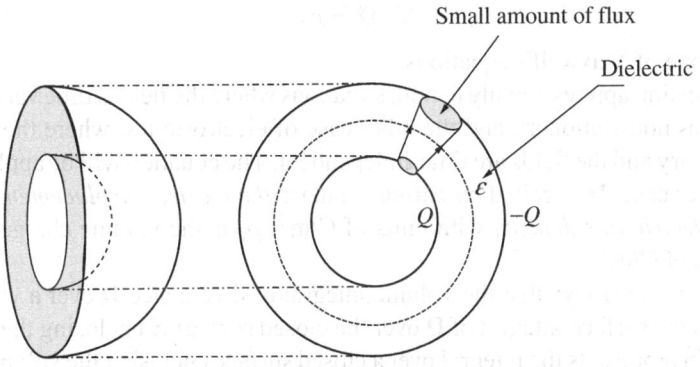

Small amount of flux

Dielectric

Fig. 4.27 Faraday's concentric spheres

The flux and displacement density are related by

$$\Delta \Psi = \mathbf{D} \cdot \Delta \mathbf{S} \tag{4.52}$$

and

$$\Psi = \iint \mathbf{D} \cdot d\mathbf{S} \tag{4.53}$$

After a little reflection one could visualise that Eqn (4.51) can be written in vector form as

$$\mathbf{D} = \frac{Q}{4\pi r^2} \mathbf{a}_r \tag{4.54}$$

4.7 | Gauss's Law

Carl Friedrich Gauss formulated, in 1835, a law applying to electrostatic fields which related arbitrary charge distributions to the electric fields produced. Gauss's law states that

The electric flux through any closed surface is equal to the enclosed electric charge.

In mathematical terms, the electric flux

$$\Psi = \oiint d\Psi = \oiint_S \mathbf{D} \cdot d\mathbf{S} \quad (\text{From Eq. 4.53}) \tag{4.55}$$

$$= Q \quad (\text{Enclosed charge}) \tag{4.56}$$

Using the divergence theorem Eqn (4.87)

$$\iiint_V \nabla \cdot \mathbf{D} \, dV = \oiint_S \mathbf{D} \cdot d\mathbf{S} \tag{4.57}$$

and using the results of Eqn (4.2),

$$Q = \iiint_V \rho_v \, dV$$

Hence, we have

$$\iiint_V \nabla \cdot \mathbf{D} \, dV = \iiint_V \rho_v \, dV \tag{4.58}$$

or

$$\nabla \cdot \mathbf{D} = \rho_v \tag{4.59}$$

which is one of Maxwell's equations.

The equation applies equally to both situations where the field is time varying and the charges non-stationary and the other case of electrostatics, where the charges are stationary and the fields are time-independent. The equation will be applied here to the latter case. $\mathbf{D}$ is called by various names: *the electric displacement density* and *the electric flux density,* with units of C/m^2. ρ_v is the volume charge density with units of C/m^3.

Equation (4.57) says that the volume integral of divergence $\mathbf{D}$ over a volume $\mathcal{V}$ is equal to the surface integral of $\mathbf{D}$ over the closed surface S enclosing the volume $\mathcal{V}$. $\oiint \ldots d\mathbf{S}$ represents the integral over a closed surface enclosing the volume. Such a volume–surface combination is shown in Fig. 4.28.

If we examine Fig. **??** and apply Gauss's law to it, then the volume integral of the divergence of **D** is equal to the total charge enclosed, that is,

$$\iiint_V \nabla \cdot \mathbf{D}\, dV = \iiint_V \rho_v\, dV \qquad (4.60)$$

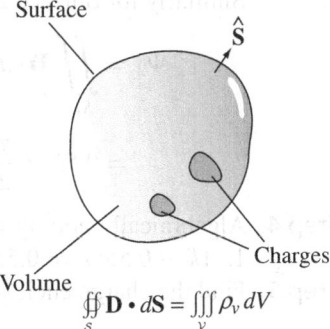

$$\oiint_s \mathbf{D} \cdot d\mathbf{S} = \iiint_v \rho_v\, dV$$

Fig. 4.28 Gauss's law

We can perform these integrations provided we know both **D** and ρ_v. Though only the volume charge, ρ_v, has been specifically mentioned it includes the cases of surface, linear, and point charges. The cases of point charges, surface charges, and linear charges can be converted into volume charges by means of the *Dirac delta function*. For example, a uniform surface of ρ_s C/m^2 covering the x-y plane can be written as $\rho_s\, \delta\,(z)$ C/m^3.

D and **E** fields are closely related. The relation between the two is given by

$$\mathbf{D} = \varepsilon\, \mathbf{E} \qquad (4.61)$$

at every point in space where ε is the permittivity of the medium being considered. If the medium is not vacuum, then the permittivity for a medium with relative dielectric constant ε_r is given by

$$\varepsilon = \varepsilon_0\, \varepsilon_r \qquad (4.62)$$

To be specific

$$\mathbf{D} = \varepsilon_0\, \mathbf{E} \quad \text{for free space} \qquad (4.63)$$

Example 4.21   The electric field in a region of space is $\mathbf{D} = (k/r)\, \mathbf{a_r}, 0 \le r \le 1$ (spherical coordinates). Find the flux leaving the region $\mathcal{R}$ bounded by $0 \le \phi \le \pi/2$, $\pi/4 \le \theta \le \pi/2$, and $0.25 \le r \le 0.5$. Also find the total charge inside this region.

Solution

Step 1. First draw and understand the bounding surfaces. $\mathcal{R}$ is bounded by six surfaces:

   1. and 2. $0 \le \phi \le \pi/2$, $\pi/4 \le \theta \le \pi/2$ with $r = 0.25$ or $r = 0.5$ with $d\mathbf{S} \propto \mathbf{a_r}$
   3. and 4. $\pi/4 \le \theta \le \pi/2$ and $0.25 \le r \le 0.5$ with $\phi = 0$ or $\phi = \pi/2$ with $d\mathbf{S} \propto \mathbf{a_\phi}$
   5. and 6. $0 \le \phi \le \pi/2$ and $0.25 \le r \le 0.5$ with $\theta = \pi/4$ or $\theta = \pi/2$ with $d\mathbf{S} \propto \mathbf{a_\theta}$

Step 2. Out of these surfaces *only surfaces 1 and 2 have flux coming in or going out*, since the surface vector $d\mathbf{S} \propto \mathbf{a_r}$. Surfaces 3-6 have $d\mathbf{S} \propto \mathbf{a_\phi}$ or $\mathbf{a_\theta}$.

Step 3. So for Surface 1 (flux entering is negative)

$$\Psi_1 = -\iint_{S_1} \mathbf{D} \cdot d\mathbf{S} = -\iint_{S_1} \left[\left(\frac{k}{r}\mathbf{a_r}\right) \cdot \mathbf{a_r} r^2 \sin\theta \right]_{r=0.25} d\phi\, d\theta$$

$$= -0.25k \times \frac{\pi}{2} \times \frac{1}{\sqrt{2}} = -0.556k$$

Similarly for Surface 2 (flux leaving is positive)

$$\Psi_2 = \iint\limits_{S_1} \mathbf{D} \cdot d\mathbf{S} = \iint\limits_{S_2} \left[\left(\frac{k}{r} \mathbf{a_r} \right) \cdot \mathbf{a}_r r^2 \sin\theta \right]_{r=0.5} d\phi \, d\theta$$

$$= 0.5k \times \frac{\pi}{2} \times \frac{1}{\sqrt{2}} = 1.112k$$

Step 4. Algebraically adding the two fluxes, the total flux leaving the region is $1.11k - 0.556k = 0.556k$.

Step 5. Find the charge enclosed. We need to first find $\nabla \cdot \mathbf{D}$:

$$\nabla \cdot \mathbf{D} = \frac{1}{r^2} \frac{\partial \left(r^2 D_r \right)}{\partial r}$$

$$= \frac{k}{r^2} = \rho_v$$

Then charge enclosed in the region is

$$\iiint\limits_{\mathcal{R}} \nabla \cdot \mathbf{D} dV = \iiint \left(\frac{k}{r^2} \right) r^2 \sin\theta \, d\theta \, d\phi \, dr$$

$$= k \times \frac{\pi}{2} \times \frac{1}{\sqrt{2}} \times 0.25 = 0.556k$$

It can be seen that the total flux leaving is equal to the enclosed charge.

Going back to the flux Eqn (4.55), we know from examining it, that the total flux leaving a closed surface is equal to the total charge inside it. Figure 4.29 illustrates this point. The volume charge, ρ_v, produces a **D** field all around it, a small portion of which is shown in the figure. Two Gaussian spheres are drawn, one which does not enclose the charge, and the other encloses it. A Gaussian surface is a closed surface in which a volume is enclosed.

The differential flux $\Delta \Psi$ ($= \mathbf{D} \cdot \Delta \mathbf{S}$), through the differential surface A is equal to that through surface B. This implies that the small flux entering b on the surface of Sphere 2 leaves through the small surface c. In this way *all* the flux *entering* the

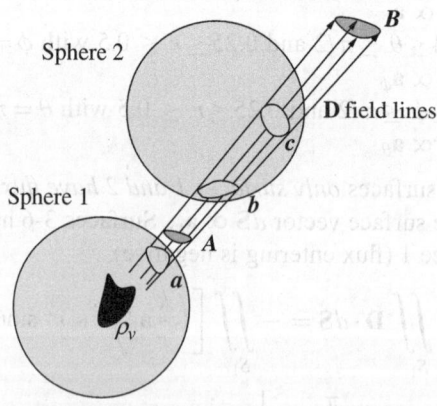

Fig. 4.29 The relationship of flux to charge

surface of Sphere 2 *leaves* through some other part of the surface of the sphere. If we give a positive sign to the flux leaving the surface then the flux entering it is negative, and the net flux leaving is zero. This is not the case with Sphere 1. Here, the flux only leaves the surface and none enters it and the net flux leaving the sphere is equal to the total charge inside the sphere.

The most important fact to remember from Gauss's law is that the flux density diverges *from positive charges!*

4.8 | Gauss's Law Applied to Cases of Spherical Symmetry

4.8.1 Gauss's Law Applied to a Point Charge

Gauss's law is best understood by applying it to various situations. This results in obtaining spectacular results with much less effort.

Let us apply Gauss's law to Fig. 4.30. We draw a closed (or Gaussian) surface which is 'equidistant' from the point charge, viz., a sphere with radius r_0. We do this from the point of view of the following reasoning:

Since all points are equidistant from the charge,

1. We expect that the **D** field will be equal at all points on the surface.
2. We also expect that **D** will have only a radial field component.

Therefore in the spherical coordinate system, (r, θ, ϕ)

$$\mathbf{D} = D_r \mathbf{a}_r \tag{4.64}$$

To calculate $\mathbf{D} \cdot d\mathbf{S}$, an element of area on the sphere is (see Fig. 4.9)

$$d\mathbf{S} = \mathbf{a}_r (r_0 \sin\theta \, d\phi)(r_0 d\theta) \tag{4.65}$$

so that

$$\mathbf{D} \cdot d\mathbf{S} = (D_r \mathbf{a}_r) \cdot (\mathbf{a}_r r_0^2 \sin\theta \, d\phi \, d\theta)$$

$$= D_r r_0^2 \sin\theta \, d\phi \, d\theta \tag{4.66}$$

Integrating this over the sphere. (Note that $\int_0^\pi \sin x \, dx = 2$ and D_r and r_0 are constants)

$$\oiint_{\text{Sphere}} \mathbf{D} \cdot d\mathbf{S} = \int_{\phi=0}^{2\pi} \int_{\theta=0}^{\pi} D_r r_0^2 \sin\theta \, d\phi \, d\theta$$

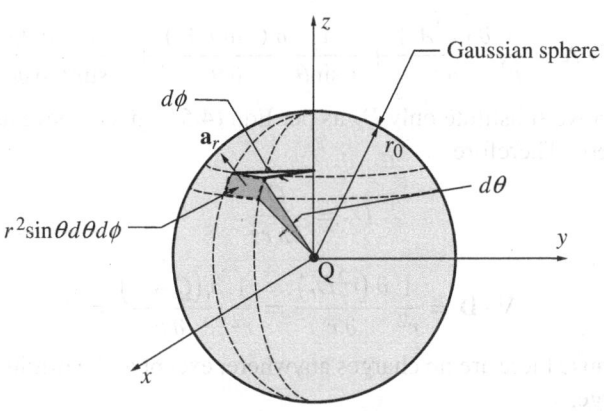

Fig. 4.30 Gauss's law applied to a point charge

$$= D_r r_0^2 \int_{\phi=0}^{2\pi} d\phi \int_{\theta=0}^{\pi} \sin\theta \, d\theta$$

$$= D_r r_0^2 (2\pi)(2)$$

$$= 4\pi \, D_r r_0^2 \tag{4.67}$$

To find the total charge enclosed

$$\iiint_{\text{Sphere}} \rho_v \, dV = \text{Charge enclosed} = Q \tag{4.68}$$

Equating the two results

$$4\pi \, D_r r_0^2 = Q$$

$$D_r = \frac{Q}{4\pi r_0^2} \tag{4.69}$$

But r_0 can assume any value, so letting r_0 be replaced by the more general value r

$$D_r = \frac{Q}{4\pi r^2} \tag{4.70}$$

If we compare the electric field at the same point in the same situation. (Note: $r_0' = 0$ and $r = r$ in Eqn (4.16).)

$$E = \frac{Q}{4\pi \varepsilon_0 \, r^2} a_r \tag{4.71}$$

which also gives

$$D = \varepsilon_0 \, E \tag{4.72}$$

Example 4.22   Compute $\nabla \cdot D$ for all values of r for the case of a point charge. Do not include the point $r = 0$. Explain your result.

Solution   For a point charge, D is (in spherical coordinates)

$$D = \frac{Q}{4\pi r^2} a_r$$

and $\nabla \cdot A$ in spherical coordinates for any general vector is

$$\nabla \cdot A = \frac{1}{r^2} \frac{\partial (r^2 A_r)}{\partial r} + \frac{1}{r \sin\theta} \frac{\partial (\sin\theta \, A_\theta)}{\partial \theta} + \frac{1}{r \sin\theta} \frac{\partial A_\phi}{\partial \phi}$$

In this formula we substitute only D_r as per Eqn (4.51) given above as D_θ and D_ϕ are equal to zero. Therefore

$$D_r = \frac{Q}{4\pi r^2}$$

and

$$\nabla \cdot D = \frac{1}{r^2} \frac{\partial (r^2 D_r)}{\partial r} = \frac{1}{r^2} \frac{\partial (Q/4\pi)}{\partial r} = 0$$

This is so because there are no charges anywhere, except at the origin, where there is a point charge.

4.8.2 Gauss's Law Applied to a Charged Sphere

Let us apply Gauss's law to yet another case, that of a hollow sphere whose surface is uniformly charged. See Fig. 4.31. The charged sphere has a radius R_0 while the Gaussian sphere has a radius r_0. The Gaussian sphere may be expanded or contracted at will, to consider different situations, by changing the value of r_0. The total charge on the surface of the sphere is considered to be Q. Two cases can be clearly distinguished.

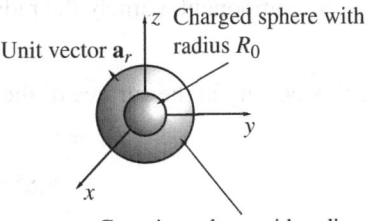

Fig. 4.31 Uniformly charged hollow sphere with radius R_0

Case 1. In this case, the Gaussian sphere has a radius *less than* the radius of the charged sphere $r_0 < R_0$.

$$\oiint_{S \equiv \text{Gaussian sphere}} \mathbf{D} \cdot d\mathbf{S} = 4\pi \, D_r r_0^2 \tag{4.73}$$

Since the Gaussian sphere has a radius less than the charged sphere, no charge is enclosed. So

$$\iiint_{\text{Gauss sphere}} \rho_v \, dV = 0 \tag{4.74}$$

By equating the two results, $D_r = 0$, $\mathbf{D} = 0$. It means that there is no field inside the sphere.

Case 2. The Gaussian sphere has a radius which is larger than the charged sphere. In this case

$$\oiint_{S} \mathbf{D} \cdot d\mathbf{S} = 4\pi \, D_r r_0^2 \qquad \text{when } r_0 > R_0 \tag{4.75}$$

and

$$\iiint_{V} \rho_v \, dV = \text{total charge enclosed} = Q \tag{4.76}$$

so

$$D_r = \frac{Q}{4\pi r_0^2} \equiv \frac{Q}{4\pi r^2} \tag{4.77}$$

Here we have replaced r_0 by the more general value of r. Using r, $\mathbf{D} = (Q/4\pi r^2)\, \mathbf{a}_r$. To sum up

$$\mathbf{D} = \begin{cases} 0, & \text{inside the charged sphere } r < R_0 \\ \dfrac{Q}{4\pi r^2}\mathbf{a}_r, & \text{outside the charged sphere } r > R_0 \end{cases} \tag{4.78}$$

Gauss's law can be applied successfully whenever there is a symmetry about the problem.

 See `Charged.Surf.Sphere.m` in Chapter 4

Example 4.23 An electric field exists at the surface of the earth which is about 110–150 V/m, radially downward. What is (a) the total charge dispersed on the surface of the earth? and (b) the charge density at the surface?

Solution   Using the spherical coordinate system (r, θ, ϕ), the electric field has only one component, namely the radial component. Therefore

$$E_r = -110 \text{ V/m}$$

The flux density at the surface of the earth is

$$D_r = \varepsilon_0 E_r$$
$$= 8.854 \times 10^{-12} \times (-110)$$
$$= -9.7394 \times 10^{-10} \text{ C/m}^2$$

where we have taken the lower value of the field. The radius of the earth is 6350 km, so

$$-9.7394 \times 10^{-10} = \frac{Q}{4\pi r_E^2}$$

where r_E is the earth's radius. Solving for Q

$$Q = -439502 \text{ C}$$

Moreover, the surface charge density is

$$\rho_s = \frac{Q}{4\pi r_E^2} = -9.7394 \times 10^{-10} \text{ C/m}^2$$

Notice that the surface charge density, ρ_s, is equal to the D_r!

Practice Problem 4.5
Find $\nabla \cdot \mathbf{D}$ everywhere for the hollow sphere with radius R_0 and surface charge ρ_s. Explain your result.

Example 4.24   A sphere of radius R_0 is uniformly charged, throughout its entire volume with a charge density ρ_v. Find the electric flux density field for all values of r in spherical coordinates, by using a Gaussian sphere.
Solution

Step 1. The charged sphere and the Gaussian sphere are shown in Fig. 4.32. Since the problem has spherical symmetry, we expect only one component of **D**, namely

$$\mathbf{D} = D_r(r)\, \mathbf{a}_r$$

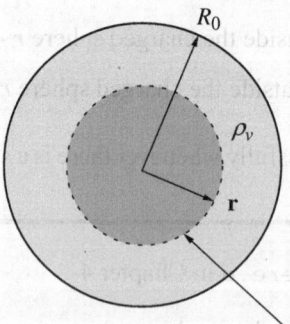

Fig. 4.32 Uniformly charged sphere

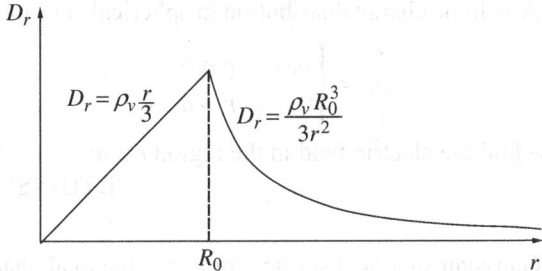

Fig. 4.33 D_r versus r for a uniformly charged sphere

Therefore applying

$$\oiint_{\text{Gaussian sphere}} \mathbf{D} \cdot d\mathbf{S} = \iiint_{\mathcal{V}} \rho_v \, dV$$

to a Gaussian sphere $r \leq R_0$ we have

$$D_r \oiint dS = \rho_v \iiint dV$$

or

$$D_r(4\pi r^2) = \frac{4\rho_v}{3}\pi r^3$$

$$D_r = \rho_v \frac{r}{3}$$

Now looking at the problem when $r \geq R_0$, we have

$$\oiint_{\text{Gaussian sphere}} \mathbf{D} \cdot d\mathbf{S} = D_r(4\pi r^2)$$

and

$$\rho_v \iiint dV = \rho_v \frac{4}{3}\pi R_0^3$$

so

$$D_r = \frac{\rho_v R_0^3}{3r^2}$$

The plot of D_r versus r is shown in Fig. 4.33.

Example 4.25 How will you check whether the answer in the previous example is correct?

Solution We know that $\nabla \cdot \mathbf{D}$ should be equal to ρ_v inside the charged sphere and zero outside.

Inside the sphere, $r < R_0$

and

$$\nabla \cdot \mathbf{D} = \frac{1}{r^2}\frac{\partial \left(r^2 D_r\right)}{\partial r}$$

$$= \frac{1}{r^2}\frac{\partial \left[r^2 \left(\rho_v \frac{r}{3}\right)\right]}{\partial r} = \rho_v$$

Outside the charged sphere, $r > R_0$

and

$$\nabla \cdot \mathbf{D} = \frac{1}{r^2}\frac{\partial \left(r^2 D_r\right)}{\partial r} = \frac{1}{r^2}\frac{\partial}{\partial r}\left[r^2 \left(\frac{\rho_v R_0^3}{3r^2}\right)\right] = 0$$

The above check proves the correctness of the answer.

Example 4.26 A volume charge distribution in spherical coordinates is given by

$$\rho_v = \begin{cases} \rho_0\, r, & r < a \\ 0, & r > a \end{cases}$$

Using Gauss's law find the electric field in the region $r < a$.

(DTU-NSIT, March 2008)

Solution

Step 1. Using a Gaussian spherical surface for $r < a$, the total charge enclosed is

$$Q = \iiint \rho_v\, dV$$

$$= \iiint_{r=0,\theta=0,\phi=0}^{r,\theta=\pi,\phi=2\pi} \rho_0\, r\,(r d\theta\ r\sin\theta\, d\phi\ dr)$$

$$= \rho_0\, 4\pi \int r^3\, dr$$

$$= \rho_0\, 4\pi\, \frac{r^4}{4}$$

$$= \rho_0\, \pi\, r^4$$

Important formula!

$$\int\int_{\theta=0,\phi=0}^{\theta=\pi,\phi=2\pi} d\theta\ \sin\theta\, d\phi = 4\pi$$

Step 2. From symmetry considerations, the electric field has a radial component only which is constant on the surface of a sphere. So

$$\oiint_{\text{Sphere}} \mathbf{D} \cdot d\mathbf{S} = D_r \times 4\pi r^2$$

$$= Q$$

$$= \rho_0\, \pi\, r^4$$

Step 3. So

$$D_r = \frac{\rho_0\, \pi\, r^4}{4\pi r^2}$$

$$= \frac{\rho_0\, r^2}{4}$$

4.9 | Gauss's Law Applied to Cases of Cylindrical Symmetry

Let us consider another set of problems which exhibits symmetry. The first of these is the infinite line charge coincident with the z-axis and with a linear charge density of ρ_l (C/m) as shown in Fig. 4.34. The obvious thing to do here is to choose a cylindrical Gaussian surface whose radius is ρ_0 and whose end surfaces lie between $z = z_0$ and $z = z_1$. From the symmetry of the figure it is clear that the **D** field will be streaming radially away. In a cylindrical coordinate system (ρ, ϕ, z), **D** will be

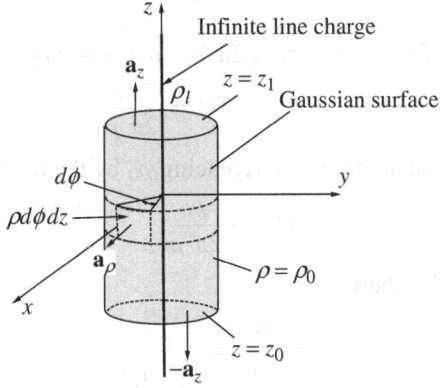

Fig. 4.34 An infinite line charge

$\mathbf{a}_\rho D_\rho$ and there will be no $\mathbf{a}_\phi$ or $\mathbf{a}_z$ component of $\mathbf{D}$. Applying Gauss's law to the surface we find that the surface can be split up into three parts: the top, the bottom, and the side. The integration over the top of the cylinder is zero since

$$\iint_{Top} \mathbf{D} \cdot d\mathbf{S} = \iint (\mathbf{a}_\rho D_\rho) \cdot (\mathbf{a}_z \rho \, d\phi \, d\rho) = 0 \qquad (\mathbf{a}_\rho \cdot \mathbf{a}_z = 0) \qquad (4.79)$$

Similar arguments apply to the bottom of the cylinder

$$\iint_{Bottom} \mathbf{D} \cdot d\mathbf{S} = \iint (\mathbf{a}_\rho D_\rho) \cdot (-\mathbf{a}_z \rho \, d\phi \, d\rho) = 0 \qquad (4.80)$$

Therefore, the contribution from the side of the cylinder is the total contribution to the flux.

$$\iint_{Side} \mathbf{D} \cdot d\mathbf{S} = \iint (\mathbf{a}_\rho D_\rho) \cdot (\mathbf{a}_\rho \rho \, d\phi \, dz)$$

$$= \int_{z=z_0}^{z_1} \int_{\phi=0}^{2\pi} D_\rho \rho_0 \, d\phi \, dz$$

$$= D_\rho \rho_0 \int_{z=z_0}^{z_1} \int_{\phi=0}^{2\pi} d\phi \, dz$$

$$= D_\rho \rho_0 \left(z_1 - z_0\right) \left(2\pi\right) \qquad (4.81)$$

Therefore

$$\oiint_{Cylinder} \mathbf{D} \cdot d\mathbf{S} = 2\pi \, D_\rho \rho_0 \left(z_1 - z_0\right) \qquad (4.82)$$

Now considering the right-hand side of Gauss's law

$$\iiint_{Cylinder} \rho_v \, dV = \int_{z=z_0}^{z_1} \rho_l \, dz$$

$$= \rho_l \int_{z=z_0}^{z_1} dz \qquad (4.83)$$

$$= \rho_l \left(z_1 - z_0\right) \qquad (4.84)$$

$$= \text{Total charge enclosed}$$

Equating these terms

$$D_\rho \rho_0 (z_1 - z_0) (2\pi) = \rho_l (z_1 - z_0) \tag{4.85}$$

$$D_\rho = \frac{\rho_l}{2\pi \rho_0} \tag{4.86}$$

But, as earlier, ρ_0 can be anything, so replacing ρ_0 by the more general value of ρ

$$D_\rho = \frac{\rho_l}{2\pi \rho} \tag{4.87}$$

And the $\mathbf{E}$ field is $\mathbf{a}_\rho E_\rho$ where

$$E_\rho = \frac{D_\rho}{\varepsilon_0} = \frac{\rho_l}{2\pi \varepsilon_0 \rho} \tag{4.88}$$

Example 4.27 Find the electric field everywhere for the case of an infinite hollow cylindrical tube of radius ρ_0 (see Fig. 4.35) with a surface charge of ρ_s C/m². Use a cylindrical Gaussian surface.

Solution We know from the symmetry of the figure that the $\mathbf{D}$ field will have only a ρ component

$$\mathbf{D} = D_\rho(\rho)\, \mathbf{a}_\rho$$

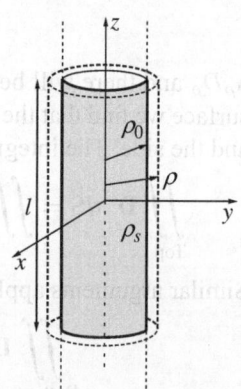

Fig. 4.35 An infinite cylindrical hollow tube of radius ρ_0 with surface charge ρ_s

and using the Gaussian cylindrical surface shown dotted in the figure for the case $\rho > \rho_0$ we apply Gauss's law

$$\iint_{\phi,z} D_\rho \rho\, d\phi\, dz (= 2\pi\, \rho \times l \times D_\rho)$$

$$= \text{Charge enclosed} = \rho_s \times l \times 2\pi\, \rho_0$$

then

$$D_\rho = \rho_s \frac{\rho_0}{\rho}, \quad \rho > \rho_0$$

On the other hand, when $\rho < \rho_0$ the charge enclosed is zero and

$$D_\rho = 0, \quad \rho < \rho_0$$

4.10 | Gauss's Law Applied to Cases of Rectangular Symmetry

Let us apply Gauss's law to a case of an infinite sheet charge. Figure 4.36 shows an infinite sheet of charge, coincident with the x-y plane extending in all directions to infinity. The Gaussian surface is a rectangular box extending along the x-axis from $x = x_0$ to $x = x_1$; along the y-axis from $y = y_0$ to $y = y_1$; and centred along the z-axis from $z = -z_0$ to $z = +z_0$. The charge on the sheet is a constant $= \rho_s$ C/m². From symmetry, on the walls of the Gaussian surface, the $\mathbf{D}$ field is assumed to have the form

$$\mathbf{D} = \begin{cases} [0, 0, D_z] & \text{for } z > 0 \\ [0, 0, -D_z] & \text{for } z < 0 \\ D_z = \text{constant} & \text{on the walls of the top and bottom} \\ & \text{of the Gaussian surface} \end{cases} \tag{4.89}$$

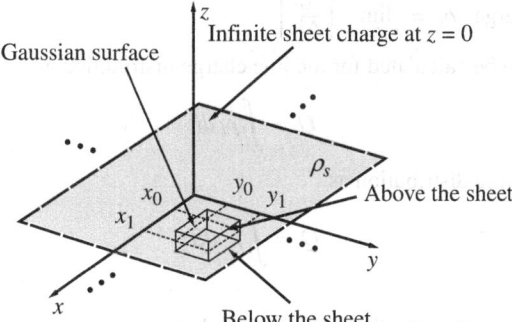

Fig. 4.36 An infinite sheet of charge

With this structure of **D**, the flux out of the sides of the box will be zero, while the flux from the top and bottom will be

$$\Psi_{top} = (x_1 - x_0)(y_1 - y_0) D_z \qquad (4.90)$$
$$\Psi_{bottom} = (x_1 - x_0)(y_1 - y_0) D_z \qquad (4.91)$$

so the total flux will be

$$\Psi_{total} = \Psi_{top} + \Psi_{bottom} = 2(x_1 - x_0)(y_1 - y_0) D_z \qquad (4.92)$$

The total charge enclosed by the Gaussian surface is

$$Q_{total} = \rho_s (x_1 - x_0)(y_1 - y_0) \text{ C} \qquad (4.93)$$

Equating the right-hand sides of the previous two equations

$$2(x_1 - x_0)(y_1 - y_0) D_z = \rho_s (x_1 - x_0)(y_1 - y_0) \qquad (4.94)$$

or

$$D_z(x,y,z_0) = \frac{\rho_s}{2} \qquad (4.95)$$

and

$$D_z(x,y,-z_0) = -D_z(x,y,z_0) = -\frac{\rho_s}{2} \qquad (4.96)$$

Therefore the electric field will be:

$$E_z = \begin{cases} \dfrac{\rho_s}{2\varepsilon_0}, & \text{for } z > 0 \\[2ex] -\dfrac{\rho_s}{2\varepsilon_0}, & \text{for } z < 0 \end{cases} \qquad (4.97)$$

the other field components being zero.

POINTS TO REMEMBER

- There are various types of charges such as the idealised point charge, Q; the line charge, ρ_l; surface charge, ρ_s; and volume charge, ρ_v.
- The SI unit of charge is Coulomb which is also 1 A-s.
- Distributed charges are calculated as below:

  – The line charge: $\rho_l = \lim\limits_{\Delta l \to 0} \left[\dfrac{\Delta q}{\Delta l} \right]$

  – The surface charge: $\rho_s = \lim\limits_{\Delta S \to 0} \left[\dfrac{\Delta q}{\Delta S} \right]$

- The volume charge: $\rho_v = \lim_{\Delta V \to 0} \left[\frac{\Delta q}{\Delta V} \right]$
- The net charge can be calculated for the line charge distribution as

$$Q = \int \rho_l \, dl$$

for the surface charge distribution as

$$Q = \int \rho_s \, dS$$

and so on.

- Coulomb's law states, that the force $\mathbf{F}$ felt by a charge q placed at $\mathbf{r}$ due to the presence of Q_1 at $\mathbf{r}_1$ is given by

$$\mathbf{F} = \frac{1}{4\pi\varepsilon_0} \left[\frac{Q_1 q \, (\mathbf{r} - \mathbf{r}_1)}{|\mathbf{r} - \mathbf{r}_1|^3} \right]$$

- The electric field at a point $\mathbf{r}$ (where the charge q is placed) due to Q_1 at $\mathbf{r}_1$ is

$$\mathbf{E}_1 = \frac{\mathbf{F}}{q} = \frac{1}{4\pi\varepsilon_0} \left(\frac{Q_1 \, (\mathbf{r} - \mathbf{r}_1)}{|\mathbf{r} - \mathbf{r}_1|^3} \right)$$

- The force $\mathbf{F}_T$ felt by a charge q placed at $\mathbf{r}$ due to N charges, $Q_1, \ldots, Q_N$ located at $\mathbf{r}_1, \ldots, \mathbf{r}_N$ is given by

$$\mathbf{F}_T = \frac{q}{4\pi\varepsilon_0} \left[\frac{Q_1 q \, (\mathbf{r} - \mathbf{r}_1)}{|\mathbf{r} - \mathbf{r}_1|^3} + \frac{Q_1 q \, (\mathbf{r} - \mathbf{r}_2)}{|\mathbf{r} - \mathbf{r}_2|^3} + \cdots \frac{Q_1 q \, (\mathbf{r} - \mathbf{r}_N)}{|\mathbf{r} - \mathbf{r}_N|^3} \right]$$

- The total electric field at the observation point $\mathbf{r}$ due to Q_1 at $\mathbf{r}'_1$, Q_2 at $\mathbf{r}'_2 \ldots$ and Q_N at $\mathbf{r}'_N$ is a superposition of all the fields produced by individual charges

$$\mathbf{E}(\mathbf{r}) = \mathbf{E}_1 + \mathbf{E}_2 + \cdots + \mathbf{E}_N$$

$$= \sum_{i=1}^{i=N} \mathbf{E}_i$$

or

$$\mathbf{E}(\mathbf{r}) = \sum_{i=1}^{i=N} \left(\frac{Q_i}{4\pi\varepsilon_0} \right) \frac{\mathbf{R}_i}{R_i^3}, \qquad \mathbf{R}_i = \mathbf{r} - \mathbf{r}'_i$$

- The electric field due a continuous charge distribution is given by

$$\mathbf{E}_{\text{at } \mathbf{r}} = \int \frac{1}{4\pi\varepsilon_0} \times \frac{dQ_{\text{at } \mathbf{r}'} \, |\mathbf{r} - \mathbf{r}'|}{|\mathbf{r} - \mathbf{r}'|^3}$$

where

$$dQ = \begin{cases} \rho_l \, (\mathbf{r}') \, dL', & \text{for a linear charge distribution} \\ \rho_s \, (\mathbf{r}') \, dS', & \text{for a surface charge distribution} \\ \rho_v \, (\mathbf{r}') \, dV', & \text{for a volume charge distribution} \end{cases}$$

- Gauss's law may be expressed as

$$\nabla \cdot \mathbf{D} = \rho_v \quad \text{(in point form)}$$

$$\oiint \mathbf{D} \cdot d\mathbf{S} = \iiint \rho_v \, dV \quad \text{(in integral form)}$$

Gauss's law is applied to cases of spherical, cylindrical, and rectangular symmetries.

- Gauss's law states that the flux leaving a region of space is equal to the charge enclosed:

$$\Psi = Q \text{ (C)}$$

$$\Psi = \iiint \rho_v \, dV$$

$$\Psi = \iint \mathbf{D} \cdot d\mathbf{S}$$

- The relationship between $\mathbf{D}$ and $\mathbf{E}$ is $\mathbf{D} = \varepsilon \, \mathbf{E}$ where ε is the permittivity of the medium $\varepsilon = \varepsilon_r \, \varepsilon_0$.

SELF ASSESSMENT

Objective Type Questions

In the following questions one or more choices may be correct. Sometimes choices may be separated with an 'or' and sometimes with an 'and'.

1. The electric field E_1 at point $\mathbf{r}$ due to Q_1 at point $\mathbf{r}_1$ is given by
    (a) $E_1 = (1/4\pi \varepsilon_0) Q_1 \mathbf{r}_1/|\mathbf{r}_1|^3$
    (b) $E_1 = (1/4\pi \varepsilon_0) Q_1 (\mathbf{r} - \mathbf{r}_1)/|\mathbf{r} - \mathbf{r}_1|^3$
    (c) $E_1 = (1/4\pi \varepsilon_0) Q_1 (\mathbf{r} - \mathbf{r}_1)/|\mathbf{r}_1|^3$
    (d) $E_1 = 1/4\pi \varepsilon_0 Q_1 \mathbf{r}/(|\mathbf{r} - \mathbf{r}_1|^3)$

2. A force of attraction exists between
    (a) two positively charged bodies
    (b) two negatively charged bodies
    (c) two oppositely charged bodies
    (d) when one body has a charge which is twice that of the first

3. What is the unit of $1/4\pi \varepsilon_0$?
    (a) $[\text{F/m}]^{-1}$ (b) $\text{m}[\text{C/V}]^{-1}$ (c) $[\text{mF}]^{-1}$ (d) None of these

4. A uniform volume charge with density of $0.2 \ \mu \ \text{C/m}^2$ is present throughout the spherical shell extending from $r = 3$ to $r = 5$ cm and $\rho = 0$ everywhere else. The total charge present will be
    (a) 41.05 pC   (b) 257.92 pC   (c) 82.1 pC   (d) 129.6 pC

5. A dipole having $Qd/4\pi \varepsilon_0 = 100$ SI units is located at origin in free space and aligned so that its moment is in the $\mathbf{a}_z$ direction. The electric field at point ($r = 1$, $\theta = 45°$, $\phi = 0°$) is (see Section 4.4)
    (a) 158.11 V/m   (b) 194.21 V/m   (c) 146.21 V/m   (d) 167.37 V/m

6. A cylindrical surface $\rho = 8$ cm contains the surface charge density $\rho_s = 5e^{-20|z|}$ (nC/m^2). The flux that leaves the surface $\rho = 8$ cm ;1 cm $< z < 5$ cm and $30° < \phi < 90°$ is
    (a) 270.07 pC   (b) 9.45 pC   (c) 270.7 nC   (d) 9.45nC

7. The unit of charge is
    (a) m   (b) F   (c) C   (d) FV

8. For the Dirac delta function, $\int_{x=-1 \, z=-1}^{x=1 \, z=1} \delta(x)\, \delta(y)\, dx dz$ is equal to
    (a) 1   (b) 0   (c) $2\delta(y)$ (d) $2\delta(z)$

9. If for the spherical coordinate system $\rho_v = 10\delta(r)$. Then this represents
    (a) a volume charge   (b) a surface charge
    (c) no charge at all   (d) it is meaningless

10. For two electrons placed 1 cm apart
    (a) The electrical force of attraction is more than a million times the gravitational attraction

(b) The electrical force of repulsion is more than a million times the gravitational attraction

(c) The electrical force of attraction is more than a million times the gravitational repulsion

(d) None of these

Short-Answer Questions

1. State and explain the divergence theorem.
2. State Gauss's law. Using the divergence theorem express it in differential form.

(Mumbai University, May 2009)

3. If two charges q_1 (>0) and q_2 (<0) of opposite polarity are placed a distance d apart, will there be a point on the straight line joining them where the electric field is zero?
4. A charged circular disk of radius a is placed on the x-y plane, with its axis coinciding with the z-axis and having a charge density ρ_s. On intuitive grounds what should be the electric field at $0 < z \ll a$ (very small) and $z \gg a$ (very large)?
5. A charge Q is placed at the centre of a thin (thickness $= \delta$) Bakelite sphere of outside radius a. What would be the electric field at a radius of $2a$? A charge $-Q$ is placed on the surface of the Bakelite sphere. What would be the electric field now?
6. Two hollow cylindrical charged infinite cylinders with radii $0 < a < b$ with surface charge densities ρ_{sa} and ρ_{sb} C/m^2 are placed concentrically. Find the electric field at $\rho = (a+b)/2$. For a hollow cylinder of radius ρ_0 and surface charge ρ_s

$$D_\rho = \rho_s \frac{\rho_0}{\rho}, \quad \rho > \rho_0$$

$$D_\rho = 0, \quad \rho < \rho_0$$

using superposition
$$\mathbf{D} = \rho_{sa} \frac{2a}{a+b} \mathbf{a}_\rho$$

7. Find the electric field everywhere due to a spherical shell of inner radius a and outer radius b of uniform charge density ρ_v (C/m^3).

[**Hint:** Use Gauss's law]

8. Find the electric field everywhere due to a cylindrical shell of uniform charge density ρ_v (C/m^3) having inner radius a and outer radius b. The cylindrical shell is infinite in length.

[**Hint:** Use Gauss's law]

9. By using the results of disk of charge with surface charge density $\rho_s = \rho_v \delta z$, find the electric field of a uniformly charged sphere.

[**Hint:** See Fig. 4.37]

Review Questions

1. Explain Gauss's law (a) in integral form (b) in differential form.

(DTU-NSIT, March 2008)

2. Write a short note on Gauss's law and the significance of electric flux.
3. Write a short note on Coulomb's law.
4. What is the relationship between the force on a point charge and the electric field?
5. Why is Gauss's law only applied to cases of symmetry?
6. For non-symmetrical cases which equations should be used to calculate the electric field?
7. What is the importance of the analysis of dipoles in electromagnetic theory?

[**Hint:** Research the modelling of dielectrics]

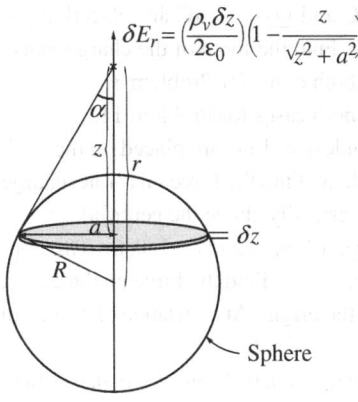

$$\delta E_r = \left(\frac{\rho_v \delta z}{2\varepsilon_0}\right)\left(1 - \frac{z}{\sqrt{z^2 + a^2}}\right)$$

Fig. 4.37

8. Can any three charges placed on a straight line be in equilibrium so that each charge feels no net force due to the action of other two charges?

   [**Hint:** Use Coulomb's law]

Numerical Problems

1. A 2 µC point charge is located at $A(4, 3, 5)$ in free space. Compute the electric field at $P(8, 12, 2)$ in cylindrical coordinates.

2. The charge density on the z-axis varies as $10^{-8}/(z^2 + 1)$ C/m for $-\infty < z < \infty$. How much charge lies on the z axis from $-10 \leq z \leq 10$?

3. The charge density on the z-axis varies as

$$\rho_l(z) = \begin{cases} 10^{-8}z^2, & \text{for } -1 < z < 1 \text{ C/m} \\ 0, & \text{otherwise} \end{cases}$$

   How much charge lies on the z-axis from $-10 \leq z \leq 10$?

4. The charge density on the z-axis varies as $10^{-8}/(z^2 + 5)$ C/m for $-\infty < z < \infty$. How much charge lies on the z-axis from $-\infty \leq z \leq \infty$?

5. The charge density on the z-axis varies as $10^{-8}/(4z^2 + 25)$ C/m for $-\infty < z < \infty$. How much charge lies on the z-axis from $-\infty \leq z \leq \infty$?

6. The surface charge density in cylindrical coordinates varies as

$$\rho_s(\rho = 1, \phi, z) = 1/(z^2 + 1) \text{ (C/m}^2), \text{ for } -\infty < z < \infty$$

   How much charge lies on an infinite cylinder?

7. The surface charge density in cylindrical coordinates varies as

$$\rho_s(\rho, \phi = 0, z) = e^{-|z|}/(\rho^2 + 1) \text{ (C/m}^2)$$

   How much charge lies on the half plane?

8. Equal charges of 1 nC are placed at point $1 = (0, 0, 0)$ and point $2 = (1, 0, 0)$. Find the force of the charge placed at 1 on 2 and vice versa.

9. Two charges, $Q_1 = 1$ nC and $Q_2 = 1$ µC are placed at point $1 = (0, 0, 0)$ and point $2 = (1, 0, 0)$ respectively. Find the force of the charge placed at 1 on 2 and vice versa.

10. Two charges, $Q_1 = 1$ nC and $Q_2 = 1$ µC are placed at point $1 = (0, 0, 0)$ and point $2 = (1/\sqrt{2}, 1/\sqrt{2}, 0)$ respectively. Find the force of the charge placed at 1 on 2 and vice versa.

11. Two charges, $Q_1 = 1$ nC and $Q_2 = 1$ μC are placed at point $1 = (0,0,0)$ and point $2 = (3,4,0)$ respectively. Find the force of the charge placed at 1 on 2 and vice versa.

12. Find the electric field in both cases for Problem 8.

13. Find the electric field in both cases for Problem 11.

14. Equal charges of magnitude $q = 1$ nC are placed on the corners of a square of side 1 m coincident with the x-y plane. Find the force on a 1 nC charge placed on the (a) centroid of the square, (b) 0.5 m vertically above the centroid.

15. 10 nC charges are placed at $(1,0,0)$ and $(-1,0,0)$. Two opposite charges $(-10$ nC$)$ are placed at $(0,1,0)$ and $(0,-1,0)$. Find the force on the charges at $(1,0,0)$ and $(0,1,0)$.

16. A charge Q_0 is placed at the origin. At a distance of 1 m from it, the electric field is 100 V/m. Find Q_0.

17. Two charges, q, are placed at $(a,0,0)$ and $(-a,0,0)$. (a) Where in all 3-D space is $\mathbf{E} = \mathbf{0}$? (b) $E_x = 0$?

18. Four charges q are placed at the corners of a square in the x-y plane, two opposite corners of which are $(a,0,0)$ and $(-a,0,0)$. Find the electric field $\mathbf{E}$ at $(2a,0,0)$.

19. Two equal charges of 10 nC are placed at $(1,2,3)$ and $(-3,-2,-1)$. Find the electric field at the origin.

20. Two equal charges of 10 nC are placed at $(r = 1, \theta = \pi/2, \phi = 3)$ and $(r = 3, \theta = \pi/2, \phi = -1)$. Find the electric field at the origin.

21. In cylindrical coordinates, two equal charges (q) are placed at $(\rho = 1, \phi = \pi/2, z = -3)$ and $(\rho = 1, \phi = \pi/2, z = 3)$. Find the electric field at the origin.

22. Calculate the divergence of $\mathbf{E} = r\mathbf{a}_r + r^2 \sin\theta \, \mathbf{a}_\theta + 5\mathbf{a}_\phi$ at $(r = 1, \theta = \pi/2, \phi = 0)$. If $\varepsilon_r = 1$ at that point then what is the charge density, ρ_v, there?

23. On the surface of a sphere of radius a, the electric field is given by $\mathbf{E} = a\mathbf{a}_r + a^2 \sin\theta \, \mathbf{a}_\theta + 5\mathbf{a}_\phi$, find the charge enclosed.

24. Find the electric field of an infinite sheet of charge of charge density ρ_s C/m^2.

(DTU-NSIT, May 2009)

Answers

Objective Type Questions

1. (b) 2. (c) 3. (a) and (b) 4. (c) 5. (a) 6. (b)
7. (c) and (d) 8. (c) 9. (a) and (b) 10. (b)

Short-Answer Questions

3. Suppose there is such a point. Place a small charge $\Delta q > 0$ at that point. Now q_1 will repel it towards q_2 and q_2 will attract it towards itself. Therefore, Δq will *always* feel a force towards q_2. Hence there will be no such point.

4. The electric field on the axis of the disk is given by

$$E_z = \left(\frac{\rho_s}{2\varepsilon_0}\right)\left(1 - \frac{z}{\sqrt{z^2 + a^2}}\right)$$

For $0 < z \ll a$, z^2 can be neglected compared to a^2 so

$$E_z = (\rho_s/2\varepsilon_0)\left(1 - \frac{z}{a}\right)$$

When z is very large, we expect

$$E_z = \frac{Q}{4\pi\varepsilon_0 z^2} = \frac{\pi a^2 \rho_s}{4\pi\varepsilon_0 z^2} = \rho_s \frac{a^2}{4\varepsilon_0 z^2}$$

5. The **D** field at $2a$ would be given by

$$\mathbf{D} = \frac{Q}{4\pi \, (2a)^2} \mathbf{a}_r$$

so

$$\mathbf{E} = \frac{Q}{4\pi \, \varepsilon_0 \, (2a)^2} \mathbf{a}_r$$

when $-Q$ is placed on the Bakelite sphere, the total charge enclosed in the sphere of radius $2a$ is zero. So $\mathbf{D} = \mathbf{E} = 0$.

Numerical Problems

1. $\mathbf{E} = 159.7\mathbf{a}_\rho + 27.4\mathbf{a}_\phi - 49.4\mathbf{a}_z$
2. $2 \times 10^{-8} \tan^{-1}(10)$ C
3. $(2/3) \times 10^{-8}$ C
4. $\pi/\sqrt{5}$ C
5. $\pi \times 10^{-7}$ C
6. $2\pi^2$ C
7. π C
8. $8.9875 \times 10^{-9} \, \mathbf{a}_x$ N and $-8.9875 \times 10^{-9} \mathbf{a}_x$ N
9. $8.9875 \times 10^{-6} \mathbf{a}_x$ N and $-8.9875 \times 10^{-6} \mathbf{a}_x$ N
10. $8.9875 \times 10^{-6} \, (\mathbf{a}_x + \mathbf{a}_y)/\sqrt{2}$ N and $-8.9875 \times 10^{-6} \, (\mathbf{a}_x + \mathbf{a}_y)/\sqrt{2}$ N
11. $0.3590 \times 10^{-6} \, (3\mathbf{a}_x + 4\mathbf{a}_y)/5$ N and $-0.3590 \times 10^{-6} \, (3\mathbf{a}_x + 4\mathbf{a}_y)/5$ N
12. $8.9875 \, \mathbf{a}_x$ V/m and $-8.9875\mathbf{a}_x$ V/m
13. $0.3590 \times 10^3 \, (3\mathbf{a}_x + 4\mathbf{a}_y)/5$ V/m and $-0.3590 \, (3\mathbf{a}_x + 4\mathbf{a}_y)/5$ V/m
14. (a) 0 (b) $-5.084 \times 10^{-8}\mathbf{a}_z$ N
15. $-4.1082 \times 10^{-07}\mathbf{a}_x$ N and $4.1082 \times 10^{-07}\mathbf{a}_y$ N
16. 1.113×10^{-8} C
17. (a) $(0, 0, 0)$ (b) yz plane
18. $(1.4202q) / (4\pi \, \varepsilon_0 \, a^2) \, \mathbf{a}_x$ V/m
19. $3.431(\mathbf{a}_x - \mathbf{a}_z)$ V/m
20. $83.58\mathbf{a}_x - 4.28\mathbf{a}_y$ V/m
21. $-5.6839 \times 10^8 q\mathbf{a}_y + 1.7052 \times 10^9 q\mathbf{a}_z$ V/m
22. $\nabla \cdot \mathbf{E} = 3 + 2r \cos\theta$; $\rho_v = 3\varepsilon_0$
23. $4\pi \varepsilon \, a^3$ C

5 | Energy and Potential

> *Let early education be a sort of amusement.*
> *You will then be better able to find out the natural bent*
> — Plato

CHAPTER OBJECTIVES

To enable the students to understand the following:

- Work done on a charge immersed in an electric field and to move it from point A to B
- Potential difference between points A and B
- Concept of the potential at infinity for a point charge
- Relation between scalar potential and the electric field

- Importance of equipotential surfaces
- Concept of potential energy
- Calculation of the potential at a point due to a number of point charges
- Calculation of the potential due to continuous charge distributions.

5.1 | Potential Due to a Point Charge

We know that if a charge Q is immersed in an electric field $\mathbf{E}$, it experiences a force $\mathbf{F}$ given by

$$\mathbf{F} = Q\mathbf{E} \tag{5.1}$$

If we tend to move this charge by the application of an external force $(-\mathbf{F})$, we need to work against this force. The incremental work done for displacement $d\mathbf{l}$ is given by

$$dW = -\mathbf{F} \cdot d\mathbf{l} = -Q\mathbf{E} \cdot d\mathbf{l}$$

The negative sign tells us that the force applied is working against the field, if dW is to be positive. The work done by the external force to move the charge from $\mathbf{r}_A$ to $\mathbf{r}_B$ will be

$$W = \int_{\mathbf{r}_A}^{\mathbf{r}_B} -Q\mathbf{E} \cdot d\mathbf{l} = -Q \int_{\mathbf{r}_A}^{\mathbf{r}_B} \mathbf{E} \cdot d\mathbf{l} \tag{5.2}$$

Example 5.1 In a region of space the electric field $\mathbf{E} = xy^2\mathbf{a}_x - xz^2\mathbf{a}_y + xyz\mathbf{a}_z$. Find the work done to move a charge of 1 nC from $(1, 1, 1)$ to $(0, 1, 1)$.

Solution

Step 1. Draw a sketch of the path of integration. The work done is

$$W = -Q \int_P \mathbf{E} \cdot d\mathbf{l}$$

where P is the integration path.

Step 2. It will be seen from the sketch that the path of integration is along a line parallel to the x-axis as y and z are fixed with $y = z = 1$.

Step 3. Put $y = z = 1$ and carry out the integration. The work done is

$$W = -Q \int \mathbf{E} \cdot d\mathbf{l} = -Q \int_{(1,1,1)}^{(0,1,1)} (xy^2\mathbf{a}_x - xz^2\mathbf{a}_y + xyz\mathbf{a}_z) \cdot \mathbf{a}_x dx$$

$$= -Q \int_{(1,1,1)}^{(0,1,1)} x\,dx \quad (\text{Putting } y = z = 1)$$

$$= -1 \times 10^{-9} \left. \frac{x^2}{2} \right|_1^0 = 0.5 \times 10^{-9} \text{ J}$$

Example 5.2 For the electric field of Example 5.1, find the work done to move a charge of 1 nC from $(3, 5, 6)$ to $(0, 0, 0)$ along a straight line.

Solution

Step 1. Draw a sketch of the straight line on the coordinate system.

Step 2. Find the parametric equation of the straight line joining the two points. The vector joining $\mathbf{r}_0 = (3, 5, 6)$ to $\mathbf{r}_1 = (0, 0, 0)$ is $\mathbf{R}_{01} = (-3, -5, -6)$. The parametric equation of the straight line joining these two points is

$$\mathbf{r} = \mathbf{r}_0 + t\mathbf{R}_{01}, \quad 0 \le t \le 1$$

which results in the three equations as

$$x = 3(1 - t)$$
$$y = 5(1 - t)$$
$$z = 6(1 - t)$$

Step 3. The work done is

$$W = -Q \int_{r_0}^{r_1} \mathbf{E} \cdot d\mathbf{l}$$

$$= -Q \int_{r_0}^{r_1} (xy^2\mathbf{a}_x - xz^2\mathbf{a}_y + xyz\mathbf{a}_z) \cdot (dx\mathbf{a}_x + dy\mathbf{a}_y + dz\mathbf{a}_z)$$

$$= -Q \int_{r_0}^{r_1} xy^2 dx + xz^2 dy + xyz dz$$

$$= -Q \int_{t=0}^{t=1} [75(1 - t)^3 (-3dt) + 108(1 - t)^3 (-5dt) + 90(1 - t)^3 (-6dt)]$$

$$= 1305\, Q \int_{t=0}^{t=1} (1 - t)^3 \, dt = 435 \text{ nJ}$$

The concept of the electric potential difference between two points (V_{AB})[1] in the electrostatic field springs directly from the concept of work. Using Eqn (5.2), we have

$$V_{AB} = \left(\frac{W}{Q}\right) = -\int_{r_A}^{r_B} \mathbf{E} \cdot d\mathbf{l}$$

This is the potential difference between points A and B situated in the region of electric field $\mathbf{E}$.

Example 5.3 For a point charge Q located at the origin, find the potential difference between two points $\mathbf{r}_A = (r_A, \pi/2, 0)$ (initial point) and $\mathbf{r}_B = (r_B, \pi/2, 0)$ (final point) in spherical coordinates.

Solution

Step 1. Draw a sketch of the problem. It will be observed that we have to integrate only in the r-direction.

Step 2. The electric field produced by the charge located at the origin is given by

$$\mathbf{E} = \frac{Q}{4\pi\varepsilon_0 r^2}\mathbf{a}_r$$

$d\mathbf{l} = dr\mathbf{a}_r$. Since θ and ϕ are constants along the path of integration ($d\theta = d\phi = 0$). So

$$V_{AB} = -\int_{r_A}^{r_B} \mathbf{E} \cdot d\mathbf{l} = -\int_{r_A}^{r_B} \frac{Q}{4\pi\varepsilon_0 r^2}dr = \left.\frac{Q}{4\pi\varepsilon_0 r}\right|_{r_A}^{r_B}$$

$$= \frac{Q}{4\pi\varepsilon_0}\left(\frac{1}{r_B} - \frac{1}{r_A}\right)$$

We now discover a property of the potential difference between two points $\mathbf{r}_A$ (initial point) and $\mathbf{r}_B$ (final point) lying on a curve as shown in Fig. 5.1. Let the curve be parametrised as $(r(t), \theta(t), \phi(t))$ with $\mathbf{r}_A = (r_A, \theta_A, \phi_A)$ and $\mathbf{r}_B = (r_B, \theta_B, \phi_B)$. Then

$$V_{AB} = -\int_A^B \mathbf{E} \cdot d\mathbf{l} = -\int_{r_A}^{r_B}\left(\frac{Q}{4\pi\varepsilon_0 r^2}\mathbf{a}_r\right) \cdot (dr\mathbf{a}_r + d\theta\,\mathbf{a}_\theta + d\phi\,\mathbf{a}_\phi)$$

$$= -\int_{r_A}^{r_B}\frac{Q dr}{4\pi\varepsilon_0 r^2} = \frac{Q}{4\pi\varepsilon_0}\left(\frac{1}{r_B} - \frac{1}{r_A}\right)$$

or $$V_{AB} = \frac{Q}{4\pi\varepsilon_0}\left(\frac{1}{r_B} - \frac{1}{r_A}\right) \tag{5.3}$$

It reveals the fact that in an electrostatic field of a point charge, *the potential difference is independent of the path of integration*. To explain it further we see that the integration path is independent of the parameter t and depends only on the r coordinate. The path may be anything, but the result obtained is the same.

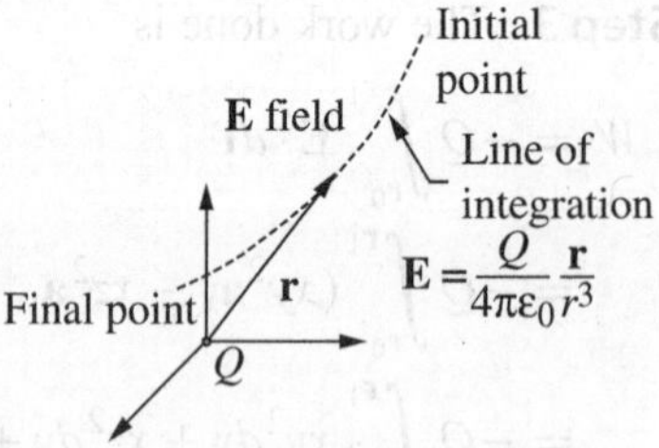

Furthermore, if $\mathbf{r}_A$ is a point at infinity $r_A \to \infty$ and $\mathbf{r}_B \to \mathbf{r}$ then from Eqn (5.3)

Fig. 5.1 Potential difference for a point charge

$$V_{AB} = \frac{Q}{4\pi\varepsilon_0}\left(\frac{1}{r} - \frac{1}{\infty}\right)$$

[1] In this book, the variable V is used both for the volume as well as for the electric potential, but this should not cause confusion.

or
$$V_r = \frac{Q}{4\pi \varepsilon_0 r} \tag{5.4}$$

Thus the point at infinity has zero potential:
$$V_\infty = 0 \tag{5.5}$$

and
$$V_{AB} = V_B - V_A = V_{\text{fin}} - V_{\text{init}} \tag{5.6}$$

Another important consequence of this property is that if we go from A to B along one path and come back from B to A along another path, then for the round trip
$$V_{A \to A} = V_{AB} + V_{BA} = (V_B - V_A) + (V_A - V_B) = 0 \tag{5.7}$$

which in terms of the line integral may be expressed as
$$\oint_L \mathbf{E} \cdot d\mathbf{l} = 0 \quad (L \text{ is any closed path}) \tag{5.8}$$

which by Stokes's theorem is
$$\oint_L \mathbf{E} \cdot d\mathbf{l} = \underset{\text{Enclosed surface}}{\iint} (\nabla \times \mathbf{E}) \cdot d\mathbf{S} = 0 \tag{5.9}$$

or
$$\nabla \times \mathbf{E} = 0 \tag{5.10}$$

Referring back to Section 4.2.3, one knows that *when the curl of a vector field is zero then it must be the gradient of a scalar field.*

As
$$dV = -\mathbf{E} \cdot d\mathbf{l} = -E_x dx - E_y dy - E_z dz \tag{5.11}$$

and
$$dV = \frac{\partial V}{\partial x} dx + \frac{\partial V}{\partial y} dy + \frac{\partial V}{\partial z} dz \tag{5.12}$$

from this, we get
$$E_x = -\frac{\partial V}{\partial x}, \qquad E_y = -\frac{\partial V}{\partial y}, \qquad E_z = -\frac{\partial V}{\partial z} \tag{5.13}$$

Hence
$$E = E_x \mathbf{a}_x + E_y \mathbf{a}_y + E_z \mathbf{a}_z = \frac{-\partial V}{\partial x} \mathbf{a}_x + \frac{-\partial V}{\partial y} \mathbf{a}_y + \frac{-\partial V}{\partial z} \mathbf{a}_z$$

or
$$\mathbf{E} = -\nabla V \tag{5.14}$$

which shows that *the electric field* $\mathbf{E}$ *is the gradient of the potential.* This equation may be used to find the units of V.

$$\text{Units of V(r)} = \text{Units of } \{E \cdot l\}$$
$$\overset{U}{=} \text{V/m} \times \text{m} \overset{U}{=} V$$

which is volts.

 See `PotDiffPtch.m` in Chapter 5

Example 5.4   The electric potential in a region of space is $V = kx^2 yz$. Find the electric field and charge in that region. The region is air.
Solution
Step 1. The electric field $\mathbf{E}$ is given by
$$\mathbf{E} = -\nabla V = -\frac{\partial V}{\partial x} \mathbf{a}_x - \frac{\partial V}{\partial y} \mathbf{a}_y - \frac{\partial V}{\partial z} \mathbf{a}_z = -2kxyz \mathbf{a}_x$$
$$- kx^2 z \mathbf{a}_y - kx^2 y \mathbf{a}_z \ (\text{V/m})$$

Step 2. From the **E** field, obtain the **D** field:

$$\mathbf{D} = \varepsilon_0\,\mathbf{E} = -k\varepsilon_0\,(\,2xyz\mathbf{a}_x + x^2 z\mathbf{a}_y + x^2 y\mathbf{a}_z\,)\ \text{C/m}^2$$

Step 3. Find the charge

$$\rho_v = \nabla\cdot\mathbf{D} = \frac{\partial D_x}{\partial x} + \frac{\partial D_y}{\partial y} + \frac{\partial D_z}{\partial z} = -2k\varepsilon_0\,yz\ \text{C}$$

Example 5.5 The charge density in the region $\rho(x)\ -L \le x \le L$ is shown in Fig. 5.2. Find the electric field and potential as a function of x. The medium has a dielectric constant ε_r.

Solution

Step 1.

$$\nabla\cdot\mathbf{D} = \frac{\partial D_x}{\partial x} + \frac{\partial D_y}{\partial y} + \frac{\partial D_z}{\partial z} = \rho$$

so

$$\frac{dD_x}{dx} = \rho\,(x)$$

Divide the region into four parts as $-L \le x \le x_1$; $-x_1 \le x \le 0$; $0 \le x \le x_2$; and $x_2 \le x \le L$

Step 2. Integrate this equation

for $-L \le x \le x_1$ $\quad D_x(x) = \displaystyle\int_{-L}^{x} \rho\,(x)\,dx = 0$

for $x_1 \le x \le 0$ $\quad D_x(x) = \displaystyle\int_{x_1}^{x} \rho_m\left(\frac{x_2}{x_1^2}\right)x\,dx = \frac{\rho_m}{2}\left(\frac{x_2}{x_1^2}\right)\left(x^2 - x_1^2\right)$

for $0 \le x \le x_2$ $\quad D_x(x) = -\dfrac{\rho_m}{2}x_2 + \displaystyle\int_{x_1}^{x} \rho_m\left(\frac{x}{x_2}\right)dx = -\dfrac{\rho_m}{2}x_2 + \dfrac{\rho_m}{2x_2}(x^2)$

for $x_2 \le x \le L$ $\quad D_x = 0$

Step 3. From the flux density, we can get the electric field

$$\mathbf{E} = \frac{\mathbf{D}}{\varepsilon_0\,\varepsilon_r} = \frac{1}{\varepsilon_0\,\varepsilon_r} \times \begin{cases} 0, & -L \le x \le x_1 \\[2mm] \dfrac{\rho_m}{2}\left(\dfrac{x_2}{x_1^2}\right)\left(x^2 - x_1^2\right)\mathbf{a}_x, & x_1 \le x \le 0 \\[2mm] \dfrac{\rho_m}{2x_2}\left(x^2 - x_2^2\right)\mathbf{a}_x, & 0 \le x \le x_2 \\[2mm] 0, & x_2 \le x \le L \end{cases}$$

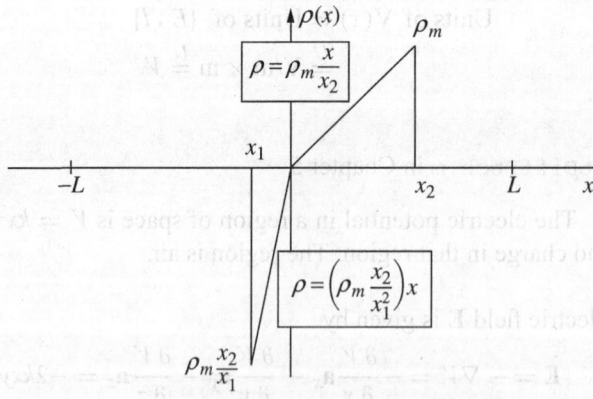

$$\rho = \rho_m\frac{x}{x_2}$$

$$\rho = \left(\rho_m\frac{x_2}{x_1^2}\right)x$$

Fig. 5.2 $\rho(x)$ for Example 5.5

Step 4. Obtain potential from electric field. Since there is only one component E_x and

$$\frac{dV}{dx} = -E_x$$

$$V(x) = -\int_{-L}^{x} E_x dx$$

It may be noted that *the absolute potential is unknown to the extent of an additive integration constant V_0, but the functional form of the potential is correct.* Generally, V_0 is taken to be zero.

Now we would like to find the potential for a point charge placed at an arbitrary point $\mathbf{r}'$. Referring to Fig. 5.3, the potential at the point (X, Y, Z) due to a point charge Q placed at the origin of the (X, Y, Z) coordinate system is

$$V(\mathbf{R}) = V(X, Y, Z)$$

$$= \frac{Q}{4\pi \varepsilon_0 R}$$

$$= \frac{Q}{4\pi \varepsilon_0 \sqrt{X^2 + Y^2 + Z^2}} \tag{5.15}$$

The point $\rho(X, Y, Z)$ is (x, y, z) with respect to origin at $(0,0,0)$ in (x, y, z) coordinate system.

In the (x, y, z) coordinate system, the same potential $V(\mathbf{R})$ is

$$V(\mathbf{r}) = \frac{Q}{4\pi \varepsilon_0 R} = \frac{Q}{4\pi \varepsilon_0 |\mathbf{r} - \mathbf{r}'|} = \frac{Q}{4\pi \varepsilon_0 [(x - x')^2 + (y - y')^2 + (z - z')^2]^{1/2}}$$

since $\mathbf{R} = \mathbf{r} - \mathbf{r}'$.

If we compute $-\nabla V (= \mathbf{E})$ of the above expression

$$-\nabla V = (\partial V/\partial x)\, \mathbf{a}_x + (\partial V/\partial y)\, \mathbf{a}_y + (\partial V/\partial z)\, \mathbf{a}_z$$

$$= \frac{Q[(x - x')\, \mathbf{a}_x + (x - y')\, \mathbf{a}_y + (x - z')\, \mathbf{a}_z]}{4\pi \varepsilon_0 [(x - x')^2 + (y - y')^2 + (z - z')^2]^{3/2}} \tag{5.16}$$

$$= \frac{Q(\mathbf{r} - \mathbf{r}')}{4\pi \varepsilon_0 |\mathbf{r} - \mathbf{r}'|^3} \tag{5.17}$$

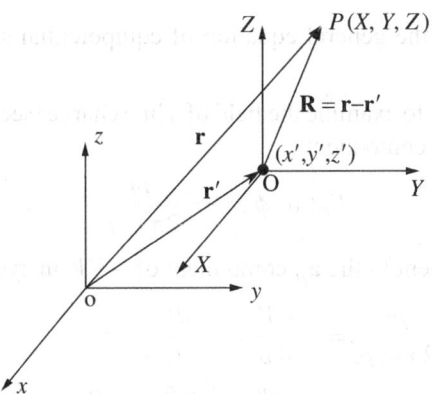

Fig. 5.3 Potential due to a point charge placed at $\mathbf{r}'$

which is the electric field at an observation point due to a charge placed at (x', y', z'). So we can conclude that the potential $V(\mathbf{r})$ due a point charge placed at $\mathbf{r}' = (x', y', z')$ *is indeed*

$$V(\mathbf{r}) = \frac{Q}{4\pi\varepsilon_0 |\mathbf{r} - \mathbf{r}'|} \tag{5.18}$$

5.2 | Equipotential Surfaces

Equipotential surfaces are those surfaces on which the potential is constant. An important attribute of an equipotential surface is that the field is normal to the surface at all points on it. This was explained in Section 4.2. If we examine the field for the three cases; the point charge, the infinite line charge, and the infinite plane charge, we can evaluate the equipotential surfaces for these three cases to give us an insight on equipotential surfaces in general.

Example 5.6 Find the general equation of equipotential surfaces for a point charge.

Solution For the point charge at (x'_0, y'_0, z'_0) the potential at (x, y, z) is

$$V(\mathbf{r}) = \frac{Q}{4\pi\varepsilon_0 \left[(x - x'_0)^2 + (y - y'_0)^2 + (z - z'_0)^2 \right]^{1/2}} \tag{5.19}$$

equating V to a constant

$$\frac{Q}{4\pi\varepsilon_0 \left[(x - x'_0)^2 + (y - y'_0)^2 + (z - z'_0)^2 \right]^{1/2}} = k_1 \tag{5.20}$$

$$\left[(x - x'_0)^2 + (y - y'_0)^2 + (z - z'_0)^2 \right]^{1/2} = \frac{Q}{4\pi\varepsilon_0 k_1} = k \tag{5.21}$$

where k is another constant. Squaring both sides, we get

$$(x - x'_0)^2 + (y - y'_0)^2 + (z - z'_0)^2 = k^2 \tag{5.22}$$

which is the equation of a sphere with radius k and centre (x'_0, y'_0, z'_0). Since the electric field is normal to this surface at all points, it is streaming radially out of $\mathbf{r}'_0$.

Example 5.7 Find the general equation of equipotential surfaces for a line of charge.

Solution Proceeding to examine the field of a line charge (see Fig. 4.34). The electric field has only one component

$$E_\rho(\rho, \phi, z) = \frac{\rho_l}{2\pi\varepsilon_0 \rho} \tag{5.23}$$

Equating this component to the $\mathbf{a}_\rho$ component of $-\nabla V$ in cylindrical coordinates

$$\frac{\rho_l}{2\pi\varepsilon_0 \rho} = -\frac{\partial V}{\partial \rho} = -\frac{dV}{d\rho} \tag{5.24}$$

or

$$V(\rho) = -\frac{\rho_l}{2\pi\varepsilon_0} \int \frac{d\rho}{\rho} = \frac{\rho_l}{2\pi\varepsilon_0} \ln \frac{1}{\rho} \tag{5.25}$$

This is the potential function. It may be seen that the function is zero at $\rho = 1$. This is the point with which all other values V are compared. The equipotential surfaces are computed by

$$\frac{\rho_l}{2\pi \varepsilon_0} \ln \frac{1}{\rho} = k_1 \tag{5.26}$$

$$\ln \frac{1}{\rho} = k_1 \frac{2\pi \varepsilon_0}{\rho_l} = k_2$$

giving

$$\rho = e^{-k_2} = k \tag{5.27}$$

where k_1, k_2, and k are constants. These equations represent a family of cylinders whose axes coincide with the z-axis, and their cross-sections are circles with centres $(0, 0, z)$. On any one such surface, the electric field is normal to the surface and equal at all points on it. The electric field streams out radially away from the z-axis.

Example 5.8 Find the general equation of equipotential surfaces for an infinite charged plane.

Solution Refer Fig. 4.36. The electric field is given by

$$E = \begin{cases} \dfrac{\rho_s}{2\varepsilon_0} a_z, & \text{when } z > 0 \\[2mm] -\dfrac{\rho_s}{2\varepsilon_0} a_z, & \text{when } z < 0 \end{cases} \tag{5.28}$$

for $z > 0$

$$-\frac{dV}{dz} = \frac{\rho_s}{2\varepsilon_0} \tag{5.29}$$

Integrating

$$V = -\frac{\rho_s}{2\varepsilon_0} z \tag{5.30}$$

$V = $ constant gives $z = k$ which are equipotential surfaces being planes parallel to the x-y plane ($z = 0$ is the x-y plane).

Example 5.9 For the potential function

$$V = x^2 - y^2$$

find the equipotential surfaces.

Solution The surface $V = x^2 - y^2$ is shown in Fig. 5.4.

Equipotential surfaces are those surfaces on which $V = $ constant which gives $x^2 - y^2 = K$ (constant).

$V = $ constant contours are shown in Fig. 5.4.

Each of these contours may be imagined to be a surface which extends from $z = -\infty$ to $z = \infty$.

5.3 | Potential Energy

Next we will investigate the physical meaning of the line integral of the electric field. We know that the line integral of the force

$$\int F \cdot dl$$

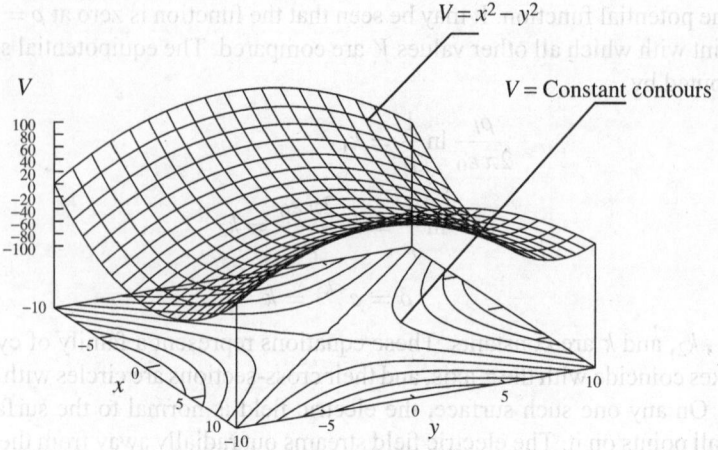

Fig. 5.4 The surface $V = x^2 - y^2$

is the work done. Let us find the work done on a test charge Q_t when moving it in an electric field due to a point charge Q placed at the origin. First let the test charge be far away at $r = \infty$. We further assume that both Q and Q_t are positive charges, and we move Q_t towards Q along a straight line. See Fig. 5.5.

Since both Q and Q_t are both of the same polarity, there is resistance to the motion since the charges repel each other. The force on the test charge is

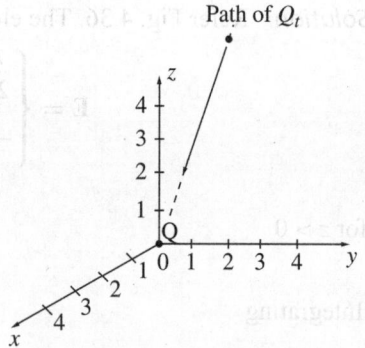

Fig. 5.5 Work done when moving Q_t towards Q

$$\mathbf{F} = Q_t \mathbf{E}_{\text{principle charge}}$$
$$= Q_t \frac{Q}{4\pi\varepsilon_0 r^2} \mathbf{a}_r$$

The work done to move the test charge from $r = \infty$ to $r = r_0$ is given by

$$\int_{r=\infty}^{r_0} \mathbf{F} \cdot d\mathbf{l} = -\int_{r=\infty}^{r_0} Q_t \mathbf{E} \cdot d\mathbf{l} = Q_t \left(-\int_{r=\infty}^{r_0} \mathbf{E} \cdot d\mathbf{l} \right)$$
$$= Q_t(V_{r_0} - V_\infty) = Q_t(V_{r_0} - 0) = Q_t V_{r_0} \qquad (5.31)$$

where $V(r_0)$ is the potential due to the charge Q at the point $r = r_0$. We have done work on the charge, but where has the 'work done' gone? In fact, Q_t has gained equivalent potential energy. If we release Q_t, it will move towards $r = \infty$ with increasing velocity and gain kinetic energy at the expense of its potential energy.

Thus   Work done = Gain in PE

and   Gain in potential energy $= Q_t[V(\text{Final position}) - V(\text{Initial position})]$
$$\qquad (5.32)$$

We may also write that

$$\text{Work done on the charge} = -Q_t \int_A^B \mathbf{E} \cdot d\mathbf{l}$$

Example 5.10 An electron has an initial velocity of $v = 2 \times 10^7 a_x$ m/sec when it is at the coordinate point $[1, 0, 0]$ m. The electron is moving in the electric field of a charge $q = 3 \times 10^{-6}$ C placed at the origin of a coordinate system. Describe the motion of the electron.

Solution Let us first draw a diagram of the configuration.

Step 1. The potential at $[1, 0, 0]$ due to the point charge is[2]

$$V_{in} = q/(4\pi\varepsilon_0 r_{in})$$
$$= (3 \times 10^{-6})/(4\pi\varepsilon_0)$$
$$= 2.696 \times 10^3 \text{ V}$$

Step 2. The initial potential energy P_{in} of the electron is

$$P_{in} = eV_{in}$$
$$= -1.6 \times 10^{-19} \text{ C} \times 2.696 \times 10^3 \text{ V}$$
$$= -4.31411 \times 10^{-15} \text{ J}$$

as $\qquad e = -1.6 \times 10^{-19}$ C, the charge of an electron

As the electron moves it slows down due to attraction to the charge at the origin and ultimately comes to rest.

Step 3. Its initial kinetic energy is

$$K_{in} = \frac{1}{2}m_e v_{in}^2$$
$$= 1.6398 \times 10^{-15} \text{ J}$$

Step 4. The initial total energy is

$$E_T = P_{in} + K_{in}$$
$$= -2.6743 \times 10^{-15} \text{ J}$$

Step 5. The final kinetic energy is zero, i.e., $K_{fin} = 0$. Hence, $E_{Tfin} = P_{fin} + 0 = P_{fin}$.

Now $\qquad\qquad\qquad E_{Tin} = E_{Tfin} = E_T$

Therefore $\qquad\qquad P_{fin} = E_T = -2.6743 \times 10^{-15}$ J

But P_{fin} is also given by

$$P_{fin} = \frac{eq}{4\pi\varepsilon_0 r_{fin}}$$

where r_{fin} is the final position of the electron.

Computing $\qquad\qquad\qquad r_{fin} = 1.6131$ m

From this position the electron reverses motion and begins moving towards the charge at the origin.

Example 5.11 In the previous example, beyond which velocity does the electron escape from the 'clutches' of the 3×10^{-6} C charge (Fig. 5.6)?

---

[2]The subscript *in* has been used for *initial* quantities while *fin* has been used for *final* quantities.

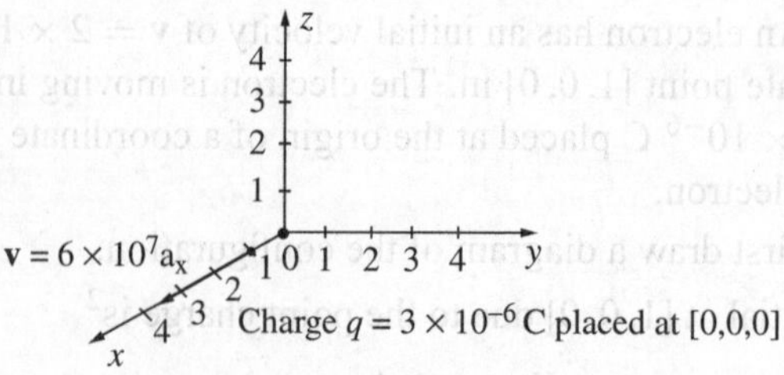

z

4
3
2
1

$v = 6 \times 10^7 a_x$ -1 0 1 2 3 4 y

x

Charge $q = 3 \times 10^{-6}$ C placed at [0,0,0]

Fig. 5.6

Solution

Step 1. The initial potential energy of the electron is

$$P_{in} = eV_{in} = -4.31411 \times 10^{-15} \text{ J}$$

Step 2. When the electron reaches $r = \infty$ its final potential energy and its kinetic energy both become zero. This means that *its total energy must be zero initially itself.*

Step 3. Solve the equation $E_T = K_{in} + P_{in}$

$$= K_{fin} + P_{fin}$$

$$= 0$$

Setting $E_T = 0$, we have

$$\frac{1}{2} m_e v_{in}^2 + (-4.31411 \times 10^{-15}) = 0$$

Solving for v_{in} $\quad v_{in} = 9.713231 \times 10^7$ m/sec

The above result can be stated as follows: if the electron has an initial velocity greater than 9.713231×10^7 m/sec then the electron does not return towards the charge at the origin.

Electron Microscope

The electron microscope (as the name implies) uses a beam of electrons to illuminate the specimen and create a magnified image of it. From quantum theory, electrons are dual in nature being waves or particles; the wavelength of an electron may be calculated according to De Broglie equation, $\lambda = h/(m_e v)$, where h is the Planck's constant and v is the velocity of the electron. Obviously the greater the velocity of an electron, the smaller its wavelength, so very small wavelengths through very high velocities are achieved. By this means wavelengths about 100,000 times shorter than light are possible which gives magnifications of up to about 2,000,000×. (Microscopes using light are limited to about 2000× magnification.) Electron microscopes use electrostatic 'lenses' to focus the electron beam. Typically, electrons are emitted by an electron gun (something like an electron gun in a TV tube) and then accelerated through a potential difference of 40 to 400 keV and transmitted through the specimen. The electrons which emerge are used to form an image. Figure 5.7 shows one such image.

Fig. 5.7 Electron microscope picture of a spider (photo taken taken from a website of University of Minnesota, http://umn.edu, 2010)

It is seen that potential energy of a charge may be negative! Hence we should rather be concerned about the *differences in potential and potential energy than their absolute values*.

The potential difference between two points $A \equiv A(x_0, y_0, z_0)$ and $B \equiv B(x_1, y_1, z_1)$ in space is equal to

$$\Delta V = \int_A^B (-\mathbf{E}) \cdot d\mathbf{l} = \int_A^B \nabla V \cdot d\mathbf{l}$$
$$= V(B) - V(A) \equiv V(\text{later point}) - V(\text{former point}) \qquad (5.33)$$

Thus due to a charge of 2 pC placed at the origin, the potentials at the points $A \equiv [1, 1, 1]$ and $B \equiv [2, 2, 2]$ are

$$V(A) = 0.0179755 \text{ V}$$
$$V(B) = 0.0051891 \text{ V}$$
$$V_B - V_A = -0.012786 \text{ V}$$
$$V(A) > V(B)$$

The electric field is directed from A to B. If a positive charge Q_t is placed at A and no restrictions are placed on it then it will feel *a force from A to B, and tend to move from A to B, losing potential energy* $= Q_t(-0.012786)$ J. On the other hand, a negative charge placed at B will feel a force towards A.

If the charge at the origin is a -2 pC charge instead of 2 pC.

then
$$V(A) = -0.0179755 \text{ V}$$
$$V(B) = -0.0051891 \text{ V}$$

and
$$V(B) > V(A)$$

The electric field is now directed from B to A. In this case if a positive charge is placed at B and no restrictions are placed on it then it will feel *a force from B to A*. On the other hand, a negative charge placed at A will feel a force towards B.

We come to two important conclusions.

1. The electric field is directed from a higher potential to a lower potential.

2. Positive charges feel forces in the direction of the electric field, and when free to move, they go from a higher potential towards a lower potential, losing potential energy. Negative charges feel forces in a direction against the electric field. They move from a lower potential to a higher potential again losing potential energy.

Example 5.12 For the case of two concentric spheres, the inner sphere has a radius r_{in} with charge Q on it, while the outer sphere has radius r_{out} with charge $-Q$. An electron leaves the outer sphere and rushes towards the inner sphere. Find KE of the electron when it reaches the inner sphere.

Solution The potential of the outer sphere is

$$V_{out} = -\frac{Q}{4\pi\varepsilon_0\, r_{out}}$$

Similarly the potential of the inner sphere is

$$V_{in} = \frac{Q}{4\pi\varepsilon_0\, r_{in}}$$

The difference in their PE is

$$\frac{eQ}{4\pi\varepsilon_0}\left(\frac{1}{r_{out}} - \frac{1}{r_{in}}\right)$$

Loss in PE = Gain in KE = $m_e V^2/2 - 0$

Hence

$$m_e v^2/2 = \frac{eQ}{4\pi\varepsilon_0}\left(\frac{1}{r_{out}} - \frac{1}{r_{in}}\right)$$

5.4 | Potential Due to a System of Point Charges

We know that the potential at $\mathbf{r} = (x, y, z)$ due to a single charge Q_1 placed at the point $\mathbf{r}'_1 = (x'_1, y'_1, z'_1)$ is given by

$$V(\mathbf{r}) = \frac{Q_1}{4\pi\varepsilon_0\, |\mathbf{r} - \mathbf{r}'_1|}$$

$$= \frac{Q_1}{4\pi\varepsilon_0\, \sqrt{(x - x'_1)^2 + (y - y'_1)^2 + (z - z'_1)^2}} \qquad (5.34)$$

As usual we use superposition to determine the potential due to more than one charges. There are two point charges Q_1 and Q_2 placed at $\mathbf{r}'_1 = (x'_1, y'_1, z'_1)$ and $\mathbf{r}'_2 = (x'_2, y'_2, z'_2)$. The potential at $\mathbf{r} = (x, y, z)$ due to these charges is

$$V(\mathbf{r}) = V_1(\mathbf{r}) + V_2(\mathbf{r})$$

$$= \sum_{i=1}^{2} \frac{Q_i}{4\pi\varepsilon_0\, \sqrt{(x - x'_i)^2 + (y - y'_i)^2 + (z - z'_i)^2}}$$

Similarly, if charges $Q_1, Q_2, \ldots, Q_N$ are placed at $\mathbf{r}'_1, \mathbf{r}'_2, \ldots, \mathbf{r}'_N$ respectively, then each of these charges produces a potential $V_i(\mathbf{r})$ at the field point $\mathbf{r}$ (Fig. 5.8). The total potential, therefore, due to all these charges placed at all these positions is

$$V(\mathbf{r}) = \sum_{i=1}^{i=N} V_i(\mathbf{r})$$

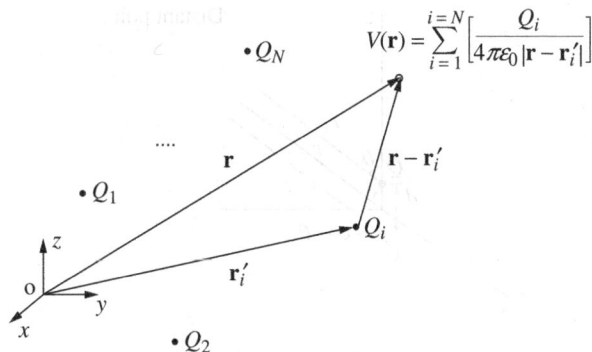

$$V(\mathbf{r}) = \sum_{i=1}^{i=N} \left[\frac{Q_i}{4\pi\varepsilon_0 |\mathbf{r} - \mathbf{r}_i'|} \right]$$

Fig. 5.8 The potential for a system of point charges

or

$$V(\mathbf{r}) = \sum_{i=1}^{i=N} \left[\frac{Q_i}{4\pi\varepsilon_0 \sqrt{(x - x_i')^2 + (y - y_i')^2 + (z - z_i')^2}} \right] \qquad (5.35)$$

And then we can evaluate the electric field by

$$\mathbf{E} = -\nabla V \qquad (5.36)$$

So, in rectangular coordinates

$$\mathbf{E} = -\nabla V$$

$$= -\left(\frac{\partial}{\partial x} \mathbf{a}_x + \frac{\partial}{\partial y} \mathbf{a}_y + \frac{\partial}{\partial z} \mathbf{a}_z \right) V(x', y', z', x, y, z)$$

$$= -\left(\frac{\partial V}{\partial x} \mathbf{a}_x + \frac{\partial V}{\partial y} \mathbf{a}_y + \frac{\partial V}{\partial z} \mathbf{a}_z \right) \qquad (5.37)$$

It may be noted that differentiation must be carried out only on the (x, y, z) trio, and *not* on the (x', y', z') set of variables.

5.4.1 Far Fields for an Electric Dipole

Example 5.13 Find the far field for the case of a dipole as shown in Fig. 5.9.

Solution

Step 1. Using the formula of Section 5.4, the potential due to the two identical charges is

$$V(\mathbf{r}) = \frac{Q}{4\pi\varepsilon_0} \left[\frac{1}{\sqrt{\left(z - \frac{d}{2}\right)^2 + y^2 + x^2}} - \frac{1}{\sqrt{\left(z + \frac{d}{2}\right)^2 + y^2 + x^2}} \right] \qquad (5.38)$$

Step 2. Going over to spherical coordinates

$$x = r \sin\theta \cos\phi, \quad y = r \sin\theta \sin\phi, \quad z = r\cos\theta \qquad (5.39)$$

and simplifying

$$V(\mathbf{r}) = \frac{Q}{2\pi\varepsilon_0} \left[\frac{1}{\sqrt{4r^2 - 4\,d\,r\cos\theta + d^2}} - \frac{1}{\sqrt{4r^2 + 4\,d\,r\cos\theta + d^2}} \right] \qquad (5.40)$$

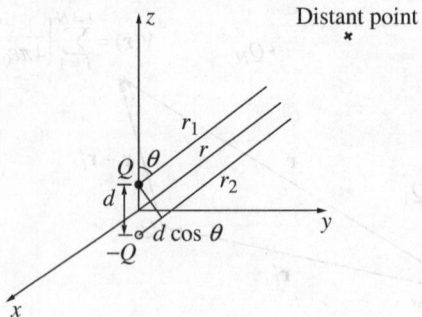

Fig. 5.9 A dipole aligned along the z-axis

Step 3. We now write the potential when $r \gg d$:

$$V(\mathbf{r}) = \frac{Q}{2\pi\varepsilon_0} \left[\frac{1}{\sqrt{4r^2 - 4\,d\,r\cos\theta + d^2}} - \frac{1}{\sqrt{4r^2 + 4\,d\,r\cos\theta + d^2}} \right]$$

$$\approx \frac{Q}{2\pi\varepsilon_0} \left[\frac{1}{\sqrt{4r^2 - 4\,d\,r\cos\theta}} - \frac{1}{\sqrt{4r^2 + 4\,d\,r\cos\theta}} \right]$$

$$= \frac{Q}{4\pi\varepsilon_0 r} \left[\frac{1}{\sqrt{1 - (d/r)\cos\theta}} - \frac{1}{\sqrt{1 + (d/r)\cos\theta}} \right]$$

Step 4. Using $\dfrac{1}{\sqrt{1 \pm x}} = 1 \mp \dfrac{x}{2} + \cdots$

we get $\quad V = \dfrac{Qd\cos\theta}{4\pi\varepsilon_0 r^2} = \dfrac{\mathbf{p}\cdot\mathbf{a}_r}{4\pi\varepsilon_0 r^2}, \qquad \mathbf{p} = Qd\mathbf{a}_z$

Step 5. Now compute the electric field

$$E_r \approx -\frac{\partial V}{\partial r} = \frac{Qd\cos\theta}{2\pi\varepsilon_0 r^3} \tag{5.41}$$

Because of the absence of ϕ there is a symmetry of the E field about the z-axis

$$E_\theta \approx -\frac{1}{r}\frac{\partial V}{\partial \theta} = \frac{Qd\sin\theta}{4\pi\varepsilon_0 r^3} \tag{5.42}$$

and $$E_\phi = 0 \tag{5.43}$$

@ See `PotDipole.m` in Chapter 5

Practice Problem 5.1
Show all the steps related to Eqn (5.40), going from rectangular to spherical coordinates.

5.5 | Potential Due to Continuous Charge Distributions

In Section 4.5 we obtained the electric field due to any continuous charge distribution. We now obtain the scalar electric potential due to a continuous charge distribution.

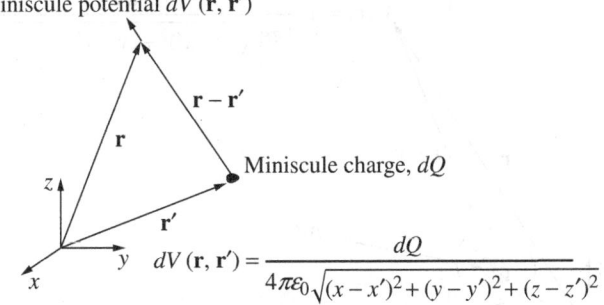

Fig. 5.10 $dV(r, r')$ due to a minuscule charge $dQ(r')$

To do so we find the infinitesimal potential dV at a point $\mathbf{r}$ in space, due to an elemental charge placed at $\mathbf{r}'$ as shown in Fig. 5.10.

$$dV(\mathbf{r}, \mathbf{r}') = \frac{dQ(\mathbf{r}')}{4\pi\varepsilon_0 \sqrt{(x - x')^2 + (y - y')^2 + (z - z')^2}} \tag{5.44}$$

where $\mathbf{r}$ is the position vector of the field point and $\mathbf{r}'$ is the position vector of the charge dQ. Notice that this is the same formula as that for a point charge, except that the point charge Q has been replaced by dQ. Now to find the potential due to a charge distribution we integrate the above equation over the region where the charge is present.

$$V'(\mathbf{r}) = \iiint_{\mathcal{R}'} dV'(\mathbf{r}) \tag{5.45}$$

$$V' = \iiint_{\mathcal{R}'} \frac{dQ(\mathbf{r}')}{4\pi\varepsilon_0 \sqrt{(x - x')^2 + (y - y')^2 + (z - z')^2}} \tag{5.46}$$

where V' is the region occupied by the charge. Now dQ will be different for different charge distributions.

$$dQ = \begin{cases} \rho_v(\mathbf{r}')\, dV', & \text{for a volume charge distribution} \\ \rho_s(\mathbf{r}')\, dS', & \text{for a surface charge distribution} \\ \rho_l(\mathbf{r}')\, dL', & \text{for a linear charge distribution} \end{cases} \tag{5.47}$$

Therefore

$$V(\mathbf{r}) = \begin{cases} \displaystyle\iiint_{V'} \frac{\rho_v(\mathbf{r}')\, dV'}{4\pi\varepsilon_0 \sqrt{(x - x')^2 + (y - y')^2 + (z - z')^2}} & \text{volume charge in } V' \\[18pt] \displaystyle\iint_{S'} \frac{\rho_s(\mathbf{r}')\, dS'}{4\pi\varepsilon_0 \sqrt{(x - x')^2 + (y - y')^2 + (z - z')^2}} & \text{surface charge in } S' \\[18pt] \displaystyle\int_{L'} \frac{\rho_l(\mathbf{r}')\, dL'}{4\pi\varepsilon_0 \sqrt{(x - x')^2 + (y - y')^2 + (z - z')^2}} & \text{line charge in } L' \end{cases} \tag{5.48}$$

Figure 5.11 shows the various regions for a volume charge distributions and the integrations are performed on V'.

Example 5.14 Obtain the potential and the electric field for the case of a pair of line charges shown in Fig. 5.12.

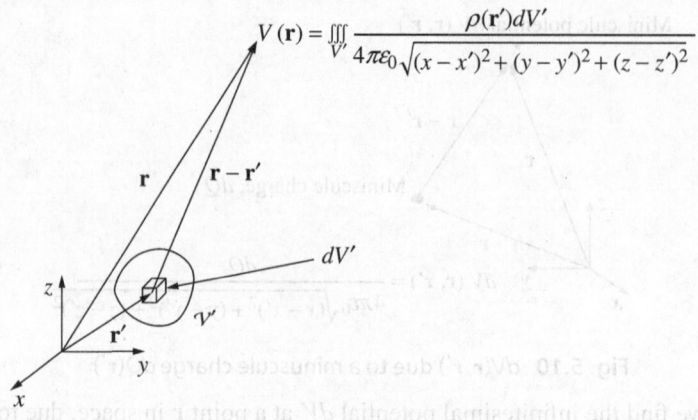

$$V(\mathbf{r}) = \iiint_{V'} \frac{\rho(\mathbf{r}')dV'}{4\pi\varepsilon_0 \sqrt{(x-x')^2 + (y-y')^2 + (z-z')^2}}$$

Fig. 5.11 Potential calculation for a volume charge distribution

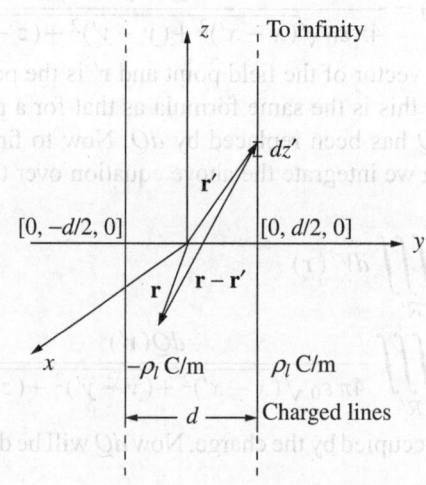

Fig. 5.12 Potential for a pair of parallel charged lines with charge ρ_l and $-\rho_l$

Solution Figure 5.12 shows a couple of charged lines piercing the x-y plane at $[0, -d/2, 0]$ and $[0, -d/2, 0]$. The lines are charged with a constant line charge of $\rho_l(z')$ and $-\rho_l(z')$ C/m. From symmetry considerations, we compute the potential only on the x-y plane, as the fields on any other plane would be the same.

$$\mathbf{r} = [x, y, 0] \tag{5.49}$$

$$\mathbf{r}' = [0, d/2, z'] \tag{5.50}$$

$$\mathbf{r}' = [0, -d/2, z'] \tag{5.51}$$

Hence

$$V(\mathbf{r}) = \int_{L'} \frac{\rho_l(\mathbf{r}')\,dL'}{4\pi\varepsilon_0 \sqrt{(x-x')^2 + (y-y')^2 + (z-z')^2}}$$

$$= \int_{-\infty}^{\infty} \left[\frac{\rho_l dz'}{4\pi\varepsilon_0 \sqrt{x^2 + (y-d/2)^2 + z'^2}} + \frac{-\rho_l dz'}{4\pi\varepsilon_0 \sqrt{x^2 + (y+d/2)^2 + z'^2}} \right] \tag{5.52}$$

The left integral is over the positively charged line while the right integral is over the negatively charged line. The integral may be written in a slightly different way, that is integrating from $-z_0$ to z_0 and letting $z_0 \rightarrow \infty$.

$$V(\mathbf{r}) = \lim_{z_0 \rightarrow \infty} \int_{-z_0}^{z_0} \left[\frac{\rho_l \, dz'}{4\pi\varepsilon_0 \sqrt{x^2 + (y - d/2)^2 + z'^2}} \right.$$

$$\left. + \frac{-\rho_l \, dz'}{4\pi\varepsilon_0 \sqrt{x^2 + (y + d/2)^2 + z'^2}} \right]$$

$$= \frac{\rho_l}{4\pi e_0} \lim_{z_0 \rightarrow \infty} \left[\sinh^{-1}\left(\frac{z'}{\sqrt{\left(y - \frac{d}{2}\right)^2 + x^2}} \right) \right.$$

$$\left. \left. - \sinh^{-1}\left(\frac{z'}{\sqrt{\left(y - \frac{d}{2}\right)^2 + x^2}} \right) \right] \right|_{z'=-z_0}^{z'=z_0}$$

$$= \lim_{z_0 \rightarrow \infty} \frac{\rho_l}{2\pi e_0} \left[\sinh^{-1}\left(\frac{2z_0}{\sqrt{4y^2 - 4dy + 4x^2 + d^2}} \right) \right.$$

$$\left. - \sinh^{-1}\left(\frac{2z_0}{\sqrt{4y^2 + 4dy + 4x^2 + d^2}} \right) \right]$$

where

$$\sinh^{-1}(p) = \log\left(\sqrt{p^2 + 1} + p \right) \tag{5.53}$$

See Appendix B, integral 22.

After taking the limit $z_0 \rightarrow \infty$ the potential in the x-y plane becomes

$$V(x,y) = \frac{\rho_l}{2\pi\varepsilon_0} \log\left(\frac{\sqrt{(y + d/2)^2 + x^2}}{\sqrt{(y - d/2)^2 + x^2}} \right) \tag{5.54}$$

If we wish to compute the equipotential surfaces for this example

$$V(x,y) = \frac{\rho_l}{2\pi\varepsilon_0} \log\left(\frac{\sqrt{(y + d/2)^2 + x^2}}{\sqrt{(y - d/2)^2 + x^2}} \right) = V_1 \text{ (a constant)} \tag{5.55}$$

Hence,

$$\frac{\sqrt{(y + d/2)^2 + x^2}}{\sqrt{(y - d/2)^2 + x^2}} = \exp\frac{V_1 \times 2\pi\varepsilon_0}{\rho_l} = k \tag{5.56}$$

where k is another constant. This equation simplifies to

$$\left[(y + d/2)^2 + x^2 \right] - k^2 \left[(y - d/2)^2 + x^2 \right] = 0$$

which is $\quad (1 - k^2)x^2 + (1 - k^2)y^2 + (1 + k^2)yd + \dfrac{d^2}{4}(1 - k^2) = 0$

or

$$x^2 + y^2 + \frac{(1+k^2)}{(1-k^2)}yd + \left[\frac{(1+k^2)}{2(1-k^2)}d\right]^2 = \frac{d^2(k^2-1)}{4(1-k^2)} + \left[\frac{(1+k^2)}{2(1-k^2)}d\right]^2$$

$$x^2 + \left[y + \frac{d(1+k^2)}{2(1-k^2)}\right]^2 = \frac{d^2k^2}{(1-k^2)^2} \tag{5.57}$$

These are equations of circles with radii

$$r = \frac{dk}{(1-k^2)} = r_k$$

and centres

$$\left[0, y_k = -\frac{d(1+k^2)}{2(1-k^2)}\right]$$

and are valid for $0 < k < 1$. If we substitute $1/k$ for $k < 1$ then we arrive at a family of complementary circles with the same radii

$$r = \frac{dk}{(1-k^2)} = r_{1/k}$$

but centres

$$\left[0, y_{1/k} = +\frac{d(1+k^2)}{2(1-k^2)}\right]$$

These plots are shown in Fig. 5.13.

 See `PotInfChgedLines.m` in Chapter 5

We can compute the electric field for this case by

$$\mathbf{E} = -\nabla V \tag{5.58}$$

Though the above expressions are quite complex, but their plot is particularly simple, as shown in Fig. 5.14. The electric field streamlines are circles which cut

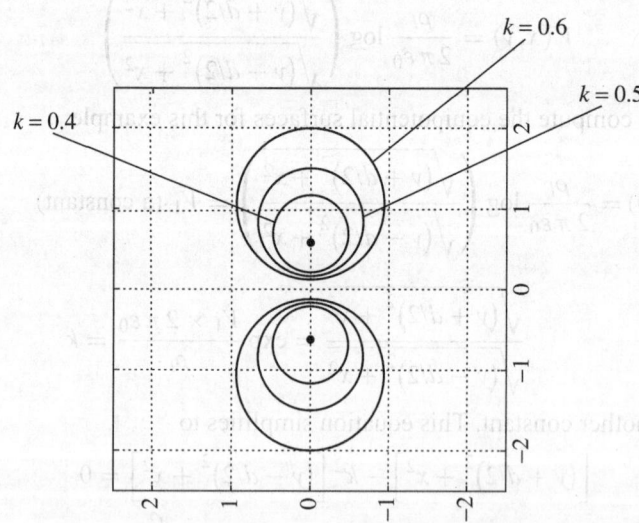

Fig. 5.13 Lines of constant electric potential for a pair of infinite line charges

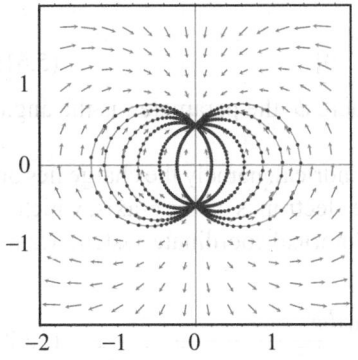

Fig. 5.14 Electric field streamlines for two parallel infinite line charges

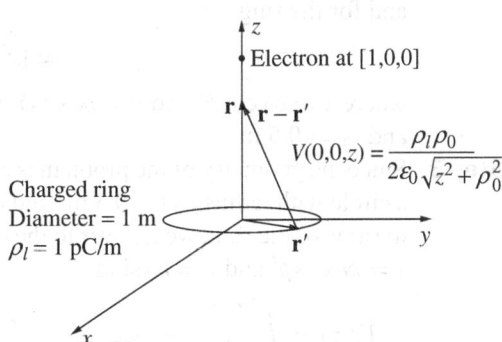

Fig. 5.15 Charged ring and electron

the equipotential circles at right angles. *In fact in any electrostatic case, the electric field lines and the equipotential lines always cut each other at right angles.*

Example 5.15 Find the potential due to an infinite line charge with charge density ρ_l along the z-axis ($z = \pm\infty$).

Solution

Step 1. The electric field due to such a configuration is given by Eqn (5.45).

$$\mathbf{E} = \frac{\rho_l}{2\pi\varepsilon_0\,\rho}\,\mathbf{a}_\rho$$

Step 2. Hence

$$V_{\rho_1\,to\,\rho_2} = -\int_{\rho_1}^{\rho_2}\mathbf{E}\cdot d\mathbf{l} = -\int_{\rho_1}^{\rho_2}\frac{\rho_l d\rho}{2\pi\varepsilon_0\,\rho} = \frac{\rho_l}{2\pi\varepsilon_0}\ln\frac{\rho_1}{\rho_2}$$

Step 3. If we make the datum $\rho = \rho_1 = 1$ and replace ρ_2 by ρ then

$$V(\rho) = -\frac{\rho_l}{2\pi\varepsilon_0}\ln\rho$$

Example 5.16 An electron at rest is released from the position of 1 m along the axis of a ring of diameter 1 m which is charged with a linear charge density of $\rho_l = 1$ pC/m as shown in Fig. 5.15. Describe the motion of the electron.

Solution

Step 1. The motion of the electron is best described by computing the potential function of the charged ring. The potential $V(\mathbf{r})$ is given by

$$V(\mathbf{r}) = \int_{Ring}\frac{\rho_l dL'}{4\pi\varepsilon_0\,|\mathbf{r} - \mathbf{r}'|}$$

$$= \int_{Ring}\frac{\rho_l dL'}{4\pi\varepsilon_0\,\sqrt{(z - z')^2 + (y - y')^2 + (x - x')^2}} \tag{5.59}$$

where ρ_l is the charge density of 1 pC/m. Both $\mathbf{r}$ and $\mathbf{r}'$ are clearly shown in the figure.

Step 2. Find the potential on the z-axis only. So

$$\mathbf{r} = [z, 0, 0] \tag{5.60}$$

and for the ring

$$\mathbf{r}' = [x', y', 0] \tag{5.61}$$

where $x' = \rho_0 \cos\phi'$ and $y' = \rho_0 \sin\phi'$ where ϕ, the parameter, is the angle and $\rho_0 = 0.5$ m.

Step 3. Since the geometry of the problem is cylindrical, namely the charge lies on a circle with radius ρ_0 (= 0.5 m) and the electron is on the z-axis, which is the axis of the ring, we go over to the cylindrical coordinate system. Using $x' = \rho_0 \cos\phi'$ and $y' = \rho_0 \sin\phi'$,

$$V(\mathbf{r}) = \int_0^{2\pi} \frac{\rho_l \rho_0 \, d\phi'}{4\pi\varepsilon_0 \sqrt{z^2 + (y - \rho_0 \sin\phi')^2 + (x - \rho_0 \cos\phi')^2}} \tag{5.62}$$

Step 4. Since the potential function needs to be calculated only on the z-axis, x and y are both equated to zero, then

$$V(z) = \int_0^{2\pi} \frac{\rho_l \rho_0 \, d\phi'}{4\pi\varepsilon_0 \sqrt{z^2 + (\rho_0 \sin\phi')^2 + (\rho_0 \cos\phi')^2}}$$

$$= \int_0^{2\pi} \frac{\rho_l \rho_0 \, d\phi'}{4\pi\varepsilon_0 \sqrt{z^2 + \rho_0^2}}$$

$$= \frac{\rho_l \rho_0}{2\varepsilon_0 \sqrt{z^2 + \rho_0^2}} \tag{5.63}$$

Step 5. Now we come to the electron. The initial kinetic energy of the electron is zero. Therefore, its initial potential energy and total energy is

$$K_{in} + P_{in} = 0 + P_{in}$$

Now

$$P_{in} = \left. \frac{\rho_l \rho_0 \, e}{2\varepsilon_0 \sqrt{z^2 + \rho_0^2}} \right|_{z=1 \text{ m}}$$

$$= -4.0458 \times 10^{-21} \text{ J} \tag{5.64}$$

Hence

$$E_T = -4.0458 \times 10^{-21} \text{ J} \tag{5.65}$$

The electron feels a strong attractive force towards the centre of the ring. As the election moves, its kinetic energy increases while its potential energy decreases keeping its total energy constant, which is equal to the initial value of total energy.

$$E_T = \text{Kinetic energy} + \text{Potential energy}$$

$$= K + P \tag{5.66}$$

$$= K_{in} + P_{in}$$

$$= -4.0458 \times 10^{-21} \text{ J} \tag{5.67}$$

Step 6. When the electron is at a distance z from the origin its potential energy is

$$P(z) = \frac{\rho_l \rho_0 \, e}{2\varepsilon_0 \sqrt{z^2 + \rho_0^2}} \tag{5.68}$$

and its kinetic energy is

$$K = \frac{1}{2}m_e v^2 \qquad (5.69)$$

so

$$E_T = K + P = \frac{\rho_l \rho_0\, e}{2\,\varepsilon_0\,\sqrt{z^2 + \rho_0^2}} + \frac{1}{2}m_e v^2$$

$$= -4.0458 \times 10^{-21} \qquad (5.70)$$

Solving for **v** (which requires some algebraic manipulations)

$$\mathbf{v} = -\frac{\sqrt{2\,\varepsilon_0\,E_T\,\sqrt{z^2 + \rho_0^2} - e\,\rho_l \rho_0}}{\left[m_e^2 \varepsilon_0^2\,(z^2 + \rho_0^2)\right]^{\frac{1}{4}}}\mathbf{a}_z \qquad (5.71)$$

To check this result, we substitute the values of E_T, ρ_0, e, m_e, ε_0, and $z = 1$, which gives a value of zero.

If we want to find the velocity of the electron when it reaches the centre of the ring then we substitute $z = 0$ in Eqn (5.71), then

$$\mathbf{v}(z = 0) = -1.047 \times 10^5 \mathbf{a}_z \text{ m/s} \qquad (5.72)$$

 See `PotChgRing.m` in Chapter 5

POINTS TO REMEMBER

- The work done by the external force to move the charge Q from $\mathbf{r}_A$ to $\mathbf{r}_B$ in region of electric field E is

$$W = -Q \int_{\mathbf{r}_A}^{\mathbf{r}_B} \mathbf{E} \cdot d\mathbf{l}$$

- The potential difference between points A and B is

$$V_{AB} = \left(\frac{W}{Q}\right) = -\int_{\mathbf{r}_A}^{\mathbf{r}_B} \mathbf{E} \cdot d\mathbf{l}$$

- The potential difference between two points, $\mathbf{r}_A$ and $\mathbf{r}_B$, in the field created by a point charge Q placed at the origin is

$$V_{AB} = \frac{Q}{4\pi\,\varepsilon_0}\left(\frac{1}{r_B} - \frac{1}{r_A}\right)$$

- The absolute potential at a point $\mathbf{r}$ in the field created by a point charge Q placed at the origin is

$$V(\mathbf{r}) = \frac{Q}{4\pi\,\varepsilon_0\,r} \text{ as } V(\infty) = 0$$

- The potential difference between two points A and B in general is

$$V_{AB} = V_B - V_A = V_{\text{fin}} - V_{\text{init}}$$

- The line integral of the electric field on a closed path is zero, i.e.

$$\oint_{\mathcal{L}} \mathbf{E} \cdot d\mathbf{l} = 0 \quad (\mathcal{L} \text{ is any closed path})$$

- Maxwell's equation for electrostatic fields is $\nabla \times \mathbf{E} = \mathbf{0}$.

- The relationship between the electric field **E** and the potential V is

$$\mathbf{E} = -\nabla V$$

- The potential at **r** for a point charge Q located at the point **r**′ is

$$V(\mathbf{r}) = \frac{Q}{4\pi\varepsilon_0 |\mathbf{r} - \mathbf{r}'|} = \frac{Q}{4\pi\varepsilon_0 \left[(x - x')^2 + (y - y')^2 + (z - z')^2 \right]^{1/2}}$$

- An equipotential surface is one where the potential at all points on the surface is constant.
- The electric field is always perpendicular to an equipotential surface.
- The work done to move a charge Q from $r = \infty$ ($V(\infty) = 0$) to $r = r_0$ in an electric field is given by

$$\int_{r=\infty}^{r_0} \mathbf{F} \cdot d\mathbf{l} = QV(r_0) \tag{5.73}$$

- Gain in potential energy for a charge Q is

$$\text{Gain in potential energy} = Q\left[V(\text{Final position}) - V(\text{Initial position}) \right]$$

- The potential at a point **r** due to N charges Q_i at points **r**′$_i$, $i = 1, \ldots, N$ in rectangular coordinates is

$$V(\mathbf{r}) = \sum_{i=1}^{i=N} \left[\frac{Q_i}{4\pi\varepsilon_0 \sqrt{(x - x_i')^2 + (y - y_i')^2 + (z - z_i')^2}} \right]$$

- If a dipole consisting of charges $(Q, -Q)$ with a separation d, oriented along the z-axis, is located at the origin. Then the potential $V(\mathbf{r})$ for $r \gg d'$ is

$$V = \frac{Qd\cos\theta}{4\pi\varepsilon_0 r^2}$$

and the electric field for such a dipole is

$$E_r = \frac{Qd\cos\theta}{2\pi\varepsilon_0 r^2}, \quad E_\theta = \frac{Qd\sin\theta}{4\pi\varepsilon_0 r^2}, \quad E_\phi = 0$$

- The potential at the point **r** due to continuous distribution of charges is

$$V(\mathbf{r}) = \iiint_{\mathcal{R}'} \frac{dQ(\mathbf{r}')}{4\pi\varepsilon_0 |\mathbf{r} - \mathbf{r}'|}$$

where $dQ = \begin{cases} \rho_v(\mathbf{r}')\,dV', & \text{for a volume charge distribution} \\ \rho_s(\mathbf{r}')\,dS', & \text{for a surface charge distribution} \\ \rho_l(\mathbf{r}')\,dL', & \text{for a linear charge distribution} \end{cases}$

- The potential at any point (x, y, z) for two infinitely long lines with charge distribution ρ_l and $-\rho_l$ is

$$V(x, y) = \frac{\rho_l}{2\pi\varepsilon_0} \log\left(\frac{\sqrt{(y + d/2)^2 + x^2}}{\sqrt{(y - d/2)^2 + x^2}} \right)$$

SELF ASSESSMENT

Objective Type Questions

In the following questions one or more choices may be correct. Sometimes choices may be separated with an 'or' and sometimes with an 'and'.

1. Which of the following sentences is *incorrect*?
   (a) The line of force or flux lines are always normal to equipotential surfaces.
   (b) The conductivity of metals generally decreases with temperature.
   (c) Work done in moving a charge along a closed path in electrostatic fields is zero.
   (d) None of the above.

2. In an electrostatic field, which of the following statements are correct?
   Work done in moving a charge Q from point A to B is
   (a) dependent on the path chosen   (b) dependent on the initial and final potentials
   (c) equal to $Q \int_A^B \mathbf{E} \cdot d\mathbf{l}$ (d) equal to $Q \int_A^B \nabla V \cdot d\mathbf{l}$

3. When moving a small charge $q = 1$ pC in the field of a charge $Q = 1$ μC located at the origin (in air) along the path $(x = 1, y = 1, z = 1)$ to $(1, 2, 1)$ to $(1, 2, 2)$ and back to $(1, 1, 1)$ (all in meters), the work done is
   (a) 10 pJ   (b) 10 nJ   (c) 10 μJ   (d) 0 J   (e) None of the above

4. For a negative point charge $-Q(Q > 0)$ the point which is taken as $V = 0$ is
   (a) not defined   (b) at infinity
   (c) at 1 m from the charge   (d) none of the above

5. For a negative point charge $-Q(Q > 0)$ at a distance of 1 m from the charge
   (a) V is negative   (b) $V = -Q/4\pi\varepsilon_0$ (c) V is positive   (d) V is zero

6. For two equal point charges Q, the potential is V everywhere, and
   (a) the potential at the mid-point between the two charges may be taken as zero
   (b) $\nabla V = 0$
   (c) the potential is the sum of the potentials of each of the charges
   (d) none of the above

7. Work done in moving a charge Q in a straight line in an electrostatic field which is varying with coordinates is
   (a) Force × distance   (b) $\int \mathbf{F} \cdot d\mathbf{l}$ (c) $-\int \mathbf{F} \cdot d\mathbf{l}$ (d) $-Q \int \mathbf{E} \cdot d\mathbf{l}$

8. The electric lines are
   (a) directed from a higher potential to a lower potential
   (b) directed from a lower potential to a higher potential
   (c) perpendicular to equipotential surfaces
   (d) None of the above

9. When a free electron is placed in an electrostatic field it moves
   (a) from a higher potential to a lower potential
   (b) from a lower potential to a higher potential
   (c) perpendicular to equipotential surfaces
   (d) None of the above

10. Equipotential surfaces
    (a) may intersect at right angles   (b) never intersect
    (c) may intersect at any angle   (d) ∇V is perpendicular to them

Short-Answer Questions

1. Give expressions of (a) electric energy density (w_e), (b) electric energy stored in the electric field.   (NTU-NSIT, March 2008)
2. Can electric field lines of electrostatic fields be closed curves? Research this topic.
   [**Hint:** Use $\mathbf{E} = -\nabla V$ and the concept of equipotential surfaces]
3. A potential field exists in a region where $\varepsilon = f(x)$. Find the value of $\nabla^2 V$ if $\rho_v = 0$ throughout the region.
4. If $V = (10/r^2) \sin\theta$ V in free space then what is ρ_v at point P $(r = 2, \theta = 30°, \phi = 0°)$?

5. Derive an expression for the potential difference between two points on a curve due to a point charge.

Review Questions

1. Explain how work has to be done to move a charge from point A to B in an electric field. If the work is positive what does it mean? If it is negative what is the implication?
2. Explain why electric field lines are perpendicular to equipotential surfaces?
3. Left alone in a static electric field with no external force applied, an electron gains momentum. What happens to its potential energy?
4. If we calculated the line integral of a static electric field along a closed contour, what should be the answer? Justify.
5. If we calculate the potential at a point in the presence of a number of point charges what is the 'law' we use and why?
6. Derive an expression for the potential at a point due to a point charge.

Numerical Problems

If the medium is not mentioned assume free space.

1. Show that the electric field for a line charge and sheet charge satisfies the equation $\nabla \times \mathbf{E} = \mathbf{0}$.
2. The electric field near the the origin of a Cartesian coordinate system is $\mathbf{E} = -(yz + 2xy^2z^2)\,\mathbf{a}_x - (xz + 2yx^2z^2)\,\mathbf{a}_y - (yx + 2zy^2x^2)\,\mathbf{a}_z$. Find the work done on a 10 nC charge to move it by 6 mm from $(1,1,1)$ in the x, y, and z directions respectively.
3. For the E field of Problem 2, find the work done to move a 10 nC charge by 6 mm from $(1,1,1)$ in the $\mathbf{a}_x + \mathbf{a}_y + \mathbf{a}_z$ direction.
4. For the E field of Problem 2, find the work done to move a 10 nC charge from $(0,0,0)$ to $(1,1,1)$ along a straight line.
5. For the E field of Problem 2, find the work done to move a 10 nC charge from $(0,0,0)$ to $(1,1,1)$ along the following zig-zag path $(0,0,0)$ to $(1,0,0)$ to $(1,1,0)$ to $(1,1,1)$. What do you infer from the answer?
6. In free space $V = x^2y^2(z + 3)$. Find $\mathbf{E}$ at $(3, 4, -6)$.
7. Let $V = xy^2z$, calculate the energy expended in transferring a $-2\mu C$ point charge from $(1, -1, 2)$ to $(2, 1, -3)$.
8. Determine the electric field for the following potentials
   (a) $V = x + y^2 + 2z^2$ (b) $V = (x^2 + y^2 + z^2)^{1/2}$
9. Determine the electric field for the potential, $V = \rho^2 z \sin\phi$.
10. Determine the electric field for the potential, $V = e^{-r}\sin\theta\,\cos\phi$.
11. Find the work done in carrying a 5 C charge from $P(1, 2, -2)$ to $R(0, 0, 0)$ in an electric field $\mathbf{E} = -2xz\mathbf{a}_x - x^2\mathbf{a}_z$ V/m.
12. The electric field in cylindrical coordinates is

$$\mathbf{E} = -\sin(\phi)\,za_\rho - \cos(\phi)\,za_\phi - \sin(\phi)\,\rho\,a_z$$

   Determine the work done in moving a 4 nC charge from
   (a) $A(1,0,0)$ to $B(4,0,0)$ (b) $B(4,0,0)$ to $C(4,30°,0)$
   (c) $C(4,30°,0)$ to $C(4,30°,2)$ (d) A to D
13. In free space, $V = x^2yz$ V. Find
   (a) $\mathbf{E}$ at $(3, 4, -6)$, (b) the charge within the cube $0 < x, y, z < 1$.
14. If there is a potential field $V = ax + by + cz$ where a, b, c are constants, then show that the electric field is constant, and find the direction in which the electric field points.

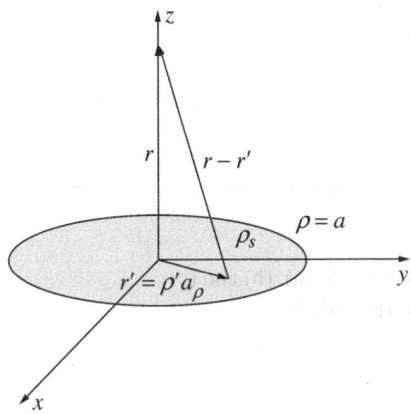

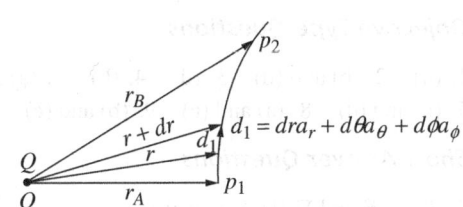

Fig. 5.16 Charged disk

Fig. 5.17 Calculation of potential difference

[Hint: Use $\mathbf{E} = -\nabla V$]

15. Three point charges $Q_1 = 1$ mC, $Q_2 = -2$ mC , $Q_3 = 3$ mC are located at $(0, 1, 4)$, $(-2, 5, 2)$, and $(3, -6, 6)$ respectively.

    (a) Find the potential V_p at $P(-1, 1, 1)$, (b) calculate potential difference V_{PQ} if Q is $(1, 2, 3)$.

16. To verify that $\mathbf{E} = [-2xy\,(z + 3), -x^2\,(z + 3), -x^2 y]$ is truly an electrostatic field, show that

    (a) $\nabla \times \mathbf{E} = 0$, (b) $\int_{\mathcal{L}} \mathbf{E} \cdot d\mathbf{l} = 0$

    where $\mathcal{L}$ is the closed loop of $(0, 0, 2)$, $(3, 0, 2)$, $(3, 3, 2)$, $(0, 3, 2)$.

17. For a disk of charge with constant surface charge ρ_s and radius a, find the potential along the axis of the disk.

18. For a disk of charge with constant surface charge ρ_s, with inner radius b, and outer radius a, find the potential along the axis of the disk.

19. For a disk of charge with surface charge ρ_s and radius a, find the potential along the axis of the disk.

20. For a single charge $\nabla \times \mathbf{E} = 0$. From this concept and the principle of super-position, show that $\nabla \times \mathbf{E}$ applies to fields generated by all kinds of charges.

$$\left[\textbf{Hint: Consider}\quad \mathbf{E}_{\text{at } \mathbf{r}} = \frac{1}{4\pi\varepsilon_0} \int_{V'} \frac{dQ_{\text{at } \mathbf{r}'}\,|\mathbf{r} - \mathbf{r}'|}{|\mathbf{r} - \mathbf{r}'|^3} \right.$$

$$\left. \nabla \times \mathbf{E}_{\text{at } \mathbf{r}} = \frac{1}{4\pi\varepsilon_0} \int_{V'} \nabla \times \left\{ \frac{dQ_{\text{at } \mathbf{r}'}\,|\mathbf{r} - \mathbf{r}'|}{|\mathbf{r} - \mathbf{r}'|^3} \right\} \right]$$

    where the $\nabla \times$ acts only on the $\mathbf{r}$ coordinate.

21. Find the potential along the axis of a circular disk with surface charge ρ_s and radius a.

$$V(\mathbf{r}) = \int_{Disk} \frac{\rho_s dS'}{4\pi\varepsilon_0 |\mathbf{r} - \mathbf{r}'|}$$

$$= \int_{Disk} \frac{\rho_s dS'}{4\pi\varepsilon_0 \sqrt{(z - z')^2 + (y - y')^2 + (x - x')^2}}$$

$$= \int_{\rho'} \int_{\phi'} \frac{\rho_s d\rho'\,\rho'\,d\phi'}{4\pi\varepsilon_0 \sqrt{z^2 + y'^2 + x'^2}}$$

$$= \int_{\rho'} \int_{\phi'} \frac{\rho_s d\rho' \, \rho' \, d\phi'}{4\pi\varepsilon_0 \, \sqrt{z^2 + \rho'^2}}$$

$$= \frac{\rho_s}{2\varepsilon_0} \left(\sqrt{z^2 + a^2} - z \right)$$

Answers

Objective Type Questions

1. (d) 2. (b) and (d) 3. (d) 4. (b) 5. (a) and (b) 6. (a), (b) and (c)
7. (c) and (d) 8. (a) and (c) 9. (b) and (c) 10. (b) and (d)

Short-Answer Questions

3. $\mathbf{D} = \varepsilon \, \mathbf{E}$ and $\nabla \cdot \mathbf{D} = \rho_v = 0$

$$\nabla \cdot \mathbf{D} = \nabla \cdot (-f(x) \nabla V)$$
$$= -\nabla f \cdot \nabla V - f \nabla \cdot \nabla V$$
$$= -[(df/dx) \, \partial V/\partial x + f(x) \nabla^2 V] = 0$$

So, $\nabla^2 V = -[1/f(x)] \, (df/dx) \, (\partial V/\partial x)$

4. Use $\nabla^2 V = -\rho_v / \varepsilon_0$;

$$\mathbf{E} = -\nabla V$$
$$= \left[\frac{20 \sin(\theta)}{r^3}, -\frac{10 \cos(\theta)}{r^3}, 0 \right]$$

$$\nabla \cdot (\varepsilon_0 \, \mathbf{E}) = \rho_v = -\varepsilon_0 \frac{10}{r^4 \sin(\theta)}$$

and at the point under consideration one can evaluate the expression.

5. Let the curve be parametrised as $(r(t), \theta(t), \phi(t))$ with $\mathbf{r}_A = (r_A, \theta_A, \phi_A)$ and $\mathbf{r}_B = (r_B, \theta_B, \phi_B)$. Then $d\mathbf{l} = dr\mathbf{a}_r + d\theta \, \mathbf{a}_\theta + d\phi \, \mathbf{a}_\phi$

$$V_{AB} = -\int_A^B \mathbf{E} \cdot d\mathbf{l} = -\int_{r_A}^{r_B} \left(\frac{Q}{4\pi\varepsilon_0 \, r^2} \mathbf{a}_r \right) \cdot (dr\mathbf{a}_r + d\theta \, \mathbf{a}_\theta + d\phi \, \mathbf{a}_\phi)$$

$$= -\int_{r_A}^{r_B} \frac{Q dr}{4\pi\varepsilon_0 \, r^2} = \frac{Q}{4\pi\varepsilon_0} \left(\frac{1}{r_B} - \frac{1}{r_A} \right)$$

or

$$V_{AB} = \frac{Q}{4\pi\varepsilon_0} \left(\frac{1}{r_B} - \frac{1}{r_A} \right)$$

Numerical Problems

2. $\approx 10^{-8} \times 0.018$ (W) in all three cases
3. $\approx 10^{-8} \times 0.031$ (W)
4. 2×10^{-8} (W)
5. 2×10^{-8}. Probably the $\mathbf{E}$ field is conservative. $\nabla \times \mathbf{E} = 0$
6. $288\mathbf{a}_x + 864\mathbf{a}_y - 144\mathbf{a}_z$
7. $16\,\mu$J

8. (a) $[-1, -2y, -4z]$ (b) $\left[\frac{-x}{\sqrt{z^2 + y^2 + x^2}}, \frac{-y}{\sqrt{z^2 + y^2 + x^2}}, \frac{-z}{\sqrt{z^2 + y^2 + x^2}} \right]$

9. $[-2 \sin(\phi) \, \rho \, z, \ -\cos(\phi) \, \rho \, z, \ -\sin(\phi) \, \rho^2]$

10. $\left[\cos(\phi) \, e^{-r} \sin(\theta), \ -\frac{\cos(\phi) \, e^{-r} \cos(\theta)}{r}, \ \frac{\sin(\phi) \, e^{-r}}{r} \right]$

11. 5 W

12. (a) 0   (b) 0   (c) 16 nJ   (d) 16 nJ

13. (a) $72\mathbf{a}_x + 54\mathbf{a}_y - 36\mathbf{a}_z$ (b) $-\varepsilon_0/2$ C

15. (a) 1.45×10^6 V, (b) 1.18×10^6 V

17. $\left(\rho_s/2\varepsilon_0\right)\left(\sqrt{z^2 + a^2} - z\right)$

18. $\left(\rho_s/2\varepsilon_0\right)\left(\sqrt{z^2 + a^2} - \sqrt{z^2 + b^2}\right)$

19. $\left(\rho_s/2\varepsilon_0\right)\left(a - \tan^{-1}\left(\dfrac{a}{z}\right)z\right)$

6

The Electric Field and Material Media

> *No matter how good teaching may be,*
> *each student must take the responsibility for his own education.*
> — John Carolus S.J.

CHAPTER OBJECTIVES

To enable the students to understand the following:

- Current and current density
- Relation between charge and current
- Velocity of mobile charge carriers and its connection with the current density
- Mobile charge density
- Continuity equation
- The criteria for classification of materials as conductors, semiconductors, and dielectrics

- Ohm's law, $\mathbf{E} = \sigma\,\mathbf{J}$, in point form
- Relation between capacitance and resistance
- Boundary conditions for electrostatic fields
- Energy stored in the electric field

6.1 | Current and Current Density

Broadly there are three types of media which interact with electric fields giving rise to different types of results. These are conductors, semiconductors, and dielectrics. In conductors and semiconductors, electric fields give rise to the flow of charge or electric currents. In conductors the charges are predominantly electrons, while in semiconductors charge flow is due to both electrons and holes. In dielectrics, the charges are not free to move and the electric field polarises the molecules comprising the nucleus and valence electrons. First we will consider conductors which support currents, and, as a prelude to charge flow we need to understand some key concepts.

Charge flow whether of the positive or negative variety gives rise to currents. Current is defined as the total charge flow per second through the cross-section of a conductor. To be specific, consider a straight conductor of cross-sectional area A through which a charge dQ passes in time dt. Then the current I is given by

$$I = dQ/dt \tag{6.1}$$

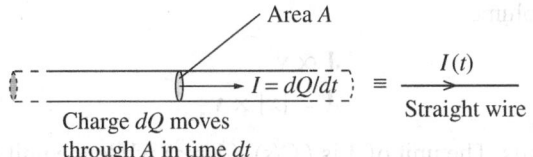

Area A

$I = dQ/dt$

$\equiv$

$I(t)$

Straight wire

Charge dQ moves
through A in time dt

Fig. 6.1 Definition of current

The equivalent straight wire carrying a current is also shown in Fig. 6.1.

Example 6.1 A wire carries a constant current of 1 A. Find the total charge passing a plane perpendicular to the cross-section of the wire in 1 s.

Solution

Step 1. The charge crossing any cross-section of the wire in time dt is given by
$$dQ = Idt.$$
Step 2. So if Q is the total charge passing the cross-section in 1 s then

$$Q = \int_{t_0}^{t_0+1} I \, dt = I \int_{t_0}^{t_0+1} dt = 1 \, \text{C}$$

Example 6.2 A wire carries a current of $I = 1 \exp\{-3t\}$ (A). Find the total charge passing a plane perpendicular to the cross-section of the wire from $t = 0$ to $t = 3$ s.

Solution

Step 1. The charge crossing any cross-section of the wire in time dt is given by
$$dQ = Idt.$$
Step 2. So if Q is the total charge passing the cross-section, it is given by

$$Q = \int_0^3 Idt = \int_0^3 1\exp\{-3t\}dt = \frac{1 - e^{-9}}{3} = 0.33 \, \text{C}$$

Since current is calculated from the movement of charge, and the cross-section of the conductor, we define a vector field $\mathbf{J}$ which is a current density in A/m^2. The small current which passes through the small surface element $\Delta \mathbf{S}$ is

$$\Delta I = \mathbf{J} \cdot \Delta \mathbf{S} \tag{6.2}$$
$$= \mathbf{J} \cdot \hat{\mathbf{n}} \Delta S \tag{6.3}$$
$$= \left(\mathbf{J}_{\parallel \text{ to } \hat{\mathbf{n}}} + \mathbf{J}_{\perp \text{ to } \hat{\mathbf{n}}} \right) \cdot \hat{\mathbf{n}} \Delta S \tag{6.4}$$
$$= \mathbf{J}_{\parallel \text{ to } \hat{\mathbf{n}}} \cdot \hat{\mathbf{n}} \Delta S \tag{6.5}$$

The relationship between $\Delta I, \mathbf{J}$, and $\Delta \mathbf{S}$ are shown in Fig. 6.2. It is important to note here that there may be current in a non-conducting medium which may be due to motion of charge between two points, say in air (such as in a Van De Graaff generator), which may also lead to a current, which is termed as a *convection current*.

Furthermore, when the average velocity of a small volume of charge is $\mathbf{v}$, then the current density should be proportional to the instantaneous

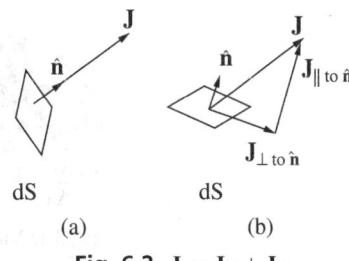

(a)   (b)

Fig. 6.2 $\mathbf{J} = \mathbf{J}_{\parallel} + \mathbf{J}_{\perp}$

velocity of that volume

$$\mathbf{J} \propto \mathbf{v} \tag{6.6}$$
$$\mathbf{J} = [x] \times \mathbf{v}$$

Looking at the units. The unit of $\mathbf{J}$ is (C/s) .($1/m^2$) while the unit of $\mathbf{v}$ is m/s so

$$C/(\sec.m^2) \overset{U}{=} [x](m/s)$$

where $[x]$ are the units of some quantity. Then

$$[x] \overset{U}{=} C/m^3$$

But C/m^3 are the units of the charge density. So $[x] = \rho_m$. Hence

$$\mathbf{J} = \rho_m \mathbf{v} \tag{6.7}$$

where ρ_m is the mobile charge density. The equation says that at any point, the current density $\mathbf{J}$ is equal to the product of the charge density and the velocity with which the charge is moving at that point. This equation applies to both convection as well as conduction currents.

Example 6.3 In a straight wire of diameter 1 mm the current is 1 A. Find the magnitude of the current density if the charges are uniformly distributed across the cross-section of the wire.

Solution

Step 1. If I is the current, A the area of cross-section, and $\mathbf{J}$ the current density, then

$$I = \int_{\text{Area of CS}} \mathbf{J} \cdot d\mathbf{S} = JA$$

Step 2. So $J = I/A = 1/(\pi \times (0.001/2)^2) = 1.2732 \times 10^6$ (A/m^2)

We now refer to Fig. 6.3, to understand some important concepts concerning ρ_m, the mobile charge density; n, the density of charge carriers, and $\mathbf{J}$ the current density. Though the explanation given below is an extremely simplified one, it would help the reader to understand the concepts involved. We visualise a situation when a conducting medium containing mobile charge carriers is under the influence of an external electric field in the $\mathbf{a}_x$ direction. We concentrate on a small rectangular parallelepiped, with sides $\Delta x, \Delta y$, and Δz buried in the medium. Referring to Fig. 6.3, which shows such a parallelepiped, we can see that a small number of charge carriers, N, are enclosed by it. For a metal, the mobile charge carriers would be electrons.

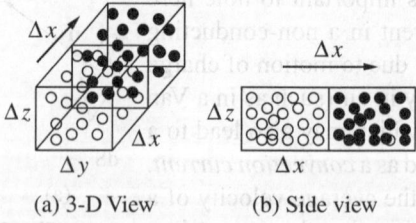

(a) 3-D View   (b) Side view

Fig. 6.3 Concepts concerning charge transport

Now if each carrier caries a charge q then the total charge in the parallelepiped would be equal to Nq. It is the charge enclosed by volume, $\Delta V = \Delta x \Delta y \Delta z$.

Hence
$$\Delta Q = Nq \tag{6.8}$$

so the density of charge carriers would be

$$n = \frac{N}{\Delta V} \tag{6.9}$$

and charge density would be

$$\rho_m = \frac{\Delta Q}{\Delta V} = nq \tag{6.10}$$

Please note that ρ_m must not be confused by the *excess* charge density ρ_v. For example, in the case of a metal, the medium is electrically neutral. That is, the negative charge contribution due to the number of mobile free electrons is equal to the positive charge contribution due to the metal nuclei, making the medium neutral. In this case ρ_m is the density of the charge contributed by the conduction band electrons. Note that generally no electric field is produced by ρ_m, but it may so happen that these mobile charge carriers may accumulate somewhere and cause excess charges, and in that case an electric field will be produced. Therefore, in most cases

$$\nabla \cdot \mathbf{D} \neq \rho_m \tag{6.11}$$

The parallelepiped is placed in such a way that the side Δx is oriented along the direction of motion of the carriers as shown. Assuming that all the charge carriers are moving in the same direction, the average velocity is

$$\mathbf{v} \cong \mathbf{a}_x \Delta x / \Delta t \tag{6.12}$$

as in small time Δt, charges get displaced by a distance Δx.

Since the charge carriers move, the total (elemental) current through the face with sides $\Delta y, \Delta z$ is

$$\Delta I = \mathbf{J} \cdot \Delta \mathbf{S} \tag{6.13}$$

Integrating
$$I = \iint \mathbf{J} \cdot d\mathbf{S} \tag{6.14}$$

where
$$\mathbf{J} = \rho_m \frac{\Delta x}{\Delta t} \mathbf{a}_x = \rho_m \mathbf{v} \tag{6.15}$$

and
$$\Delta \mathbf{S} = \Delta y \Delta z \mathbf{a}_x \tag{6.16}$$

 See JFromNAndV.m in Chapter 6

Example 6.4 Estimate the mobile charge density ρ_m of silver.
Solution
Step 1. The density of silver is 10.5×10^3 kg/m^3; silver has an atomic weight of 107.9. So 107.9 g have $N_A = 6.022 \times 10^{23}$ atoms. (Avogadro's number)
Step 2. The number of atoms per kg is
$$(6.022 \times 10^{23}/107.9) \times 10^3 = 5.58 \times 10^{24} \text{ atoms/kg} \tag{6.17}$$
Step 3. Multiplying the number of atoms/kg by the density of silver (kg/m^3), we get the number of atoms/m^3,
$$(5.58 \times 10^{24}) \times (10.5 \times 10^3) = 5.86 \times 10^{28} \text{ atoms/m}^3 \tag{6.18}$$

Step 4. The valency of silver is 1. So if each atom contributes a maximum of 1 electron to the conduction band,

$$\rho_m = (5.86 \times 10^{28}) \times (-1.602 \times 10^{-19}) \tag{6.19}$$

$$= -9.39 \times 10^9 \ \text{C/m}^3 \tag{6.20}$$

which is the maximum value of mobile charge density.

Example 6.5 Using ρ_m for silver, and J from Example 6.3, estimate the magnitude of the velocity v of the mobile charge carriers.

Solution

Step 1. We use Eqn (6.7), $J = \rho_m v = -9.39 \times 10^9 v$

Step 2. From Example 6.3, $J = 1.2732 \times 10^6$

So $v = 1.36 \times 10^{-4}$ (m/s)

6.2 | Continuity Equation

Let us apply the divergence theorem to **J**. Referring to Fig. 6.4, in a region where there are currents,

$$\iiint_V \nabla \cdot \mathbf{J} dV = \oiint_S \mathbf{J} \cdot d\mathbf{S} \tag{6.21}$$

But

$$\oiint_S \mathbf{J} \cdot d\mathbf{S} = \text{Current leaving the volume} = I = -\frac{dQ}{dt} \tag{6.22}$$

The negative sign indicates that the charge inside the volume is decreasing.
But $Q = \iiint \rho_v \, dV$. So

$$\iiint \nabla \cdot \mathbf{J} dV = -\frac{\partial}{\partial t} \iiint \rho_v \, dV \tag{6.23}$$

Since space coordinates and time are independent of each other we can change the order of differentiation and integration to get

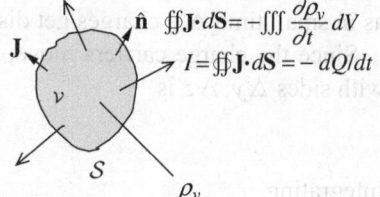

$$\oiint \mathbf{J} \cdot d\mathbf{S} = -\iiint \frac{\partial \rho_v}{\partial t} dV$$

$$I = \oiint \mathbf{J} \cdot d\mathbf{S} = -dQ/dt$$

Fig. 6.4 Illustration for the continuity equation

$$\frac{\partial}{\partial t} \iiint \rho_v \, dV = \iiint \frac{\partial \rho_v}{\partial t} dV \tag{6.24}$$

Using the last result:

$$\iiint \nabla \cdot \mathbf{J} dV = -\iiint \frac{\partial \rho_v}{\partial t} dV \tag{6.25}$$

which gives

$$\nabla \cdot \mathbf{J} = -\frac{\partial \rho_v}{\partial t} \tag{6.26}$$

This is known as the *continuity equation*. It connects the current density with the rate of change of charge in any region of space. Its integral form is

$$\oiint \mathbf{J} \cdot d\mathbf{S} = -\iiint \frac{\partial \rho_v}{\partial t} dV \tag{6.27}$$

The importance of the continuity equation is that it tells us that charge is *conserved*. That is charge is neither created nor destroyed.

Example 6.6   Obtain the continuity equation from Maxwell's equations.

Solution

Step 1. One of Maxwell's equations is

$$\nabla \times \mathbf{H} = \partial \, \mathbf{D}/\partial \, t + \mathbf{J} \tag{6.28}$$

The equation says that the curl of $\mathbf{H}$ is equal to the rate of change of the flux density plus the current density.

Step 2. Take the divergence of this equation

$$\nabla \cdot (\nabla \times \mathbf{H}) = \nabla \cdot (\partial \, \mathbf{D}/\partial \, t + \mathbf{J}) \tag{6.29}$$

Step 3. Using the identity $\nabla \cdot (\nabla \times \mathbf{A}) = 0$ and $\nabla \cdot (\partial \, \mathbf{D}/\partial \, t) = \partial \, /\partial \, t(\nabla \cdot \mathbf{D})$ (since space and time are independent of each other) the previous equation becomes

$$
\begin{aligned}
\nabla \cdot \mathbf{J} &= -\frac{\partial}{\partial \, t}\nabla \cdot \mathbf{D} \\
&= -\frac{\partial \rho_v}{\partial \, t}
\end{aligned}
\tag{6.30}
$$

where we have used Gauss's law, namely, $\nabla \cdot \mathbf{D} = \rho_v$.

Example 6.7   Apply the continuity equation to a straight wire carrying a current I and having circular cross-sectional area A.

Solution

Step 1. Let the wire be oriented in the z direction.

Step 2. Considering a Gaussian surface: *lower and upper surfaces: $z = z_0, z_1$; side surface $\rho = a \, (= \sqrt{A/\pi})$*

Step 3. Observing Fig. 4.34, we find that

$$\oiint \mathbf{J} \cdot d\mathbf{S} = \underbrace{\iint \mathbf{J} \cdot d\mathbf{S}}_{\text{Lower surface}} + \underbrace{\iint \mathbf{J} \cdot d\mathbf{S}}_{\text{Upper surface}} + \underbrace{\iint \mathbf{J} \cdot d\mathbf{S}}_{\text{Side}}$$

Step 4. Now $\underset{\text{Side}}{\iint} \mathbf{J} \cdot d\mathbf{S} = 0$ (No current leaves through the side)

$$\underset{\text{Upper}}{\iint} \mathbf{J} \cdot d\mathbf{S} = -\underset{\text{Lower}}{\iint} \mathbf{J} \cdot d\mathbf{S} = IA$$

Step 5. Since there is no accumulation of charge with time

$$\frac{dQ}{dt} = 0$$

$$Q = \text{constant} = 0$$

Step 6. So $\oiint \mathbf{J} \cdot d\mathbf{S} = -\iiint \frac{\rho_v}{dt}dV = \frac{dQ}{dt}$

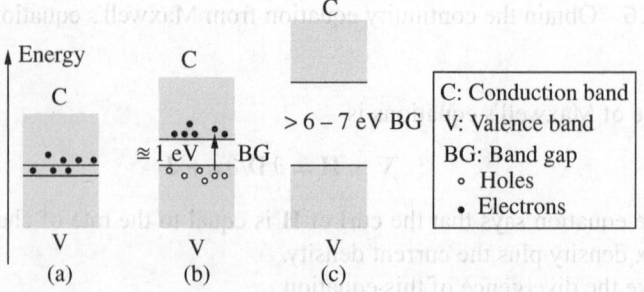

Fig. 6.5 The energy levels of the outermost shell of materials—the valence and conduction band in (a) metals, (b) semiconductors, and (c) dielectrics

6.3 | Conductors, Semiconductors, and Dielectrics

The band theory of materials postulates that the outermost shell of atoms have two types of energy bands. Referring to Fig. 6.5,

1. A valence band, where the electrons are tightly bound to the nucleus. These electrons do not take part in charge transport. Under the influence of an external electric field, the nuclei and the valence band electrons take part in formation of minuscule dipoles.
2. A conduction band whose lower edge is generally higher than the upper edge of the valence band. The electrons is this band take part in current formation under the influence of external fields.

In conductors, such as metals, the conduction band overlaps the valence band. This is shown in part (a) of the figure. In semiconductors (part (b) of the figure) the two bands do not overlap and there is about a 1 eV gap between the upper edge of the valence band and lower edge of the conduction band and which is $\approx 40kT$. Note that the *average* energy of electrons in the material is at most only a few kT. Normally the conduction band is empty but at room temperature some electrons acquire enough energy to migrate from the valence band to the conduction band giving rise to two types of charge carriers: conduction band electrons and valence band holes. In the case of dielectrics (part (c) of the figure) the band gap is of the order of 6–7 eV. In this case electrons never acquire enough energy to move from the valence band to the conduction band.

6.3.1 Conductors

Referring to Fig. 6.6, we all know Ohm's law is

$$V = IR \qquad (6.31)$$

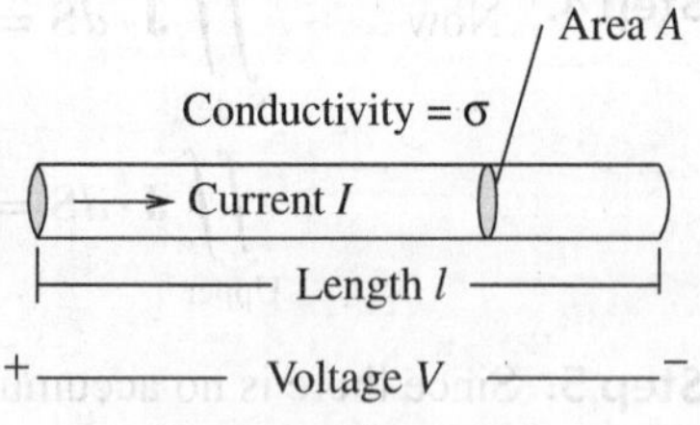

where V and I are the voltage across the conductor and the current through it, while R is its resistance. If l is the length of the conductor, A the area of cross-section, and σ the conductivity of the material

Fig. 6.6 Ohm's law

$$R = \frac{l}{\sigma A} \qquad (6.32)$$

Going over to field quantities, the electric field is

$$E = V/l \qquad (6.33)$$

The *conduction current density* is

$$J = I/A \tag{6.34}$$

then

$$El = JA\left(\frac{l}{\sigma A}\right)$$

$$E = \frac{J}{\sigma}$$

or

$$E\sigma = J$$

The two quantities E and J are actually vectors while σ is a scalar. The vector equation is

$$\mathbf{J} = \sigma\,\mathbf{E} \tag{6.35}$$

This is Ohm's law in vector form. In a good conductor, σ can have very large values.

We can look upon a metal in a microscopic form. Due to the presence of an electric field, electrons are accelerated. However, on acceleration they are slowed down by collisions with the atoms of the lattice. So

$$\langle v \rangle = a\tau \tag{6.36}$$

where $\langle v \rangle$ is the average velocity of electrons, a is the acceleration, and τ is the mean free time between collisions. As we know,

$$a = \frac{eE}{m_e} \tag{6.37}$$

so

$$\langle v \rangle = \frac{e\tau\,E}{m_e} \tag{6.38}$$

or

$$J = \rho_m \langle v \rangle = \frac{\rho_m\,e\tau\,E}{m_e} \tag{6.39}$$

Hence,

$$\sigma = \frac{\rho_m\,e\tau}{m_e} \tag{6.40}$$

 See `JFromSigmaAndE.m` in Chapter 6

Example 6.8   Find the current density in copper for an electric field of 1 V/m. What is the current due to this field in a wire of 1 mm diameter?
Solution
Step 1. Copper has a conductivity of $58.14 \times 10^6 \mho/m$. So an electric field of $E = 1$ V/m gives $J = \sigma E = 58.4 \times 10^6$ A/m^2
Step 2. A 1 mm diameter wire has an area of

$$A = \pi\left(\frac{d}{2}\right)^2 = 1.96 \times 10^{-7} \text{ m}^2$$

Step 3. The current in the wire is therefore $I = JA = 11.46$ A

A perfect conductor is one whose conductivity $\dot{\sigma}$ tends to infinity. Copper practically satisfies the definition.

Example 6.9   Find the resistance of the non-uniform resistor whose geometry is shown in Fig. 6.7. The resistor consists of deposited material of conductivity σ and thickness t in a coaxial mode.

Solution

Step 1. The current enters the resistor at $\rho = a$ and leaves at $\rho = b$. Consider the miniscule resistor between ρ and $\rho + d\rho$, its resistance dR is

$$dR = \frac{d\rho}{\sigma \rho \theta t}$$

Step 2. Since all these miniscule resistors are in series along the path of the current, they have to be added together,

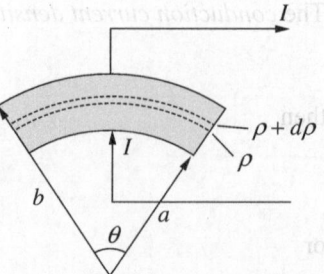

Fig. 6.7 A non-uniform resistor

so

$$R \approx \sum_{\text{infinite number}} \frac{d\rho}{\sigma \rho \theta t}$$

or

$$R = \frac{1}{\sigma \theta t} \int_a^b \frac{d\rho}{\rho} = \frac{1}{\sigma \theta t} \ln \frac{b}{a} \qquad (6.41)$$

6.3.2 Relaxation Time for Conductors

Suppose in a conductor charge accumulates somewhere in the interior of a conductor, for some reason. Applying Gauss's law, a **D** field develops. Also since we are working in a conductor, the charge dissipates. From Eqn (6.28),

$$\nabla \times \mathbf{H} = \partial \mathbf{D}/\partial t + \mathbf{J}$$

Since there is no magnetic field ($\mathbf{H} = 0$) then at any point

$$\partial \mathbf{D}/\partial t + \mathbf{J} = 0 \qquad (6.42)$$

Also at that point, $\mathbf{D} = \varepsilon_0 \mathbf{E} = (\varepsilon_0/\sigma) \mathbf{J}$

$$\frac{\partial \mathbf{J}}{\partial t} + \left(\frac{\sigma}{\varepsilon_0}\right) \mathbf{J} = 0$$

or

$$\mathbf{J} = \mathbf{J}_0 e^{-\frac{\sigma}{\varepsilon_0} t} \qquad (6.43)$$

Equation (6.43) says that after the accumulation of charge at any point, a current density develops which dissipates the charge till no charge remains. The rate at which the charge dissipates is proportional to $\exp\{-\sigma/\varepsilon_0 t\}$.

Example 6.10 A charge of 1 C accumulates in copper. Find the time it takes to dissipate to 10^{-18} C.

Solution

Step 1. Since $\mathbf{J} = \mathbf{J}_0 e^{-\frac{\sigma}{\varepsilon_0} t}$

Step 2. Therefore, $\rho_v(t) = \rho_{v0} e^{-\frac{\sigma}{\varepsilon_0} t} = 1 e^{-\frac{\sigma}{\varepsilon_0} t}$

Step 3. For copper $\sigma/\varepsilon_0 = 6.6 \times 10^{18}$

Step 4. Taking natural logarithms on both sides

$$\ln[\rho_v(t_f)] = -6.6 \times 10^{18} t_f$$

$$-41.44 = -6.6 \times 10^{18} t_f$$

$$t_f = 6.3 \times 10^{-18} \text{ s}$$

we can see that even if a 1 C charge accumulates in copper, it dissipates in the twinkling of an eye!

There is an electric field in a conductor due to accumulation of charge. Observing Eqn (6.43) and using $\mathbf{J} = \sigma \mathbf{E}$, the electric field dissipates at the same rate

$$\mathbf{E} = \mathbf{E}_0 e^{-(\sigma/\varepsilon_0)t} \tag{6.44}$$

The mechanism by which actual conduction takes place is as follows. Electrons are under constant thermal motion. As they move about in the conductor lattice they impinge upon the stationary metal atoms, rebound, then hit another atom, and so on, somewhat as shown in Fig. 6.8. Under the influence of an external field, they continue to move in the manner discussed, but along with that they constantly 'drift' slowly against the field. The total motion of such a large number of these mobile electrons constitutes a current. The relationship between the drift velocity, $\mathbf{v}_d$, and the electric field is a linear one for any particular material

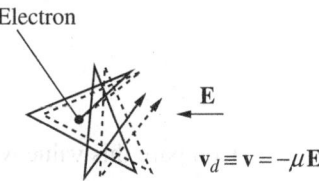

Electron

→ Normal thermal motion

--► Motion when E field is present

Fig. 6.8 The motion of electrons under the influence of an external field

$$\mathbf{v}_d = -\mu_e \mathbf{E} \tag{6.45}$$

where μ_e is a constant called the mobility. The negative sign is there since the motion of electrons are in the opposite direction of the field.

Example 6.11 Estimate the drift velocity of electrons for silver in a conductor of diameter 1 mm carrying a current of 1 A and compare the result with the velocities attained due to thermal motion. Also estimate the mobility for silver.

Solution

Step 1. The conductor carries a current of 1 A. So the current density J is

$$J = \frac{I}{\pi r^2}$$
$$= 1.27 \times 10^6 \text{ A/m}^2 \tag{6.46}$$

Step 2. Recall that

$$J = \rho_m v_d \tag{6.47}$$

and the value for $\rho_m = 9.39 \times 10^9$ C/m³ for silver from Example 6.4. So the drift velocity $\qquad v_d \approx 1.36 \times 10^{-4}$ m/s $\qquad$ (6.48)

Step 3. The electric field is related to J by

$$J = \sigma E \tag{6.49}$$

Step 4. The experimental value of σ for silver is $\sigma = 6.25 \times 10^7$ ℧/m, which gives the value of the electric field to be

$$E \approx 0.02032 \text{ V/m} \tag{6.50}$$

Step 5. Since $v_d = \mu_e E$ so the mobility for silver is

$$\mu_e \approx 6.65 \times 10^{-3} \text{ m}^2/(\text{V s}) \tag{6.51}$$

Step 6. The kinetic energy of electron under normal thermal motion is about a few kT.

Step 7. Let us estimate the velocity of an electron whose kinetic energy is exactly $1\ kT$, just to get an idea of the velocities involved. For $T = 300°K$,

$$\frac{1}{2}m_e v_{th}^2 = kT$$

$$v_{th} = \sqrt{\frac{2kT}{m_e}}$$

$$\approx 9.5 \times 10^4 \text{ m/s}$$

Compare this value with $v_d \approx 1.36 \times 10^{-4}$ m/s

Suppose a conductor is suddenly immersed in an electric field (Fig. 6.9(a)). Initially there would be a field inside the conductor, which would set up large currents. The electrons would move very quickly *against* the direction of the electric field that is to the left side. As a result right side would get positively charged due to the stationary nuclei bereft of electrons. The migration of the electrons to the left would result in a zero resultant field *inside* the conductor as shown in Fig. 6.9(b). Thus if there is a field inside the conductor there would be a perpetual migration of charges producing heat which is not seen in practice. On the surface of the conductor, for the same reason, there would be no tangential electric field as otherwise the charges would migrate on the surface perpetually. There would be a charge on the surface distributed in such a way that only a normal resultant electric field would be present. The surface of the conductor would be a equipotential surface. These results are summarised as

1. No electrostatic field is present inside a conductor
2. On the surface of the conductor, electric fields are only normal to the surface
3. Charges are present on the surface but distribute themselves so as to make the surface an equipotential one

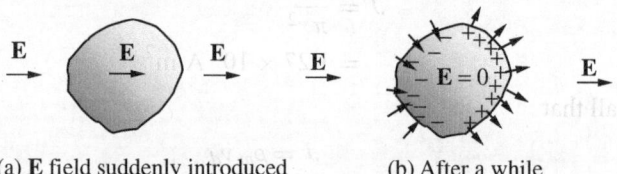

(a) **E** field suddenly introduced   (b) After a while

Fig. 6.9 Conductor in the presence of an external electrostatic field

Example 6.12 A spherical metal shell with inner radius r_0 and outer radius r_1 has a very small hole in it through which a positive charge Q is introduced as shown in Fig. 6.10. Find the fields everywhere and discuss. (b) what happens if the shell is grounded?

Solution Part (a)

Step 1. Neglecting the small hole as it will *not* change the fields greatly, we proceed to apply the case of a charge Q enclosed by a spherical shell. Using a spherical Gaussian surface with the charge Q as centre and radius $r < r_0$, the electric field is given by

$$\mathbf{E} = \frac{Q}{4\pi\varepsilon_0 r^2}\mathbf{a}_r \quad r < r_0 \text{ V/m} \tag{6.52}$$

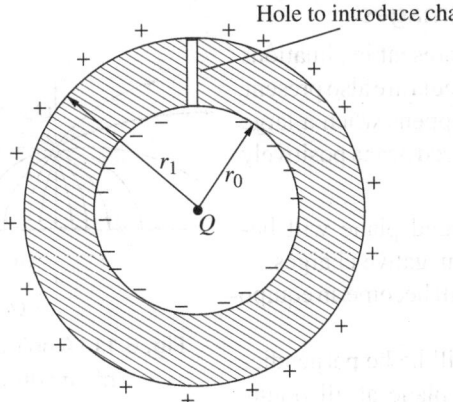

Fig. 6.10 A charge Q enclosed by a spherical shell

Step 2. We now increase the radius of the Gaussian sphere. When $r_1 > r > r_0$ the electric field inside the metal is zero. So, by Gauss's law, the total charge enclosed must be zero. It means that the inside surface of metal shell must have a uniform surface charge whose total value is $-Q$. The surface charge

$$\sigma_0 \, (r = r_0) = \frac{\text{Total charge}}{\text{Total surface area}} = -\frac{Q}{4\pi r_0^2} \text{ C/m}^2 \qquad (6.53)$$

Step 3. The field must satisfy all the conditions discussed above, that is the field inside the metal is zero; it is normal to the metal surface $r = r_0$, and the surface $r = r_0$ is an equipotential surface.

Step 4. Mobile electrons have moved to the inner surface of the shell. As a result the outer surface develops a surface charge Q which is positive consisting of the immobile nuclei whose outer electrons have migrated to the inner surface of the shell. Now on increasing the radius of the Gaussian surface still further, $r > r_1$, the total charge enclosed is once again $Q : Q$ at the centre, $-Q$ on the inner surface, and Q on the outer surface.

Step 5. The electric field then becomes

$$\mathbf{E} = \frac{Q}{4\pi \varepsilon_0 \, r^2} \mathbf{a}_r, \quad r > r_0 \text{ V/m} \qquad (6.54)$$

and the surface charge on the outer surface is

$$\sigma_1 \, (r = r_1) = \frac{Q}{4\pi r_1^2} \text{ C/m}^2 \qquad (6.55)$$

Part (b)

Step 1. As soon as we ground the outer surface which is positively charged, negative charge rushes in from ground and neutralises all the positive charge on the outer surface. The inner surface is still negatively charged because of the presence of the charge at the centre, and the electric field inside the shell is unchanged. Outside the inner surface there is no electric field anywhere. Hence,

$$\mathbf{E} = \begin{cases} \dfrac{Q}{4\pi \varepsilon_0 \, r^2} \mathbf{a}_r, & r < r_0 \\ 0, & r > r_0 \end{cases} \qquad (6.56)$$

6.3.3 Method of Images

Normally charges are present in situations where other material media are also present. Let us discuss what happens when a large ground plane is introduced near a positively charged point charge.

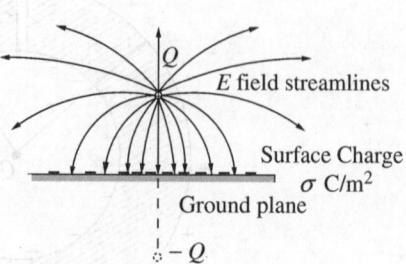

1. The surface of ground plane will become charged with negative charges.
2. The ground plane will become an equipotential surface.
3. The electric field will be be perpendicular to the ground plane at all points on the surface of the ground plane.

Fig. 6.11 A point charge near an infinite ground plane

In fact whenever a conductor is introduced near any charge distribution, then what has been outlined as points 1 to 3 above will always take place. The situation will be somewhat as shown in Fig. 6.11.

If we recall the electric field of a dipole the fields look similar—not only is the electric field similar, but the field above the ground plane is indeed the same. The solution of such types of problems is done by a well known method, namely, the method of images. In this method,

1. We start with some set of charges which give us the electric field and the potential everywhere.
2. If some equipotential surface is replaced by a thin metal plane then the electric field and the potential field are not disturbed because the metal acts like an equipotential surface.
3. We will then have a new problem which consists of the same set of charges and the metal plane which we have just put in place
4. The new problem has the solution as in *1.*

Example 6.13   Apply the method of images to the dipole problem.

Solution

Step 1. Using the formula of Section 5.4, the potential due to the two identical charges is

$$V(\mathbf{r}) = \frac{Q}{4\pi\varepsilon_0}\left[\frac{1}{\sqrt{\left(z - \frac{d}{2}\right)^2 + y^2 + x^2}} - \frac{1}{\sqrt{\left(z + \frac{d}{2}\right)^2 + y^2 + x^2}}\right] \quad (6.57)$$

Step 2. In this equation let $z = 0$. Immediately we can see that $V(\mathbf{r})|_{z=0} = 0$ which is an equipotential surface.

Step 3. At half the distance between the positive and negative charges, the potential surface $V = 0$ is an infinite plane.

Step 4. An infinite metal plane can now be introduced right there. The fields everywhere are the same as before. The lower charge can be removed, and the fields above the plane are still the same, while the fields below the plane vanish.

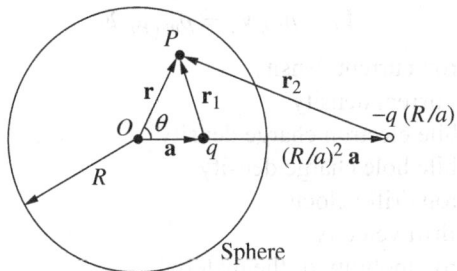

Fig. 6.12 Method of images applied to a single charge
and a perfectly conducting sphere

Example 6.14 Apply the method of images to the problem of a single charge placed inside a sphere as shown in Fig. 6.12.

Solution The method of images may be applied to a sphere as analysed by Tikhonov (1963) who considers the case of a point charge in a hollow metallic sphere.

Step 1. Referring to the figure, we would like to find the potential $V(\mathbf{r})$ inside a sphere of radius R, centred at O, due to a point charge q inside the sphere at position $\mathbf{a}$.

Step 2. *The stipulation* is that the image of this charge with respect to the first charge for the sphere is placed at $(R/a)^2\,\mathbf{a}$. It has a charge of $-qR/a$.

Step 3. Hence for the configuration of the two charges shown,

$$V(\mathbf{r}) = \frac{1}{4\pi\varepsilon_0}\left[\frac{q}{|\mathbf{r}_1|} + \frac{(-qR/a)}{|\mathbf{r}_2|}\right] \qquad (\text{anywhere})$$

$$= \frac{1}{4\pi\varepsilon_0}\left[\frac{q}{\sqrt{r^2 + a^2 - 2ar\cos\theta}} - \frac{(qR/a)}{\sqrt{r^2 + \frac{R^4}{a^2} - \frac{2R^2}{a^2}ar\cos\theta}}\right]$$

$$= \frac{1}{4\pi\varepsilon_0}\left[\frac{q}{\sqrt{r^2 + a^2 - 2ar\cos\theta}} - \frac{q}{\sqrt{\frac{r^2a^2}{R^2} + R^2 - 2ar\cos\theta}}\right] \qquad (6.58)$$

Step 4. In the above expression if we put $r = R$, then

$$V(R) = \frac{1}{4\pi\varepsilon_0}\left[\frac{q}{\sqrt{R^2 + a^2 - 2aR\cos\theta}} - \frac{q}{\sqrt{a^2 + R^2 - 2aR\cos\theta}}\right] = 0$$

Step 5. We can therefore see that $V = 0$ on the sphere.

Step 6. Therefore, the potential inside the sphere for a single charge q is given by Eqn (6.58).

6.3.4 Semiconductors

A semiconductor behaves just like a conductor does, except that there are two kinds of charge carriers. This is so because valence electrons acquire enough energy to enter the conduction band, and they interact with the electric field and give rise to currents. On the other hand atoms bereft of electrons in the valence band become positively charged and they also become charge carriers. These are called holes. Since there are two types of charge carriers,

$$\mathbf{J}_e = \rho_{me}\,\mathbf{v}_e = -\rho_{me}\,\mu_e\,\mathbf{E} \qquad (6.59)$$

and
$$\mathbf{J}_h = \rho_{mh}\,\mathbf{v}_h = \rho_{mh}\,\mu_h\,\mathbf{E} \tag{6.60}$$

where $\mathbf{J}_e$ is the electron current density
$\mathbf{J}_h$ is the hole current density
ρ_{me} is the mobile electron charge density
ρ_{mh} is the mobile hole charge density
$\mathbf{v}_e$ is the electron drift velocity
$\mathbf{v}_h$ is the hole drift velocity
μ_e is the electron mobility of the material
μ_h is the hole mobility of the material and
$\mathbf{E}$ is the electric field within the semiconductor

In any conducting material,
$$\rho_m = nq \tag{6.61}$$

where ρ_m being the mobile charge carrier density; n the carrier concentration (number/m^3); and q is the value of the charge: $q = e$ for electrons and $q = -e$ for holes. Using these expressions
$$\mathbf{J} = \mathbf{J}_h + \mathbf{J}_e = -n_h e \mathbf{v}_h + n_e e \mathbf{v}_e \tag{6.62}$$

where n_e is the electron concentration and n_h is the hole concentration. The velocity of the holes is in the direction of the electric field, while the velocity of the electrons is in the direction opposite to the field,

$$\mathbf{v}_h = \mu_h\,\mathbf{E} \tag{6.63}$$
$$\mathbf{v}_e = -\mu_e\,\mathbf{E} \tag{6.64}$$
$$\mathbf{J} = -n_h e \mu_h\,\mathbf{E} - n_e e \mu_e\,\mathbf{E} = \sigma\,\mathbf{E} \tag{6.65}$$

where
$$\sigma = -n_h e \mu_h - n_e e \mu_e \tag{6.66}$$

σ will be positive since e is negative,

$$\sigma = n_h |e| \mu_h + n_e |e| \mu_e \tag{6.67}$$

Example 6.15   A sample of silicon is doped with a type 3 element (having 3 outermost electrons) with an effective concentration of 3×10^{23} atoms per m^3. Find the conductivity of the material: $\mu_e = 0.14$ and $\mu_p = 0.05$ m^2/(V s). $e = -1.6 \times 10^{-19}$ C.

Solution

Step 1. The sample is silicon, a type 4 element and a semiconductor.
Step 2. The impurity is type 3 so the doped material is a p-type semiconductor.
$$n_h = 3 \times 10^{23}.$$
Step 3. Since $n_h \gg n_e$ (due to the presence of an impurity)
$$\sigma \approx n_h \mu_h |e| = 2400\ \mho/m$$

John Bardeen–Scientist and Inventor

John Bardeen (1908–1991), an American physicist and engineer, is the only person to have won the Nobel Prize in Physics twice. First with W. Shockley and W. Brattain in 1956 for the invention of the transistor, and then in 1972 when he proposed a fundamental theory of superconductivity (in collaboration with others.)

The transistor as we all know has fundamentally changed society with the development of micro-miniaturised electronic components and has made available many modern electronic devices: the telephone, television, and computers among many others. Bardeen's proposed theory is also used in MRIs [a].

---

[a] Magnetic Resonance Imaging

6.3.5 Dielectrics

Referring to Fig. 6.5, dielectric materials are those materials which have a minimum band gap of 6–7 eV between the valence band and conduction band. For conduction to take place, electrons must be present in the conduction band. If one remembers that 1 eV corresponds to an energy difference of about $40\,kT$ at room temperature, a quick calculation based on the Maxwell-Boltzmann model shows that, very very few electrons have enough energy to enter the conduction band. As a result, practically all the electrons are present in the valence band and are tightly bound to the nucleus.

A detailed treatment of dielectrics will not be given here, but the basic idea is that a dielectric polarises under the influence of an external field into dipole moments at the atomic/molecular levels. How this takes place is shown in Fig. 6.13.

The figure shows the two cases, the first case (a) when no field is present. An electron cloud exists and spherical symmetry is maintained. In the second case (b), the electron cloud shifts to the *left* as the field is directed towards the right. The dipole that is created has a dipole moment

$$\mathbf{p} = q\mathbf{d}\,\mathrm{C\,m} \tag{6.68}$$

where $\mathbf{d}$ is the position vector directed from the negative charge to the positive charge.

Since we are talking at an atomic level we can talk of a polarisation density $\mathbf{P}$, which is the number of dipole moments/m^3, but which have been added vectorially. If in a very small volume ΔV there are N dipoles then

$$\mathbf{P} = \lim_{\Delta V \to 0} \left(\frac{1}{\Delta V} \sum_{i=1}^{N} \mathbf{p}_i \right) \tag{6.69}$$

Then, it turns out that under the influence of an external field $\mathbf{E}$, the relationship between $\mathbf{E}$ and $\mathbf{P}$ is a linear one

$$\mathbf{P} = \varepsilon_0 \, \chi_e \, \mathbf{E} \tag{6.70}$$

where χ_e is called the electric susceptibility. The flux density in the dielectric material is related to $\mathbf{E}$ and $\mathbf{P}$ by the relation

$$\mathbf{D} = \varepsilon_0 \, \mathbf{E} + \mathbf{P} \tag{6.71}$$

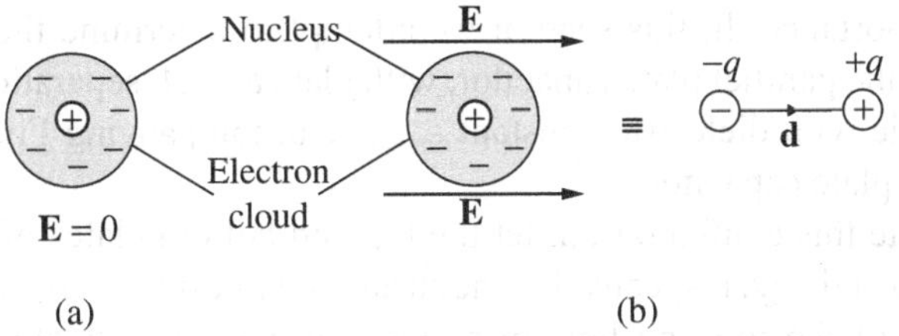

Fig. 6.13 Polarisation of a single molecule under the influence of an external field. (a) molecule when the field is absent, (b) molecule when the field is present

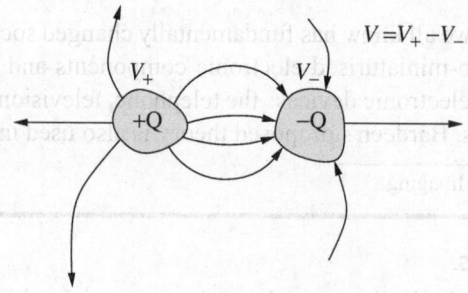

Fig. 6.14 Two metal bodies representing a capacitor

$$= \varepsilon_0 \mathbf{E} \left(1 + \chi_e\right) \qquad (6.72)$$

$$= \varepsilon_0 \varepsilon_r \mathbf{E} \qquad (6.73)$$

where $\varepsilon_r = 1 + \chi_e$ is the relative dielectric constant.

Example 6.16 Find the polarisation density in a dielectric material of $\varepsilon_r = 1.5$ immersed in an electric field of 1 V/m.

Solution

Step 1. First $\chi_e = \varepsilon_r - 1 = 0.5$

Step 2. $\mathbf{P} = \varepsilon_0 \chi_e \mathbf{E} = \varepsilon_0 \times 0.5 \text{ C/m}^2$

6.4 | Capacitance

If we connect two metal bodies by an emf source like a battery we find that current flows from one body to the other, through the source. As the current flows one body becomes negatively charged while the other body becomes positively charged. Since each of the two bodies is a conductor, each becomes an equipotential surface, with potentials which we designate as V_+ and V_-. It is obvious that $V = V_+ - V_-$, where V is the emf of the battery. When this potential difference is reached, the current stops flowing and charge is established on each body, according to the 'capacity' of the system. The capacitance of this configuration is defined as

$$C = \frac{Q}{V} \qquad (6.74)$$

6.4.1 Parallel Plate Capacitor

The capacitance in various configurations plays an important role in the field of electrical engineering. In a very large frequency range (from dc all the way up to hundreds of megahertz) in circuits, capacitors are required and their design is of paramount importance. In this section we attempt to determine the approximate capacitance of the parallel plate capacitor, with plate area A, separation d and filled with a dielectric with dielectric constant ε_r. The accompanying Fig. 6.15, shows such a parallel plate capacitor.

To investigate this configuration, let the top and bottom plates of the capacitor have a charge Q and $-Q$, respectively. The surface charge density ρ_s and $-\rho_s$ on the inner surfaces of the top and bottom plates may be approximated by Q/A and $-Q/A$, respectively. We draw a Gaussian surface which would be a rectangular parallelepiped with sides $a = $ depth, $b = $ width, and height $= 2h$. Let only half

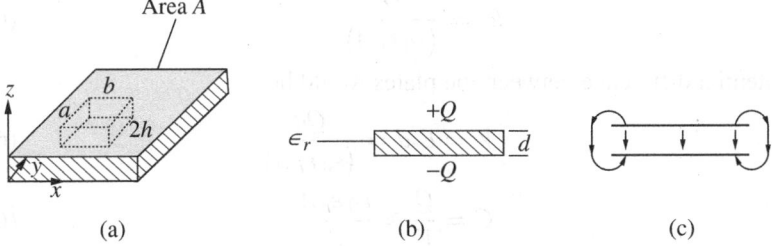

Fig. 6.15 Geometry and fields of the parallel plate capacitor (a) 3-D view, (b) cross-section, (c) rough sketch of the electric field

the height h, be towards the bottom plate and the other half above the top plate. The electric field lines would run from the inner surface of the top plate to the inner surface of the bottom plate. Applying Gauss's theorem to the parallelepiped

$$\oiint \mathbf{D} \cdot d\mathbf{S} = \underbrace{\iint \mathbf{D} \cdot d\mathbf{S}}_{\text{Sides}} + \underbrace{\iint \mathbf{D} \cdot d\mathbf{S}}_{\text{Top surface}} + \underbrace{\iint \mathbf{D} \cdot d\mathbf{S}}_{\text{Bottom surface}} \qquad (6.75)$$

Recalling the case of a sheet of charge, we expect that to a first approximation on the inner surface of the upper plate $\mathbf{D}$ will have the form $\mathbf{D} = [0, 0, -D_z]$. On the upper surface of the top plate, however, $\mathbf{D}$ will be approximately zero.

$$\underbrace{\iint \mathbf{D} \cdot d\mathbf{S}}_{\text{Sides}} = \underbrace{\iint D_y \cdot (-\mathbf{a}_y) \, dydz}_{\text{Left side}} + \underbrace{\iint D_y \cdot \mathbf{a}_y dydz}_{\text{Right side}}$$

$$+ \underbrace{\iint D_x \cdot (-\mathbf{a}_x) \, dydz}_{\text{Behind}} + \underbrace{\iint D_x \cdot \mathbf{a}_x dydz}_{\text{Front}} \qquad (6.76)$$

Now from the formulation, D_x and D_y on the sides will be approximately zero. Hence

$$\iint_{\text{Sides}} \mathbf{D} \cdot d\mathbf{S} \approx 0$$

The field on the upper surface of the top plate would also be zero. So

$$\iint_{\text{Top surface}} \mathbf{D} \cdot d\mathbf{S} \approx 0$$

Only on the bottom surface we can expect the integral to be non-zero.

$$\iint_{\text{Bottom surface}} \mathbf{D} \cdot d\mathbf{S} = \int_{x=x_0}^{x=x_0+a} \int_{y=y_0}^{y=y_0+b} (-D_z)(-\mathbf{a}_z) \, dxdy$$

$$= D_z ab$$

The charge enclosed is $\rho_s \, ab$ where ρ_s is the surface charge density on the inner surface. Recall that $\rho_s = Q/A$.

$$D_z ab = \rho_s \, ab = \left(\frac{Q}{A}\right) ab$$

$$D_z = \varepsilon_0 \, \varepsilon_r \, E_z = Q/A$$

$$E_z = \frac{Q}{(\varepsilon_0 \, \varepsilon_r \, A)} \tag{6.77}$$

The potential difference between the plates would be

$$V = E_z d = \frac{Qd}{(\varepsilon_0 \, \varepsilon_r \, A)} \tag{6.78}$$

and

$$C = \frac{Q}{V} \approx \frac{\varepsilon_0 \, \varepsilon_r \, A}{d} \tag{6.79}$$

which is the approximate value of the capacitance of a parallel plate capacitor. In this formulation, we have neglected the fringing fields at the extreme ends of the capacitor plates, and the fields on the outer surfaces of both the top and the bottom plates.

6.4.2 Coaxial Line

We look at another example, that of a coaxial line. The importance of the capacitance (per meter) of a coaxial line is important in transmission line theory. Much of the television cables laid by cable companies consist of coaxial lines. Computer cables also consist of two- and four-wire lines.

In this section we investigate the capacitance of the coaxial line shown in Fig. 6.16. The inner and outer conductors have a radii of a and b respectively. Between the two conductors there is a dielectric of dielectric constant ε_r. The surface charge density on the inner surface is ρ_s C/m^2.

We assume that the fields are radial, that is, they *diverge* from the positive charge on the inner conductor and converge on the negative charge on the outer conductor. Therefore, the flux density field **D** is assumed to have a structure,

$$\mathbf{D} = [D_\rho, 0, 0]$$

in the cylindrical coordinate system. The field is also assumed to be uniform throughout the length of the coaxial line. A Gaussian surface is drawn as shown in Fig. 6.16. The surface is cylindrical with a radius ρ, $a < \rho < b$ and the length of the cylinder is h. Integrating on the Gaussian surface

$$\oiint \mathbf{D} \cdot d\mathbf{S} = \underbrace{\iint \mathbf{D} \cdot d\mathbf{S}}_{\text{Sides}} + \underbrace{\iint \mathbf{D} \cdot d\mathbf{S}}_{\text{Cylindrical surface}}$$

$$= \underbrace{\iint \mathbf{D} \cdot d\mathbf{S}}_{\text{Cylindrical surface}}$$

$$= \underbrace{\iint D_\rho \rho \, d\phi \, dz}_{\text{Cylindrical surface}}$$

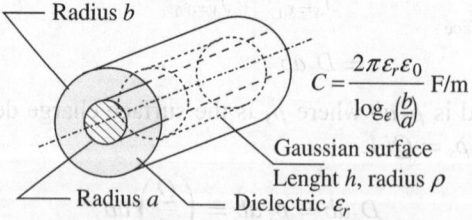

$$C = \frac{2\pi \varepsilon_r \varepsilon_0}{\log_e\left(\frac{b}{a}\right)} \text{ F/m}$$

Radius b

Radius a

Gaussian surface
Lenght h, radius ρ
Dielectric ε_r

Fig. 6.16 Capacitance of a coaxial line

$$= \int_{z_0}^{z_0+h} \int_0^{2\pi} D_\rho \, \rho \, d\phi \, dz$$

$$= 2\rho h\pi D_\rho \tag{6.80}$$

The total charge enclosed is

$$Q = \int\!\!\int \rho \, dS dz = \int_{z_0}^{z_0+h} \int_0^{2\pi} \rho_s \, a \, d\phi \, dz$$

$$= h \times \rho_s \, a 2\pi \tag{6.81}$$

Equating these two equations, we get

$$2\rho h\pi D_\rho = h\rho_s \, a 2\pi = Q \tag{6.82}$$

$$D_\rho = \frac{\rho_s \, a}{\rho} \tag{6.83}$$

and

$$E_\rho = \frac{\rho_s \, a}{\rho \, \varepsilon_r \, \varepsilon_0} \tag{6.84}$$

The potential difference between the inner and outer conductor is

$$V_{ab} = \int_a^b \mathbf{E} \cdot d\mathbf{l} = \frac{\rho_s \, a}{\varepsilon_r \, \varepsilon_0} \log_e \rho \Big|_a^b = \frac{\rho_s \, a}{\varepsilon_r \, \varepsilon_0} \log_e(b/a) \tag{6.85}$$

Hence

$$C = \frac{Q}{V_{ab}} = \frac{h\rho_s \, a 2\pi}{(\rho_s \, a/\varepsilon_r \, \varepsilon_0) \log_e(b/a)} = \frac{2\pi \varepsilon_r \, \varepsilon_0 \, h}{\log_e(b/a)} \tag{6.86}$$

This is the capacitance of a length h of the structure. For $h = 1\,\mathrm{m}$, the capacitance/meter is

$$C = \frac{2\pi \varepsilon_r \, \varepsilon_0}{\log_e(b/a)} \ \mathrm{F/m} \tag{6.87}$$

 See `CapOfCoaxLine.m` in Chapter 6

6.4.3 Two Conductor Line

Let us look at the example of a two conductor line illustrated in Fig. 6.17. We find the capacitance of such a structure using a slightly convoluted approach to a structure

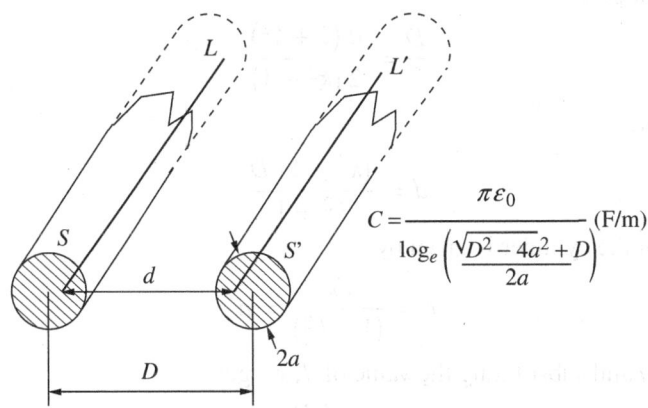

Fig. 6.17 A two conductor line

which we have tackled earlier. Each conductor has radius a and their centres are separated by a distance D. The surface charge density on the left and right conductors are such that the the total charge per meter on each is ρ_l and $-\rho_l$ Coulomb/m, respectively.

To solve this problem we consider another problem, that of the dual line charge considered in Example 5.14. We know that the equipotential surfaces are cylinders. Using the method of images (Section 6.3.3) we can replace each of the equipotential cylinders by a metal surface. Referring to Fig. 6.17, lines L and L' are the two line charges, with lines charges ρ_l and $-\rho_l$ Coulomb/m; the equipotential surfaces are S and S', which are replaced by metal cylinders.

Using the analysis outlined in Section 6.5, if V_1 is potential on the left equipotential cylinder, the centre line of the cylinder is located at

$$\left[x_k = \frac{d\left(1 + k^2\right)}{2\left(k^2 - 1\right)}, y = 0, z = z \right]$$

where

$$k = \exp \frac{V_1 \times 2\pi\varepsilon_0}{\rho_l} = \exp \frac{V_L \times 2\pi\varepsilon_0}{\rho_l}$$

$$V_L = \frac{\rho_l}{2\pi\varepsilon_0} \log_e k$$

where V_L is the potential on the left conductor. Similarly, the potential on the right conductor is

$$1/k = \exp \frac{V_2 \times 2\pi\varepsilon_0}{\rho_l} = \exp \frac{V_R \times 2\pi\varepsilon_0}{\rho_l}$$

$$V_R = \frac{\rho_l}{2\pi\varepsilon_0} \log_e \left(1/k\right) = -\frac{\rho_l}{2\pi\varepsilon_0} \log_e k$$

$$V_L - V_R = \frac{\rho_l}{\pi\varepsilon_0} \log_e k$$

The capacitance/meter is

$$C = \frac{\rho_l}{V_L - V_R} = \frac{\pi\varepsilon_0}{\log_e k}$$

To calculate k we proceed as follows. The x-coordinate of the centre of the left conductor translates to

$$\frac{D}{2} = \frac{d\left(1 + k^2\right)}{2\left(k^2 - 1\right)} \tag{6.88}$$

Solving for d,

$$d = \frac{\left(k^2 - 1\right) D}{k^2 + 1} \tag{6.89}$$

The radius of the cylinder is given by

$$r = \frac{dk}{\left(1 - k^2\right)} = a \tag{6.90}$$

Solving for a and substituting the value of d, we get

$$a = \frac{kD}{k^2 + 1} \tag{6.91}$$

Solving for k,

$$k = \frac{\sqrt{D^2 - 4a^2} + D}{2a}$$

using this value of k, we get

$$C = \frac{\pi \varepsilon_0}{\log_e k} = \frac{\pi \varepsilon_0}{\log_e \left(\frac{\sqrt{D^2 - 4a^2} + D}{2a} \right)} \quad \text{(F/m)} \qquad (6.92)$$

 See CapTwoParlellCond.m in Chapter 6

6.4.4 Capacitances of Concentric Spheres

Example 6.17 Find the capacitance of two concentric spheres and from there find also the capacitance of a single sphere (Fig. 6.18).

Solution The electric field between the two spheres is

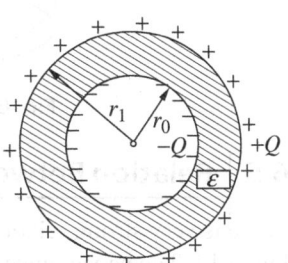

Fig. 6.18 A charge Q enclosed by a spherical shell

$$\mathbf{E} = -\frac{Q}{4\pi \varepsilon r^2} \mathbf{a}_r, \qquad r_0 < r < r_1$$

and the potential difference is

$$V = \int_{r_1}^{r_0} \left(-\frac{Q}{4\pi \varepsilon r^2} \right) dr$$

$$= \frac{Q}{4\pi \varepsilon r} \Big]_{r_1}^{r_0}$$

$$= \frac{Q}{4\pi \varepsilon} \left(\frac{1}{r_0} - \frac{1}{r_1} \right)$$

hence the capacitance of the configuration is

$$C = Q/V = \frac{4\pi \varepsilon}{\frac{1}{r_0} - \frac{1}{r_1}} \qquad \text{(F)}$$

if the outer sphere is removed with $r_1 \to \infty$, then the capacitance of a sphere of radius r_0 (in air) is

$$C = 4\pi \varepsilon_0 r_0$$

Leyden Jar

Records indicate that a German scientist named E. G. Von Kleist invented the capacitor in 1745. However, about the same time P. van Musschenbroek, from the University of Leyden, Holland, invented a capacitor in the form of a glass jar which became famous as the Leyden jar (see Fig. 6.19).

The Leyden jar consisted of a glass jar lined on the inside and outside with metal foil, and partially filled with water. The water really had no part to play, though at that time it was thought to play an important role. The glass in fact was the dielectric, and the charge was stored on the metal foils. To convey the charge to the inner foil, there was a metal chain connecting the inner foil to the metal rod of the top of the jar.

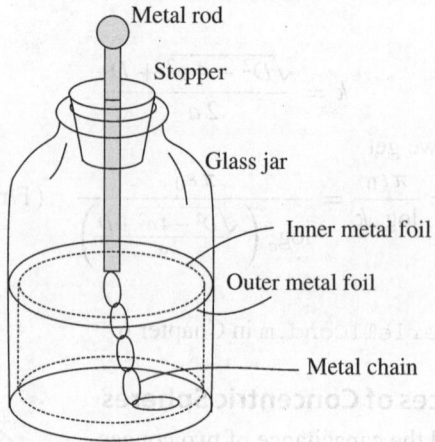

Fig. 6.19 Construction of the Leyden jar

6.5 | Relation Between Capacitance and Resistance

There is very strong relationship between capacitance and resistance. Let us study Fig. 6.14. The capacitance is given by

$$C = \frac{Q}{V} = \frac{\iint_A \mathbf{D} \cdot d\mathbf{S}}{\int_L \mathbf{E} \cdot d\mathbf{l}} = \frac{\varepsilon \iint_A \mathbf{E} \cdot d\mathbf{S}}{\int_L \mathbf{E} \cdot d\mathbf{l}} \tag{6.93}$$

where L is the integration along any line connecting one conductor to the other; the line may be curved or straight. A is the area of the surface of any one of the conductors; the integral

$$\iint_A \mathbf{D} \cdot d\mathbf{S}$$

is equal to the total charge residing on the conductor, and ε is the permittivity of the medium. Let us consider the same geometry but with the case where the medium has a conductivity σ. In this case the current I which flows along the electric field lines is given by

$$I = \iint_A \mathbf{J} \cdot d\mathbf{S} = \sigma \iint_A \mathbf{E} \cdot d\mathbf{S}$$

The definition of voltage difference is exactly the same as before. So

$$R = \frac{V}{I} = \frac{\int_L \mathbf{E} \cdot d\mathbf{l}}{\iint_A \mathbf{J} \cdot d\mathbf{S}} = \frac{\int_L \mathbf{E} \cdot d\mathbf{l}}{\sigma \iint_A \mathbf{E} \cdot d\mathbf{S}} \tag{6.94}$$

The integrations being performed are identical. Therefore,

$$RC = \frac{\int_L \mathbf{E} \cdot d\mathbf{l}}{\sigma \iint_A \mathbf{E} \cdot d\mathbf{S}} \times \frac{\varepsilon \iint_A \mathbf{E} \cdot d\mathbf{S}}{\int_L \mathbf{E} \cdot d\mathbf{l}} = \frac{\varepsilon}{\sigma} \tag{6.95}$$

or

$$R = \frac{\varepsilon}{\sigma C} \tag{6.96}$$

Example 6.18 Find the resistance of length l of a coaxial line (geometry as shown in Fig. 6.20) where the current moves from the inner conductor to the outer conductor of the line. The inner and outer conductors have radii a and b respectively. The medium between the inner and outer conductors is a material of conductivity σ.

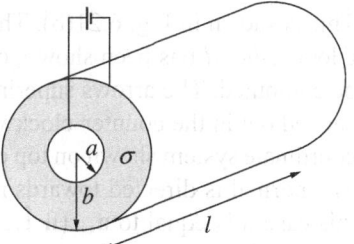

Fig. 6.20 A coaxial resistor

Solution

Step 1. The capacitance per meter of a coaxial line is [Eqn (6.87)]

$$C = \frac{2\pi \varepsilon_r \varepsilon_0}{\log_e (b/a)}$$

Step 2. The capacitance for length l is $c = Cl$

Step 3. The resistance for length l is given by (Eqn (6.96))

$$R = \frac{\varepsilon}{\sigma c} = \frac{\varepsilon_r \varepsilon_0}{\sigma l \left[2\pi \varepsilon_r \varepsilon_0 / \log_e (b/a)\right]} = \frac{\log_e (b/a)}{2\pi \sigma l}$$

Step 4. Compare with a similar situation, in Example 6.9, Eqn (6.41).

6.6 | Boundary Conditions for Electrostatic Fields

What happens when there is an electric field in a region where more than one media are present? Suppose there is an electric field present in a region comprising air, dielectrics, and metal bodies.

In this section we will examine the behaviour of the field on the very boundary of two dissimilar media. The cases of interest are the dielectric–dielectric boundary and the dielectric–metal boundary.

A dielectric (region 2) is immersed in an external electric field as illustrated in Fig. 6.21. The region outside the dielectric (region 1) consists of a medium with permittivity ε_1 (e.g. if the medium is air then $\varepsilon_1 \equiv \varepsilon_0$). The immersed dielectric has a permittivity of ε_2. We concentrate on a very small region R on the boundary of the two dielectrics shown in Fig. 6.21(a) and apply the integral form of the electrostatic Maxwell's equation to the interface,

$$\oint \mathbf{E} \cdot d\mathbf{l} = 0 \qquad (6.97)$$

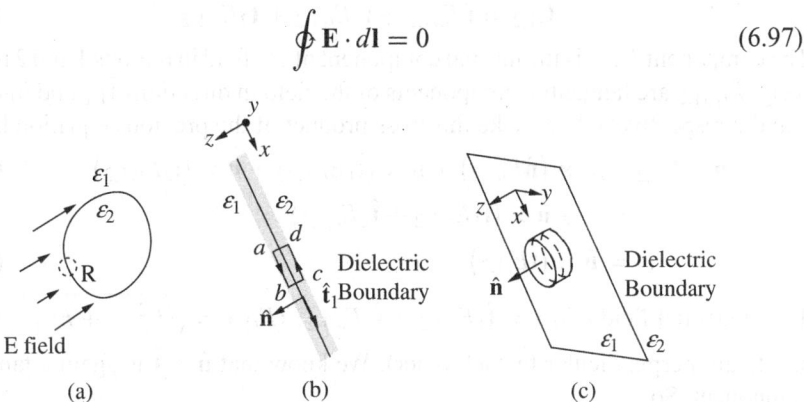

(a) (b) (c)

Fig. 6.21 Behaviour of the electric field near a dielectric–dielectric boundary

This is shown in Fig. 6.21(b). The figure depicts a close-up of the interface where a loop a-b-c-d has been shown, on which the line integral of the electric field is to be computed. The arrows superimposed on the loop show that integration is being carried out in the counter-clockwise sense as per the requirement. From the small coordinate system shown on top of the figure, $\hat{n} \equiv a_z$ is the normal to the boundary. The normal is directed towards region 1. $\hat{t}_1$ is the tangent to the boundary and in this case it is equal to a_x. $(\hat{n}, \hat{t}_1, \hat{t}_2)$ in that order, form a right-handed coordinate system. We assume that d-a and b-c are vanishingly small, while a-b and c-d are small enough so that that the field is approximately constant over these lengths.

$$\text{length } (b - c) = \text{length } (d - a) \approx 0$$

$$\text{length } (a - b) = \text{length } (c - d) \approx |\Delta x|$$

Applying Eqn (6.97) to the loop a-b-c-d,

$$\mathbf{E}_1 \cdot a_x |\Delta x| + \mathbf{E}_2 \cdot (-a_x) |\Delta x| = 0$$

where $\mathbf{E}_1$ and $\mathbf{E}_2$ are the electric fields in regions 1 and 2, respectively. The integrations over the line lengths b-c and d-a are assumed to be negligible and so have been dropped from the equation. The previous equation reduces to

$$E_{x1} - E_{x2} = 0 \tag{6.98}$$

In the same manner by considering a similar loop going into the page the longer sides of which lie in medias 1 and 2, while the shorter side cuts across the two media, we perform a similar line integral over it to get

$$E_{y1} - E_{y2} = 0 \tag{6.99}$$

which can be summarised as

$$E_{t1} - E_{t2} = 0 \tag{6.100}$$

The subscript t is used to signify the *tangential* component. $E_{t1,2}$ are the tangential components of the electric field next to the boundary but in media 1 and 2 respectively.

$$E_{t1,2} = \sqrt{E_{x1,2}^2 + E_{y1,2}^2}$$

The tangential component of the electric field is continuous across a dielectric boundary. Considering the electric field as consisting of a normal and tangential components where $\hat{n} \equiv a_z$, $\hat{t}_1 \equiv a_x$ $\hat{t}_2 \equiv a_y$ the field is

$$\mathbf{E}_{1,2} = \hat{n}E_{n1,2} + \hat{t}_1 E_{t_1 1,2} + \hat{t}_2 E_{t_2 1,2} \tag{6.101}$$

The component $E_{n1,2}$ is the normal component of the field in regions 1 and 2 respectively; $E_{t_1,2 1,2}$ are tangential components of the field in directions $\hat{t}_{1,2}$ and in regions 1 and 2 respectively. If we take the cross product of the previous equation by $\hat{n}$

$$\hat{n} \times \mathbf{E}_{1,2} = \hat{n} \times (\hat{n}E_{n1,2}) + \hat{n} \times (\hat{t}_1 E_{t_1 1,2}) + \hat{n} \times (\hat{t}_2 E_{t_2 1,2}) \tag{6.102}$$

$$= 0 + \hat{n} \times (\hat{t}_1 E_{t_1 1,2} + \hat{t}_2 E_{t_2 1,2})$$

$$= \hat{n} \times (\hat{t}E_{t1,2}) \tag{6.103}$$

The tangential field $\hat{t}E_{t1,2} = \hat{t}_1 E_{t_1 1,2} + \hat{t}_2 E_{t_2 1,2}$. ($E_{t1,2} = \sqrt{E_{t_1 1,2}^2 + E_{t_2 1,2}^2}$ since $\hat{t}_1$ and $\hat{t}_2$ are perpendicular to each other). We know that $\hat{n} \times \hat{t}$ is again a tangential component. So

$$\hat{n} \times (\mathbf{E}_1 - \mathbf{E}_2) = 0 \tag{6.104}$$

This equation is written in this form because it is independent of coordinate notation and because of the importance of its comparison with

$$\nabla \times \mathbf{E} = 0$$

We will come back to this kind of notation when we consider other boundary conditions.

Let us now consider the normal component of the field. If we apply Gauss's law

$$\oiint \mathbf{D} \cdot d\mathbf{S} = \rho_v$$

to the configuration shown in Fig. 6.21(c). The figure shows a close up of region R in Fig. 6.21(a). We draw a small pill-box as shown in (c) whose height is negligible and the top and bottom areas are so small that the field is assumed not to change much in those regions

$$\text{Height of pill-box} \approx 0$$

$$\text{Area of the top and bottom} \approx \Delta A$$

Applying Gauss's law to this pill-box with the knowledge that there is no accumulated charge on the interface ($\rho_s = 0$)

$$\mathbf{D}_1 \cdot \mathbf{a}_z \Delta A + \mathbf{D}_2 \cdot (-\mathbf{a}_z) \Delta A = 0 \tag{6.105}$$

$$D_{z1} - D_{z2} = 0 \tag{6.106}$$

or in other words, *the normal component of the flux density is continuous across a dielectric–dielectric boundary*,

$$D_{n1} - D_{n2} = 0$$

$$\hat{\mathbf{n}} \cdot (\mathbf{D}_1 - \mathbf{D}_2) = 0 \tag{6.107}$$

Compare this last equation with $\nabla \cdot \mathbf{D} = 0$.

Let us investigate the behaviour of the field near a dielectric–metal boundary. We know from our earlier discussion (Section 6.3.1) that there can be no electric field inside a conductor. In the presence of a field, a surface charge develops and makes the surface an equipotential one. Figure 6.22(a) shows a metal body of conductivity σ immersed in a dielectric of permittivity ε_1. We once again apply Maxwell's Eqn (6.97) to the dielectric–metal body

$$\oint \mathbf{E} \cdot d\mathbf{l} = 0$$

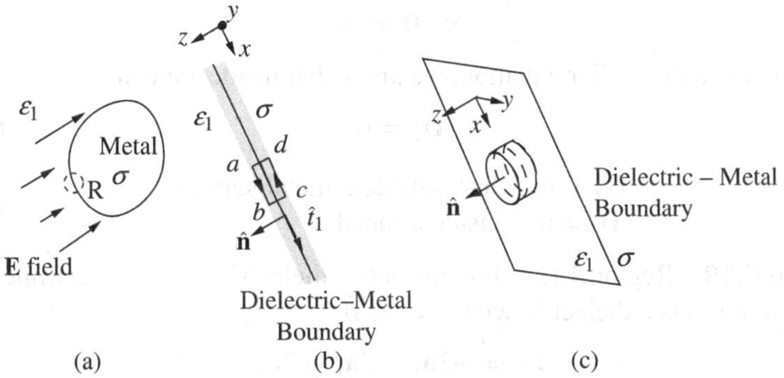

(a)   (b)   (c)

Fig. 6.22 Behaviour of the electric field near dielectric–metal boundary

Applying this equation to the loop *a-b-c-d* in Fig. 6.22(b) and, using the same arguments as earlier in this section (disregarding the fact that we are dealing with a metal) we have

$$\hat{n} \times (\mathbf{E}_1 - \mathbf{E}_2) = 0$$

$$E_{t1} = E_{t2} \tag{6.108}$$

but there can be no field inside the metal; that is $E_{t2} = 0$. Therefore, just outside the metal, the tangential electric field must be zero.

$$\begin{aligned} E_t &= 0 \quad \text{Just outside a metal surface} \\ \mathbf{E} &= 0 \quad \text{inside a metal} \end{aligned} \tag{6.109}$$

Similarly, applying Gauss's law to the pill-box shown on the boundary interface of Fig. 6.22(c).

$$\oiint \mathbf{D} \cdot d\mathbf{S} = \rho_v \tag{6.110}$$

$$\text{Height of pill-box} \approx 0$$
$$\text{Area of the top and bottom} \approx \Delta A$$

then, applying Gauss's law to this pill-box, whose volume is ΔV

$$\mathbf{D}_1 \cdot \mathbf{a}_z \Delta A + \mathbf{D}_2 \cdot (-\mathbf{a}_z) \Delta A = \rho_v \Delta V \tag{6.111}$$

where ρ_v is the volume charge density accumulated near the surface. However, we know from earlier arguments that the charge exists *only on the surface*

$$\rho_v \Delta V = \rho_s \Delta A$$

where ρ_s is the surface charge density

$$\mathbf{D}_1 \cdot \mathbf{a}_z \Delta A + \mathbf{D}_2 \cdot (-\mathbf{a}_z) \Delta A = \rho_s \Delta A \tag{6.112}$$

$$D_{z1} - D_{z2} = \rho_s \tag{6.113}$$

and so in general, we have independent of coordinate notation

$$\hat{n} \cdot (\mathbf{D}_1 - \mathbf{D}_2) = \rho_s \tag{6.114}$$

where $\hat{n}$ is the normal to the surface but directed towards the outside from the metal, that is medium 1; ρ_s is the charge on the surface of the metal; and $\mathbf{D}_{1,2}$ is the flux density outside and inside the metal, respectively. That this equation is indeed true is borne out by observing the units. Note the comparison with

$$\nabla \cdot \mathbf{D} = \rho_v$$

Coming back to the earlier equation, we know that inside a metal

$$\mathbf{D}_2 = 0 \tag{6.115}$$

so
$$\begin{aligned} D_n &= \rho_s \quad \text{Just outside a metal surface} \\ \mathbf{D} &= 0 \quad \text{inside a metal} \end{aligned} \tag{6.116}$$

Example 6.19 Region 1 ($z < 0$) represents a dielectric with $\varepsilon_r = 2$ while region 2 ($z > 0$) represents a dielectric with $\varepsilon_r = 4$. If

$$\mathbf{E}_1 = -3\mathbf{a}_x + 5\mathbf{a}_y + 7\mathbf{a}_z$$

Find $\mathbf{D}_2$, $\mathbf{E}_2$, and $\mathbf{P}_2$.   (NTU-NSIT, March 2008)

Solution

Step 1. We know that the tangential electric field is continuous and the normal electric flux density is continuous. Therefore

$$\mathbf{E}_{2(\tan)} = \mathbf{E}_{1(\tan)} = -3\mathbf{a}_x + 5\mathbf{a}_y$$

Step 2. And therefore $\quad \mathbf{D}_{2(\tan)} = \varepsilon_2\,\mathbf{E}_{2(\tan)} = 4\varepsilon_0\left(-3\mathbf{a}_x + 5\mathbf{a}_y\right)$

Step 3. Also $\qquad\qquad \mathbf{D}_{2(n)} = \mathbf{D}_{1(n)} = 2\varepsilon_0\,(7\mathbf{a}_z)$

Step 4. So $\qquad\qquad \mathbf{D}_2 = 4\varepsilon_0\left(-3\mathbf{a}_x + 5\mathbf{a}_y\right) + 2\varepsilon_0\,(7\mathbf{a}_z)$

$$\mathbf{E}_2 = \mathbf{D}_2/4\varepsilon_0 = -3\mathbf{a}_x + 5\mathbf{a}_y + (7/2)\,\mathbf{a}_z$$

Step 5. Also $\chi_{e1} = \varepsilon_{r1} - 1 = 2 - 1 = 1$ and $\chi_{e2} = 4 - 1 = 3$. So

$$\mathbf{P}_2 = \varepsilon_0\,\chi_{e2}\,\mathbf{E}_2 = 3\varepsilon_0\left[-3\mathbf{a}_x + 5\mathbf{a}_y + (7/2)\,\mathbf{a}_z\right]$$

Example 6.20 A uniform electric field $\mathbf{E}_1$ meets a dielectric with dielectric constant ε_r filling the half space as shown in Fig. 6.23. Find the field inside the dielectric.

Solution

Step 1. The field is split into two components, tangential and normal to the dielectric boundary, $\mathbf{E}_1 = \mathbf{E}_{1t} + \mathbf{E}_{1n}$.

Step 2. Quantifying these components

$$E_{1t} = E_1\cos\theta\,, \quad E_{1n} = E_1\sin\theta$$

Also $\qquad\qquad D_{1t} = \varepsilon_0\,E_1\cos\theta\,, \quad D_{1n} = \varepsilon_0\,E_1\sin\theta$

Step 3. The tangential component at the boundary is continuous, so

$$E_{2t} = E_{1t} = E_1\cos\theta$$

Step 4. The normal component of the $\mathbf{D}$ is continuous, hence

$$D_{2n} = D_{1n} = \varepsilon_0\,E_1\sin\theta$$

Step 5. The normal component of the electric field is (in the dielectric) is therefore

$$E_{2n} = \frac{D_{2n}}{\varepsilon_0\,\varepsilon_r} = \frac{E_1\sin\theta}{\varepsilon_r}$$

Step 6. Find the magnitude of the field in the dielectric

$$|\mathbf{E}_2| = \sqrt{E_{2t}^2 + E_{2n}^2} = \sqrt{E_1^2\cos^2\theta + E_1^2\frac{\sin^2\theta}{\varepsilon_r^2}} = E_1\sqrt{\cos^2\theta + (\sin\theta/\varepsilon_r)^2}$$

The magnitude of the electric field in the dielectric is less than that in air.

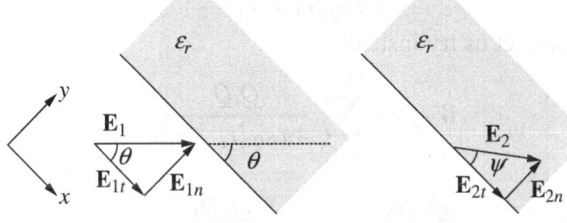

Fig. 6.23 Electric field in the presence of a dielectric occupying a half-space

Step 7. Visualising the electric field in the dielectric to be $\mathbf{E}_2 = \mathbf{a}_x E_{2t} + \mathbf{a}_y E_{2n}$, then its angle from the x-axis is

$$\psi = \tan^{-1}\left\{\frac{E(\sin\theta/\varepsilon_r)}{E\cos\theta}\right\} = \tan^{-1}(\tan\theta/\varepsilon_r)$$

which is less than θ.

6.7 | Energy Stored in the Electric Field

Let us investigate what happens as we move point charges into a region of space. Figure 6.24 shows how charges from infinity are moved into a region of space. The figure shows charges Q_i, $i = 1, 2, \ldots, N$ being moved to positions given by $\mathbf{r}_i$, $i = 1, 2, \ldots, N$ from infinity. For example, the first charge of magnitude Q_1 is moved to the position $\mathbf{R}_1 \equiv (\mathbf{0})$ (the origin) from $R = \infty$. The amount of work done is zero, since no other charges are present to produce an electric field. The work done to move charge number 2 from infinity to $\mathbf{r}_2$ is equal to the potential at $\mathbf{r}_2$ due to Q_1 multiplied by the magnitude of the charge Q_2

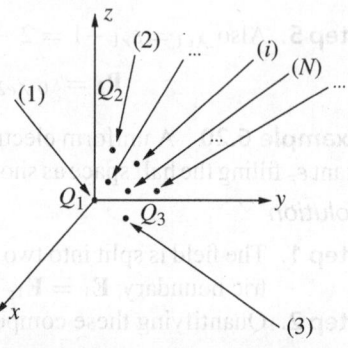

Fig. 6.24 Moving charges from infinity into a finite region of space

$$W_2 = \frac{Q_1 Q_2}{4\pi\varepsilon_0|\mathbf{r}_2 - \mathbf{r}_1|}\ (\mathrm{J})$$

To move charge number 3, the work required is

$$W_3 = \frac{Q_1 Q_3}{4\pi\varepsilon_0|\mathbf{r}_3 - \mathbf{r}_1|} + \frac{Q_2 Q_3}{4\pi\varepsilon_0|\mathbf{r}_3 - \mathbf{r}_2|}$$

And the total work ($W_{T,3}$ for three charges) done is

$$W_{T,3} = W_2 + W_3 = \frac{Q_1 Q_2}{4\pi\varepsilon_0|\mathbf{r}_2 - \mathbf{r}_1|} + \frac{Q_1 Q_3}{4\pi\varepsilon_0|\mathbf{r}_3 - \mathbf{r}_1|} + \frac{Q_2 Q_3}{4\pi\varepsilon_0|\mathbf{r}_3 - \mathbf{r}_2|}$$

How to represent this mathematically? Let us try

$$W_{T,3} = \sum_{i=1}^{i=3}\sum_{j=1}^{j=3}\frac{Q_i Q_j}{4\pi\varepsilon_0|\mathbf{r}_i - \mathbf{r}_j|}$$

The very first term

$$\frac{Q_1 Q_1}{4\pi\varepsilon_0|\mathbf{r}_1 - \mathbf{r}_1|}$$

is wrong. Therefore, let us try instead

$$W_{T,3} = \sum_{\substack{i,j=1 \\ i\neq j}}^{i,j=3}\frac{Q_i Q_j}{4\pi\varepsilon_0|\mathbf{r}_i - \mathbf{r}_j|}$$

This would be $W_{T,3} = \dfrac{Q_1 Q_2}{4\pi\varepsilon_0|\mathbf{r}_2 - \mathbf{r}_1|} + \dfrac{Q_2 Q_1}{4\pi\varepsilon_0|\mathbf{r}_1 - \mathbf{r}_2|} + \cdots$

that is every term would occur twice. Therefore, the correct formula would be

$$W_{T,3} = \frac{1}{2} \sum_{\substack{i,j=1 \\ i \neq j}}^{i,j=3} \frac{Q_i Q_j}{4\pi\varepsilon_0 |\mathbf{r}_i - \mathbf{r}_j|}$$

In the same way, for N charges, the work done would be

$$W_{T,N} = \frac{1}{2} \sum_{\substack{i,j=1 \\ i \neq j}}^{i,j=N} \frac{Q_i Q_j}{4\pi\varepsilon_0 |\mathbf{r}_i - \mathbf{r}_j|} \tag{6.117}$$

Let us play around with this equation to get a significant result. The equation rewritten is

$$W_{T,N} = \frac{1}{2} \underbrace{\sum_{i=1}^{i=N} Q_i \times \sum_{j=1}^{j=N} \left[\frac{Q_j}{4\pi\varepsilon_0 |\mathbf{r}_i - \mathbf{r}_j|} \right]}_{i \neq j}$$

the term in the square brackets is the potential at $\mathbf{r}_i$ due to all the other charges. If we write the square bracket summation as

$$V_i = \sum_{j=1}^{j=N} \frac{Q_j}{4\pi\varepsilon_0 |\mathbf{r}_i - \mathbf{r}_j|}, \qquad \text{with } j \neq i \tag{6.118}$$

then
$$W_{T,N} = \frac{1}{2} \sum_{i=1}^{i=N} Q_i V_i \tag{6.119}$$

Now instead of large charges, let the charges accumulated be small, that is, each Q_i is replaced by ΔQ_i and N can made very large

$$W_{T,N} = \frac{1}{2} \sum_{i=1}^{i=N} (\Delta Q_i) V_i$$

this accumulation can done by replacing ΔQ_i by $\rho_{vi} \Delta v_i$ where $\rho_{vi} \Delta v_i$ is a very small volume of charge which is moved into the region where the accumulation is taking place. Note that here we have used v_i instead of V_i for the differential volume so as not to confuse it with the potential V_i. Then

$$W_{T,N} = \frac{1}{2} \sum_{i=1}^{i=N} (\rho_{vi} \Delta v_i) V_i$$

Replacing the summation by an integration

$$W_T = \frac{1}{2} \iiint_V (\rho_v \, dv) V \tag{6.120}$$

The volume V is over the ρ_v. That is where ρ_v is present, there is a contribution to the volume integral, and where there is no charge density, there is no contribution. We can therefore safely integrate over all space.

$$W_T = \frac{1}{2} \iiint_{\text{All space}} (\rho_v \, dv) V \tag{6.121}$$

Now we know that
$$\nabla \cdot \mathbf{D} = \rho_v$$
and
$$\nabla \cdot (\mathbf{D}V) = V\nabla \cdot \mathbf{D} + \mathbf{D} \cdot \nabla V$$
therefore

$$W_T = \frac{1}{2} \iiint\limits_{\text{All space}} (\nabla \cdot (\mathbf{D}V) - \mathbf{D} \cdot \nabla V) \, dv$$

$$= \frac{1}{2} \iiint\limits_{\text{All space}} \nabla \cdot (\mathbf{D}V) \, dv + \frac{1}{2} \iiint\limits_{\text{All space}} (-\mathbf{D} \cdot \nabla V) \, dv$$

$$= \frac{1}{2} \oiint\limits_{\text{Surface at } \infty} (\mathbf{D}V) \cdot d\mathbf{S} + \frac{1}{2} \iiint\limits_{\text{All space}} \mathbf{D} \cdot (-\nabla V) \, dv \quad (\text{Divergence Theorem})$$

$$= 0 + \frac{1}{2} \iiint\limits_{\text{All space}} \mathbf{D} \cdot (-\nabla V) \, dv \quad (\text{both } \mathbf{D} \text{ and } \nabla V \text{ are zero at infinity})$$

$$= \frac{1}{2} \iiint\limits_{\text{All space}} \mathbf{D} \cdot \mathbf{E} dv \quad (\mathbf{E} = -\nabla V) \tag{6.122}$$

A surprising result. The work done on the charges has been stored in the electrostatic field!

$$W_e = \frac{1}{2} \iiint\limits_{\text{All space}} \mathbf{D} \cdot \mathbf{E} dV = \frac{1}{2} \iiint\limits_{\text{All space}} \varepsilon |\mathbf{E}|^2 \, dV \quad (\text{J}) \tag{6.123}$$

The term $w_e = \varepsilon |\mathbf{E}|^2$ is the *energy density* in J/m^3 at all points permeated by the electric field.

Example 6.21   Find the approximate energy stored in a parallel plate capacitor of capacitance $C = \varepsilon \, (A/d)$ where A is the area of the plates and d is the separation between the plates.

Solution   The electric field between the capacitor plates is given by

$$|\mathbf{E}| \approx \frac{V}{d}$$

where V is the potential difference across the capacitor plates. The total stored energy is therefore

$$\frac{1}{2} \varepsilon |\mathbf{E}|^2 \times (\text{volume between the capacitor plates}) \approx \frac{\varepsilon V^2}{2d^2} \times (Ad)$$

$$= \frac{V^2}{2} \left(\varepsilon \frac{A}{d} \right)$$

$$= \frac{1}{2} V^2 C$$

POINTS TO REMEMBER

- The current is related to the charge in a wire by $I = dQ/dt$ where dQ/dt is evaluated at any cross-section of the wire.

- The current density $\mathbf{J}$ and the current I are connected by the relation

$$I = \iint_{cs} \mathbf{J} \cdot d\mathbf{S}$$

where cs is the cross-section of the wire.
- At any point in a conducting medium, if the charge carriers have a velocity $\mathbf{v}$ and the charge density of these mobile carriers is ρ_m then $\mathbf{J} = \rho_m \mathbf{v}$
- If the number of charge carries per unit volume is n (No./m³) and each charge has a charge of q (C), then the charge density ρ is $\rho = nq$.
- The *continuity equation* states that

$$\nabla \cdot \mathbf{J} = -\frac{\partial \rho}{\partial t}$$

- The *continuity equation* in integral form says that the amount of current leaving a closed surface is equal to the total time rate of *depletion* of charge in the volume enclosed

$$\oiint \mathbf{J} \cdot d\mathbf{S} = -\iiint \frac{\partial \rho}{\partial t} \, dV$$

- Ohm's law is $V = IR$
  where V is the voltage or potential difference across the resistor of value R, and I is the current through it.
- Ohm's law in point form is $\mathbf{J} = \sigma \mathbf{E}$
  where $\mathbf{J}$ is the current density at a point and $\mathbf{E}$ is the electric field at that point and σ is the conductivity.
- The resistance of a piece of material is given by

$$R = \frac{l}{\sigma A}$$

where l is the length of the material (the direction in which the current will travel), A is the area of cross-section and σ is the conductivity.
- The relaxation time τ for a material is the amount of time, charge takes to dissipate in a material from its initial value (Q_0) to $Q_0 e^{-1}$. τ for conductors is $(\sigma/\varepsilon)^{-1}$.
- The mobility (μ) in a conductor or semiconductor is defined by the relation

$$\mathbf{v} = \mu \mathbf{E}$$

where $\mathbf{v}$ is the drift velocity of charge carriers in the material and $\mathbf{E}$ is the electric field.
- The dipole moment $\mathbf{p}$ of two charges q and $-q$ is given by $\mathbf{p} = q\mathbf{d}$
  where the vector $\mathbf{d}$ is a position vector directed from the negative charge to the positive charge.
- The polarisation density $\mathbf{P}$, the number of dipole moments/m³ in a dielectric is defined by

$$\mathbf{P} = \lim_{\Delta V \to 0} \left(\frac{1}{\Delta V} \sum_{i=1}^{N} \mathbf{p}_i \right)$$

where $\mathbf{p}_i$ are the dipole moments in the dielectric and ΔV is a small volume while N is the number of dipole moments in ΔV.
- Under the influence of an external field $\mathbf{E}$, the relationship between $\mathbf{E}$ and $\mathbf{P}$ is a linear one, $\mathbf{P} = \varepsilon_0 \chi_e \mathbf{E}$ where χ_e is called the electric susceptibility.
- For a dielectric $\chi_e = \varepsilon_r - 1$
- In a dielectric $\mathbf{D} = \varepsilon_0 \mathbf{E} + \mathbf{P}$

- The capacitance of a capacitor formed by two metallic surfaces is

$$C = \frac{Q}{V}$$

where Q is the charge residing on one of the surfaces and V is the potential difference between them.

- The approximate value of capacitance for a parallel plate capacitor is

$$C \approx \frac{\varepsilon_0 \varepsilon_r A}{d}$$

where A is the area of the plates and d is the separation between them.

- The capacitance per metre of a coaxial line is

$$C = \frac{2\pi \varepsilon_r \varepsilon_0}{\log_e(b/a)} \tag{6.124}$$

where a, b are the inner and outer radii of the line respectively and ε_r is the dielectric constant of the material filling the region between the conductors.

- The capacitance per metre of two conductors which are in parallel is

$$C = \frac{\pi \varepsilon_0}{\log_e\left(\frac{\sqrt{D^2 - 4a^2} + D}{2a}\right)} \tag{6.125}$$

where D is the distance between the centres of the two conductors and a is the radius of each conductor.

- The capacitance of two concentric spheres is

$$C = \frac{4\pi\varepsilon}{(1/r_0) - (1/r_1)}$$

where r_0 and r_1 are the radii of the inner and outer sphere respectively and ε is the permittivity of the material filling the region between the spheres.

- The capacitance of a single sphere in air is $C = 4\pi\varepsilon_0 r_0$ where r_0 is the radius of the sphere.

- The capacitance between two conductors is given by

$$C = \frac{Q}{V} = \frac{\iint_A \mathbf{D} \cdot d\mathbf{S}}{\int_L \mathbf{E} \cdot d\mathbf{l}} = \frac{\varepsilon \iint_A \mathbf{E} \cdot d\mathbf{S}}{\int_L \mathbf{E} \cdot d\mathbf{l}}$$

where L is the integration along any line connecting one conductor to the other; the line may be curved or straight. A is the area on the surface of any one of the conductors.

- The resistance between two conductors is given by

$$R = \frac{V}{I} = \frac{\int_L \mathbf{E} \cdot d\mathbf{l}}{\iint_A \mathbf{J} \cdot d\mathbf{S}} = \frac{\int_L \mathbf{E} \cdot d\mathbf{l}}{\sigma \iint_A \mathbf{E} \cdot d\mathbf{S}}$$

where the integrations performed are the same as in the previous bullet, and

$$RC = \frac{\varepsilon}{\sigma}$$

- At a dielectric–dielectric boundary, the tangential E fields and the normal D fields are continuous

$$E_{1t} = E_{2t}, \quad D_{1n} = D_{2n}$$

- At a dielectric–metal boundary, the tangential E field is zero and the normal D field is equal to the surface charge density

$$E_t = 0, \quad D_n = \rho_s$$

no fields can exist inside a metal.

- The energy, $W_{T,N}$ stored in an electrostatic field for N charges, $Q_1 \ldots Q_N$ at position vectors $\mathbf{r}_1 \ldots \mathbf{r}_N$ is

$$W_{T,N} = \frac{1}{2} \sum_{\substack{i,j = 1 \\ i \neq j}}^{i,j=N} \frac{Q_i Q_j}{4\pi \varepsilon_0 |\mathbf{r}_i - \mathbf{r}_j|}$$

- The energy stored in the electrostatic field in general is

$$W_e = \frac{1}{2} \iiint_{\text{All space}} \mathbf{D} \cdot \mathbf{E} dV = \frac{1}{2} \iiint_{\text{All space}} \varepsilon |\mathbf{E}|^2 dV$$

SELF ASSESSMENT

Objective Type Questions

1. If the current in a wire is $I(t)$ A, then the charge passing through the cross-section of the wire in 1 s is
   (a) I C   (b) $1/I$ C   (c) $\int_0^1 I(t)\, dt$ (d) none of the above

2. If $\mathbf{J} = e^{-ax^2 - by^2 - cz^2} \mathbf{a}_x$, how much current crosses the x-y plane?
   (a) 0   (b) πa (c) $1/\pi a$ (d) none of the above

3. If $\mathbf{J} = e^{-ax^2 - by^2 - cz^2} \mathbf{a}_x$, is charge accumulating at $(0,0,0)$?
   (a) yes   (b) no   (c) increasing at a constant rate   (d) decreasing at a constant rate

4. If $\mathbf{J} = e^{-ax^2 - by^2 - cz^2} \mathbf{a}_x$ in a region of space and the mobile charge density is -1 mC/m^3 at $(0,0,0)$, what is the drift velocity of the charge?
   (a) 1×10^6 m/s   (b) 1×10^4 m/s   (c) 1×10^3 m/s   (d) none of the above

5. If no charge has to accumulate at a point in space, then
   (a) $\partial \rho_v / \partial t = t$ (b) $\partial \rho_v / \partial t = 0$ (c) none of the above

6. The mobile charge density in a metal is
   (a) directly proportional to the number of valence electrons
   (b) inversely proportional to the number of valence electrons
   (c) there is no connection with the number of valence electrons
   (d) none of the above

7. The continuity equation says that
   (a) matter is neither created nor destroyed
   (b) the current density is created through the presence of an electric field
   (c) charge is neither created nor destroyed
   (d) the total current leaving a closed surface is equal to the rate of depletion of charge in that surface

8. Which statements are true?
   (a) The continuity equation can be obtained from Maxwell's equations
   (b) One of Maxwell's equations can be obtained from the continuity equation
   (c) The continuity equation has nothing to do with Maxwell's equations
   (d) None of the above

9. In a semiconductor,
   (a) holes are present in the conduction band which contribute to the current
   (b) electrons are present in the conduction band which contribute to the current
   (c) electrons and holes are present in the conduction band which contribute to the current
   (d) holes are present in the valence band which contribute to the current
10. σ/ε has the units of
    (a) charge   (b) length   (c) mass   (d) time
11. $(\varepsilon/2)\,\mathbf{E}\cdot\mathbf{E}$
    (a) has the units of $\mathrm{J\,m}^{-3}$
    (b) is the energy density of the electric field
    (c) is a nonsensical quantity
    (d) has meaning but is nothing to do with energy density

Short-Answer Questions

1. Derive the expression to calculate the the energy stored in the electrostatic field of a section of coaxial cable of length L.   (DTU-NSIT, 2010)
2. A coaxial line of length L with inner radius a and outer radius b (Fig. 6.16) is filled with a material with dielectric constant ε_r and conductivity σ. If a battery of V V with source resistance R_s Ω is connected between the inner and outer conductors, find the current as a function of time (Fig. 6.25).
3. State and explain the boundary conditions for electrostatics.
        (Mumbai University, December 2008)
4. Explain polarisation.   (Mumbai University, May 2009)
5. In a circular wire of radius a if the current density across the wire is

$$\mathbf{J} = J_{z0}\frac{1 - e^{\rho/a}}{1 - e}\mathbf{a}_z$$

calculate the resistance of length l.
6. Find the capacitance of a capacitor with three layers of dielectrics (Fig. 6.26).
7. Find the capacitance of two concentric spheres (Fig. 6.27) with the dielectric between the spheres being a function of r. The radii of the inner and outer spheres is r_0 and r_1, respectively. The value of the dielectric is $\varepsilon_0\,\varepsilon_{r1}$ at the surface of the outer sphere and is ε_0 at the surface of the inner sphere and varies linearly with r.

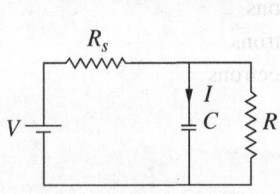

Fig. 6.25

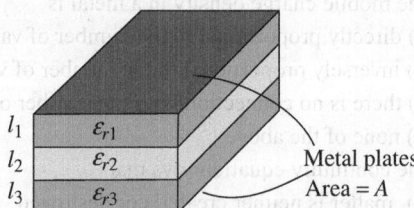

Fig. 6.26 Three layer dielectric construction in series

Review Questions

1. Explain the concept of flux of a vector field out of a surface and link it with current density $\mathbf{J}$ and the current I.
2. Explain why the current density $\mathbf{J}$ is directly proportional to the velocity of the charge carriers ($\mathbf{v}$) and also the mobile charge density, ρ_m.

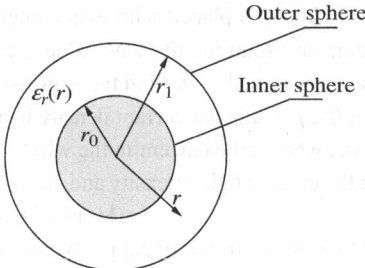

Fig. 6.27 Concentric spheres with dielectric filling with $\varepsilon = \varepsilon_0\, \varepsilon_r\,(r)$

3. Explain the concept of charge neutrality in a metal.
4. Write a short note on the continuity equation and explain how does it predict the conservation of charge.
5. Write a short note on conductors, dielectrics, and semiconductors.
6. Why is it that in a conductor, electrons in the presence of an electric field, are only described as moving with constant drift velocity? (Ordinarily electric fields accelerate electrons.) How is this related to heat dissipated in a conductor?
7. Why is it that metals immersed in an electrostatic electric field have no fields in their interior?
8. If a current is present above an infinite ground plane, explain how this combination may be replaced by the original current and an image current.
9. Explain the concept of polarisation density **P** and explain how it is related to the electric field.
10. Define the terms (a) polarisation, (b) electric susceptibility. (NTU-NSIT, March 2008)
11. How is energy stored in a capacitor?
12. Write a short note on the boundary conditions of electrostatic fields for dielectrics and metals.

Numerical Problems

1. Given the current density $\mathbf{J} = e^{-2y}\mathbf{a}_x + e^{-2y}\mathbf{a}_y$ A/m², (a) find the total current crossing the plane $y = 0.5$ in the $\mathbf{a}_y$ direction in the region $-1 < x < 1$, $-2 < z < 2$, (b) find the total current crossing the plane $x = 0.5$ in the $\mathbf{a}_x$ direction in the region $-1 < y < 1$, $-2 < z < 2$, (c) find the total current crossing the plane $z = 0.5$ in the $\mathbf{a}_z$ direction in the region $-1 < y < 1$, $-2 < x < 2$.
2. Given the current density $\mathbf{J} = e^{-2y}\mathbf{a}_x + e^{-2y}\mathbf{a}_y$ A/m². Find the total current leaving the region $0 < x$, $y < 0.5$, $2.5 < z < 3$ by (a) integrating $\mathbf{J} \cdot d\mathbf{S}$ over the surface of the cube; (b) employing the divergence theorem.
3. Let the current density be $\mathbf{J} = 2\rho\, \mathbf{a}_\rho - \sin\phi\, \mathbf{a}_\phi$ A/m² within a region of space. Find the total current I crossing the surface (a) $\rho = 2$, $0 < \phi < 2\pi$, $0 < z < 5$, in the $\mathbf{a}_\rho$ direction. (b) Evaluate $\nabla \cdot \mathbf{J}$ at $P(\rho = 2.4, \phi = 0.08, z = 6.05)$.
4. In spherical coordinates, $\mathbf{J} = 20\sin\theta\, \cos\phi\, \mathbf{a}_r + \dfrac{1}{r}\mathbf{a}_\phi$ A/m²
   (a) Find the total current flowing through the surface $r = 1$, (b) find the total current flowing through the surface $r = 1$, $0 < \theta < \pi/2$, $0 < \phi < \pi/2$.
5. If a beam of electrons occupying a region of space with density 5×10^{20} (No/cc) are moving with uniform velocity $\mathbf{v} = 2 \times 10^6 \mathbf{a}_x$ (m/s) (a) what is the charge density ρ_v? (b) the value of **J**? If the beam is passing through a circular pipe of 1 cm radius, what is the current?

6. A circular metal wire of radius a is placed with axis coinciding with the z-axis. The current density is maximum on circumference with value $J_0 \mathbf{a}_z$ and decays exponentially along the radius at a rate proportional to α. Find the expression for the current density.

7. If $\mathbf{J} = 20e^{1000(\rho-a)} \mathbf{a}_z$ for $0 < \rho < a$ is the current density in a circular wire with radius a oriented along the z-axis, what is the current in the wire?

8. Derive an expression for the electric field intensity and electric potential due to a dipole.

(Mumbai University, December 2008)

9. Find the resistance per meter of ten strands of copper twisted together with the diameter of each strand equal to 0.05 mm.

10. Find the resistance of 1 m of wire with the current density given in Problem 7.

11. A current $I = I_0 e^{-\alpha t}$ $t > 0$, charges one plate of a capacitor of value C. Find the charge on the capacitor plate, and the voltage across the plates as a function of time.

12. A plane glass is coated with a resistive material of $\sigma = 10^5$ S/m. The thickness of the coating is $10\,\mu\text{m}$ and the width is 1 mm. Find the resistance of 1 mm of this conductor.

13. The potential in a region of space is $V = \ln(\sqrt{x^2 + y^2})$. The point $(1, 0)$ is on a conducting surface. Find the $\mathbf{E}$ field and ρ_s at that point.

14. The potential in a region of space $(\varepsilon_r = 1)$ is $V = x^5 + xy^4 - 10x^3y^2$. The point $(0, 0)$ is on a conducting surface, $z = 0$. Find the $\mathbf{E}$ field and ρ_s at that point and the equation of the surface in the neighbourhood of $(0, 0)$.

15. Two point charges each with charge 1 pC are embedded in a dielectric of dielectric constant ε_r and placed 1 m apart. Find the force of repulsion when ε_r is equal to (a) 1, (b) 2, and (c) 10.

16. Calculate the average shift in the electron cloud of a single molecule when oxygen gas O_2 (at STP) is placed in an electric field of 1 V/m. ε_r for oxygen is 1.0005 at 0 °C. [**Hint:** 16 g of Oxygen at STP occupies 22.4 l.16 g of Oxygen also contains 1 Avagadro number of molecules. Calculate the number of molecules per cubic meter.]

17. A sphere of radius 10 cm is charged with 1 μC of charge. Another sphere with 5 cm radius without any charge on it is connected by a wire to the first sphere. Find the charges on both spheres.

18. A slab of dielectric of large dimensions (width and height very much greater than the thickness, $t = 1$ cm) is placed at an angle of 45° with respect an electric field of 1 V/m, as shown in Fig. 6.28. Find the electric field in the dielectric $(\varepsilon_r = 10)$ and on the other side.

19. A liquid of unknown conductivity is placed in a large trough of glass as shown in Fig. 6.29. Two rods are immersed in the liquid. The dimensions of the rods are: the diameter of

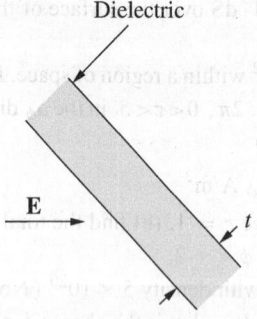

Dielectric

E

t

Fig. 6.28

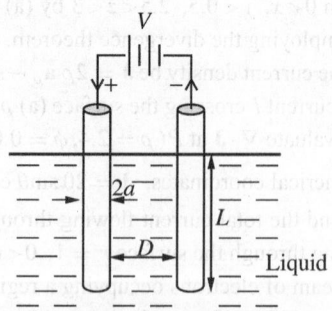

V

$2a$

L

D

Liquid

Fig. 6.29 Measurement of σ for a liquid conductor

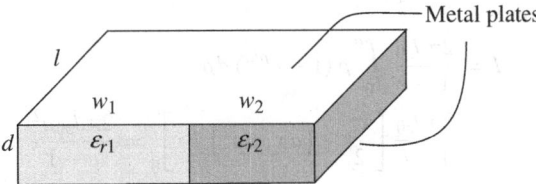

Fig. 6.30 A capacitor with two dielectrics in parallel

each rod, 1 cm; the distance between the rods, 15 cm, and the depth of immersion is 15 cm. If the voltage applied is 10 V and the current is found to be 100 mA, find the conductivity of the liquid.

20. For Problem 14, find the energy stored in the electric field in the region $0<x<1$, $0<y<1$ and $0<z<1$. The region is air.

21. For a capacitor with two dielectrics calculate the capacitance with respect to the parameters given in Fig. 6.30.

Answers

Objective Type Questions

1. (c)   2. (a)   3. (b)   4. (c)   5. (b)   6. (a)   7. (c) and (d)   8. (a)
9. (b) and (d)   10. (d)   11. (a) and (b)

Short-Answer Questions

2. The capacitance of the coaxial line is given by (refer Eqn (6.87))

$$C = \frac{2\pi \varepsilon_r \varepsilon_0 L}{\log_e(b/a)}$$

on the other hand the resistance of the line is (Eqn (6.96))

$$R = \frac{\varepsilon}{\sigma C} = \frac{\varepsilon}{\sigma}\left[\frac{2\pi \varepsilon L}{\log_e(b/a)}\right]^{-1}$$

This resistance and capacitance are in parallel. The circuit is as shown in Fig. 6.25. When the battery is connected the capacitor acts like a short circuit and the initial current is $I(0) = V/R_s$. The final current is zero, and the time constant is $R_{||}C$ where

$$R_{||} = R||R_s = \frac{R_s R}{R_s + R}$$

so

$$I(t) = I_C(t) = \frac{V}{R_s}e^{-t/CR_{||}}$$

now the current through the resistance is zero at $t = 0$ and $V/(R_s + R)$ at $t = \infty$ so the current as a function of time is

$$I_R = \frac{V}{R_s + R}(1 - e^{-t/R_{||}})$$

the total current is therefore

$$I_T(t) = \frac{V}{R_s}e^{-t/CR_{||}} + \frac{V}{R_s + R}(1 - e^{-t/R_{||}})$$

5. In a tubular section of wire between ρ and $\rho + d\rho$ the current that flows is

$$dI = 2\pi \rho d\rho J_{z0} \frac{1 - e^{\rho/a}}{1 - e}$$

so the total current is

$$I = \frac{2\pi J_{z0}}{1-e} \int_0^a \rho \left(1 - e^{\rho/a}\right) d\rho$$

$$= \frac{2\pi J_{z0}}{1-e} \left[\frac{x^2}{2} - \left(ax - a^2\right) e^{\frac{x}{a}} \right]_0^a = \frac{2\pi J_{z0} a^2}{e-1}$$

The heat dissipated in the wire is ($\mathbf{E} = \mathbf{J}/\sigma$)

$$\iiint \mathbf{E} \cdot \mathbf{J}\, dV$$

$$= \frac{2\pi J_{z0}^2 l}{\sigma (1-e)^2} \int_0^a \rho \left(1 - e^{\rho/a}\right)^2 d\rho$$

$$= \frac{2\pi J_{z0}^2 l}{\sigma (1-e)^2} \left[\frac{(2ax - a^2)\, e^{\frac{2x}{a}}}{4} - 2\left(ax - a^2\right) e^{\frac{x}{a}} + \frac{x^2}{2} \right]_0^a$$

$$= \frac{2\pi J_{z0}^2 l}{\sigma (1-e)^2} (0.5969 a^2)$$

if we divide the total power dissipated by the square of the current,

$$R = \frac{\left\{ (2\pi J_{z0}^2 l) / [\sigma (1-e)^2] \right\} (0.5969 a^2)}{\left(2\pi J_{z0} a^2/(e-1)\right)^2} = \frac{0.5969 l}{2\pi \sigma\, a^2}$$

6. Each capacitor has a capacitance

$$C_i = \frac{\varepsilon_i l_i}{\sigma A}$$

and the capacitance C is

$$\frac{1}{C} = \sum_i \frac{1}{C_i}$$

7. The dielectric can be modelled as

$$\varepsilon_r (r) = ar + b$$

$$a = \frac{\varepsilon_0 - \varepsilon_0\, \varepsilon_{r1}}{r_0 - r_1}$$

$$b = \frac{\varepsilon_0\, \varepsilon_{r1}\, r_0 - \varepsilon_0\, r_1}{r_0 - r_1}$$

Note that in these expressions, if ε_{r1} is set equal to 1 then $a = 0$ and $b = \varepsilon_0$. Now if the total charge on the outer sphere is Q and inner sphere is $-Q$, so the surface charge densities on the outer and inner spheres is

$$\frac{Q}{4\pi r_1^2} \quad \text{and} \quad \frac{-Q}{4\pi r_0^2}$$

The $\mathbf{D}$ field between the two spheres is

$$\mathbf{D} = \frac{Q}{4\pi r^2} (-\mathbf{a}_r)$$

and the electric field is

$$\mathbf{E} = \frac{Q}{4\pi \varepsilon (r)\, r^2}$$

The energy density is

$$\frac{1}{2} \iiint \mathbf{E} \cdot \mathbf{D}\, dV$$

where the integration is to be performed over the volume between the spheres.

$$E = \frac{1}{2} \iiint \left(\frac{Q}{4\pi \varepsilon (r) r^2} \right) \left(\frac{Q}{4\pi r^2} \right) r \sin\theta \, d\phi \, r\theta \, dr$$

$$= \frac{4\pi Q^2}{2(4\pi)^2} \int_{r_0}^{r_1} \frac{r^2 dr}{\varepsilon (r) r^4} \qquad \text{(integrate by partial fractions)}$$

$$= \frac{Q^2}{8\pi} \left[\frac{a \ln (ar + b)}{b^2} - \frac{a \ln(r)}{b^2} - \frac{1}{br} \right]_{r_0}^{r_1} \qquad \text{(check by differentiation)}$$

$$E = \frac{Q^2}{8\pi} \left[\frac{a}{b^2} \ln \frac{r_0(ar_1 + b)}{r_1(ar_0 + b)} - \frac{1}{b} \left(\frac{1}{r_1} - \frac{1}{r_0} \right) \right]$$

setting the energy stored on a capacitor to $Q^2/2C$

$$\frac{Q^2}{2C} = \frac{Q^2}{8\pi} \left[\frac{a}{b^2} \ln \frac{r_0(ar_1 + b)}{r_1(ar_0 + b)} - \frac{1}{b} \left(\frac{1}{r_1} - \frac{1}{r_0} \right) \right]$$

$$C = \frac{4\pi}{\left[\frac{a}{b^2} \ln \frac{r_0(ar_1+b)}{r_1(ar_0+b)} - \frac{1}{b} \left(\frac{1}{r_1} - \frac{1}{r_0} \right) \right]}$$

in these expressions if $a = 0$ and $b = \varepsilon_0$, then the capacitance reduces to that two concentric spheres.

Numerical Problems

1. (a) 2.943 (b) 14.5074 (c) 0
2. -0.15803 A for both
3. (a) 0 (b) $\left[\dfrac{4\rho - \cos(\varphi)}{\rho} \right]_P = 3.585$
4. (a) 0 (b) 5π
5. (a) 80 C/cc (b) 1.6×10^{14} A/m^2 (c) 1.256×10^{10} A
6. $J_0 e^{-\alpha(a-\rho)}$
7. (a) Current in the wire $\approx 40\pi \, a \times 10^{-3}$ A (b) $a - 1/1000 \lesssim \rho < a$ for $1000a \gg 1$
9. 0.88Ω/m
10. Approximately $R \approx \dfrac{125000(2000 \, a - 1)}{\pi \, (1000a - 1)^2}$
11. $V = \dfrac{I_0}{\alpha C} \left(1 - e^{-\alpha t} \right), \, q = \dfrac{I_0}{\alpha} \left(1 - e^{-\alpha t} \right)$
12. 10Ω
13. $\mathbf{E} = [-2, 0, 0]$, $\rho_s = 2\varepsilon_0$ C/m^2
14. $\mathbf{E} = -[-y^4 + 30x^2y^2 - 5x^4, \, 20x^3y - 4xy^3, \, 0]$; $\rho_s = 0$; Equation: $x^5 + xy^4 - 10x^3y^2 = 0$
15. (a) $\dfrac{2.49 \times 10^{-25}}{\pi \, \varepsilon_0}$-N (b) $\dfrac{2.49 \times 10^{-25}}{2\pi \, \varepsilon_0}$ N (c) $\dfrac{2.49 \times 10^{-25}}{10\pi \, \varepsilon_0}$-N
16. 2.29×10^{-22} m
17. Charge on the bigger sphere is 2/3 μC
18. If $\mathbf{a}_x$ is the direction of the electric field in region 1, then in the dielectric, $\mathbf{D}_2 = \dfrac{\varepsilon_0}{2} \times$ ($\mathbf{a}_x + \mathbf{a}_y$) $+ 5\varepsilon_0$ ($\mathbf{a}_x - \mathbf{a}_y$); on the other side the electric field is the same as in region 1
19. $\sigma = 0.0016$
20. $(160/9) \, \varepsilon \, 0$ W
21. $C = \dfrac{lw_1}{d} \varepsilon_0 \varepsilon_{r1} + \dfrac{lw_2}{d} \varepsilon_0 \varepsilon \, r_2$

7

Laplace's and Poisson's Equations

The origin of thinking is some perplexity, confusion or doubt.
— John Dewey

CHAPTER OBJECTIVES

To enable the students to understand the following:

- Derivation of Laplace's and Poisson's equations from Maxwell's equations
- Uniqueness theorem
- One-dimensional solutions to Laplace's and Poisson's equations
- Two-dimensional solutions to Laplace's and Poisson's equations
- Separation of variables technique
- Solution of Laplace's equation by Numerical Techniques

7.1 | Introduction

In many situations, in electrostatic problems one may know only the boundary conditions of some engineering problems. For example, one may know the voltage on some boundary as in the form of a voltage on a set of plates and is required to find the potential field in the region enclosed. Or one may be given the charge distribution in a region of space and it is required to obtain the electric field in that region.

In such situations one may have to set up a partial differential equation to extract the solution. The partial differential equations and their solution are well known in mathematics as boundary value problems:

To set up such a partial differential equation, we start with Gauss's law

$$\nabla \cdot \mathbf{D} = \rho_v$$

Using $\mathbf{D} = \varepsilon \mathbf{E}$, ε being the permittivity of the (homogeneous) medium, we can write the equation in terms of the electric field,

$$\nabla \cdot \mathbf{E} = \frac{\rho_v}{\varepsilon} \tag{7.1}$$

To get a equation involving the potential V we use the gradient relationship between the electric field and the potential, $\mathbf{E} = -\nabla V$

$$\nabla \cdot \nabla V = \nabla^2 V = -\frac{\rho_v}{\varepsilon} \tag{7.2}$$

this is Poisson's equation. In a charge-free region of space this becomes Laplace's equation

$$\nabla^2 V = 0 \tag{7.3}$$

The operator ∇^2 is called the Laplacian.

Laplace's Equation

The solutions of Laplace's equation are called harmonic functions. Apart from electromagnetism, Laplace's equation is important in the areas of astronomy, fluid dynamics, and steady state heat conduction. Surprisingly, Laplace's equation appears in a modified form (the Young Laplace equation) in the theory of respiration.

7.2 | Uniqueness Theorem

A very important point in solving partial differential equations is that once we have found one solution of the equation for given boundary conditions then should we continue to look for other solutions?

Let us take Laplace's equation. Suppose two solutions namely, $V_1(\mathbf{r})$ and $V_2(\mathbf{r})$ exist to Laplace's equation in the region $\mathcal{V}$. At the boundary, $\mathcal{S}$,

$$V_{1b} = V_{2b} = V(\mathbf{r})\,|_{\mathbf{r}=\mathbf{r}_b} \tag{7.4}$$

where $\mathbf{r}_b$ is the position vector of the boundary of the surface $\mathcal{S}$. Then let a function ψ be defined as the difference of these two solutions:

$$\psi(\mathbf{r}) = V_1 - V_2 \tag{7.5}$$

$$\psi(\mathbf{r} = \mathbf{r}_b) = V_{1b} - V_{2b} = 0 \tag{7.6}$$

Now consider the identity

$$\nabla \cdot (\psi \nabla \psi) = \psi \nabla^2 \psi + (\nabla \psi)^2 \tag{7.7}$$

If we take a look at $\nabla^2 \psi$

$$\nabla^2 \psi = 0 \qquad (\text{since } \underbrace{\nabla^2 V_1}_{=0} - \underbrace{\nabla^2 V_2}_{=0} = 0) \tag{7.8}$$

as both terms on the right are solutions to Laplace's equation. Hence Eqn (7.7) becomes

$$\nabla \cdot (\psi \nabla \psi) = (\nabla \psi)^2 \tag{7.9}$$

Integrating this equation over the volume $\mathcal{V}$ (note we are using v for volume)

$$\iiint_{\mathcal{V}} \nabla \cdot (\psi \nabla \psi)\, dv = \iiint_{\mathcal{V}} (\nabla \psi)^2\, dv \tag{7.10}$$

And also integrating the left of Eqn (7.9) over the surface $\mathcal{S}$ enclosing $\mathcal{V}$ and applying Divergence theorem

$$\iiint_{\mathcal{V}} \nabla \cdot (\psi \nabla \psi)\, dv = \oiint \psi \nabla \psi \cdot d\mathbf{S} \tag{7.11}$$

So

$$\oiint_{\mathcal{S}} \psi \nabla \psi \cdot d\mathbf{S} = \iiint_{\mathcal{V}} (\nabla \psi)^2\, dv \tag{7.12}$$

Now we know that ψ on S is zero: $\psi|_S = 0$. Therefore

$$\oiint_S \psi \nabla \psi \cdot d\mathbf{S} = 0$$

and so

$$\iiint_V (\nabla \psi)^2 \, dv = 0$$

but $(\nabla \psi)^2$ can never be negative. So

$$\nabla \psi = 0 \tag{7.13}$$
$$\nabla (V_1 - V_2) = 0 \tag{7.14}$$
$$V_1 - V_2 = \text{constant} = k \tag{7.15}$$

but at the boundary

$$V_{1b} = V_{2b} \tag{7.16}$$

So the constant k is zero. Therefore

$$V_1(\mathbf{r}) = V_2(\mathbf{r}) \qquad \text{everywhere} \tag{7.17}$$

From this it is clear that once a solution of Laplace's equation is obtained, it is the only solution. The same reasoning applies to Poisson's equation. The only place where there is a departure is in Eqn (7.8) which should be written as

$$\nabla^2 \psi = 0 \qquad (\text{since } \underbrace{\nabla^2 V_1}_{=-\rho_v/\varepsilon} - \underbrace{\nabla^2 V_2}_{=-\rho_v/\varepsilon} = 0)$$

Example 7.1 If the potential function in a region of space is

$$V = e^{-x} \cos y$$

find whether this is a solution to Laplace's equation.

Solution

Step 1. We find the vector ∇V

$$\nabla V = [-e^{-x} \cos(y), -e^{x} \sin(y), 0]$$

Step 2. Taking the divergence of this vector

$$\nabla \cdot \nabla V = \frac{\partial \left[-e^{-x} \cos(y)\right]}{\partial x} + \frac{\partial \left[-e^{-x} \sin(y)\right]}{\partial y}$$
$$= e^{-x} \cos(y) - e^{-x} \cos(y)$$
$$= 0$$

Hence the potential function does satisfy Laplace's equation.

Example 7.2 Check whether the potential function $V = xz$ is a solution of Laplace's equation.

Solution

Step 1. We find the vector ∇V:

$$\nabla V = [z, 0, x]$$

Step 2. Taking the divergence of this vector

$$\nabla \cdot \nabla V = \frac{\partial z}{\partial x} + \frac{\partial x}{\partial z} = 0$$

Hence the potential function does satisfy Laplace's equation.

Example 7.3   Check whether the potential function $V = A \ln \rho + B$ in cylindrical coordinates is a solution of Laplace's equation. A and B are constants.

Solution

$$V = A \ln \rho + B$$

$$\nabla^2 V = \frac{1}{\rho} \frac{\partial}{\partial \rho} \left(\rho \frac{\partial V}{\partial \rho} \right) + \frac{1}{\rho^2} \frac{\partial^2 V}{\partial \phi^2} + \frac{\partial^2 V}{\partial z^2} \quad (\text{in cylindrical coordinates})$$

$$= \frac{1}{\rho} \frac{\partial}{\partial \rho} \left[\rho \frac{\partial}{\partial \rho} (A \ln \rho + B) \right]$$

$$= \frac{1}{\rho} \frac{\partial}{\partial \rho} (A)$$

$$= 0$$

Hence the function $V = A \ln \rho + B$ satisfies Laplace's equation.

7.3 | Laplace's Equation—Applications

7.3.1   One-dimensional Solutions

Out of the two Eqns (7.2) and (7.3) the simpler of the two is Laplace's equation, because $\rho \to \rho_v$, the charge density is absent. In Cartesian coordinates, Laplace's equation becomes

$$\nabla^2 V = \frac{\partial^2 V}{\partial x^2} + \frac{\partial^2 V}{\partial y^2} + \frac{\partial^2 V}{\partial z^2} = 0 \qquad (7.18)$$

7.3.1.1   Laplace's Equation Applied to Infinite Parallel Planes

Example 7.4   Find a solution of Laplace's equation for two infinite parallel planes separated by a distance d with the lower plane at $V = 0$ and the upper plane at $V = V_0$.

Solution   Let us consider the simplest of all cases: let V be a function of only one dimension, namely, x. That is $V \equiv V(x)$. This implies that there is no variation in either the y or z directions. Laplace's equation then becomes

$$\frac{\partial^2 V}{\partial x^2} = 0$$

performing two integrations

$$\int \frac{d^2 V}{dx^2} \, dx = \frac{dV}{dx} = k_1 \qquad (7.19)$$

and

$$\int \frac{dV}{dx} \, dx = V(x) = k_1 x + k_2 \qquad (7.20)$$

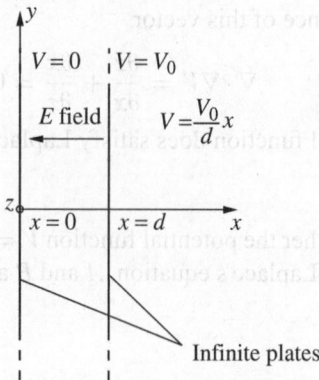

Fig. 7.1 Laplace's equation applied to two infinite plates

where k_1 and k_2 are constants which will be determined by the boundary conditions. Looking at the solution carefully we realise that this solution would apply to two infinite plates with a potential difference of V_0 as in the case of the parallel plate capacitor as shown in Fig. 7.1. To check it let us compute the **E**-field

$$\mathbf{E} = -\nabla V = -k_1 \qquad \text{a constant}$$

The electric field would be a constant only in the case where there would be two parallel plates! So we apply the solution just obtained above to Fig. 7.1. We set $V = 0$ at $x = 0$ and $V = V_0$ at $x = d$. From the first condition

$$k_1 x + k_2|_{x=0} = 0$$

$$k_2 = 0$$

$$k_1 x|_{x=d} = V_0$$

$$k_1 = \frac{V_0}{d}$$

so the total solution is

$$V = \frac{V_0}{d}x \tag{7.21}$$

$$\mathbf{E} = -\nabla V = -\frac{V_0}{d} \tag{7.22}$$

we can see that we have obtained the solution to Laplace's equation for one dimension in the Cartesian coordinate system.

To check the answer, we put $x = 0$ and $x = d$ in the first of the above equations and we can see that we have obtained the correct solution to Laplace's equation for one dimension in the Cartesian coordinate system.

Example 7.5 Use the solution to Laplace's equation to find the capacitance of a parallel plate capacitor.

Solution

Step 1. The parallel plate capacitor consists of two plates, each of area A placed in parallel, with a separation of d and filled with a dielectric ε.

Step 2. Suppose the upper and lower plates have charges of Q and $-Q$, respectively, with voltages of V_0 of the upper plate and 0 of the lower plate.

Step 3. Since we have two plates, placing x coordinate on the lower plate, the equation of the voltage is

$$V = \frac{V_0}{d} x$$

which satisfies the boundary condition, $V = 0$ at $x = 0$ and $V = V_0$ at $x = d$. The equation was obtained from the analysis conducted earlier.

Step 4. The upper and lower plates each have surface charges, $\rho_s = \pm Q/A$.

Step 5. Recall that at the metal boundary, the normal component of the electric flux density (**D**), is given by

$$D_n = \rho_s$$

and

$$D_n = \varepsilon E_n$$

so

$$\rho_s = \frac{Q}{A} = \varepsilon E_n$$

$$Q = \varepsilon A E_n$$

and ρ_s is the charge on any plate per m^2.

Step 6. The normal electric field is given by

$$E_n = -\frac{dV}{dx} = -\frac{V_0}{d}$$

Step 7. Therefore, the capacitance is

$$C = \frac{Q}{V_0} = \frac{(\varepsilon A V_0/d)}{V_0} = \frac{\varepsilon A}{d} \qquad (7.23)$$

Let us examine Laplace's equation in other coordinate systems and apply it to some specific configurations. In the cylindrical and spherical coordinate systems, Laplace's equation is

$$\nabla^2 V = \frac{1}{\rho}\frac{\partial}{\partial \rho}\left(\rho\frac{\partial V}{\partial \rho}\right) + \frac{1}{\rho^2}\frac{\partial^2 V}{\partial \phi^2} + \frac{\partial^2 V}{\partial z^2} = 0 \qquad \text{(Cylindrical)} \qquad (7.24)$$

$$\nabla^2 V = \frac{1}{r^2}\frac{\partial}{\partial r}\left(r^2\frac{\partial V}{\partial r}\right) + \frac{1}{r^2 \sin\theta}\frac{\partial}{\partial \theta}\left(\sin\theta\frac{\partial V}{\partial \theta}\right)$$

$$+ \frac{1}{r^2 \sin^2\theta}\frac{\partial^2 V}{\partial \phi^2} = 0 \qquad \text{(Spherical)} \qquad (7.25)$$

7.3.1.2 Laplace's Equation Applied to Concentric Cylinders

Example 7.6 Find the general solution to Laplace's equation for concentric cylinders.

Solution

Step 1. Using Laplace's equation in cylindrical coordinates,

$$\nabla^2 V = \frac{1}{\rho}\frac{\partial}{\partial \rho}\left(\rho\frac{\partial V}{\partial \rho}\right) + \frac{1}{\rho^2}\frac{\partial^2 V}{\partial \phi^2} + \frac{\partial^2 V}{\partial z^2} = 0$$

Step 2. We want the solution V such that $V = 0$ on the inner conductor and $V = V_0$ on the outer conductor of two coaxial cylinders with a and b as the inner and outer radii

Step 3. Let $V \equiv V(\rho)$ and $\partial V/\partial \phi = \partial V/\partial z = 0$. We want a solution where there is no dependence on the ϕ and z coordinates.

$$\nabla^2 V = \frac{1}{\rho} \frac{d}{d\rho} \left(\rho \frac{dV}{d\rho} \right) = 0 \qquad (7.26)$$

Step 4. Integrating this twice

$$\int \frac{d}{d\rho} \left(\rho \frac{dV}{d\rho} \right) d\rho = \rho \frac{dV}{d\rho} = k_1$$

$$\frac{dV}{d\rho} = \frac{k_1}{\rho}$$

$$\int \frac{dV}{d\rho} d\rho = V = k_1 \ln \rho + k_2$$

Step 5. Using the boundary condition, $V = 0$ on the inner conductor ($\rho = a$) and $V = V_0$ on the outer conductor ($\rho = b$),

$$V|_{\rho=b} = k_1 \ln \rho + k_2|_{\rho=b} = V_0$$

$$k_1 \ln b + k_2 = V_0 \qquad (1)$$

$$V|_{\rho=a} = k_1 \ln \rho + k_2|_{\rho=a} = 0$$

$$k_1 \ln a + k_2 = 0 \qquad (2)$$

Subtracting Eqn (2) from Eqn (1) gives

$$k_1 \ln \frac{b}{a} = V_0$$

$$k_1 = \frac{V_0}{\ln(b/a)}$$

$$k_2 = -k_1 \ln b = \frac{-V_0 \ln b}{\ln(b/a)}$$

Step 6. Therefore the solution is

$$V = \frac{V_0}{\ln(b/a)} \ln \rho - \frac{V_0}{\ln(b/a)} \ln a$$

$$= \frac{V_0}{\ln(b/a)} \ln \frac{\rho}{a}$$

Step 7. By the uniqueness theorem, this solution is the only one and correct one.

Step 8. The electric field is

$$E_\rho = -\frac{dV}{d\rho} = -\frac{V_0}{\ln(b/a) \rho}$$

The reader can use the above solution to obtain an expression for the capacitance per meter of the coaxial line.[1]

 See VOfCoaxLine.m and CapOfCoaxLine.m in Chapter 7

Example 7.7 Two long coaxial cylinders of radii 2 and 6 cm are maintained at 60 V and 20 V, respectively. Calculate the voltage at $r = 4$ cm.

(DTU-NSIT, March 2008)

[1] The reader is encouraged to apply his common sense to check the correctness of this solution.

Solution

Step 1. The potential V as a function of coordinates is

$$V = A \ln \rho + B$$

where A and B are constants.

Step 2. Using the boundary conditions,

$$60 = A \ln(2) + B$$
$$20 = A \ln(6) + B$$

Step 3. Solving these equations, we get

$$A = -\frac{40}{\ln(2/6)} = -36.4, \qquad B = \frac{20 \ln(2) - 60 \ln(6)}{\ln(2/6)} = 85.2$$

Step 4. At $\rho = 4$ cm

$$V = A \ln 4 + B = 34.73 \text{ V}$$

Practice Problem 7.1

Show that the capacitance per meter of the coaxial line with inner radius equal to a and outer radius equal to b is given by

$$C = \frac{2\pi \varepsilon}{\ln(b/a)} \text{ (F/m)}$$

7.3.1.3 Laplace's Equation Applied to Concentric Spheres

Example 7.8   Obtain one-dimensional solution to the Laplace's equation in r for spherical coordinates, apply it two concentric metal spheres of radii a and b, $a < b$. Let the inner sphere be at a potential V_0 and the outer one at a potential $V = 0$. Also devise a method to calculate the capacitance of this combination.

Solution

Step 1. In spherical coordinates,

$$\nabla^2 V = \frac{1}{r^2} \frac{\partial}{\partial r} \left(r^2 \frac{\partial V}{\partial r} \right) + \frac{1}{r^2 \sin \theta} \frac{\partial}{\partial \theta} \left(\sin \theta \frac{\partial V}{\partial \theta} \right) + \frac{1}{r^2 \sin^2 \theta} \frac{\partial^2 V}{\partial \phi^2}$$

Step 2. Since the configuration is symmetric, V is a function of r alone.
$$V \equiv V(r)$$

Step 3. Laplace's equation becomes

$$\frac{1}{r^2} \frac{\partial}{\partial r} \left(r^2 \frac{\partial V}{\partial r} \right) = 0$$

$$\frac{\partial}{\partial r} \left(r^2 \frac{\partial V}{\partial r} \right) = 0$$

Integrating
$$\left(r^2 \frac{\partial V}{\partial r} \right) = k_1$$

where k_1 is a constant.

Step 4.

$$\frac{\partial V}{\partial r} = \frac{k_1}{r^2}$$

Integrating
$$V = -\frac{k_1}{r} + k_2$$

where k_2 is another constant.

Step 5. Solving for k_1, k_2 after setting the boundary condition at $r = a$ and $r = b$

$$-\frac{k_1}{a} + k_2 = V_0$$

$$-\frac{k_1}{b} + k_2 = 0$$

so
$$k_1 = \frac{ab V_0}{a - b}; \qquad k_2 = \frac{a V_0}{a - b}$$

Step 6. Hence the potential function is

$$V(r) = \frac{1}{b - a}\left[\frac{abV_0}{r} - aV_0\right]$$

Step 7. Check the boundary condition and also whether the function satisfies Laplace's equation. (Left for the student as an exercise)

Step 8. The electric field is

$$\mathbf{E} = -\nabla \cdot V$$

$$= -\frac{\partial}{\partial r}\left\{\frac{1}{b - a}\left[\frac{abV_0}{r} - aV_0\right]\right\}$$

$$= -\frac{1}{b - a}\left(-\frac{abV_0}{r^2}\right)$$

$$= \frac{abV_0}{(b - a)\,r^2}\mathbf{a}_r$$

Step 9. At $r = a$, the electric field and electric flux density are

$$\mathbf{E}(r = a) = \frac{bV_0}{(b - a)\,a}\mathbf{a}_r, \qquad \mathbf{D}(r = a) = \frac{\varepsilon bV_0}{(b - a)\,a}\mathbf{a}_r$$

Step 10. The surface charge density is

$$D_n = \rho_s = \frac{\varepsilon bV_0}{(b - a)\,a}$$

Step 11. And therefore, the total charge

$$Q = 4\pi a^2 \rho_s = \frac{4\pi a^2 \varepsilon bV_0}{(b - a)\,a}$$

Step 12. Hence the capacitance is

$$C = \frac{Q}{V_0} = \frac{4\pi \varepsilon ba}{(b - a)}$$

Step 13. Compare with Example 6.17.

 See VOfConcntricSpheres.m in Chapter 7

7.3.1.4 Laplace's Equation Applied to Two Coaxial Cones

Example 7.9 Find the solution to Laplace's equation for the configuration shown in Fig. 7.2. The outer cone is maintained at $V = 0$ and the inner one at $V = V_0$.

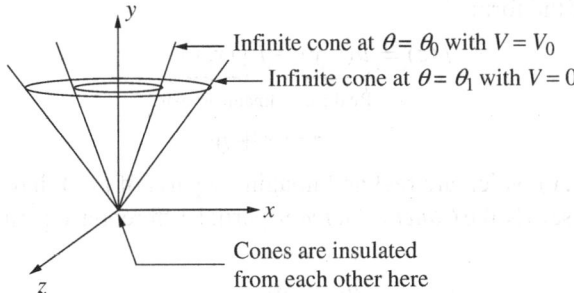

Fig. 7.2

Solution

Step 1. In spherical coordinates,

$$\nabla^2 V = \frac{1}{r^2}\frac{\partial}{\partial r}\left(r^2\frac{\partial V}{\partial r}\right) + \frac{1}{r^2 \sin\theta}\frac{\partial}{\partial\theta}\left(\sin\theta\frac{\partial V}{\partial\theta}\right) + \frac{1}{r^2 \sin^2\theta}\frac{\partial^2 V}{\partial\phi^2}$$

Step 2. Since the configuration is variable with respect to the θ coordinate, $V \equiv V(\theta)$.

Step 3. Therefore

$$\frac{1}{r^2 \sin\theta}\frac{\partial}{\partial\theta}\left(\sin\theta\frac{\partial V}{\partial\theta}\right) = 0$$

$$\frac{\partial}{\partial\theta}\left(\sin\theta\frac{\partial V}{\partial\theta}\right) = 0$$

Integrating
$$\sin\theta\frac{\partial V}{\partial\theta} = k_1$$

$$\frac{\partial V}{\partial\theta} = \frac{k_1}{\sin\theta}$$

Step 4. Integrating again

$$V = k_1 \ln[\tan(\theta/2)] + k_2$$

The integral is also equal to $-\ln(\csc\theta + \cot\theta)$

Step 5. If
$$f(\theta) = \ln[\tan(\theta/2)]$$
$$V = k_1 f(\theta) + k2$$

and enforcing the boundary conditions at $\theta = \theta_0$ ($V = V_0$) and $\theta = \theta_1$ ($V = 0$), then

$$k_1 = \frac{V_0}{f(\theta_0) - f(\theta_1)}; \qquad k_2 = -\frac{f(\theta_1)\, V_0}{f(\theta_0) - f(\theta_1)}$$

Step 6. And
$$V = \frac{V_0}{f(\theta_0) - f(\theta_1)}[f(\theta) - f(\theta_1)]$$

 See VOfConcbtricCones.m in Chapter 7

7.3.2 Two-dimensional Solutions to Laplace's Equation

7.3.2.1 Analytic Functions

One of the simplest ways of solving Laplace's equation, which is indirect in nature, is to use an analytic function of a complex variable. Analytic functions of the complex variable z are of the form

$$f(z) = \underbrace{u(x,y)}_{\text{Real part}} + j \underbrace{v(x,y)}_{\text{Imaginary part}} \qquad (7.27)$$

where

$$z = x + jy \qquad (7.28)$$

$u(x,y)$ and $v(x,y)$ which are real and imaginary parts of $f(z)$, have the property that they have to satisfy the *Cauchy-Riemann* partial differential equations (Narayan, 2001)

$$\frac{\partial u}{\partial x} = \frac{\partial v}{\partial y} \qquad (7.29)$$

$$\frac{\partial u}{\partial y} = -\frac{\partial v}{\partial x} \qquad (7.30)$$

differentiating the first of these equations with respect to x and the second one with respect to y, we get

$$\frac{\partial^2 u}{\partial x^2} = \frac{\partial^2 v}{\partial x \partial y} \qquad (7.31)$$

$$\frac{\partial^2 u}{\partial y^2} = -\frac{\partial^2 v}{\partial y \partial x} \qquad (7.32)$$

by adding the two equations given above, we get

$$\frac{\partial^2 u}{\partial x^2} + \frac{\partial^2 u}{\partial y^2} = \frac{\partial^2 v}{\partial x \partial y} - \frac{\partial^2 v}{\partial y \partial x} \qquad (7.33)$$

but at a point which is analytic

$$\frac{\partial^2 v}{\partial x \partial y} = \frac{\partial^2 v}{\partial y \partial x} \qquad (7.34)$$

we get Laplace's equation in two dimensions

$$\frac{\partial^2 u}{\partial x^2} + \frac{\partial^2 u}{\partial y^2} = 0 \qquad (7.35)$$

similarly we can show that

$$\frac{\partial^2 v}{\partial x^2} + \frac{\partial^2 v}{\partial y^2} = 0 \qquad (7.36)$$

therefore, the real and imaginary parts of every analytic function, $f(z)$, can be a potential source of a solution to an engineering problem. Let us take some concrete examples.

Example 7.10 Plot the real and imaginary plots of the function $\sin(z)$ where $z = x + jy$.

Solution Take the case of the function

$$f(z) = \sin(z) \qquad (7.37)$$

$$= \sin(x + jy)$$

$$= \sin(x)\cosh(y) + j\cos(x)\sinh(y) \qquad (7.38)$$

where we have used the two relations

$$\sin(jy) = j\sinh(y)$$

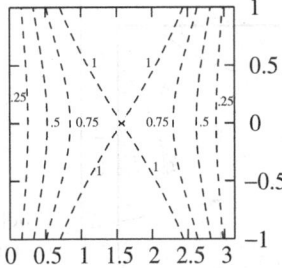

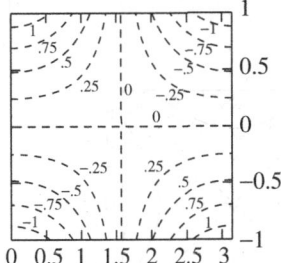

Fig. 7.3 Contour plot of the real part of $\sin z$, $z = x + jy$

Fig. 7.4 Contour plot of the imaginary part of $\sin z$, $z = x + jy$

$$\cos(jy) = \cosh(y)$$

If we plot the real and imaginary parts of this function

$$u = \sin(x)\cosh(y) \text{ (the real part)}$$
$$v = \cos(x)\sinh(y) \text{ (the imaginary part)}$$

as contour plots

$$u(x,y) = 1.0, \ 0.75, \ 0.5, \ 0.25 \text{ and } 0$$
$$v(x,y) = \pm 1.0, \ \pm 0.75, \ \pm 0.5, \ \pm 0.25 \text{ and } 0$$

for the region $0 \le x \le \pi$ and $-1 \le y \le 1$, then some very interesting observations follow.

Both Figs 7.3 and 7.4, represent a possible potential field configuration. The usefulness of these plots is that they give us a feel of what the potential looks like in complicated two-dimensional cases. For example, if we were to place a metal curved plane specially shaped and infinite in extent along the z-direction along the wedge-shaped $u = 1$ curve (the centre of which is at $x = \pi/2$, $y = 0$) and an infinite plane along the $u = 0$ at $x = \pi$ an engineer gets an idea of how the potential changes near a wedge placed at potentials of $V = 1$ and $V = 0$, respectively. From the right-hand graph, we can see how the potential changes near a right-angled corner.

Example 7.11   Plot the real and imaginary plots of the function z^2 where $z = x + jy$.

Solution   Another interesting function is $f(z) = z^2$. Both the real and imaginary parts of the function are shown plotted on the same graph (Fig. 7.5). The labels with 'V' apply to the real part of the function while the ordinary labels apply to the imaginary part of the function.

Analysing the function in more detail,

$$z^2 = (x + jy)^2$$
$$= (x^2 - y^2) + 2jxy$$
$$u = x^2 - y^2 \text{ (real part)}$$
$$v = 2xy \text{ (imaginary part)}$$

Both parts of the function are hyperbolas (Feynman & Sands, 2001). The plot of the real part of the function z^2 with the electric field superimposed is shown in Fig. 7.6.

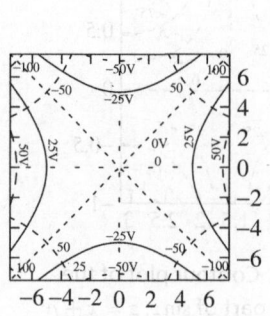

Fig. 7.5 The real and imaginary parts of the function $f(z) = z^2$

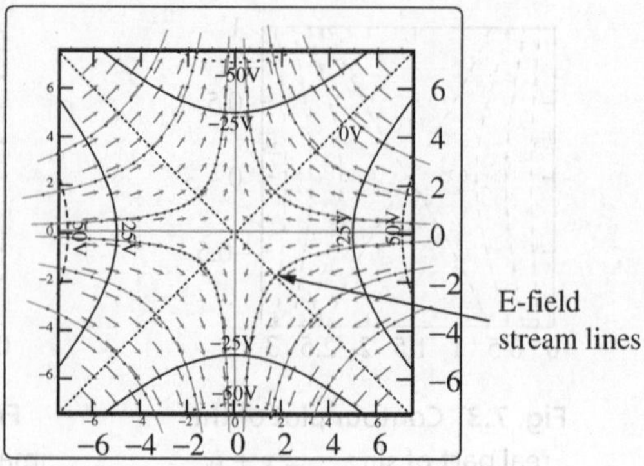

Fig. 7.6 Contour plot of $u = x^2 - y^2$ along with the electric field superimposed on the potential field

E-field stream lines

7.3.3 Separation of Variables

Another method which is a direct analytical method is the well-known separation of variables technique. Let us consider Laplace's equation in two dimensions (though the method can be applied to three-dimensional problems as well)

$$\frac{\partial^2 V}{\partial x^2} + \frac{\partial^2 V}{\partial y^2} = 0$$

Let $V = X(x)\,Y(y)$ a multiplication of two functions, X which is a function of x while Y is a function of y. Then

$$\frac{\partial^2 V}{\partial x^2} = Y\frac{d^2 X}{dx^2}$$

$$\frac{\partial^2 V}{\partial y^2} = X\frac{d^2 Y}{dy^2}$$

$$\frac{\partial^2 V}{\partial x^2} + \frac{\partial^2 V}{\partial y^2} = Y\frac{d^2 X}{dx^2} + X\frac{d^2 Y}{dy^2}$$

which gives

$$Y\frac{d^2 X}{dx^2} + X\frac{d^2 Y}{dy^2} = 0 \qquad (7.39)$$

Divide by XY, we get

$$\underbrace{\frac{1}{X}\frac{d^2 X}{dx^2}}_{\text{A function only of } x} + \underbrace{\frac{1}{Y}\frac{d^2 Y}{dy^2}}_{\text{A function only of } y} = 0 \qquad (7.40)$$

Let us deliberate on an argument: a function of x is added to a function of y to give zero!

$$f(x) + g(y) = 0$$

For it to be true, the following must be true

$$f(x) = k \text{ (a constant: could be zero, negative or positive)}$$

and

$$g(y) = -k \text{ (the same constant with a negative sign!)}$$

therefore
$$f(x) = \frac{1}{X}\frac{d^2X}{dx^2} = k$$

$$\frac{d^2X}{dx^2} = kX \qquad (7.41)$$

Hence from the theory of ordinary differential equations (for a and b constants)

$$X = \begin{cases} a\sinh(\sqrt{k}\,x) + b\cosh(\sqrt{k}\,x), & \text{for } k > 0 \\ a\sin(\sqrt{-k}\,x) + b\cos(\sqrt{-k}\,x), & \text{for } k < 0 \\ ax + b, & \text{for } k = 0 \end{cases} \qquad (7.42)$$

the reader may differentiate the above expressions to satisfy himself that they indeed satisfy the previous differential equation. One may use the following relations:

$$\frac{d}{dx}\sinh(\sqrt{k}\,x) = \sqrt{k}\cosh(\sqrt{k}\,x) \quad (k > 0)$$

$$\frac{d}{dx}\cosh(\sqrt{k}\,x) = \sqrt{k}\sinh(\sqrt{k}\,x) \quad (k > 0)$$

$$\frac{d}{dx}\sin(\sqrt{-k}\,x) = \sqrt{-k}\cos(\sqrt{-k}\,x) \quad (k < 0)$$

$$\frac{d}{dx}\cos(\sqrt{-k}\,x) = -\sqrt{-k}\sin(\sqrt{-k}\,x) \quad (k < 0)$$

Similarly, we can proceed with the y relation

$$g(y) = \frac{1}{Y}\frac{d^2Y}{dy^2} = -k$$

$$\frac{d^2Y}{dy^2} = -kY \qquad (7.43)$$

Hence (with c and d constants)

$$Y = \begin{cases} c\sin(\sqrt{k}\,y) + d\cos(\sqrt{k}\,y), & \text{for } k > 0 \\ c\sinh(\sqrt{-k}\,y) + d\cosh(\sqrt{-k}\,y), & \text{for } k < 0 \\ cy + d, & \text{for } k = 0 \end{cases} \qquad (7.44)$$

so $V = XY$

$$= \begin{cases} \left\{a\sinh(\sqrt{k}\,x) + b\cosh(\sqrt{k}\,x)\right\}\left\{c\sin(\sqrt{k}\,y) + d\cos(\sqrt{k}\,y)\right\}, & \text{for } k > 0 \\ \left\{a\sin(\sqrt{-k}\,x) + b\cos(\sqrt{-k}\,x)\right\}\left\{c\sinh(\sqrt{-k}\,y) + d\cosh(\sqrt{-k}\,y)\right\}, & \text{for } k < 0 \\ (ax + b)(cy + d), & \text{for } k = 0 \end{cases}$$
$$(7.45)$$

Instead of hyperbolic functions, we may instead use exponential functions. An alternative formulation is

$V = XY$

$$= \begin{cases} \left\{a\exp(\sqrt{k}\,x) + b\exp(-\sqrt{k}\,x)\right\}\left\{c\sin(\sqrt{k}\,y) + d\cos(\sqrt{k}\,y)\right\}, & \text{for } k > 0 \\ \left\{a\sin(\sqrt{-k}\,x) + b\cos(\sqrt{-k}\,x)\right\}\left\{c\exp(\sqrt{-k}\,y) + d\exp(-\sqrt{-k}\,y)\right\}, & \text{for } k < 0 \\ (ax + b)(cy + d), & \text{for } k = 0 \end{cases}$$
$$(7.46)$$

Let us apply these relations to an actual problem.

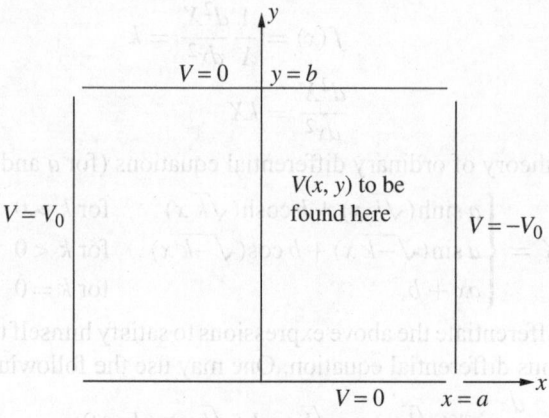

Metal boundaries with corners insulated

Fig. 7.7 Laplace's equation applied to a rectangular region

Example 7.12 We apply Laplace's equation to the two-dimensional layout shown in Fig. 7.7. The figure shows a rectangular region of width $2a$ and height b, bounded by metal plates which are maintained at constant potential. The lower and upper plates at $y = 0$ and $y = b$ are maintained at a potential of $V = 0$, while the two side plates at $x = -a$ and $x = a$ are maintained at $V = 20$ V and $V = -20$ V, respectively.

Solution To analyse this configuration, the potential function can be as that given in Eqn set (7.45).

Since the bottom plate and top plate are both at $V = 0$ the only function which fits the y direction is $\sin(\alpha y)$ where α is to be determined so as to make the top plate have a potential of zero. Note that at $y = 0$ the function is zero. That is

$$\sin(\alpha y)\,|_{y=b} = \sin(\alpha b) = 0 \tag{7.47}$$

or
$$\alpha b = m\pi \qquad m = 1,\ 2,\ 3,\ldots$$

$$\alpha = \frac{m\pi}{b} \qquad m = 1,\ 2,\ 3,\ldots \tag{7.48}$$

Observing the other boundary condition we must use $\sinh \alpha b$ as the function since it is an *odd* function and the boundary condition is 20 V at the left plate and -20 V at the right plate. So the complete solution is

$$V(x,y) = \sum_{m=1}^{m=\infty} d_m \sin(\alpha_m y) \sinh(\alpha_m x) \tag{7.49}$$

$$\alpha_m = \frac{m\pi}{b}$$

$$d_m \qquad \text{yet to be determined}$$

Let us see whether the function above fits the bill

1. It satisfies Laplace's equation (most important). Let us apply the Laplacian to Eqn (7.49).

$$\nabla^2 \left[\sum_{m=1}^{m=\infty} d_m \sin(\alpha_m y) \sinh(\alpha_m x) \right]$$

$$= \left(\frac{\partial^2}{\partial x^2} + \frac{\partial^2}{\partial y^2} \right) \left[\sum_{m=1}^{m=\infty} d_m \sin(\alpha_m y) \sinh(\alpha_m x) \right]$$

$$= \sum_{m=1}^{m=\infty} \left(\frac{\partial^2}{\partial x^2} + \frac{\partial^2}{\partial y^2} \right) [d_m \sin(\alpha_m y) \sinh(\alpha_m x)]$$

$$= \sum_{m=1}^{m=\infty} \left\{ \left[\frac{\partial^2}{\partial x^2} d_m \sin(\alpha_m y) \sinh(\alpha_m x) \right] \right.$$

$$\left. + \left[\frac{\partial^2}{\partial y^2} d_m \sin(\alpha_m y) \sinh(\alpha_m x) \right] \right\}$$

$$= \sum_{m=1}^{m=\infty} \left\{ \left[\alpha_m^2 d_m \sin(\alpha_m y) \sinh(\alpha_m x) \right] \right.$$

$$\left. + \left[-\alpha_m^2 d_m \sin(\alpha_m y) \sinh(\alpha_m x) \right] \right\}$$

$$= 0$$

2. It satisfies the lower and upper boundaries since the sine function is zero on these boundaries.
3. d_m are to be determined to satisfy the left boundary. *If it satisfies the left boundary, it will satisfy the right boundary since* $\sinh(\cdots)$ *is an odd function.*

So we must proceed to satisfy the left or right boundary.

$$V(x,y) = \sum_{m=1}^{m=\infty} d_m \sin(\alpha_m y) \sinh(\alpha_m x) \bigg|_{x=a} \tag{7.50}$$

$$= \sum_{m=1}^{m=\infty} d_m \sin(\alpha_m y) \sinh(\alpha_m a) \tag{7.51}$$

$$= -V_0 \quad \text{(for all values of } y) \tag{7.52}$$

How are we to ensure the last equality?

$$\sum_{m=1}^{m=\infty} d_m \sin(\alpha_m y) \sinh(\alpha_m a) = -V_0 \tag{7.53}$$

A little thought tells us that if we multiply both sides of the equation by $\sin(\alpha_n y)$ and integrate both sides in the interval $[0, b]$ then the orthogonality property of the sine function will give us d_m. Lets try this: multiply both sides by $\sin(\alpha_n y)$ with $n = 1, 2, \ldots m, \ldots$

$$\sin(\alpha_n y) \times \left[\sum_{m=1}^{m=\infty} d_m \sin(\alpha_m y) \sinh(\alpha_m a) \right] = \sin(\alpha_n y) \times (-V_0)$$

taking this term into the summation sign

$$\sum_{m=1}^{m=\infty} d_m \sin(\alpha_n y) \sin(\alpha_m y) \sinh(\alpha_m a) = -V_0 \sin(\alpha_n y)$$

integrating both sides in the limits $[0, b]$

$$\sum_{m=1}^{m=\infty} d_m \sinh(\alpha_m a) \underbrace{\left[\int_0^b \sin(\alpha_n y) \sin(\alpha_m y) \, dy \right]}_{=0 \text{ for } m \neq n} = -V_0 \int_0^b \sin(\alpha_n y) \, dy$$

we find that when $n \neq m$, the integral

$$\int_0^b \sin(\alpha_n y) \sin(\alpha_m y) \, dy = 0$$

so every term of the infinite sum becomes zero on the left-hand side. Now for the $m = n$

$$d_n \sinh(\alpha_n a) \left[\int_0^b \sin^2(\alpha_n y) \, dy \right] = -V_0 \int_0^b \sin(\alpha_n y) \, dy$$

the integral

$$\int_0^b \sin^2(\alpha_n y) \, dy = \int_0^b \left(\frac{1 - \cos 2\alpha_n}{2} \right) dy$$

$$= \frac{b}{2} - \frac{b \sin (2 n\pi)}{2 n\pi}$$

$$= \frac{b}{2}$$

and the integral

$$\int_0^b \sin(\alpha_n y) \, dy = \frac{b}{\pi n} - \frac{b \cos(\pi n)}{\pi n}$$

using these results

$$d_n \left(\frac{b}{2} \right) \sinh(\alpha_n a) = -V_0 \left[\frac{b}{\pi n} - \frac{b \cos(\pi n)}{\pi n} \right]$$

$$d_n = \frac{-2V_0 \left[1 - \cos(\pi n) \right]}{\pi n \sinh(\alpha_n a)}, \qquad n = 1, 2, 3, \ldots$$

we can see that for $n = 2, 4, 6, \ldots$, $\cos(\pi n) = 1$ and for $n = 1, 3, 5, \ldots$, $\cos(\pi n) = -1$ so

$$d_n = \frac{-4V_0}{\pi n \sinh(\alpha_n a)} \text{ for } n = 1, 3, 5, \ldots$$

and $d_n = 0$ for $n = 2, 4, 6, \ldots$

using these values of d_n

$$V(x, y) = \sum_{n=1,3,\cdots}^{n=\infty} d_n \sin(\alpha_n y) \sinh(\alpha_n x)$$

$$\alpha_n = \frac{n\pi}{b}$$

$$d_n = \frac{-4V_0}{\pi n \sinh(\alpha_n a)} \qquad n = 1, 3, 5 \ldots$$

anticipating the next section, let $a = 3$, $b = 7$ cm, and $V_0 = 20$ V. Let us calculate d_n for $n = 1, 3, \ldots$

n	1	3	5	7
d_n	-14.213	-0.2991	-0.0121	-0.0006

Let us calculate the potential at ($x = 1$ cm, $y = 5$ cm). The results are given in table below. The top row shows the number of terms in the summation, while the bottom row shows how fast the sum converges.

n, number of terms	1 term	2 terms	3 terms	4 terms
$V_n(V)$	-5.1562	-5.3887	-5.3337	-5.3337

Example 7.13 Use the separation of variables technique to obtain the general solution of Laplace's equation in rectangular coordinates for three dimensions.

Solution

Step 1. In three dimensions,

$$\frac{\partial^2 V}{\partial x^2} + \frac{\partial^2 V}{\partial y^2} + \frac{\partial^2 V}{\partial z^2} = 0$$

Step 2. We set

$$V = X(x)\, Y(y)\, Z(z) \tag{7.54}$$

where X, Y, and Z are solely functions of x, y, and z respectively.

Step 3. Substituting in Laplace's equation and dividing by XYZ (as given earlier in this section for the case of two variables)

$$\underbrace{\frac{1}{X}\frac{d^2X}{dx^2}}_{\text{A function only of } x} + \underbrace{\frac{1}{Y}\frac{d^2Y}{dy^2}}_{\text{A function only of } y} + \underbrace{\frac{1}{Z}\frac{d^2Z}{dy^2}}_{\text{A function only of } z} = 0$$

Step 4. Since

$$f(x) + g(y) + h(z) = 0$$

where $f(x) = (1/X)\,(d^2X/dx^2)$, etc. Therefore, each of these functions is equal to a constant

$$\frac{1}{X}\frac{d^2X}{dx^2} = k_x$$

$$\frac{1}{Y}\frac{d^2Y}{dx^2} = k_y$$

$$\frac{1}{Z}\frac{d^2Z}{dx^2} = k_z$$

where k_x, k_y, and k_z are constants which may be real, imaginary, or complex, but

$$k_x + k_y + k_z = 0$$

Step 5. The importance of these solutions is that they are combined as infinite sums to match any boundary condition as we saw in the example earlier in this chapter.

Example 7.14 Use the separation of variables technique to solve Laplace's equation in cylindrical coordinates for three dimensions, for various cases.

Solution

Step 1. In cylindrical coordinates, Laplace's equation is

$$\nabla^2 V = \frac{1}{\rho} \frac{\partial}{\partial \rho} \left(\rho \frac{\partial V}{\partial \rho} \right) + \frac{1}{\rho^2} \frac{\partial^2 V}{\partial \phi^2} + \frac{\partial^2 V}{\partial z^2} = 0$$

Step 2. We use a two step process: let $V \equiv F(\rho, \phi) Z(z)$

then
$$\frac{Z}{\rho} \frac{\partial}{\partial \rho} \left(\rho \frac{\partial F}{\partial \rho} \right) + \frac{Z}{\rho^2} \frac{\partial^2 F}{\partial \phi^2} + F \frac{\partial^2 Z}{\partial z^2} = 0$$

Step 3. Dividing by FZ and using earlier arguments

$$\frac{1}{\rho} \frac{\partial}{\partial \rho} \left(\rho \frac{\partial F}{\partial \rho} \right) + \frac{1}{\rho^2} \frac{\partial^2 F}{\partial \phi^2} = -m^2 F \qquad (7.55)$$

$$\frac{\partial^2 Z}{\partial z^2} = m^2 Z \qquad (7.56)$$

Step 4. The solution to Eqn (7.56) has been dealt with earlier, while for Eqn (7.55), we let $F \equiv R(\rho) \Phi(\phi)$. Then

$$\frac{\Phi}{\rho} \frac{\partial}{\partial \rho} \left(\rho \frac{\partial R}{\partial \rho} \right) + \frac{R}{\rho^2} \frac{\partial^2 \Phi}{\partial \phi^2} = -m^2 R \Phi$$

Step 5. Dividing by $R\Phi$ and multiplying by ρ^2

$$\frac{\rho}{R} \frac{\partial}{\partial \rho} \left(\rho \frac{\partial R}{\partial \rho} \right) + \frac{1}{\Phi} \frac{\partial^2 \Phi}{\partial \phi^2} = -\rho^2 m^2$$

$$\frac{\rho}{R} \frac{\partial}{\partial \rho} \left(\rho \frac{\partial R}{\partial \rho} \right) - \rho^2 m^2 + \frac{1}{\Phi} \frac{\partial^2 \Phi}{\partial \phi^2} = 0$$

Step 6. *Case 1.* We can set these equations to constants in the usual manner

$$\frac{\rho}{R} \frac{\partial}{\partial \rho} \left(\rho \frac{\partial R}{\partial \rho} \right) + \rho^2 m^2 = n^2 \qquad (7.57)$$

$$\frac{1}{\Phi} \frac{\partial^2 \Phi}{\partial \phi^2} = -n^2 \qquad (7.58)$$

Step 7. We know how to solve Eqn (7.58), but Eqn (7.57) becomes

$$\rho \frac{\partial}{\partial \rho} \left(\rho \frac{\partial R}{\partial \rho} \right) + \left(\rho^2 m^2 - n^2 \right) R = 0$$

or,
$$\rho^2 \frac{\partial^2 R}{\partial \rho^2} + \rho \frac{\partial R}{\partial \rho} + \left(\rho^2 m^2 - n^2 \right) R = 0 \qquad (7.59)$$

Step 7a. As an aside, set $m = 0$ (no z variation), then $R = \rho^n$ and $R = \rho^{-n}$ are solutions to Eqn (7.59).

Step 8. To solve Eqn (7.59), we compare this equation with the following equation (Bowman, 1968)

$$x^2 \frac{d^2 y}{dx^2} + (2p + 1) x \frac{dy}{dx} + (\alpha^2 x^{2r} + \beta^2) x = 0 \qquad (7.60)$$

whose solution is

$$y = x^{-p}\left[C_1 J_{q/r}\left(\frac{\alpha}{r}x^r\right) + C_2 Y_{q/r}\left(\frac{\alpha}{r}x^r\right)\right]$$

where J and Y are the Bessel functions of the first and second kind, and $q = \sqrt{p^2 - \beta^2}$. Then for Eqn (7.60), $p = 0, r = 1, \alpha = m, \beta = jn$, and $q = n$

$$R = C_1 J_n(m\rho) + C_2 Y_n(m\rho)$$

or
$$= C_1' I_n(m\rho) + C_2' K_n(m\rho)$$

Step 9. A typical solution will have three types of functions (Bowman, 1968).

$$\sum \begin{bmatrix} \sin(mz) \text{ and } \cos(mz) \\ \sinh(mz) \text{ and } \cosh(mz) \\ Az + B \end{bmatrix} \times \begin{bmatrix} I_n(m\rho) \text{ and } K_n(m\rho) \\ J_n(m\rho) \text{ and } Y_n(m\rho) \\ \rho^n \text{ and } \rho^{-n} \end{bmatrix}$$
$$\times \begin{bmatrix} \sin(n\phi) \text{ and } \cos(n\phi) \end{bmatrix}$$

where I and K are the modified Bessel functions of the first and second kind respectively.

7.3.4 Numerical Techniques

With the advent of the digital computer, most applications involving partial differential equations and boundary-value problems are solved numerically, and Laplace's equation is no exception.

The basis of the numerical solution of Laplace's equation, is the following. At a point (x, y) and its neighbourhood $(x+h, y), (x-h, y), (x, y+h)$, and $(x, y-h)$ the potential function $V(x, y)$ takes on the values $V(x+h, y), V(x-h, y), V(x, y+h)$, and $V(x, y-h)$. The potential function at a neighbourhood point, say $(x+h, y)$ is given by the Taylor series expansion

$$V(x+h, y) = V(x, y) + h\frac{\partial V}{\partial x} + \frac{h^2}{2!}\frac{\partial^2 V}{\partial x^2} + \frac{h^3}{3!}\frac{\partial^3 V}{\partial x^3} + O(h^4) \qquad (7.61)$$

This is the standard expansion where y is treated like a constant. Similarly, at the point $(x-h, y)$

$$V(x-h, y) = V(x, y) - h\frac{\partial V}{\partial x} + \frac{h^2}{2!}\frac{\partial^2 V}{\partial x^2} - \frac{h^3}{3!}\frac{\partial^3 V}{\partial x^3} + O(h^4) \qquad (7.62)$$

adding these two equations

$$V(x+h, y) + V(x-h, y) = 2V(x, y) + 2\frac{h^2}{2!}\frac{\partial^2 V}{\partial x^2} + O(h^4) \qquad (7.63)$$

we can similarly obtain an equation for the y-neighbourhood

$$V(x, y+h) + V(x, y-h) = 2V(x, y) + 2\frac{h^2}{2!}\frac{\partial^2 V}{\partial y^2} + O(h^4) \qquad (7.64)$$

adding the above and previous equations

$$V(x+h, y) + V(x-h, y) + V(x, y+h)$$
$$+ V(x, y-h) \approx 4V(x, y) + h^2\left(\frac{\partial^2 V}{\partial x^2} + \frac{\partial^2 V}{\partial x^2}\right) \qquad (7.65)$$

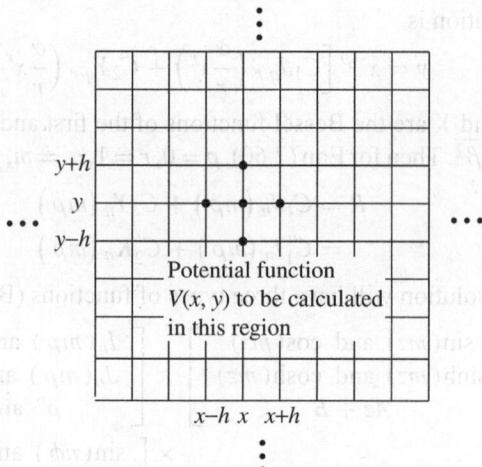

Fig. 7.8 Grid applied to a region of two-space where Laplace's has to be solved

but

$$\left(\frac{\partial^2 V}{\partial x^2} + \frac{\partial^2 V}{\partial x^2}\right) = 0$$

therefore

$$V(x+h,y) + V(x-h,y) + V(x,y+h) + V(x,y-h) \cong 4V(x,y) \qquad (7.66)$$

or $\quad V(x,y) \cong \dfrac{1}{4}[V(x+h,y) + V(x-h,y) + V(x,y+h) + V(x,y-h)] \quad (7.67)$

To apply this equation, the region where Laplace's is to be solved, is divided into a large number of grid points, taking care that the grid points coincide with the boundaries where the potential is specified (Fig. 7.8).

Example 7.15 Apply the 'grid' method just discussed to a rectangular region of width 6 cm and height 7 cm as shown in Fig. 7.9, with the left plate held at a potential

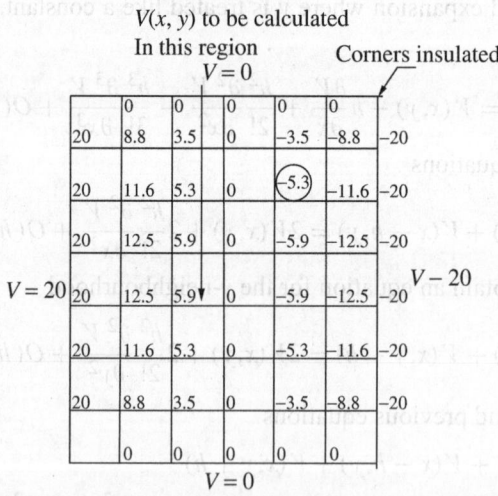

Fig. 7.9 The 'grid' method applied to a rectangular domain of width of 6 cm and height 7 cm

of 20 V, the right plate at -20 V and the lower and upper plates held at 0 V each.

Solution The problem is the same one which was discussed in Ex. 7.12 onward with $2a = 6$ cm and $b = 7$ cm. The method was applied to this problem with a grid of 6×7 as shown in Fig. 7.9. After a number of iterations the values stabilised to the ones shown in the figure. The results have been shown after rounding off to the nearest tenth of a volt. At the grid point $x = 1$ cm, $y = 5$ cm, shown circled ($V = -5.3$ V which is -5.26 rounded off), the comparison with the accurate value of -5.3337 V (from the theoretical treatment) is encouraging.

7.4 | Poisson's Equation

As already discussed, Eqn (7.2) in Section 7.1 is the Poisson's equation

$$\nabla \cdot \nabla V = \nabla^2 V = -\frac{\rho_v}{\varepsilon} \tag{7.68}$$

The equation states that the potential function $V(\mathbf{r})$ satisfies Eqn (7.68), given a set of charges $\rho_v(\mathbf{r})$. The electric field $\mathbf{E}$ maybe obtained everywhere in that region by

$$\mathbf{E} = -\nabla V \tag{7.69}$$

7.4.1 One-Dimensional Solutions

We can apply Poisson's equation to the simplest cases, where only a single dimension is involved. Let us take an example.

Example 7.16 Apply Poisson's equation to the region between a pair of infinite plates separated by a distance d and filled with a charge $\rho_v = c$, a constant.
Solution

Step 1. The geometry of the problem suggests using rectangular coordinates just as in the case of the parallel plate capacitor problem. In Cartesian coordinates, Poisson's equation becomes

$$\nabla^2 V = \frac{\partial^2 V}{\partial x^2} + \frac{\partial^2 V}{\partial y^2} + \frac{\partial^2 V}{\partial z^2} = \frac{\rho_v}{\varepsilon} \tag{7.70}$$

The boundary conditions are that the lower plate is at a potential $V = V_0$ and the upper plate is at $V = V_1$. From the geometry of the problem, let V be a function of only one dimension, namely, x. That is $V \equiv V(x)$. This implies that there is no variation in either the y or z directions.

Step 2. Poisson's equation then becomes

$$\frac{\partial^2 V}{\partial x^2} = -\frac{c}{\varepsilon}$$

Step 3. Performing integration twice, we get

$$\int \frac{d^2 V}{dx^2}\, dx = \frac{dV}{dx} = -\frac{c}{\varepsilon}x + k_1 \tag{7.71}$$

and

$$\int \frac{dV}{dx}\, dx = V(x) = -c\frac{x^2}{2\varepsilon} + k_1 x + k_2 \tag{7.72}$$

where k_1 and k_2 are constants which will be determined by the boundary conditions.

Step 4. At $x = 0$ when we apply $V = V_0$, we get $k_2 = V_0$. Looking further, at $x = d$

$$V_1 = -c\frac{d^2}{2\varepsilon} + k_1 d + V_0 \qquad (7.73)$$

or

$$k_1 = \frac{V_1 - V_0}{d} + \frac{cd}{2\varepsilon} \qquad (7.74)$$

hence the potential is

$$V(x) = -c\frac{x^2}{2\varepsilon} + \left(\frac{V_1 - V_0}{d} + \frac{cd}{2\varepsilon}\right)x + V_0 \qquad (7.75)$$

Step 5. As a check, put $x = 0$ and $x = d$ to see that the boundary conditions are satisfied. (Left for the student to do)

Example 7.17 Find the surface charge on the plates on the previous example. Assume $\varepsilon = \varepsilon_0$.

Solution

Step 1. The electric field is obtained by using the potential function:

$$\mathbf{E} = -\nabla V = -\mathbf{a}_x \left[-\frac{cx}{\varepsilon_0} + \left(\frac{V_1 - V_0}{d} + \frac{cd}{2\varepsilon_0}\right) \right]$$

Step 2. At $x = 0$ and $x = d$, the electric field is

$$\mathbf{E} = -\mathbf{a}_x \left(\frac{V_1 - V_0}{d} + \frac{cd}{2\varepsilon_0}\right) \qquad (\text{at } x = 0)$$

$$\mathbf{E} = -\mathbf{a}_x \left(\frac{V_1 - V_0}{d} - \frac{cd}{2\varepsilon_0}\right) \qquad (\text{at } x = d)$$

Step 3. The surface charge is $|D_n| = |\rho_s|$. At the lower and upper plates, the electric field is directed from the upper plate to the lower plate

$$\rho_{sl} = -\varepsilon_0 \left(\frac{V_1 - V_0}{d} + \frac{cd}{2\varepsilon_0}\right) \ \text{C/m}^2 \qquad (\text{lower plate } x = 0)$$

$$\rho_{su} = \varepsilon_0 \left(\frac{V_1 - V_0}{d} - \frac{cd}{2\varepsilon_0}\right) \ \text{C/m}^2 \qquad (\text{upper plate } x = d)$$

Step 4. Note that the two surface charges are unequal. Also note that $\rho_{su} + \rho_{sl} = -cd$. Why are the signs chosen as they are?

POINTS TO REMEMBER

- Laplace's equation is $\nabla^2 V = 0$
- Poisson's equation $\nabla^2 V = -\dfrac{\rho_v}{\varepsilon}$
- Once a solution is obtained which satisfies the boundary conditions, that solution is the only unique solution for both Laplace and Poisson's equations.
- In rectangular coordinates,

$$\nabla^2 V = \frac{\partial^2 V}{\partial x^2} + \frac{\partial^2 V}{\partial y^2} + \frac{\partial^2 V}{\partial z^2}$$

- In cylindrical coordinates,

$$\nabla^2 V = \frac{1}{\rho}\frac{\partial}{\partial \rho}\left(\rho \frac{\partial V}{\partial \rho}\right) + \frac{1}{\rho^2}\frac{\partial^2 V}{\partial \phi^2} + \frac{\partial^2 V}{\partial z^2}$$

- In spherical coordinates,

$$\nabla^2 V = \frac{1}{r^2}\frac{\partial}{\partial r}\left(r^2 \frac{\partial V}{\partial r}\right) + \frac{1}{r^2 \sin\theta}\frac{\partial}{\partial \theta}\left(\sin\theta \frac{\partial V}{\partial \theta}\right) + \frac{1}{r^2 \sin^2\theta}\frac{\partial^2 V}{\partial \phi^2}$$

- In separation of variables

$$V = X(x)\,Y(y)\,Z(z) \qquad \text{(rectangular coordinates)}$$
$$V = F(\rho,\phi)\,Z(z) \qquad \text{(cylindrical coordinates)}$$
$$V = F(r,\theta)\,\Phi(\phi) \qquad \text{(spherical coordinates)}$$

- Two dimensional solutions can be obtained using the theory of complex variables.
- When solving Laplace's equation by a numerical technique,

$$V(x,y) \approx \frac{1}{4}\,[V(x+h,y) + V(x-h,y) + V(x,y+h) + V(x,y-h)]$$

SELF ASSESSMENT

Objective Type Questions

1. The function $f = e^{-kx}e^{-k^2y}$
   (a) satisfies Laplace's equation   (b) Does not satisfy Laplace's equation
   (c) satisfies Poisson's equation   (d) none of these
2. To solve Laplace's equation between two conducting cylinders of radii a and b with $b > a$, the potential between the cylinders has a coordinate dependence of
   (a) $V = \rho\cos\phi$ (b) $V = 1/(\rho\cos\phi)$
   (c) $V = 1/z$ (d) $V = \ln\rho$
3. To solve Laplace's equation between two conducting spheres of radii a and b with $b > a$, the potential between the spheres has a coordinate dependence of
   (a) $V = r\cos\phi$ (b) $V = 1/(r\cos\phi)$
   (c) $V = 1/r$ (d) none of these
4. $-\varepsilon\,\nabla V$ is
   (a) the electric field, $\mathbf{E}$ (b) the electric flux density, $\mathbf{D}$,
   (c) the magnetic field $\mathbf{H}$ (d) none of the above
5. In cylindrical coordinates, $\dfrac{1}{\rho}\dfrac{\partial}{\partial \rho}\left(\rho \dfrac{\partial V}{\partial \rho}\right) + \dfrac{1}{\rho^2}\dfrac{\partial^2 V}{\partial \phi^2} = 0$ is

   (a) Laplace's equation   (b) Poisson's equation
   (c) Euler's equation   (d) none of the above
6. In cylindrical coordinates, $\dfrac{1}{\rho}\dfrac{\partial}{\partial \rho}\left(\rho \dfrac{\partial V}{\partial \rho}\right) + \dfrac{1}{\rho^2}\dfrac{\partial^2 V}{\partial \phi^2} = 0$ is

   (a) Laplace's equation with no z-dependence
   (b) Poisson's equation with no z-dependence
   (c) Euler's equation with no z-dependence
   (d) none of the above
7. The solution to $y''(x) + 3y(x) = 0$ is
   (a) $y = 0$ (b) $y = -5\sin 3x$
   (c) $-5\sin\sqrt{3}x$ (d) $-5\sin\sqrt{3}x + 8\cos\sqrt{3}x$

8. The solution to $y''(x) - 3y(x) = 0$ is
   (a) $y = 0$
   (b) $y = -5 \sinh 3x$
   (c) $-15 \cosh \sqrt{3}x$
   (d) $-5 \sin \sqrt{3}x + 8 \cos \sqrt{3}x$

9. In spherical coordinates, the equation

$$\frac{1}{r^2} \frac{\partial}{\partial r}\left(r^2 \frac{\partial V}{\partial r}\right) + \frac{1}{r^2 \sin\theta} \frac{\partial}{\partial\theta}\left(\sin\theta \frac{\partial V}{\partial\theta}\right) + \frac{1}{r^2 \sin^2\theta} \frac{\partial^2 V}{\partial\phi^2} = 10$$

   is
   (a) Laplace's equation
   (b) Poisson's equation
   (c) Mitter's equation
   (d) none of the above

10. In spherical coordinates, the equation

$$\frac{1}{r^2} \frac{\partial}{\partial r}\left(r^2 \frac{\partial V}{\partial r}\right) + \frac{1}{r^2 \sin\theta} \frac{\partial}{\partial\theta}\left(\sin\theta \frac{\partial V}{\partial\theta}\right) + \frac{1}{r^2 \sin^2\theta} \frac{\partial^2 V}{\partial\phi^2} = 10 \sin\phi$$

   is
   (a) Laplace's equation
   (b) Poisson's equation
   (c) Perseval's relation
   (d) none of the above

Short-Answer Questions

1. Write Laplace's equation in all three coordinate systems.
2. Discuss the separation of variables technique for solving Laplace's equation.
3. Derive Laplace's equation in the rectangular coordinate system.
4. Derive Laplace's equation in the cylindrical coordinate system.
5. Derive Laplace's equation in the spherical coordinate system.
6. Two coaxial cylindrical conductors of radii 2 m and 5 m are maintained at 100 V and 20 V, respectively. Calculate the potential at $\rho = 3$ m. (DTU-NSIT, May 2009)
7. $V = 0$ V for $r = 0.1$ m and $V = 100$ V for $r = 2.0$ m in spherical co-ordinates. Assuming free space between the concentric spherical shells, find **E** and **D** using Laplace's equation. (Mumbai University, December 2007)

Review Questions

1. Explain why Gauss's law cannot solve problems solved by Laplace's equation.
2. Explain why Gauss's law can only solve some of the problems solved by Poisson's equation.
3. Explain why the uniqueness theorem is important in solutions of Laplace's and Poisson's equations.
4. Why is the solution $V = 5x + 2$ the only solution to the problem of the potential between two infinite plates at $x = 0$ and $x = 1$ held at potentials of $V = 2$ V and $V = 7$ V, respectively?
5. Why is that a solution of Laplace's equation will always satisfy Poisson's equation?
6. Derive Poisson's and Laplace's equations and give their applications in electromagnetics. (DTU-NSIT, May 2009)
7. Derive Laplace's equation in all three coordinate systems. (Mumbai University, May 2009)

Numerical Problems

Note: Wherever the medium is not mentioned assume $\varepsilon = \varepsilon_0$.

1. Calculate the Laplacian for the following functions:
   (a) $f = x + y + z$

(b) $f = xyz$

(c) $f = \sqrt{x^2 + y^2}$

(d) $f = \rho \cos\phi + \rho^2 \sin\phi$

(e) $f = r \cos\phi \sin\theta$

2. Find the gradient of the scalar field given, and then the Laplacian.
   (a) $f = r^2$, (b) $f = \rho + \cos\phi$.

3. For the potential field $V = xy$, find the equipotential surfaces with a potential of $V = 0$ and $V = 10$.

4. Does the potential field $V = xy + yz + xz$ satisfy Laplace's equation? At $P = (1, 2, 3)$ what is the potential?

5. Does the function $f = e^{-kx} \sin(ky)$, for any value of k, satisfy Laplace's equation?

6. Does the potential field in cylindrical coordinates

$$V = \frac{10 \sin(2\phi)}{\rho^2}$$

   satisfy Laplace's equation? Find the potential at $\rho = 1$, $\phi = \pi/3$. Give the equation of the equipotential surface at this point.

7. Does the potential field in cylindrical coordinates

$$V = \frac{20 \cos(2\phi)}{\rho^2}$$

   satisfy Laplace's equation? Find the potential at $\rho = 5$, $\phi = \pi/6$. What is electric field at this point?

8. Does the potential function in spherical coordinates satisfy Laplace's equation?

$$V = K \frac{\ln\left(r^2 \sin(\theta)^2\right)}{2}$$

9. For the potential field $V = xy + yz + xz$ at $P = (1, 2, 3)$ what is the equipotential surface? If a metallic surface with the shape of the equipotential surface computed above be placed at P, will the potential function change due to the presence of the metal?

10. For Problem 9, find the direction of the electric field and surface charge at P. Is the point $(2, 2, 7/4)$ on this surface? Find the surface charge at $(2, 2, 7/4)$.

11. A point charge is placed at a height h above an infinite plane. Compute the potential field above the plane and show that it satisfies Laplace's equation.

12. A point charge is placed at a height h above an infinite plane. Compute the distribution of the surface charge on the plane.

13. Two infinite plates at angle of θ between them as shown in Fig. 7.10, are maintained at potentials of $V = 0$ and V_0. (a) Which coordinate system is to be used? (b) Find the potential field between the plates.

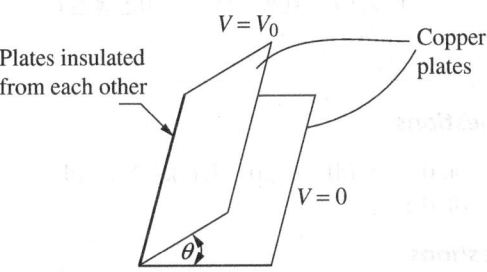

Fig. 7.10 Two plates at angle θ

14. Verify that the expression obtained for the potential due to an electric dipole satisfies the Laplace's equation.

15. The potential due to an infinitely long charged wire with charge ρ_l C/m is given by

$$-\frac{\rho_l}{2\pi\varepsilon_0}\ln\rho$$

and show that it satisfies Laplace's equation.

16. Find the potential function and the electric field intensity for the region between two concentric right circular cylinders, where $V = 0$ at $r = 1$ mm and $V = 150$ V at $r = 20$ mm: Neglect fringing. (Mumbai University, June 2008)

17. The potential due to two parallel infinitely long oppositely charged wires as considered in Example 6.14.

$$V(x,y) = \frac{\rho_l}{2\pi\varepsilon_0}\log\left(\frac{\sqrt{(y+d/2)^2 + x^2}}{\sqrt{(y-d/2)^2 + x^2}}\right)$$

Show that it is a solution of the Laplace's equation.

18. Two metal plates are placed as shown in Fig. 7.11 and held at potentials of 0 and 5 V. The plates are infinite in the direction perpendicular to the plane of the page. *Only formulate* the potential as a function of x and y in the region enclosed by the plates.

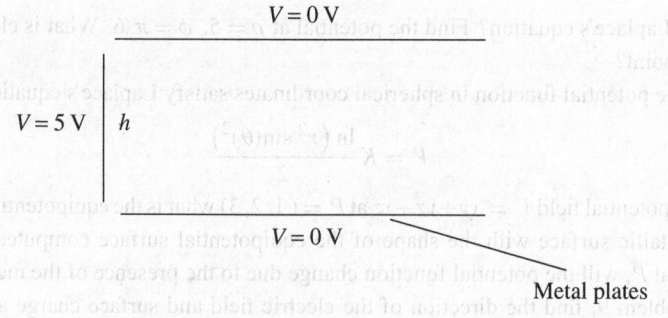

Fig. 7.11 Metal plates insulated from each other in a rectangular shape. The top and bottom plates are infinite in extent

19. The plate $y = x$ is grounded and the plate $y = x - 2$ is kept at a potential of 10 V. Find the potential between the plates.

20. Solve $\nabla^2 V = 0$ in the semi-infinite strip $0 < x < a$, $y > 0$ with the boundary condition

$$\left.\begin{array}{l} V(0,y) = 0 \\ V(a,y) = 0 \end{array}\right\} y > 0$$

$$V(x,0) = x(a-x), \qquad 0 \le x \le a$$

Answers

Objective Type Questions

1. (b)   2. (d)   3. (c)   4. (b)   5. (d)   6. (a)   7. (a), (c) and (d)
8. (a) and (c)   9. (b)   10. (b)

Short-Answer Questions

6. $V = -87.3\ln\rho + 160.5$. At $\rho = 3$, $V = 64.6$ V

7. $V = -10.53/r + 105.26$; $\mathbf{E} = -\left(10.53\mathbf{a}_r\right)/r^2$; $\mathbf{D} = -\left(10.53\mathbf{a}_r\varepsilon_0\right)/r^2$

Numerical Problems

1. (a) 0 (b) 0 (c) $1/\sqrt{x^2 + y^2}$ (d) $3\sin\phi$ (e) 0
2. (a) 6 (b) $(\rho - \cos\phi)/\rho^2$
3. For $V = 0$ $xy = 0$; for $V = 10$, $xy = 10$.
4. Yes it satisfies Laplace's equation. The potential at that point is 11.
5. Yes
6. Yes it satisfies Laplace's equation. $5\sqrt{3}$
7. Yes it satisfies Laplace's equation. $\mathbf{E}\rho = 0.16\mathbf{a}_\rho + 0.277\mathbf{a}_\phi$
8. Yes $V = K\dfrac{\ln\left(r^2\sin(\theta)^2\right)}{2}$
9. $xy + yz + xz = 11$. No the field will not change
10. Yes, the point $(2, 2, 7/4)$ is on the equipotential surface. The surface charge at P is $-7.07\ \varepsilon_0$ C/m^2. The surface charge at $(2, 2, 7/4)$ is $-6.643\ \varepsilon_0$ C/m^2
11. $V = \dfrac{Q}{4\pi\varepsilon_0}\left(\dfrac{1}{\sqrt{(z-h)^2 + y^2 + x^2}} - \dfrac{1}{\sqrt{(z+h)^2 + y^2 + x^2}}\right)$
12. $\rho_s = \dfrac{hQ}{2\pi(y^2 + x^2 + h^2)^{\frac{3}{2}}}$
13. (a) Use the cylindrical coordinate system, (b) $V = (V_0/\theta)\phi$
16. $V = 50.07\ln\rho$ where ρ is in mm. $\mathbf{E} = -50.07/\rho$
18. $\sum a_n \sin(n\pi y/h)\exp(-n\pi x/h)$
19. $V = -5(y - x)$
20. $V = \displaystyle\sum_{n=1}^{\infty} b_n \sin(n\pi x/a)\,e^{-n\pi y/a}$ where $b_n = \dfrac{4a^2[1-(-1)^n]}{\pi^3 n^3}$

PART III

Magnetostatics

8

The Steady Magnetic Field

Learning is a treasure that will follow its owner everywhere.

— Chinese Proverb

CHAPTER OBJECTIVES

To enable the students to understand the following:

- Biot-Savart law and different types of currents
- Ampere's law and its applications
- Use of the right-hand thumb rule to ascertain the direction of the magnetic field
- The magnetic scalar potential

- The magnetic flux density, and its analogy with the electric flux density
- Derivation of the Biot-Savart law using the magnetic vector potential
- Far-field approximation

8.1 | Introduction

The earliest knowledge of magnetism can be traced back to the Chinese around the year 1000, who discovered naturally occurring magnets called load stone made up of iron-rich ore. This state of affairs existed till about 1820.

It was believed around that time that the earth was a giant load stone, magnetised in the same way, and the greatest puzzle then was the slow variation of the earth's magnetic field. That the earth's magnetic field was changing was confirmed by the fact that at any one place the direction of the compass needle slowly shifted over time.

Around 1820, Hans Christian Oersted noted the connection between varying currents and the movement of a compass needle, but was not able to explain the phenomenon. Experimental investigations by Andre-Marie Ampere in France on two parallel wires carrying current showed that they interacted magnetically. Parallel wires carrying currents in the same direction attracted and anti-parallel currents repelled each other. Ampere further showed that the force between two long straight parallel currents was (a) inversely proportional to the distance between them and (b) proportional to the intensity of the current flowing in each. It was up to Maxwell to connect the two types of forces—electric and magnetic. He neatly tied up both mathematically into the now famous Maxwell's equations.

8.2 | Biot-Savart Law

The 'source' of magnetic fields are electric currents. The word 'source' has been put in inverted commas, because currents are only moving charges, and simply put— when charges move, they produce magnetic fields. How are currents quantitatively related to magnetic fields? We already know from Coulomb's law that electric fields produced by charges involve an inverse square law. That is

$$E \propto \frac{q}{R^2}$$

where R is the distance from the *point* charge q.

Similarly, magnetic fields are also inversely proportional to the distance from a 'point current'. What is this 'point current'? Let us write the Biot-Savart law to understand this. Referring to Fig. 8.1, we see that in a conductor carrying a current there is

- a minuscule current element $I'd\mathbf{l}'$,
- located at a position vector vector $\mathbf{r}'$
- where I' is the current and
- $d\mathbf{l}'$ is a minuscule part of the current carrying conductor,
- This current element produces a minuscule *magnetic field* $d\mathbf{H}$
- At the point $P(\mathbf{r})$ (at a position vector $\mathbf{r}$), whose value is given by

$$d\mathbf{H} = \frac{I'd\mathbf{l}' \times \hat{\mathbf{R}}}{4\pi|\mathbf{R}|^2} \tag{8.1}$$

- $I'd\mathbf{l}'$, which is the source term, has been set in the *'prime'* notation.
- $\hat{\mathbf{R}}$ is the unit vector in the direction of the vector $\mathbf{R} = \mathbf{r} - \mathbf{r}'$.

Let us proceed to evaluate this expression in Cartesian coordinates. In general with no restriction whatsoever,

$$d\mathbf{l}' = \mathbf{a}_x dx' + \mathbf{a}_y dy' + \mathbf{a}_z dz'$$

$$\mathbf{r} = [x, y, z]$$

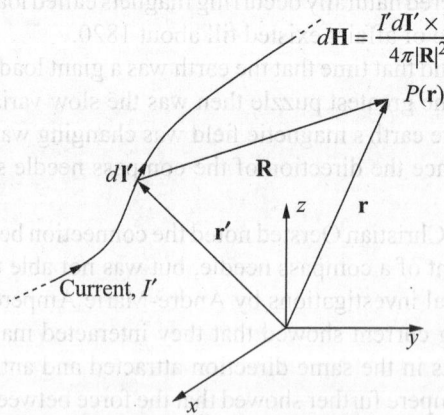

Fig. 8.1 The Biot-Savart law

$$\mathbf{r'} = [x', y', z']$$

$$\mathbf{R} = [x - x', y - y', z - z']$$

$$\hat{\mathbf{R}} = \frac{[x - x', y - y', z - z']}{\sqrt{(x - x')^2 + (y - y')^2 + (z - z')^2}} \qquad (8.2)$$

Using these results, the three differential components of the magnetic field are given below in the rectangular coordinate system:

$$dH_x = \frac{I'[dy'(z - z') - dz'(y - y')]}{4\pi\left[\sqrt{(x - x')^2 + (y - y')^2 + (z - z')^2}\right]^3}$$

$$dH_y = \frac{I'[dz'(x - x') - dx'(z - z')]}{4\pi\left[\sqrt{(x - x')^2 + (y - y')^2 + (z - z')^2}\right]^3}$$

$$dH_z = \frac{I'[dx'(y - y') - dy'(x - x')]}{4\pi\left[\sqrt{(x - x')^2 + (y - y')^2 + (z - z')^2}\right]^3} \qquad (8.3)$$

These expressions are fairly complicated, but to give us a better understanding we apply these expressions to specific cases in Section 8.5, for specific current distributions. We will integrate these expressions to get the magnetic field.

8.2.1 Biot-Savart Law Applied to a Tiny Filamentary Current

Example 8.1 Apply these equations to a z-directed filamentary current placed at the origin of a coordinate system as shown in Fig. 8.2.

Solution

Step 1. The figure depicts a filamentary current $I'dz'\mathbf{a}_z$ placed at the origin of a coordinate system.

Step 2. As the filamentary current is at the origin, $\mathbf{r'} = 0$.

since $$\mathbf{r'} = 0 \text{ therefore } \begin{cases} x' = 0 \\ y' = 0 \\ z' = 0 \end{cases}$$

and $$d\mathbf{l'} = \mathbf{a}_z dz'$$

so $$\begin{cases} dx' = 0 \\ dy' = 0 \end{cases}$$

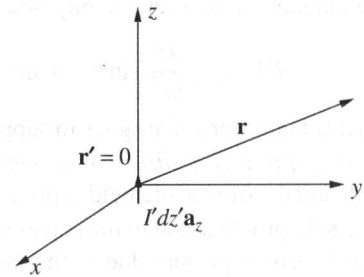

Fig. 8.2 z-directed filamentary current at the origin

Plugging these values in Eqn (8.3)

$$dH_x = \frac{I'(-dz'y)}{4\pi \left[\sqrt{x^2 + y^2 + z^2}\right]^3}$$

$$dH_y = \frac{I'(dz'x)}{4\pi \left[\sqrt{x^2 + y^2 + z^2}\right]^3}$$

$$dH_z = 0 \qquad (8.4)$$

Step 3. If we go over to the spherical coordinate system, then

$$x = r \sin\theta \cos\phi$$
$$y = r \sin\theta \sin\phi$$
$$r = \sqrt{x^2 + y^2 + z^2}$$

and
$$\mathbf{a}_x = \mathbf{a}_r \sin\theta \cos\phi + \mathbf{a}_\theta \cos\theta \cos\phi - \mathbf{a}_\phi \sin\phi$$
$$\mathbf{a}_y = \mathbf{a}_r \sin\theta \sin\phi + \mathbf{a}_\theta \cos\theta \sin\phi + \mathbf{a}_\phi \cos\phi$$

then
$$d\mathbf{H} = \frac{I'(-dz'y)}{4\pi \left[\sqrt{x^2 + y^2 + z^2}\right]^3}\mathbf{a}_x + \frac{I'(dz'x)}{4\pi \left[\sqrt{x^2 + y^2 + z^2}\right]^3}\mathbf{a}_y$$

$$= \frac{I'(-dz'y)}{4\pi r^3}\mathbf{a}_x + \frac{I'(dz'x)}{4\pi r^3}\mathbf{a}_y$$

$$= \frac{I'dz'}{4\pi r^3}(-y\mathbf{a}_x + x\mathbf{a}_y)$$

Now $-y\mathbf{a}_x = -r\sin\theta\sin\phi\,(\mathbf{a}_r \sin\theta\cos\phi + \mathbf{a}_\theta \cos\theta\cos\phi - \mathbf{a}_\phi\sin\phi)$

$$= -\left\{\frac{1}{2}\mathbf{a}_r r\sin^2\theta\sin 2\phi + \frac{1}{4}\mathbf{a}_\theta r\sin 2\theta\sin 2\phi - \mathbf{a}_\phi r\sin\theta\sin^2\phi\right\}$$

$x\mathbf{a}_y = r\sin\theta\cos\phi\,(\mathbf{a}_r \sin\theta\sin\phi + \mathbf{a}_\theta \cos\theta\sin\phi + \mathbf{a}_\phi\cos\phi)$

$$= \frac{1}{2}\mathbf{a}_r r\sin^2\theta\sin 2\phi + \frac{1}{4}\mathbf{a}_\theta r\sin 2\theta\sin 2\phi + \mathbf{a}_\phi r\sin\theta\cos^2\phi$$

adding these two terms

$$\frac{I'}{4\pi r^3}(-y\mathbf{a}_x + x\mathbf{a}_y) = \mathbf{a}_\phi\frac{I'}{4\pi r^3}r\sin\theta$$

going back to the magnetic field there is only one component, namely,

$$dH_\phi = \frac{I'dz'}{4\pi r^2}\sin\theta \quad \text{(A/m)} \qquad (8.5)$$

the other components being zero. This goes to support the important result that *z-directed currents produce ϕ-directed magnetic fields*. The field 'curls' around the current, and the minuscule field is proportional to the current I' and $\sin\theta$, and inversely proportional to the square of the distance from the source $d\mathbf{l}$. The $\sin\theta$ term is present due to the fact that there is a cross-product in the Biot-Savart law.

The units of the magnetic field can be established from the Biot-Savart law. Rewriting the law,

$$d\mathbf{H} = \frac{I'd\mathbf{l}' \times \hat{\mathbf{R}}}{4\pi R^2}$$

$$\text{Units of } d\mathbf{H} \overset{U}{=} \frac{\text{A} \times \text{m}}{\text{m}^2} \quad (\hat{\mathbf{R}} \text{ has no units})$$

$$\overset{U}{=} \text{A/m}$$

8.3 | Types of Currents

In electromagnetic theory, one will meet with several types of idealisations of the current. Referring to Fig. 8.3, minuscule amounts of these currents are $I'dl'$, $\mathbf{J}'_s ds'$, and $\mathbf{J}'dV'$. We can write alternate forms of the Biot-Savart law using these currents. We have already written out this law for the term I'. In some cases the current exists on the surface of a conductor. The idealisation then is that the current is infinite thin, but covering the surface. $\mathbf{J}'_s$ is the notation we will use for the surface current idealisation; it has the units of A/m and is the surface current density. We can change the type of current in the Biot-Savart law by

$$\mathbf{J}' \ldots dV' = \mathbf{J}_s \ldots dS' = I' \ldots d\mathbf{l} \tag{8.6}$$

Therefore using Eqn (8.6), the Biot-Savart law using $\mathbf{J}'_s$ is

$$d\mathbf{H} = \frac{\mathbf{J}'_s \times \hat{\mathbf{R}}}{4\pi R^2} dS'$$

$$= \frac{\mathbf{J}'_s \times \mathbf{R}}{4\pi R^3} dS' \tag{8.7}$$

And the magnetic field using this equation is

$$\mathbf{H} = \iint_{S'} \frac{\mathbf{J}'_s \times \mathbf{R}}{4\pi R^3} dS' \quad \text{(A/m)} \tag{8.8}$$

In this equation, $\mathbf{J}'_s \equiv \mathbf{J}'_s(\mathbf{r}')$ is the source surface current, $\mathbf{r}'$ is the position vector of the source, $\mathbf{r}$ is the position vector of the field point, $\mathbf{R} = \mathbf{r} - \mathbf{r}'$, dS' is a differential element at the source, and $\mathbf{H}$ is evaluated at the field point $\mathbf{r}$.

In the same manner, when the current is a volume current density, $\mathbf{J}'(\text{A/m}^2)$, then

$$d\mathbf{H} = \frac{\mathbf{J}' \times \hat{\mathbf{R}}}{4\pi R^2} dV'$$

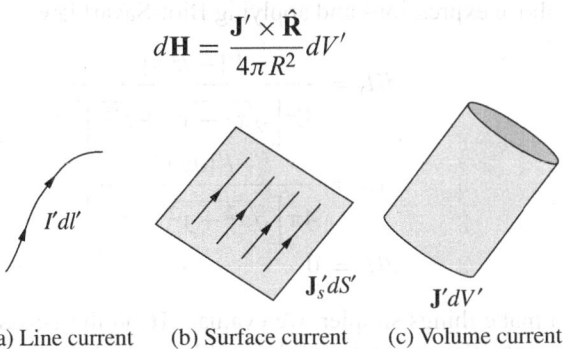

(a) Line current   (b) Surface current   (c) Volume current

Fig. 8.3 Types of currents

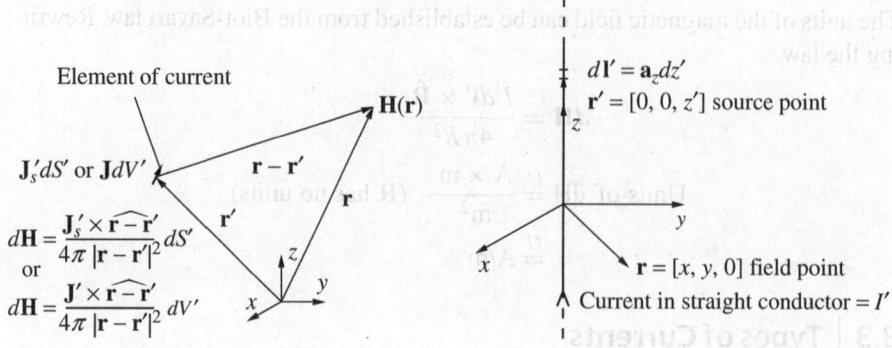

Fig. 8.4 The Biot-Savart law

Fig. 8.5 Applying the Biot-Savart law to a current carrying straight conductor

or

$$H = \iint_{V'} \frac{J' \times R}{4\pi R^3} dV' \qquad (8.9)$$

r and r' are shown in Fig. 8.4.

8.3.1 Biot-Savart Law Applied to an Infinitely Long Straight Wire

Example 8.2 Calculate the magnetic field of a straight infinite wire shown in Fig. 8.5 using the Biot-Savart law.

Solution

Step 1. Observing Fig. 8.5,

$$d\mathbf{l}' = \mathbf{a}_z dz'$$
$$\mathbf{r} = [x, y, 0]$$
$$\mathbf{r}' = [0, 0, z']$$
$$\mathbf{r} - \mathbf{r}' = \mathbf{R} = [x, y, -z']$$
$$\hat{\mathbf{R}} = \frac{[x, y, -z']}{\sqrt{x^2 + y^2 + z'^2}} \qquad (8.10)$$

Step 2. Using above expressions and applying Biot-Savart law,

$$dH_x = \frac{I'[-dz'y]}{4\pi \left[\sqrt{x^2 + y^2 + z'^2}\right]^3}$$

$$dH_y = \frac{I'[dz'x]}{4\pi \left[\sqrt{x^2 + y^2 + z'^2}\right]^3}$$

$$dH_z = 0 \qquad (8.11)$$

Step 3. Now to make things simpler, we evaluate **H** on the x-axis ($y = z = 0$), remembering the symmetry of the problem. We have to keep x constant, while integrating over z' from $-\infty$ to ∞.

Let us first integrate (by formula 29 of Appendix B.6.2)

$$\Gamma(x, z') = \int \frac{dz'}{(z'^2 + x^2)^{\frac{3}{2}}} = \frac{z'}{x^2 \sqrt{z'^2 + x^2}} \qquad (8.12)$$

Using the value of $\Gamma(x, z' \to \pm\infty)$

$$\Gamma(x, y, z' \to \infty) = \frac{1}{x^2}$$

$$\Gamma(x, y, z' \to -\infty) = -\frac{1}{x^2} \qquad (8.13)$$

$$H_x = 0 \quad (\text{since } y = 0)$$

$$H_y = \frac{I'x}{4\pi} \left[\frac{2}{x^2} \right]$$

$$H_z = 0 \qquad (8.14)$$

therefore

$$H_y = \frac{I'}{2\pi x}$$

Step 4. We now remember the symmetry of the problem, and realise that H_y is actually H_ϕ ($a_y \to a_\phi$), and x the distance from the z-axis is ρ. Therefore,

$$\mathbf{H} = \frac{I'}{2\pi\rho} a_\phi$$

Writing out the fields in (ρ, ϕ, z) coordinates

$$H_\rho = 0$$

$$H_\phi = \frac{I'}{2\pi\rho}$$

$$H_z = 0 \qquad (8.15)$$

What does the magnetic field for a straight conductor look like? Scientists and engineers have thought of a simple way called the right-hand rule to visualise the field pattern. If one holds the conductor in the right-hand with the thumb pointing in the direction of the current, then the magnetic field lines have a direction as that of the fingers, as shown in Fig. 8.10.

8.3.2 Magnetic Field Lines of a Long Straight Wire

We can also plot the field lines of the magnetic field. The field line equations are

$$\frac{dx}{H_x} = \frac{dy}{H_y}$$

$$\frac{dx}{-\frac{I'y}{2\pi(y^2+x^2)}} = \frac{dy}{\frac{I'x}{2\pi(y^2+x^2)}}$$

$$x\,dx = -y\,dy$$

Integrating both sides,

$$\int x\,dx = -\int y\,dy$$

$$x^2 = -y^2 + c^2$$

$$x^2 + y^2 = c^2$$

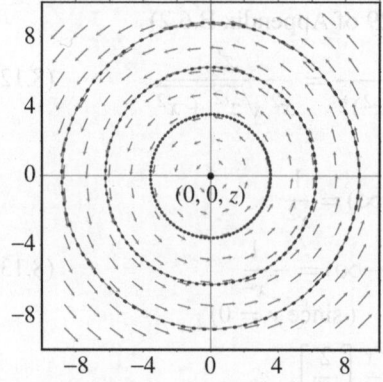

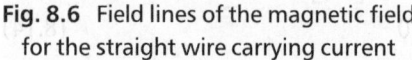

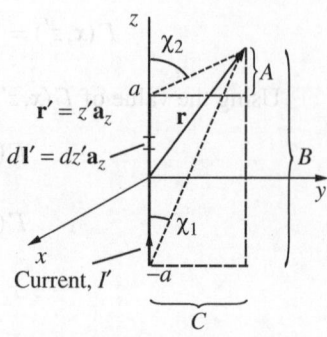

Fig. 8.6 Field lines of the magnetic field for the straight wire carrying current

Fig. 8.7 Applying the Biot-Savart law to a current carrying straight conductor

where c is a constant. Hence, the field lines consist of a family of circles with centre $[0, 0, z]$: These are shown in Fig. 8.6.

Some important points may be noted here for the comparison of field lines of the electric and magnetic fields

1. Electric field lines start at some charge or start at infinity, or some metal surface.
2. Electric field lines end at a charge, at infinity or a metal surface.
3. Magnetic field lines form closed loops.
4. If a magnetic field line starts at infinity then it also ends at infinity.

8.3.3 Biot-Savart Law Applied to a Short Straight Wire

Example 8.3 Find the magnetic field for the case of filament of current flowing in the z direction which lies from $z = -a$ to $z = a$ as shown in Fig. 8.7.

Solution
Step 1. As earlier we use Eqn set 8.3 with $dy' = dx' = 0$, $x' = y' = 0$.

$$dH_x = \frac{I'[-dz'y]}{4\pi\left[\sqrt{x^2 + y^2 + (z - z')^2}\right]^3}$$

$$dH_y = \frac{I'[dz'x]}{4\pi\left[\sqrt{x^2 + y^2 + (z - z')^2}\right]^3}$$

$$dH_z = 0 \tag{8.16}$$

Step 2. Writing the equations for $\mathbf{r}$ in cylindrical coordinates

$$dH_x = \frac{-\rho \sin \phi \, I' dz'}{4\pi\left[\sqrt{\rho^2 + (z - z')^2}\right]^3}$$

$$dH_y = \frac{\rho \cos \phi \, I' dz'}{4\pi\left[\sqrt{\rho^2 + (z - z')^2}\right]^3}$$

$$dH_z = 0 \tag{8.17}$$

Step 3. Noting that there is no H_z component (keep in mind Example 8.1). Then (using formula 29, Appendix B.6.2)

$$H_x = -\frac{\rho \sin \phi}{4\pi} \int_{-a}^{a} \frac{-\rho \sin \phi \, I' dz'}{4\pi \left[\sqrt{\rho^2 + (z - z')^2} \right]^3}$$

$$= -\left(\frac{I' \sin \phi}{4\pi\rho} \right) \left[\frac{(z + a)}{\sqrt{(z + a)^2 + \rho^2}} - \frac{(z - a)}{\sqrt{(z - a)^2 + \rho^2}} \right]$$

$$H_y = \left(\frac{I' \cos\phi}{4\pi\rho} \right) \left[\frac{(z + a)}{\sqrt{(z + a)^2 + \rho^2}} - \frac{(z - a)}{\sqrt{(z - a)^2 + \rho^2}} \right]$$

Step 4. After some simple manipulations (see Example 8.1)

$$\mathbf{H} = \mathbf{a}_\phi \frac{I'}{4\pi\rho} \left[\frac{(z + a)}{\sqrt{(z + a)^2 + \rho^2}} - \frac{(z - a)}{\sqrt{(z - a)^2 + \rho^2}} \right] \qquad (8.18)$$

Example 8.4 Find the magnetic field in the case of filament of current flowing in the z direction which lies from $z = -a$ to $z = b$ as shown in Fig. 8.7 where the upper limit is $z = b$.

Solution

Step 1. From Fig. 8.7, we can realise that $A = z - a$, $B = z + a$, and $C = \rho$, and therefore (see Sadiku, 2006)

$$\mathbf{H} = \mathbf{a}_\phi \frac{I'}{4\pi\rho} [\cos\chi_1 - \cos\chi_2]$$

Step 2. We can generalise these results: if the wire were to exist from $-a$ to b $(a, b > 0)$ then

$$\mathbf{H} = \mathbf{a}_\phi \frac{I'}{4\pi\rho} \left[\frac{(z + a)}{\sqrt{(z + a)^2 + \rho^2}} - \frac{(z - b)}{\sqrt{(z - b)^2 + \rho^2}} \right] \qquad (8.19)$$

 See `Chapter8d.m` and `FilamntWireATOB.m` in Chapter 8

Biot-Savart Law

In 1819, Oersted found that a compass needle was deflected by a current carrying wire when the needle was brought near the wire. Based on the results of Oersted, Jean-Baptiste Biot and Felix Savart performed experiments in 1920s to quantify the force exerted on the compass needle by the current, which resulted in the Biot-Savart law.

8.4 | Ampere's Law

Note that before we read this section, we need to brush up our concepts on the curl (Section 3.4.3) and the line integral (Section 3.2.1).

The Biot-Savart law gives an equation from which we can compute the magnetic field at any point in space and the geometry of the source currents and current densities is quite arbitrary, but Ampere's law is more of a generalisation in terms of development of Maxwell's equations. Ampere's law states that if a magnetic field exists in some region of space, then if we were to take a closed contour (or loop) and find the line integral of the magnetic field over that loop in the counter-clockwise sense, then the line integral would be equal to the total current passing through the loop. The direction of the integration is given by the direction of the curled fingers as per right-hand thumb rule (refer Fig. 1.12) and the direction of the thumb gives the direction of the current. That is

$$\oint_{\mathcal{L}} \mathbf{H} \cdot d\mathbf{l} = I_{\text{enclosed}} \qquad (8.20)$$

where $\mathcal{L}$ is the closed loop and $\oint$ is the integral taken in the counter-clockwise sense the situation is shown in Fig. 8.8b. This is Ampere's law in integral form.

If a law has to be accurate, it has to apply to all situations. Let us look at Eqn (8.20) once again. The left-hand side of the equation is

$$\oint_{\mathcal{L}} \mathbf{H} \cdot d\mathbf{l}$$

Let us recall Stokes's theorem. From Eqn (3.73) which is Stokes's theorem for any vector field, $\mathbf{A}$, the line integral of $\mathbf{A}$ taken in the clockwise sense over a closed loop $\mathcal{L}$ is equal to the surface integral of the curl of $\mathbf{A}$ over the surface $\mathcal{S}$, enclosed by $\mathcal{L}$. This is shown in Fig. 3.19,

$$\oint_{\mathcal{L}} \mathbf{A} \cdot d\mathbf{l} = \iint_{S} (\nabla \times \mathbf{A}) \cdot d\mathbf{S}$$

Therefore, if we substitute $\mathbf{H}$ instead of $\mathbf{A}$ then

$$\oint_{\mathcal{L}} \mathbf{H} \cdot d\mathbf{l} = \iint_{S} (\nabla \times \mathbf{H}) \cdot d\mathbf{S} \qquad (8.21)$$

where $\mathcal{S}$ is the surface enclosed by $\mathcal{L}$. Now let us examine the right-hand side of Eqn (8.20), which is I_{enc}. We know that

$$I_{\text{enc}} = \iint_{S} \mathbf{J} \cdot d\mathbf{S} \qquad (8.22)$$

where $\mathbf{J}$ is the current density in $\mathcal{S}$. So

$$\iint_{S} (\nabla \times \mathbf{H}) \cdot d\mathbf{S} = \iint_{S} \mathbf{J} \cdot d\mathbf{S} \qquad (8.23)$$

and this is generally true. Hence $\quad \nabla \times \mathbf{H} = \mathbf{J} \qquad (8.24)$

This is Ampere's law in differential form. It is also Maxwell's equation for the steady magnetic field. The law states that the curl of the magnetic field is equal to the current density at a point in space. If at that point, no current density exists, then the curl of the magnetic field is zero. Let us examine Fig. 8.8. Notice that the figure has two parts (a) and (b). In (a) part of the figure, the surface $\mathcal{S}$ bulges out and in part (b) of the figure, $\mathcal{S}$ is flat. In both cases, $\mathcal{L}$ is counter-clockwise and encloses the $\mathcal{S}$. In both cases, the Eqn (8.23) given above applies. As we take various examples, the application of Ampere's law will become clearer.

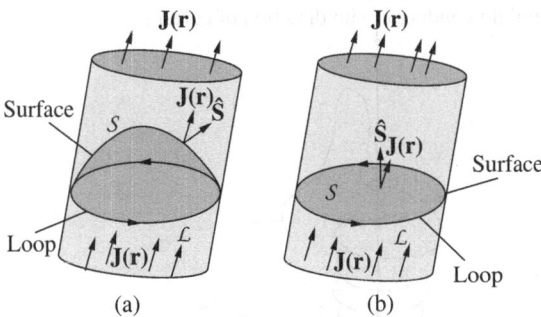

Fig. 8.8 Ampere's law

8.4.1 Ampere's Law Applied to a Long Straight Wire

Example 8.5 Apply Ampere's law to a thin straight wire of infinite length.
Solution Ampere's law is like Gauss's law. If applied properly to certain symmetrical configurations; it can give us quick results with the minimum of effort. Ampere's law also gives a very sound 'feel' for how the magnetic field is produced by current carrying conductors.

One of the most popular applications of Ampere's law is its application to the straight infinite wire, which we consider now.

Figure 8.9 shows an infinite wire extending from $z = -\infty$ to $z = +\infty$ and carrying a current I'. To correctly apply Ampere's law, a closed contour must be chosen to the configuration in question. The line integral of the magnetic field must be equated to the current passing through the enclosed surface. To obtain the current, we may have to do a surface integration.

Various loops (or contours) are shown in Fig. 8.9 which may be considered for application of the law. These will now be considered in turn.

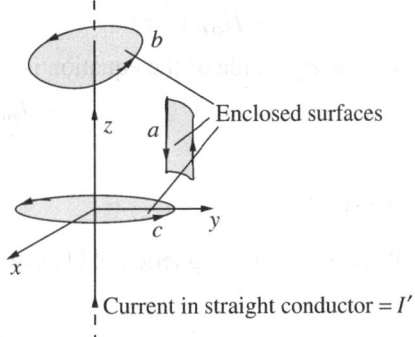

Fig. 8.9 Choosing various contours for Ampere's law

Loop *a* is not good enough because it encloses no current and application of the law will yield no result. The loop *b* is not symmetrical though it does enclose the current. Loop *c* serves our purpose. It has the required symmetry. Loop *c* is a circle lying on a plane parallel to the *x-y* plane with radius ρ. On this loop we expect the magnetic field to have the same value everywhere because of the symmetry of the loop. Using the integral definition of Ampere's law

$$\oint \mathbf{H} \cdot d\mathbf{l} = I_{enclosed} = \iint \mathbf{J} \cdot d\mathbf{S}$$

In cylindrical coordinates, which are the natural coordinates to use in this situation, the left side of the equation leads to

$$\oint \mathbf{H} \cdot d\mathbf{l} = \oint (H_\rho d\rho + H_z dz + H_\phi \rho d\phi)$$

$$= \oint H_\phi \rho \, d\phi \quad (dz \text{ and } d\rho \text{ are both zero})$$

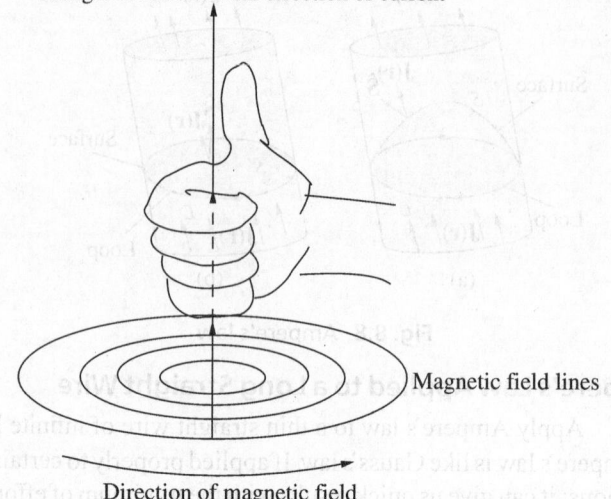

Straight conductor with direction of current

Magnetic field lines

Direction of magnetic field

Fig. 8.10 The direction of the magnetic field vis-a-vis a straight conductor carrying current

$$= H_\phi \rho \int_0^{2\pi} d\phi \quad (H_\phi \text{ and } \rho \text{ are both constant on the loop}) \quad (8.25)$$

$$= H_\phi \rho \, (2\pi) \quad (8.26)$$

and the right side of the equation is

$$I_{enclosed} = I' \quad (8.27)$$

so

$$2\pi H_\phi \rho = I' \quad$$

hence

$$H_\phi = \frac{I'}{2\pi\rho} \quad (8.28)$$

We see that the magnetic field has only one component, namely, the ϕ component

$$\mathbf{H} = [0, H_\phi, 0]$$
$$= \mathbf{a}_\phi H_\phi \quad (8.29)$$

Re-examining the magnetic field, since $H_\phi \propto 1/\rho$,

$$H_\phi \to \infty \quad \text{as} \quad \rho \to 0$$

and
$$H_\phi \to 0 \quad \text{as} \quad \rho \to \infty$$

The magnetic field 'CURLS' *around the current* as shown in Fig. 8.10 as per the right-hand thumb rule.

Example 8.6 Find the curl of **H** in the case of a straight wire carrying a current.
Solution The differential statement of Ampere's law states that the curl of **H** at a point is equal to volume current density, **J**, at that point. For example for the straight wire, if we were to find the curl of **H** at any point except the line $\rho = 0$, then (using the equations for the curl in cylindrical coordinates).

$$(\nabla \times \mathbf{H})_\rho = \left(\underbrace{\frac{1}{\rho} \frac{\partial H_z}{\partial \phi}}_{H_z = 0} - \underbrace{\frac{\partial H_\phi}{\partial z}}_{H_\phi(\rho)} \right) = 0$$

$$(\nabla \times \mathbf{H})_\phi = \left(\underbrace{\frac{\partial H_\rho}{\partial z}}_{H_\rho = 0} - \underbrace{\frac{\partial H_z}{\partial \rho}}_{H_z = 0} \right) = 0$$

$$(\nabla \times \mathbf{H})_z = \frac{1}{\rho} \left\{ \underbrace{\frac{\partial (\rho H_\phi)}{\partial \rho}}_{=0} - \underbrace{\frac{\partial H_\rho}{\partial \phi}}_{H_\rho = 0} \right\} = 0 \qquad (8.30)$$

The first equation above is zero because H_z is zero and H_ϕ is not a function of z. The second equation is zero because H_ρ and H_z are both zero. The third equation is zero because ρH_ϕ is a constant and so

$$\frac{\partial (\rho H_\phi)}{\partial \rho} = \frac{\partial \left(\rho \frac{I'}{2\pi\rho} \right)}{\partial \rho} = 0$$

The second part of the third equation is zero because $H_\rho = 0$.

Why is the curl of the magnetic field zero everywhere? It is zero because everywhere (except along the z-axis, where the current exists) the current density is zero! There is a magnetic field, but its curl is zero.

On the other hand if we were to consider the integral form of Ampere's law and apply it to an infinite wire carrying a constant current I'

$$\oint \mathbf{H} \cdot d\mathbf{l} = \iint \mathbf{J} \cdot d\mathbf{S} = I_{enclosed}$$

$$\oint \mathbf{H} \cdot d\mathbf{l} = \oint \left[H_\rho d\rho + H_\phi (\rho d\phi) + H_z dz \right]$$

$$= \oint H_\phi \rho d\phi \quad (\text{because } H_\rho \text{ and } H_z \text{ are zero})$$

$$= \int_0^{2\pi} \left(\frac{I'}{2\pi\rho} \right) \rho d\phi$$

$$= \int_0^{2\pi} \frac{I'}{2\pi} d\phi$$

$$= I' \quad (\text{which is } = I_{enclosed}) \qquad (8.31)$$

Example 8.7 Find the curl of $\mathbf{H}$ in the case of a short straight wire carrying a current as considered in Example 8.3.

Solution

Step 1. Consider the fields produced by a short straight wire which extends from $-a$ to a. The field is

$$\mathbf{H} = \mathbf{a}_\phi \frac{I'}{4\pi\rho} \left[\frac{(z+a)}{\sqrt{(z+a)^2 + \rho^2}} - \frac{(z-a)}{\sqrt{(z-a)^2 + \rho^2}} \right]$$

Step 2. Applying the curl to $\mathbf{H}$:

$$\nabla \times \mathbf{H} = -\frac{\partial H_\phi}{\partial z} \mathbf{a}_\rho + \frac{1}{\rho} \frac{\partial (\rho H_\phi)}{\partial \rho} \mathbf{a}_z$$

where the H_z and H_ρ fields are equated to zero. Taking the curl

$$(\nabla \times \mathbf{H})_\rho = \frac{1}{\rho}\left(\frac{1}{\sqrt{(z+a)^2+\rho^2}} - \frac{(z+a)^2}{\left((z+a)^2+\rho^2\right)^{\frac{3}{2}}}\right.$$
$$\left. - \frac{1}{\sqrt{(z-a)^2+\rho^2}} + \frac{(z-a)^2}{\left((z-a)^2+\rho^2\right)^{\frac{3}{2}}}\right)$$

$$(\nabla \times \mathbf{H})_z = \frac{(z-a)}{\left((z-a)^2+\rho^2\right)^{\frac{3}{2}}} - \frac{(z+a)}{\left((z+a)^2+\rho^2\right)^{\frac{3}{2}}}$$

which is not zero even at a point where no current density exists. Why is this so? The reason is that the wire does not form a closed loop and there is no physical case where a wire of this type can exist in reality!

8.4.2 Ampere's Law Applied to a Wire of Radius a

Example 8.8 Apply Ampere's law to a current carrying wire of radius a which is infinite in length.

Solution We now apply Ampere's law to straight conductor of circular cross-section of radius a and of infinite length as shown in Fig. 8.11. The wire caries a current I'.

Step 1. Given these conditions, the volume current density is

$$\mathbf{J}' = \frac{I'}{\pi a^2}\mathbf{a}_z \tag{8.32}$$

inside the conductor.

Step 2. To apply Ampere's law in accordance with Eqn (8.20), we choose a circular contour of radius $b < a$ along which to carry out the line integration as shown in the figure. Then

$$\oint_\mathcal{L} \mathbf{H} \cdot d\mathbf{l} = \int_0^{2\pi} H_\phi b \, d\phi$$
$$= H_\phi b \int_0^{2\pi} d\phi$$
$$= 2\pi b H_\phi \tag{8.33}$$

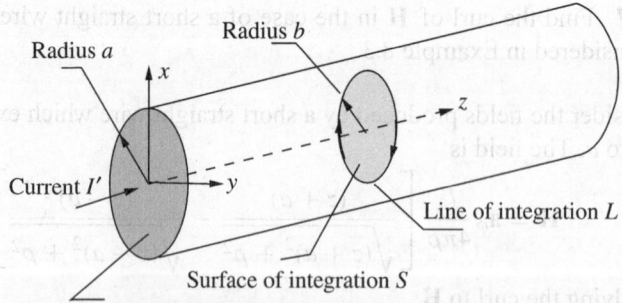

Radius b

Radius a

x

Current I'

y

z

Line of integration L

Surface of integration S

Cross-section of conductor

Fig. 8.11 Straight wire of radius a carrying a current I'

H_ϕ is assumed to be constant on the contour, from symmetry considerations. The surface integration is now carried out

$$\iint_S \mathbf{J}' \cdot d\mathbf{S} = \int\int_{\rho=0,\phi=0}^{\rho=b,\phi=2\pi} \left(\frac{I'}{\pi a^2}\mathbf{a}_z\right) \cdot (\rho\, d\phi\, d\rho\, \mathbf{a}_z)$$

$$= \frac{I'}{\pi a^2} \int\int_{\rho=0,\phi=0}^{\rho=b,\phi=2\pi} \rho\, d\phi\, d\rho$$

$$= \frac{I'b^2}{a^2} \tag{8.34}$$

Step 3. Equating these two results

$$2\pi b H_\phi = \frac{I'b^2}{a^2}$$

$$H_\phi = \frac{I'b}{2\pi a^2} \tag{8.35}$$

Since b is actually a variable ($\equiv \rho$)

$$H_\phi = \frac{I'\rho}{2\pi a^2} \tag{8.36}$$

When $b > a$ then

$$\oint_\mathcal{L} \mathbf{H} \cdot d\mathbf{l} = \int_0^{2\pi} H_\phi b\, d\phi$$

$$= H_\phi b \int_0^{2\pi} d\phi$$

$$= 2\pi b H_\phi \tag{8.37}$$

and the surface integral is

$$\iint_S \mathbf{J}' \cdot d\mathbf{S} = \int\int_{\rho=0,\phi=0}^{\rho=a,\phi=2\pi} \left(\frac{I'}{\pi a^2}\mathbf{a}_z\right) \cdot (\rho\, d\phi\, d\rho\, \mathbf{a}_z)$$

$$= \frac{I'}{\pi a^2} \int\int_{\rho=0,\phi=0}^{\rho=a,\phi=2\pi} \rho\, d\phi\, d\rho$$

$$= \frac{I'a^2}{a^2}$$

$$= I' \tag{8.38}$$

In the above equations the upper limit in the integral sign is $\rho = a$ since the current is only present in the conductor, which is defined by the region $0 \le \phi \le 2\pi$ and $0 \le \rho \le a$. In addition, it is not surprising that the result is equal to I' since the total current crossing the surface is I'. Equating the two equations as earlier, we get

$$2\pi b H_\phi = I'$$

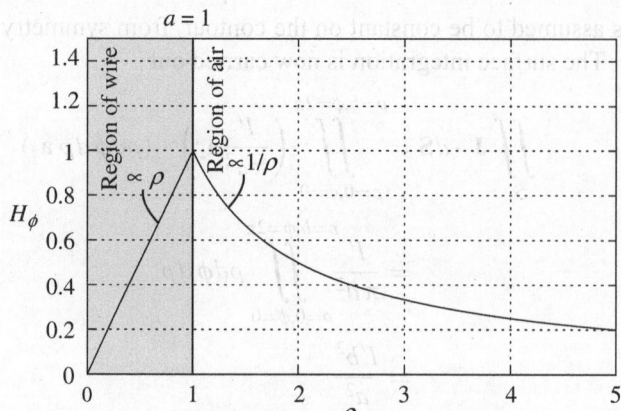

Fig. 8.12 Normalised plot of H_ϕ vs ρ for $a = 1$

or
$$H_\phi = \frac{I'}{2\pi b}$$

but b may be treated as a variable, which is ρ, in the above equation, which then gives

$$H_\phi = \frac{I'}{2\pi \rho} \tag{8.39}$$

Outside the conductor, the result is the same as in the case of an infinitely thin wire.

so $\quad \mathbf{H} = H_\phi \mathbf{a}_\phi = \begin{cases} \dfrac{I'\rho}{2\pi a^2}\mathbf{a}_\phi, & \text{inside the conductor} \\[2mm] \dfrac{I'}{2\pi \rho}\mathbf{a}_\phi, & \text{outside the conductor} \\[2mm] \dfrac{I'}{2\pi a}\mathbf{a}_\phi, & \text{at the boundary} \end{cases}$

Notice that at the boundary, the field is continuous. The normalised plot of H_ϕ is shown in Fig. 8.12.

8.4.3 Ampere's Law Applied to an Infinite Solenoid

Next we apply Ampere's law to the case of an infinite solenoid each turn of which carries a current I'.

Example 8.9 Find the field inside and outside an infinite solenoid as shown in Fig. 8.13.

Solution We next apply Ampere's law to the infinite helix or solenoid shown in Fig. 8.13. The radius of the cross-section of the helix is r and a current I' flows through it. To apply Ampere's law, we draw a closed rectangular loop, namely, a-b-c-d which will be the path of integration in the counter-clockwise sense. The points a,b,c,d in cylindrical coordinates are a is $(\rho_0, \phi_0, -z_0)$, b is (ρ_0, ϕ_0, z_0), c is (ρ_1, ϕ_0, z_0) and the point d is $(\rho_1, \phi_0, -z_0)$.

Applying the right-hand thumb rule to the path of integration we come to the conclusion that the current enclosed by the loop is the total current going *into* the

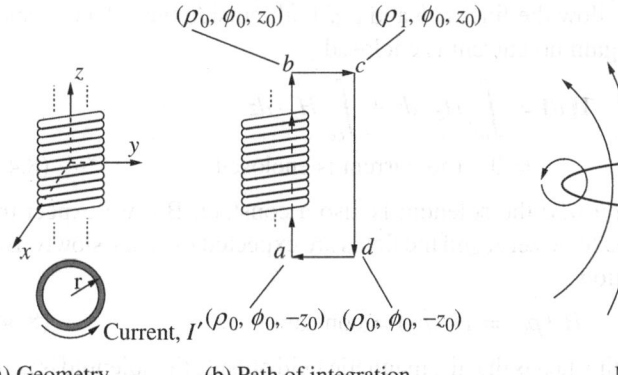

(a) Geometry   (b) Path of integration

Fig. 8.13 The infinite helix

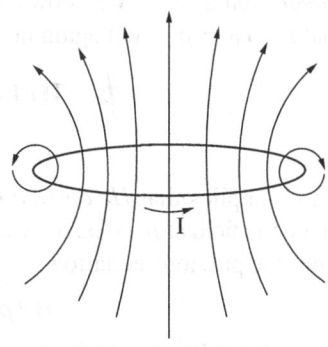

Magnetic field lines

Fig. 8.14 Magnetic field lines for a single loop

plane of the paper. If the helix current is I' and the number of turns in the loop is n then nI' is the total current enclosed. Then if the number of turns/m of the helix be N, then the total number of turns enclosed is $2Nz_0$ turns (because $2z_0$ is the length of the loop) and therefore the total current enclosed is $2Nz_0I'$ (A). The line integral

$$\oint_{abcd} \mathbf{H}\, d\mathbf{l} = \int_{ab} H_z \cdot dz + \int_{bc} H_\rho d\rho + \int_{cd} H_z \cdot dz + \int_{da} H_\rho d\rho \qquad (8.40)$$

$$= 2Nz_0I' \quad \text{(a constant)} \qquad (8.41)$$

We can now play with this line integral. Let us have an idea of the fields existing inside and outside the solenoid. We can get an idea of the fields by imagining that one holds one turn of the solenoid with the right hand with the thumb pointing in the direction of the current. The magnetic field *curls* around the turn. Now as we go all round the turn, we can draw a rough sketch of the fields. Such a sketch is shown in Fig. 8.14 for one turn.

From the sketch it is clear that the fields in the *plane of the loop* will be the fields which will be present in the interior of the infinite solenoid. This is so because for the infinite number of turns, the field configuration in the plane of the loop will be repeated again and again. H_ϕ should be zero because of symmetry considerations. H_ρ will also be zero inside and outside the solenoid because there is no H_ρ in the plane of the loop for a single turn. Only H_z should be present everywhere inside, and perhaps outside. It is also clear that H_z must be a function of ρ only.

Integration 1. Referring to Fig. 8.13, we make line *c-d* to be inside the helix itself, that is $\rho_0 < \rho_1 < r$ (where r is the radius of the solenoid). Using this path, no current is enclosed

$$\oint_{abcd} \mathbf{H}\, d\mathbf{l} = \int_{ab} H_z \cdot dz + \int_{bc} H_\rho d\rho + \int_{cd} H_z \cdot dz + \int_{da} H_\rho d\rho$$

$$= \int_{ab} H_z \cdot dz + \int_{cd} H_z \cdot dz \quad \text{since } H_\rho = 0$$

$$= 0 \quad \text{(no current is enclosed)}$$

which implies that

$$H_z(\rho_0) = H_z(\rho_1) = c \quad \text{(a constant)} \qquad (8.42)$$

This is the magnetic field inside the solenoid.

Integration 2. Now we allow the lines *a-b* and *c-d* both outside the solenoid, such that $r < \rho_0 < \rho_1$, and again no current is enclosed

$$\oint_{abcd} \mathbf{H}\,d\mathbf{l} = \int_{ab} H_z \cdot dz + \int_{cd} H_z \cdot dz$$

$$= 0 \quad \text{(no current is enclosed)} \tag{8.43}$$

which implies that H_z *outside* the solenoid is also a constant. But we expect the magnetic field at $\rho \to \infty$ to be zero, and the fields are expected to decay slowly. But from the previous equation

$$H_z(\rho_0) = H_z(\rho_1) = \text{constant} \tag{8.44}$$

The only explanation to this fact is that the magnetic field outside the solenoid is zero. Hence

$$H_z(\rho_0) = H_z(\rho_1) = \text{constant} = 0 \tag{8.45}$$

Integration 3. Now we allow *a-b* to be inside and *c-d* to be outside the solenoid, that is $\rho_0 < r < \rho_1$, so that

$$\oint_{abcd} \mathbf{H}\,d\mathbf{l} = \int_{ab} H_z \cdot dz + \int_{cd} H_z \cdot dz$$

$$= \int_{ab} H_z \cdot dz$$

$$= 2z_0 H_z(\rho_0)$$

But the right-hand side must be equal to the total current enclosed

$$\text{Current enclosed} = 2Nz_0 I'$$

From the previous equations

$$2z_0 H_z(\rho_0) = 2Nz_0 I'$$

$$H_z(\rho_0) = \frac{2Nz_0 I'}{2z_0}$$

$$= NI'$$

This is the only component of the magnetic field, the other components being zero. Hence

$$\mathbf{H} = \begin{cases} [0, 0, NI'], & \text{for } 0 < \rho < r \\ 0, & \text{for } r < \rho < \infty \end{cases} \tag{8.46}$$

8.4.4 Ampere's Law Applied to a Winding Around a Torus

Example 8.10 Apply Ampere's law to a torus wound with wire carrying a constant current as shown in Fig. 8.15.

Solution

Step 1. Ampere's law applied to a torus is

$$\oint \mathbf{H} \cdot d\mathbf{l} = NI \tag{8.47}$$

where the integration is to be performed along the path shown. N is the total number of turns and I is the current. Note that the current enters into surface enclosed by the path.

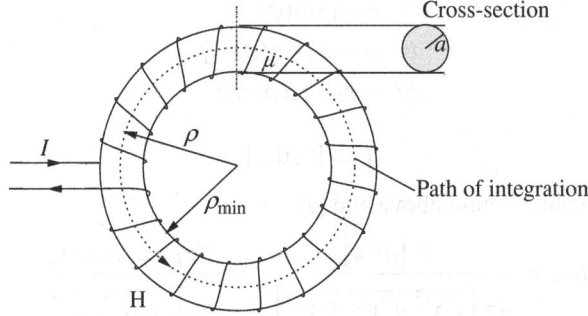

Fig. 8.15 A torus with a winding

Step 2.
$$\oint H_\phi \rho d\phi = NI$$
$$H_\phi = \frac{NI}{2\pi\rho} \quad \rho_{min} < \rho < \rho_{min} + 2a \tag{8.48}$$

Step 3. If the circular closed path is chosen outside ($\rho > \rho_{min} + 2a$) or inside the torus ($\rho < \rho_{min}$), no current crosses the surface and the field in these regions is zero.

8.5 | The Magnetic Field

8.5.1 Loop of Wire Carrying a Current

We now compute the magnetic field for a loop of current.

Example 8.11 Find the magnetic field on the axis of a ring carrying a constant current.

Solution In this example we apply our Biot-Savart law formulae to find the magnetic field along the axis of a current loop shown in Fig. 8.16. The current I' flows in the wire and the loop is placed flat on the x-y plane. The radius of the loop is ρ_0. Scrutinising the figure, it is clear that

$$\mathbf{r}' = [x', y', 0]$$
$$x' = \rho_0 \cos(\phi)$$

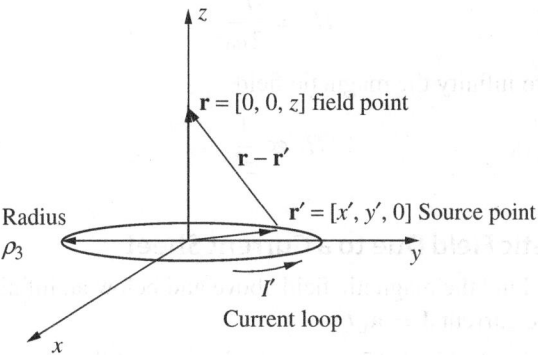

Fig. 8.16 Magnetic field on the central axis due to a current loop

$$y' = \rho_0 \sin(\phi)$$
$$dx' = -\rho_0 \sin(\phi)\, d\phi$$
$$dy' = \rho_0 \cos(\phi)\, d\phi \tag{8.49}$$

with
$$\mathbf{r} = [0, 0, z] \tag{8.50}$$

We plug in the values found above and get

$$dH_x = \frac{I'\left[dy'z\right]}{4\pi\left[\sqrt{x'^2 + y'^2 + z^2}\,\right]^3} = \frac{I'z\rho_0 \cos(\phi)\, d\phi}{4\pi\left[\sqrt{\rho_0^2 + z^2}\,\right]^3}$$

$$dH_y = \frac{I'\left[-dx'z\right]}{4\pi\left[\sqrt{x'^2 + y'^2 + z^2}\,\right]^3} = \frac{I'z\rho_0 \sin(\phi)\, d\phi}{4\pi\left[\sqrt{\rho_0^2 + z^2}\,\right]^3}$$

$$dH_z = \frac{I'\left[dx'(-y') - dy'(-x')\right]}{4\pi\left[\sqrt{x'^2 + y'^2 + z^2}\,\right]^3} = \frac{I'\rho_0^2\, d\phi}{4\pi\left[\sqrt{\rho_0^2 + z^2}\,\right]^3} \tag{8.51}$$

Integrating these equations

$$H_x = \int_0^{2\pi} \frac{I'z\rho_0 \cos(\phi)\, d\phi}{4\pi\left[\sqrt{\rho_0^2 + z^2}\,\right]^3} = 0$$

$$H_y = \int_0^{2\pi} \frac{I'z\rho_0 \sin(\phi)\, d\phi}{4\pi\left[\sqrt{\rho_0^2 + z^2}\,\right]^3} = 0$$

$$H_z = \int_0^{2\pi} \frac{I'\rho_0^2\, d\phi}{4\pi\left[\sqrt{\rho_0^2 + z^2}\,\right]^3} = \frac{I'\rho_0^2}{2\left[\sqrt{\rho_0^2 + z^2}\,\right]^3} \tag{8.52}$$

The magnetic field along the axis of the loop has only a single component, namely, the z component, and the field line starts at $-\infty$ and proceeds to ∞. Note that at $z = 0$ the loop produces a field

$$H_z = \frac{I'}{2\rho_0}$$

while as z tends to infinity the magnetic field

$$H_z \propto \frac{1}{z^3}$$

8.5.2 Magnetic Field Due to a Current Sheet

Example 8.12 Find the magnetic field above and below an infinite current sheet carrying a surface current $\mathbf{J} = \mathbf{a}_x J_{sx}$.

Solution Referring to Fig. 8.17 we can apply some of the above concepts to the case of the magnetic field produced by a current sheet. The figure shows a current

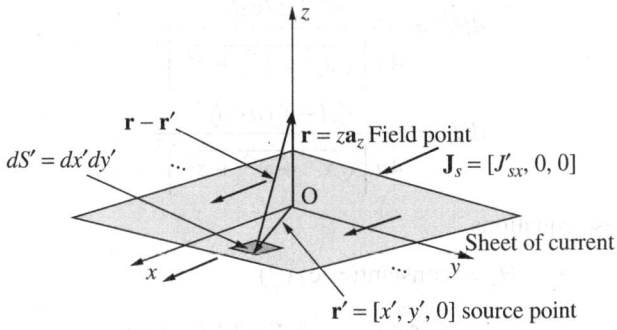

Fig. 8.17 A current sheet with an x directed J_s

flowing in the $\mathbf{a}_x$ direction with $\mathbf{J}'_s = \left[J'_{sx}, 0, 0\right]$. If we use Eqn (8.8) the various terms of this equation in the most general case, in rectangular coordinates, are

$$\mathbf{J}'_s = \mathbf{a}_x J'_{sx} + \mathbf{a}_y J'_{sy} + \mathbf{a}_z J'_{sz}$$
$$\mathbf{r} = [x, y, z]$$
$$\mathbf{r}' = [x', y', z']$$
$$\mathbf{r} - \mathbf{r}' == \mathbf{R} = [x - x', y - y', z - z']$$
$$\hat{\mathbf{R}} = \frac{[x - x', y - y', z - z']}{\sqrt{(x - x')^2 + (y - y')^2 + (z - z')^2}} \quad (8.53)$$

Hence

$$dH_x = \frac{\left[J'_{sy}(z - z') - J'_{sz}(y - y')\right] dS'}{4\pi \left[\sqrt{(x - x')^2 + (y - y')^2 + (z - z')^2}\,\right]^3}$$

$$dH_y = \frac{\left[J'_{sz}(x - x') - J'_{sx}(z - z')\right] dS'}{4\pi \left[\sqrt{(x - x')^2 + (y - y')^2 + (z - z')^2}\,\right]^3}$$

$$dH_z = \frac{\left[J'_{sx}(y - y') - J'_{sy}(x - x')\right] dS'}{4\pi \left[\sqrt{(x - x')^2 + (y - y')^2 + (z - z')^2}\,\right]^3} \quad (8.54)$$

From Fig. 8.17,

$$\mathbf{J}'_s = \mathbf{a}_x J'_{sx}$$
$$\mathbf{r} = [0, 0, z]$$
$$\mathbf{r}' = [x', y', 0]$$
$$\mathbf{r} - \mathbf{r}' = [x - x', y - y', z - z']$$
$$= [-x', -y', z]$$
$$\hat{\mathbf{R}} = \frac{[-x', -y', z]}{\sqrt{x'^2 + y'^2 + z^2}}$$
$$dS' = dx'\, dy' \quad (8.55)$$

Here $\mathbf{r} = [0, 0, z]$ since the field on the z-axis will be the same as the field anywhere. Using these equations

$$dH_x = 0$$

$$dH_y = \frac{-J'_{sx}z\,dx'\,dy'}{4\pi\left[\sqrt{x'^2+y'^2+z^2}\,\right]^3}$$

$$dH_z = \frac{J'_{sx}(-y')\,dx'\,dy'}{4\pi\left[\sqrt{x'^2+y'^2+z^2}\,\right]^3} \qquad (8.56)$$

Integrating these equations

$$H_x = \text{constant}(= c)\;(?)$$

$$H_y = \int\!\!\int_{x,y=-\infty}^{x,y=\infty} \frac{-J'_{sx}z\,dx'\,dy'}{4\pi\left[\sqrt{x'^2+y'^2+z^2}\,\right]^3}$$

$$H_z = \int\!\!\int_{x,y=-\infty}^{x,y=\infty} \frac{J'_{sx}(-y')\,dx'\,dy'}{4\pi\left[\sqrt{x'^2+y'^2+z^2}\,\right]^3} \qquad (8.57)$$

In the very first equation a question mark appears. That is so because the integration is a definite integral and *the definite integration of zero is zero*. The third integral is zero because the integrand is an odd function of y'. So we are left with the second integration

$$H_y = \int\!\!\int \frac{-J'_{sx}z\,dx'\,dy'}{4\pi\left[\sqrt{x'^2+y'^2+z^2}\,\right]^3} \qquad (8.58)$$

Integrating with respect to x', using the integral

$$\int \frac{-J'_{sx}z\,dx'}{4\pi\left[\sqrt{x'^2+y'^2+z^2}\,\right]^3} = \frac{-J'_{sx}zx'}{4\pi\left[(z^2+y'^2)\,\sqrt{z^2+y'^2+x'^2}\,\right]} \qquad (8.59)$$

so
$$H_y = \int_{y'=-\infty}^{y'=\infty} dy'\left[\frac{-J'_{sx}zx'}{4\pi\left[(z^2+y'^2)\,\sqrt{z^2+y'^2+x'^2}\,\right]}\right]_{x'=-\infty}^{x'=\infty}$$

$$= \int_{y'=-\infty}^{y'=\infty} \frac{-J'_{sx}z\,dy'}{2\pi\left[(z^2+y'^2)\right]} \qquad (8.60)$$

Performing the second integration, we get

$$H_y = -\frac{J'_{sx}}{2\pi} \times \left[\arctan\left(\frac{y'}{z}\right)\Big|_{y'=-\infty}^{y'=\infty}\right]$$

$$= -\frac{J'_{sx}}{2\pi}(\pi), \qquad \text{for } z > 0$$

$$= -\frac{J'_{sx}}{2\pi}(-\pi), \qquad \text{for } z < 0 \qquad (8.61)$$

or more compactly

$$H_x = 0$$

$$H_y = \begin{cases} -\dfrac{J'_{sx}}{2}, & \text{for } z > 0 \\ \dfrac{J'_{sx}}{2}, & \text{for } z < 0 \end{cases}$$

$$H_z = 0 \tag{8.62}$$

If we consider the region above the sheet, the unit vector of the surface towards the region above is $\mathbf{a}_z$. That is, this is the unit vector of the surface towards the upper region. Then

$$\mathbf{a}_z \times \left(\mathbf{H}_{\text{above}} - \mathbf{H}_{\text{below}}\right) = \mathbf{a}_z \times \left[\left(-\frac{J_{sx}}{2}\right)\mathbf{a}_y - \left(\frac{J_{sx}}{2}\right)\mathbf{a}_y\right]$$

$$= \mathbf{a}_z \times \left[(-J_{sx})\,\mathbf{a}_y\right]$$

$$= J_{sx}\mathbf{a}_x$$

$$= \mathbf{J}_s \tag{8.63}$$

This result is significant as will be clear when we consider boundary conditions.

8.5.3 Magnetic Field in the Interior of an Infinite Solenoid

Example 8.13 Find the magnetic field at the origin for an infinite solenoid shown in Fig. 8.18. The figure shows a cross-section of the solenoid with current coming out of the plane of the paper in the upper conductors and going into the plane of the paper in the lower conductors.

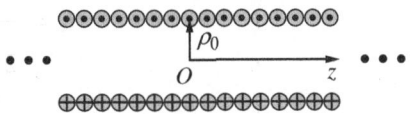

Fig. 8.18 Field at the origin of an infinite solenoid

Solution For a single loop, the magnetic field on the axis of a loop (from Example 8.11) is

$$H_z = \frac{I'\rho_0^2}{2\left[\sqrt{\rho_0^2 + z^2}\right]^3}$$

and if there are N turns per metre along an infinite solenoid, then at the origin, the z-directed magnetic field is given by the integral (by Appendix B.6.2, formula 29)

$$H_z = \int_{-\infty}^{\infty} \frac{NI'\rho_0^2\, dz}{2\left[\sqrt{\rho_0^2 + z^2}\right]^3} = \left[\frac{NI'\,z}{2\sqrt{z^2 + \rho_0^2}}\right]_{-\infty}^{\infty} = 2\left(\frac{NI'}{2}\right) = NI'$$

This calculation corroborates the result of Example 8.9.

8.5.4 Magnetic Field in the Interior of a Finite Solenoid

Example 8.14 Find the magnetic field on the axis of a finite solenoid shown in Fig. 8.19. The figure shows a cross-section of the solenoid with current coming out of the plane of the paper in the upper conductors and going into the plane of the paper in the lower conductors. The solenoid is of length L

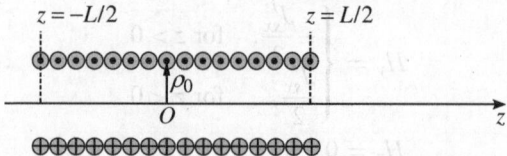

Fig. 8.19 Field at position z for a finite solenoid of length L with N turns/m

Solution

Step 1. We start with equation of the magnetic field for a single loop in flush with the xy plane with its centre at the origin and carrying a current I', Eqn (8.52)

$$H_z = \frac{I'\rho_0^2}{2\left[\sqrt{\rho_0^2 + z^2}\,\right]^3}$$

Step 2. Now from the structure of the Biot-Savart law equations, we can straight away see that if the centre of the loop was to be shifted to z', the magnetic field at $(0, 0, z)$ would be

$$H_z = \frac{I'\rho_0^2}{2\left[\sqrt{\rho_0^2 + (z - z')^2}\,\right]^3}$$

Step 3. Now for a current $NI'dz'$, the infinitesimal magnetic field would become

$$dH_z = \frac{NI'\rho_0^2\, dz'}{2\left[\sqrt{\rho_0^2 + (z - z')^2}\,\right]^3}$$

where N is the number of turns per metre and Ndz' are the differential number of turns. Integrating this expression using integral 29 of Appendix B.6.2,

$$H_z = \frac{NI'\rho_0^2}{2}\left[\frac{(z' - z)}{\rho_0^2\left(\sqrt{(z' - z)^2 + \rho_0^2}\right)}\right]_{-L/2}^{L/2}$$

$$= \frac{NI'}{2}\left[\frac{(L/2 - z)}{\left(\sqrt{(L/2 - z)^2 + \rho_0^2}\right)} + \frac{(L/2 + z)}{\left(\sqrt{(L/2 + z)^2 + \rho_0^2}\right)}\right] \quad (8.64)$$

if we plot only the bracketed part ($NI'/2 = 1$) the plot of H_z versus z is shown in Fig. 8.20. The dimensions chosen are of course unrealistic but they are chosen just to show a representative plot.

@ See FiniteSolenoid.m in Chapter 8

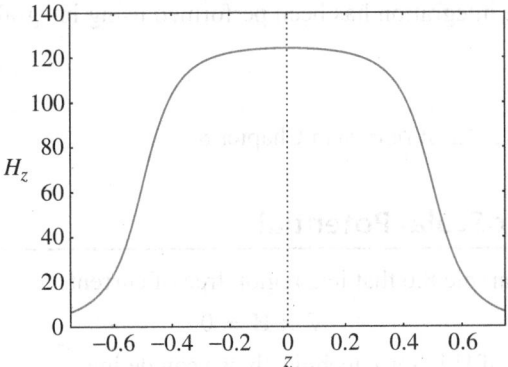

Fig. 8.20 Plot of H_z vs z for a solenoid of length $L = 1$ m and radius $\rho_0 = 0.125$ m

8.5.5 Magnetic Field on the Axis of a Rotating Charged Disk

Example 8.15 A charged disk with surface charge ρ_s and radius a is rotated with an angular velocity ω rad/s. Find the magnetic field at a distance z from the centre of the disk as shown in Fig. 8.21.

Solution

Step 1. The current density at a distance ρ along the surface of the disk is

$$\mathbf{J}_s = \rho_s \mathbf{v} = \rho_s \omega \rho \mathbf{a}_\phi$$

where ρ_s is the surface charge density, $\mathbf{v}$ is the velocity at that point, and ω is the angular velocity. The total convection current flowing there in the form of a ring is

$$I'(\rho)\, d\rho = \rho_s \omega \rho \, d\rho$$

Step 2. Using the results of a ring of current, for small ring of width $d\rho$ carrying a current $\rho_s \omega \rho$, by Example 8.11,

$$dH_z = \frac{I'(\rho)\rho^2 d\rho}{2\left[\sqrt{\rho^2 + z^2}\,\right]^3} = \frac{\rho_s \omega \rho^3 d\rho}{2\left[\sqrt{\rho^2 + z^2}\,\right]^3}$$

$$H_z = 2\rho_s \omega \left[\frac{z^2}{\sqrt{z^2 + \rho^2}} + \frac{\rho^2}{\sqrt{z^2 + \rho^2}}\right]_{\rho=0}^{\rho=a}$$

$$= 2\rho_s \omega \left(\sqrt{z^2 + a^2} - z\right)$$

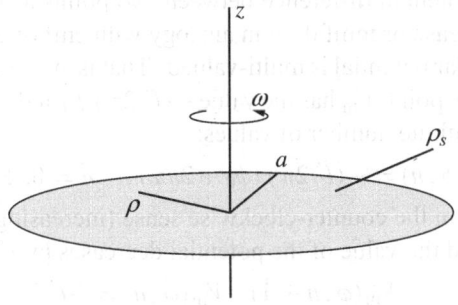

Fig. 8.21 Rotating disk of charge

where the integration has been performed using integral 32 of Appendix B.6.2.

 See `MagFIdChRotDsc.m` in Chapter 8

8.6 | Magnetic Scalar Potential

We know from Example 8.6 that in a region free of currents,

$$\nabla \times \mathbf{H} = 0$$

then since the curl of H is zero, technically we can define

$$\mathbf{H} = -\nabla V_\mathrm{m} \qquad (8.65)$$

where V_m is the magnetic scalar potential. Let us take the case of a wire with radius a. The magnetic field inside and outside the wire is

$$\mathbf{H} = \begin{cases} \dfrac{I' \rho}{2\pi a^2} \mathbf{a}_\phi, & \text{inside the conductor} \\[2mm] \dfrac{I'}{2\pi \rho} \mathbf{a}_\phi, & \text{outside the conductor} \end{cases}$$

Observing these equations we realise that inside the conductor we cannot define a magnetic scalar potential since

$$\nabla \times \mathbf{H} \neq 0$$

since there is a current density present. Outside the conductor, however, we may define a magnetic scalar potential: taking the line integral of **H** outside the conductor along a circle whose centre lies on the z-axis,

$$\mathbf{H} = -\nabla V_\mathrm{m}$$

$$\frac{I'}{2\pi \rho} \mathbf{a}_\phi = -\left(\frac{\partial V_\mathrm{m}}{\partial \rho} \mathbf{a}_\rho + \frac{1}{\rho} \frac{\partial V_\mathrm{m}}{\partial \phi} \mathbf{a}_\phi + \frac{\partial V_\mathrm{m}}{\partial z} \mathbf{a}_z \right)$$

$$\frac{I'}{2\pi \rho} = -\frac{1}{\rho} \frac{\partial V_\mathrm{m}}{\partial \phi}$$

or

$$V_\mathrm{m} = -\int_0^\phi \frac{I'}{2\pi \rho} \rho \, d\phi = -\frac{I'}{2\pi} \phi$$

Let us take a moment and examine this equation carefully.

1. The unit of the potential is amperes.
2. We can call the potential difference between two points as the 'magnetomotive force' (mmf) increase or mmf drop in analogy with emf of electric circuits.
3. The magnetic scalar potential is multi-valued. That is at $\phi = \phi_0$ and $\phi = \phi_0 + 2\pi$ which are the same point, V_m has the value $-(I'/2\pi) \phi_0$ and $-(I'/2\pi)(\phi_0 + 2\pi)$. In fact it has an infinite number of values:

$$V_\mathrm{m}(\phi, n) = -(I'/2\pi)(\phi_0 + 2n\pi), \quad n = 0, \pm 1, \ldots$$

4. When one circuit in the counter-clockwise sense (increasing value of ϕ) of the circle is completed the value of the potential decreases by I'

$$V_\mathrm{m}(\phi, n+1) - V_\mathrm{m}(\phi, n) = -I'$$

so the mmf decrease for one circuit is I'.

5. Magnetic field lines flow from higher potentials to lower potentials. Therefore, for this case the field flows along the ϕ-direction.

6. If apply Laplace's equation to V_m then

$$\nabla^2 V_m = \frac{1}{\rho}\frac{\partial}{\partial \rho}\left(\rho\frac{\partial V_m}{\partial \rho}\right) + \frac{1}{\rho^2}\frac{\partial^2 V_m}{\partial \phi^2} + \frac{\partial^2 V_m}{\partial z^2}$$

$$= \frac{1}{\rho^2}\frac{\partial^2 V_m}{\partial \phi^2} \quad (\text{cancelled terms as } V_m \text{ not function of } \rho \text{ and } z)$$

$$= \frac{I'}{2\pi\rho^2}\frac{\partial^2 \phi}{\partial \phi^2} = 0$$

One way to remove difficulty given under point 3 above is to define V_m by

$$V_m = -\frac{I'}{2\pi}\phi, \quad \text{for } 0 < \phi < 2\pi$$

We will look at these points again when we consider magnetic circuits.

8.6.1 Scalar Potential in the Interior of an Infinite Solenoid

Example 8.16 Find the magnetic scalar potential for the field inside and outside of an infinite solenoid.

Solution The field inside a solenoid is

$$\mathbf{H} = NI'\mathbf{a}_z$$

where N is the number of turns/metre and I' is the current in the winding. Then

$$\mathbf{H} = -\left(\frac{\partial V_m}{\partial \rho}\mathbf{a}_\rho + \frac{1}{\rho}\frac{\partial V_m}{\partial \phi}\mathbf{a}_\phi + \frac{\partial V_m}{\partial z}\mathbf{a}_z\right)$$

$$NI' = -\frac{\partial V_m}{\partial z}$$

$$V_m = -NI'z$$

In this case the function is single-valued. Outside the solenoid, there is no field, so there

$$V_m = c \quad \text{(a constant)}$$

Note that just like the case of electric fields, *magnetic field lines flow from higher potentials to lower potentials.*

8.7 | Vector Potential and Magnetic Flux Density

In the history of the study of electromagnetics, the magnetic field has never been known to possess an isolated pole. For the electric field, the electric flux density vector, $\mathbf{D}$, is known to obey Gauss's law, namely

$$\nabla \cdot \mathbf{D} = \rho_v \quad \text{and} \quad \iiint_V \mathbf{D} \cdot d\mathbf{S} = Q_{\text{enclosed}}$$

where Q_{enclosed} and ρ_v are the sources of the field. On the other hand, Ampere's law states that the magnetic field is produced by currents which, of course, are only moving charges.

On the basis of these ideas, we define an analogous vector, **B**, which is the magnetic flux per metre squared. The magnetic flux Ψ_m is measured in Webers, and so **B** is measured in webers per metre squared (which is also called Tesla, T). So

$$\oiint_S \mathbf{B} \cdot d\mathbf{S} = 0 \quad \text{(no magnetic sources)}$$

where S is a closed surface. Then by using the divergence theorem

$$\iiint_V \nabla \cdot \mathbf{B} \, dV = \oiint_S \mathbf{B} \cdot d\mathbf{S} = 0 \tag{8.66}$$

which leads to another of Maxwell's equations, namely,

$$\nabla \cdot \mathbf{B} = 0 \tag{8.67}$$

where **B** is the magnetic flux density. The flux density and magnetic field are linked by a constant, namely

$$\mathbf{B} = \mu \mathbf{H} \tag{8.68}$$

where μ is the permeability of a medium. For vacuum or air $\mu = \mu_0 = 4\pi \times 10^{-7}$ (H/m). Equation (8.66) says that the surface integral of **B** over the closed surface S is zero, or the total *flux* leaving the surface is always zero (see Fig. 8.22). *Since the total magnetic flux out of any closed surface is always zero, therefore there are no magnetic charges!* (or mono-poles)

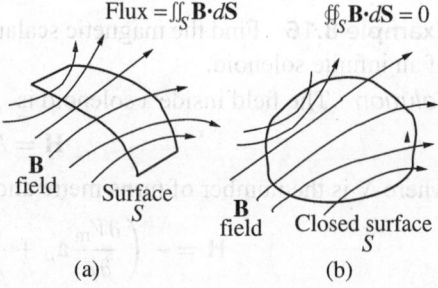

Flux $= \iint_S \mathbf{B} \cdot d\mathbf{S}$ $\oiint_S \mathbf{B} \cdot d\mathbf{S} = 0$

(a)   (b)

Fig. 8.22 Magnetic flux

Going back to Eqn (8.67) we know from our knowledge of vector analysis that when the divergence of a vector is zero, that vector field must be the curl of some other vector. In other words, if

$$\nabla \cdot \mathbf{B} = 0$$

then

$$\mathbf{B} = \nabla \times \mathbf{A} \tag{8.69}$$

where **A**, in theory can be any vector field. Thus given any vector field **A** it will, by the previous equation give us a possible **B** field. **A** is called the vector potential. Can two vector potentials give us the same **B** field? Let us investigate this.

8.7.1 Calculation of the Vector Potential

Let two fields $\mathbf{A}_1$ and $\mathbf{A}_2$ give us the same **B** field

$$\mathbf{B} = \nabla \times \mathbf{A}_1 \tag{8.70}$$
$$= \nabla \times \mathbf{A}_2 \tag{8.71}$$

therefore

$$\nabla \times (\mathbf{A}_1 - \mathbf{A}_2) = 0$$

recall that when the curl of a vector field is zero then that vector must be the gradient of a scalar. Therefore

$$\mathbf{A}_1 - \mathbf{A}_2 = \nabla \phi \tag{8.72}$$
$$\mathbf{A}_1 = \mathbf{A}_2 + \nabla \phi \tag{8.73}$$

So, vector potential is not unique: two different **A** fields can give us the same **B** field. The difference between the two possible vector potentials is the gradient of a scalar.

How do we get a *practical or real* vector potential? That is, given a set of currents, how do we proceed to obtain the vector potential? To do this let us look at another of Maxwell's equations, namely Ampere's law in magnetostatics

$$\nabla \times \mathbf{H} = \mathbf{J}$$

or
$$\nabla \times \mathbf{B} = \mu_0 \mathbf{J} \tag{8.74}$$

substituting Eqn (8.69) in the above equation and using

$$\nabla \times (\nabla \times \mathbf{A}) = \nabla (\nabla \cdot \mathbf{A}) - (\nabla \cdot \nabla) \mathbf{A} \tag{8.75}$$

we get
$$\nabla \times (\nabla \times \mathbf{A}) = \mu_0 \mathbf{J}$$
$$\nabla (\nabla \cdot \mathbf{A}) - (\nabla \cdot \nabla) \mathbf{A} = \mu_0 \mathbf{J} \tag{8.76}$$

Any vector is not fully determined until both its curl and divergence are both specified. We choose (for a reason)

$$\nabla \cdot \mathbf{A} = 0 \tag{8.77}$$

so that the previous equation becomes

$$-(\nabla \cdot \nabla) \mathbf{A} = \mu_0 \mathbf{J}$$
$$\nabla^2 \mathbf{A} = -\mu_0 \mathbf{J} \tag{8.78}$$

On examination of the above equation, we find that we are dealing with three equations

$$\nabla^2 A_x = -\mu_0 J_x$$
$$\nabla^2 A_y = -\mu_0 J_y$$
$$\nabla^2 A_z = -\mu_0 J_z \tag{8.79}$$

Each of these equations are mathematically similar to the potential equation

$$\nabla^2 V = -\rho_v / \varepsilon_0 \tag{8.80}$$

The solution to the above equation is

$$V(\mathbf{r}) = \frac{1}{4\pi \varepsilon_0} \iiint_{\mathcal{V}'} \frac{\rho_v (\mathbf{r}') \, dv'}{|\mathbf{r} - \mathbf{r}'|} \tag{8.81}$$

where $V(\mathbf{r})$ is the potential at the field point $\mathbf{r}$ due to charges $\rho_v (\mathbf{r}')$ in the volume $\mathcal{V}'$ whose position vector is specified by the $\mathbf{r}'$.[1] Using the above result we can assert that[2]

$$A_i(\mathbf{r}) = \frac{\mu_0}{4\pi} \iiint_{\mathcal{V}'} \frac{J_i(\mathbf{r}') \, dV'}{|\mathbf{r} - \mathbf{r}'|} \quad i = x, y, z \tag{8.82}$$

where $A_i(\mathbf{r})$ is the ith component ($i = x, y$ or z) of the *vector* potential at the field point $\mathbf{r}$ due to currents $J_i(\mathbf{r}')$ in the volume $\mathcal{V}'$ whose position vector is specified by the $\mathbf{r}'$. These three equations can be recombined into

$$\mathbf{A}(\mathbf{r}) = \frac{\mu_0}{4\pi} \iiint_{\mathcal{V}'} \frac{\mathbf{J}(\mathbf{r}') \, dV'}{|\mathbf{r} - \mathbf{r}'|} \tag{8.83}$$

---

[1] Note that dv' is used instead of dV' to avoid confusion with V, the potential.
[2] Note that instead of ε_0 in the denominator there is μ_0 in the numerator.

Figure 8.23 depicts the various variables. We can formulate the above equations a bit differently. Instead of the term $\mathbf{J}_v dV'$, we can use other formulations (as given in Section 8.3)

$$\mathbf{A}(\mathbf{r}) = \frac{\mu_0}{4\pi} \iint_{S'} \frac{\mathbf{J}_s(\mathbf{r}')\,dS'}{|\mathbf{r}-\mathbf{r}'|} \quad \text{for surface currents}$$

(8.84)

$$\mathbf{A}(\mathbf{r}) = \frac{\mu_0}{4\pi} \int_{\mathcal{L}'} \frac{I(\mathbf{r}')\,d\mathbf{l}'}{|\mathbf{r}-\mathbf{r}'|} \quad \text{for line currents}$$

(8.85)

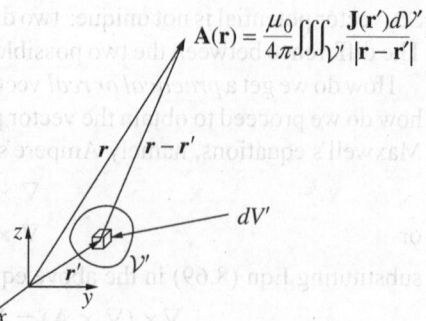

$$\mathbf{A}(\mathbf{r}) = \frac{\mu_0}{4\pi} \iiint_{V'} \frac{\mathbf{J}(\mathbf{r}')dV'}{|\mathbf{r}-\mathbf{r}'|}$$

Fig. 8.23 The geometry of how the vector potential is calculated

8.7.2 Vector Potential of a Circular Loop

Example 8.17 Find the magnetic vector potential, far away from the origin, for a loop of current of radius ρ_0 with current I' as shown in Fig. 8.16.

Solution Since $d\mathbf{l} = \rho_0\,d\phi\,\mathbf{a}_\phi$ and $I(\mathbf{r}')$ is a constant, and using the results of Appendix B.1.3

$$\mathbf{A}(\mathbf{r}) = \frac{\mu_0 I'}{4\pi} \int \frac{\rho_0\,d\phi'\,\mathbf{a}_\phi}{\sqrt{r'^2 + r^2 - 2r\,r'\cos(\phi-\phi')\sin(\theta)\sin(\theta') - 2\,r\,r'\cos(\theta)\cos(\theta')}}$$

Since the loop lies on the xy plane, $\theta' = \pi/2$ and $r' = \rho_0$. Also we evaluate the integral for the field point $\phi = 0$ or the xz plane:

$$\mathbf{A}(\mathbf{r}) = \frac{\mu_0 I'}{4\pi} \int \frac{\rho_0\,d\phi'\,\mathbf{a}_\phi}{\sqrt{\rho_0^2 + r^2 - 2r\,r'\cos\phi'\,\sin(\theta)}}$$

$$\approx \frac{\mu_0 I'}{4\pi r} \int \frac{\rho_0\,d\phi'\,(-\sin\phi\,\mathbf{a}_x + \cos\phi\,\mathbf{a}_y)}{\sqrt{1 - 2\,(\rho_0/r)\cos\phi'\,\sin(\theta)}}$$

$$\approx \frac{\mu_0 I'}{4\pi r} \int_{\phi'} \rho_0\,(-\sin\phi\,\mathbf{a}_x + \cos\phi\,\mathbf{a}_y)\left[1 + (\rho_0/r)\cos(\phi-\phi')\sin(\theta)\right]d\phi'$$

where we have used $\mathbf{a}_x$ and $\mathbf{a}_y$ since these are constant vectors, otherwise integration with vectors which are always changing direction is impossible. Evaluating this integral on the xz plane where $\phi = 0$

$$A_x = \frac{\mu_0 I'\rho_0}{4\pi r} \int_{\phi'} (-\sin\phi')\,[1 + (\rho_0/r)\cos(\phi')\sin(\theta)]d\phi' = 0 \quad (A_\rho = 0)$$

$$A_y = \frac{\mu_0 I'\rho_0}{4\pi r} \int_{\phi'} (\cos\phi')\,[1 + (\rho_0/r)\cos(\phi')\sin(\theta)]d\phi'$$

$$= \frac{\mu_0 \pi I'\rho_0^2 \sin\theta}{4\pi} \frac{1}{r^2} \quad (A_\phi)$$

$$A_z = 0$$

Note that A_y is A_ϕ.

Therefore in spherical coordinates,

$$\mathbf{A} = \left[0, 0, \frac{\mu_0\pi I'\rho_0^2 \sin\theta}{4\pi}\frac{1}{r^2}\right]$$

(8.86)

hence
$$\mathbf{B} = \nabla \times \mathbf{A}$$

$$= \frac{\mu_0 \pi I' \rho_0^2}{4\pi} \left[\frac{2 \cos(\theta)}{r^3}, \frac{\sin(\theta)}{r^3}, 0 \right] \qquad (8.87)$$

8.8 | Biot-Savart Law-Revisited

Let us take the curl of the vector potential at a field point,

$$\nabla_{(r)} \times \mathbf{A}(\mathbf{r}) = \frac{\mu_0}{4\pi} \nabla_{(r)} \times \left[\iiint_{V'} \frac{\mathbf{J}(\mathbf{r}') \, dV'}{|\mathbf{r} - \mathbf{r}'|} \right]$$

$$= \frac{\mu_0}{4\pi} \iiint_{V'} \nabla_{(r)} \times \left[\frac{\mathbf{J}(\mathbf{r}') \, dV'}{|\mathbf{r} - \mathbf{r}'|} \right]$$

$$= \frac{\mu_0}{4\pi} \iiint_{V'} \left[\nabla_{(r)} \left(\frac{1}{|\mathbf{r} - \mathbf{r}'|} \right) \times \mathbf{J}(\mathbf{r}') \, dV' \right] \qquad (8.88)$$

the nomenclature $\nabla_{(r)}$ is the nabla operator operating on the $\mathbf{r}$ coordinates only, (which is where the field point is) and *not* the $\mathbf{r}'$ coordinates (which are the coordinates of where the currents are). In the last equation we have used

$$\nabla \times (a\mathbf{A}) = \nabla a \times \mathbf{A} + a \nabla \times \mathbf{A}$$

Now
$$|\mathbf{r} - \mathbf{r}'|^{-1} = \left[(x - x')^2 + (y - y')^2 + (z - z')^2 \right]^{-(1/2)}$$

$$\nabla_{(r)} \left(|\mathbf{r} - \mathbf{r}'|^{-1} \right) = -\frac{\left[(x - x') \, \mathbf{a}_x + (y - y') \, \mathbf{a}_y + (z - z') \, \mathbf{a}_z \right]}{\left[(x - x')^2 + (y - y')^2 + (z - z')^2 \right]^{(3/2)}}$$

$$= -\frac{(\mathbf{r} - \mathbf{r}')}{|\mathbf{r} - \mathbf{r}'|^3}$$

therefore
$$\nabla_{(r)} \times \mathbf{A}(\mathbf{r}) = \frac{\mu_0}{4\pi} \iiint_{V'} \left[-\frac{(\mathbf{r} - \mathbf{r}')}{|\mathbf{r} - \mathbf{r}'|^3} \right] \times \mathbf{J}(\mathbf{r}') \, dV'$$

$$\mathbf{B}(\mathbf{r}) = \frac{\mu_0}{4\pi} \iiint_{V'} \frac{\mathbf{J}(\mathbf{r}') \times (\mathbf{r} - \mathbf{r}')}{|\mathbf{r} - \mathbf{r}'|^3} \, dV'$$

which is the Biot-Savart law.

8.9 | Vector Potential

The important aspect about the vector potential is that we can take the results of the scalar potential and *apply them directly*. Let us take some examples.

8.9.1 Vector Potential for a Current Carrying Straight Conductor

Example 8.18 Let us look at a straight wire along the z-axis of a coordinate system, carrying a steady current I.

Solution The potential of such a wire with a charge density ρ_l is given by

$$V(x, y) = -\frac{\rho_l}{2\pi\varepsilon_0} \ln \sqrt{y^2 + x^2}$$

since the vector potential is proportional to the current, and the directions to be properly chosen, we can straight away write ($\rho_l \Rightarrow I \mathbf{a}_z$ and $1/\varepsilon_0 \Rightarrow \mu_0$)

hence
$$A_z = -\frac{I\mu_0}{2\pi} \ln \sqrt{y^2 + x^2}$$

$$= -\frac{I\mu_0}{2\pi} \ln \rho$$

Note that in this case only one component of $\mathbf{A}$, namely A_z is present since the current is z-directed and the vector potential is in the direction of the current (Examine equation (8.85)) the other components being zero. Therefore using cylindrical coordinates

$$\nabla \times \mathbf{A} = \underbrace{\left(\frac{1}{\rho}\frac{\partial A_z}{\partial \phi} - \frac{\partial A_\phi}{\partial z}\right)}_{0} \mathbf{a}_\rho + \underbrace{\left(\frac{\partial A_\rho}{\partial z} - \frac{\partial A_z}{\partial \rho}\right)}_{} \mathbf{a}_\phi + \frac{1}{\rho}\underbrace{\left\{\frac{\partial (\rho A_\phi)}{\partial \rho} - \frac{\partial A_\rho}{\partial \phi}\right\}}_{0} \mathbf{a}_z$$

$$= -\frac{\partial A_z}{\partial \rho}\mathbf{a}_\phi$$

$$\mathbf{B} = \frac{I\mu_0}{2\pi\rho}\mathbf{a}_\phi \qquad (8.89)$$

which is the same result which we obtained using Ampere's law.

8.9.2 Two Current Carrying Straight Conductors

Example 8.19   We can proceed along the same lines and find the field for two infinite wires parallel to the z-axis placed on the x-y plane at $(x = 0, y = d/2)$ and $(x = 0, y = -d/2)$.

Solution   The potential from electrostatics is

$$V(x,y) = \frac{\rho_l}{2\pi\varepsilon_0} \log\left(\frac{\sqrt{(y + d/2)^2 + x^2}}{\sqrt{(y - d/2)^2 + x^2}}\right) \qquad (8.90)$$

where the wires are charged ρ_l and $-\rho_l$. Using $\rho_l \Rightarrow I\mathbf{a}_z$, $-\rho_l \Rightarrow -I\mathbf{a}_z$ and $1/\varepsilon_0 \Rightarrow \mu_0$ the analogous vector potential is

$$A_z = \frac{I\mu_0}{2\pi} \log\left(\frac{\sqrt{(y + d/2)^2 + x^2}}{\sqrt{(y - d/2)^2 + x^2}}\right)$$

$$A_x = 0$$

$$A_y = 0 \qquad (8.91)$$

then
$$B_x = \partial_y A_z = \frac{I\mu_0}{2\pi}\left(\frac{y + d/2}{\sqrt{(y + d/2)^2 + x^2}} - \frac{y - d/2}{\sqrt{(y - d/2)^2 + x^2}}\right)$$

$$B_y = -\partial_x A_z = -\frac{I\mu_0}{2\pi}\left(\frac{x}{\sqrt{(y + d/2)^2 + x^2}} - \frac{x}{\sqrt{(y - d/2)^2 + x^2}}\right)$$

$$B_z = 0 \qquad (8.92)$$

We can see that we have obtained quite complicated expressions in a simple manner.

8.10 | Far Field Approximation

Very often we would like to find the magnetic field far from where the currents are placed—the far field. In such cases it is much easier to find the field using the vector potential. Figure 8.24 shows the geometry of the region where the currents lie, along with the region of the field point.

The concerned accurate equation for calculation of the vector potential is

$$\mathbf{A}(\mathbf{r}) = \frac{\mu_0}{4\pi} \iiint\limits_{\mathcal{V}'} \frac{\mathbf{J}(\mathbf{r}')\, dV'}{|\mathbf{r} - \mathbf{r}'|}$$

the term

$$|\mathbf{r} - \mathbf{r}'|^{-1} = \frac{1}{\sqrt{r^2 - 2r'\cos\chi\, r + r'^2}}$$

by the cosine law. χ is the angle between $\mathbf{r}$ and $\mathbf{r}'$. The far field is defined by $r \gg r'$.

$$T = \frac{1}{\sqrt{r^2 - 2r'\cos\chi\, r + r'^2}} = \frac{1}{r\sqrt{1 - 2\left(\frac{r'}{r}\right)\cos\chi + \left(\frac{r'}{r}\right)^2}}$$

since r'/r is much smaller than 1, therefore

$$T \approx \frac{1}{r}\left(\frac{1}{\sqrt{1 - 2\left(\frac{r'}{r}\right)\cos\chi}}\right) \qquad \left(\text{neglecting } (r'/r)^2\right)$$

$$\approx \frac{1}{r}\left[1 + \left(\frac{r'}{r}\right)\cos\chi\right] \qquad (8.93)$$

where we have used the Taylor series expansion of $1/\sqrt{1 - 2\,(r'/r)\cos\chi} \cong 1 + (r'/r)\cos\chi$.

$$\frac{1}{\sqrt{r^2 - 2r'\cos\chi\, r + r'^2}} \approx \underbrace{\frac{1}{r}}_{\text{First term}} + \underbrace{\frac{r'\cos\chi}{r^2}}_{\text{Second term}}$$

using this expansion we write

$$\mathbf{A}(\mathbf{r}) = \frac{\mu_0}{4\pi} \iiint\limits_{\mathcal{V}'} \mathbf{J}(\mathbf{r}') \left(\frac{1}{r} + \frac{r'\cos\chi}{r^2}\right) dV'$$

$$\mathbf{A}(\mathbf{r}) = \frac{\mu_0}{4\pi r^2} \iiint_{\mathcal{V}'} \mathbf{J}(\mathbf{r})\, (\hat{\mathbf{r}} \bullet \hat{\mathbf{r}}')r' dV'$$

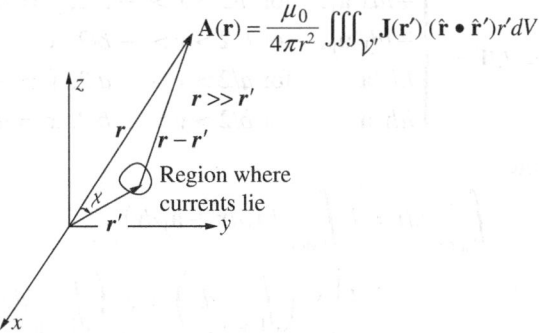

Fig. 8.24 The far field approximation to the vector potential

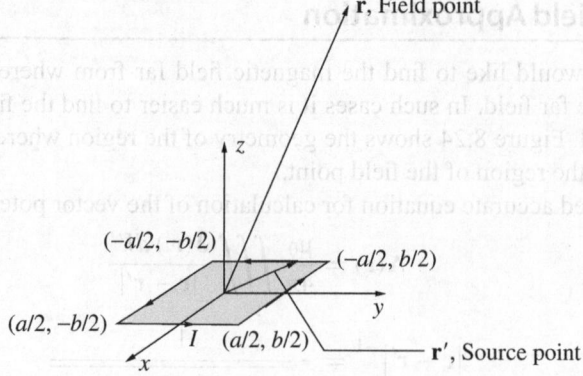

Fig. 8.25 Magnetic field due to a square loop of current

generally we use only the first two terms. However, the term

$$\iiint_{\mathcal{V}'} \mathbf{J}(\mathbf{r}') \, dV' = 0$$

since the steady current always flows in closed loops.[3] The second term always contributes

$$A(\mathbf{r}) \approx \frac{\mu_0}{4\pi r^2} \iiint_{\mathcal{V}'} \mathbf{J}(\mathbf{r}') \, r' \cos\chi \, dV'$$

but $\cos\chi = \hat{\mathbf{r}} \cdot \hat{\mathbf{r}}'$. Therefore

$$A(\mathbf{r}) \approx \frac{\mu_0}{4\pi r^2} \iiint_{\mathcal{V}'} \mathbf{J}(\mathbf{r}') \, (\hat{\mathbf{r}} \cdot \hat{\mathbf{r}}') \, r' dV' \qquad (8.94)$$

 See `MagFldRectLoop.m` in Chapter 8

8.10.1 Square Current Loop and Magnetic Dipole

Example 8.20 Consider a square loop of steady current as shown in Fig. 8.25.
Solution Mathematically, the current can be described by the set of equations on the xy plane.

$$\mathbf{J}'dV' = Id\mathbf{l}' = \begin{cases} -Idx'\mathbf{a}_x, & \text{for } a/2 > x > -a/2, \ y = b/2 \\ -Idy'\mathbf{a}_y, & \text{for } b/2 > y > -b/2, \ x = -a/2 \\ Idx'\mathbf{a}_x, & \text{for } a/2 > x > -a/2, \ y = -b/2 \\ Idy'\mathbf{a}_y, & \text{for } b/2 > y > -b/2, \ x = a/2 \end{cases}$$

Let us first calculate

$$\iiint \mathbf{J}dV = \int_{\text{Loop}} Id\mathbf{l} = I \int_{\text{Loop}} (\mathbf{a}_x dx + \mathbf{a}_y dy)$$

$$= I \left[\mathbf{a}_x \left(\int_{\text{Loop}} dx \right) + \mathbf{a}_y \left(\int_{\text{Loop}} dy \right) \right] = 0$$

---

[3]This point will be clear when we take applications of this formula.

Let us now calculate the second integral

$$\iiint_{V'} \mathbf{J}(\mathbf{r}')(\hat{\mathbf{r}}\cdot\hat{\mathbf{r}}')r'dV' = \mathbf{a}_x \iiint_{V'} J_x(\mathbf{r}')(\hat{\mathbf{r}}\cdot\hat{\mathbf{r}}')r'dV' + \mathbf{a}_y \iiint_{V'} J_y(\mathbf{r}')(\hat{\mathbf{r}}\cdot\hat{\mathbf{r}}')r'dV'$$

For the first integral on the right, $\mathbf{r} = x\mathbf{a}_x + y\mathbf{a}_y + z\mathbf{a}_z$, $\mathbf{r}' = x'\mathbf{a}_x \pm (b/2)\,\mathbf{a}_y$. So

$$\hat{\mathbf{r}}\cdot\hat{\mathbf{r}}' = \frac{xx' \pm y(b/2)}{r\sqrt{x'^2 + (b/2)^2}}$$

$$r' = \sqrt{x'^2 + (b/2)^2}$$

$$\mathbf{a}_x \iiint_{V'} J_x(\mathbf{r}')\,(\hat{\mathbf{r}}\cdot\hat{\mathbf{r}}')\,r'dV' = \frac{\mathbf{a}_x}{r}\left\{ \int_{-a/2}^{a/2} (-I)\,[xx' + y(b/2)\,]dx' \right.$$

$$\left. + \int_{-a/2}^{a/2} (I)\,[xx' - y(b/2)\,]\,dx' \right\}$$

$$= \frac{\mathbf{a}_x}{r}(-Iyab) \quad \text{(cancelled terms of opposite signs)}$$

Similarly for the second integral on the right $\mathbf{r} = x\mathbf{a}_x + y\mathbf{a}_y + z\mathbf{a}_z$, $\mathbf{r}' = \pm(a/2)\,\mathbf{a}_x + y'\mathbf{a}_y$

$$\hat{\mathbf{r}}\cdot\hat{\mathbf{r}}' = \frac{x(a/2) \pm yy'}{r\sqrt{(a/2)^2 + y'^2}}$$

$$r' = \sqrt{(a/2)^2 + y'^2}$$

hence

$$\mathbf{a}_y \iiint_{V'} J_y(\mathbf{r}')\,(\hat{\mathbf{r}}\cdot\hat{\mathbf{r}}')\,r'dV' = \frac{\mathbf{a}_y}{R}\left\{ \int_{-b/2}^{b/2} (-I)\,[x(-a/2) + yy']\,dy' \right.$$

$$\left. + \int_{-b/2}^{b/2} (I)\,[x(a/2) + yy']\,dy' \right\}$$

$$= \frac{\mathbf{a}_y}{R}(Ixab)$$

or

$$\mathbf{A} = \frac{\mu_0 Iab}{4\pi r^3}\left[-\mathbf{a}_x y + \mathbf{a}_y x\right]$$

$$= \frac{\mu_0 Iab}{4\pi r^3}\,(\mathbf{a}_z \times \mathbf{r})$$

$$= \frac{\mu_0 Iab}{4\pi r^3}r\,(\mathbf{a}_z \times \hat{\mathbf{r}})$$

$$= \frac{\mu_0 Iab}{4\pi r^2}\left[\mathbf{a}_z \times (\mathbf{a}_z \cos\theta + \mathbf{a}_\rho \sin\theta)\right]$$

$$= \frac{\mu_0 Iab}{4\pi r^2}\mathbf{a}_\phi \sin\theta \tag{8.95}$$

We can of course obtain the same result by the long method, which is good exercise for the reader.

$$\mathbf{a}_x = \mathbf{a}_r \cos\phi \sin\theta + \cos\theta \cos\phi\, \mathbf{a}_\theta - \mathbf{a}_\phi \sin\phi$$

$$\mathbf{a}_y = \mathbf{a}_r \sin\phi \sin\theta + \cos\theta \sin\phi\, \mathbf{a}_\theta + \mathbf{a}_\phi \cos\phi$$

$$x = r \sin \theta \cos \phi$$
$$y = r \sin \theta \sin \phi$$

using these results

$$-\mathbf{a}_x y = -(\mathbf{a}_r \cos \phi \sin \theta + \cos \theta \cos \phi \, \mathbf{a}_\theta - \mathbf{a}_\phi \sin \phi)(r \sin \theta \sin \phi)$$

$$= -(\mathbf{a}_r r \cos \phi \sin \phi \sin^2\theta + \mathbf{a}_\theta r \sin \theta \cos \theta \sin \phi \cos \phi - \mathbf{a}_\phi r \sin \theta \sin^2\phi)$$

$$\mathbf{a}_y x = (\mathbf{a}_r \sin \phi \sin \theta + \cos \theta \sin \phi \, \mathbf{a}_\theta + \mathbf{a}_\phi \cos \phi)(r \sin \theta \cos \phi)$$

$$= \mathbf{a}_r r \cos \phi \sin \phi \sin^2\theta + \mathbf{a}_\theta r \sin \theta \cos \theta \sin \phi \cos \phi + \mathbf{a}_\phi r \sin \theta \cos^2\phi$$

$$\mathbf{a}_y x - \mathbf{a}_x y = \mathbf{a}_\phi r \sin \theta$$

therefore

$$A_\phi = \frac{\mu_0 I a b \sin \theta}{4\pi r^2} \tag{8.96}$$

the other components being zero.

Based on this vector potential, we can obtain the **B** field

$$
\mathbf{B} = \frac{1}{r \sin \theta} \left\{ \frac{\partial (\sin \theta \, A_\phi)}{\partial \theta} - \frac{\partial A_\theta}{\partial \phi} \right\} \mathbf{a}_r + \frac{1}{r} \left\{ \frac{1}{\sin \theta} \frac{\partial A_r}{\partial \phi} - \frac{\partial (r A_\phi)}{\partial r} \right\} \mathbf{a}_\theta
$$
$$
+ \frac{1}{r} \left\{ \frac{\partial (r A_\theta)}{\partial r} - \frac{\partial A_r}{\partial \theta} \right\} \mathbf{a}_\phi \tag{8.97}
$$
$$
= \frac{1}{r \sin \theta} \left\{ \frac{\partial (\sin \theta \, A_\phi)}{\partial \theta} \right\} \mathbf{a}_r + \frac{1}{r} \left\{ -\frac{\partial (r A_\phi)}{\partial r} \right\} \mathbf{a}_\theta
$$
$$
= \frac{\mu_0 I a b \cos \theta}{2\pi r^3} \mathbf{a}_r + \frac{\mu_0 I a b \sin \theta}{4\pi r^3} \mathbf{a}_\theta \tag{8.98}
$$

These expressions are identical with the expressions obtained for the electric field for an electric dipole given in Section 6.4.1, except that Qd, the dipole moment is replaced by the expression Iab, which is the current multiplied by the area of the loop. Because of this similarity, Iab is called the *magnetic dipole moment*. Or in terms of vector notation

$$\mathbf{m} = I\mathbf{S} \tag{8.99}$$

where **m** is the magnetic dipole moment, I is the current in the loop and **S** is the vector area of the loop. This is not an isolated result, if we were to examine the far fields of a circular loop, we get a similar result as Example 8.7 shows.

Practice Problem 8.1

Using worked out far field expression (Eqn 8.94) for the vector potential, obtain the far field magnetic field for a current loop of radius a. Show that the expressions for the field are the same ones which we obtained above except that ab is replaced by πa^2.

POINTS TO REMEMBER

- The Biot-Savart law for the three types of current is

$$\mathbf{H} = \int \frac{I'd\mathbf{l}' \times \hat{\mathbf{R}}}{4\pi |\mathbf{R}|^2} \qquad \text{(line current)}$$

$$H = \iint_{S'} \frac{J'_s \times R}{4\pi R^3} dS' \quad (\text{surface current})$$

and

$$H = \iint_{S'} \frac{J' \times R}{4\pi R^3} dS' \quad (\text{volume current})$$

where the primed coordinates represent the source terms.

- Ampere's law is $\quad \oint_{\mathcal{L}} H \cdot dl = I_{\text{enclosed}}$
- Maxwell's equation for steady magnetic fields is

$$\nabla \times H = J$$

- The magnetic scalar potential is defined through the relation

$$H = -\nabla V_m$$

in a current free region ($J = 0$).
- The magnetic fields obey the two laws equivalent to each other

$$\left. \begin{array}{c} \oiint_S B \cdot dS = 0 \\ \\ \nabla \cdot B = 0 \end{array} \right\} \quad (\text{no magnetic charges})$$

- The magnetic field and the magnetic flux density are related by

$$B = \mu H$$

- The vector potential is given by the equation

$$B = \nabla \times A$$

where A is the vector potential.
- The vector potential obeys the equation

$$\nabla^2 A = -\mu J$$

with the solution

$$A(r) = \frac{\mu}{4\pi} \iiint_{V'} \frac{J(r') \, dV'}{|r - r'|}$$

- In the far-field, the vector potential is calculated from

$$A(r) \approx \frac{\mu_0}{4\pi r^2} \iiint_{V'} J(r') \, (\hat{r} \cdot \hat{r}') \, r' dV'$$

where r is the position vector of the field point, and r' is the position vector of the source point.
- For a flat loop, the magnetic dipole is given by

$$m = IS$$

SELF ASSESSMENT

Objective Type Questions

1. The Biot-Savart law is used
   (a) to calculate the electric field generated by a current
   (b) to calculate the electric field due to a point charge
   (c) to calculate the magnetic field produced by a steady current
   (d) None of the above

2. According to the Biot-Savart law, the magnetic field is proportional to
   (a) R (b) R^2 (c) $1/R^3$ (d) $1/R^2$
   where R is the distance from source point to the field point.

3. A magnetostatic field in a region which is free of currents is
   (a) conservative   (b) non-conservative   (c) None of these

4. A magnetostatic field in a region which is free of currents but where static charges are present is
   (a) conservative   (b) non-conservative   (c) None of these

5. A magnetostatic field in a region where currents are present is
   (a) conservative   (b) non-conservative   (c) None of these

6. Ampere's law states that the line integral of
   (a) $\mathbf{H}$ from point A to point B is independent of the path
   (b) $\mathbf{H}$ from point A to point B and back from point B to A is zero
   (c) $\mathbf{H}$ from point A to point B and back from point B to A is equal to the current enclosed by the path
   (d) None of the above

7. In a region where the current density is of the order of 10 kA/m^2, the value of $|\nabla \times \mathbf{H}|$ will be of the order of
   (a) A   (b)   mA   (c) kA   (d) μA

8. If the current, I, in a wire is flowing in direction $d\mathbf{l}$ with position vector $\mathbf{r'}$, and we wish to calculate the magnetic field contribution from this current element at the field point $\mathbf{r}$ (with $\mathbf{R} = \mathbf{r} - \mathbf{r'}$) then the magnetic field $d\mathbf{H}$ is proportional to
   (a) $Id\mathbf{l}$ (b) $1/r^2$ (c) $1/R^2$ (d) $1/r'^2$

9. If the current, I, in a wire is flowing in direction $d\mathbf{l}$ with position vector $\mathbf{r'}$, and we wish to calculate the magnetic field contribution from this current element at the field point $\mathbf{r}$ (with $\mathbf{R} = \mathbf{r} - \mathbf{r'}$) then the magnetic field $d\mathbf{H}$ is proportional to
   (a) $Id\mathbf{l} \times \hat{\mathbf{R}}/r^2$ (b) $Id\mathbf{l} \cdot \hat{\mathbf{R}}$ (c) $Id\mathbf{l} \times \hat{\mathbf{R}}/R^2$ (d) $1/r'^2$

10. The magnet scalar potential has significance where
    (a) the region is free of static charges   (b) the region is free of steady currents
    (c) the region has static charges   (d) the region has steady currents

Short-Answer Questions

1. Explain the Biot-Savart law.   (Mumbai University, December 2008).
2. Starting with the Biot-Savart law, calculate the divergence of the magnetic induction vector $\mathbf{B}$.   (DTU-NSIT, June 2010)
3. What are the magnetic scalar and vector potentials. What are their importance?
                (DTU-NSIT, June 2010)
4. Define the current density vector, $\mathbf{J}$. Using Ampere's circuital law derive the relationship between $\mathbf{B}$ and $\mathbf{J}$.   (Mumbai University, May 2009)
5. Define (a) magnetic field strength (b) magnetomotive force.
6. Write a short note on Biot-Savart law.
7. State and explain Amperes law and derive Maxwell's equation (in magnetostatics).
8. Explain why the magnetic scalar potential is not a true potential like the electric scalar potential.

Review Questions

1. When would we apply the Biot-Savart law?
2. Discuss the similarities and dissimilarities between Biot-Savart law and Coulomb's law.

3. Discuss the three types of current which were discussed in the chapter.

4. When applying Biot-Savart law why do we need to pay special attention to the three types of current?

5. What is the connection between the surface integral of the curl of a vector and line integral of the vector?

6. Discuss Ampere's law and its significance with respect to Maxwell's equation for magnetostatics.

7. Why do we apply Ampere's law to symmetrical applications?

8. Where can we work with the magnetic scalar potential?

9. Why is the magnetic flux density equal to curl[A]?

10. What is the advantage of using A?

11. In the derivation of Eqn (8.82) why do we choose div[A] = 0?

12. Why do we call $\mathbf{m} = I\mathbf{S}$ the magnetic dipole moment?

Numerical Problems

If the medium is not specified, assume that $\mu = \mu_0$, $\varepsilon = \varepsilon_0$

1. Given that current density $\mathbf{J} = -10(\sin 2\phi\, e^{-2z}\mathbf{a}_\phi + \cos 2\phi\, e^{-2z}\mathbf{a}_z)$ A/m^2

    (a) find the total current crossing the plane $\phi = \pi/2$ in the $\mathbf{a}_\phi$ direction in the region $0 < \rho < 1$, $0 < z < 1$.

    (b) find the total current crossing the plane $\phi = \pi/4$ in the $\mathbf{a}_\phi$ direction in the region $0 < \rho < 1$, $0 < z < 1$.

2. A current I, flows along a wire located on the x-axis from $x = \infty$ to the origin and then along the y-axis from the origin to $y = \infty$ (see Fig. 8.26). Find the magnetic field at $(1, 1, 0)$ using the Biot-Savart law.

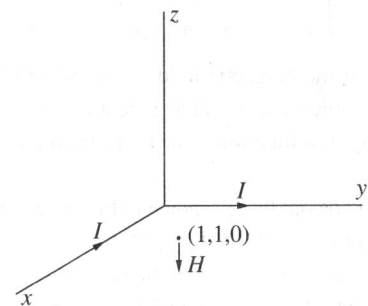

Fig. 8.26

3. A current I, flows along a wire located on the the y-axis from $y = -\infty$ to the origin and then along the x-axis from the origin to $x = -\infty$ (see Fig. 8.27). Find the magnetic field at $(1, 1, 0)$ using the Biot-Savart law.

4. Using Biot-Savart law derive an expression for $\mathbf{H}$ due to an infinitely long straight filament carrying a current of I amperes.

    (Mumbai University, June 2008)

5. A square filamentary loop of 2 m side is placed in $z = 0$ plane with its centre at the origin. If current of 10 A is passing through loop, find $\mathbf{H}$ at origin.

    (Mumbai University, December 2007)

6. An infinite long current filament is placed along z-axis. The magnetic field intensity at point P $(3, 4, 0)$ is $10(-0.8\mathbf{a}_x + 0.6\mathbf{a}_y)$ A/m. Find the current through the filament.

    (Mumbai University, December 2008)

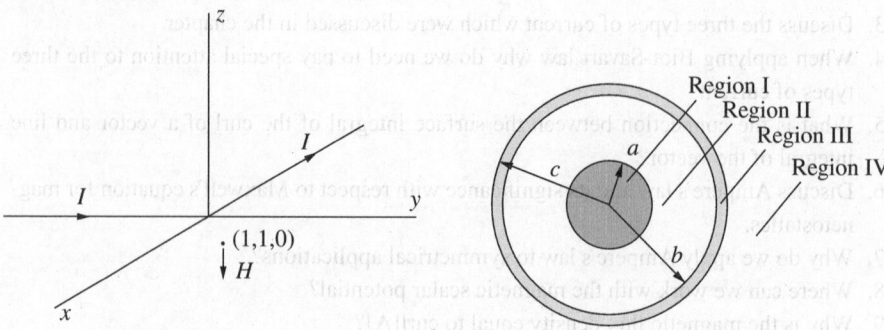

Fig. 8.27 Problem 3   **Fig. 8.28** A practical coaxial line

7. Find the magnetic field at the centre of a square loop of side $2L$ carrying a current I. The current flows in an anti-clockwise sense.

8. In a certain region the material has $\mu_r = 10$, $\mathbf{H} = [2y+1, -2z-2y, 2z-2y+2x-1]$, find the flux density at $(1, 1, 1)$; find the flux density everywhere and show that $\nabla \cdot \mathbf{B} = 0$

9. In a certain region, $\mathbf{H} = [2y + 1, -2z - 2y, 2z - 2y + 2x - 1]$. Find $\mathbf{J}$ at all points of that region.

10. Apply Stokes's theorem to the magnetic field $\mathbf{H}$ of Problem 9. Let the circuit be $(0, 0, 0)$ to $(1, 0, 0)$ to $(1, 1, 0)$ to $(0, 1, 0)$ and back to $(0, 0, 0)$.

11. Using Ampere's circuital law in integral form, find $\mathbf{H}$ at all points due to the following current density in cylindrical coordinates

$$\mathbf{J} = \begin{cases} 0, & 0 < \rho < a \\ k\rho \, \mathbf{a}_z, & a < \rho < b \\ 0, & b < \rho < \infty \end{cases} \qquad \text{(DTU-NSIT, May 2009)}$$

12. Develop an expression for the magnetic field intensity both inside and outside a solid cylindrical conductor of radius a carrying a current I with uniform density. Sketch the variation of field intensity as a function of distance from the conductor axis.

(DTU-NSIT, June 2010)

13. For the case of a practical coaxial line as shown in Fig. 8.28, calculate the magnetic field $\mathbf{H}$ using Ampere's law in Regions I, II, III, and IV if the current in the inner conductor is I going into the page and I coming out of the page in the outer conductor.

14. Given that the vector potential is $\mathbf{A} = 10 \sin \theta \, \mathbf{a}_r$, find the magnetic field $\mathbf{H}$.

15. If the vector potential in a certain region is $\mathbf{A} = x^2yz\mathbf{a}_x + y^2zx\mathbf{a}_y + z^2xy\mathbf{a}_z$. Find the magnetic flux density, $\mathbf{B}$.

16. For the vector potential $\mathbf{A} = r^2 \sin \theta \, \mathbf{a}_r + r^2 \sin \theta \cos \phi \, \mathbf{a}_\theta$, find the magnetic flux density in a region with $\mu_r = 11$.

17. In a region where $\mathbf{B} = 0$, show that the vector potential, if non-zero, can only be the gradient of a scalar field.

18. Given the magnetic flux density, $\mathbf{B} = 6 \cos(10^6 t) \sin(0.01x) \, \mathbf{a}_z$ mT, find (a) the magnetic flux passing through the surface $z = 0$; $0 < x < 20$ m; $0 < y < 3$ m at $t = 1$ μs.

(DTU-NSIT, June 2010)

19. Given that the vector potential is $\mathbf{A} = 10 \sin \theta \, \mathbf{a}_\theta$, find the magnetic field $\mathbf{H}$. Also show that $\nabla \cdot \mathbf{B} = 0$.

20. In a certain region of space, $(x, y, z \geq 0)$ the vector $\mathbf{H}$ is given by

$$\mathbf{H} = [0, 0, -2 \, e^{-2y}]$$

(a) Find the current density at $(2, 5, 9)$ and

(b) Show that $\nabla \cdot \mathbf{H} = 0$, in the region $(x, y, z \geq 0)$.

Answers

Objective Type Questions

1. (c)   2. (d)   3. (a)   4. (a)   5. (b)   6. (c)   7. (d)   8. (a) and (d)   9. (c)   10. (b)

Short-Answer Questions

5. (a) A magnetic field strength of 1 A/m is produced at a perpendicular distance of 1 m due to a steady current of 1 A in a very long wire carrying the current. (b) The magnetomotive force increase of 1 A is found when we traverse a closed circuit around a very long wire carrying a current of 1 A. Magnetic field lines flow from lower potentials to higher potentials.

6. **Hint:** See Section 8.2.

7. **Hint:** See Section 8.4.

8. The magnetic scalar potential is essentially applicable to regions where $\nabla \times \mathbf{H} = 0$. The potential may be multi-valued. In regions where $\mathbf{J} = 0$ it does obey Laplace's equation. For more explanation, see Section 8.6.

Numerical Problems

1. (a) $I = 0$ (b) $I = -5(1 - e^{-2})$

2. $H_z = -\dfrac{2(\sqrt{2} + 1)I}{2^{\frac{5}{2}}\pi}$

3. $H_z = -\dfrac{2(\sqrt{2} - 1)I}{2^{\frac{5}{2}}\pi}$

4. See Section 8.3.1

5. $H_z = \dfrac{40}{\pi\sqrt{2}}$

6. $I = 100\pi\,\text{A}$

7. $H = \dfrac{4I}{\sqrt{2\pi}\,L}$

8. $\mathbf{B}$ at $(1, 1, 1)$ is $10\mu_0\,[3, -4, 1]$

9. $J = [0, 2, 2]$

10. 2

11. $H_\phi = 0$ for $0 < \rho < a$; $H_\phi = (k/3\rho)(\rho^3 - a^3)$ for $a < \rho < b$;
    $H_\phi = (k/3\rho)(b^3 - a^3)$ for $b < \rho < \infty$

12. See Section 8.4.2

13. Region I: $H_\phi = -(I\rho)/(2\pi a^2)$ Region II: $H_\phi = -I/(2\pi\rho)$;
    Region III: $H_\phi = -I/(2\pi\rho) + [I(\rho^2 - b^2)]/[2\pi\rho(c^2 - b^2)]$; Region IV: 0

14. $\mathbf{H} = (1/\mu_0)[\cos(\theta/r)]a_\phi$

15. $\mathbf{B} = [xy^2 - xz^2, \ yz^2 - x^2y, \ x^2z - y^2z]$

16. $\mathbf{B} = [\sin(\phi)\,r, \ 0, \ \{3\cos(\phi)\,r^2\sin(\theta) - r^2\cos(\theta)\}/r]$; $\mathbf{H} = \mathbf{B}/(11\mu_0)$

18. 19.386

19. $\mathbf{H} = (1/\mu_0)(\sin(\theta)/r)\,\mathbf{a}_\phi$

20. Calculate $\nabla \times \mathbf{H}$. Then $\nabla \times \mathbf{H} = \mathbf{J}\ (= [-4e^{-10}, \ 0, \ 0])$

The Steady Magnetic Field | 279

9

Magnetic Forces, Inductance, and Magnetisation

An educational system isn't worth a great deal
if it teaches young people how to make a living
but doesn't teach them how to make a life.
— Anonymous

CHAPTER OBJECTIVES

To enable the students to understand the following:

- Lorentz force
- Electron moving in a steady magnetic field
- Straight wire carrying a current in a magnetic field, **B**
- Force between two current carrying parallel conductors
- Loop carrying a current in a constant magnetic field
- Magnetic dipole

- Force between two current carrying elements
- Inductance
- Inductance of a coaxial line
- Mutual inductance between inductors
- Magnetisation and the magnetic susceptibility
- Magnetic circuits

9.1 | Lorentz Force

Forces appear in electrodynamics in two varieties. The first one consists of the force on a charged particle due to the presence of an electric field. The law here is

$$\mathbf{F} = q\mathbf{E} \qquad (9.1)$$

where q is the charge, **E** is the electric field in which the charge is located, and **F** is the force felt by the charge. The first point to be noted is that the force on the charge is in the direction of the electric field. The second point is that the force is proportional to the magnitude of the charge and the magnitude of the electric field.

The second kind of force, experimentally verified, is the force on a *moving* charge. Referring to Fig. 9.1, if a charge q is moving with a velocity $\mathbf{v}$ in a magnetic field with flux density $\mathbf{B}$ then the magnitude of the force experienced by the charge is given by

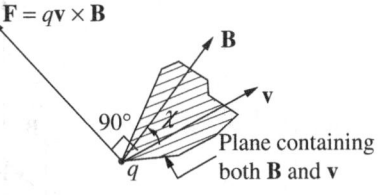

Fig. 9.1 $\mathbf{v} \times \mathbf{B}$ force

$$F = qvB \sin\chi \qquad (9.2)$$

where χ is the angle between $\mathbf{v}$ and $\mathbf{B}$. The *direction* of $\mathbf{F}$ is perpendicular to both the vectors $\mathbf{v}$ and $\mathbf{B}$ and is in the direction of the thumb when the fingers are curled from $\mathbf{v}$ to $\mathbf{B}$ as given by the right-hand rule. This is expressed in vector notation as

$$\mathbf{F} = q\mathbf{v} \times \mathbf{B} \qquad (9.3)$$

The total force on a charge in presence of both the electric and the magnetic fields is given by

$$\mathbf{F} = q(\mathbf{E} + \mathbf{v} \times \mathbf{B}) \qquad (9.4)$$

this force is called the *Lorentz force*.

Maglev Trains

In the case of a maglev train—short for magnetic levitation—powerful electromagnets are used. These trains float over a 'track' using magnetic forces, and these trains are accelerated and decelerated by means of magnetic forces. Maglev trains are capable to go at over 500 km/h.

9.2 | Electron Moving in a Steady Magnetic Field

Let us think of the application of this new law ($\mathbf{v} \times \mathbf{B}$ force). First we require a magnetic field. Then we require some moving charges so that we can see the effect of the force. We start with a single electron moving perpendicular to a magnetic field. We move it perpendicularly, so that $\mathbf{v} \times \mathbf{B}$ is simplified.

Example 9.1 Find the equation of motion of an electron moving in a plane perpendicular to a steady magnetic field.

Solution

Step 1. The situation is depicted in Fig. 9.2 which shows

$$\mathbf{B} = B_0\mathbf{a}_z \qquad (9.5)$$

$$\mathbf{v} = v_0\mathbf{a}_y \qquad (9.6)$$

$$x = \text{some value which we specify later}$$

$$y = 0$$

$$z = 0 \qquad (9.7)$$

$$\mathbf{F} = e\mathbf{v} \times \mathbf{B}$$

$$= ev_0 B_0\mathbf{a}_x \quad \text{at } t = 0 \qquad (9.8)$$

Step 2. The equations of motion are obtained from $\mathbf{F} = ev_y B_z\mathbf{a}_x - ev_x B_z\mathbf{a}_y$.

$$m_e \frac{dv_x}{dt} = ev_y B_z \qquad (9.9)$$

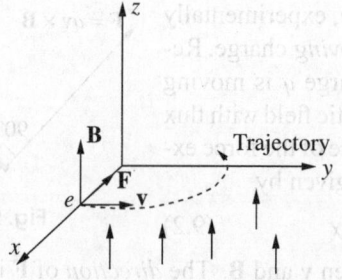

Uniform magnetic field $\mathbf{B} = B_0 \mathbf{a}_z$

Fig. 9.2 Single electron moving perpendicularly to a steady magnetic field

$$m_e \frac{dv_y}{dt} = -ev_x B_z \tag{9.10}$$

$$m_e \frac{dv_z}{dt} = 0 \tag{9.11}$$

Step 3. The third equation integrates to

$$v_z = 0 \text{ or } c \text{ (a constant)} \tag{9.12}$$

and since the initial value of $v_z = 0$ therefore $c = 0$

Step 4. Using Laplace transforms with the notation

$$a(t) \overset{L}{\Longleftrightarrow} A(s)$$

the first two equations can be written as

$$m_e[sV_x(s)] = eV_y(s)\, B_0 \tag{9.13}$$

$$m_e[sV_y(s) - v_0] = -eV_x(s)\, B_0 \tag{9.14}$$

Step 5. From Eq. (9.13)

$$V_x(s) = \frac{eV_y(s)\, B_0}{m_e s}$$

substituting this value in Eq. (9.14)

$$m_e[sV_y(s) - v_0] = -eB_0 \left[\frac{eV_y(s)\, B_0}{m_e s} \right]$$

$$sV_y(s) - v_0 = -\left(\frac{eB_0}{m_e} \right)^2 \frac{V_y(s)}{s}$$

$$sV_y(s) + \left(\frac{eB_0}{m_e} \right)^2 \frac{V_y(s)}{s} = v_0$$

$$V_y(s) \left[s^2 + \left(\frac{eB_0}{m_e} \right)^2 \right] = sv_0$$

or

$$V_y(s) = \frac{sv_0}{\left[s^2 + \left(\frac{eB_0}{m_e} \right)^2 \right]}$$

Step 6. Taking the inverse Laplace transform

$$v_y(t) = v_0 \cos(\omega_c t) \tag{9.15}$$

(Check: at $t = 0$ the value of $v_y = v_0$)

where $$\omega_c = -\frac{eB_0}{m_e} \tag{9.16}$$

Step 7. Note that the right-hand side is positive since e is negative. ω_c is the cyclotron frequency. v_x can now be calculated,

since $$m_e \frac{dv_y}{dt} = -ev_x B_0$$

therefore $$v_x = \frac{1}{\omega_c} \frac{dv_y}{dt} = -v_0 \sin\left(\omega_c t\right) \tag{9.17}$$

Notice that the magnitude of the velocity is constant

as $$\sqrt{v_x^2 + v_y^2} = v_0 \quad \text{(the initial velocity)} \tag{9.18}$$

which means that the kinetic energy of the electron *neither increases nor decreases and no work is done by this force.* Let us find the trajectory of the electron

$$v_x = -v_0 \sin(\omega_c t)$$

or $$\frac{dx}{dt} = -v_0 \sin(\omega_c t)$$

Integrating $$x = \frac{v_0}{\omega_c} \cos(\omega_c t) + x_0 \quad \text{where} \quad \frac{v_0}{\omega_c} + x_0 \text{ is } x \text{ at } t = 0$$

$$v_y = v_0 \cos(\omega_c t)$$

or $$\frac{dy}{dt} = v_0 \cos(\omega_c t)$$

Integrating $$y = \frac{v_0}{\omega_c} \sin(\omega_c t) \quad y = 0 \text{ at } t = 0$$

To obtain the geometry of the trajectory of the electron

$$x - x_0 = \frac{v_0}{\omega_c} \cos(\omega_c t)$$

$$y = \frac{v_0}{\omega_c} \sin(\omega_c t) \tag{9.19}$$

so $$(x - x_0)^2 + y^2 = \left(\frac{v_0}{\omega_c}\right)^2 \tag{9.20}$$

which is the equation of a circle. Hence the electron is moving in a circle with radius v_0/ω_c, and centre $(x_0, 0)$ with a frequency of ω_c.

Example 9.2 An electron with a velocity equal to 4×10^4 m/s (which corresponds to an energy of kT) is released perpendicular to the direction of a magnetic field, approximately equal to that of the earth, i.e., $B = 0.4 \times 10^{-4}$ T. Find the radius of the circle and the frequency with which the electron goes round in circles.
Solution From the previous example, the radius of the circle is

$$R = \frac{v}{\omega_c}$$

where
$$\omega_c = -\frac{eB}{m_e}$$

putting $e = -1.6 \times 10^{-19}$ C, $m_e = 9.1 \times 10^{-31}$ kg, and $B = 0.4 \times 10^{-4}$ T, in above relations

$$\omega_c = 7 \times 10^6 \text{ r/s}$$

which gives a frequency of 1.12 MHz and the radius of the circle is

$$R = 4 \times 10^4 / 7 \times 10^6 = 5.7 \text{ mm}$$

Example 9.3 Show that no work is done by the $\mathbf{v} \times \mathbf{B}$ force.

Solution Let us take a look at the general case. Let a charge q move with a velocity v in a magnetic field with flux density $\mathbf{B}$. Then the differential work done in time dt is

$$dW = \underbrace{q(\mathbf{v} \times \mathbf{B})}_{\text{Force}} \cdot \underbrace{\frac{d\mathbf{r}}{dt} dt}_{\text{Displacement}} \qquad (9.21)$$

where $\mathbf{r}$ is the position vector of the charge. But

$$\frac{d\mathbf{r}}{dt} = \mathbf{v} \qquad (9.22)$$

therefore
$$dW = q \underbrace{(\mathbf{v} \times \mathbf{B}) \cdot \mathbf{v}}_{\text{Scalar triple product}} dt = 0 \qquad (9.23)$$

Hence *charges moving in magnetic fields neither gain nor loss kinetic energy.*

Example 9.4 Find the force, $\mathbf{F}$, on a charge of 1 nC placed in a magnetic field of $\mathbf{H} = 1\mathbf{a}_x$ A/m when (a) $\mathbf{v} = 0$, (b) $\mathbf{v} = 2\mathbf{a}_x + 3\mathbf{a}_y$ m/s.

Solution The force on the charge is

$$\mathbf{F} = 1 \times 10^{-9}(\mathbf{v} \times \mathbf{B})$$

for (a) since $\mathbf{v} = 0$, so $\mathbf{F} = 0$.

for (b)
$$\mathbf{F} = 1 \times 10^{-9} \left[(2\mathbf{a}_x + 3\mathbf{a}_y) \times (1/\mu_0)\,\mathbf{a}_x \right]$$
$$= -\frac{3 \times 10^{-9}}{\mu_0} \mathbf{a}_z \text{ N}$$

 See `LorentzForce.m` in Chapter 9

9.3 | A Straight Wire Carrying a Current in a Magnetic Field

We next consider the case of a long straight metallic wire immersed in a uniform magnetic field of magnetic flux density $\mathbf{B} = B_0 \mathbf{a}_z$ shown in Fig. 9.3. To get a better feel for the problem, we will analyse the problem from the micro level, and then move on to the macro level.

Example 9.5 Find the force on a current carrying conductor as shown in Fig. 9.3.

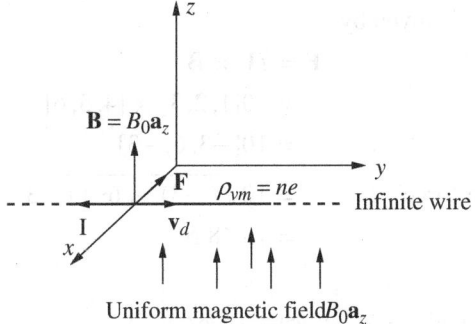

Fig. 9.3 A long straight wire in a steady magnetic field

Solution In a metal conductor the mobile charge carriers are electrons. If the density of electrons are n electrons/m^3 then the mobile volume charge density is

$$\rho_{vm} = ne \ (C/m^3) \tag{9.24}$$

the total mobile charge per meter of line is again

$$\rho_{lm} = \rho_{vm} \times A \ (m^2)$$
$$= \rho_{vm} A \ (C/m) \ (A \text{ is the area of cross-section of the wire}) \tag{9.25}$$

Let $\mathbf{v}_d = v_d \mathbf{a}_y$ be the drift velocity, then the force on the wire per meter $\mathbf{F}_u$ is

$$\mathbf{F}_u = \rho_{lm} \mathbf{v}_d \times \mathbf{B}$$
$$= \rho_{lm} v_d B_0 \mathbf{a}_x \ (N/m) \tag{9.26}$$

But the current is related to the line mobile charge and the drift velocity through the equation

$$I = -\rho_l v_d \tag{9.27}$$

the negative sign occurs since the conventional current direction and the *electron* current direction are of opposite sign. Therefore

$$\mathbf{F}_u = -IB_0 \mathbf{a}_x \ (N/m)$$

and the force on a length of straight line L (in vector notation, I is in the direction $\mathbf{L}$) is

$$\mathbf{F} = -ILB_0 \mathbf{a}_x$$
$$= I\mathbf{L} \times \mathbf{B} \quad \text{(confirm this from the figure)} \tag{9.28}$$

A simpler manner in which we can approach this problem is to start from the equation

$$\mathbf{F} = q\mathbf{v} \times \mathbf{B}$$

Since the wire is straight, $q\mathbf{v}$ may be replaced by $I\mathbf{L}$. So

$$\mathbf{F} = I\mathbf{L} \times \mathbf{B} \tag{9.29}$$

Example 9.6 A wire extends from $A = [0, 0, 0]$ to $B = [1, 2, 3]$ and is placed in a uniform magnetic field $\mathbf{B} = [4, 5, 6]$ with a notation using rectangular coordinates. If the current in the wire is 10 A flowing from A to B, find the force on the wire.

Solution The force is given by

$$\mathbf{F} = I\mathbf{L} \times \mathbf{B}$$
$$= 10[1, 2, 3] \times [4, 5, 6]$$
$$= 10[-3, 6, -3]$$

the magnitude of the force
$$= 10\sqrt{(-3)^2 + 6^2 + (-3)^2}$$
$$= 73.48 \text{ N}$$

 See `ForceWireBFld.m` in Chapter 9

9.3.1 Force Between Two Current Carrying Parallel Conductors

We now consider the application of $\mathbf{v} \times \mathbf{B}$ law in determining the force between two current carrying conductors.

Example 9.7 Two long thin wires are separated by 1 cm in air and carry currents of 100 A in opposite directions. Find the force vector on 5 m on one wire.

(Mumbai University, May 2009)

Solution Placing the z-axis (current flowing along the positive z-axis) along one of the wires, and the other wire at $y = 0.01$ m, the magnetic field $\mathbf{H}$ produced by the wire along z-axis is

$$\mathbf{H} = -\frac{100\mathbf{a}_x}{2\pi(0.01)} = -\frac{10^4 \mathbf{a}_x}{2\pi} \text{ (A/m)}$$

Hence
$$\mathbf{B} = \mu_0 \mathbf{H} = -\mu_0 \frac{10^4 \mathbf{a}_x}{2\pi} \tag{9.30}$$

Then the force per unit length on the other wire placed in the magnetic field of former wire is

$$\frac{\mathbf{F}}{l} = \underbrace{100}_{\text{Current}} \underbrace{(-\mathbf{a}_z)}_{\text{Direction}} \times \mathbf{B}$$

$$= 100\mathbf{a}_z \times \left(\mu_0 \frac{10^4 \mathbf{a}_x}{2\pi} \right)$$

$$= 100 \times \mu_0 \frac{10^4}{2\pi} \mathbf{a}_y$$

$$= 100 \times (4\pi \times 10^{-7}) \frac{10^4}{2\pi} \mathbf{a}_y = 0.2\mathbf{a}_y$$

so the force on 5 m of wire is

$$\mathbf{F}_{5\text{ m}} = 1\mathbf{a}_y \text{ (N)}$$

the force is *repulsive*.

Note: If the two infinitely long parallel wires carry currents of I_1 and I_2 and are separated by a distance a, then the force per unit length on any of them is given by

$$\frac{F}{l} = \frac{\mu_0 I_1 I_2}{2\pi a} \tag{9.31}$$

the force is attractive, if the currents flow in the same direction and repulsive if they flow in opposite directions.

9.4 | Other Formulations

We can manipulate the basic equation

$$\mathbf{F} = q\mathbf{v} \times \mathbf{B}$$

to suit our needs as below

$$d\mathbf{F} = dQ \, \mathbf{v} \times \mathbf{B} \tag{9.32}$$

$$= (\rho_v \, dV) \, \mathbf{v} \times \mathbf{B} \quad (dQ = \rho_v \, dV) \tag{9.33}$$

$$= \mathbf{J} \times \mathbf{B} dV \quad\quad (\mathbf{J} = \rho_v \, \mathbf{v}) \tag{9.34}$$

In the above equations, $d\mathbf{F}$ is the elemental force which the elemental charge dQ feels; ρ_v is the charge density; dV is a volume element; $\mathbf{J}$ is the current density, and $\mathbf{B}$ is the external magnetic flux density. If we use a current sheet, then

$$d\mathbf{F} = \mathbf{J}_s \times \mathbf{B} dS$$

Another formulation is the force felt by a differential current element $Id\mathbf{l}$

$$d\mathbf{F} = Id\mathbf{l} \times \mathbf{B} \quad (dQ\mathbf{v} = Id\mathbf{l}) \tag{9.35}$$

9.5 | Loop Carrying a Current in a Constant Magnetic Field

We consider now a loop immersed in a constant magnetic field as shown in Fig. 9.4.

Example 9.8 Let us consider yet another case: that of a current carrying loop in a uniform magnetic flux density $\mathbf{B} = B_0\mathbf{a}_z$, shown in Fig. 9.4.

Solution In this case we use the formulation

$$d\mathbf{F} = Id\mathbf{l} \times \mathbf{B}$$

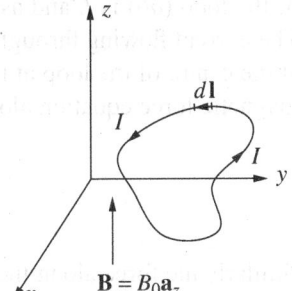

so

$$\mathbf{F} = I\left(\oint_{\mathcal{L}} d\mathbf{l} \times \mathbf{B}\right) \quad (I \text{ is a constant})$$

$$= I\left(\oint_{\mathcal{L}} d\mathbf{l}\right) \times \mathbf{B} \quad (\mathbf{B} \text{ is constant})$$

$$= 0 \tag{9.36}$$

since

$$\oint_{\mathcal{L}} d\mathbf{l} = 0 \tag{9.37}$$

Fig. 9.4 A loop carrying current in a constant magnetic field

This can be proved as follows:

$$\oint_{\mathcal{L}} d\mathbf{l} = \oint_{\mathcal{L}} (\mathbf{a}_x dx + \mathbf{a}_y dy + \mathbf{a}_z dz)$$

since we are integrating on a loop, let the parametric equations for x, y, z be $x(t), y(t), z(t)$. Then

$$\oint_{\mathcal{L}} d\mathbf{l} = \int_{t=t_i}^{t=t_f} (\mathbf{a}_x x' dt + \mathbf{a}_y y' dt + \mathbf{a}_z z' dt) \quad (x' = dx/dt, \text{ etc.})$$

$$= \left|\left[\mathbf{a}_x x(t) + \mathbf{a}_y y(t) + \mathbf{a}_z z(t)\, t\right]\right|_{t=t_i}^{t=t_f}$$

$$= \mathbf{a}_x \left[x(t_f) - x(t_i)\right] + \mathbf{a}_y \left[y(t_f) - y(t_i)\right] + \mathbf{a}_z \left[z(t_f) - z(t_i)\right]$$

but since we are considering a loop, $x(t_f) = x(t_i)$; $y(t_f) = y(t_i)$, and $z(t_f) = z(t_i)$. Therefore

$$\oint_L d\mathbf{l} = 0 \tag{9.38}$$

Therefore no force is felt by a loop carrying a steady current I and immersed in a steady magnetic field.

 See `TorqueOnLoop.m` in Chapter 9

9.6 | Torque in a Loop Carrying a Current in a Constant Magnetic Field

The total force on a loop carrying a current is zero, but that does not mean that the loop does not feel a torque, as the following example shows.

Example 9.9   Take the example shown in Fig. 9.5. The figure shows a square loop *abcd* immersed in a constant magnetic field **B** and it is allowed to rotate about a central axis as shown. Find the torque on the loop.

Solution   The magnetic field is directed from the north pole of the magnet to the south pole, with the flux density modelled as $\mathbf{B} = -\mathbf{a}_z B_0$. The length of the side of the loop (ab) is L and as a vector it is $\mathbf{L} = \mathbf{a}_y L$. The other side is of length d. The current flowing through the loop is I. The coordinate system chosen is placed at the centre of the loop at the point o with ab along the y-axis. Applying now the magnetic force equation along the side of the loop ab:

$$\mathbf{F}_{ab} = I\mathbf{L} \times \mathbf{B}$$
$$= I(\mathbf{a}_y L) \times (-\mathbf{a}_z B_0)$$
$$= -\mathbf{a}_x I L B_0 \tag{9.39}$$

similarly the force along the side cd is

$$\mathbf{F}_{cd} = I\mathbf{L} \times \mathbf{B}$$

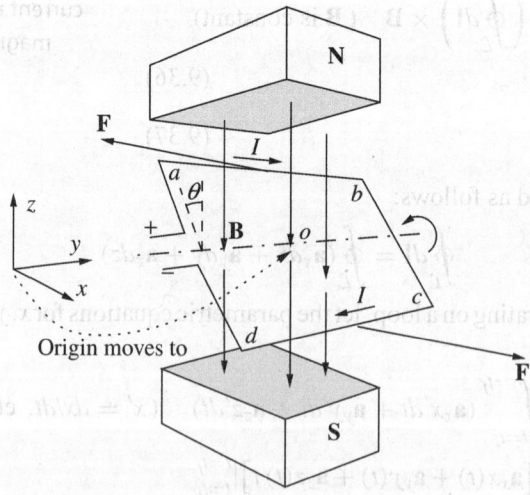

Fig. 9.5 Torque in a loop carrying a current

$$= I(-\mathbf{a}_y L) \times (-\mathbf{a}_z B_0)$$

$$= \mathbf{a}_x ILB_0 \qquad (9.40)$$

both these forces tend to rotate the loop in the counter-clockwise sense when viewed from the side ad. The torque about the axis of rotation due to $\mathbf{F}_{ab}$ is

$$\mathbf{T}_{ab} = \mathbf{r} \times \mathbf{F}_{ab}$$

$$= \frac{d}{2}(\mathbf{a}_z \cos\theta - \mathbf{a}_x \sin\theta) \times (-\mathbf{a}_x ILB_0)$$

$$= -\frac{d}{2}\mathbf{a}_y \cos\theta \, ILB_0 \qquad (9.41)$$

and torque due to $\mathbf{F}_{cd}$ is

$$\mathbf{T}_{cd} = \mathbf{r} \times \mathbf{F}_{cd}$$

$$= \frac{d}{2}(-\mathbf{a}_z \cos\theta + \mathbf{a}_x \sin\theta) \times (\mathbf{a}_x ILB_0)$$

$$= -\frac{d}{2}\mathbf{a}_y \cos\theta \, ILB_0 \qquad (9.42)$$

hence the total torque is

$$\mathbf{T} = -\mathbf{a}_y ILdB_0 \cos\theta$$

$$= -\mathbf{a}_y IAB_0 \cos\theta$$

where $A = Ld$ is the area of the loop. Notice that the forces on sides bc and da are equal and opposite and pass through the centre of the loop. This formulation can be further simplified by using the concept of a magnetic dipole.

9.6.1 Magnetic Dipole and Torque on an Arbitrary Loop

The magnetic dipole of a current carrying loop is defined as

$$\mathbf{m} = I\mathbf{S} \qquad (9.43)$$

where I is the loop current, S is the area of the loop and

$$\mathbf{S} = S\hat{\mathbf{S}} \qquad (9.44)$$

is the vector normal to the area S. This is shown in Fig. 9.6. If we take a look at the expression of the torque on a current loop as considered in Section 9.6, we find that

$$\mathbf{T} = -\mathbf{a}_y IAB_0 \cos\theta$$

$$= -\mathbf{a}_y mB_0 \cos\theta$$

which gives us the important formula

$$\mathbf{T} = \mathbf{m} \times \mathbf{B} \qquad (9.45)$$

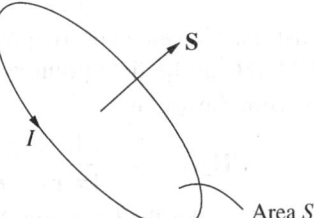

Fig. 9.6 Magnetic moment definition

where $\mathbf{m}$ is the magnetic dipole.

If we consider the far fields of a magnetic dipole using the far field approximation, the magnetic flux density $\mathbf{B}$ is given by (see Eqn (8.98)),

$$\mathbf{B} \approx \frac{\mu_0 \, m}{4\pi r^3}\left(2\cos\theta \, \mathbf{a}_r + \sin\theta \, \mathbf{a}_\theta\right) \qquad (9.46)$$

where $m = Iab = IS$ is the magnitude of the magnetic dipole of the square loop. $S = ab$ is the surface area of the loop. Similar results are obtained for a circular loop. If we take a look at the far fields of an electric dipole, the electric field is

$$\mathbf{E} \approx \frac{p}{4\pi\varepsilon_0 r^3}\left(2\cos\theta\,\mathbf{a}_r + \sin\theta\,\mathbf{a}_\theta\right) \tag{9.47}$$

$p = qd$ being the magnitude of the electric dipole. A comparison of Eqns (9.46) and (9.47) shows why the magnetic dipole is so named.

Example 9.10 Derive a general formula for the torque on an arbitrary loop, $\mathcal{L}$ immersed in an arbitrary but steady field $\mathbf{B}(\mathbf{r})$.

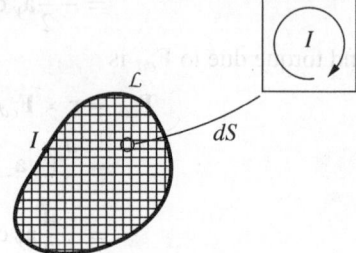

Solution Let loop be as shown in Fig. 9.7. Divide the loop into a grid of loops, each carrying a current I. Since each of the tiny loops carries the same current, the current in neighbouring loops cancel, except for the current in the loop $\mathcal{L}$. The tiny torque on each tiny loop is given by

$$d\mathbf{T} = d\mathbf{m} \times \mathbf{B}$$

In this formula, the field $\mathbf{B}$ may be assumed constant over the differential loop.

Fig. 9.7 Torque on an arbitrary loop placed in a magnetic field

Hence

$$\mathbf{T} = I\iint_S d\mathbf{S} \times \mathbf{B} \tag{9.48}$$

9.7 | Force between Two Current Elements

We now proceed to compute the force which a current element exerts on another one. Ampere, was the first to explain this effect: he showed that two parallel wires carrying current, attracted each other if the currents flowed in the same direction and repelled each other if the currents flowed in opposite directions.

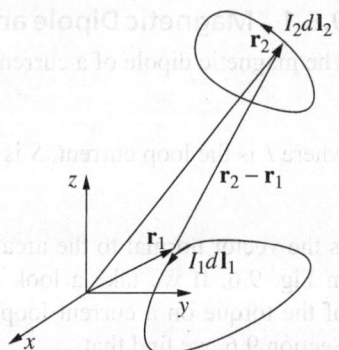

Consider Fig. 9.8 where two current elements exist. The first element $I_1 d\mathbf{l}_1$ produces a magnetic field $d\mathbf{H}_2$ at the field point $\mathbf{r}_2$ which is given by the Biot-Savart law:

$$d\mathbf{H}_2(\mathbf{r}_2) = \frac{I_1 d\mathbf{l}_1 \times \widehat{\mathbf{r}_2 - \mathbf{r}_1}}{4\pi|\mathbf{r}_2 - \mathbf{r}_1|^2} \tag{9.49}$$

Fig. 9.8 Calculation of the force between two current elements

Therefore, the field at point 2 due to the circuit 1 would be an integration of the above equation:

$$\mathbf{H}_{(2)} = \oint_{\mathcal{L}_1} \frac{I_1 d\mathbf{l}_1 \times \widehat{\mathbf{r}_2 - \mathbf{r}_1}}{4\pi|\mathbf{r}_2 - \mathbf{r}_1|^2}$$

Normally, if at point 2 there is a magnetic field $\mathbf{H}_{(2)}$ then the current element $I_2 d\mathbf{l}_2$ would feel a minuscule force

$$d\mathbf{F}_2 = \mu_0 I_2 d\mathbf{l}_2 \times \mathbf{H}_{(2)} \tag{9.50}$$

Therefore, the minuscule element on circuit 2 feels a force

$$d\mathbf{F}_2 = \mu_0 \, I_2 d\mathbf{l}_2 \times \mathbf{H}_{(2)}$$

$$= \mu_0 \, I_2 d\mathbf{l}_2 \times \left(\oint_{L_1} \frac{I_1 d\mathbf{l}_1 \times \widehat{\mathbf{r}_2 - \mathbf{r}_1}}{4\pi |\mathbf{r}_2 - \mathbf{r}_1|^2} \right)$$

And the total force that circuit 2 feels due to the current flow in circuit 1 is

$$\mathbf{F}_2 = \mu_0 \, I_2 \oint_{L_2} d\mathbf{l}_2 \times \left(\oint_{L_1} \frac{I_1 d\mathbf{l}_1 \times \widehat{\mathbf{r}_2 - \mathbf{r}_1}}{4\pi |\mathbf{r}_2 - \mathbf{r}_1|^2} \right)$$

$$= \mu_0 \frac{I_2 I_1}{4\pi} \oint_{L_2} \oint_{L_1} \frac{d\mathbf{l}_2 \times \left(d\mathbf{l}_1 \times \widehat{\mathbf{r}_2 - \mathbf{r}_1} \right)}{|\mathbf{r}_2 - \mathbf{r}_1|^2}$$

9.8 | Inductance

As in the case of non-dissipative electric circuit elements which store energy, namely, capacitances, the analogous magnetic elements which are non-dissipative and store magnetic energy are inductances. Thus in the case of inductances as we try to increase the current in an inductor, a back emf is formed opposing the change. The back emf, v_L is given by

$$v_L = L \frac{dI}{dt} \tag{9.51}$$

where L is the inductance of the inductor, and I is the current through it.

We know that the inductor does not dissipate energy, so the power $v_L I$ that is being supplied to the inductor is being stored in the magnetic field as magnetic energy. The magnetic energy, W_m, stored at any time is

$$W_m = \int_{t=0}^{t} \underbrace{\left(L \frac{dI}{dt} \right)}_{v_L} I dt$$

$$= L \int_{t=0}^{t} I \frac{dI}{dt} dt$$

$$= L \int_{i=0}^{i} I dI$$

$$= L \frac{I^2}{2} \tag{9.52}$$

so,
$$L = \frac{2 W_m}{I^2} \tag{9.53}$$

While the current is increasing at the field level we find that the magnetic field is also on the increase, and energy as calculated above is being stored in the field. Generally an inductor consists of N turns, and each turn links a certain amount of magnetic flux, Ψ_m, given by (see Fig. 9.9)

$$\Psi_m = \iint_{\text{Cross-section}} \mathbf{B} \cdot d\mathbf{S} \tag{9.54}$$

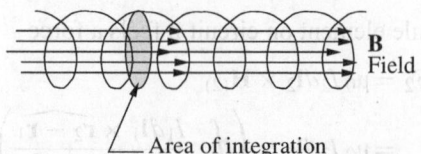

Fig. 9.9 Area of integration for calculating the flux linked by a single turn, Ψ_m

therefore the total flux linked by N turns (λ) is

$$\lambda = N\Psi_m \tag{9.55}$$

The inductance of the inductor is

$$L = \frac{\lambda}{I} = \frac{N\Psi_m}{I} \tag{9.56}$$

9.8.1 Inductance of a Coil

Example 9.11 Find approximate inductance of a coil of length l and area of cross-section A which has n turns/m.

Solution As a start let us do a quick calculation for an inductance in the form of a coil. Referring to Section 9.4.3 we know that a solenoid carrying a current I and having n turns per meter produces a uniform magnetic field **H** in its core, given by

$$\mathbf{H} = nI \text{ A/m} \tag{9.57}$$

Now consider an inductor whose length is l with N turns wound on a cylinder with area A. Then $n(\text{turns/m}) \simeq N(\text{turns})/l(\text{m})$. The magnetic field in the core of the inductance is therefore (approximately)

$$\mathbf{H} \simeq \frac{NI}{l} \tag{9.58}$$

The magnetic flux density, **B** is related to the magnetic field by

$$\mathbf{B} = \mu\,\mathbf{H} \simeq \mu_0 \frac{NI}{l} \tag{9.59}$$

The flux linked per turn is

$$\Psi_m \simeq \mu_0 \frac{NIA}{l} \tag{9.60}$$

The total flux linked is

$$\lambda = N\Psi_m \simeq \mu_0 \frac{N^2 IA}{l} \tag{9.61}$$

therefore the inductance is

$$L = \frac{\lambda}{I} \simeq \mu_0 \frac{N^2 A}{l} \tag{9.62}$$

This value of inductance is only approximate, since the magnetic field was borrowed from that of an infinitely long solenoid where the field in the core is uniform, but the approximation is a fairly good one and can be relied upon as the starting point for an inductance calculation.

Example 9.12 Find a better approximation than Example 9.11 for the inductance of a coil of length l and area of cross-section A which has n turns/m.

Solution Using the z-directed magnetic field for a finite length solenoid as derived in Eqn (8.64)

$$H_z = \frac{nI'}{2}\left[\frac{(l/2-z)}{\left(\sqrt{(l/2-z)^2 + A/\pi}\right)} + \frac{(l/2+z)}{\left(\sqrt{(l/2+z)^2 + A/\pi}\right)}\right]$$

$$= \frac{nI'}{2}\frac{l}{\left(\sqrt{(l/2)^2 + A/\pi}\right)} \quad (\text{for } z = 0)$$

Using $N = nl$ as earlier where N is the number of turns,

$$H_z = \frac{NI'}{\left(2\sqrt{(l/2)^2 + A/\pi}\right)}$$

assuming that the field at the centre will be only z-directed from symmetry considerations. Then

$$\Psi_m = A\mu_0 \frac{NI'}{\left(2\sqrt{(l/2)^2 + A/\pi}\right)}$$

and

$$\lambda = A\mu_0 \frac{N^2 I'}{\left(2\sqrt{(l/2)^2 + A/\pi}\right)}$$

So

$$L = \frac{\lambda}{I'} = A\mu_0 \frac{N^2}{\left(2\sqrt{(l/2)^2 + A/\pi}\right)}$$

Example 9.13 For a coil of diameter 1 cm and length 1 inch and with number of turns $N = 20$, find the inductance of the coil by the two methods outlined in Examples 9.11 and 9.12.

Solution By the method of Example 9.11,

$$L = \mu_0 \frac{N^2 A}{l}$$

$$= \mu_0 \frac{400 \times (\pi \times 0.005^2)}{0.0254}$$

$$= 1.237\mu_0 \ (\text{H}) = 1.55 \ \mu\text{H}$$

By the method of Example 9.12,

$$L = A\mu_0 \frac{N^2}{\left(2\sqrt{(l/2)^2 + A/\pi}\right)}$$

$$= \mu_0 \times 7.854 \times 10^{-5} \times \frac{400}{2\sqrt{(0.0254/2)^2 + A/\pi}}$$

$$= 1.151\mu_0 = 1.45 \ \mu\text{H}$$

If we compute the percent difference, then the second answer is within 7% of the first answer.

 See `MagFldCentLoop.m` in Chapter 9

9.8.1.1 Wheeler's Formula

A great deal of work has been done in the area of inductance calculation and one of the simplest but fairly accurate formula is

$$L(\mu H) = \frac{r^2 N^2}{(9r + 10l)}$$

where r is the radius of the coil and l is the length, both in inches.

Example 9.14 For the coil of dimensions given in Example 9.13, use Wheeler's formula to calculate the inductance.

Solution Using Wheeler's formula

$$L(\mu H) = \frac{r^2 N^2}{(9r + 10l)} = \frac{400(0.5/2.54)^2}{[9(0.5/2.54) + 10 \times 1]}$$
$$= 1.32 \, \mu H$$

9.8.2 Inductance of a Coaxial Line

Let us do some more inductance calculations, this time the inductance per meter of a coaxial line. The configuration is shown in Fig. 9.10, which also shows the area of integration to calculate the flux Ψ_m.

Example 9.15 Find the inductance per meter of a coaxial line as shown in Fig. 9.10.
Solution For a current I flowing into the centre conductor, the magnetic field between the inner and outer conductor is given by

$$\mathbf{H} = \frac{I}{2\pi \rho} \mathbf{a}_\phi \tag{9.63}$$

in a coordinate system with the z-axis pointed down the line and whose origin is placed at the centre of the inner conductor. $\mathbf{B}$ is therefore

$$\mathbf{B} = \mu_0 \frac{I}{2\pi \rho} \mathbf{a}_\phi$$

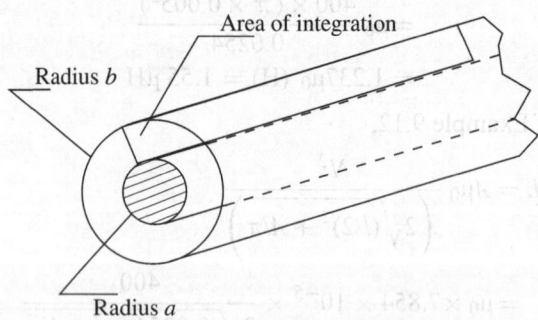

Fig. 9.10 Inductance of a coaxial line

and (note that z goes from 0 to 1 m)

$$\Psi_m = \int\limits_{\rho=a,\,z=0}^{\rho=b,\,z=1}\int \mu_0 \frac{I}{2\pi\rho} d\rho\,dz$$

$$= \frac{\mu_0 I}{2\pi} \ln\left(\frac{b}{a}\right) \text{ (Wb)} \tag{9.64}$$

and hence the inductance per meter is

$$L = \frac{\Psi_m}{I} = \frac{\mu_0}{2\pi} \ln\left(\frac{b}{a}\right) \text{ (H/m)} \tag{9.65}$$

9.8.3 Magnetic Energy

In Section 6.7 we studied that the energy stored in the electric field in a region of space V is given by

$$W_e = \iiint\limits_V \frac{1}{2} \mathbf{D} \cdot \mathbf{E}\, dV \text{ (J)} \tag{9.66}$$

where

$$w_e = \frac{1}{2} \mathbf{D} \cdot \mathbf{E} \text{ (J/m}^3) \tag{9.67}$$

is the electrical energy density at any point in space. The electrical energy stored in a capacitor is given by

$$W_e = \frac{1}{2} CV^2 \tag{9.68}$$

where V was the potential difference across the capacitor plates. In a similar manner we now would like to investigate the energy stored in an inductor from the field point of view. To do this, we take the the coaxial line as a case in point.

Taking a small slice of the coaxial line of length dz along the z-axis (see Fig. 9.11), the magnetic energy stored in this small section is

$$dW_m = \frac{1}{2} dL\, I^2$$

$$= \frac{1}{2} \frac{d\Psi_m}{I} I^2$$

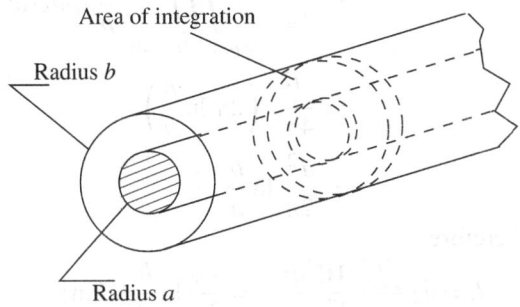

Fig. 9.11 Calculation of inductance from the field point of view

$$= \frac{1}{2} \underbrace{\left(\int_{\rho=a}^{\rho=b} B_\phi d\rho dz \right)}_{d\Psi_m} I$$

$$= \frac{1}{2} \int_\rho B_\phi d\rho dz \underbrace{\left(\int_{\phi=0}^{\phi=2\pi} H_\phi \rho d\phi \right)}_{\text{Ampere's law}}$$

$$= \frac{\mu_0}{2} \left(\int\!\!\!\int_{\rho=a,\, \phi=0}^{\rho=b,\, \phi=2\pi} H_\phi^2 \rho d\rho d\phi \right) dz \; \text{(J)} \tag{9.69}$$

where dL is the infinitesimal inductance of that part; in the last step, the two integrations are taken together. Therefore

$$W_m = \frac{\mu_0}{2} \int\!\!\!\int\!\!\!\int_{\rho=a,\, \phi=0,\, z=0}^{\rho=b,\, \phi=2\pi,\, z=1} H_\phi^2 \rho d\rho d\phi \, dz \; \text{(J/m)} \tag{9.70}$$

Generalising this result,

$$W_m = \frac{\mu_0}{2} \int\!\!\!\int\!\!\!\int_V \mathbf{H} \cdot \mathbf{H} \, dV \; \text{(J/m)} \tag{9.71}$$

where $\rho d\rho d\phi \, dz = dV$ and $H_\phi^2 \to \mathbf{H} \cdot \mathbf{H}$. The magnetic energy density is therefore

$$w_m = \frac{1}{2} \mathbf{B} \cdot \mathbf{H} \; \text{(J/m}^3) \tag{9.72}$$

Example 9.16 Find the inductance per meter of a coaxial line using stored magnetic energy concepts.

Solution We now calculate the inductance from the field point of view for a coaxial line. The inductance is

$$L = \frac{2W_m}{I^2} = \mu_0 \frac{\int\!\!\int\!\!\int |\mathbf{H}|^2 \, dV}{I^2} \; \text{(H)} \tag{9.73}$$

Applying this formula to a coaxial line, $\mathbf{H} = I/2\pi \rho \; \mathbf{a}_\phi$.

So

$$\int\!\!\int\!\!\int |\mathbf{H}|^2 \, dV = \int\!\!\int\!\!\int \frac{I^2}{4\pi^2 \rho^2} \rho d\rho d\phi \, dz$$

$$= \frac{I^2}{4\pi^2} \int\!\!\int\!\!\int_{\rho=a,\, \phi=0,\, z=0}^{\rho=b,\, \phi=2\pi,\, z=1} \frac{1}{\rho} d\rho d\phi \, dz$$

$$= \frac{I^2}{4\pi^2} \left(2\pi \ln \frac{b}{a} \right)$$

$$= \frac{I^2}{2\pi} \ln \frac{b}{a} \tag{9.74}$$

the inductance is therefore

$$L = \mu_0 \frac{\int\!\!\int\!\!\int |\mathbf{H}|^2 dV}{I^2} = \frac{\mu_0}{2\pi} \ln \frac{b}{a} \; \text{(H/m)} \tag{9.75}$$

which is the same result as earlier. See Eqn (9.63).

From Eqn (9.53), we can get another formula for the inductance. Consider

$$\frac{1}{2}\iiint \mathbf{B} \cdot \mathbf{H}\, dV = \frac{1}{2}\iiint (\nabla \times \mathbf{A}) \cdot \mathbf{H}\, dV \tag{9.76}$$

where we have used $\mathbf{B} = \nabla \times \mathbf{A}$, $\mathbf{A}$ being the vector potential

and from
$$\nabla \cdot (\mathbf{A} \times \mathbf{H}) = \mathbf{H} \cdot (\nabla \times \mathbf{A}) - \mathbf{A} \cdot (\nabla \times \mathbf{H}) \tag{9.77}$$

so
$$\mathbf{H} \cdot (\nabla \times \mathbf{A}) = \nabla \cdot (\mathbf{A} \times \mathbf{H}) + \mathbf{A} \cdot (\nabla \times \mathbf{H}) \tag{9.78}$$

Equation (9.76) becomes

$$\iiint \mathbf{B} \cdot \mathbf{H}\, dV = \iiint [\nabla \cdot (\mathbf{A} \times \mathbf{H}) + \mathbf{A} \cdot (\nabla \times \mathbf{H})]\, dV$$

$$= \oiint_{S} \mathbf{A} \times \mathbf{H} \cdot d\mathbf{S} + \iiint_{V} \mathbf{A} \cdot \mathbf{J}\, dV \tag{9.79}$$

the surface integral on the right is zero since we may take the surface S at infinity where the fields are zero, so

$$W_m = \frac{1}{2}\iiint_{V} \mathbf{B} \cdot \mathbf{H}\, dV = \frac{1}{2}\iiint_{V_C} \mathbf{A} \cdot \mathbf{J}\, dV \tag{9.80}$$

where V_C is the volume where the currents are.

Example 9.17 Find the inductance per meter of a coaxial line with inner diameter 1 mm and outer diameter 4.7 mm (RG6/U standard coaxial cable).

Solution The inductance per meter is given by

$$L = \frac{\mu_0}{2\pi}\ln(b/a)$$

$$= \frac{4\pi \times 10^{-7}}{2\pi}\ln(4.7)$$

$$= 30.95\ \mu\,\mathrm{H}$$

9.8.4 Inductance of a Circular Loop

Example 9.18 Calculate the approximate inductance of a single circular loop.

Solution Consider a loop of radius R as shown in Fig. 9.12 of circular cross-section, where a is the radius of the wire. The vector potential $\mathbf{A}$ near the wire is given by Eqn (9.82) to be

$$A_i(\mathbf{r}) = \frac{\mu_0}{4\pi}\iiint_{V'} \frac{J_i(\mathbf{r}')\, dV'}{|\mathbf{r} - \mathbf{r}'|} \qquad i = x, y, z$$

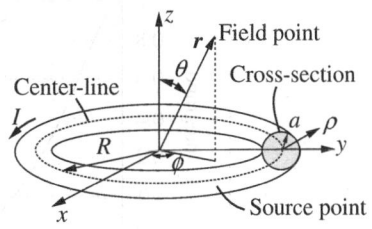

Fig. 9.12 A circular loop of wire carrying a current I

taking a point in the wire we can see that there is only component of $\mathbf{J}$, namely J_ϕ and

$$\mathbf{J} = J_\phi\left(-\mathbf{a}_x \sin\phi' + \mathbf{a}_y \cos\phi'\right) \tag{9.81}$$

where J_ϕ is the magnitude of the current density which is uniform throughout the cross-section. Considering a point on the x–z plane, where $\phi = 0$

$$|\mathbf{r} - \mathbf{r}'| = \sqrt{(x - x')^2 + (y - y')^2 + (z - z')^2}$$

$$= \sqrt{R^2 + r^2 - 2Rr \sin\theta \cos\phi'} \qquad (9.82)$$

and (Jackson, 1999, Section 5.5)

$$A_x = \left(\frac{\mu_0}{4\pi}\right) IR \int_0^{2\pi} \frac{(-\sin\phi')\,d\phi'}{\sqrt{R^2 + r^2 - 2Rr \sin\theta \cos\phi'}} = 0$$

$$A_y = \left(\frac{\mu_0}{4\pi}\right) IR \int \frac{(\cos\phi')\,d\phi'}{\sqrt{R^2 + r^2 - 2Rr \sin\theta \cos\phi'}}$$

$$A_z = 0 \qquad (9.83)$$

the first equation is zero since the integrand has odd symmetry about ϕ, the second equation is not zero and is A_ϕ and the third equation is zero since $J_z = 0$. The A_x and A_y integrands are shown in Fig. 9.13. The second equation can be expressed in terms of the complete elliptic integrals K and E,

$$A_\phi = \left(\frac{\mu_0}{4\pi}\right)\left(\frac{4IR}{\sqrt{R^2 + r^2 + 2Rr \sin\theta}}\right)\left[\frac{(2 - k^2)\,K(k) - 2E(k)}{k^2}\right] \qquad (9.84)$$

where

$$k^2 = \frac{4Rr \sin\theta}{R^2 + r^2 + 2Rr \sin\theta} \qquad (9.85)$$

for $r = R$, $\theta = \pi/2$ we find that $k \approx 1$. On closer examination, we discover that $K(k)$ becomes infinity but $E(k)$ remains finite and ≈ 1. The functions near this

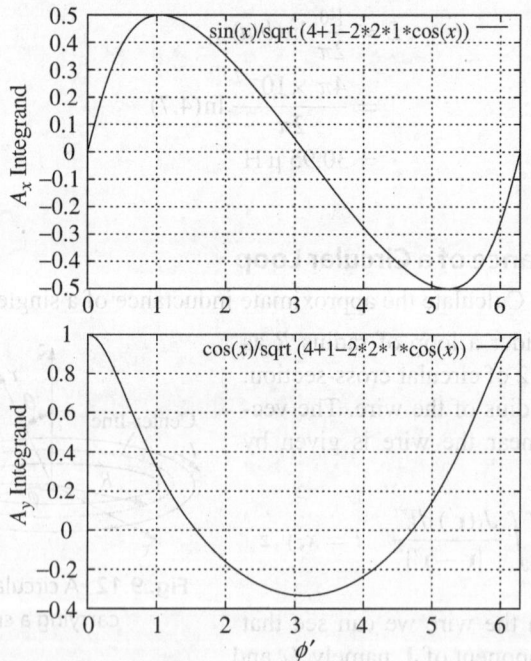

Fig. 9.13 A_x and A_y integrands for $R = 2$, $r = 1$, $\theta = \pi/2$

point[1] can be approximated by

$$K(z)\,|_{z \to 1} \approx -\frac{1}{2}\ln(1-z)\left(1 - \frac{z-1}{4} + O((z-1)^2)\right) + \ln 4$$
$$+ \frac{1}{4}(1 - \ln 4)(z-1) + O((z-1)^2) \tag{9.86}$$

$$E(z)\,|_{z \to 1} \approx 1 + \frac{1}{2}\ln(1-z)\left(\frac{z-1}{4}\right) + \frac{z-1}{4}\left(-2\ln 4 + 1 + O((z-1)^2)\right) \tag{9.87}$$

we can see from these equations that as $z \to 1$ the largest terms in these expansions are

$$K(z)\,|_{z \to 1} \approx -\frac{1}{2}\ln(1-z) + \ln 4 \tag{9.88}$$

$$E(z)\,|_{z \to 1} \approx 1 \tag{9.89}$$

which we introduce into Eqn (9.84) to get

$$A_\phi \approx \left(\frac{\mu_0}{4\pi}\right) 2I \left(\frac{1}{2}\ln\frac{16}{(1-k)} - 2\right) \tag{9.90}$$

where we have substituted $r \approx R$ in the loop. We now proceed to evaluate $1-k$ on the inside of the wire. We allow $r = R + \rho$ where $\rho < a$ (the radius of the loop). Here we find the Taylor series expansion of k is

$$k \approx 1 - \frac{\rho^2}{4R^2} + \frac{\rho^3}{4R^3} + \cdots \tag{9.91}$$

substituting in Eqn (9.90), and taking only the ρ^2 term, $1-k \approx (\rho/2R)^2$

so,
$$A_\phi \approx \left(\frac{\mu_0 I}{2\pi}\right)\left(\ln\frac{8R}{\rho} - 2\right) \tag{9.92}$$

Integrating first only on the area of cross-section

$$\iint A_\phi J_\phi \rho'\, d\phi'\, d\rho' = J_\phi \iint A_\phi \rho'\, d\phi'\, d\rho'$$
$$= \left(\frac{\mu_0 I}{2\pi}\right)\left(\frac{I}{\pi a^2}\right) \iint \left(\ln\frac{8R}{\rho'} - 2\right)\rho'\, d\phi'\, d\rho'$$
$$= \left(\frac{\mu_0 I}{2\pi}\right)\left(\frac{I}{\pi a^2}\right)(2\pi)\int\left(\ln\frac{8R}{\rho'} - 2\right)\rho'\, d\rho'$$

since
$$\int [x\ln(8R/x) - 2x]\, dx = \frac{x^2\ln(8R/x)}{2} - \frac{3x^2}{4}$$

therefore
$$\frac{1}{2}\iint A_\phi J_\phi \rho'\, d\phi'\, d\rho' = \left(\frac{\mu_0 I}{4\pi}\right)\frac{I}{\pi a^2}(2\pi)\frac{a^2}{2}\left[\ln(8R/a) - \frac{3}{2}\right]$$
$$= \left(\frac{\mu_0 I^2}{4\pi}\right)\left[\ln(8R/a) - \frac{3}{2}\right] \tag{9.93}$$

---

[1] http://functions.wolfram.com/EllipticIntegrals/EllipticK/introductions/
CompleteEllipticIntegrals/ShowAll.html or any other reference dealing with elliptic integrals.

integrating along the wire

$$W_m = 2\pi R \times \left(\frac{\mu_0 I^2}{4\pi}\right) \left[\ln(8R/a) - \frac{3}{2}\right]$$

$$= \left(\frac{\mu_0 I^2}{2}\right) R \left[\ln(8R/a) - \frac{3}{2}\right] \qquad (9.94)$$

$$= L\frac{I^2}{2}$$

which gives the inductance of a single turn to be

$$L = \mu_0 R \left[\ln(8R/a) - \frac{3}{2}\right] \qquad (9.95)$$

There are various other results given in the literature where instead of the constant 3/2 they have either 2 or 7/8. This difference is chiefly due to the level of approximation of the K and E functions.

Example 9.19 Find the inductance of a single loop of wire of radius $R = 0.5$ cm and the radius of cross-section $a = 0.1$ cm
Solution Using the formula, we get

$$L = \mu_0 R \left[\ln(8R/a) - \frac{3}{2}\right]$$

$$= \mu_0 (0.005) \left[\ln(8 \times 0.5/0.1) - 1.5\right]$$

$$= 0.011\mu_0 = 13.75 \text{ nH}$$

If we use 2 as the constant instead of 3/2, the inductance is 10.61 nH, and if we use 7/8 then $L = 17.7$ nH. The formulae tend to give the same result when $R \gg a$.

9.8.5 Mutual Inductance

If a loop or coil produces a magnetic field $\mathbf{B}_1$ and a second loop or coil links the field of the first one, then the flux linked by the second loop or coil may be called Ψ_{m21} and read as the flux linked in coil (loop) 2 due to field 1 is

$$\Psi_{m21} = \iint_{S_2} \mathbf{B}_1 \cdot d\mathbf{S}_2 \qquad (9.96)$$

where S_2 is the region where the integration is performed. Then the *mutual inductance* of the second loop or coil is given by

$$M_{21} = \frac{N_2 \Psi_{m21}}{I_1} \qquad (9.97)$$

where N_2 are the number of turns in the second coil and I_1 is the current in the first coil or loop. Similarly, if the second loop has a current I_2 in it which produces a field $\mathbf{B}_2$ then the mutual inductance of coil 1 due to current 2 is given by

$$M_{12} = \frac{N_1 \Psi_{m12}}{I_2} \qquad (9.98)$$

Due to reciprocity,

$$M_{12} = M_{21} \qquad (9.99)$$

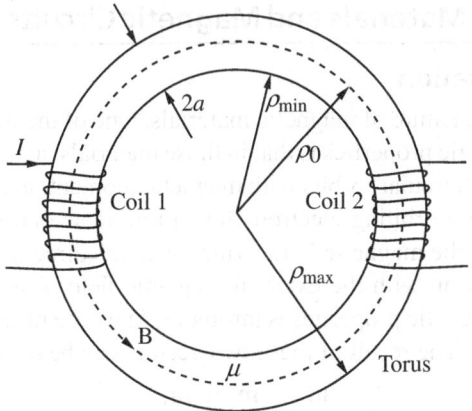

Fig. 9.14 Mutual inductance of two coils

Example 9.20 Find the mutual inductance of the two coils with turns N_1 and N_2 wound on a torus as shown in Fig. 9.14. Find also the self inductance of coil 1.

Solution From Eqn (9.48), if the current I is flowing in coil 1 then

$$H_{\phi 1} \approx \frac{N_1 I}{2\pi \rho} \qquad \rho_{min} < \rho < \rho_{min} + 2a$$

The result of Eqn (9.48) applies even though the torus is partially wound since for $\mu \gg \mu_0$ the magnetic field is confined entirely to the inside of the torus.

so

$$B_{\phi 1} = \mu H_{\phi 1} = \frac{\mu N_1 I}{2\pi \rho_0} \qquad \rho_0 = \rho_{min} + a$$

then the flux linked by a single turn of coil 2 is

$$\Psi_{m1} = B_{\phi 1} \times \text{Area of cross-section} = \frac{\mu N_1 I}{2\pi \rho_0} \times \pi a^2$$

and the total flux linked is

$$N_2 \Psi_{m1} = \frac{\mu N_1 N_2 I a^2}{2\rho_0}$$

so the mutual inductance is

$$M_{21} = \frac{N_2 \Psi_{m1}}{I} = \frac{\mu N_1 N_2 a^2}{2\rho_0}$$

By reciprocity, $\qquad M_{12} = M_{21}$

To calculate the (self) inductance of coil 1, the flux linked per turn is again

$$\Psi_{m1} = \frac{\mu N_1 I a^2}{2\rho_0}$$

so the total flux linked in coil 1 is

$$N_1 \Psi_{m1} = \frac{\mu N_1^2 I a^2}{2\rho_0}$$

hence the inductance of coil 1 is

$$L_1 = \frac{N_1 \Psi_{m1}}{I} = \frac{\mu N_1^2 a^2}{2\rho_0}$$

9.9 | Magnetic Materials and Magnetic Circuits

9.9.1 Magnetisation

We now consider the nature of magnetic materials. One of the ways in which a material exhibits magnetic properties is that in those materials unpaired electrons orbit around the nuclei in 'circular' orbits with magnetic moment, $\mathbf{m}_o$. In the presence of a magnetic field these orbiting electrons feel a torque as outlined in Eqn (9.45), which tends to align the magnetic field (which is in the direction of $\mathbf{m}_o$) produced by the orbiting electron with the external magnetic field. Another way by which electrons exhibit magnetic properties is through spin whose magnetic moments can be designated as $\mathbf{m}_s$. The result of these two vectors can be denoted as $\mathbf{m}_r$ with

$$\mathbf{m}_r = \mathbf{m}_o + \mathbf{m}_s \tag{9.100}$$

Most materials can be considered as one of three types: *diamagnetic, paramagnetic or ferromagnetic.*

For *diamagnetic* materials, $\mathbf{m}_r = 0$. These are weakly repelled by magnetic fields; Since $\mathbf{m}_r$ is zero, external magnetic fields have little effect on them. Most elements in the periodic table, including copper, silver, and gold, are diamagnetic.

In the second case, $\mathbf{m}_r \cong 0$. These substances are weakly attracted to magnetic fields and external fields are enhanced by their presence. These are *paramagnetic* substances. An example of a paramagnetic element is potassium. After the field is removed, the material also does not exhibit any magnetic properties.

Finally we consider *ferromagnetic* materials. In these materials

$$|\mathbf{m}_s| \gg |\mathbf{m}_o| \tag{9.101}$$

$$\mathbf{m}_r \cong \mathbf{m}_s \gg 0 \tag{9.102}$$

In ferromagnetic materials, adjacent atoms crystallise in similar arrangements in regions which are known as magnetic domains. These materials are strongly attracted by external magnetic fields. And after the field is removed some of these materials retain their magnetism.

In a magnetic material immersed in an external field in a small volume ΔV, let there be N small magnetic dipole moments. Then the *magnetisation* $\mathbf{M}$ is given by

$$\mathbf{M} = \lim_{\Delta V \to 0} \sum_{i=1}^{N} \frac{\mathbf{m}_i}{\Delta V} \tag{9.103}$$

and has the same units as the magnetic field $\mathbf{H}$. Without going into greater detail we can show that

$$\mathbf{B} = \mu_0 \left(\mathbf{H} + \mathbf{M} \right) \tag{9.104}$$

$$= \mu_0 \mu_r \mathbf{H} \tag{9.105}$$

where $\mathbf{H}$ is the magnetic field in the absence of the material. Assuming a linear model, we can say that

$$\mathbf{M} = \chi_m \mathbf{H} \tag{9.106}$$

where χ_m is the magnetic susceptibility.

$$\chi_m > 0, \qquad \text{for paramagnetic materials}$$
$$\chi_m \gg 0, \qquad \text{for ferromagnetic materials and}$$
$$\chi_m < 0, \qquad \text{for diamagnetic materials}$$

using Eqns (9.105) and (9.106), we have

$$\mu = \mu_0 \mu_r \tag{9.107}$$
$$\mu_r = 1 + \chi_m \tag{9.108}$$

9.9.2 Magnetostatic Boundary Conditions

What is the behaviour of the magnetic field at the boundary of various media? The steady magnetic field is connected with those media where there is a change of permeability, μ, rather than changes of permittivity, ε. Similarly, changes in the conductivity, σ, does not have any effect on the magnetic field.

Figure 9.15, shows a magnetic field, $\mathbf{H}$, in a region where there is a change in media. There are two media, μ_1 and μ_2. The $\mathbf{B}$ field in the two media is $\mathbf{B}_1 = \mu_1 \mathbf{H}_1$ and $\mathbf{B}_2 = \mu_2 \mathbf{H}_2$. There are two magnetostatic laws that govern the fields, namely

$$\nabla \times \mathbf{H} = \mathbf{J}$$

and
$$\nabla \cdot \mathbf{B} = 0 \tag{9.109}$$

The reader is advised to read Section 6.6 to understand the detailed treatment of how equations such as Eqn set (9.109) are applied to boundaries.

A magnetic material (region 2) is immersed in an external magnetic field as illustrated in Fig. 9.15. The region outside the magnetic material (region 1) consists of a medium with permeability μ_1 (e.g. if the medium is air then $\mu_1 \equiv \mu_0$). The immersed magnetic material has a permeability of μ_2. We concentrate on a very small region R on the boundary of the two magnetic materials shown in Fig. 9.15(a) and apply the integral form of the magnetostatic Maxwell's equation (the first equation of Eqn set (9.109)) to the interface.

This is shown in Fig. 9.15(b). The figure depicts a close-up of the interface where a loop a-b-c-d has been shown, on which the line integral of the magnetic field is to be computed. The arrows superimposed on the loop show that integration is being carried out in the counter-clockwise sense. From the coordinate system shown on top of the figure, $\hat{\mathbf{n}} = \mathbf{a}_z$ is the normal to the boundary. The normal is directed towards region 1. $\hat{\mathbf{t}}_1$ is the tangent to the boundary and in this case it is equal to $\mathbf{a}_x$. We assume that d-a and b-c are very small, while a-b and c-d are small enough so

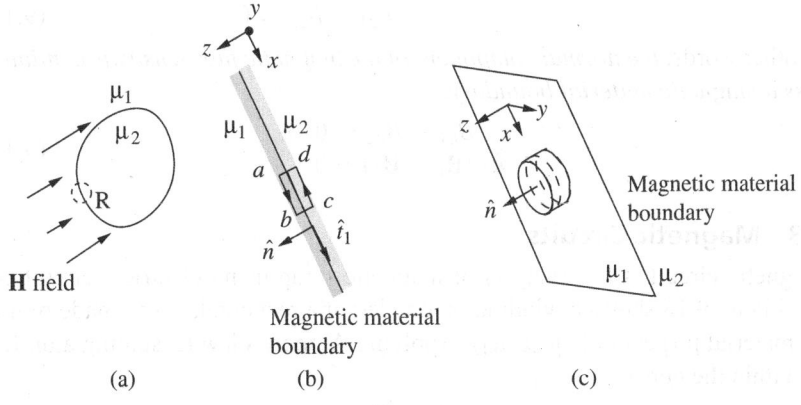

Fig. 9.15 Magnetic boundary conditions

that that the field is approximately constant over them. In other words,

$$\text{Length } (b - c) = \text{Length } (d - a) \cong 0$$

$$\text{Length } (a - b) = \text{Length } (c - d) \cong |\Delta x|$$

Applying $\nabla \times \mathbf{H} = \mathbf{J}$ to the loop a-b-c-d,

$$\mathbf{H}_1 \cdot \mathbf{a}_x |\Delta x| + \mathbf{H}_2 \cdot (-\mathbf{a}_x) |\Delta x| = J_{sy2}$$

where $\mathbf{H}_1$ and $\mathbf{H}_2$ are the magnetic fields in regions 1 and 2 respectively and J_{sy2} is the y-directed surface current at the boundary. The surface current has to be taken into account if the conductivity σ_2 of region 2 is almost infinity and if a surface current is present. The integrations over the line lengths b-c and d-a are assumed to be negligible and so have been dropped from the equation. The previous equation now becomes

$$H_{x1} - H_{x2} = J_{sy2} \tag{9.110}$$

In the same manner by considering a similar loop going into the page the longer sides of which lie in media 1 and 2, while the shorter side cuts across the two media, we perform a similar line integral over it to get

$$H_{y1} - H_{y2} = -J_{sx2} \tag{9.111}$$

which can be summarised as

$$\hat{\mathbf{n}} \times (\mathbf{H}_1 - \mathbf{H}_2) = \mathbf{J}_{st} \tag{9.112}$$

The subscript t is used to signify the *tangential* component.

Let us now consider the normal component of the field. We apply

$$\oiint \mathbf{B} \cdot d\mathbf{S} = 0 \tag{9.113}$$

to the configuration shown in Fig. 9.15(c). The figure is a close up of region R in Fig. 9.15(a). We draw a small pill-box as shown in (c) whose height is negligible and the top and bottom areas are so small that the field is assumed not to change much in those regions

$$\text{Height of pill-box} \cong 0$$

$$\text{Area of the top and bottom} \cong \Delta A$$

Applying Eqn (9.113) to this pill-box

$$\mathbf{B}_1 \cdot \mathbf{a}_z \Delta A + \mathbf{B}_2 \cdot (-\mathbf{a}_z) \Delta A = 0 \tag{9.114}$$

$$B_{z1} - B_{z2} = 0 \tag{9.115}$$

or in other words, *the normal component of the magnetic flux density is continuous across a magnetic material boundary.*

$$B_{n1} - B_{n2} = 0$$
$$\hat{\mathbf{n}} \cdot (\mathbf{B}_1 - \mathbf{B}_2) = 0 \tag{9.116}$$

9.9.3 Magnetic Circuits

A magnetic circuit is the analysis of magnetic setup using electrical circuit concepts. Figure 9.16 shows a winding on one leg of a rectangular core made of magnetic material μ (generally $\mu \gg \mu_0$). Applying Ampere's law (assuming a uniform field within the core)

$$NI = Hl \tag{9.117}$$

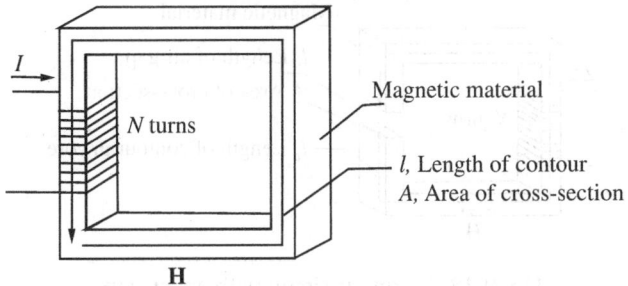

Fig. 9.16 A coil wound round a core

where I is the current in the winding; N is the number of turns of the winding; H is the magnetic field in the core and l is the length of the contour. H is therefore

$$H = \frac{NI}{l} \tag{9.118}$$

the flux Ψ_m is given by

$$\Psi_m = BA = \mu \frac{NI}{l} A \tag{9.119}$$

In this formulation,

$$NI = \left(\frac{l}{\mu A}\right) \Psi_m \tag{9.120}$$

if

$$\mathcal{F} = NI \quad \text{and} \quad \mathcal{R} = \left(\frac{l}{\mu A}\right) \tag{9.121}$$

then

$$\mathcal{F} = \mathcal{R}\Psi_m \tag{9.122}$$

where

- $\mathcal{F}$ is called the *magnetomotive force* or mmf (analogous to emf);
- $\mathcal{R}$ is the reluctance (analogous to resistance).
  Observe: the formula for resistance is $l/\sigma A$ and reluctance is $l/\mu A$
- and Ψ_m is analogous to the current.

So the equivalent magnetic circuit is shown in Fig. 9.17.

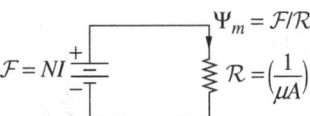

Fig. 9.17 A magnetic circuit

Example 9.21 Let us analyse another magnetic circuit, one with an air-gap in it, as shown in Fig. 9.18.

Solution If we employ Ampere's law and apply it on the circuit around the core and air-gap, we have

$$NI = H_a l_a + H_c l_c \tag{9.123}$$

where

- H_a is the magnetic field in the gap (which we assume to be constant)

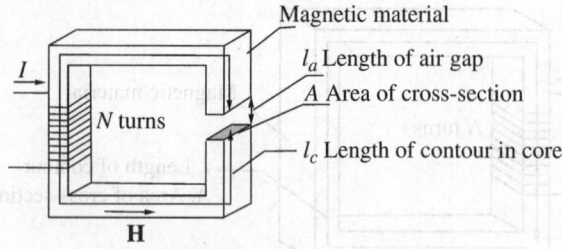

Fig. 9.18 Magnetic circuit with an air-gap

- l_a is the length of the air-gap
- H_c is the magnetic field in the core (which also we assume to be constant) and
- l_c is the length of the contour in the core
- The **B** field in the two regions is

$$B_a = \mu_0 H_a, \qquad B_c = \mu_c H_c \qquad (9.124)$$

- where B_a and B_c are the flux densities in the air and core, respectively
- and μ_0 and μ_c are the permeabilties of air and core, respectively

From the conditions on the boundary between the core and air: the normal component of B is continuous

$$B_a = B_c (\equiv B) \qquad (9.125)$$

so re-writing Eq. (9.122)

$$\underbrace{NI}_{\mathcal{F}} = \frac{B}{\mu_0} l_a + \frac{B}{\mu_c} l_c$$

$$= \underbrace{BA}_{\Psi_m} \underbrace{\left(\frac{l_a}{\mu_0 A}\right)}_{\mathcal{R}_a} + BA \underbrace{\left(\frac{l_c}{\mu_c A}\right)}_{\mathcal{R}_c}$$

$$\mathcal{F} = \Psi_m (\mathcal{R}_a + \mathcal{R}_c) \qquad (9.126)$$

the same result may be obtained by applying the 'Magnetic Ohm's law' to the circuit shown in Fig. 9.19.

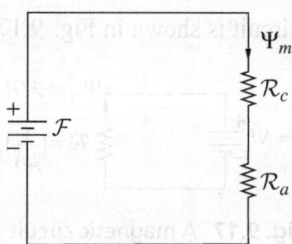

Fig. 9.19 Equivalent circuit of a magnetic core with an air-gap

POINTS TO REMEMBER

- The total force on a charge in the presence of both the electric as well as the magnetic fields is given by

$$F = q(\mathbf{E} + \mathbf{v} \times \mathbf{B})$$

this force is called the *Lorentz force*.

- The cyclotron frequency of an electron moving perpendicular to a magnetic field B_0 is

$$\omega_c = -\frac{eB_0}{m_e}$$

where e is the electronic charge, m_e is the mass of the electron
- The force on a length of straight line L (in vector notation, I is in the direction **L**) immersed in a steady magnetic field **B** is

$$\mathbf{F} = I\mathbf{L} \times \mathbf{B}$$

- No force is felt by a loop carrying a steady current I and immersed in a steady magnetic field.
- The torque **T** felt by a loop in a constant magnetic field **B** is

$$\mathbf{T} = \mathbf{m} \times \mathbf{B}$$

where $\mathbf{m} = I\mathbf{S}$ is the magnetic dipole.
- In the general case, where **B** varies from point to point but is steady, the torque is given by

$$\mathbf{T} = I \iint_{S} d\mathbf{S} \times \mathbf{B}$$

- The value of inductance of an inductor is given by

$$L = \frac{2W_m}{I^2}$$

where W_m is the total magnetic energy stored and I is the current through the inductor.
- The inductance of the inductor of N turns is

$$L = \frac{N\Psi_m}{I}$$

where Ψ_m is the flux linked by one turn and I is the current.
- The inductance of a coil of N turns and cross-sectional area A and length l is

$$L \simeq \mu_0 \frac{N^2 A}{l}$$

- The inductance per meter of a coaxial line is

$$L = \frac{\mu_0}{2\pi} \ln\left(\frac{b}{a}\right) \quad \text{(H/m)} \tag{9.127}$$

where b and a are the outer and inner radii of the line.
- The magnetic energy density is

$$w_m = \frac{1}{2}\mathbf{B} \cdot \mathbf{H} \quad \text{(J/m}^3) \tag{9.128}$$

- The inductance of a single circular turn is

$$L = \mu_0 R \left[\ln(8R/a) - \frac{3}{2}\right] \tag{9.129}$$

where R is the radius of the loop and a is the radius of the wire.
- If N_2 are the number of turns in coil 2 and I_1 is the current in coil 1 producing a field $\mathbf{B}_1$ then the mutual inductance of coil 2 is

$$M_{21} = \frac{N_2 \Psi_{m21}}{I_1}$$

where Ψ_{m21} is the flux linked by a single turn of coil 2. Similarly, if the coil 2 has a current I_2 in it which produces a field $\mathbf{B}_2$ then the mutual inductance of coil 1 due to current 2 is given by

$$M_{12} = \frac{N_1 \Psi_{m12}}{I_2}$$

Also due to reciprocity, $\qquad M_{12} = M_{21}$

- The magnetic flux density, $\mathbf{B}$ is related to the magnetic field, $\mathbf{H}$ and magnetisation, $\mathbf{M}$ through the relation

$$\mathbf{B} = \mu_0 \, (\mathbf{H} + \mathbf{M}) = \mu_0 \mu_r \, \mathbf{H}$$

Assuming a linear model we can say that

$$\mathbf{M} = \chi_m \, \mathbf{H}$$

where χ_m is the magnetic susceptibility.

- The magnetic susceptibility

$$\chi_m > 0, \qquad \text{for paramagnetic materials}$$
$$\chi_m \gg 0, \qquad \text{for ferromagnetic materials and}$$
$$\chi_m < 0, \qquad \text{for diamagnetic materials}$$

- Magnetic field boundary conditions are as follows:

$$\hat{n} \times (\mathbf{H}_1 - \mathbf{H}_2) = \mathbf{J}_{st}$$

which says that the tangential magnetic field is discontinuous by the amount of the surface current. And the normal component of the magnetic flux density is continuous across a magnetic material boundary.

$$B_{n1} - B_{n2} = 0$$
$$\hat{n} \cdot (\mathbf{B}_1 - \mathbf{B}_2) = 0$$

- In a magnetic circuit, if

$$\mathcal{F} = NI \text{ (mmf)} \qquad \text{and} \qquad \mathcal{R} = \left(\frac{l}{\mu A} \right) \text{ (reluctance)}$$

then $$\mathcal{F} = \mathcal{R} \Psi_m$$

where Ψ_m is the flux

SELF ASSESSMENT

Objective Type Questions

1. In an isotropic medium, $\mathbf{B}$ the magnetic flux density and $\mathbf{H}$ at the same point in space
   (a) are parallel   (b) are perpendicular
   (c) are parallel but differ in magnitude   (d) None of the above
2. In an isotropic medium, with χ_m, $\mathbf{B}$ the magnetic flux density and $\mathbf{H}$ at the same point in space have the relation
   (a) $\mathbf{B} = \mu_0 \, \mathbf{H}$ (b) $\mathbf{B} = \mathbf{H}/\mu_0$
   (c) $\mathbf{B} = \mu_0 \, \mu_r \, \mathbf{H}$ (d) $\mu_r = 1 + \chi_m$
3. In an anisotropic medium, $\mathbf{B}$ the magnetic flux density and $\mathbf{H}$ at the same point in space
   (a) are parallel   (b) are perpendicular
   (c) are parallel but differ in magnitude   (d) are at some angle to each other

4. What is the Lorentz force
   (a) $q\mathbf{v} \times \mathbf{B}$ (b) $q(\mathbf{E} + \mathbf{v} \times \mathbf{B})$
   (c) $q\mathbf{E}$ (d) None of these
5. Force due to a magnetic field on a point charge is directed
   (a) in the same plane and in the direction of the charge motion
   (b) in the same plane but in the opposite direction of the charge motion
   (c) in a direction perpendicular to charge motion
   (d) none of the above
6. The kinetic energy of a charge moving in a constant magnetic field
   (a) increases linearly with time   (b) decreases linearly with time
   (c) remains constant with time   (d) increases as the square root of he time
7. For a charge moving from $[0, 0, 0]$ to $[1, 1, 1]$ with constant velocity in a $\mathbf{B}$ field given
   by $[3, 4, 5]$ the direction of the acceleration is
   (a) $[0.41, 0.41, 0.82]$ (b) $[0.41, -0.82, 0.41]$
   (c) $[-0.82, 0.41, 0.41]$ (d) None of these
8. A charged particle of charge q and mass m moves with a velocity $\mathbf{v}$ and acceleration $\mathbf{a}$
   in a region where there is a magnetic field only
   (a) the acceleration is in the direction of $\mathbf{B}$
   (b) the acceleration is perpendicular to $\mathbf{B}$
   (c) the acceleration is in the direction of $\mathbf{v}$
   (d) the acceleration is perpendicular to $\mathbf{v}$
9. Two parallel wires, carrying currents I_1 and I_2 are separated by a distance a
   (a) The force on each wire is proportional to I_1
   (b) The force on each wire is proportional to I_2
   (c) The force on each wire is proportional to a
   (d) The force on each wire is inversely proportional to a^2
10. A loop carrying a steady current in a steady and spatially constant magnetic field $\mathbf{B}$
    (a) may feel a force   (b) may feel a torque
    (c) will always feel a force   (d) will always feel a torque

Short-Answer Questions

1. Give examples of an isotropic medium which is lossless, an isotropic medium which is lossy, and an anisotropic medium.
2. Define permeability, magnetisation, and susceptibility.
3. For a current (I) carrying straight conductor of length l in direction $\hat{\mathbf{I}}$, immersed in a constant magnetic field $\mathbf{B}$, find an expression of the force.
4. Find the inductance of a single loop of wire (wire diameter is 1 mm) whose loop radius is 1 cm.
5. Does a loop carrying a steady current feel a force or a torque in the presence of the earth's magnetic field?

Review Questions

1. What is the nature of the $\mathbf{v} \times \mathbf{B}$ force? Why does the kinetic energy of a moving electron subjected only to this force remain constant?
2. If a current element carrying a current I is oriented in a direction $d\mathbf{l}$, and is immersed in a magnetic field $\mathbf{B}$ then what is the direction of the force $d\mathbf{F}$?
3. Does a loop of current feel a force in a constant magnetic field? Can a loop of current feel a force when the magnetic field is not constant?

4. Give reasons why a small circular current carrying loop is called a magnetic dipole.
5. Give one engineering application of the use of magnetic forces.

Numerical Problems

1. A point charge $q = 1$ nC with mass $m = 1$ µg is in the presence of external field $\mathbf{E} = 1\mathbf{a}_z$ V/m; find the equation of motion for the particle at rest at $(x, y, z) = (1, 0, 0)$.
2. A point charge $q = 1$ nC with mass $m = 1$ µg is in the presence of external field $\mathbf{E} = 1\mathbf{a}_z$ V/m; find the equation of motion for the particle moving with an initial velocity $\mathbf{v} = (0.5, 0, 0)$ at $(x, y, z) = (1, 0, 0)$.
3. A point charge $q = 1$ nC with mass $m = 1$ µg is in the presence of external field $\mathbf{E} = 1\mathbf{a}_z$ V/m; find the equation of motion for the particle moving with an initial velocity $\mathbf{v} = (0, 0, 0.5)$ at $(x, y, z) = (1, 0, 0)$.
4. A point charge $q = 1$ nC with mass $m = 1$ µg is in the presence of external field $\mathbf{E} = 1\mathbf{a}_z$ V/m and $\mathbf{B} = 1$ T; find by superposition the equation of motion for the particle moving with an initial velocity $\mathbf{v} = (0.5, 0, 0.5)$ at $(x, y, z) = (1, 0, 0)$.
5. Find the force on a current filament of magnitude I_2 of length L placed at a distance a from an infinite wire as shown in Fig. 9.20.
6. Find the force on a current filament of magnitude I_2 of length L whose end is placed at a distance a from an infinite wire as shown in Fig. 9.21.
7. Using the results of Problems 5 and 6, find the force on the rectangular loop as shown in Fig. 9.22.
8. Derive the magnitude and direction of the force per meter felt by a sheet of current given by

$$\mathbf{J}_s = 10\delta(z)\,\mathbf{a}_y, \qquad -1 \le x \le 1$$

in a magnetic field $\mathbf{B} = 10^{-6}\mathbf{a}_z$ T.
9. Derive the torque $\mathbf{T}$ on a square loop of current of side a placed in a constant magnetic field $\mathbf{B}_0$.
10. For the configuration of Fig. 9.22, calculate the torque on the loop.
11. For the magnetic field $\mathbf{B} = xy\mathbf{a}_x - y^2\mathbf{a}_y$ find the torque on a square loop of side a placed on the xy plane with its centre coinciding with the origin. The loop carries a current I.

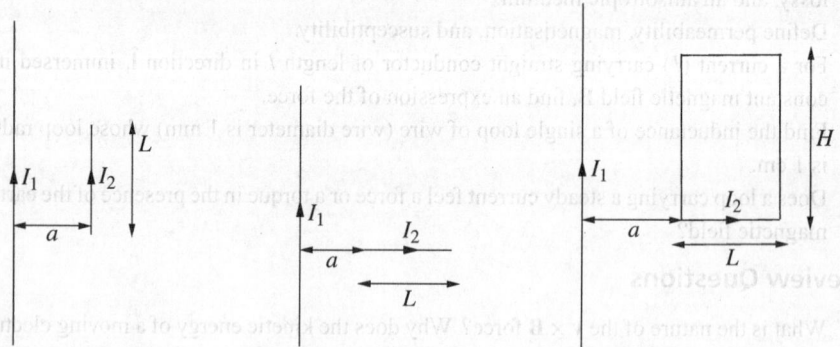

Fig. 9.20 A current filament of length L placed parallel to an infinite line

Fig. 9.21 A current filament of length L placed perpendicularly near an infinite line

Fig. 9.22 A rectangular loop of width L and height H placed near an infinite line

12. For the magnetic field $\mathbf{B} = xyz\mathbf{a}_x - zy^2\mathbf{a}_y$, find the torque on a square loop of side a placed parallel to the yz plane with its centre coinciding with the origin. The loop carries a current I.

13. Show that for a planar loop carrying a current I_1 immersed in a constant external magnetic field $\mathbf{B}_0$, the torque felt by it is such that it aligns the magnetic field produced by the current $\mathbf{B}_1$, with $\mathbf{B}_0$.

14. Explain qualitatively the force per meter felt by two infinitely long parallel wires separated by a distance a each carrying a current I.

15. Calculate the force per meter felt by an infinitely long wire carrying a current I_1 by another parallel wire at a distance a carrying a current I_2.

16. Find the approximate inductance of length 1 in. of cross-section 3 cm² and with 10 turns.

17. Calculate the approximate number of turns required to make an inductance of 10 μH on a circular bobbin of length 1 in. and cross-section 3 cm².

18. Using Wheeler's formula, calculate the number of turns required to make an inductance of 10 μH on a circular bobbin of length 1 in. and cross-section 3 cm².

19. Calculate the inductance of
    (a) A 10 m length of coaxial cable with a filling of $\varepsilon_r = 18$, $\sigma = 0$, and $\mu_r = 80$ having dimensions of $a = 1$ mm and $b = 4$ mm.
    (b) A toroid formed by surfaces $\rho = 3$ cm, $\rho = 5$ cm, $z = 0$, and $z = 1.5$ cm wrapped with 5000 turns of wire and consisting of magnetic material with $\mu_r = 6$.
    (c) A solenoid 8 cm in length, 2 cm in radius having $\mu_r = 100$ with 900 turns of wire.
    (DTU-NSIT, June 2010)

20. For the configuration of Fig. 9.14, if coil 1 has 10 turns and coil 2 has 100 turns, with $\mu_r = 5000$, find the self inductance and mutual inductances of each coil. Dimensions of the torus is $\rho_0 = 3$ cm and $a = 0.2$ cm.

21. For the configuration of Fig. 9.14, if coil 1 has 15 turns with a current of 1 A and coil 2 has 30 turns with a current of 2 A but whose flux opposes that of coil 1. If $\mu_r = 5000$, find the equivalent magnetic circuit consisting of the flux, reluctance and mmf. Dimensions of the torus is $\rho_0 = 3$ cm and $a = 0.2$ cm.

22. The region $y < 0$ (Region 1) is a air and $y > 0$ (Region 2) has $\mu_r = 10$. If there is a uniform magnetic field $\mathbf{H} = 5\mathbf{a}_x + 6\mathbf{a}_y + 7\mathbf{a}_z$ A/m in Region 1, find $\mathbf{B}$ and $\mathbf{H}$ in Region 2.

23. The region $z < 0$ (Region 1) is a air and $z > 0$ (Region 2) has $\mu_r = 10$. If the magnetic field $\mathbf{B} = 5\mathbf{a}_x + 6\mathbf{a}_y + 7\mathbf{a}_z$ T in Region 2, find $\mathbf{B}$ and $\mathbf{H}$ in Region 1.

24. For the configuration of Fig. 9.18 obtain the magnetic circuit. The number of turns is 100, the area of cross-section is 1 cm², the length of the core is 6 cm, the air gap is 0.25 cm, $\mu_r = 5000$, and the current $I = 1$ A. What are the values of the mmf, the flux in the core, reluctance of the core and reluctance of the air-gap?

Answers

Objective Type Questions

1. (a) and (c) 2. (c) and (d) 3. (d) 4. (b) 5. (c)
6. (c) 7. (b) 8. (b) and (d) 9. (a) and (b) 10. (b)

Short-Answer Questions

4. 36.2 nH

Numerical Problems

1. $x = 1$, $y = 0$, $z = t^2/2$

2. $x = 1 + 0.5t$, $y = 0$, $z = t^2/2$

3. $x = 1$, $y = 0$, $z = t^2/2 + 0.5t$

4. $z = t^2/2 + 0.5t$; $\rho(t) = 1$; $\phi(t) = \omega_0 t$; $\omega_0 = 1$

5. $F = (\mu_0 I_1 I_2 L)/(2\pi a)$, the force is attractive

6. $F = [(\mu_0 I_1 I_2 L)/(2\pi)] \ln[(a+L)/a]$. The force is in the direction of I_1

7. $F = (\mu_0 I_1 I_2 H) \{1/(2\pi a) - 1/[2\pi(a+L)]\}$ the force 3 is repulsive

8. $\mathbf{F} = 2 \times 10^{-5} \mathbf{a}_x \, \text{N/m}$

9. See Section 9.6.1

10. 0

11. $[(Ia^4)/12] \mathbf{a}_x$

12. 0

15. See Example 9.7

16. 1.47×10^{-6} H

17. 26 turns

18. 30 turns

19. (a) $L = 176.5 \mu_0$ (b) $L = 3.66 \mu_0$ (c) $L = 1.2723 \times 10^6 \mu_0$

20. Mutual inductance $= 333.3 \mu_0$; self inductance of coil $1 = 33.3 \mu_0$; self inductance of coil $2 = 3333.3 \mu_0$

21. The mmf source due to coil 1 is $\mathcal{F}_1 = 15$ A. The mmf source due to coil 2 is $\mathcal{F}_2 = 60$ A. These sources oppose each other. The reluctance $\mathcal{R} = 3/\mu_0$

22. $\mathbf{B}_2 = \mu_0 (50\mathbf{a}_x + 6\mathbf{a}_y + 70\mathbf{a}_z)$

23. $\mathbf{B}_1 = 0.5\mathbf{a}_x + 0.6\mathbf{a}_y + 7\mathbf{a}_z$

24. The mmf $= 600$ A, $\mathcal{R}_{\text{core}} = 0.12/\mu_0$, $\mathcal{R}_{\text{air gap}} = 25/\mu_0$, $\Psi_m = 23.88\mu_0$ Wb

PART IV

Time Varying Fields, Radiation, and Propagation

10 | Time-Dependant Fields

It is a thousand times better to have common sense without education than to have education without common sense.
— Robert G. Ingersoll

CHAPTER OBJECTIVES

To enable the students to understand the following:

- Faraday's law
- Maxwell's equation from Faraday's law
- Displacement current density
- Maxwell's equations in point and integral forms
- Time and frequency domain wave equations
- Use of phasors

10.1 | Faraday's Law

So far we have worked with static electromagnetic fields. We now turn our attention to time varying phenomena. In 1831, Michael Faraday (1791–1867) and Joseph Henry (1797–1878) independently came to the conclusion that when a changing magnetic field was linked by a coil of wire connected to a galvanometer, it was found that the needle moved, thus showing that an emf was generated in the coil. The conclusion was that a changing magnetic field linking a circular wire creates an emf.

> **Michael Faraday**
>
> Faraday, who was the son of a blacksmith, was educated only up to the school level. Even though he had this limited academic background, he was considered one of the greatest experimentalist of his time due to his path-breaking work in the fields of electromagnetism and electrochemistry. His contributions to electricity and magnetism were of such great importance that a unit of charge (faraday) and a unit of capacitance (farad) have been named after him.

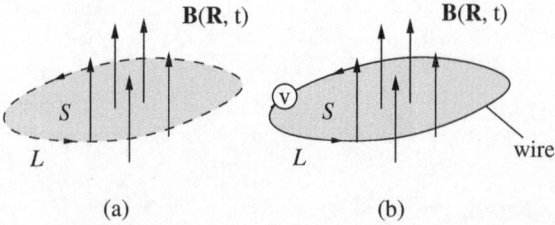

Fig. 10.1 Stokes's theorem applied to $\nabla \times \mathbf{E} = -\partial \mathbf{B}/\partial t$. (a) Applied to a closed curve, (b) Applied to a wire coinciding with the closed curve giving Faraday's laws

Faraday's law

The induced electromotive force, $\mathcal{E}$, in any closed circuit is equal to the time rate of change of the magnetic flux through the circuit.

Faraday's law is a general statement of the fact that an emf is induced in a closed circuit where a time varying magnetic field is present as shown in Fig. 10.1(a).

With this in mind we concentrate on the set-up shown in Fig. 10.1(b) which consists of a wire in the shape of the closed curve $\mathcal{L}$ and having a voltmeter in the circuit. The voltmeter shows a voltage which is given by *Faraday's* and *Lenz's* laws.

$$\text{emf generated} = \mathcal{E} = -\frac{\partial \Psi_m}{\partial t} \tag{10.1}$$

Lenz's law gives us the direction of the current in the closed loop and the sign of the emf produced.

Lenz's law

The induced current flows in a direction such that the current opposes the change that induced it.

The law of electromagnetic induction is used in the design of generators, where the motion of the conductors causes the flux enclosed by a closed loop to change and thereby cause an induced emf (Heinrich Lenz formulated the law in 1834).

Example 10.1 For the configuration shown in Fig. 10.2, the magnetic flux density is given by the equation

$$\mathbf{B} = \mathbf{a}_z B_0 f(t) = \begin{cases} \mathbf{a}_z B_0 t u(t), & \text{for } t = 0 \text{ to } 1 \text{ s} \\ \mathbf{a}_z B_0, & \text{for } t \geq 1 \text{ s} \end{cases}$$

Find the induced voltage and direction of current in the loop of wire. $u(t)$ is the step function.

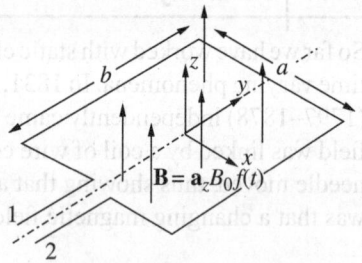

Fig. 10.2 Induced current in a time-varying magnetic field

Solution The flux in the loop is equal to

$$\Psi_m(t) = \underset{\text{Area of loop}}{\iint} [\mathbf{a}_z B_0 f(t)] \cdot \mathbf{a}_z dS = ab B_0 f(t)$$

where

$$f(t) = \begin{cases} tu(t), & \text{for } t = 0 \text{ to } 1 \text{ s} \\ 1, & \text{for } t \geq 1 \text{ s} \end{cases}$$

The induced emf $\mathcal{E}$ is given by

$$\mathcal{E} = -\frac{\partial \Psi_m}{\partial t} = \begin{cases} -abB_0, & \text{for } t = 0 \text{ to } 1 \text{ s} \\ 0, & \text{for } t \geq 1 \text{ s} \end{cases}$$

Example 10.2 For the configuration shown in Fig. 10.2, the magnetic flux density is given by the equation

$$\mathbf{B} = \mathbf{a}_z B_0 f(t) = \begin{cases} \mu_0 \sin(2\pi \times 10^7 t), & \text{for } 0 < t < 1 \text{ s} \\ 0, & \text{for } 1 < t \text{ s} \end{cases}$$

Find the induced voltage and direction of current in the loop of wire. $u(t)$ is the step function.

Solution The flux in the loop is equal to

$$\Psi_m(t) = \iint\limits_{\text{Area of loop}} \mathbf{B} \cdot \mathbf{a}_z dS = \mu_0 \, ab \sin(2\pi \times 10^7 t)$$

The induced emf $\mathcal{E}$ is given by

$$\mathcal{E} = -\frac{\partial \Psi_m}{\partial t} = -2\pi \times 10^7 \mu_0 \, ab \cos(2\pi \times 10^7 t)$$

Using similar arguments as earlier in the previous problem, the potential at terminal 1 is higher than the potential at terminal 2, and therefore the current flows from 1 to 2. Notice that the induced current produces a B field in the $-\mathbf{a}_z$ direction, which opposes the change as per Lenz's law.

It must be noted that when there is a coil of N turns instead of just a single turn then

$$\mathcal{E} = -N\frac{\partial \Psi_m}{\partial t} \tag{10.2}$$

Next we apply Faraday's law to the case where the conductor itself is moving.

 See `InducedEMFGen.m` in Chapter 10

Example 10.3 Find the emf generated for the case of Fig. 10.3 across the terminals marked '+' and '−'.

Solution Referring to Fig. 10.3, we find that we have a complicated arrangement where a single loop of wire is being turned in the anti-clockwise sense (as seen from the side ad) immersed in a uniform and *constant* magnetic field, $\mathbf{B} = -\mathbf{a}_z B_0$. The loop is being rotated at an angular speed $\omega = d\theta/dt$. To apply Faraday's law, we need to compute the vector normal to the plane $abcd$. Two unit vectors, which are perpendicular to each other, and which lie on the plane are

$$\hat{\mathbf{u}}_1 = \mathbf{a}_y \quad \text{and} \quad \hat{\mathbf{u}}_2 = \mathbf{a}_z \cos\theta - \mathbf{a}_x \sin\theta \tag{10.3}$$

So the vector normal to the plane is

$$\hat{\mathbf{n}} = \mathbf{a}_y \times \hat{\mathbf{u}}_2 = \mathbf{a}_x \cos\theta + \mathbf{a}_z \sin\theta \tag{10.4}$$

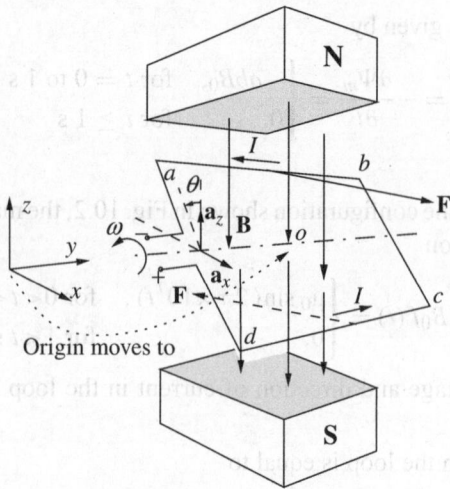

Fig. 10.3 Faraday's law applied to generators

The flux through this plane is

$$\Psi_m = \iint\limits_{abcd} \mathbf{B} \cdot d\mathbf{S} = \iint\limits_{abcd} [-\mathbf{a}_z B_0] \cdot (\mathbf{a}_x \cos\theta + \mathbf{a}_z \sin\theta) \, dS = -B_0 A \sin\theta \quad (10.5)$$

where A is the area of the loop (as a check: note that for $\theta = 0$ there is no flux linked). The induced emf is therefore

$$\mathcal{E} = -\frac{d}{dt}(-B_0 A \sin\theta) = B_0 A \cos\theta \, \frac{d\theta}{dt} = B_0 A \omega \cos\theta \ \text{(V)} \quad (10.6)$$

The induced emf drives a current in the loop which develops a torque (due to the two forces **F** on the wires ab and cd) to turn the loop in the opposite direction of the angular motion (Lenz's law).

In the case under discussion, the magnetic field is constant and the conductors are moving, causing the flux linked by the loop to change, which causes the emf to be induced.

Another way of looking at the same problem is to view it from the point of view of the Lorentz force. The conductors, which move in the constant magnetic field, contain charges which feel a force given by

$$\mathbf{F} = q\mathbf{v} \times \mathbf{B} = q\mathbf{E}_m \quad (10.7)$$

The vector $\mathbf{E}_m$ can be looked upon as a *motional* electric field. The charges feeling the force then set up currents.

 See `CurrCoilWire.m` in Chapter 10

Example 10.4 A circular loop of 10 turns of conducting wire of radius $r = 5$ cm and resistance $R = 10 \, \Omega$ is placed in a slowly varying uniform magnetic field. The magnetic field makes an angle of 45° with respect to the direction of the surface of the loop. The magnetic field magnitude is given by

$$B = \cos(2\pi t) \ \text{(T)}$$

Find the emf and the current generated in a wire.

Solution The emf generated is given by

$$\mathcal{E} = -N\frac{\partial \Psi_m}{\partial t}$$

where $N = 10$, and

$$\Psi_m = BA\cos 45° = \cos(2\pi t) \times \pi \times (.05)^2 \cos 45° = 5.9722 \times 10^{-3} \cos(2\pi t)$$

so the emf generated is

$$|\mathcal{E}| = 10 \times 2\pi \times 5.9722 \times 10^{-3} \cos(2\pi t) = 0.37524\cos(2\pi t) \text{ (V)}$$

and the current is $0.037524\cos(2\pi t)$ mA.

An Application of Faraday's Law

The ignition system of a car uses the concepts developed above to create a spark in the spark plug. Referring to Fig. 10.4, we see that when the contact **A** is opened, the magnetic field of the primary winding slumps to zero. Due to rate of change of flux, the secondary winding develops very high voltages (about 40,000 V) which in turn causes a spark in the spark plug.

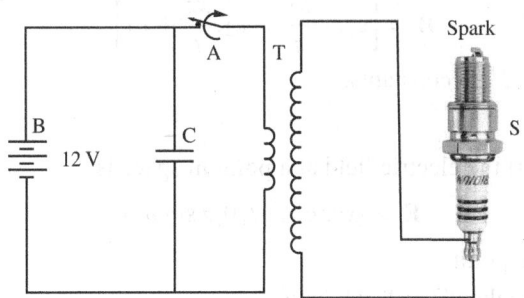

Fig. 10.4 Ignition system of a car

10.2 | Maxwell Equation from Faraday's Law

Equation (10.1) can be used to derive an equation in vector calculus. The emf generated in a closed loop is (see Fig. 10.1) given by

$$\text{emf generated} = \mathcal{E} = \oint_{\mathcal{L}} \mathbf{E} \cdot d\mathbf{l} \tag{10.8}$$

which by Stokes's theorem becomes

$$\mathcal{E} = \iint_S \nabla \times \mathbf{E} \cdot d\mathbf{S} \tag{10.9}$$

The flux Ψ_m through that loop is

$$\Psi_m = \iint_S \mathbf{B} \cdot d\mathbf{S} \tag{10.10}$$

while the rate of change of flux is

$$\frac{\partial \Psi_m}{\partial t} = \frac{\partial}{\partial t}\iint \mathbf{B} \cdot d\mathbf{S} = \iint_S \frac{\partial \mathbf{B}}{\partial t} \cdot d\mathbf{S} \tag{10.11}$$

using the above equation and Eqn (10.9), we get

$$\iint\limits_S \nabla \times \mathbf{E} \cdot d\mathbf{S} = -\iint\limits_S \frac{\partial \mathbf{B}}{\partial t} \cdot d\mathbf{S}$$

or

$$\nabla \times \mathbf{E} = -\frac{\partial \mathbf{B}}{\partial t}$$

which is one of Maxwell's equations.

Example 10.5 If the electric field at a point in space is

$$\mathbf{E} = \left[\frac{xyz}{t^2}, 0, \frac{z}{e^t} \right], \quad t > 1$$

Calculate **B** at that point.

Solution

$$\partial\mathbf{B}/\partial t = -\nabla \times \mathbf{E}$$

$$= -\left[0, -\frac{xy}{t^2}, \frac{xz}{t^2} \right]$$

By integrating, we get

$$\mathbf{B} = \left[c_1, -\frac{xy}{t} + c_2, \frac{xz}{t} + c_3 \right]$$

where c_i, $i = 1, 2, 3$ are constants.

Example 10.6 If the electric field at a point in space is

$$\mathbf{E} = [xyz \cos\omega t, 0, z \sin\omega t]$$

Calculate **B** at that point.

Solution We calculate $\nabla \times \mathbf{E}$ which is

$$\nabla \times \mathbf{E} = [0, -xy \cos(\omega t), xz \cos(\omega t)]$$

So

$$\frac{\partial \mathbf{B}}{\partial t} = -\nabla \times \mathbf{E}$$

$$= [0, xy \cos(\omega t), -xz \cos(\omega t)]$$

By integrating, we get

$$\mathbf{B} = [0, (xy/\omega) \sin(\omega t), -(xz/\omega) \sin(\omega t)]$$

10.3 | Displacement Current Density

We have studied Ampere's law as applied to *steady* magnetic fields produced by steady currents. What about cases where the current is time varying? Take the case of a resistor–capacitor series combination connected to a battery as shown in Fig. 10.5. In part (a) of the figure, we apply Ampere's law:

$$\oint_{\mathcal{L}} \mathbf{H} \cdot d\mathbf{l} = I$$

or

$$\nabla \times \mathbf{H} = \mathbf{J} \tag{10.12}$$

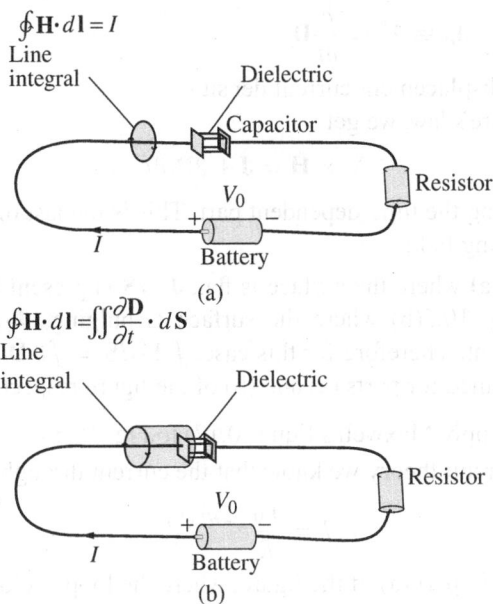

$$\oint \mathbf{H} \cdot d\mathbf{l} = I$$

Line integral

Dielectric

Capacitor

V_0

Resistor

Battery

(a)

$$\oint \mathbf{H} \cdot d\mathbf{l} = \iint \frac{\partial \mathbf{D}}{\partial t} \cdot d\mathbf{S}$$

Line integral

Dielectric

Resistor

V_0

Battery

(b)

Fig. 10.5 Setup to show how the displacement current comes into play

and we find that the line integral shown indeed gives us the enclosed current, but when we apply Ampere's law to part (b) of the figure (the surface being inside the dielectric), we find that the line integral gives us a zero result (there being no current between the plates of the capacitor). Though the line integral is the same as earlier, the two results are different. So obviously something is missing.

Let us now approach Ampere's law from a different manner: taking the divergence of Eqn (10.12),

$$\nabla \cdot (\nabla \times \mathbf{H}) = \nabla \cdot \mathbf{J} \tag{10.13}$$

but div(curl($\mathbf{H}$)) = 0 for any vector and $\nabla \cdot \mathbf{J} = -\partial \rho_v / \partial t$, the continuity equation, giving

$$-\frac{\partial \rho_v}{\partial t} = 0!$$

which is an absurd result!

Let us therefore modify Eqn (10.12) to include another current density, $\mathbf{J}^*$.

$$\nabla \times \mathbf{H} = \mathbf{J} + \mathbf{J}^* \tag{10.14}$$

taking the divergence of this equation, we get

$$\nabla \cdot \nabla \times \mathbf{H} = \nabla \cdot (\mathbf{J} + \mathbf{J}^*) = 0 \tag{10.15}$$

or

$$\nabla \cdot \mathbf{J} = -\nabla \cdot \mathbf{J}^*$$

$$-\frac{\partial \rho_v}{\partial t} = -\nabla \cdot \mathbf{J}^*$$

$$\frac{\partial}{\partial t}(\nabla \cdot \mathbf{D}) = \nabla \cdot \mathbf{J}^* \quad \text{(using Gauss's law)}$$

$$\nabla \cdot \left(\frac{\partial}{\partial t}\mathbf{D}\right) = \nabla \cdot \mathbf{J}^* \quad \text{(interchanging differentiations)}$$

which gives
$$\mathbf{J}_d \equiv \mathbf{J}^* = \frac{\partial}{\partial t}\mathbf{D} \qquad (10.16)$$

which we call the displacement current density.

Rewriting Ampere's law, we get
$$\nabla \times \mathbf{H} = \mathbf{J} + \partial \mathbf{D}/\partial t \qquad (10.17)$$

the second part being the time-dependent part. This is the last of Maxwell's equations for time-varying field.

Note: In Fig. 10.5(a) where the surface is flat, $\mathbf{J} \cdot d\mathbf{S}$ is present but $\partial \mathbf{D}/\partial t \cdot d\mathbf{S}$ is absent, and in Fig. 10.5(b) where the surface is bulging, $\mathbf{J} \cdot d\mathbf{S}$ is absent but $\partial \mathbf{D}/\partial t \cdot d\mathbf{S}$ is present. Therefore for this case, $\oint \mathbf{J} \cdot d\mathbf{S} = \oint \partial \mathbf{D}/\partial t \cdot d\mathbf{S}$ where the integrals are performed for parts (a) and (b) of the figure respectively.

Example 10.7 Apply Maxwell's Eqn (10.17) to Fig. 10.5.

Solution From circuit theory we know that the current through the circuit is
$$I = \frac{V_0}{R}e^{-t/RC} \qquad (10.18)$$

So the line integral in part (a) of the figure, where the loop encloses current I is
$$\oint \mathbf{H} \cdot d\mathbf{l} = I \qquad (10.19)$$

but in part (b) of the figure, the loop apparently does not enclose current but the line integral is the same as before. That is
$$\oint \mathbf{H} \cdot d\mathbf{l}$$

is the same in both figures. Let us pause a moment and calculate the charge accumulated on the capacitor plates
$$Q(t) = \int_{t=0}^{t} \frac{V_0}{R}e^{-t/RC}\,dt = -CV_0 e^{-t/RC}\Big|_0^t = CV_0(1 - e^{-t/RC}) \qquad (10.20)$$

The charge produces a voltage across the plates
$$V(t) = Q(t)/C = V_0(1 - e^{-t/RC}) \qquad (10.21)$$

which produces an electric field
$$E = V/d = \frac{V_0}{d}(1 - e^{-t/RC}) \qquad (10.22)$$

where d is the distance between the plates. The D field is therefore
$$D = \varepsilon E = \frac{\varepsilon V_0}{d}\left(1 - e^{-t/RC}\right) \qquad (10.23)$$

and
$$\iint (\partial \mathbf{D}/\partial t) \cdot d\mathbf{S} = \frac{\varepsilon V_0}{d} \times \frac{1}{RC}e^{-t/RC} \times A \qquad (10.24)$$

where A is the area of the capacitor plates. Recalling that $C = \varepsilon A/d$ the expression given above becomes
$$\iint (\partial \mathbf{D}/\partial t) \cdot d\mathbf{S} = \frac{\varepsilon V_0 A}{Rd(\varepsilon A/d)}e^{-t/RC} = \frac{V_0}{R}e^{-t/RC} = I(t)! \qquad (10.25)$$

The current is the same as given in Eqn (10.18). The analysis given above is rather crude, but the idea is correct.

So in summary, in part (a) of the figure

$$\oint \mathbf{H} \cdot d\mathbf{l} = I \qquad (10.26)$$

and in part (b)

$$\oint \mathbf{H} \cdot d\mathbf{l} = \iint (\partial \mathbf{D}/\partial t) \cdot d\mathbf{S} = I \qquad (10.27)$$

the term $(\partial \mathbf{D}/\partial t)$ is the displacement current density, $\mathbf{J}_d$.

Example 10.8 The electric field in a dielectric with $\varepsilon_r = 15$ is given by $\mathbf{E} = [10\mathbf{a}_x, 20\mathbf{a}_y + 30\mathbf{a}_z] \cos(2\pi \times 10^6 t + \pi/6)$. Find the displacement current density.

Solution The displacement current density is given by $\partial \mathbf{D}/\partial t$.

$$\mathbf{D} = \varepsilon\, \mathbf{E} = 15\varepsilon_0 \,[10\mathbf{a}_x, 20\mathbf{a}_y + 30\mathbf{a}_z] \cos(2\pi \times 10^6 t + \pi/6)$$

$$\frac{\partial \mathbf{D}}{\partial t} = -30\pi \times 10^6 \times \varepsilon_0 \,[10\mathbf{a}_x, 20\mathbf{a}_y + 30\mathbf{a}_z] \sin(2\pi \times 10^6 t + \pi/6)$$

 See `CurrentRCCkt.m` in Chapter 10

10.4 | Time-dependent Maxwell's Equations

10.4.1 Point Form of the Equations

So far we have studied the time-independent electromagnetic fields. The relevant equations (i.e., Maxwell's equations) including the time term are given below

$$\nabla \cdot \mathbf{D} = \rho_v \qquad (10.28)$$

$$\nabla \times \mathbf{E} = \boxed{-\partial \mathbf{B}/\partial t} \qquad (10.29)$$

$$\nabla \cdot \mathbf{B} = 0 \qquad (10.30)$$

$$\nabla \times \mathbf{H} = \mathbf{J} + \boxed{\partial \mathbf{D}/\partial t} \qquad (10.31)$$

In the second and fourth equations given above, there are time varying terms which are shown boxed. The first equation is Gauss's law, the last is Ampere's law, the second equation is Faraday's law and the third equation generated the vector potential and tells us that there are no magnetic charges. The two boxed terms are the time dependent terms which add to the static equations and complete all of Maxwell's equations.

Example 10.9 If $\mathbf{A}$ is the vector potential and V is the scalar potential then another form of Faraday's law is

$$\mathbf{E} = -\frac{\partial \mathbf{A}}{\partial t} - \nabla V$$

Since

$$\nabla \times \mathbf{E} = -\frac{\partial \mathbf{B}}{\partial t} = -\frac{\partial \nabla \times \mathbf{A}}{\partial t}$$

or

$$\nabla \times \left(\mathbf{E} + \frac{\partial \mathbf{A}}{\partial t} \right) = 0$$

hence

$$\mathbf{E} = -\frac{\partial \mathbf{A}}{\partial t} - \nabla V \quad (\text{since curl(grad}(\dots)) = 0$$

10.5 | Integral Form of Maxwell's Equations

We can write Maxwell's equations in integral form from the point form of the equations.

1. Integrating Eqn (10.28) over a volume $\mathcal{V}$ and using the divergence theorem, we get

$$\iiint_{\mathcal{V}} \nabla \cdot \mathbf{D} dV = \oiint_{S} \mathbf{D} \cdot d\mathbf{S} = \iiint_{\mathcal{V}} \rho_v \, dV \qquad (10.32)$$

where S is the surface enclosing the volume $\mathcal{V}$.

2. Next we integrate Eqn (10.29) over a surface S enclosed by the closed curve $\mathcal{L}$ and using Stokes's theorem, we get

$$\iint_{S} \nabla \times \mathbf{E} \cdot d\mathbf{S} = \oint_{\mathcal{L}} \mathbf{E} \cdot d\mathbf{l} = -\iint_{S} \frac{\partial \mathbf{B}}{\partial t} \cdot d\mathbf{S} \qquad (10.33)$$

the counter-clockwise direction on the line integral shows us the direction of the line integral.

3. Integrating the third equation:

$$\iiint_{\mathcal{V}} \nabla \cdot \mathbf{B} = \oiint \mathbf{B} \cdot d\mathbf{S} = 0 \qquad (10.34)$$

4. And the fourth

$$\iint_{S} \nabla \times \mathbf{H} \cdot d\mathbf{S} = \oint_{\mathcal{L}} \mathbf{H} \cdot d\mathbf{l} = \iint_{S} \mathbf{J} \cdot d\mathbf{S} + \iint_{S} \frac{\partial \mathbf{D}}{\partial t} \cdot d\mathbf{S} \qquad (10.35)$$

10.6 | Fundamental Equations of Radiation and Propagation

In any situation where we have to communicate between a source and a receiver we usually use high frequency signals and either have wires or cables connecting the two, or we radiate energy from the one to the other. Figure 10.6 illustrates such a communication setup where antennas are used to transfer the signal.

The figure displays a transmitter and a receiver. The transmitter is connected to a transmitting antenna through a transmission line, which sends information to a distant receiver in the form of radiated signals. A typical communication link consists of an information source, consisting of electrical signals, suitably modulated—for example using AM or FM—sent via the antenna. The receiver has the same configuration as that of the transmitter.

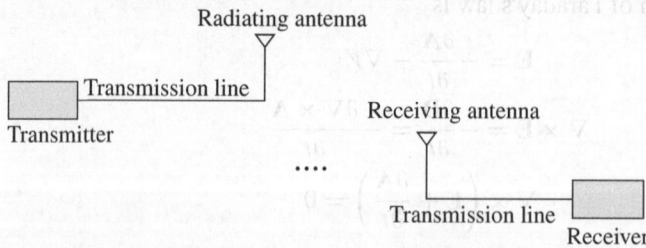

Fig. 10.6 Typical communication setup with transmitting and receiving antennas

To properly design the transmission lines or the antennas, which are the parts of the equipment concerning the electromagnetic part, we need to understand the concepts of electromagnetic waves, whether these travel through transmission lines or through space itself. Surprisingly, both these phenomena are governed by the same kind of equations. In this chapter, we will examine these equations in greater detail.

10.7 | Time Domain Wave Equation

To begin with, since we are dealing with electromagnetic phenomena, we must take Maxwell's equations into account. Remember that *all* electromagnetic phenomena are governed by Maxwell's equations whether these phenomena be time-independent or time-dependent. Keeping this in mind, let us then begin with the basic electromagnetic equations of Maxwell introduced and used frequently and work with them to get wave equations. The equations, this time, are the complete equations which include the time-dependent part. Maxwell's equations are reproduced here for the convenience of the reader.

$$\nabla \cdot \mathbf{D} = \rho_v \tag{10.36}$$
$$\nabla \times \mathbf{E} = -\partial \mathbf{B}/\partial t \tag{10.37}$$
$$\nabla \cdot \mathbf{B} = 0 \tag{10.38}$$
$$\nabla \times \mathbf{H} = \partial \mathbf{D}/\partial t + \mathbf{J} \tag{10.39}$$

To recapitulate: $\mathbf{D}$ is the electric flux density, ρ_v is the volume charge density, and these two are connected by the first equation, which formulates that the divergence of $\mathbf{D}$ is equal to the volume charge density. $\mathbf{E}$ is the electric field and $\mathbf{B}$ is the magnetic flux density. The curl of $\mathbf{E}$ is the negative of the rate of change of $\mathbf{B}$. Next we have the equation, the divergence of $\mathbf{B}$ is always zero, and finally the curl of $\mathbf{H}$, the magnetic field, is equal to the displacement current density $\partial \mathbf{D}/\partial t$ plus $\mathbf{J}$, the current density.

The relations between $\mathbf{D}$, $\mathbf{E}$ on one hand and $\mathbf{B}$, $\mathbf{H}$ on the other are

$$\mathbf{D} = \varepsilon \, \mathbf{E} \tag{10.40}$$
$$\mathbf{B} = \mu \, \mathbf{H} \tag{10.41}$$

where the constants of proportionality ε and μ are the permittivity and permeability respectively, of the medium in which the waves travel. Some of these vector fields may or may not be needed depending on the physical situation.

To obtain the wave equation of the $\mathbf{E}$ field, we take the curl of Eqn (10.37):

$$\nabla \times (\nabla \times \mathbf{E}) = -\nabla \times (\partial \mathbf{B}/\partial t) = -\partial (\nabla \times \mathbf{B})/\partial t \tag{10.42}$$

In the second equality, $\partial/\partial t$ and $\nabla \times$ were transposed since the two operators are independent of each other and their transposition gives the same result. Using the vector identity

$$\nabla \times (\nabla \times \mathbf{A}) = \nabla(\nabla \cdot \mathbf{A}) - (\nabla \cdot \nabla) \mathbf{A}$$

and applying it to the left-hand side of the previous equation, we get

$$\nabla (\nabla \cdot \mathbf{E}) - (\nabla \cdot \nabla) \mathbf{E} = -\frac{\partial}{\partial t} (\nabla \times \mathbf{B})$$

$$= -\frac{\partial}{\partial t} [\nabla \times (\mu \, \mathbf{H})]$$

$$= -\mu \frac{\partial}{\partial t} (\nabla \times \mathbf{H}) \quad (\mu \text{ is a constant})$$

$$= -\mu \frac{\partial}{\partial t} (\partial \mathbf{D}/\partial t + \mathbf{J}) \tag{10.43}$$

In the second equation, of the above equations, we have used Eqn (10.41). In the last equation, we have substituted Eqn (10.39) in the right-side of the equation. In a region where the wave is propagating, we assume that there are no charges or currents, $\rho_v = 0$ and $\mathbf{J} = \mathbf{0}$. So that $\nabla \cdot \mathbf{D} = \rho_v = 0$; which in turn implies that $\nabla \cdot \mathbf{E} = 0$ (remember that $\mathbf{D} = \varepsilon \mathbf{E}$). This simplifies the time domain wave equation

$$\nabla (\nabla \cdot \mathbf{E}) - (\nabla \cdot \nabla) \mathbf{E} = -\mu \frac{\partial}{\partial t} (\partial \mathbf{D}/\partial t + \mathbf{J})$$

$$-(\nabla \cdot \nabla) \mathbf{E} = -\mu \frac{\partial}{\partial t} (\partial \mathbf{D}/\partial t + \mathbf{J}) \quad (\nabla \cdot \mathbf{E} = 0 \text{ as no free charges})$$

$$= -\mu \frac{\partial}{\partial t} (\partial \mathbf{D}/\partial t) \quad (\mathbf{J} = 0 \text{ as no currents present})$$

$$= -\mu \frac{\partial^2}{\partial t^2} \mathbf{D}$$

$$= -\mu \frac{\partial^2}{\partial t^2} (\varepsilon \mathbf{E}) \quad (\mathbf{D} = \varepsilon \mathbf{E})$$

$$= -\mu \varepsilon \frac{\partial^2 \mathbf{E}}{\partial t^2} \tag{10.44}$$

Let us find the units of $1/\sqrt{\mu \varepsilon}$

$$\text{Units of } \frac{1}{\sqrt{\mu \varepsilon}} = \frac{1}{\sqrt{(\text{H/m}) \cdot (\text{F/m})}}$$

$$= \frac{m}{\sqrt{\text{H} \cdot \text{F}}}$$

$$= \frac{m}{\sqrt{(\Omega \cdot s) \, (\mho \cdot s)}}$$

$$= \frac{m}{s}$$

which we denote by velocity v. A simple calculation of $1/\sqrt{\mu_0 \, \varepsilon_0}$ gives 2.998×10^8 m/s, which we recognise to be c, the velocity of light in free space. Remember that μ_0 and ε_0 are the permeability and the permittivity of free space. Using this notation, the wave equation just derived becomes

$$\nabla^2 \mathbf{E} - [1/(v^2)] \partial^2 \mathbf{E}/\partial t^2 = 0 \tag{10.45}$$

In Cartesian coordinates, the operator

$$\nabla \cdot \nabla \equiv \nabla^2$$

$$= \left(\frac{\partial}{\partial x} \mathbf{a}_x + \frac{\partial}{\partial y} \mathbf{a}_y + \frac{\partial}{\partial z} \mathbf{a}_z \right) \cdot \left(\frac{\partial}{\partial x} \mathbf{a}_x + \frac{\partial}{\partial y} \mathbf{a}_y + \frac{\partial}{\partial z} \mathbf{a}_z \right)$$

$$= \frac{\partial^2}{\partial x^2} + \frac{\partial^2}{\partial y^2} + \frac{\partial^2}{\partial z^2} \tag{10.46}$$

is the Laplacian. Applied to the electric field **E**

$$\nabla^2 \mathbf{E} = \frac{\partial^2 \mathbf{E}}{\partial x^2} + \frac{\partial^2 \mathbf{E}}{\partial y^2} + \frac{\partial^2 \mathbf{E}}{\partial z^2}$$

$$= [1/(v^2)]\partial^2 \mathbf{E}/\partial t^2 \tag{10.47}$$

This equation is a vector equation. It is actually three equations in E_x, E_y, and E_z. In Cartesian coordinates, these equations are

$$\frac{\partial^2 E_x}{\partial x^2} + \frac{\partial^2 E_x}{\partial y^2} + \frac{\partial^2 E_x}{\partial z^2} - [1/(v^2)]\frac{\partial^2 E_x}{\partial t^2} = 0 \tag{10.48}$$

$$\frac{\partial^2 E_y}{\partial x^2} + \frac{\partial^2 E_y}{\partial y^2} + \frac{\partial^2 E_y}{\partial z^2} - [1/(v^2)]\frac{\partial^2 E_y}{\partial t^2} = 0 \tag{10.49}$$

and

$$\frac{\partial^2 E_z}{\partial x^2} + \frac{\partial^2 E_z}{\partial y^2} + \frac{\partial^2 E_z}{\partial z^2} - [1/(v^2)]\frac{\partial^2 E_z}{\partial t^2} = 0 \tag{10.50}$$

Example 10.10 Show that

$$E_z(x, t) = E_{x0}\cos(\omega t - \beta x) \tag{10.51}$$

with $v = \omega/\beta$ satisfies the wave equation, just derived.

Solution

$$\nabla^2 E_z = -E_{x0}\beta^2 \cos(\omega t - \beta x)$$

and

$$[1/(v^2)]\frac{\partial^2 E_z}{\partial t^2} = -(\omega/v)^2 E_{x0}\cos(\omega t - \beta x)$$

$$= -\beta^2 E_{x0}\cos(\omega t - \beta x)$$

substituting both these terms into the wave equation, we find that it is satisfied.

These partial differential equations look quite complex. In order to gain a picture of the solution to these equations, we do two things. (a) We look at an analogous scalar equation, and (b) we reduce the number of independent variables. To get a scalar equation, we first concentrate on just one of these three equations: the very first one in E_x and replace E_x by ϕ. To reduce the number of independent variables we set $\partial^2/\partial y^2$ and $\partial^2/\partial z^2$, equal to zero (which is the equivalent of saying that ϕ is not a function of either y or z). Then we get

$$\frac{\partial^2 \phi}{\partial x^2} - [1/(v^2)]\frac{\partial^2 \phi}{\partial t^2} = 0 \tag{10.52}$$

This is the scalar wave equation in one coordinate and time. We now go over to the work done in the past to find the solution of such an equation. Mathematicians had studied this equation and found that *any* function ϕ whose argument is of the type $x - vt$, i.e., $\phi \equiv \phi(x - vt)$, satisfies the equation. Let us verify it. Let the variable $a \equiv a(x, t) = x - vt$. Then

$$\partial a/\partial t = -v \quad \text{and} \quad \partial a/\partial x = 1$$

We know that $\qquad \phi(x - vt) = \phi(a)$

Using the notation that $\phi' = d\phi/da$,

$$\frac{\partial}{\partial x}\phi(a) = \frac{d\phi}{da}\frac{\partial a}{\partial x} = \phi' \times 1 = \phi'$$

$$\frac{\partial^2}{\partial x^2}\phi = \frac{\partial}{\partial x}\frac{\partial}{\partial x}\phi$$

$$= \frac{\partial}{\partial x}\phi' \quad \text{(from the previous equation)}$$

$$= \frac{d\phi'}{da}\frac{\partial a}{\partial x}$$

$$= \phi'' \times 1$$

$$= \phi''$$

Similarly, $$\frac{\partial}{\partial t}\phi\,(a) = \frac{d\phi}{da}\frac{\partial a}{\partial t} = \phi' \cdot (-v) = -v\phi'$$

$$\frac{\partial^2}{\partial t^2}\phi = \frac{\partial}{\partial t}\frac{\partial}{\partial t}\phi$$

$$= \frac{\partial}{\partial t}(-v\phi') \quad \text{(from the previous equation)}$$

$$= -v\frac{d\phi'}{da}\frac{\partial a}{\partial t}$$

$$= (-v)\ \phi'' \times (-v)$$

$$= v^2\phi''$$

Substituting these relations into the scalar wave equation,

$$\frac{\partial^2 \phi}{\partial x^2} - [1/v^2]\frac{\partial^2 \phi}{\partial t^2} = \phi'' - [1/v^2][v^2]\phi'' = 0$$

we find that the scalar wave equation is satisfied.

 See `TravellingWave.m` in Chapter 10

Example 10.11 Let $\phi\,(x - vt) = \sin[(2\pi/\lambda)\,(x - vt)\,]$, where λ is the wavelength and v the velocity. Show that this function satisfies the wave equation.
Solution Taking the first derivative

$$\phi' = \frac{2\pi}{\lambda}\cos\left[\frac{2\pi}{\lambda}(x - vt)\right] \tag{10.53}$$

and taking the second derivative

$$\phi'' = -\left(\frac{2\pi}{\lambda}\right)\sin\left[\frac{2\pi}{\lambda}(x - vt)\right] \tag{10.54}$$

Now, $\partial^2 \phi/\partial x^2$ and $\partial^2 \phi/\partial t^2$ are obtained in a similar manner

$$\frac{\partial}{\partial x}\phi = \frac{2\pi}{\lambda}\cos\left[\frac{2\pi}{\lambda}(x - vt)\right] = \phi'$$

$$\frac{\partial^2}{\partial x^2}\phi = -\left(\frac{2\pi}{\lambda}\right)^2\sin\left[\frac{2\pi}{\lambda}(x - vt)\right] = \phi''$$

$$\frac{\partial}{\partial t}\phi = -\frac{2\pi}{\lambda}v\cos\left[\frac{2\pi}{\lambda}(x - vt)\right] = -v\phi'$$

$$\frac{\partial^2}{\partial t^2}\phi = -v^2\left(\frac{2\pi}{\lambda}\right)^2\sin\left[\frac{2\pi}{\lambda}(x - vt)\right] = v^2\phi''$$

Substituting these terms into the scalar wave equation, we find that it is identically satisfied.

The next question that we ask is that why is this a wave equation? The solution to the wave equation should be a wave. And what is a wave? A wave is a signal carrying energy which *travels*. Observe the signal $\phi(x - vt)$, which is a solution of the wave equation. At $t = 0$, the signal has a spatial profile given by $\phi(x)$. Now observe the signal at time t. The spatial profile of the signal is now $\phi(x - vt)$. Recall that if $f(x)$ is compared to $f(x - x_0)$, then the second function is shifted to the right

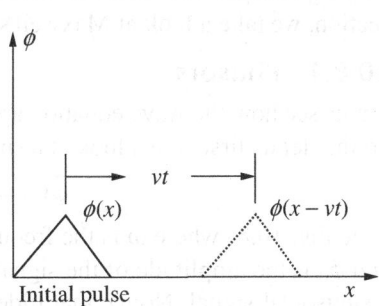

Fig. 10.7 A travelling wave

by x_0. Applying this argument, $\phi(x - vt)$ is $\phi(x)$ shifted to the right by vt. But t is increasing all the time, so $\phi(x)$ is *continuously shifting to the right*. In effect $\phi(x)$ is *travelling* to the right, a distance of vt in time t. And the velocity with which the wave travels is v. This situation is pictured in Fig. 10.7, where a triangular pulse is shown travelling to the right with a velocity v.

A wave travelling towards the right (the positive x direction) is a function of the type $\phi(x - vt)$, as discussed, while a wave travelling in the $-x$ direction is any function of the type $\phi \equiv \phi(x + vt)$. This can be logically understood by applying the reasoning given earlier. The reader can substitute this function into the wave equation and verify that this function is also a solution to the wave equation.

Example 10.12 Verify that the function $\phi(x + vt)$ is indeed a solution of the wave equation. Also verify that this is a wave travelling to the *left*.

Solution Substituting the function in

$$\frac{\partial^2}{\partial t^2}\phi = v^2\phi''$$

Similarly

$$\frac{\partial^2}{\partial x^2}\phi = \phi''$$

so the wave equation

$$\frac{\partial^2 \phi}{\partial x^2} - [1/v^2]\frac{\partial^2 \phi}{\partial t^2} = 0$$

Since the argument of ϕ is $x + vt$ where there is a '+' sign, the wave has to be travelling in the $-x$ direction.

What about the wave equation for the magnetic field? We find that we can obtain it in a similar fashion to give

$$\nabla^2 \mathbf{H} - [1/(v^2)]\partial^2 \mathbf{H}/\partial t^2 = 0 \tag{10.55}$$

Practice Problem 10.1
Obtain the wave equation for the magnetic field from first principles.

10.8 | Frequency Domain Wave Equation

Generally, engineering applications require sources which produce sinusoidally varying voltages and currents, which in turn produce electromagnetic fields. In this section, we take a look at Maxwell's equations when the fields vary sinusoidally.

10.8.1 Phasors

Let us see how the wave equation appears when using sinusoidal quantities, and to do that let us first take a look at a sinusoidally varying quantity such as

$$\tilde{\phi}(t) = \phi_0 \cos(\omega t + \theta) \tag{10.56}$$

(see Fig. 10.8) where ω is the frequency in rad/s, θ is a constant phase in radians and ϕ_0 is the amplitude of the signal. The tilde sign ($\sim$) signifies that the entity is a sinusoidal signal. Notice that tilde looks like a sine wave. This equation can be written as

$$\tilde{\phi}(t) = \Re\,[\phi_0\, e^{j(\omega t + \theta)}]$$
$$= \Re\,[\phi_0\, e^{j\theta}\, e^{j\omega t}]$$
$$= \Re\,[\Phi\, e^{j\omega t}] \tag{10.57}$$

Here the symbol $\Re$ means 'the real part of' and Φ is a complex number given a special name—phasor—corresponding to $\phi(t)$

$$\Phi = \phi_0\, e^{j\theta} \tag{10.58}$$

The relation between the sinusoidally varying quantity, in time domain, and the phasor Φ is Eqn (10.57). The two can be symbolically shown connected as

$$\tilde{\phi}(t) \xrightarrow{\text{Ph}} \Phi \tag{10.59}$$

Differentiating Eqn (10.56) with respect to time t

$$\tilde{\phi}'(t) = -\phi_0\, \omega \sin(\omega t + \theta)$$

Is this result the same as differentiating Eqn (10.57)? Let us differentiate this equation

$$\frac{d}{dt}\tilde{\phi}(t) = \frac{d}{dt}\Re[\phi_0\, e^{j(\omega t + \theta)}]$$
$$\overset{?}{=} \Re\left\{\frac{d}{dt}[\phi_0\, e^{j\,(\omega t + \theta)}]\right\}$$
$$= \Re\{j\omega\, \phi_0\, e^{j\,(\omega t + \theta)}\}$$

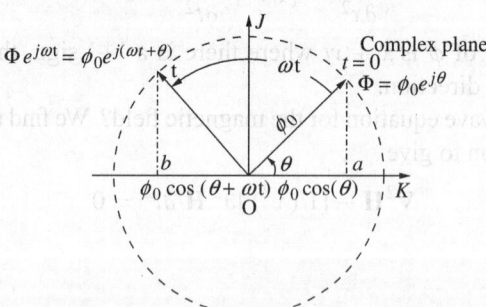

Fig. 10.8 Showing the relation between time-domain and frequency-domain entities

$$= \Re \{j\omega \phi_0 [\cos(\omega t + \theta) + j \sin(\omega t + \theta)]\}$$
$$= \Re \{j\omega \phi_0 \cos(\omega t + \theta) - \omega \phi_0 \sin(\omega t + \theta)\}$$
$$= -\omega \phi_0 \sin(\omega t + \theta) \qquad (10.60)$$

There is a question mark against the second equality. Can we interchange $\Re$ and d/dt? The answer is yes, because $d\phi/dt$ as calculated above and that calculated by the previous equation are indeed equal. To get it more clearly, when $\phi(t)$ is a sinusoidal function, and Φ is the corresponding phasor,

$$\frac{d}{dt}\tilde{\phi} = \frac{d}{dt}\Re(\Phi e^{j\omega t})$$

$$= \Re\left[\frac{d}{dt}(\Phi e^{j\omega t})\right]$$

$$= \Re[j\omega \Phi e^{j\omega t}] \qquad (10.61)$$

In the same manner, we can show that

$$\int \tilde{\phi}(t)\ dt = \int \phi_0 \cos(\omega t + \theta)\ dt$$

$$= \int \Re\left[\phi_0 e^{j(\omega t + \theta)}\right]\ dt$$

$$= \Re\left[\int \phi_0 e^{j(\omega t + \theta)}\ dt\right]$$

$$= \Re\left[\frac{\Phi e^{j\omega t}}{j\omega}\right]$$

Example 10.13 Find the phasors for (a) $10\cos(200t)$, (b) $20\sin(2\pi \times 10^6 t)$, (c) $13\cos(10^5 t + 1.5)$, (d) $25\sin(100t + 45)$.
Solution
(a) The term $10\cos(200t) = \Re[10e^{(200t+0)}]$

that is $\theta = 0$. So the phasor is $10e^{j0} = 10\angle 0$
(b) The term

$$20\sin(2\pi \times 10^6 t) = 20\cos(\pi/2 - 2\pi \times 10^6 t)$$
$$= 20\cos(2\pi \times 10^6 t - \pi/2)$$
$$= \Re[20e^{j\ (2\pi \times 10^6 t - \pi/2)}]$$

so the phasor is $20e^{j(-\pi/2)} = 20\angle(-\pi/2)$.
(c) $13\cos(10^5 t + 1.5)$ is by inspection $= \Re[13e^{j(10^5 t + 1.5)}]$, which gives the phasor $13e^{j1.5}$.
(d) The phasor is by inspection $25e^{j\ (45 - \pi/2)}$.

Example 10.14 Convert the phasors (a) $10\angle 10$, (b) $1e^{j15}$ to their real-time counterparts.
Solution
(a) $10\angle 10 = 10e^{j10}$. The conversion to time domain is $10e^{j(10+\omega t)}$. The real-time function is therefore

$$\Re[10e^{j(10+\omega t)}] = 10\cos(10+\omega t)$$

(b) $1e^{j15}$ converts to $1\cos(15+\omega t)$.

10.9 | Wave Equation

Though we have considered the real part of the complex phasor, we could have worked with the imaginary part ($\Im(\ldots)$) as well. Applying this knowledge to sinusoidally varying fields, let $\tilde{\mathbf{E}}(\mathbf{R}, t)$ be the fields in time but sinusoidally varying, and $\mathbf{E}(\mathbf{R})$ be the corresponding (complex) phasor, then from definition

$$\tilde{\mathbf{E}}(\mathbf{R}, t) = \Re[\mathbf{E}e^{j\omega t}]$$

The other field vectors are similarly written. A single Maxwell's equation may be written in the phasor notation as

$$\nabla \times \tilde{\mathbf{E}}(\mathbf{R}, t) = -\frac{\partial \tilde{\mathbf{B}}(\mathbf{R}, t)}{\partial t}$$

$$\nabla \times \Re[\mathbf{E}(\mathbf{R})\,e^{j\omega t}] = -\frac{\partial}{\partial t}\Re[\mathbf{B}(\mathbf{R})\,e^{j\omega t}] \qquad (10.62)$$

$$\nabla \times \Re[\mathbf{E}(\mathbf{R})\,e^{j\omega t}] = -\Re\left\{\frac{\partial}{\partial t}[\mathbf{B}(\mathbf{R})\,e^{j\omega t}]\right\}$$

$$\Re\{\nabla \times [\mathbf{E}(\mathbf{R})\,e^{j\omega t}]\} = -\Re[j\omega\,\mathbf{B}(\mathbf{R})\,e^{j\omega t}]$$

$$\Re\{\nabla \times [\mathbf{E}(\mathbf{R})\,e^{j\omega t}]\} + \Re[j\omega\,\mathbf{B}(\mathbf{R})\,e^{j\omega t}] = 0$$

$$\Re\{\nabla \times [\mathbf{E}(\mathbf{R})\,e^{j\omega t}] + [j\omega\,\mathbf{B}(\mathbf{R})\,e^{j\omega t}]\} = 0$$

$$\nabla \times [\mathbf{E}(\mathbf{R})\,e^{j\omega t}] + [j\omega\,\mathbf{B}(\mathbf{R})\,e^{j\omega t}] = 0$$

$$\nabla \times \mathbf{E} = -j\omega\,\mathbf{B} \qquad (10.63)$$

In the last equation we have *dropped* $e^{j\omega t}$ and this factor is understood to be there.

> Note that all of Maxwell's equations remain the same except that $\partial/\partial t$ is replaced by $j\omega$.

Equations (10.36) to (10.39) become

$$\nabla \cdot \mathbf{D} = \rho_v \qquad (10.64)$$
$$\nabla \times \mathbf{E} = -j\omega\,\mu\,\mathbf{H} \qquad (10.65)$$
$$\nabla \cdot \mathbf{B} = 0 \qquad (10.66)$$
$$\nabla \times \mathbf{H} = j\omega\,\varepsilon\,\mathbf{E} + \mathbf{J} \qquad (10.67)$$

Practice Problem 10.2
Show that Maxwell's equations in phasor notation maybe written as in Eqns (10.64) to (10.67). Every term in these equations is a phasor.

Note: We will make no distinction between the notation (i.e., for example $\mathbf{E}$) used for sinusoidal and time-domain mathematical quantities since the discussion itself will clearly indicate whether the reference is towards the first set or the second.

To obtain the wave equation, we take the curl of Eqn (10.65):

$$\nabla \times (\nabla \times \mathbf{E}) = -j\omega\,\mu\,\nabla \times \mathbf{H} = -j\omega\,\mu\,(j\omega\,\varepsilon\,\mathbf{E} + \mathbf{J}) \qquad (10.68)$$

Using the identity

$$\nabla \times (\nabla \times \mathbf{A}) = \nabla(\nabla \cdot \mathbf{A}) - (\nabla \cdot \nabla)\,\mathbf{A}$$

we have

$$\nabla(\nabla \cdot \mathbf{E}) - (\nabla \cdot \nabla)\,\mathbf{E} = -j\omega\,\mu\,(j\omega\,\varepsilon\,\mathbf{E} + \mathbf{J}) \qquad (10.69)$$

In a region where the wave is propagating, we assume that there are no charges or currents. Therefore, there $\rho_v = 0$ (which implies that $\nabla \cdot \mathbf{E} \equiv 0$) and $\mathbf{J} = 0$.

$$-(\nabla \cdot \nabla)\,\mathbf{E} = -j\omega\mu\,(j\omega\varepsilon)\,\mathbf{E}$$

Furthermore,

$$-(j\omega\mu)\,(j\omega\varepsilon) = \omega^2\,\mu\varepsilon$$

Therefore,

$$-(\nabla \cdot \nabla)\,\mathbf{E} = \omega^2\,\mu\varepsilon\,\mathbf{E}$$

The phase velocity v_p of the wave is related to μ and ε by

$$v_p^2 = 1/(\mu\varepsilon) \tag{10.70}$$

Hence, $\omega^2\,\mu\varepsilon = \omega^2\,[1/(1/\mu\,\varepsilon)] = \omega^2/v_p^2 = (2\pi f)^2/(f\lambda)^2 = (2\pi/\lambda)^2 = k^2$ (10.71)

where λ is the wavelength of the wave in meters and k is the free-space propagation constant in rad/m. Using these relations, Eqn (10.69) becomes

$$-(\nabla \cdot \nabla)\,\mathbf{E} = k^2\mathbf{E} \tag{10.72}$$

which is the wave equation. The wave equation for sinusoidal functions is called the *Helmholtz's equation*.

This equation is in fact three different equations (as shown earlier): which are

$$\frac{\partial^2 E_x}{\partial x^2} + \frac{\partial^2 E_x}{\partial y^2} + \frac{\partial^2 E_x}{\partial z^2} + k^2 E_x = 0 \tag{10.73}$$

$$\frac{\partial^2 E_y}{\partial x^2} + \frac{\partial^2 E_y}{\partial y^2} + \frac{\partial^2 E_y}{\partial z^2} + k^2 E_y = 0 \tag{10.74}$$

$$\frac{\partial^2 E_z}{\partial x^2} + \frac{\partial^2 E_z}{\partial y^2} + \frac{\partial^2 E_z}{\partial z^2} + k^2 E_z = 0 \tag{10.75}$$

Similarly, the frequency domain scalar wave equation in one variable is

$$\frac{d^2\phi}{dz^2} + k^2\phi = 0 \tag{10.76}$$

where $k = \omega\sqrt{\mu\varepsilon}$. In the above equation we have replaced E_x by ϕ in the first of the three previous equations and $\partial/\partial y$, $\partial/\partial x$ are equated to zero, that is ϕ is not a function of these coordinates. If we observe Eqn (10.76) we note that its solution is of the sinusoidal type of function [i.e., $\sin(\ldots)$ or $\cos(\ldots)$ etc.]

If

$$\phi = \phi_0 \sin kz$$

then

$$\frac{\partial^2\phi}{\partial z^2} = -k^2\phi$$

is satisfied. Let us look at a specific function of the type, i.e.,

$$\phi = Ae^{-jkz}$$

where A is a constant. This function satisfies the wave equation. If we pause a moment and further examine the nature of the full function $A\exp\{j(\omega t - kz)\}$, we realise that it must be a travelling wave. Since we are looking at a phasor, in real time it must be

$$\phi\,(z,t) = \Re\{Ae^{j(\omega t - kz)}\} = A\cos(\omega t - kz)$$

Denoting the phase of this travelling wave by $\theta = \omega t - kz$, then at some time $t = t_0$ and position $z = z_0$,

$$\theta\,|_{t=t_0,\,x=x_0} = \theta_0 = \omega t_0 - kz_0$$

As in the case of the time domain scalar wave which we examined earlier, we allow a different set of values of z and t but changed in such a way that the phase remains the same, that is we travel with the phase

$$\theta_0 = \omega\,(t_0 + \Delta t) - k(z_0 + \Delta z)$$
$$= \omega\,t_0 + \omega\,\Delta t - kz_0 - k\Delta z$$
$$= (\omega\,t_0 - kz_0) + \omega\,\Delta t - k\Delta z$$
$$= \theta_0 + \omega\,\Delta t - k\Delta z \tag{10.77}$$

or,
$$0 = \omega\,\Delta t - k\Delta z \tag{10.78}$$

which gives,
$$\frac{\Delta z}{\Delta t} = \frac{\omega}{k} = v_p \tag{10.79}$$

Notice that v_p is the *phase velocity* since it is the velocity with which the *phase* travels. We can derive the magnetic field Helmholtz's equation in a similar way

$$-(\nabla \cdot \nabla)\,\mathbf{H} = k^2\mathbf{H} \tag{10.80}$$

Example 10.15   A scalar wave function $\phi = \phi_0 \cos(2\pi \times 10^7 t - 100 \times 2\pi x)$. Find the direction in which the wave is travelling and find its velocity.

Solution   Since the argument of the wave function is

$$2\pi \times 10^7 t - 100 \times 2\pi x$$
$$\uparrow$$

therefore the wave is travelling in the positive x-direction due to the negative sign indicated by the arrow. In this expression,

$$\omega = 2\pi \times 10^7$$
$$k = 2\pi/\lambda = 100 \times 2\pi$$

therefore
$$v = \omega/k = 10^5 \text{ m/s}$$

POINTS TO REMEMBER

- Faraday's law: emf generated $= \mathcal{E} = -\dfrac{\partial \Psi_m}{\partial t}$

- Faraday's law for an N turn coil: emf generated $= \mathcal{E} = -N\dfrac{\partial \Psi_m}{\partial t}$

- Maxwell's equations in point form:

$$\nabla \cdot \mathbf{D} = \rho_v$$
$$\nabla \times \mathbf{E} = -\partial\,\mathbf{B}/\partial t$$
$$\nabla \cdot \mathbf{B} = 0$$
$$\nabla \times \mathbf{H} = \mathbf{J} + \partial\mathbf{D}/\partial t$$

- Maxwell's equations in integral form:

$$\oiint_S \mathbf{D} \cdot d\mathbf{S} = \iiint \rho_v \, dV$$

$$\oint_L \mathbf{E} \cdot d\mathbf{l} = -\iint_S \frac{\partial \mathbf{B}}{\partial t} \cdot d\mathbf{S}$$

$$\oiint_S \mathbf{B} \cdot d\mathbf{S} = 0$$

$$\oint_L \mathbf{H} \cdot d\mathbf{l} = \iint_S \mathbf{J} \cdot d\mathbf{S} + \iint_S \frac{\partial \mathbf{D}}{\partial t} \cdot d\mathbf{S}$$

- The time domain wave equation(s):

$$\nabla^2 \mathbf{E} = \frac{\partial^2 \mathbf{E}}{\partial x^2} + \frac{\partial^2 \mathbf{E}}{\partial y^2} + \frac{\partial^2 \mathbf{E}}{\partial z^2} = [1/(v^2)] \partial^2 \mathbf{E}/\partial t^2$$

$$\frac{\partial^2 E_x}{\partial x^2} + \frac{\partial^2 E_x}{\partial y^2} + \frac{\partial^2 E_x}{\partial z^2} - [1/(v^2)] \frac{\partial^2 E_x}{\partial t^2} = 0$$

$$\frac{\partial^2 E_y}{\partial x^2} + \frac{\partial^2 E_y}{\partial y^2} + \frac{\partial^2 E_y}{\partial z^2} - [1/(v^2)] \frac{\partial^2 E_y}{\partial t^2} = 0$$

$$\frac{\partial^2 E_z}{\partial x^2} + \frac{\partial^2 E_z}{\partial y^2} + \frac{\partial^2 E_z}{\partial z^2} - [1/(v^2)] \frac{\partial^2 E_z}{\partial t^2} = 0$$

- Frequency domain Maxwell's equations in point form:

$$\nabla \cdot \mathbf{D} = \rho_v$$

$$\nabla \times \mathbf{E} = -j\omega \mu \mathbf{H}$$

$$\nabla \cdot \mathbf{B} = 0$$

$$\nabla \times \mathbf{H} = j\omega \varepsilon \mathbf{E} + \mathbf{J}$$

- Frequency domain wave equation(s):

$$(\nabla \cdot \nabla) \mathbf{E} = -k^2 \mathbf{E} \quad (k = \omega \sqrt{\mu \varepsilon})$$

$$\frac{\partial^2 E_x}{\partial x^2} + \frac{\partial^2 E_x}{\partial y^2} + \frac{\partial^2 E_x}{\partial z^2} + k^2 E_x = 0$$

$$\frac{\partial^2 E_y}{\partial x^2} + \frac{\partial^2 E_y}{\partial y^2} + \frac{\partial^2 E_y}{\partial z^2} + k^2 E_y = 0$$

$$\frac{\partial^2 E_z}{\partial x^2} + \frac{\partial^2 E_z}{\partial y^2} + \frac{\partial^2 E_z}{\partial z^2} + k^2 E_z = 0$$

- Frequency domain wave equation of the vector potential with currents as sources:

$$\frac{\partial^2 \mathbf{A}}{\partial x^2} + \frac{\partial^2 \mathbf{A}}{\partial y^2} + \frac{\partial^2 \mathbf{A}}{\partial z^2} - k^2 \mathbf{A} = -\mu \mathbf{J}$$

$$\frac{\partial^2 A_x}{\partial x^2} + \frac{\partial^2 A_x}{\partial y^2} + \frac{\partial^2 A_x}{\partial z^2} - k^2 A_x = -\mu J_x$$

$$\frac{\partial^2 A_y}{\partial x^2} + \frac{\partial^2 A_y}{\partial y^2} + \frac{\partial^2 A_y}{\partial z^2} - k^2 A_y = -\mu J_y$$

$$\frac{\partial^2 A_z}{\partial x^2} + \frac{\partial^2 A_z}{\partial y^2} + \frac{\partial^2 A_z}{\partial z^2} - k^2 A_z = -\mu J_z$$

SELF ASSESSMENT

Objective Type Questions

1. If a loop is placed in a steady magnetic field which is perpendicular to the loop, then
   (a) a current flows in the counter clockwise sense in the loop
   (b) a current flows in the clockwise sense in the loop
   (c) No current flows in the loop
   (d) None of the above
2. Faraday's law states that "The induced electromotive force, $\mathcal{E}$, in any closed circuit is equal to the time rate of change of the magnetic field through the circuit."
   (a) This is a statement of Faraday's law
   (b) This is a statement of Lenz's law
   (c) None of the above
3. The induced electromotive force, $\mathcal{E}$, in any closed circuit is equal to the time rate of change of the magnetic flux through the circuit.
   (a) This is a statement of Faraday's law
   (b) This is a statement of Lenz's law
   (c) None of the above
4. The equation

$$\text{emf generated} = \mathcal{E} = -\frac{\partial \Psi_m}{\partial t}$$

   is the emf generated by Faraday's law where
   (a) Ψ_m is the magnetic field through the loop
   (b) Ψ_m is the magnetic flux through the loop
   (c) Ψ_m is the electric flux through the loop
   (d) None of the above
5. The induced current flows in a direction such that the current opposes the change that induced it.
   (a) This is a statement of Faraday's law
   (b) This is a statement of Lenz's law
   (c) None of the above
6. The ignition system of a car uses _____ to create a spark in the spark plug. In the blank we may write
   (a) Faraday's law   (b) Lenz's law
   (c) Faraday's and Lenz's laws   (d) These concepts
7. In the equation $\nabla \times \mathbf{E} =$ _____, in the blank we may fill
   (a) $\partial \mathbf{D}/\partial t$ (b) $\partial \mathbf{B}/\partial t$
   (c) $-\partial \mathbf{B}/\partial t$ (d) $-\partial \mathbf{D}/\partial t$
8. $\nabla \cdot \mathbf{B} = 0$ is
   (a) an equation which has not been studied
   (b) one of Maxwell's equations
   (c) a statement of the fact that there are no magnetic charges
   (d) none of the above
9. In the function $\phi\,(x - vt)$ the unit of v is
   (a) m   (b) s   (c) s/m   (d) m/s
10. The unit of $\partial \mathbf{D}/\partial t$ is
    (a) $C/(m^2 s)$ (b) A   (c) A/m   (d) A/m^2

Short-Answer Questions

1. Figure 10.9 consists of a loop of metallic wire immersed in a constant **B**-field. Does a current flow in the metal hoop as shown in part (a) of the figure?
2. Does a current flow in the metal hoop as shown in part (b) of Fig. 10.9?

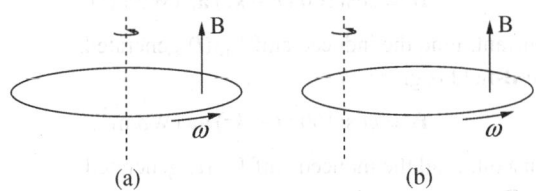

(a)   (b)

Fig. 10.9 Rotating hoop immersed in a constant **B**-field

3. The magnet shown in Fig. 10.10 is moved as shown in part (a) of the figure. In which direction does the current go in the upper part of the coil? Does the current go into the plane of the paper or come out of it?
4. The magnet shown in Fig. 10.10 is moved as shown in part (b) of the figure. In which direction does the current go in the upper part of the coil? Does the current go into the plane of the paper or come out of it?

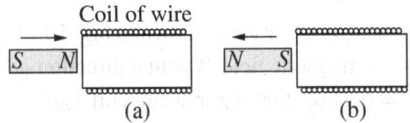

Coil of wire

(a)   (b)

Fig. 10.10 Moving magnet and stationary coil

5. Does $\phi\,(x/v - t)$ satisfy the wave equation? v is the velocity of the wave.
6. Is $\varepsilon\,\partial\mathbf{E}/\partial t$ the displacement current?
7. What is the angle of the phasors (a) j, (b) 1, (c) $1 + j1$?
8. In which quadrant does $1 - j1$ lie? What is its angle?

Review Questions

1. State Faraday's law.
2. State Lenz's law.
3. Explain what is displacement current density and why is it neccessary to include it as a term in Ampere's law.
4. Does current flow between the plates of a capacitor? Is the displacement current density a movement of charges?
5. Write out the point form and integral forms of Maxwell's equations.
6. Explain why the function $\phi\,(x - vt)$ where v is a constant, is a travelling wave.

Numerical Problems

1. In Fig. 10.2, the **B**-field is given by

$$\mathbf{B} = t^2 u(t)\, \mathbf{a}_z \ (\text{Wb/m}^2), \quad 0 \le t \le 2\ (\text{s})$$
$$= 4\ (\text{Wb/m}^2) \quad t \ge 4\ (\text{s})$$

Find the induced emf $V_{12}(t)$ generated.

2. In Fig. 10.2, the **B**-field is given by

$$\mathbf{B} = \cos(100\pi t)\,\mathbf{a}_z \ (\text{Wb/m}^2)$$

Find the induced emf $V_{12}(t)$ generated.

3. In Fig. 10.2, the **B**-field is given by

$$\mathbf{B} = \cos(100\pi t - ky)\,\mathbf{a}_y \ (\text{Wb/m}^2)$$

where k is a constant. Find the induced emf $V_{12}(t)$ generated.

4. In Fig. 10.2, the **B**-field is given by

$$\mathbf{B} = \cos(100\pi t - kz)\,\mathbf{a}_x \ (\text{Wb/m}^2)$$

where k is a constant. Find the induced emf $V_{12}(t)$ generated.

5. In Fig. 10.2, the **B**-field is given by

$$\mathbf{B} = \cos(100\pi t - kz)\,xy\,(\mathbf{a}_x + \mathbf{a}_y + \mathbf{a}_z) \ (\text{Wb/m}^2)$$

where k is a constant and $a = 1$, $b = 2$. Find the induced emf $V_{12}(t)$ generated.

6. The magnetic field $\mathbf{H}(x,y,z,t)$ of a plane wave is $H_0 \sin(\omega t - kx)\,\mathbf{a}_z$ and is incident on the 'antenna' as shown in Fig. 10.2. Find the magnitude of the emf $|V(t)|$ between terminals 1 and 2. Show that it is

$$\left| H_0\mu_0 \ \frac{2\omega \sin\left(\frac{ak}{2}\right)\cos(\omega t)}{k} \ b \right|$$

7. A slider moves on rails with a velocity v as shown in Fig. 10.11. The whole arrangement is immersed in a constant magnetic field **B** with a direction going into the plane of the paper as shown. If $R_2 = 0$, show that the induced emf V_{ab} is

$$V_{ab} = BvL$$

8. In Fig. 10.11, If $B = 0.1$ (T), the position of the slider in the y direction is given by $y(t) = 0.1t$ (m), and $L = 0.1$ (m), find the emf generated in the loop.

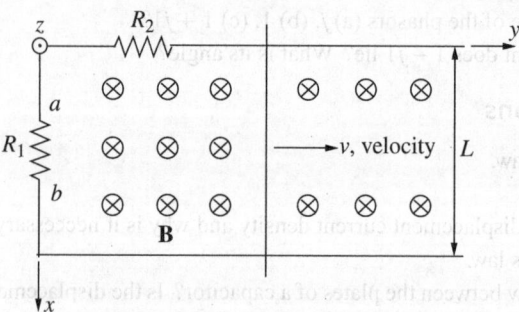

Fig. 10.11 Moving slider on rails

9. In problem 8, if $R_1 = 10 \ \Omega$ and $R_2 = 20 \ \Omega$, find the voltage drops across the two resistors and the current.

10. In Fig. 10.11, if $B = 0.1$ (T), the position of the slider in the y direction is given by $y(t) = 0.1t + .05t^2$ (m) and $L = 0.1$ (m), if $R_2 = 0$ and $R_1 = 1 \ \Omega$, find the emf generated and the current in the 1 Ω resistor.

11. In Fig. 10.11, If $B(x,y) = 0.1xy$ (T), the position of the slider in the y-direction is given by $y(t) = 0.1t$ (m) and $L = 0.1$ (m), if $R_2 = 0$ and $R_1 = 1 \ \Omega$, find the emf generated and the current in the 1 Ω resistor when the y-position of the slider is 1 m.

12. From Maxwell's equation in integral form, Eqn (10.33), show how both Faraday's law as well as Lenz's laws can be obtained.

13. In a region of free space $\varepsilon = \varepsilon_0$ and $\mu = \mu_0$, the electric field is a constant, $\mathbf{E} = E_0 \mathbf{a}_z$. What is the displacement current density?

14. In a region of free space $\varepsilon = \varepsilon_0$ and $\mu = \mu_0$, the electric field is $\mathbf{E} = E_0 \cos(110t - kx) \mathbf{a}_z$. What is the displacement current density? What is the conduction current density?

15. In a region $\varepsilon = 10^{-6}$, $\mu = 10^{-5}$, $\sigma = 10^{-5}$ the electric field is $\mathbf{E} = E_0 \cos(110t - kx) \mathbf{a}_z$. What is the displacement current density? What is the conduction current density?

16. If $V = 10 \sin(\omega t + 30°)$, find the phasor of this sinusoid.

17. If $V = 30\angle 45°$ is a phasor of a sinusoid at frequency 1 GHz, find the time-domain function.

18. By using the arguments given in Section 10.3 show that for a parallel plate capacitor with a dielectric filling of ε, σ the input current to the capacitor is equal to the displacement current plus the conduction current within the capacitor.

19. If $P(x) = \begin{cases} 1, & 0 < x < 1 \\ 0, & \text{otherwise} \end{cases}$

    Plot $P(x - 0.5t)$ for $t = 0, 1, 5$ s.

20. If $\phi(x) = \begin{cases} \sin(2\pi x), & 0 < x < 1 \\ 0, & \text{otherwise} \end{cases}$

    Plot $\phi(x - 0.5t)$ for $t = 0, 1, 5$ s.

21. The function $10 \sin(\omega t - \beta x)$ represents a wave travelling in air. If the frequency is 100 MHz, find ω and β. Plot the advance of the wave for $t = 0.1T$, $0.3T$, and $0.5T$ where T is the time period.

Answers

Objective Type Questions

1. (c) 2. (c) 3. (a) 4. (b) 5. (b) 6. (a) 7. (c)
8. (b) and (c) 9. (d) 10. (a) and (d)

Short-Answer Questions

1. No
2. No
3. Into the plane of the paper
4. Into the plane of the paper
5. Yes
6. No
7. (a) $\pi/2$ (b) 0 (c) $\pi/4$
8. 4th quadrant; $-\pi/4$

Numerical Problems

1. $V_{12} = t^2 u(t) ab$ for $0 < t < 4$; $V_{12} = 0$ for $t > 4$
2. $V_{12} = -100\pi ab \sin(100\pi t)$
3. $V_{12} = 0$
4. $V_{12} = 0$
5. $V_{12} = 0$
8. 10^{-3} V

9. $I = 33.3\,\mu A$; $V_{R_1} = 0.33$ mV; $V_{R_2} = 0.66$ mV
10. $10^{-2}t + 10^{-3}$ V
11. Emf $= 5t\,\mu V$
13. 0
14. $-110\varepsilon_0 E_0 \sin(110t - kx)\,\mathbf{a}_z\,A/m^2$
15. $\partial \mathbf{D}/\partial t = -1.1 \times 10^{-4} E_0 \sin(110t - kx)\,a_z$; $\mathbf{J} = 10^{-5}\mathbf{E}$
16. $10\angle - 60°$
17. $10\cos(2\pi \times 10^9 t + 45°)$
19. At $t = 0$, the function is

$$
\begin{array}{ll}
0, & \text{for } -\infty < x < 0 \\
1, & \text{for } 0 < x < 1 \\
0, & \text{for } 1 < x < \infty
\end{array}
$$

at $t = 1$, the function is

$$
\begin{array}{ll}
0, & \text{for } -\infty < x < 1 \\
1, & \text{for } 0.5 < x < 1.5 \\
0, & \text{for } 1.5 < x < \infty
\end{array}
$$

etc.

21. $\omega = 6.28 \times 10^8$ r/s, $\beta = 2.09$ r/m.

11

Electromagnetic Waves

Education should be exercise;
it has become massage.
— Martin H. Fischer

CHAPTER OBJECTIVES

To enable the students to understand the following:

- Uniform plane wave
- Electromagnetic spectrum
- Wave polarisation—linear, circular, and elliptic polarisations
- Wave propagation in conducting media and lossy dielectrics
- Reflection and refraction of waves
- Poynting Vector and the flow of power

11.1 | Introduction

When we set up a communication link between a transmitter and receiver without the help of cables we usually set up two antennas, a transmitting antenna and a receiving antenna. The transmitting antenna transmits electromagnetic waves which are essentially waves whose nature is that of light waves but at very lower frequency. These waves are generally sinusoidal, described by frequency and the velocity of propagation, which in turn depend on the medium in which the waves propagate. Thus if the waves propagate in a dielectric, then the medium is described by its permittivity, $\varepsilon = \varepsilon_r \varepsilon_0$. If the medium is a conductor, then the conductivity σ comes into play, and so on.

In this chapter we concentrate on the fundamental properties of electromagnetic waves as they propagate in various types of media, and their interaction with matter.

11.2 | Uniform Plane Wave

In the very beginning we consider the simplest of waves, namely, the uniform plane wave. Let a sinusoidally time-varying wave travel in the $+z$ direction but with no variation in either x or y directions. Figure 11.1 depicts how the wave

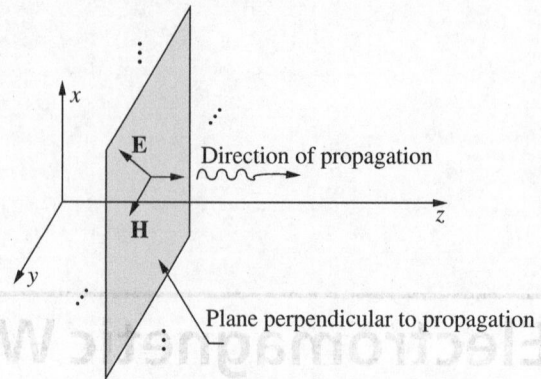

Fig. 11.1 The geometry of the uniform plane wave

travels. The shaded region shows a single phase front, perpendicular to the direction of propagation.

The procedure outlined below is standard for the analysis of most wave phenomena.

• Since we are considering a wave, the electric and magnetic fields must satisfy the wave equation.

• Since the nature of the wave is oscillatory, having a $\exp(j\omega t)$ time dependence, the fields must satisfy Helmholtz's equation (see Section 10.9 for a better understanding of this point). The Helmholtz's equation for the electric field is Eqn (11.72)

$$\nabla^2 \mathbf{E} + k^2 \mathbf{E} = 0$$

where

$$k = \omega \sqrt{\mu \varepsilon}$$

• If we recall the results of the last chapter, a travelling wave must have a functional dependence,

$$\tilde{\mathbf{E}}(\mathbf{R}, t) = \Re\{\mathbf{E}_0 e^{j(\omega t - \beta z)}\} \tag{11.1}$$

where β is the propagation constant of the wave and ω is the frequency of oscillation. Here β may or may not be equal to k. In the case of a uniform plane wave, $\mathbf{E}_0$ is a constant but real phasor,

$$\mathbf{E}_0 = \mathbf{a}_x E_{x0} + \mathbf{a}_y E_{y0} + \mathbf{a}_z E_{z0} \quad (E_{x0}, E_{y0}, E_{z0} \text{ are real and constant}) \tag{11.2}$$

Let us first satisfy the Helmholtz equation

$$\nabla^2 \mathbf{E} = -k^2 \mathbf{E}$$

$$\mathbf{E}_0 \nabla^2 e^{j(\omega t - \beta z)} \stackrel{?}{=} -k^2 \mathbf{E}_0 e^{j(\omega t - \beta z)}$$

$$\mathbf{E}_0 e^{j\omega t} \frac{\partial^2}{\partial z^2} e^{-j\beta z} \stackrel{?}{=} -k^2 \mathbf{E}_0 e^{j(\omega t - \beta z)}$$

$$\mathbf{E}_0 e^{j\omega t} (-\beta^2 e^{-j\beta z}) \stackrel{?}{=} -k^2 \mathbf{E}_0 e^{j(\omega t - \beta z)}$$

$$-\beta^2 \mathbf{E} = -k^2 \mathbf{E}$$

which is satisfied if $\beta = k$. The term $\exp\{j(\omega t - kz)\}$ ensures that the wave is a travelling wave, travelling in the $+z$ direction. The phase of the wave at time t_0 is given by

$$\omega t_0 - kz = \text{constant} = K$$

which is the equation of a *plane* in three dimensions and which is parallel to the $z = 0$ plane.

$$z = (\omega t_0 - K)/k = \text{another constant} = K_1$$

Now working only with the phasor

$$\mathbf{E} = \mathbf{E}_0 e^{-jkz} \tag{11.3}$$

The sinusoidal fields in time are

$$\Re(\mathbf{E}e^{j\omega t}) = \Re(\mathbf{E}_0 e^{-jkz} e^{j\omega t})$$
$$= \mathbf{E}_0 \cos(\omega t - kz) \tag{11.4}$$

The term $\mathbf{E}_0$ can be pulled out of the bracket since it is real and constant. If this electric field is to represent an actual wave, it must satisfy Maxwell's equations. The first Maxwell equation which has to be satisfied is

$$\nabla \cdot \mathbf{E} = \frac{\rho_v}{\varepsilon}$$

But we know that the wave is travelling in a region free of charges. So ρ_v must be zero.

$$\nabla \cdot \mathbf{E} = 0$$

$$\frac{\partial}{\partial x}(E_{x0}e^{-jkz}) + \frac{\partial}{\partial y}(E_{y0}e^{-jkz}) + \frac{\partial}{\partial z}(E_{z0}e^{-jkz}) = 0$$

$$0 + 0 - jkE_{z0} = 0$$

$$E_{z0} = 0 \tag{11.5}$$

In other words there is no electric field in the direction of propagation; there are only transverse components.

Hence

$$\mathbf{E} = \mathbf{a}_x E_{x0} e^{-jkz} + \mathbf{a}_y E_{y0} e^{-jkz} \tag{11.6}$$

The second Maxwell equation which the electric and magnetic fields have to satisfy is

$$\nabla \times \mathbf{E} = -j\omega \mu \mathbf{H} \tag{11.7}$$

which is Eqn (11.65). Writing out the various components starting with the x-component

$$(\nabla \times \mathbf{E})_x = -j\omega \mu H_x$$

$$\frac{\partial E_z}{\partial y} - \frac{\partial E_y}{\partial z} = -j\omega \mu H_x$$

$$0 + jkE_{y0}e^{-jkz} = -j\omega \mu H_x$$

Similarly, we can get the other components

$$(\nabla \times \mathbf{E})_y = \frac{\partial E_x}{\partial z} - \frac{\partial E_z}{\partial x} = -jkE_{x0}e^{-j\beta z} = -j\omega \mu H_y$$

$$(\nabla \times \mathbf{E})_z = \frac{\partial E_y}{\partial x} - \frac{\partial E_x}{\partial y} = 0 = -j\omega \mu H_z$$

From the last equation, we get $H_z = 0$ or the *magnetic field is also transverse to the direction of propagation*. Figure 11.1 clearly shows how both the electric and magnetic fields lie on $z=$ constant planes. From the above equations, the magnetic field components are

$$-j\omega\mu H_x = j\beta E_{y0}e^{-jkz}$$

$$H_x = \frac{j\omega\sqrt{\mu\varepsilon}}{-j\omega\mu}E_{y0}$$

$$= -\sqrt{\frac{\varepsilon}{\mu}}E_{y0} \tag{11.8}$$

and

$$-j\omega\mu H_y = -j\beta E_{x0}e^{-jkz}$$

$$H_y = \frac{-j\omega\sqrt{\mu\varepsilon}}{-j\omega\mu}E_{x0}$$

$$= \sqrt{\frac{\varepsilon}{\mu}}E_{x0} \tag{11.9}$$

Concentrating on the term $\sqrt{\varepsilon/\mu}$, we calculate its units

$$\text{Units of}\left\{\sqrt{\frac{\varepsilon}{\mu}}\right\} = \sqrt{\frac{F/m}{H/m}}$$

$$= \sqrt{\frac{F}{H}}$$

$$= \sqrt{\frac{\mho\cdot s}{\Omega\cdot s}}$$

$$= \sqrt{\mho^2}$$

$$= \mho \tag{11.10}$$

Since $\sqrt{\varepsilon/\mu}$ has the units of $\mho$, we represent $\sqrt{\mu/\varepsilon}$ by Z, a resistance

$$H_x = -\frac{1}{Z}E_y \tag{11.11}$$

$$H_y = \frac{1}{Z}E_x \tag{11.12}$$

$$Z = \sqrt{\frac{\mu}{\varepsilon}} \tag{11.13}$$

Z is called *the characteristic or intrinsic impedance of the medium.* For air or vacuum, a quick calculation of the characteristic impedance, $Z \equiv Z_0$ gives

$$Z_0 = \sqrt{\frac{\mu_0}{\varepsilon_0}} \approx 377\ \Omega \tag{11.14}$$

Since $\mu_0 = 4\pi\times10^{-7}$ and $\varepsilon_0 = (1/36\pi)\times10^{-9}$; therefore $\sqrt{\mu_0/\varepsilon_0} = \sqrt{4\times36\pi^2\times100} = 120\pi = 377\ \Omega$. The magnitude of the magnetic field is

$$H = \sqrt{H_x^2 + H_y^2} = \frac{1}{Z}\sqrt{E_y^2 + E_x^2}$$

$$= \frac{E}{Z} \tag{11.15}$$

Taking a different tack, we calculate $\mathbf{E}\times\mathbf{H}$ for the uniform plane wave.

$$\mathbf{E}\times\mathbf{H} = \mathbf{a}_x(E_yH_z - E_zH_y) + \mathbf{a}_y(E_zH_x - H_zE_x) + \mathbf{a}_z(E_xH_y - E_yH_x)$$

$$= \mathbf{a}_z(E_x H_y - E_y H_x) \quad (E_z \text{ and } H_z \text{ are both zero})$$

$$= \mathbf{a}_z \left[E_x \left(\frac{E_x}{Z} \right) - E_y \left(\frac{-E_y}{Z} \right) \right] \quad (\text{From the previous set of equations})$$

$$= \mathbf{a}_z \frac{1}{Z} \left(E_x^2 + E_y^2 \right) \quad V^2/(\Omega \, m^2) \equiv W/m^2$$

$$|\mathbf{E} \times \mathbf{H}| = \frac{1}{Z} \left(E_x^2 + E_y^2 \right) = \frac{1}{Z} |\mathbf{E}|^2 \tag{11.16}$$

It is important to note that the *direction* of $\mathbf{E} \times \mathbf{H}$ is the direction of propagation.

To compute the angle θ, between $\mathbf{E}$ and $\mathbf{H}$ we must compute the dot product between the two. Since

$$\mathbf{E} \cdot \mathbf{H} = EH \cos \theta$$

Using Eqns (11.11) and (11.12)

$$\mathbf{E} \cdot \mathbf{H} = (\mathbf{a}_x E_{x0} e^{-jkz} + \mathbf{a}_y E_{y0} e^{-jkz}) \cdot (\mathbf{a}_x H_{x0} e^{-jkz} + \mathbf{a}_y H_{y0} e^{-jkz})$$

$$= \frac{e^{-j2kz}}{Z} (\mathbf{a}_x E_{x0} + \mathbf{a}_y E_{y0}) \cdot (-\mathbf{a}_x E_{y0} + \mathbf{a}_y E_{x0})$$

$$= \frac{e^{-j2kz}}{Z} (-E_{x0} E_{y0} + E_{y0} E_{x0})$$

$$= 0 \tag{11.17}$$

Now $E \neq 0$ and $H \neq 0$, therefore $\cos \theta$ must be zero. In other words, the angle from $\mathbf{E}$ to $\mathbf{H}$ is $\pi/2$. It is an important conclusion:

 See `MagPoyntVect.m` in Chapter 11

Not only are the electric and magnetic field perpendicular to the direction of propagation, but they are perpendicular to each other; the vector $\mathbf{E} \times \mathbf{H}$ is directed towards the propagation direction.

Figure 11.2 depicts the advance of the plane wave where the electric field is drawn as the solid line while the magnetic field as the dotted line. Notice that *the magnetic field is always in phase with the electric field*; that is when the electric field increases, the magnetic field increases, and when the electric field decreases the magnetic field

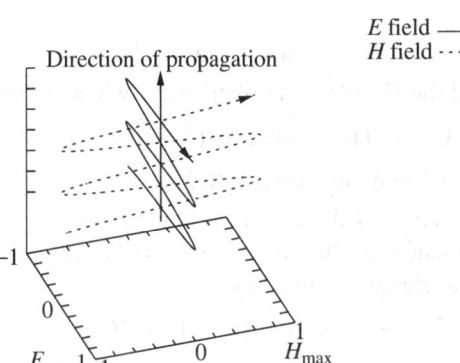

Fig. 11.2 Electric and magnetic fields of a uniform plane wave

does the same, keeping the ratio of the electric to the magnetic field constant $(= Z,$ the characteristic impedance).

Example 11.1 For a plane wave travelling in air, if $\omega = 2\pi \times 10^9$ rad/s, find the propagation constant k.

Solution The propagation constant k is given by

$$k = \frac{2\pi}{\lambda}$$

where λ is the wavelength of the wave. Also

$$f\lambda = c$$

where $f = \omega/(2\pi)$ is the frequency of the wave in Hz and c is the velocity of the wave in air $(= 3 \times 10^8$ m/s). So

$$k = \frac{2\pi f}{c}$$

$$= \frac{\omega}{c}$$

$$= \frac{2\pi \times 10^9}{3 \times 10^8}$$

$$= \frac{20\pi}{3} \text{ m}^{-1}$$

Example 11.2 If the electric field $\mathbf{E} = (10\mathbf{a}_z + 20\mathbf{a}_y) \cos(2\pi \times 10^7 t - kx)$, show that this is the electric field of a wave.

Solution If we compare the functional dependance of the electric field with Eqn (11.51), we can identify that

$$\omega = 2\pi \times 10^7 \text{ r/s}$$

and k as the propagation constant. Therefore, this is the electric field of a plane wave.

Example 11.3 Express $\mathbf{E} = (10\mathbf{a}_z + 20\mathbf{a}_y) \cos(2\pi \times 10^7 t - kx)$ as a phasor. $\omega = 2\pi \times 10^7$.

Solution
$$\mathbf{E} = \Re\{(10\mathbf{a}_z + 20\mathbf{a}_y) \, e^{j(2\pi \cdot 10^7 t - kx)}\}$$

$$= (10\mathbf{a}_z + 20\mathbf{a}_y) \, \Re\{e^{j(2\pi \cdot 10^7 t - kx)}\}$$

so the phasor is

$$\mathbf{E} = (10\mathbf{a}_z + 20\mathbf{a}_y) \, e^{-jkx}$$

Example 11.4 Find the $\mathbf{H}$ field of an plane wave whose $\mathbf{E}$ vector is given by

$$\mathbf{E} = (10\mathbf{a}_z + 20\mathbf{a}_y) \cos(2\pi \cdot 10^7 t - kx)$$

in air. Find the value of k and the wavelength λ.

Solution From inspection of the $\mathbf{E}$ field it is clear that $\omega = 2\pi \times 10^7 t (= 2\pi f)$ where f is the frequency of the wave. So $f = 10^7$ Hz. Since the velocity $c = 3 \times 10^8$ m/s the wavelength is given by

$$\lambda = c/f = 3 \times 10^8/10^7 = 30 \text{ m}$$

from λ, we can calculate k by

$$k = 2\pi/\lambda = 0.20944 \text{ rad/m}$$

Again from inspection it is clear that the wave is travelling in the $+x$ direction. So

$$\mathbf{H} = \frac{1}{Z_0}\mathbf{a}_x \times \mathbf{E} = \frac{1}{377}(-10\mathbf{a}_y + 20\mathbf{a}_z)\cos(2\pi \times 10^7 t - kx) \ \text{A/m}$$

 See `PhaseConstant.m` in Chapter 11

Example 11.5 A plane wave caries a power density of 1 MW/m². Find the magnitude of the electric and magnetic field vectors.

Solution The power density S is given by

$$S = E^2/Z_0 = 10^6 \ \text{W/m}^2$$

so

$$E = \sqrt{10^6 \times 377} = 1.9416 \times 10^4 \ \text{V/m}$$

and

$$H = E/Z_0 = 51.501 \ \text{A/m}$$

11.3 | Electromagnetic Spectrum

Electromagnetic waves in actual use for different purposes are characterised by their frequency f and wavelength, λ. These bands of frequencies have been given names by engineers, and there is an agreement all over the world about their use. Figure 11.3 shows the frequency bands, their names and their uses.

In the figure, the frequency spectrum is depicted on a logarithmic scale from 3 Hz to 3 PHz. (Read as *peta*hertz). At the top the approximate analytical technique is outlined: for example the *lumped* element approach is used from DC to a few

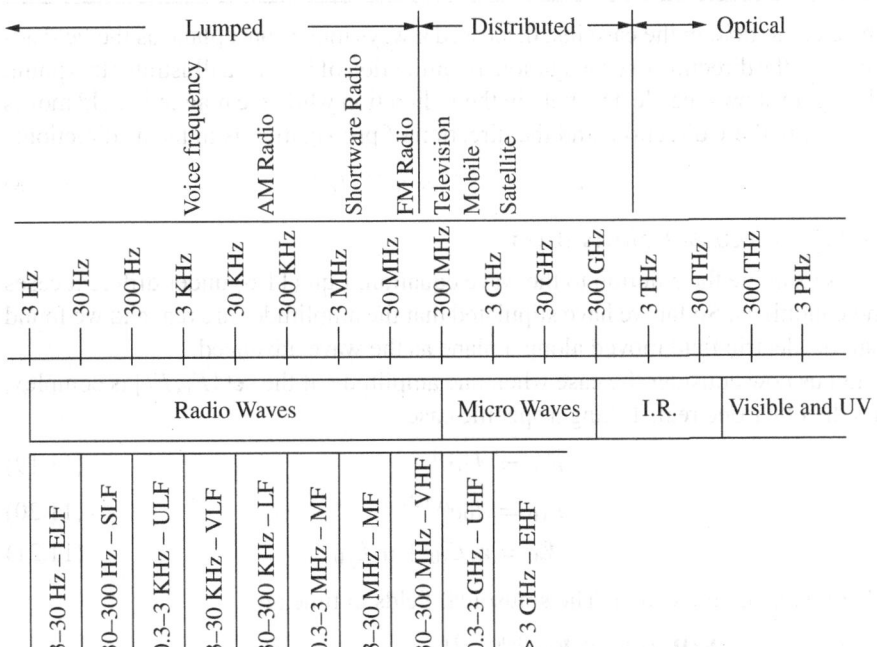

Fig. 11.3 The electromagnetic frequency spectrum

Table 11.1 IEEE microwave band designations

Band	Frequency
L	1–2 GHz
S	2–4 GHz
C	4–8 GHz
X	8–12 GHz
Ku	12–18 GHz
K	18–27 GHz
Ka	27–40 GHz
V	40–75 GHz
W	75–110 GHz
mm	110–300 GHz

hundred megahertz; the distributed element approach is used from a few hundred megahertz to about a few hundred gigahertz, and so on. Below the logarithmic spectrum is the rough demarcation of the frequency bands: radio waves, microwaves, infrared, visible, and beyond. At the bottom of the figure, the designated frequency bands are named: VLF, very low frequency; LF, low frequency; MF, medium frequency; HF, high frequency; VHF, very high frequency; UHF, ultra high frequency and EHF extremely high frequency and so on.

Apart from the radio frequency bands there is yet another set of designations which is primarily due to the US military, but now adopted worldwide. These bands are shown in Table 11.1.

11.4 | Wave Polarisation

The electric field in the case just discussed always moves on a plane as the field advances in the direction of propagation. Examination of Fig. 11.2 illustrates this point. The figure shows the electric field in the x-direction while the magnetic field moves in phase in the y direction, and the direction of propagation is in the $\mathbf{a}_z$ direction:

$$\mathbf{a}_x E_x \times \mathbf{a}_y H_y = \mathbf{a}_z |E_x H_y| \tag{11.18}$$

11.4.1 Circular Polarisation

Let us examine the solution to the wave equation, Eqn (11.6) under different cases and conditions. So far we have stipulated that the amplitudes are real and we found that the electric field moved along a plane as the wave advanced.

Let us now consider the case when one amplitude of the set $[E_x, E_y]$ is complex, and the other one real. Taking a specific case

$$E_{x0} = A_0 e^{j0} \tag{11.19}$$

$$E_{y0} = A_0 e^{j(\pi/2)} \tag{11.20}$$

$$\mathbf{E}_0 = \mathbf{a}_x E_{x0} + \mathbf{a}_y E_{y0} \tag{11.21}$$

where A_0 is of course real. The sinusoidal fields in time are

$$\tilde{\mathbf{E}}(\mathbf{R}, t) = \Re\{\mathbf{E}_0 e^{j(\omega t - kz)}\}$$

$$= \Re\{(\mathbf{a}_x E_{x0} + \mathbf{a}_y E_{y0}) e^{j(\omega t - kz)}\}$$

$$= \Re\{[\mathbf{a}_x A_0 + \mathbf{a}_y A_0 e^{j(\pi/2)}] e^{j(\omega t - kz)}\}$$

$$= A_0 \Re\{\mathbf{a}_x e^{j(\omega t - kz)} + \mathbf{a}_y e^{j(\pi/2)} e^{j(\omega t - kz)}\}$$

$$= A_0 \Re\{\mathbf{a}_x e^{j(\omega t - kz)} + \mathbf{a}_y e^{j(\omega t - kz + \pi/2)}\}$$

$$= A_0 [\mathbf{a}_x \cos(\omega t - kz) - \mathbf{a}_y \sin(\omega t - kz)] \qquad (11.22)$$

The magnitude of the field in time is

$$|\tilde{\mathbf{E}}(\mathbf{R}, t)| = \sqrt{A_0^2 [\cos^2(\omega t - kz) + \sin^2(\omega t - kz)]}$$

$$= A_0 \qquad (11.23)$$

which is constant. How do we interpret this result? Let us analyse this situation in greater detail.

The original electric field in time is

$$\tilde{\mathbf{E}} = \mathbf{a}_x \tilde{E}_x + \mathbf{a}_y \tilde{E}_y$$

$$= A_0 [\mathbf{a}_x \cos(\omega t - kz) - \mathbf{a}_y \sin(\omega t - kz)] \qquad (11.24)$$

We get rid of the z coordinate by setting it to zero. By setting z to zero we are analysing the behaviour of the electric field on the plane $z = 0$.

$$\tilde{\mathbf{E}}(x, y, z = 0, t) = A_0 [\mathbf{a}_x \cos(\omega t) - \mathbf{a}_y \sin(\omega t)] \qquad (11.25)$$

This seems simpler. Let us now proceed to calculate the electric field $\tilde{\mathbf{E}}$ at different times: at $t = 0$, $t = T/4$, $t = T/2$, and $t = 3T/4$, where T is the time period of one cycle. Since $\omega = 2\pi/T$, $\omega t = 0$, $\pi/2$, π, and $3\pi/4$ at these times.

$$\tilde{\mathbf{E}} = A_0 \mathbf{a}_x \qquad (\text{At } t = 0)$$

$$\tilde{\mathbf{E}} = -A_0 \mathbf{a}_y \qquad (\text{At } t = T/4)$$

$$\tilde{\mathbf{E}} = -A_0 \mathbf{a}_x \qquad (\text{At } t = T/2)$$

$$\tilde{\mathbf{E}} = A_0 \mathbf{a}_y \qquad (\text{At } t = 3T/4)$$

We notice that the electric field vector rotates in a direction of the fingers of the *left hand* when made into a fist and the thumb points in the direction of propagation in accordance left hand thumb rule. Figure 11.4 illustrates this point.

The next Fig. 11.5 shows how the wave advances helically in a polarised wave. The figure shows the wave at time $t = T$. The tip of the electric field vector at $t = 0$ is at point a; at $t = T/4$ it is at b. In this way it traverses a full circle in time T, going through points c and d. In time T different parts of the wave advances by varying

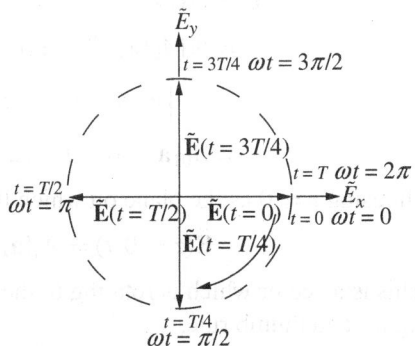

Fig. 11.4 Figure illustrating left circular polarisation

amounts. The electric field at point a advances to a' which is one wavelength away. The electric field at point b advances to b' three quarters of a wavelength away, and so on. The electric field vectors all lie parallel to the x-y plane but all their tips lie on a helix. On the x-y plane, the electric field traces a clockwise circle. This type of polarisation is called left circular polarisation (LCP) because as

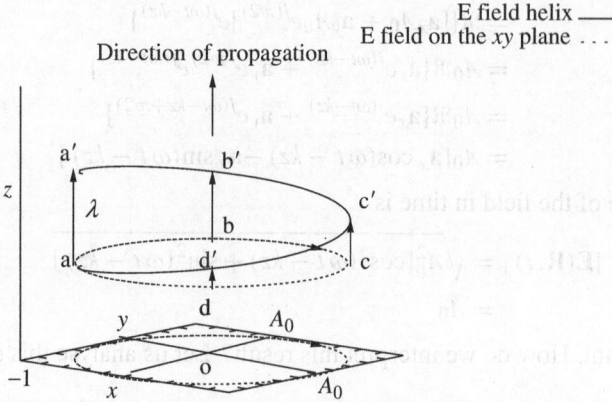

Fig. 11.5 The advance of the wave in a left circular polarisation (LCP) plane wave

the left hand is held in the form of a fist, with the thumb extended, the fingers curl in the direction of motion of the electric field while the thumb points in the direction of the propagation of the wave.

Consider the case of a right hand circularly polarised wave. For such a wave, advancing in the $+z$ direction, the electric field rotates in the anti-clockwise direction on any plane parallel to the x-y plane and the wave travels in the direction of the thumb. The electric field then can be written as

$$E_{x0} = A_0 e^{j0}$$

$$E_{y0} = A_0 e^{-j(\pi/2)}$$

$$\mathbf{E}_0 = \mathbf{a}_x E_{x0} + \mathbf{a}_y E_{y0}$$

$$\mathbf{E} = \mathbf{E}_0 e^{-j\beta z} \qquad (11.26)$$

The sinusoidal time-dependent field is

$$\begin{aligned}
\tilde{\mathbf{E}} &= \Re\{\mathbf{E}e^{j\omega t}\} \\
&= \Re\{A_0[\mathbf{a}_x e^{j0} + \mathbf{a}_y e^{-j(\pi/2)}]e^{j(\omega t - kz)}\} \\
&= A_0\left[\mathbf{a}_x \cos(\omega t - kz) + \mathbf{a}_y \cos\left(\omega t - kz - \frac{\pi}{2}\right)\right] \\
&= A_0[\mathbf{a}_x \cos(\omega t - kz) + \mathbf{a}_y \sin(\omega t - kz)] \qquad (11.27)
\end{aligned}$$

choosing $z = 0$ as the plane on which the electric field is evaluated

$$\tilde{\mathbf{E}}(z = 0, t) = A_0[\mathbf{a}_x \cos(\omega t) + \mathbf{a}_y \sin(\omega t)] \qquad (11.28)$$

This is a vector which is rotating in the counter-clockwise direction and obeys the right-hand thumb rule.

Example 11.6 If $E_x = 5\angle 45°$ and $E_y = 5\angle 45°$, find the type of polarisation.
Solution If $E_x = 5\angle 45°$ and $E_y = 5\angle 45°$ then

$$\tilde{E}_x = 5\cos(\omega t + 45°)$$

$$\tilde{E}_y = 5\cos(\omega t + 45°)$$

and

$$\tilde{E}_y = 1 \times \tilde{E}_x$$

which is the equation of a straight line. So the wave is linearly polarised. ▬▬▬

Example 11.7 If $E_x = 5\angle 45°$ and $E_y = 15\angle 45°$ find the type of polarisation.

Solution If $E_x = 5\angle 45°$ and $E_y = 15\angle 45°$ then

$$\tilde{E}_x = 5\cos(\omega t + 45°)$$
$$\tilde{E}_y = 15\cos(\omega t + 45°)$$

and
$$\tilde{E}_y = 3 \times \tilde{E}_x$$

which is the equation of a straight line. So the wave is linearly polarised.

Example 11.8 If $E_x = 5\angle 0°$ and $E_y = 5\angle 90°$, find the type of polarisation.

Solution If $E_x = 5\angle 0°$ and $E_y = 5\angle 90°$ then

$$\tilde{E}_x = 5\cos(\omega t)$$
$$\tilde{E}_y = 5\cos(\omega t + 90°)$$
$$= -5\sin \omega t$$

and
$$\tilde{E}_y^2 + \tilde{E}_x^2 = 50$$

which is the equation of a circle. So the wave is circularly polarised.

11.4.2 Elliptical Polarisation

We have studied two types of polarisations—linear and circular polarisation. Let us take a look at the most general type of polarisation, namely elliptical polarisation. Let

$$\tilde{\mathbf{E}} = \mathbf{a}_x \tilde{E}_x + \mathbf{a}_y \tilde{E}_y$$
$$= \mathbf{a}_x E_{x0} \cos(\omega t - kz) + \mathbf{a}_y E_{y0} \cos(\omega t - kz + \theta) \tag{11.29}$$

where E_{x0} and E_{y0} are the wave amplitudes in the x and y directions and θ is the phase angle with which $\tilde{E}_y$ leads $\tilde{E}_x$. Choosing the plane $z = 0$ to evaluate the polarisation

$$\tilde{\mathbf{E}}(z = 0) = \mathbf{a}_x E_{x0} \cos(\omega t) + \mathbf{a}_y E_{y0} \cos(\omega t + \theta)$$
$$= \mathbf{a}_x E_{x0} \cos \omega t + \mathbf{a}_y E_{y0}(\cos \omega t \cos \theta - \sin \omega t \sin \theta) \tag{11.30}$$

$$\cos \omega t = \frac{\tilde{E}_x}{E_{x0}} \tag{11.31}$$

$$\sin \omega t = \sqrt{1 - \left(\frac{\tilde{E}_x}{E_{x0}}\right)^2} \tag{11.32}$$

$$\frac{\tilde{E}_y}{E_{y0}} = \cos \omega t \cos \theta - \sin \omega t \sin \theta$$

$$= \frac{\tilde{E}_x}{E_{x0}} \cos \theta - \sqrt{1 - \left(\frac{\tilde{E}_x}{E_{x0}}\right)^2} \sin \theta \tag{11.33}$$

or
$$\left(\frac{\tilde{E}_y}{E_{y0}}\right)^2 + \left(\frac{\tilde{E}_x}{E_{x0}} \cos \theta\right)^2 - 2\left(\frac{\tilde{E}_y}{E_{y0}}\right)\left(\frac{\tilde{E}_x}{E_{x0}}\right) \cos \theta = \left[1 - \left(\frac{\tilde{E}_x}{E_{x0}}\right)^2\right] \sin^2 \theta \tag{11.34}$$

$$\left(\frac{\tilde{E}_y}{E_{y0}\sin\theta}\right)^2 + \left(\frac{\tilde{E}_x}{E_{x0}\sin\theta}\right)^2 - 2\left(\frac{\tilde{E}_y}{E_{y0}}\right)\left(\frac{\tilde{E}_x}{E_{x0}}\right)\frac{\cos\theta}{\sin^2\theta} = 1 \qquad (11.35)$$

Let us discuss these equations.

Case 1. $\theta = 0$ or π; E_{x0} and E_{y0} can be anything. This case gives us linear polarisation.

To illustrate this statement we use Eqn (11.34) since Eqn (11.35) is unsuitable. Letting $\sin\theta = 0$, we have

$$\left(\frac{\tilde{E}_y}{E_{y0}}\right)^2 + \left(\frac{\tilde{E}_x}{E_{x0}}\cos\theta\right)^2 - 2\left(\frac{\tilde{E}_y}{E_{y0}}\right)\left(\frac{\tilde{E}_x}{E_{x0}}\cos\theta\right) = 0$$

$$\left(\frac{\tilde{E}_y}{E_{y0}} - \frac{\tilde{E}_x}{E_{x0}}\cos\theta\right)^2 = 0$$

$$\frac{\tilde{E}_y}{E_{y0}} = \frac{\tilde{E}_x}{E_{x0}}\cos\theta$$

$$\tilde{E}_y = \begin{cases} \dfrac{E_{y0}}{E_{x0}}\tilde{E}_x & \text{for } \theta = 0 \\[2mm] -\dfrac{E_{y0}}{E_{x0}}\tilde{E}_x & \text{for } \theta = \pi \end{cases} \qquad (11.36)$$

which is a straight line between the variables $\tilde{E}_y$ and $\tilde{E}_x$. This is shown in Figs 11.6 (a) and (b).

Case 2. $\theta = \pm\pi/2$, E_{x0} and E_{y0} can be anything. This gives elliptical polarisation with the major and minor axes of the ellipse along the coordinate axes. We use

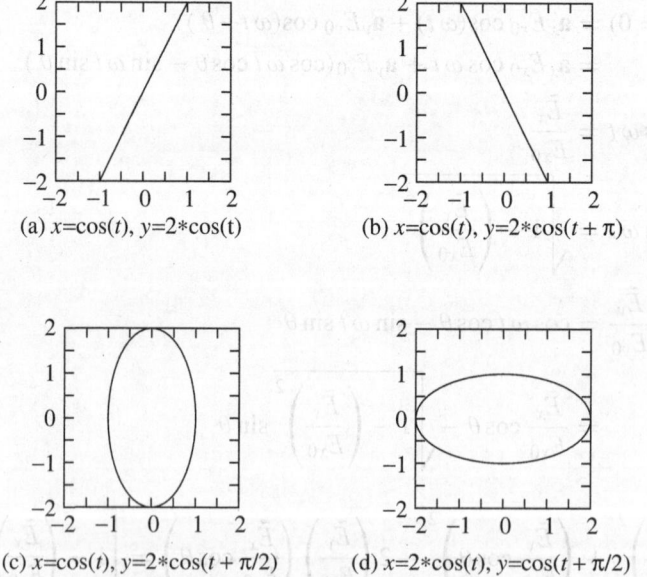

(a) $x=\cos(t)$, $y=2*\cos(t)$

(b) $x=\cos(t)$, $y=2*\cos(t+\pi)$

(c) $x=\cos(t)$, $y=2*\cos(t+\pi/2)$

(d) $x=2*\cos(t)$, $y=\cos(t+\pi/2)$

Fig. 11.6 Linear and elliptical polarisations

Eqn (11.35) with $\cos\theta = 0$ and $\sin\theta = \pm 1$

$$\left(\frac{\tilde{E}_y}{E_{y0}}\right)^2 + \left(\frac{\tilde{E}_x}{E_{x0}}\right)^2 = 1 \tag{11.37}$$

which is the equation of an ellipse with major and minor axis along the x and y directions, respectively. This is shown in Figs 11.6(c) and (d). If $E_{y0} = E_{x0}$ then we obviously get circular polarisation.

Case 3. θ, E_{y0}, E_{x0} all of which can be anything. In this general case, we have elliptical polarisation, with the major axis of the ellipse tilted in any general direction. The case for $\theta = \pi/6$, $E_{y0} = 2$ and $E_{x0} = 1.5$ is shown in Fig. 11.7.

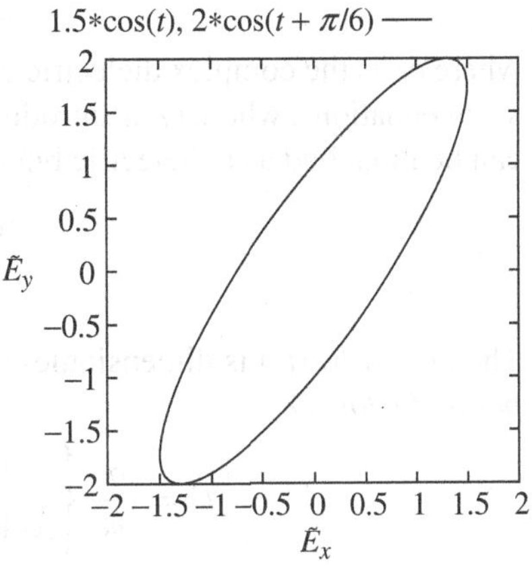

Fig. 11.7 The case for $\theta = \pi/6$, $E_{y0} = 2$ and $E_{x0} = 1.5$

Example 11.9 If $E_x = 5\angle 15°$ and $E_y = 7\angle 90°$, find the type of polarisation.

Solution We first remove the excess phase by 15°. $E_x = 5\angle 0°$ and $E_y = 7\angle 75°$. Then

$$\tilde{E}_x = 5\cos\omega t$$

$$\frac{\tilde{E}_x}{5} = \cos\omega t$$

$$\tilde{E}_y = 7\cos(\omega t + 75°)$$

$$\begin{aligned}\frac{\tilde{E}_y}{7} &= \cos\omega t\cos 75° - \sin\omega t\sin 75°\\ &= 0.2588\cos\omega t - 6.7615\sin\omega t\\ &= 0.2588\left(\frac{\tilde{E}_x}{5}\right) - 0.9659\sqrt{1 - \left(\frac{\tilde{E}_x}{5}\right)^2}\end{aligned}$$

from the previous theory we realise that this is elliptical polarisation.

11.5 Wave Propagation in Conducting Media

In conducting media, as the wave propagates, currents are set up. These currents circulate in the medium and produce heat. Thus, energy carried by the wave must necessarily get diminished. We already know that where an electric field is present in a medium with conductivity σ, there the current density generated at a point in the medium is related to the electric field through the equation

$$\mathbf{J} = \sigma\,\mathbf{E}$$

Using Maxwell's Eqn (11.67), in phasor form

$$\begin{aligned}\nabla\times\mathbf{H} &= j\omega\varepsilon\,\mathbf{E} + \mathbf{J}\\ &= j\omega\varepsilon\,\mathbf{E} + \sigma\,\mathbf{E}\end{aligned} \tag{11.38}$$

$$= j\omega \left(\varepsilon - \frac{j\sigma}{\omega} \right) \mathbf{E} \tag{11.39}$$

$$= j\omega \varepsilon \left(1 - \frac{j\sigma}{\omega \varepsilon} \right) \mathbf{E} \tag{11.40}$$

$$= j\omega \varepsilon_C \, \mathbf{E} \tag{11.41}$$

where ε_C is the complex dielectric constant which is so called because of the above set of equations, where ε_C is introduced to simplify things. The conducting medium can be modelled as a dielectric but one with a complex permittivity.

$$\varepsilon_C = \varepsilon' - j\varepsilon'' \tag{11.42}$$

$$= \varepsilon - j\frac{\sigma}{\omega} \tag{11.43}$$

The ratio $\sigma/(\omega\varepsilon)$ is dimensionless (the same as ε_r). This ratio is called the *dissipation factor*, D

$$D = \frac{\sigma}{\omega\varepsilon} \begin{cases} \ll 1, & \text{for a good dielectric} \\ \gg 1, & \text{for a good conductor} \end{cases} \tag{11.44}$$

Generally, the dissipation factor for lossy dielectrics is also called its loss tangent.

$$D = \frac{\sigma}{\omega\varepsilon} = \frac{\varepsilon''}{\varepsilon'} \tag{11.45}$$

Why it is so called may be seen from examining Fig. 11.8.

Using the notation of β_C as the *complex propagation constant*

$$\beta_C = \beta_R + j\beta_I = \beta + j\alpha$$
$$= \omega\sqrt{\mu\varepsilon_C}$$

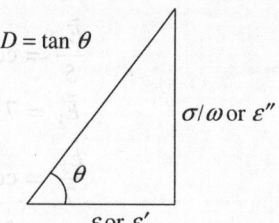

$D = \tan\theta$

σ/ω or ε''

ε or ε'

Fig. 11.8 Loss tangent for a dielectric

To compute β_C

$$\beta_C^2 = (\beta + j\alpha)^2$$
$$= \beta^2 - \alpha^2 + j2\beta\alpha$$
$$\omega^2 \mu\varepsilon_C = \beta^2 - \alpha^2 + j2\beta\alpha$$
$$\omega^2 \mu \left(\varepsilon - j\frac{\sigma}{\omega} \right) = \beta^2 - \alpha^2 + j2\beta\alpha$$

Equating the real and imaginary parts, and solving the resulting equations, we get

$$\beta = \beta_R = \frac{\omega\sqrt{\mu\varepsilon}\sqrt{\sqrt{1 + \left(\frac{\sigma}{\omega\varepsilon}\right)^2} + 1}}{\sqrt{2}} \tag{11.46}$$

$$\alpha = \beta_I = \frac{\omega\sqrt{\mu\varepsilon}\sqrt{\sqrt{1 + \left(\frac{\sigma}{\omega\varepsilon}\right)^2} - 1}}{\sqrt{2}} \tag{11.47}$$

What is the meaning of a complex propagation constant β_C? Let us examine a wave travelling in the z-direction, with the electric field in the x-direction. The wave in complex notation is then

$$E_x = E_{x0}e^{-j\beta_C z}$$

$$= E_{x0}e^{-j(\beta - j\alpha)z}$$
$$= E_{x0}e^{-\alpha z}e^{-j\beta z} \qquad (11.48)$$

Using sines and cosines, the wave in the real world is

$$\tilde{E}_x = \Re[E_{x0}e^{-\alpha z}e^{-j\beta z}e^{j\omega t}]$$
$$= E_{x0}\Re[e^{-\alpha z}e^{j(\omega t - \beta z)}]$$
$$= E_{x0}e^{-\alpha z}\cos(\omega t - \beta z) \qquad (11.49)$$

Observing the previous equation we can see that as the wave progresses in the z direction, the electric field is still oscillating (the cosine factor), but its amplitude decays (the exponential factor). This is shown in Fig. 11.9. The figure shows a snapshot at some time instant of a wave incident from air into a conductive medium. As the wave progresses into the medium, the following happens

1. The wavelength of the wave decreases. This is so because

$$\beta_{\text{Cond Med}} = \frac{2\pi}{\lambda_C} = \frac{\omega\sqrt{\mu\varepsilon}\sqrt{\sqrt{1+\left(\frac{\sigma}{\omega\varepsilon}\right)^2}+1}}{\sqrt{2}} > \omega\sqrt{\mu\varepsilon} \quad \left(=\frac{2\pi}{\lambda}\right)$$

where λ_C is the wavelength in the conducting medium.

2. The wave amplitude decays as $\exp(-\alpha z)$. The term $\exp(-\alpha z)$ forms an envelope of the cosine term.

@ See `AttenuationConstant.m` and `DissipationFactor.m` in Chapter 11

Concentrating on the decaying term, the amplitude of the wave falls by $E_{x0}\exp(-1)$ in a distance given by

$$z_{\text{skin depth}} = \delta = 1/\alpha$$
$$E_{x0}e^{-\delta\alpha} = E_{x0}e^{-1}$$
$$= 0.3679E_{x0}$$

δ is called the *skin depth* and depending on the value of α, the skin depth varies. Let us obtain the value of the magnetic field. Using the notation $\partial x \equiv \partial/\partial x$ etc.,

$$\mathbf{H} = -\frac{1}{j\omega\mu}\nabla \times \mathbf{E}$$
$$= -\frac{1}{j\omega\mu}[\mathbf{a}_z(\partial_x E_y - \partial_y E_x) + \mathbf{a}_x(\partial_y E_z - \partial_z E_y) + \mathbf{a}_y(\partial_z E_x - \partial_x E_z)]$$

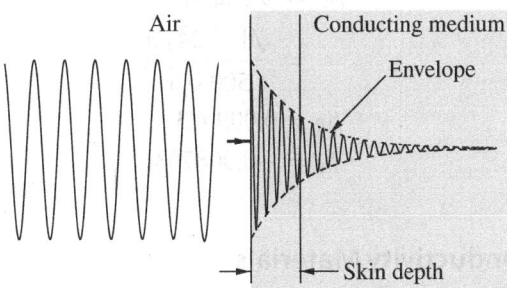

Fig. 11.9 Profile of a wave propagating from air to a conducting medium

$$= -\frac{1}{j\omega\mu}\left[\mathbf{a}_z\underbrace{(\partial_x E_y - \partial_y E_x)}_{\partial_x = 0,\ \partial_y = 0} + \mathbf{a}_x\underbrace{(\partial_y E_z - \partial_z E_y)}_{\partial_y = 0,\ E_y = 0} + \mathbf{a}_y\underbrace{(\partial_z E_x - \partial_x E_z)}_{\partial_z = -j\beta_C,\ E_z = 0}\right]$$

In the first factor the E-field is not a function of either x or y. The same argument applies to the next factor taking into account the additional fact that $E_y = 0$. In the last bracket, differentiation with respect to z is the same as multiplication by $-j\beta_C$. Therefore,

$$\mathbf{H} = -\frac{1}{j\omega\mu}[\mathbf{a}_y(-j\beta_C E_x)]$$

$$= \mathbf{a}_y\frac{\beta_C}{\omega\mu}E_x$$

$$= \mathbf{a}_y\sqrt{\frac{\varepsilon_C}{\mu}}E_x \qquad (11.50)$$

which means that the characteristic impedance is complex

$$Z_C = \sqrt{\frac{\mu}{\varepsilon_C}} = \sqrt{\frac{\mu}{\varepsilon' - j\varepsilon''}} = \sqrt{\frac{\mu}{\varepsilon - j(\sigma/\omega)}}$$

The meaning of a complex characteristic impedance is that the electric and magnetic fields are not in phase with each other. Thus, in a propagating wave, propagating in the $\hat{\mathbf{n}}$ direction if the electric field is $\mathbf{E}$ then the magnetic field is $(1/Z)\hat{\mathbf{n}} \times \mathbf{E}$. Let us apply what we have learnt to different materials.

Example 11.10 A wave of frequency $f = 1$ MHz travels in a material with $\sigma = 2 \times 10^{-5}$ ℧/m and $\varepsilon_r = 15$. Compute the complex propagation constant.
Solution For the material

$$\varepsilon = \varepsilon_0\,\varepsilon_r$$

$$= [15/(36\pi)] \times 10^{-12}$$

$$= [1/(24\pi)] \times 10^{-11}$$

and

$$\varepsilon_C = \varepsilon\left(1 - \frac{j\sigma}{\omega\varepsilon}\right)$$

$$= \left(\frac{1 - 24j}{24\pi}\right) \times 10^{-11}$$

The formula for β is

$$\beta_C = \omega\sqrt{\mu_0\,\varepsilon_C}$$

$$= \frac{\sqrt{1 - 24j}\,\pi}{500\sqrt{6}}$$

with

$$\beta_R = 0.009073$$

$$\beta_I = -0.008703$$

11.5.1 Low Conductivity Materials

Low conductivity materials are generally lossy dielectrics with a small dissipation factor. Consider such a dielectric. In the following equations, we have used the

Taylor series expansions, where $x \ll 1$

$$\sqrt{1+x} \approx 1 + \frac{x}{2} \qquad (11.51)$$

$$\frac{1}{1+x} \approx 1 - x \qquad (11.52)$$

Reproducing Eqns (11.46) and (11.47) with

$$\frac{\sigma}{\omega\varepsilon} = \frac{\varepsilon'}{\varepsilon''} \ll 1$$

$$\beta = \frac{\omega\sqrt{\mu\varepsilon}\sqrt{\sqrt{1+\left(\frac{\sigma}{\omega\varepsilon}\right)^2}+1}}{\sqrt{2}}$$

$$\approx \frac{\omega\sqrt{\mu\varepsilon}\sqrt{\left[1+\frac{1}{2}\left(\frac{\sigma}{\omega\varepsilon}\right)^2\right]+1}}{\sqrt{2}} \qquad \text{(Taylor's expansion of sq. root)}$$

$$= \omega\sqrt{\mu\varepsilon}\sqrt{1+\frac{1}{4}\left(\frac{\sigma}{\omega\varepsilon}\right)^2}$$

$$\approx \omega\sqrt{\mu\varepsilon}\left[1+\frac{1}{8}\left(\frac{\sigma}{\omega\varepsilon}\right)^2\right] \qquad \text{(Taylor's expansion of sq. root)} \qquad (11.53)$$

$$\alpha = \frac{\omega\sqrt{\mu\varepsilon}\sqrt{\sqrt{1+\left(\frac{\sigma}{\omega\varepsilon}\right)^2}-1}}{\sqrt{2}}$$

$$\approx \frac{\omega\sqrt{\mu\varepsilon}\sqrt{\left[1+\frac{1}{2}\left(\frac{\sigma}{\omega\varepsilon}\right)^2\right]-1}}{\sqrt{2}}$$

$$= \left(\frac{\omega\sqrt{\mu\varepsilon}}{2}\right)\left(\frac{\sigma}{\omega\varepsilon}\right) \qquad (11.54)$$

The characteristic impedance

$$Z = \sqrt{\frac{\mu}{\varepsilon_C}}$$

$$= \sqrt{\frac{\mu}{\varepsilon}} \times \frac{1}{\sqrt{1-j(\sigma/\omega\varepsilon)}}$$

$$= \sqrt{\frac{\mu}{\varepsilon}}\left(1+j\frac{\sigma}{2\omega\varepsilon}\right) \qquad \text{(applying the Taylor series expansion)}$$

Example 11.11 Apply these results to a case of a wave travelling in a dielectric with a loss tangent of 0.05 and a dielectric constant of 2. Let the frequency of the wave be 10 GHz. The loss tangent is $0.05 \ll 1$.

Solution The permeability $\mu \equiv \mu_0$; the permittivity $\varepsilon = \varepsilon_r \varepsilon_0 = 2\varepsilon_0$.

$$\omega = 2\pi f = 2\pi \times 10^6 \text{ rad/s}$$

also (see Fig. 11.8)

$$\frac{\sigma}{\omega\varepsilon} = 0.05 \ll 1$$

$$\beta_0 = \omega \sqrt{\mu_0 \varepsilon} = 298.3412\pi$$

$$\beta = \beta_0 \left[1 + \frac{1}{8}\left(\frac{\sigma}{\omega\varepsilon}\right)^2\right] = 298.4345\pi$$

$$\alpha \approx \frac{\beta_0}{2}\left(\frac{\sigma}{\omega\varepsilon}\right) = 7.4585\pi$$

$$\delta = 1/\alpha = 0.04268 \text{ m} \quad (\text{Skin depth})$$

$$\sqrt{\frac{\mu_0}{\varepsilon}} = 266.6 \ \Omega$$

The characteristic impedance

$$Z = \sqrt{\frac{\mu_0}{\varepsilon_C}}$$

$$= \sqrt{\frac{\mu_0}{\varepsilon}} \times \frac{1}{\sqrt{1 - j(\sigma/\omega\varepsilon)}}$$

$$\approx \sqrt{\frac{\mu_0}{\varepsilon}}\left(1 + j\frac{\sigma}{2\omega\varepsilon}\right)$$

$$= 266.6(1 + j0.0025)$$

$$= 266.6\angle 0.14°$$

The magnetic field may be obtained from

$$\mathbf{H} = \frac{1}{Z}(\hat{\mathbf{n}} \times \mathbf{E}) = \sqrt{\frac{\varepsilon_C}{\mu_0}}(\hat{\mathbf{n}} \times \mathbf{E})$$

11.5.2 High Conductivity Materials

For the case of high conductivity materials like metals,

$$\frac{\sigma}{\omega\varepsilon} \gg 1$$

and

$$\beta = \frac{\omega\sqrt{\mu\varepsilon}\sqrt{\sqrt{1 + \left(\frac{\sigma}{\omega\varepsilon}\right)^2} + 1}}{\sqrt{2}}$$

$$\approx \frac{\omega\sqrt{\mu\varepsilon}\sqrt{\left(\frac{\sigma}{\omega\varepsilon}\right) + 1}}{\sqrt{2}} \quad \left(\text{since } \frac{\sigma}{\omega\varepsilon} \gg 1\right)$$

$$\approx \frac{\omega\sqrt{\mu\varepsilon}\sqrt{\left(\frac{\sigma}{\omega\varepsilon}\right)}}{\sqrt{2}} \quad \left(\text{since } \frac{\sigma}{\omega\varepsilon} \gg 1\right)$$

$$\approx \sqrt{\frac{\omega\mu\sigma}{2}} \tag{11.55}$$

and

$$\alpha = \frac{\omega\sqrt{\mu\varepsilon}\sqrt{\sqrt{1 + \left(\frac{\sigma}{\omega\varepsilon}\right)^2} - 1}}{\sqrt{2}}$$

$$\approx \frac{\omega\sqrt{\mu\varepsilon}\sqrt{\left(\frac{\sigma}{\omega\varepsilon}\right) - 1}}{\sqrt{2}}$$

$$\approx \frac{\omega \sqrt{\mu \varepsilon} \sqrt{\left(\frac{\sigma}{\omega \varepsilon}\right)}}{\sqrt{2}}$$

$$= \sqrt{\frac{\omega \mu \sigma}{2}} \tag{11.56}$$

Therefore, for high conductivity materials

$$\alpha = \beta = \sqrt{\frac{\omega \mu \sigma}{2}} \tag{11.57}$$

The characteristic impedance is given by

$$Z = \sqrt{\frac{\mu_0}{\varepsilon_C}}$$

$$= \sqrt{\frac{\mu_0}{\varepsilon_0 - j(\sigma/\omega)}}$$

$$= \sqrt{\frac{\mu_0}{\varepsilon_0}} \sqrt{\frac{1}{1 - j\left(\frac{\sigma}{\omega \varepsilon_0}\right)}}$$

$$\approx Z_0 \sqrt{j} \sqrt{\frac{\omega \varepsilon_0}{\sigma}}$$

$$= Z_0 \left(\frac{1+j}{\sqrt{2}}\right) \sqrt{\frac{\omega \varepsilon_0}{\sigma}}$$

The characteristic impedance of a high conductivity material is therefore

$$Z \approx Z_0 \left(\frac{1+j}{\sqrt{2}}\right) \sqrt{\frac{\omega \varepsilon_0}{\sigma}} \; \Omega$$

where Z_0 is the characteristic impedance of free space ($\approx 377 \, \Omega$).

Example 11.12 Taking the example of copper with a conductivity $\sigma = 5.814 \times 10^7$ ℧/m; $\varepsilon \equiv \varepsilon_0$ at a frequency of $f = 1$ MHz ($\omega = 6.28 \times 10^6$ rad/s), find β, the propagation constant; δ, the skin depth and λ, the wavelength.

Solution

$$\sigma /(\omega \varepsilon_0) = 1.045 \times 10^{12} \gg 1$$

and so

$$\beta = \alpha \approx \sqrt{\frac{\omega \mu \sigma}{2}} = 15150$$

The skin depth

$$\delta = 1/\alpha = 6.6 \times 10^{-5} \, \text{m} \tag{11.58}$$

The wave decays to $e^{-1} = 0.369$ in a distance of 0.066 mm. The wavelength in air is

$$\lambda_{\text{air}} = \frac{c}{f} = \frac{3 \times 10^8}{1 \times 10^6} = 300 \, \text{m}$$

the wavelength in copper at the same frequency is

$$\lambda_{\text{copper}} = \frac{2\pi}{\beta}$$

$$= \frac{2\pi}{15150}$$

$$= 4.15 \times 10^{-4} \, \text{m}$$

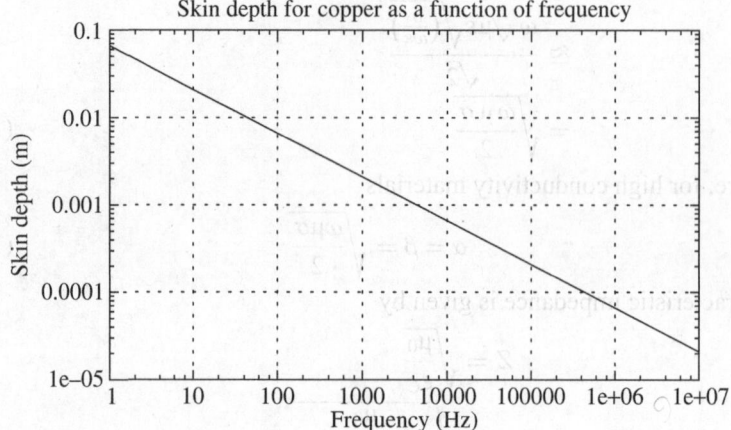

Fig. 11.10 Skin depth for copper as a function of frequency

Example 11.13 Find the skin depth as a function of frequency for copper.

Solution The skin depth by Eqn (11.58) is given by

$$\delta = 1/\alpha = \sqrt{\frac{2}{\omega \mu_0 \sigma}}$$

A typical example of the skin depth as a function of frequency for copper is shown in Fig. 11.10. As we can see from the figure, the skin depth rapidly decreases with increasing frequency, and at dc (zero frequency) the skin depth becomes infinity. On the other hand as the frequency is increased into the GHz range, the skin depth goes into the μm range. It is for this reason that in many microwave components, the conducting surface is coated with a very thin layer of gold (in which the wave penetrates) to minimise the energy dissipated. In high conductivity materials, the wavelength and skin depth are related by

$$\lambda_{\text{material}} = 2\pi \delta_{\text{material}} \tag{11.59}$$

The characteristic impedance for copper is given by

$$Z \approx Z_0 \left(\frac{1+j}{\sqrt{2}}\right) \sqrt{\frac{\omega \varepsilon_0}{\sigma}}$$
$$= 377 \times 1\angle 45° \times 9.57 \times 10^{-13}$$
$$= 3.61 \times 10^{-10} \angle 45°$$

11.6 | Boundary Conditions

When we consider electromagnetic problems we need to look at the conditions of the electromagnetic fields at the boundary between two media. Principally what we are interested in is: what are the relations between the electromagnetic field components in the two media which are (a) tangential to the surface separating the two media? and (b) normal to this surface?

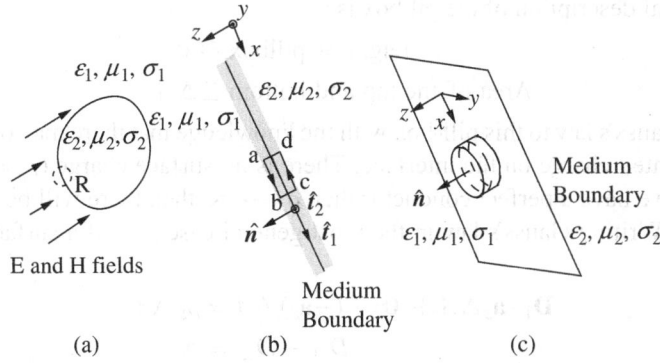

Fig. 11.11 The behaviour of electromagnetic fields near a boundary consisting of a change of medium

There is one special case which we always have to keep in mind: the case of a perfect electric conductor, the conductivity, $\sigma \to \infty$. In this case, no *time varying fields* can exist inside a conductor: $\mathbf{E}$ and $\mathbf{H}$ will both be zero. In this case there will be both a surface charge, ρ_s, as well as a surface current, $\mathbf{J}_s$, at the boundary of the two media.

To investigate these problems we take a look at the case where electromagnetic fields are present inside and outside a body with permittivity and permeability (ε_2, μ_2, σ_2). This we call region 2. The body is immersed in a region with permittivity and permeability (ε_1, μ_1, σ_1), called region 1. This configuration is shown in Fig. 11.11. In the (a) part of the figure the macro level diagram is shown where a small region labelled $\mathcal{R}$ is shown. This region is shown blown up in parts (b) and (c) of the figure.

The fields in regions 1 and 2 must satisfy Maxwell's equations.

$$\nabla \cdot \mathbf{D} = \rho_v$$

$$\nabla \times \mathbf{E} = -\frac{\partial \mathbf{B}}{\partial t}$$

$$\nabla \cdot \mathbf{B} = 0$$

$$\nabla \times \mathbf{H} = \frac{\partial \mathbf{D}}{\partial t} + \mathbf{J} \qquad (11.60)$$

These are the differential form of the equations. The integral forms are

$$\iiint \nabla \cdot \mathbf{D} dV = \oiint \mathbf{D} \cdot d\mathbf{S} = \iiint \rho_v \, dV$$

$$\iint \nabla \times \mathbf{E} \cdot d\mathbf{S} = \oint \mathbf{E} \cdot d\mathbf{l} = -\iint \frac{\partial \mathbf{B}}{\partial t} \cdot d\mathbf{S}$$

$$\iiint \nabla \cdot \mathbf{B} dV = \oiint \mathbf{B} \cdot d\mathbf{S} = 0$$

$$\iint \nabla \times \mathbf{H} \cdot d\mathbf{S} = \oint \mathbf{H} \cdot d\mathbf{l} = \iint \frac{\partial \mathbf{D}}{\partial t} \cdot d\mathbf{S} + \iint \mathbf{J} \cdot d\mathbf{S} \qquad (11.61)$$

We first take a look at the first of the above equations (Maxwell's equations in integral form). The equation is applied to the pillbox shown in part (c) of the figure. The mathematical description of the pillbox is

$$\text{Height of pillbox} \approx 0$$

$$\text{Area of the top and bottom} \approx \Delta A$$

Applying Gauss's law to this pill-box with the knowledge that there may or may not be accumulated charge on the interface. There is no surface charge ($\rho_s = 0$) if σ_2 is finite. If we have a perfect conductor then $\sigma_2 \to \infty$ then there will be a surface charge, ρ_s. Writing Gauss's law in the most general case (i.e., if a surface charge exists)

$$\mathbf{D}_1 \cdot \mathbf{a}_z \Delta A + \mathbf{D}_2 \cdot (-\mathbf{a}_z) \Delta A = \rho_s \, \Delta A \tag{11.62}$$

$$D_{z1} - D_{z2} = \rho_s \tag{11.63}$$

in other words, *the normal component of the flux density is discontinuous across a dielectric–dielectric boundary by the amount of the surface charge density.* Writing the above equation in a more general form

$$D_{n1} - D_{n2} = \rho_s$$

$$\hat{\mathbf{n}} \cdot (\mathbf{D}_1 - \mathbf{D}_2) = \rho_s \tag{11.64}$$

If there is no surface charge ($\rho_s = 0$), then the right-hand side is replaced by zero and in that case, the normal component of the **D** field is continuous.

Let us take the case where the second medium is a perfect electric conductor. Then there will be no fields there and $D_{n2} = 0$ but there will be a surface charge, ρ_s. So

$$D_{n1} = \rho_s \tag{11.65}$$

Now considering the the second of the Maxwell's equations in integral form (the second of the Eqn set (11.61)) to part (b) of Fig. 11.11. In this case

$$\text{Length } (b - c) = \text{Length } (d - a) \approx 0$$

$$\text{Length } (a - b) = \text{Length } (c - d) \approx |\Delta x|$$

Applying the line integral to the loop a-b-c-d in the anti-clockwise direction,

$$\mathbf{E}_1 \cdot \mathbf{a}_x |\Delta x| + \mathbf{E}_2 \cdot (-\mathbf{a}_x) |\Delta x| = 0$$

where $\mathbf{E}_1$ and $\mathbf{E}_2$ are the electric fields in regions 1 and 2, respectively. The right-hand side of the integral

$$\iint \frac{\partial \mathbf{B}}{\partial t} \cdot d\mathbf{S} \leqslant \iint \left| \frac{\partial \mathbf{B}}{\partial t} \right| \times |d\mathbf{S}| = 0$$

since *one side enclosing the area is vanishingly small, the area is vanishingly small,* i.e.,

$$|d\mathbf{S}| \approx \Delta x \times \text{Length } (b - c) = 0 \tag{11.66}$$

is zero. The condition on the electric field may then be summarised as

$$E_{t1} - E_{t2} = 0 \tag{11.67}$$

The subscript t is used to signify the *tangential* component. $E_{t1,2}$ are the tangential components of the electric field next to the boundary but in media 1 and 2 respectively. Writing this in a more compact form

$$\hat{\mathbf{n}} \times (\mathbf{E}_1 - \mathbf{E}_2) = 0 \tag{11.68}$$

Notice that the *normal goes from medium 2 to medium 1.*

Let us take the case where the second medium is a perfect electric conductor. Then there will be no fields there and $E_{t2} = 0$. So

$$E_{t1} = 0 \qquad (11.69)$$

Taking a look at the next Maxwell equation

$$\iiint \nabla \cdot \mathbf{B} dV = \oiint \mathbf{B} \cdot d\mathbf{S} = 0$$

we apply the same arguments as for the $\mathbf{D}$ vector and get

$$B_{n1} - B_{n2} = 0 \qquad (11.70)$$

or

$$\hat{\mathbf{n}} \cdot (\mathbf{B}_1 - \mathbf{B}_2) = 0 \qquad (11.71)$$

Let us take the case where the second medium is a perfect electric conductor. Then there will be no fields there and $B_{n2} = 0$. So

$$B_{n1} = 0 \qquad (11.72)$$

Finally we treat the last Maxwell equation

$$\iint \nabla \times \mathbf{H} \cdot d\mathbf{S} = \oint \mathbf{H} \cdot d\mathbf{l} = \iint \frac{\partial \mathbf{D}}{\partial t} \cdot d\mathbf{S} + \iint \mathbf{J} \cdot d\mathbf{S}$$

Integrating over the small loop $abcd$ in the anticlockwise sense,

$$H_{x1}|\Delta x| - H_{x2}|\Delta x| = \frac{\partial D_y}{\partial t} \Delta A + J_y \Delta A + J_{sy}|\Delta x|$$

Here $|\Delta x|$ is the longer side of the loop and ΔA is the area of the loop. J_y is the y-directed volume current while J_{sy} is the surface current just at the boundary. The line integrals of $\mathbf{H}$ over the shorter sides of the loop are equated to zero since there lengths are virtually zero. Using the same argument, ΔA is also considered to be zero. Therefore

$$H_{x1}|\Delta x| - H_{x2}|\Delta x| = J_{sy}|\Delta x|$$

or

$$H_{t_1 1} - H_{t_1 2} = J_{st_2} \qquad (11.73)$$

$\hat{\mathbf{t}}_1, \hat{\mathbf{t}}_2$, and $\hat{\mathbf{n}}$ form a right-handed orthogonal coordinate system. $\hat{\mathbf{t}}_1, \hat{\mathbf{t}}_2$, and $\hat{\mathbf{n}}$ are akin to $\mathbf{a}_x, \mathbf{a}_y$, and $\mathbf{a}_z$. Scrutinising the figure, J_{st_2} is the y- (or $\hat{\mathbf{t}}_2$-) directed surface current. Writing equation in vector notation

$$\mathbf{a}_z \times (\mathbf{H}_1 - \mathbf{H}_2) = \mathbf{J}_s \qquad (11.74)$$

or

$$\hat{\mathbf{n}} \times (\mathbf{H}_1 - \mathbf{H}_2) = \mathbf{J}_s \qquad (11.75)$$

Let us take the case where the second medium is a perfect electric conductor. Then there will be no fields there and $\mathbf{H}_2 = 0$. But in this case there will be the surface current $\mathbf{J}_s$. So

$$\hat{\mathbf{n}} \times \mathbf{H}_1 = \mathbf{J}_s \qquad (11.76)$$

note that $\hat{\mathbf{n}}$ will be from the pec (perfect electric conductor) into medium 1.

11.7 | Reflection and Refraction of Waves

11.7.1 Reflection from a Metal Surface

Electromagnetic waves have the same nature as light and therefore reflect from metallic objects and suffer refraction in the presence of dielectrics. The simplest case is one where a plane wave falls on a metal or dielectric plane surface. Let us consider first the case of a uniform plane wave obliquely incident on a air metal boundary. For the sake of convenience the metal may be considered to be a perfect one with $\sigma \to \infty$. The wave may approach the boundary in either of two configurations: where the electric field is perpendicular to the plane of incidence or when it is parallel to the plane of incidence. Both configurations are shown in Fig. 11.12.

We need to look at a plane wave travelling in a general direction whose unit vector is $\hat{\mathbf{n}}$. The equation of a plane is

$$f(x, y, z) = ax + by + cz = \text{constant} = p \qquad (11.77)$$

then the normal to this plane is

$$\nabla f = a\mathbf{a}_x + b\mathbf{a}_y + c\mathbf{a}_z \qquad (11.78)$$

(Recall that the normal to any surface $f(x, y, z)$ is ∇f) and the unit normal to the plane is

$$\hat{\mathbf{n}} = \frac{a\mathbf{a}_x + b\mathbf{a}_y + c\mathbf{a}_z}{\sqrt{a^2 + b^2 + c^2}} \qquad (11.79)$$

note that $a/\sqrt{a^2 + b^2 + c^2}$ is the direction cosine of the vector $\hat{\mathbf{n}}$ in the x-direction, and the other direction cosines are $b/\sqrt{a^2 + b^2 + c^2}$ and $c/\sqrt{a^2 + b^2 + c^2}$. The equation of the plane, Eqn (11.77), may now be written as

$$\hat{\mathbf{n}} \cdot \mathbf{r} = \frac{p}{\sqrt{a^2 + b^2 + c^2}} = p' \qquad (11.80)$$

where p' is the shortest distance of the plane from the origin. With this introduction, the equation of a uniform plane wave travelling in the $\hat{\mathbf{n}}$ direction is

$$\mathbf{E} = \mathbf{E}_0 e^{-jk\hat{\mathbf{n}} \cdot \mathbf{r}} \qquad (11.81)$$

where $\mathbf{E}_0$ is perpendicular to $\hat{\mathbf{n}}$ and

$$\hat{\mathbf{n}} \cdot \mathbf{r} = \mathbf{a}_x x + \mathbf{a}_y y + \mathbf{a}_z z$$

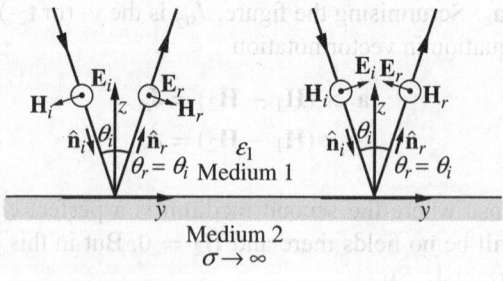

(a) Perpendicular
    polarisation

(b) Parallel
    polarisation

Fig. 11.12 A wave obliquely incident from air on a metal ($\sigma \to \infty$)

describes different planes all parallel to to each other. We can see for instance if the wave travels in the $\mathbf{a}_y$ direction then $\mathbf{a}_y \cdot \mathbf{r} = y$, then

$$\mathbf{E} = \mathbf{E}_0 e^{-jky}$$

and so on.

Example 11.14 Find the equation of the electric field with amplitude E_0 when the wave travels in the (a) x direction, (b) z direction and (c) in the direction $\mathbf{a}_x + \mathbf{a}_y + \mathbf{a}_z$.
Solution

(a) When the wave travels in the x-direction, $\hat{\mathbf{n}} = \mathbf{a}_x$, $\hat{\mathbf{n}} \cdot \mathbf{r} = x$, and $\mathbf{E} = E_0\hat{\mathbf{n}}_\perp$ $\exp\{-jkx\}$, where $\hat{\mathbf{n}}_\perp$ is a unit vector perpendicular to $\mathbf{a}_x$.
(b) When the wave travels in the z-direction, $\hat{\mathbf{n}} = \mathbf{a}_z$, $\hat{\mathbf{n}} \cdot \mathbf{r} = z$, and $\mathbf{E} = E_0\hat{\mathbf{n}}_\perp$ $\exp\{-jkz\}$, where $\hat{\mathbf{n}}_\perp$ is a unit vector perpendicular to $\mathbf{a}_z$.
(c) When the wave travels in the $\mathbf{a}_x + \mathbf{a}_y + \mathbf{a}_z$ direction, $\hat{\mathbf{n}} = (\mathbf{a}_x + \mathbf{a}_y + \mathbf{a}_z)/\sqrt{3}$, $\hat{\mathbf{n}} \cdot \mathbf{r} = (x + y + z)/\sqrt{3}$ and $\mathbf{E} = E_0\hat{\mathbf{n}}_\perp \exp\{-jk(x + y + z)/\sqrt{3}\}$, where $\hat{\mathbf{n}}_\perp$ is a unit vector perpendicular to $\hat{\mathbf{n}}$.

Referring now to Fig. 11.12(a),

$$\hat{\mathbf{n}}_i = -\mathbf{a}_z \cos\theta_i + \mathbf{a}_y \sin\theta_i \tag{11.82}$$

$$\hat{\mathbf{n}}_r = \mathbf{a}_z \cos\theta_i + \mathbf{a}_y \sin\theta_i \tag{11.83}$$

$$\mathbf{E}_i = \mathbf{a}_x E_{x0} e^{-jk(y\sin\theta_i - z\cos\theta_i)} \tag{11.84}$$

Let the reflected E-field be

$$\mathbf{E}_r = \mathbf{a}_x RE_{x0} e^{-jk(y\sin\theta_i + z\cos\theta_i)} \tag{11.85}$$

where R is the reflection coefficient which is to be determined.

At the boundary of the metal, the tangential electric field must vanish as per the previous section. The coordinate system is: the y and z coordinates as shown in the figure and the x coordinate comes out of the plane of the paper. The metal surface is defined by the equation $z = 0$. Therefore,

$$(\mathbf{E}_i + \mathbf{E}_r)_{z=0} = 0 \qquad \text{or} \tag{11.86}$$

$$\left[\mathbf{a}_x E_{x0} e^{-jk(y\sin\theta_i - z\cos\theta_i)} + \mathbf{a}_x RE_{x0} e^{-jk(y\sin\theta_i + z\cos\theta_i)} \right]_{z=0} 0 \qquad \text{or} \tag{11.87}$$

$$R = -1 \tag{11.88}$$

the reflected electric field is therefore for perpendicular incidence,

$$\mathbf{E}_r = -\mathbf{a}_x E_{x0} e^{-jk(y\sin\theta_i + z\cos\theta_i)} \tag{11.89}$$

We now analyse the case of parallel incidence where the magnetic field is perpendicular to the plane of incidence as shown in Fig. 11.12(b). The electric field is given by

$$\mathbf{E}_{0i} = Z_0 \mathbf{H}_{0i} \times \hat{\mathbf{n}}_i$$
$$= Z_0 H_{x0i}[\mathbf{a}_x \times (-\mathbf{a}_z \cos\theta_i + \mathbf{a}_y \sin\theta_i)]$$
$$= Z_0 H_{x0i}(-\mathbf{a}_y \cos\theta_i + \mathbf{a}_z \sin\theta_i) \tag{11.90}$$

where $Z_0 = \sqrt{\mu/\varepsilon_1}$, ($377\,\Omega$), the intrinsic impedance of free space. We now proceed to apply the boundary condition at the metal boundary: the tangential electric field is zero at the boundary.

$$E_{y0r} = -Z_0 H_{x0i} \cos\theta_i \tag{11.91}$$

so

$$\mathbf{E}_{0r} = Z_0 H_{x0i}(-\mathbf{a}_y \cos\theta_i + \mathbf{a}_z \sin\theta_i) \quad (11.92)$$

and the reflected magnetic field is $\hat{\mathbf{n}}_r \times \mathbf{E}_{0r}/Z_0$

$$\mathbf{H}_{0r} = (\mathbf{a}_z \cos\theta_i + \mathbf{a}_y \sin\theta_i) \times H_{x0i}(-\mathbf{a}_y \cos\theta_i + \mathbf{a}_z \sin\theta_i)$$
$$= H_{x0i}(\cos^2\theta_i + \sin^2\theta_i)\,\mathbf{a}_x$$
$$= H_{x0i}\mathbf{a}_x \quad (11.93)$$

We notice that the reflected magnetic field remains as it is at $z = 0$! The incident and reflected magnetic fields are

$$\mathbf{H}_i = \mathbf{a}_x H_{x0} e^{-jk(y\sin\theta y_i - z\cos\theta_i)} \quad (11.94)$$
$$\mathbf{H}_r = \mathbf{a}_x H_{x0} e^{-jk(y\sin\theta_i + z\cos\theta_i)} \quad (11.95)$$

11.7.1.1 Normal Incidence

From the above discussion it is clear that for normal incidence, the two polarisations become one and the same. Referring to Eqns (11.84) and (11.85), we set $\theta_i = \theta_r = 0$,

$$\mathbf{E}_i = \mathbf{a}_x E_{x0} e^{+jkz}$$
$$\mathbf{E}_r = \mathbf{a}_x R E_{x0} e^{-jkz}$$

and set $R = -1$

$$\mathbf{E}_i = \mathbf{a}_x E_{x0} e^{+jkz}$$
$$\mathbf{E}_r = -\mathbf{a}_x E_{x0} e^{-jkz}$$

Adding these two waves,

$$\mathbf{E}_T = \mathbf{E}_i + \mathbf{E}_r$$
$$= \mathbf{a}_x E_{x0} e^{+jkz} - \mathbf{a}_x E_{x0} e^{-jkz}$$
$$= 2j\mathbf{a}_x E_{x0} \sin(kz)$$

which is a standing wave.

The electric field is zero on the surface of the metal and goes through maximas and minimas as we move vertically perpendicular to the metal surface. That is

$$\sin(kz) = \begin{cases} 1, & kz_{min} = \pi/2,\ 5\pi/2,\ \dots \\ -1, & kz_{max} = 3\pi/2,\ 7\pi/2,\ \dots \\ 0, & kz_{zero} = 0,\ \pi,\ 2\pi,\ \dots \end{cases}$$

Therefore, the minimas occur at

$$z_{min} = \lambda/4,\ 5\lambda/4,\ \dots$$

the maximas occur at

$$z_{max} = 3\lambda/4,\ 7\lambda/4,\ \dots$$

and the zeros at

$$z_{zero} = 0,\ \lambda/2,\ \lambda,\ \dots$$

since the electric field in real time is $\Re[\mathbf{E}_T \exp\{j\omega t\}]$ therefore

$$\mathbf{E}_T(z,t) = 2\mathbf{a}_x \sin(kz)\cos(\omega t + \pi/2)$$

which means that every half cycle the maxima becomes a minima and the minima becomes a maxima. The zeros remain where they are.

Example 11.15 For a wave of 1 GHz normally incident on a metal surface, find the maximas and minimas and zeros of the standing wave for $t = T/4$.

Solution For 1 GHz, the wavelength is

$$\lambda = c/f = 3 \times 10^8/1 \times 10^9 = 0.3 \text{ m}$$

and the time period T is

$$T = 1/f = 1 \text{ ns}$$

The equation of the electric field at $t = T/4$ is

$$E = 2 \sin(kz) \cos(\omega T/4 + \pi/2)$$

$$= 2 \sin\left(\frac{2\pi}{\lambda}z\right) \cos\left[\frac{2\pi}{T}(T/4) + \pi/2\right]$$

$$= 2 \sin\left(\frac{2\pi}{0.3}z\right) \cos[\pi] = -2 \sin\left(\frac{2\pi}{0.3}z\right)$$

Now the maximas are defined by

$$\frac{2\pi}{0.3}z_{max} = \pi/2,\ 5\pi/2,\ 9\pi/2 \ldots$$

$$z_{max\ 1} = 0.075 \text{ m}$$

$$z_{max\ 2} = 0.375 \text{ m}$$

$$z_{max\ 3} = 0.675 \text{ m}$$

$$\vdots$$

Similarly, the minimas are defined by

$$\frac{2\pi}{0.3}z_{min} = 3\pi/2,\ 7\pi/2,\ 11\pi/2 \ldots$$

$$z_{min\ 1} = 0.225 \text{ m}$$

$$z_{min\ 2} = 0.525 \text{ m}$$

$$z_{min\ 3} = 0.825 \text{ m}$$

$$\vdots$$

and the zeros are defined by

$$\frac{2\pi}{0.3}z_{zero} = 0,\ \pi,\ 2\pi, \ldots$$

$$z_{zero1} = 0 \text{ m}$$

$$z_{zero2} = 0.15 \text{ m}$$

$$z_{zero3} = 0.3 \text{ m}$$

$$\vdots$$

11.7.2 Refraction from a Dielectric Surface

11.7.2.1 Perpendicular Polarisation

We now consider a wave obliquely incident on a dielectric surface as shown in Fig. 11.13. First let us consider the case where the electric field is perpendicular

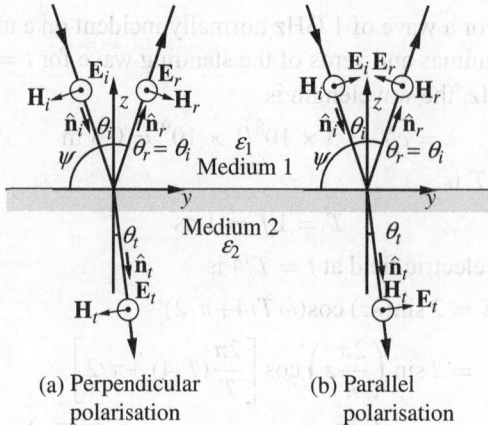

(a) Perpendicular polarisation

(b) Parallel polarisation

Fig. 11.13 A wave obliquely incident from air (ε_1) on a dielectric (ε_2)

to the plane of incidence as shown in part (a) of the figure. The incident, reflected, and transmitted electric fields are given by

$$\mathbf{E}_i = \mathbf{a}_x E_{x0i} e^{-jk(y \sin \theta_i - z \cos \theta_i)} \tag{11.96}$$

$$\mathbf{E}_r = \mathbf{a}_x E_{x0r} e^{-jk(y \sin \theta_i + z \cos \theta_i)} \tag{11.97}$$

$$\mathbf{E}_t = \mathbf{a}_x E_{x0t} e^{-jk(y \sin \theta_t - z \cos \theta_t)} \tag{11.98}$$

At the dielectric interface ($z = 0$) the tangential fields must be continuous, so we equate the sum of the fields in Region 1 to the fields in Region 2.

$$E_{x0i} + E_{x0r} = E_{x0t} \tag{11.99}$$

similarly, the incident, reflected, and transmitted magnetic fields are given by

$$\mathbf{H}_m = \frac{1}{Z_j} \hat{\mathbf{n}}_m \times \mathbf{E}_m \tag{11.100}$$

where $Z_j = \sqrt{\mu_0 / \varepsilon_j}$ characterises the medium (1 or 2) and m characterises the wave: incident, transmitted, or reflected waves. So

$$\mathbf{H}_i = \frac{E_{x0i}}{Z_1} (\mathbf{a}_y \cos \theta_i + \mathbf{a}_z \sin \theta_i) \, e^{-jk(y \sin \theta_i - z \cos \theta_i)} \tag{11.101}$$

$$\mathbf{H}_r = \frac{E_{x0r}}{Z_1} (-\mathbf{a}_y \cos \theta_i + \mathbf{a}_z \sin \theta_i) \, e^{-jk(y \sin \theta_i + z \cos \theta_i)} \tag{11.102}$$

$$\mathbf{H}_t = \frac{E_{x0t}}{Z_2} (\mathbf{a}_y \cos \theta_t + \mathbf{a}_z \sin \theta_t) \, e^{-jk(y \sin \theta_t - z \cos \theta_t)} \tag{11.103}$$

At the dielectric interface, the tangential magnetic fields must also be continuous

$$\frac{(E_{x0i} - E_{x0r}) \cos \theta_i}{Z_1} = \frac{E_{x0t} \cos \theta_t}{Z_2} \tag{11.104}$$

using Eqn (11.99),

$$\frac{(E_{x0i} - E_{x0r}) \cos \theta_i}{Z_1} = \frac{(E_{x0i} + E_{x0r}) \cos \theta_t}{Z_2} \tag{11.105}$$

or $\quad E_{x0r}(\cos \theta_i / Z_1 + \cos \theta_t / Z_2) = E_{x0i}(\cos \theta_i / Z_1 - \cos \theta_t / Z_2) \tag{11.106}$

$$\frac{E_{x0r}}{E_{x0i}} = \frac{\cos \theta_i / Z_1 - \cos \theta_t / Z_2}{\cos \theta_i / Z_1 + \cos \theta_t / Z_2} \tag{11.107}$$

$$R_\perp = \frac{\cos\theta_i - Z_1 \cos\theta_t / Z_2}{\cos\theta_i + Z_1 \cos\theta_t / Z_2} \tag{11.108}$$

$$= \frac{\cos\theta_i - \sqrt{(\varepsilon_2/\varepsilon_1)}\,\cos\theta_t}{\cos\theta_i + \sqrt{(\varepsilon_2/\varepsilon_1)}\,\cos\theta_t} \tag{11.109}$$

In these equations, $R_\perp$ is the reflection coefficient. Also from Snell's law

$$\frac{\sin\theta_i}{\sin\theta_t} = \sqrt{\frac{\varepsilon_2}{\varepsilon_1}} \tag{11.110}$$

which makes
$$\cos\theta_t = \sqrt{1 - \sin^2\theta_t} = \sqrt{1 - (\varepsilon_1/\varepsilon_2)\sin^2\theta_i} \tag{11.111}$$

which gives
$$R_\perp = \frac{E_r}{E_i} = \frac{\cos\theta_i - \sqrt{(\varepsilon_2/\varepsilon_1) - \sin^2\theta_i}}{\cos\theta_i + \sqrt{(\varepsilon_2/\varepsilon_1) - \sin^2\theta_i}} \tag{11.112}$$

Example 11.16   Prove that $R_\perp$ of Eqn (11.112) will never be zero for any angle $0 < \theta_i < \pi/2$ and $\varepsilon_2 > \varepsilon_1$.

Proof   We know that

$$\sqrt{1 - \sin^2\theta_i} = \cos\theta_i$$

therefore
$$\sqrt{(\varepsilon_2/\varepsilon_1) - \sin^2\theta_i} > \cos\theta_i, \quad \text{for } \varepsilon_2 > \varepsilon_1$$

therefore
$$\cos\theta_i - \sqrt{(\varepsilon_2/\varepsilon_1) - \sin^2\theta_i} < 0$$

11.7.2.2   Parallel Polarisation

Proceeding to analyse the case of parallel polarisation, the magnetic fields for the incident, reflected, and transmitted waves are

$$\mathbf{H}_i = \mathbf{a}_x H_{x0i} e^{-jk(y\sin\theta_i - z\cos\theta_i)} \tag{11.113}$$

$$\mathbf{H}_r = \mathbf{a}_x H_{x0r} e^{-jk(y\sin\theta_i + z\cos\theta_i)} \tag{11.114}$$

$$\mathbf{H}_t = \mathbf{a}_x H_{x0t} e^{-jk(y\sin\theta_t - z\cos\theta_t)} \tag{11.115}$$

equating the tangential magnetic fields at the boundary

$$H_{x0i} + H_{x0r} = H_{x0t} \tag{11.116}$$

we now find the corresponding electric fields

$$\mathbf{E}_i = -Z_1 H_{x0i}(\mathbf{a}_y \cos\theta_i + \mathbf{a}_z \sin\theta_i)\, e^{-jk(y\sin\theta_i - z\cos\theta_i)} \tag{11.117}$$

$$\mathbf{E}_r = -Z_1 H_{x0r}(-\mathbf{a}_y \cos\theta_i + \mathbf{a}_z \sin\theta_i)\, e^{-jk(y\sin\theta_i + z\cos\theta_i)} \tag{11.118}$$

$$\mathbf{E}_t = -Z_2 H_{x0t}(\mathbf{a}_y \cos\theta_t + \mathbf{a}_z \sin\theta_t)\, e^{-jk(y\sin\theta_t - z\cos\theta_t)} \tag{11.119}$$

equating the tangential electric fields at the boundary

$$Z_1(H_{x0i}\cos\theta_i - H_{x0r}\cos\theta_i) = Z_2 H_{x0t}\cos\theta_t \tag{11.120}$$

substituting Eqn (11.116) in this equation

$$Z_1 \cos\theta_i (H_{x0i} - H_{x0r}) = Z_2 \cos\theta_t (H_{x0i} + H_{x0r}) \tag{11.121}$$

$$H_{x0r}(Z_1\cos\theta_i + Z_2\cos\theta_t) = H_{x0i}(Z_1\cos\theta_i - Z_2\cos\theta_t) \tag{11.122}$$

$$\frac{H_{x0r}}{H_{x0i}} = \frac{(Z_1\cos\theta_i - Z_2\cos\theta_t)}{(Z_1\cos\theta_i + Z_2\cos\theta_t)} \tag{11.123}$$

from where we get using Snell's law

$$R_{\parallel} = \frac{E_r}{E_i} = \frac{(\varepsilon_2/\varepsilon_1)\cos\theta_i - \sqrt{(\varepsilon_2/\varepsilon_1) - \sin^2\theta_i}}{(\varepsilon_2/\varepsilon_1)\cos\theta_i + \sqrt{(\varepsilon_2/\varepsilon_1) - \sin^2\theta_i}} \tag{11.124}$$

Example 11.17 Prove that $R_{\parallel}$ of Eqn (11.124) *may be* zero for some angle $0 < \theta_i < \pi/2$ and $\varepsilon_2 > \varepsilon_1$.

Proof We know that

$$\sqrt{1 - \sin^2\theta_i} = \cos\theta_i$$

therefore

$$\sqrt{(\varepsilon_2/\varepsilon_1) - \sin^2\theta_i} > \cos\theta_i, \quad \text{for } \varepsilon_2 > \varepsilon_1$$

or

$$\cos\theta_i - \sqrt{(\varepsilon_2/\varepsilon_1) - \sin^2\theta_i} < 0$$

and

$$(\varepsilon_2/\varepsilon_1)\cos\theta_i - \sqrt{(\varepsilon_2/\varepsilon_1) - \sin^2\theta_i}$$

may be zero. (See next section)

Brewster Angle We now consider a case when $R_{\parallel}$ of Eqn (11.124) is zero at some angle when $\theta_i = \theta_b$ which is the Brewster angle and which implies no reflection. For $R_{\parallel}$ to be zero, the numerator must be zero,

$$(\varepsilon_2/\varepsilon_1)\cos\theta_b - \sqrt{(\varepsilon_2/\varepsilon_1) - \sin^2\theta_b} = 0$$

$$(\varepsilon_2/\varepsilon_1)\cos\theta_b = \sqrt{(\varepsilon_2/\varepsilon_1) - \sin^2\theta_b}$$

$$(\varepsilon_2/\varepsilon_1)^2 \cos^2\theta_b = (\varepsilon_2/\varepsilon_1) - \sin^2\theta_b$$

$$(\varepsilon_2/\varepsilon_1)^2 (1 - \sin^2\theta_b) = (\varepsilon_2/\varepsilon_1) - \sin^2\theta_b$$

$$\sin^2\theta_b \left[\left(\frac{\varepsilon_2}{\varepsilon_1}\right)^2 - 1\right] = \frac{\varepsilon_2}{\varepsilon_1}\left(\frac{\varepsilon_2}{\varepsilon_1} - 1\right)$$

$$\sin^2\theta_b = \frac{\varepsilon_2/\varepsilon_1}{\left(\frac{\varepsilon_2}{\varepsilon_1}\right) + 1}$$

we now compute the value of $\sin\theta_b$ and $\cos\theta_b$

$$\sin^2\theta_b = \frac{\varepsilon_2}{\varepsilon_1 + \varepsilon_2}$$

$$\cos^2\theta_b = 1 - \sin^2\theta_b = \frac{\varepsilon_1}{\varepsilon_1 + \varepsilon_2}$$

and from which we get

$$\tan\theta_b = \sqrt{\frac{\varepsilon_2}{\varepsilon_1}} \tag{11.125}$$

where θ_b is the Brewster angle, when no reflection takes place.

Example 11.18 Find the Brewster angle for fresh water with $\varepsilon_r = 80$.

Solution The Brewster angle is

$$\tan\theta_b = \sqrt{\frac{\varepsilon_2}{\varepsilon_1}}$$

$$= \sqrt{\frac{\varepsilon_0 \, \varepsilon_r}{\varepsilon_0}}$$

$$\theta_b = \tan^{-1} \sqrt{80}$$

$$= 83.62°$$

11.8 | Poynting Vector and the Flow of Power

If we stand out in the sun we know that our skin is warmed by the rays from the sun. How did the energy from the sun reach us? It is obvious that the light which falls on our skin has warmed our skin. And light is a form of electromagnetic radiation—the same waves which we have studied. But our studies have not given us a clue as to how waves carry energy and power. All we know—as a hint—is that the product

$$|\mathbf{E}| \cdot |\mathbf{H}|$$

has the units of power density:

$$(\text{V/m}) \times (\text{A/m}) = \text{W/m}^2$$

From the knowledge obtained from this chapter, we can speculate that the dot product will not do, since the electric and magnetic fields are perpendicular to each other

$$\mathbf{E} \cdot \mathbf{H} = 0 \tag{11.126}$$

Though the dot product is zero, the wave still caries power. We can consider the cross product, which denoted by $\mathbf{P}$ is

$$\mathbf{P} = \mathbf{E} \times \mathbf{H} \tag{11.127}$$

The units are right. That is, though $\mathbf{P}$ has the units of power density, but it is a vector? To understand about the concept of "Power which is a vector" we need to study the basic equations in more detail. In many books $\mathbf{S} = \mathbf{E} \times \mathbf{H}$ is used instead of $\mathbf{P}$.

The basic work on $\mathbf{P}$, called the Poynting vector, was carried out by Henry Poynting (1852–1914) in 1884. He was an English physicist, and a professor of physics at Mason Science College (now the University of Birmingham) from 1880 until his death. The Poynting vector describes the direction and magnitude of electromagnetic energy flow and is used in the Poynting theorem, a statement about energy conservation for electric and magnetic fields.

11.8.1 Poynting's Theorem

Let us concentrate on the vector identity

$$\nabla \cdot (\mathbf{E} \times \mathbf{H}) = \mathbf{H} \cdot \nabla \times \mathbf{E} - \mathbf{E} \cdot \nabla \times \mathbf{H} \tag{11.128}$$

The units of this equation is W/m^3 throughout. Now

$$\nabla \times \mathbf{E} = -\mu \frac{\partial \mathbf{H}}{\partial t} \tag{11.129}$$

$$\nabla \times \mathbf{H} = \varepsilon \frac{\partial \mathbf{E}}{\partial t} + \mathbf{J} \tag{11.130}$$

which are the standard Maxwell's equations. Substituting these equations in the previous identity

$$\nabla \cdot (\mathbf{E} \times \mathbf{H}) = \mathbf{H} \cdot \left(-\mu \frac{\partial \mathbf{H}}{\partial t}\right) - \mathbf{E} \cdot \left(\varepsilon \frac{\partial \mathbf{E}}{\partial t} + \mathbf{J}\right) \qquad (11.131)$$

Considering each term

$$\mathbf{H} \cdot \left(\mu \frac{\partial \mathbf{H}}{\partial t}\right)$$

is the power density (W/m^3), of the magnetic field at any point in space. If we integrate over any given volume we will get the total power of the magnetic field in that region. Obviously, the term

$$\mu \mathbf{H} \cdot \mathbf{H}$$

is proportional to the *energy density stored* in the magnetic field (J/m^3) at any instant of time, and at a given point in space. Similarly,

$$\mathbf{E} \cdot \left(\varepsilon \frac{\partial \mathbf{E}}{\partial t}\right)$$

is the power density of the electric field, and

$$\varepsilon \mathbf{E} \cdot \mathbf{E}$$

is proportional to the energy density stored in the electric field. The last term of Eqn (11.131),

$$\mathbf{E} \cdot \mathbf{J}$$

is the ohmic power density. Integrating Eqn (11.131) over a region in space in accordance with Fig. 11.14

$$\iiint_V \nabla \cdot (\mathbf{E} \times \mathbf{H}) \, dV = \iiint_V \left[\mathbf{H} \cdot \left(-\mu \frac{\partial \mathbf{H}}{\partial t}\right) - \mathbf{E} \cdot \left(\varepsilon \frac{\partial \mathbf{E}}{\partial t} + \mathbf{J}\right)\right] dV$$

$$-\underbrace{\oiint_S (\mathbf{E} \times \mathbf{H}) \cdot d\mathbf{S}}_{\text{Poynting Vector}} = \iiint_V \left[\underbrace{\frac{\mu}{2} \frac{\partial |\mathbf{H}|^2}{\partial t}}_{\text{Magnetic Field}} + \underbrace{\frac{\varepsilon}{2} \frac{\partial |\mathbf{E}|^2}{\partial t}}_{\text{Electric Field}} + \underbrace{\mathbf{E} \cdot \mathbf{J}}_{\text{Heat}}\right] dV$$

$$(11.132)$$

where we have used

$$\mathbf{A} \cdot \frac{\partial \mathbf{A}}{\partial t} = \frac{1}{2} \frac{\partial |\mathbf{A}|^2}{\partial t}$$

To prove this, we expand the left- and right-hand sides

$$\sum_{i=x,y,z} \left[A_i \frac{\partial A_i}{\partial t}\right] = \frac{1}{2} \sum_{i=x,y,z} \left[\frac{\partial A_i^2}{\partial t}\right]$$

$$= \frac{1}{2} \sum_{i=x,y,z} \left[2A_i \frac{\partial A_i}{\partial t}\right]$$

$$= \text{LHS}$$

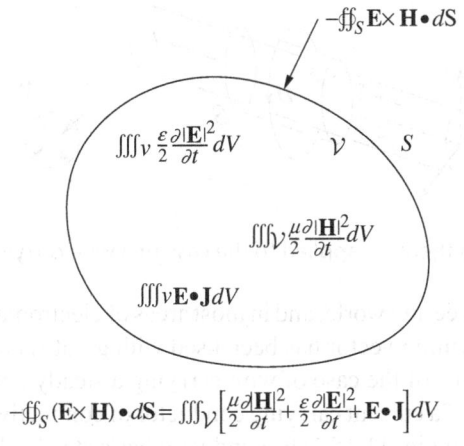

$$-\oiint_S (\mathbf{E} \times \mathbf{H}) \bullet d\mathbf{S} = \iiint_\mathcal{V} \left[\frac{\mu}{2} \frac{\partial |\mathbf{H}|^2}{\partial t} + \frac{\varepsilon}{2} \frac{\partial |\mathbf{E}|^2}{\partial t} + \mathbf{E} \bullet \mathbf{J} \right] dV$$

Fig. 11.14 Figure Illustrating Poynting's theorem

To correctly interpret Eqn (11.132) (refer to Fig. 11.14) the term on the left is the total power *entering* the surface S which encloses the volume $\mathcal{V}$. The surface integral denotes the total flux leaving the surface. The negative sign implies that the power is entering. The terms on the right are (in that order): the total power gained—(a) by the magnetic field, and (b) by the electric field. The last term is the power dissipated in ohmic losses. Hence we can write this equation as

$$\text{Power entering } S = \text{Power gained by} \{\mathbf{E} \text{ in } \mathcal{V} + \mathbf{H} \text{ in } \mathcal{V}\}$$
$$+ \text{ Heat dissipated in } \mathcal{V}$$

The two terms

$$\frac{\mu}{2} \mathbf{H} \cdot \mathbf{H}, \quad \frac{\varepsilon}{2} \mathbf{E} \cdot \mathbf{E}$$

are the *energy densities,* J/m³, stored in the electric and magnetic fields respectively.

11.8.2 Poynting Vector

The Poynting vector, $\mathbf{P}$, is associated with the flow of power. Does the vector have actual physical significance? Famous scientists have expressed doubt about its reality.

Sir James Jeans in his book *The Mathematical Theory of Electricity and Magnetism* said that

> "*The integral of the Poynting Flux over a closed surface gives the total flow of energy into or out of a surface, but it has not been proved, and we are not entitled to assume, that there is an actual flow of energy at every point equal to the Poynting Flux.*"

> "For instance, if an electrified sphere is placed near to a bar magnet, this latter assumption would require a perpetual flow of energy at every point in the field except the special points at which the electric and magnetic lines of force are tangential to one another. It is difficult to believe that this predicted circulation of energy can have any physical reality. . . "

The italicised part states that Poynting's theorem has meaning, but not the Poynting vector. That is the Poynting vector has no real physical significance whatsoever.

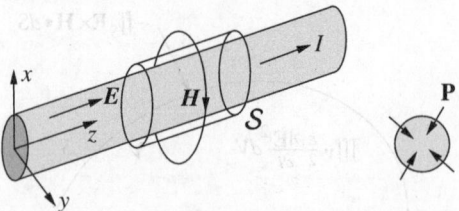

Fig. 11.15 Poynting theorem applied to the case of a wire carrying a steady current

However, in the engineering world, and in most areas of electromagnetics—especially in antennas—the Poynting vector has been used with great success. We look at the case of the energy flow of the case of wire carrying a steady current.

Consider the case of a current carrying conductor of conductivity σ which carries a current I as shown in Fig. 11.15. The conductor has a steady electric field E_z. The magnetic field in the cylindrical coordinate system, at a point (ρ, ϕ, z) outside (and just inside the conductor) is given by

$$H_\phi = \frac{I}{2\pi\rho}$$

Now $\qquad I = J_z \times \pi a^2 = \pi a^2 \sigma \, E_z \quad (J_z = \sigma \, E_z)$

where the radius of the conductor is a, and J_z is the (constant) current density inside the conductor. Therefore

$$E_z = \frac{I}{\pi a^2 \sigma}$$

$$J_z = \frac{I}{\pi a^2}$$

H at the surface of the conductor is

$$H_\phi = \frac{I}{2\pi a}$$

We apply Poynting's theorem, Eqn (11.132), to the Gaussian surface shown as a cylinder S in the figure. The Poynting vector is

$$\mathbf{P} = \mathbf{E} \times \mathbf{H}$$

$$= \begin{vmatrix} \mathbf{a}_\rho & \mathbf{a}_\phi & \mathbf{a}_z \\ 0 & 0 & E_z \\ 0 & H_\phi & 0 \end{vmatrix}$$

$$= \mathbf{a}_\rho(-E_z H_\phi)$$

It is interesting to note that the power flow is inward from the surface of the cylinder. This is due to the negative sign with the unit vector $\mathbf{a}_\rho$. The only region where the electric field and magnetic field are present together is *inside* the conductor. The Gaussian surface (shown as S in the figure) is made to coincide with the surface of the conductor, but *infinitesimally smaller* and its length is d. Integrations on the two flat surfaces contribute nothing, because the direction of the Poynting vector is parallel to those surfaces. The Pointing vector contribution is from the curved surface, an element of area which is

$$d\mathbf{S} = \mathbf{a}_\rho a\, d\phi\, dz$$

On the surface of the conductor, the $\mathbf{a}_\rho$ part of the Poynting vector is

$$P_\rho = -E_z H_\phi = -\left(\frac{I}{\pi a^2 \sigma}\right) \times \left(\frac{I}{2\pi a}\right) = -\frac{I^2}{2\pi^2 a^3 \sigma}$$

where we have used the earlier equations of E_z and H_ϕ. Now

$$-\oiint_{S} \mathbf{P} \cdot d\mathbf{S} = \int\int_{\phi=0,z=0}^{\phi=2\pi,z=d} \frac{I^2 a d\phi dz}{2\pi^2 a^3 \sigma}$$

$$= \frac{I^2 (2\pi) d}{2\pi^2 a^2 \sigma} = \frac{I^2 d}{\pi a^2 \sigma}$$

$$= \text{Total power flow into the conductor}$$

If we calculate the total $\mathbf{E} \cdot \mathbf{J}$ term

$$\iiint_{V} \mathbf{E} \cdot \mathbf{J} = E_z J_z \pi a^2 d$$

$$= \left(\frac{I}{\pi a^2 \sigma}\right)\left(\frac{I}{\pi a^2}\right) \times \pi a^2 d = \frac{I^2 d}{\pi a^2 \sigma}$$

$$= \text{Total power dissipated inside the conductor}$$

We notice that the two results are equal.

Let us look at another example, that of a uniform plane wave travelling in space in the $\mathbf{a}_z$. The electric and magnetic fields in phasor notation are given by

$$E_x = E_0 e^{-jkz}$$

$$H_y = \frac{E_0}{Z_0} e^{-jkz} \qquad (11.133)$$

where k is the notation used for the propagation constant β, for free space and Z_0 is the characteristic impedance of free space. We cannot work with phasor quantities since the Poynting theorem has been defined for real time variables. These fields in the real time are

$$\tilde{E}_x = E_0 \cos(\omega t - kz)$$

$$\tilde{H}_y = \frac{E_0}{Z_0} \cos(\omega t - kz) \qquad (11.134)$$

Then

$$\tilde{\mathbf{P}} = \tilde{\mathbf{E}} \times \tilde{\mathbf{H}} = \begin{vmatrix} \mathbf{a}_x & \mathbf{a}_y & \mathbf{a}_z \\ \tilde{E}_x & 0 & 0 \\ 0 & \tilde{H}_y & 0 \end{vmatrix}$$

$$= \mathbf{a}_z \tilde{E}_x \tilde{H}_y$$

$$= \mathbf{a}_z \frac{E_0^2}{Z_0} \cos^2(\omega t - kz)$$

$$= \mathbf{a}_z \frac{E_0^2}{Z_0} \left[\frac{1}{2} - \frac{\cos 2(\omega t - kz)}{2}\right] \qquad (11.135)$$

We notice that (a) The Poynting vector travels in the direction of wave propagation, which is intuitively satisfying; (b) the Poynting vector is always positive but pulsating, increasing from 0 to E_0^2/Z_0 and then back to 0, and so on, sinusoidally;

and (c) the average value of the $\tilde{P}$ is $E_0^2/(2Z_0)$, since the average value is given by, $(\omega T = 2\pi)$

$$\tilde{P}_{av} = \frac{1}{T}\int_0^T \frac{E_0^2}{Z_0}\left[\frac{1}{2} - \frac{\cos 2(\omega t - kz)}{2}\right]dt$$

$$= \frac{1}{T} \times \frac{E_0^2}{Z_0} \times \frac{T}{2} \quad \text{(integral of the cosine term is zero)}$$

$$= \frac{E_0^2}{2Z_0}$$

These ideas expressed above are reminiscent of circuits where the average power supplied, for example, to a load is given by

$$P = VI\cos\phi$$

where V and I are the *rms* phasor voltage across and current through the load and ϕ is the angle between them. Similarly, working with electromagnetic phasors *(but not rms values)*

$$\mathbf{P} = \frac{1}{2}(\mathbf{E} \times \mathbf{H}^*) \tag{11.136}$$

Because these are not rms values, the factor 1/2 has to be included. $\mathbf{P}$ is called the Poynting vector, and $\mathbf{E}$ and $\mathbf{H}$ are the phasor electric and magnetic fields. On the other hand the *time averaged Poynting vector* is given by

$$\mathbf{P}_{av} = \frac{1}{2}\Re\{(\mathbf{E} \times \mathbf{H}^*)\} \tag{11.137}$$

Let us find out whether we are right. Using Eqn set (11.133) and applying the definition given above, we get

$$\mathbf{P}_{av} = \frac{1}{2}\Re\{(\mathbf{E} \times \mathbf{H}^*)\}$$

$$= \frac{1}{2}\Re\left\{\begin{vmatrix} \mathbf{a}_x & \mathbf{a}_y & \mathbf{a}_z \\ E_x & 0 & 0 \\ 0 & H_y^* & 0 \end{vmatrix}\right\}$$

$$= \frac{1}{2}\Re\{(E_x H_y^*)\}$$

$$= \frac{1}{2} \times \Re\{E_0 e^{-jkz} \times H_0^* e^{+jkz}\} = \frac{|E_0|^2}{2Z_0}$$

POINTS TO REMEMBER

- The Helmholtz equation for the electric field for a travelling wave in air is

$$\nabla^2\mathbf{E} + k^2\mathbf{E} = 0$$

  where

$$k = \omega\sqrt{\mu\varepsilon}$$

- A travelling wave must have a functional dependence,

$$\tilde{\mathbf{E}}(\mathbf{R}, t) = \Re\{E_0 e^{j(\omega t - \beta z)}\} \tag{11.138}$$

where β is the propagation constant of the wave and ω is the frequency of oscillation. Here β may or may not be equal to k.

- For the uniform plane wave, there is no electric field in the direction of propagation; there are only transverse components.

$$\mathbf{E} = \mathbf{a}_x E_{x0} e^{-jkz} + \mathbf{a}_y E_{y0} e^{-jkz}$$

- The magnetic field is given by

$$H_x = -\frac{1}{Z} E_y$$

$$H_y = \frac{1}{Z} E_x$$

and

$$Z = \sqrt{\frac{\mu}{\varepsilon}}$$

- The magnitude of the magnetic field is

$$H = \sqrt{H_x^2 + H_y^2}$$

$$= \frac{1}{Z} \sqrt{E_y^2 + E_x^2}$$

$$= \frac{E}{Z}$$

- The magnitude of the Poyting vector $\mathbf{E} \times \mathbf{H}$ is given by

$$|\mathbf{E} \times \mathbf{H}| = \frac{1}{Z} \left(E_x^2 + E_y^2 \right) \quad \text{W/m}^2$$

- *Z* is called *the characteristic or intrinsic impedance of the medium.* For air or vacuum,

$$Z_0 = \sqrt{\frac{\mu_0}{\varepsilon_0}} \approx 377 \ \Omega$$

- Wave polarisation is defined as
  - *linear* when the tip of the electric field vector moves on a *line* in a plane perpendicular to the direction of propagation
  - *circular* when the tip of the electric field vector moves on a *circle* in a plane perpendicular to the direction of propagation
  - *elliptical* when the tip of the electric field vector moves on an *ellipse* in a plane perpendicular to the direction of propagation
- The complex permittivity is given by

$$\varepsilon_C = \varepsilon' - j\varepsilon'' = \varepsilon - j\frac{\sigma}{\omega}$$

- The *dissipation factor, D*

$$D = \frac{\sigma}{\omega \varepsilon} \begin{cases} \ll 1, & \text{for a good dielectric} \\ \gg 1, & \text{for a good conductor} \end{cases}$$

  for lossy dielectrics it is also called the *loss tangent.*
- Using the notation of β_C as the *complex propagation constant*

$$\beta_C = \beta_R + j\beta_I = \beta + j\alpha$$

$$= \omega \sqrt{\mu \varepsilon_C}$$

- For $\beta_C = \beta_R - j\beta_I$, the complex propagation constant (also called γ)

$$\beta = \beta_R = \frac{\omega \sqrt{\mu \varepsilon} \sqrt{\sqrt{1 + \left(\frac{\sigma}{\omega \varepsilon}\right)^2} + 1}}{\sqrt{2}}$$

$$\alpha = \beta_I = \frac{\omega \sqrt{\mu\varepsilon}\sqrt{\sqrt{1+\left(\frac{\sigma}{\omega\varepsilon}\right)^2}-1}}{\sqrt{2}}$$

- The wave in complex notation is then $E = E_0 e^{-\alpha z} e^{-j\beta z}$
- The skin depth is given by $\delta = 1/\alpha$
- For low conductivity materials,

$$\beta \approx \omega \sqrt{\mu\varepsilon}\left[1+\frac{1}{8}\left(\frac{\sigma}{\omega\varepsilon}\right)^2\right]$$

$$\alpha \approx \left(\frac{\omega\sqrt{\mu\varepsilon}}{2}\right)\left(\frac{\sigma}{\omega\varepsilon}\right)$$

$$Z \approx \sqrt{\frac{\mu}{\varepsilon}}\left(1+j\frac{\sigma}{2\omega\varepsilon}\right)$$

- For high conductivity materials,

$$\alpha = \beta = \sqrt{\frac{\omega\mu\sigma}{2}}$$

The characteristic impedance is given by

$$Z \approx Z_0 \left(\frac{1+j}{\sqrt{2}}\right)\sqrt{\frac{\omega\varepsilon_0}{\sigma}}$$

- The boundary conditions for time-varying fields between two media are
  - For **E**
  $$E_{t1} - E_{t2} = 0$$
  $$\hat{n}\times(\mathbf{E}_1 - \mathbf{E}_2) = 0$$
  - For **D**
  $$D_{n1} - D_{n2} = \rho_s$$
  $$\hat{n}\cdot(\mathbf{D}_1 - \mathbf{D}_2) = \rho_s$$
  - For **B**
  $$B_{n1} - B_{n2} = 0$$
  $$\hat{n}\cdot(\mathbf{B}_1 - \mathbf{B}_2) = 0$$
  and
  - For **H**
  $$\hat{n}\times(\mathbf{H}_1 - \mathbf{H}_2) = \mathbf{J}_s$$
  in all these cases the unit normal $\hat{n}$ extends from medium 2 to medium 1. Also if medium 2 is a perfect electric conductor, all the fields are zero.
- For perpedicular polarisation (electric field perpendicular to the plane of polarisation) the reflection coefficient is

$$R_\perp = \frac{E_r}{E_i} = \frac{\cos\theta_i - \sqrt{(\varepsilon_2/\varepsilon_1) - \sin^2\theta_i}}{\cos\theta_i + \sqrt{(\varepsilon_2/\varepsilon_1) - \sin^2\theta_i}}$$

for a wave incident from medium 1 (ε_1) to medium 2 (ε_2).
- For parallel polarisation (electric field parallel to the plane of polarisation) the reflection coefficient is

$$R_\| = \frac{E_r}{E_i} = \frac{(\varepsilon_2/\varepsilon_1)\cos\theta_i - \sqrt{(\varepsilon_2/\varepsilon_1) - \sin^2\theta_i}}{(\varepsilon_2/\varepsilon_1)\cos\theta_i + \sqrt{(\varepsilon_2/\varepsilon_1) - \sin^2\theta_i}}$$

for a wave incident from medium 1 (ε_1) to medium 2 (ε_2).
- The Brewster angle, θ_b, the angle at which $R_\| = 0$ is given by the relation

$$\tan\theta_b = \sqrt{\frac{\varepsilon_2}{\varepsilon_1}}$$

for a wave incident from medium 1 (ε_1) to medium 2 (ε_2).

- Poynting's theorem is

$$-\oiint_S (\mathbf{E} \times \mathbf{H}) \cdot d\mathbf{S} = \iiint_V \left[\frac{\mu}{2} \frac{\partial |\mathbf{H}|^2}{\partial t} + \frac{\varepsilon}{2} \frac{\partial |\mathbf{E}|^2}{\partial t} + \mathbf{E} \cdot \mathbf{J} \right] dV$$

where S is the surface enclosing a volume V. The term on the left is the power entering the surface S. The first term on the right is the power stored in the magnetic field; the second term on the right is the power stored in the electric field; and the third term on the right is the power dissipated as heat.
- The average power propagated in a plane wave is given by

$$P = \frac{|E|^2}{2Z_0}$$

where E is the magnitude of the electric field and Z_0 is the intrinsic impedence of the medium.

SELF ASSESSMENT

Objective Type Questions

1. A wave is called a plane wave because
   (a) it travels in planes
   (b) it appears like a plane
   (c) its phase is constant over a plane
   (d) none of the above
2. For a wave travelling in free space at a frequency of 1 GHz and in a material with $\sigma = 10^3$ and $\varepsilon = \varepsilon_0$ compare the velocity (v) wavelength (λ) and radian frequency (ω) in the two media.
   (a) all three are equal
   (b) two out of three are equal
   (c) one out of three are equal
   (d) none of the three are equal
3. The magnitude of the magnetic field of a circularly polarised wave travelling in space at a point z = constant is
   (a) sinusoidally varying
   (b) circularly varying
   (c) is a constant
   (d) none of the above
4. The tip of the electric field vector of a linearly polarised wave travelling in space at a point z = constant is
   (a) sinusoidally varying
   (b) circularly varying
   (c) is a constant
   (d) lies on a straight line
5. The power density in W/m^2 of a plane wave
   (a) propagates with the wave
   (b) is equal to $(1/2) |\mathbf{E} \times \mathbf{H} *|$
   (c) is equal to $E^2/(2Z_0)$
   (d) none of the above
6. Distributed components are used in circuits in the frequency range
   (a) 3 Hz to 3 kHz
   (b) 3 kHz to 300 kHz
   (c) 3 GHz to 30 GHz
   (d) none of the above
7. The complex permittivity is
   (a) $\varepsilon_0 \varepsilon_r$ where ε_r is a real number
   (b) $\varepsilon_0 \varepsilon_r (1 - j\sigma)$
   (c) $\varepsilon_0 \varepsilon_r [1 + j(\sigma/\omega)]$
   (d) $\varepsilon_0 \varepsilon_r \{1 + j[\sigma/(\omega \varepsilon_0 \varepsilon_r)]\}$
8. The unit of σ/ε_0 is
   (a) Ω (b) $\mho$ (c) dimensionless   (d) s^{-1}
9. The skin depth δ is
   (a) $1/\beta_C$ (b) $1/\alpha$ (c) $1/(\omega \varepsilon)$ (d) none of the above

10. In a metal of high conductivity
    (a) $\alpha = \beta$ (b) Z_0 is complex   (c) $\lambda_{\text{metal}} \ll \lambda$ (d) none of the above
11. The boundary condition of sinusoidally varying electric field phasor at an air–metal surface is
    (a) $\mathbf{E}_{\text{tan}} = 0$ (b) $\mathbf{E}_n \neq 0$ (c) $\mathbf{D}_n = \rho_s$ (d) none of the above

Short-Answer Questions

1. Many "Direct to Home" channels, like "Tata Sky" and "Airtel" operate in a range of frequencies. Find the downlink satellite frequency band in terms of name and numbers.
2. If the electric field of a uniform plane wave is z directed, and the wave is also travelling in the z direction, is this possible?
3. Find the electric field strength

$$E = 10\cos(\omega t - \beta z)$$

for

(a) For $t = 0$ and for all values of z
(b) For $z = 0$ and for all values of t
(c) For $\omega t = \pi/4$ and for all values of z
(d) For $\beta z = \pi/2$ and for all values of t

4. Find the magnetic field associated with

$$\mathbf{E} = E_0 \mathbf{a}_x e^{-\alpha z} e^{-j(\omega t - \beta z)}$$

5. Find the propagation constant γ, for free space at $f = 1\text{MHz}$, 10 MHz, and 1 GHz.

Review Questions

1. Why is the "plane wave" so called?
2. Write the Helmholtz equation and outline its solution in rectangular coordinates.
3. Write the the the functional dependance of the electric field which is sinusoidal in nature and travelling in the z-direction.
4. Write a short note on the complex propagation constant.
5. How would you identify RCP and LCP waves?
6. Explain why a circularly polarised wave can still be a plane wave.
7. Explain how standing waves form for normal incidence on a metal surface.

Numerical Problems

If the medium is not specified, please assume $\varepsilon = \varepsilon_0$ and $\mu = \mu_0$.

1. Write the Helmholtz equation of the electric field which is z directed. Write its solution for a wave travelling in the y direction with a velocity c and frequency ω.
2. In what range of frequencies is the wavelength in the range 1–10 cm? Which are the frequency bands involved?
3. $\tilde{\mathbf{E}} = \mathbf{E}_0 \cos(2.1 \times 10^8 t - \beta z)$ is the equation of the electric field of a wave travelling in air. What is the velocity of the wave? Find the value of (a) β, (b) λ, (c) ω
4. If the electric field expression for a plane wave is given as

$$E_x = E_0 \cos(1.5 \times 10^{10} t + 60z)$$

then determine (a) its wavelength, (b) phase velocity, and (c) direction of travel. Find the magnetic field $\mathbf{H}$ associated with this electric field.

5. A plane wave is written as

$$\tilde{\mathbf{E}}(z,t) = E_0\mathbf{a}_x \cos(\omega t - kz) + 2E_0\mathbf{a}_y \cos(\omega t - kz)$$

Express this wave in phasor form and find its polarisation.

6. A plane wave is written as

$$\tilde{\mathbf{E}}(z,t) = E_0\mathbf{a}_x \cos(\omega t - kz) - 2E_0\mathbf{a}_y \cos(\omega t - kz)$$

Find its polarisation.

7. Find the magnetic field associated with

$$\mathbf{E} = E_0\mathbf{a}_z e^{-\alpha x} e^{+j(\omega t - \beta x)}$$

8. A plane wave is travelling from free space to a medium with $\varepsilon_r = 3$ and $\mu_r = 3$. If the expression in free-space is

$$\mathbf{E} = E_0\mathbf{a}_x e^{j(\omega_0 t - \beta_0 z)}$$

write the general expression of the electric field in the medium of the same wave.

9. Show that $$E(x,t) = f_1(x - ct) + f_2(x + ct)$$

is a solution of the wave equation

$$(\nabla^2 + \omega^2 \mu\varepsilon)E = 0$$

with $k = \omega\sqrt{\mu\varepsilon}$ and $c = \omega/k$.

10. Find the intrinsic impedence of a medium at a frequency of 1 GHz with $\sigma = 5.8 \times 10^7$ ʊ/m, ε_r and $\mu_r = 4$. What happens to the impedence for the two cases of (a) $\sigma = 0$ and (b) $\sigma = \infty$ if the other parameters are constant.

11. A plane wave propagating through a medium with $\varepsilon_r = 6$ and $\mu_r = 4$ has a magnetic field given by

$$\tilde{\mathbf{H}} = 10\mathbf{a}_z e^{-z/5} \sin(2\pi \times 10^8 t + \beta x)$$

find the expression for the electric field, the wave velocity, and the wave impedance.

12. An EM plane wave travels with

$$\mathbf{E} = 20\mathbf{a}_y e^{j\left(\frac{\sqrt{3}}{2}y + \frac{1}{2}z\right)} e^{j2\pi \times 10^8 t}$$

determine (a) β, (b) the direction in which the wave travels, and (c) ω.

13. Given that $\tilde{\mathbf{E}} = \mathbf{a}_y E_0 \sin(10\pi x) \cos(6\pi \times 10^9 t - \beta z)$

Find $\mathbf{H}$ and β.

14. Specify the polarisation of the wave if its electric field is defined as

$$\mathbf{E} = E_{0x}\mathbf{a}_x \cos(\omega t - \beta z + \theta_x) + E_{0y}\mathbf{a}_y \cos(\omega t - \beta z + \theta_y)$$

for the following cases:

(a) $E_{0x} = E_{0y}$, and $\theta_x = \theta_y + 2n\pi$, and where $n = 0, 1, \ldots$
(b) $E_{0x} = E_{0y}$, and $\theta_x = \theta_y + (2n + 1)\pi$, and where $n = 0, 1, \ldots$

15. Specify the polarisation of the wave if its electric field is defined as

$$\mathbf{E} = E_{0x}\mathbf{a}_x \cos(\omega t - \beta z + \theta_x) + E_{0y}\mathbf{a}_y \cos(\omega t - \beta z + \theta_y)$$

for the following cases:

(a) $E_{0x} = E_{0y}$ and $\theta_x = \theta_y + (2n + 1)(\pi/2)$, and where $n = 0, 1, \ldots$
(b) $E_{0x} = E_{0y}$ and $\theta_y = \theta_x + (2n + 1)(\pi/2)$, and where $n = 0, 1, \ldots$

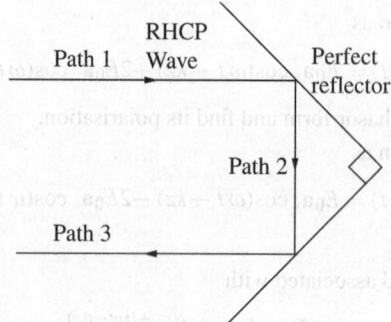

Fig. 11.16

16. An RHCP wave is incident on a perfect reflector as shown in Fig. 11.16. Specify the polarisations on Paths 2 and 3. Write equations for all cases

17. Show that any linearly polarised wave can be written as a sum of two RHCP and LHCP waves.

18. For a material having $\mu_r = 2$, $\varepsilon_r = 4$, and $\sigma = 100$ S/m, calculate the attenuation constant, phase constant, and intrinsic impedence at 10 GHz.

19. Find the skin depth δ for silver and gold at frequencies 1 MHz and 1 GHz. Also find the propagation constant and wave velocity for both cases.

20. If a material has $\sigma = 341$ nS/m, what are the frequency ranges in which the material can be considered to be (a) a low-loss dielectric and (b) a good conductor.

21. A 500 MHz plane propagates normal to a material of $\sigma = 10^5$ S/m. Calculate the skin depth and the distance where the wave amplitude decreases to 20% of its initial value.

22. Plot the dissipation diagram ($\omega - \beta$) for silver.

23. A medium has a propagation constant $\gamma = 200 + j500$/m. (Note that the dependence is exp$\{-\gamma z\}$). The intrinsic impedence $\eta = Z = 75\angle 45°$ at a frequency $f = 500$ MHz. The electric field is given as

$$E_x = 100 e^{-\alpha z} \cos(2\pi \times 5 \times 10^8 t - \beta z)$$

Calculate the magnetic field and average power/m^2 at $z = 1$ mm.

24. In free space

$$E(x, t) = 100 \mathbf{a}_y \sin(\omega t - \beta x)$$

Find the average Poynting vector and power crossing a square area of 1 m^2 in the plane $x = 3$.

25. A plane wave is incident from air on a dielectric with $\varepsilon_r = 10$. If the angle of incidence is 75°, find the reflection coefficient for both polarisations. Find also the power reflected and transmitted in both cases.

26. Find the Brewster angle for dielectric with $\varepsilon_r = 10$.

Answers

Objective Type Questions

1. (c)   2. (c)   3. (b) and (c)   4. (a) and (d)   5. (a), (b) and (c)
6. (c)   7. (d)   8. (d)   9. (b)   10. (d)   11. (a), (b) and (c)

Short-Answer Questions

1. C-band (4–8 GHz) or Ku-band (12–18 GHz)

2. No

3. (a) $E = 10 \cos(\beta z)$, (b) $E = 10 \cos(\omega t)$, (c) $E = 10 \cos(\beta z - \pi/4)$,
   (d) $E = 10 \cos(\omega t - \pi/2)$

4. $\mathbf{H} = (E_0/Z)\,\mathbf{a}_y e^{-\alpha z} e^{-j(\omega t - \beta z)}$ where Z is the intrinsic impedence of the medium.

5. $\gamma = j\beta$; At 1 MHz $\beta = 0.0209$; at 10 MHz $\beta = 0.209$; At 1 GHz $\beta = 20.9$ r/m.

Numerical Problems

1. $\nabla^2 E_z + k^2 E_z = 0$; $\mathbf{E} = E_{0z}\mathbf{a}_z e^{-j(w/c)y}$

2. $\lambda = 1$ cm corresponds to $f = 30$ GHz (Ka band); $\lambda = 10$ cm corresponds to $f = 3$ GHz (C band); bands involved are C bank, X band, Ku band, and Ka band.

3. $c = 3 \times 10^8$ m/s (a) $\beta = 0.7$ r/m, (b) $\lambda = 8.98$ m, (c) $\omega = 2.1 \times 10^8$ r/s

4. $\mathbf{H} = \mathbf{a}_y \dfrac{E_0 \times 1.2}{120\pi} \cos(1.5 \times 10^{10} t + 60z)$
   (a) $\lambda = 10.47$ cm (b) $V_p = 2.5 \times 10^8$ (c) travelling in the z-direction.

5. $\mathbf{E} = E_0\mathbf{a}_x\angle 0 + 2E_0\mathbf{a}_y\angle 0$; linear polarisation.

6. Linear polarisation.

7. $\mathbf{H} = \dfrac{E_0}{Z}\mathbf{a}_y e^{-\alpha x} e^{-j(\omega t - \beta x)}$

8. $\mathbf{E} = E_0\mathbf{a}_x e^{j(w_0 t - 3\beta_0 z)}$

10. $Z = 0.0165 + j0.0165$; (a) $Z = 376.73$ (b) $Z = 0$

11. $\tilde{\mathbf{E}} = -Z10\mathbf{a}_y e^{-z/5} \sin(2\pi \times 10^8 t + \beta x)$ where $Z = 307.8 + j6$; $\beta = 10.26$ r/m.
    $V = 6.12 \times 10^7$.

12. (a) $\beta = 1$ (b) $\hat{\mathbf{n}} = -\frac{\sqrt{3}}{2}\mathbf{a}_y - \frac{1}{2}\mathbf{a}_z$ (c) $\omega = 2\pi \times 10^8$

13. $\tilde{\mathbf{H}} = -\mathbf{a}_x(E_0/377) \sin(10\pi x) \cos(6\pi \times 10^9 t - \beta z)$ where $\beta = 62.83$ r/m

14. (a) Linear polarisation
    (b) Linear polarisation

15. (a) RHCP
    (b) LHCP

16. Path 2: LHCP; Path 3: RHCP

18. $\alpha = 2778$ Np/m, $\beta = 2840$ r/m and $Z = 28.4 + j27.8\ \Omega$

19. For silver: at 1 GHz $\delta = 2.02\ \mu$m, $\alpha = \beta = 4.935 \times 10^5$/m; at 1MHz $\delta = 64.07\ \mu$m, $\alpha = \beta = 1.56 \times 10^4$/m. For gold: at 1 GHz $\delta = 2.485\ \mu$m, $\alpha = \beta = 4.02 \times 10^5$/m; at 1 MHz $\delta = 78.6\ \mu$m, $\alpha = \beta = 1.27 \times 10^4$/m. Wave velocity for silver $= 1.27 \times 10^4$ m/s at GHz, $= 402$ m/s at 1 MHz. Wave velocity for gold $= 1.56 \times 10^4$ m/s at 1 GHz, $= 493$ m/s at 1MHz.

20. The material may be considered a low loss dielectric for $f > 0.48$ MHz Hz and a good conductor for $f < 4.8$ kHz.

21. $\delta = 7.117 \times 10^{-5}$ m. Distance at which the wave decays to 0.2 of its initial value is 1.15×10^{-4} m.

23. $H_y = \dfrac{4e^{-\alpha z} \cos(\beta z - \omega t)}{3\left(\frac{i}{\sqrt{2}} + \frac{1}{\sqrt{2}}\right)}$ average power $= 31.6$ W/m^2.

24. $\mathbf{P} = 13.26\mathbf{a}_x$ W/m^2; average power crossing 1 m^2 at $x = 3$ is 13.26 W.

25. $R_\perp = -0.8417$, $R_{||} = 0.0755$; Power reflected. $R_\perp^2 P_{\text{inc}} = 0.7085 P_{\text{inc}}$; $R_{||}^2 P_{\text{inc}} = 0.0057 P_{\text{inc}}$; Transmitted power. $P_\perp t = 0.2915 P_{\text{inc}}$; $P_{||} t = 0.9943 P_{\text{inc}}$

26. $72.5°$

12

Transmission Lines and Waveguides

Learn as much as you can while you are young,
since life becomes too busy later.
— Dana Stewart Scott

CHAPTER OBJECTIVES

To enable the students to understand the following:

- Transmission lines and their representation
- Time domain equation
- Frequency domain equation
- Frequency domain transmission line equation
- Power considerations line
- Reflection coefficient, Γ
- The standing wave ratio (SWR)

- Short wavelength UHF circuit elements
- Smith charts, their construction and use
- Transformer matching
- The fields, propagation constant, and modes of the following waveguides:
  (a) parallel plate waveguide,
  (b) two conductor waveguides,
  (c) rectangular waveguide, and
  (d) circular waveguide

12.1 | Introduction

In various communication equipment when the frequency is high (>100 MHz) and the size of our lines is comparable to the wavelength of propagation, concepts of transmission lines have to be employed. Generally, for low frequencies, the length of the wires connecting the components can be ignored, and the voltage and current along the connecting wires can be assumed to be same at all points along the line. This is the *lumped* model of electrical circuits and is normally employed in regular circuit theory. For direct current this is strictly true, but for alternating current, if the length is comparable to the wavelength, the length becomes important and connecting wires must be treated as transmission lines.

Examining Fig. 10.6 we can see that at sufficiently high frequencies, the line connecting the antenna to the receiver must be modelled as a transmission line.

To be specific, take the example of a television connected to an antenna on the roof of a house. Typically the frequency of the received signal is of the order of a hundred megahertz. So (using $\lambda = c/f$), $\lambda \sim 3$ m. The connecting line between the antenna and TV will be about 20 m which is about 7 wavelengths. Thus we see that to model the connecting line as a short lumped segment would lead to a faulty design. Let us consider another case where the frequency is only 30 kHz. Here the wavelength is about 10,000 m, so 20 m may be modelled as a 'short' segment of wire, and the lumped model may be applied without loss of accuracy. Transmission line theory must therefore be used in the design of circuits when the conditions outlined above apply.

A common rule of thumb in connection with the length of the line is that the cable or wire should be treated as a transmission line if the length of the line is greater than $\sim 1/10$ of the wavelength. At these lengths the phase delay and the interference of any reflections on the line become important and can lead to unpredictable behaviour in systems which have not been carefully designed using transmission line theory.

To analyse transmission lines it has to be treated as composed of inductances, capacitances and resistors which are *distributed* rather than lumped. A lumped component has a value like $L = 1$ mH for an inductor, $C = 10\,\mu$F for a capacitor, but distributed elements must be described in terms as $L = 1$ mH/m and $C = 10\,\mu$ F/m, that is, their values are per meter rather than lump sum. The schematic of a transmission line is shown in Fig. 12.1, while examples of actual transmission lines are shown in Fig. 12.2.

The figure shows two conductors, an upper conductor and a lower conductor with a distributed inductance of L (H/m) along the line, a distributed capacitance of C (F/m) between the upper and lower conductor, a resistance of R (Ω/m) along

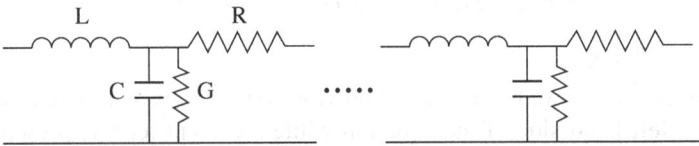

Fig. 12.1 Equivalent circuit of a transmission line

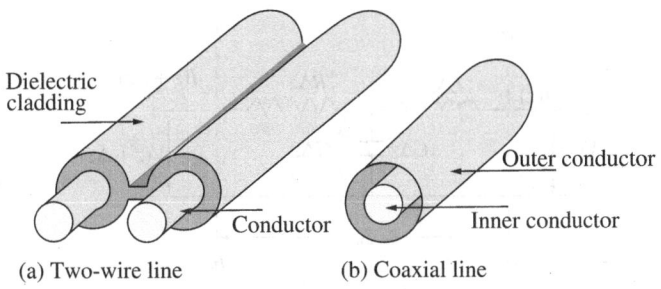

(a) Two-wire line   (b) Coaxial line

Fig. 12.2 Examples of transmission lines

the line and a conductance of G ($\mho$/m) in parallel with the distributed capacitance of dielectric cladding.

12.2 | Time Domain Equation

To analyse a transmission line we take a short section of such a line of length Δz, between two planes $a - a'$ and $b - b'$ on the line in accordance with Fig. 12.3. The plane $a - a'$ is at coordinate point z while the plane $b - b'$ is at a coordinate point $z + \Delta z$, where Δz is a very small distance as compared to the wavelength. To obtain the voltage equation and the voltage at plane $b - b'$, we proceed as follows: from the voltage $V(z, t)$, at plane $a - a'$, we subtract the two voltage drops ($= L\Delta z \times \partial I/\partial t$) of the series inductor and ($= R\Delta z \times I$) of series resistor between the two planes and thereby obtain the voltage $V(z + \Delta z, t)$ at the plane $b - b'$. The equation is

$$V(z + \Delta z, t) = \underbrace{V(z, t)}_{1} - \underbrace{(L\Delta z)\frac{\partial I(z, t)}{\partial t}}_{2} - \underbrace{(R\Delta z)\, I(z, t)}_{3} \qquad (12.1)$$

Here term 1 is the voltage at plane $a - a'$ in the figure. Term 2 is the voltage drop across the inductance, $L\Delta z$, which is the total series inductance between the two planes. The term 3, is a voltage drop across a resistance, $R\Delta z$, which is the total series resistance between the two planes.

By taking $V(z, t)$ to the left side of the equation

$$V(z + \Delta z, t) - V(z, t) = -(L\Delta z)\frac{\partial I(z, t)}{\partial t} - (R\Delta z)\, I(z, t)$$

$$\frac{\partial V(z, t)}{\partial z}\Delta z \approx -L\frac{\partial I(z, t)}{\partial t}\Delta z - RI(z, t)\,\Delta z$$

$$\frac{\partial V(z, t)}{\partial z} = -L\frac{\partial I(z, t)}{\partial t} - RI(z, t)$$

where we have used

$$V(z + \Delta z, t) - V(z, t) \cong \frac{\partial V(z, t)}{\partial z}\Delta z \qquad (12.2)$$

and cancelled the Δz term on both sides.

The concern which we have with the previous equation is that there is a voltage term on the left-hand side of the equation while a current term is present on the right hand side. Since the equation under discussion is a differential equation, we would like voltage terms to be present on both sides of the equation.

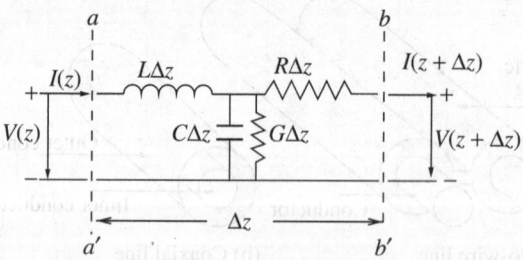

Fig. 12.3 Analysis of a transmission line

With this in mind, we obtain an analogous 'current' differential equation using the current at the two planes $a - a'$ and $b - b'$. To obtain this equation, the current at plane $b - b'$, $I(z + \Delta z, t)$, is equal to the current at plane $a - a'$, $I(z, t)$, minus the two leakage currents, one through the shunt capacitance ($= C\Delta z \partial V/\partial t$) and the other through the conductance ($= G\Delta z V$).

$$I(z + \Delta z, t) = I(z, t) - C\Delta z \frac{\partial V(z, t)}{\partial t} - G\Delta z V(z, t) \tag{12.3}$$

Here $(C\Delta z)(\partial V(z, t)/\partial t)$ is the shunt current from the upper line to the lower line due to the capacitance $C\Delta z$, and $(G\Delta z) V(z, t)$ is the total shunt current loss due to a lossy dielectric between the conductors. As earlier

$$I(z + \Delta z, t) - I(z, t) \approx \frac{\partial I(z, t)}{\partial z} \Delta z \tag{12.4}$$

gives us the current equation

$$\frac{\partial I(z, t)}{\partial z} = -C \frac{\partial V(z, t)}{\partial t} - GV(z, t) \tag{12.5}$$

To reduce the number of terms and to consider only the most important terms we neglect (a) R, the series resistance which contributes to the copper loss and (b) G, the conductance which contributes to the losses in the dielectric separating the two lines. Then the equations become

$$\frac{\partial V(z, t)}{\partial z} = -L \frac{\partial I(z, t)}{\partial t} \tag{12.6}$$

and

$$\frac{\partial I(z, t)}{\partial z} = -C \frac{\partial V(z, t)}{\partial t} \tag{12.7}$$

We have here two partial differential equations in two independent variables (z, t) and two dependent variables (V, I). Seems complicated? Let us differentiate the first equation with respect to z.

$$\frac{\partial^2 V(z, t)}{\partial z^2} = -L \frac{\partial}{\partial z} \left(\frac{\partial I(z, t)}{\partial t} \right)$$

$$= -L \frac{\partial}{\partial t} \left(\frac{\partial I(z, t)}{\partial z} \right) \quad \text{(changing the order of differentiation)}$$

$$= -L \frac{\partial}{\partial t} \left(-C \frac{\partial V(z, t)}{\partial t} \right) \quad \text{(substitution from the other equation)}$$

$$= LC \frac{\partial^2 V(z, t)}{\partial t^2} \tag{12.8}$$

Let us take a look at the units of LC. L has units of (Ω sec) /m; C has the units of ($\mho$sec) /m; so the unit of LC is (sec/m)^2. A little reflection tells us that $1/\sqrt{LC}$ has the units of velocity, v, i.e., $v = 1/\sqrt{LC}$. Equation (12.8) now becomes

$$\frac{\partial^2 V(z, t)}{\partial z^2} = LC \frac{\partial^2 V(z, t)}{\partial t^2} = (1/v^2) \left(\frac{\partial^2 V(z, t)}{\partial t^2} \right) \tag{12.9}$$

This is a wave equation in one dimension, which we can corroborate by comparing this equation with Eqn (11.52). We observe that both equations are identical, and therefore we understand that $V(z, t)$ defines a voltage wave travelling with a velocity v. This is the *most* important finding: *when a voltage is applied to one end of a transmission line it travels down the line with a velocity v.* In a similar manner we can define the governing equation for the current $I(z, t)$

$$\frac{\partial^2 I(z,t)}{\partial z^2} = LC \frac{\partial^2 I(z,t)}{\partial t^2} = (1/v^2) \left(\frac{\partial^2 I(z,t)}{\partial t^2} \right) \quad (12.10)$$

which is also a wave equation. *Therefore in a transmission line, both the voltage and current are travelling waves.*

Example 12.1   If a pulse is defined as

$$P(x) = \begin{cases} P_0, & 0 < x < 1 \\ 0, & \text{elsewhere} \end{cases}$$

and if the voltage is

$$V(x, t = 0) = P(x)$$

find the equation of the voltage travelling in a transmission line at velocity $v = 2.8 \times 10^8$ m/s.

Solution   We know that the functional dependence of a travelling wave is

$$\phi = f(x - vt)$$

therefore the travelling voltage pulse is defined as

$$V(x, t) = P(x - 2.8 \times 10^8 t)$$

12.3 | Frequency Domain Equation

The frequency domain equation for transmission lines is one where the voltages and currents are sinusoidal oscillations. As earlier, the time dependence is of the type $\exp(j\omega t)$ and we have to define distributed *impedances* and *admittances* based on inductances and capacitances. For example, the distributed series impedance is

$$Z = j\omega L + R \quad (12.11)$$

where L and R have been defined in the previous section. R is the resistance along the length of the line, contributing to copper losses. Similarly, the shunt admittance is

$$Y = j\omega C + G \quad (12.12)$$

where G is the conductance describing the dielectric loss. Then

$$V(z + \Delta z) = V(z, t) - j\omega (L\Delta z) I(z) - (R\Delta z) I(z)$$
$$V(z + \Delta z) - V(z, t) = -j\omega (L\Delta z) I(z) - (R\Delta z) I(z)$$
$$\frac{\partial V}{\partial z} \Delta z = -j\omega (L\Delta z) I(z) - (R\Delta z) I(z) \quad (12.13)$$

In these equations $\exp(j\omega t)$ is implicit, and has not been included in the equation; that is, $V(z, t) = V(z) \exp(j\omega t)$ and $I(z, t) = I(z) \exp(j\omega t)$ where $V(z, t)$ and $I(z, t)$ are the total voltage and current. Now proceeding as earlier

$$\frac{dV(z)}{dz} = -(j\omega L + R) I(z) \quad (12.14)$$

Similarly, we can obtain

$$\frac{dI(z)}{dz} = -(j\omega C + G) V(z) \tag{12.15}$$

Differentiating the previous equation with respect to z followed by substituting the above equation, we have

$$\frac{d}{dz}\left(\frac{dV(z)}{dz}\right) = -(j\omega L + R)\frac{d}{dz}(I(z)) = (j\omega L + R)(j\omega C + G) V(z) \tag{12.16}$$

which gives

$$\frac{d^2 V(z)}{dz^2} = (j\omega L + R)(j\omega C + G) V(z) \tag{12.17}$$

$$= \gamma^2 V(z) \tag{12.18}$$

In these equations, $\sqrt{(j\omega L + R)(j\omega C + G)}$ is the *complex* propagation constant

$$\gamma = \alpha + j\beta = \sqrt{(j\omega L + R)(j\omega C + G)}$$

This kind of realisation comes from comparing with the scalar Helmholtz equation (10.72). In it α is called the attenuation constant in Nepers/m and β is the propagation constant in radians/sec.

Coming back to the second equation, we get the equation for the current

$$\frac{d^2 I(z)}{dz^2} = (j\omega L + R)(j\omega C + G) I(z) \tag{12.19}$$

In these equations, if we set $R \approx G \approx 0$ for low loss lines, the two equations become

$$\frac{d^2 V(z)}{dz^2} = (j\omega L)(j\omega C) V(z) = -\omega^2 LC V(z)$$

$$\frac{d^2 I(z)}{dz^2} = (j\omega L)(j\omega C) I(z) = -\omega^2 LC I(z) \tag{12.20}$$

if we set

$$\beta^2 = \omega^2 LC \tag{12.21}$$

then these equations become

$$\frac{d^2 V(z)}{dz^2} = -\beta^2 V(z) \tag{12.22}$$

$$\frac{d^2 I(z)}{dz^2} = -\beta^2 I(z) \tag{12.23}$$

which are two one-dimensional Helmholtz's wave equations for $V(z)$ and $I(z)$, respectively. The solution to these equations which are travelling waves are

$$V(z) = V_0 e^{j(\omega t - \beta z)}$$

$$I(z) = I_0 e^{j(\omega t - \beta z)} \tag{12.24}$$

If we drop the $\exp(j\omega t)$ (this term is conventionally understood to be present) then

$$V(z) = V_0 e^{-j\beta z}$$

$$I(z) = I_0 e^{-j\beta z} \tag{12.25}$$

are the equations for waves travelling in the $+z$ direction (as discussed in earlier chapters). It is important to note that $Z_0 = \sqrt{(R + jwL)(G + jwC)}$ is called the characteristic equation of the line.

12.3.1 Lossy Transmission Lines and Distortion-less Lines

The solution to the wave equation where both R and G are not neglected is

$$V(z) = V_- e^{\gamma z} + V_+ e^{-\gamma z}$$

where

$$\gamma = \alpha + j\beta = \sqrt{(j\omega L + R)(j\omega C + G)}$$

The solution may be verified by substitution.

For a loss-less line where $R = G = 0$, $\gamma = j\omega \sqrt{LC} = j\beta$ and the functional dependence is $\exp(-j\beta z)$ or $\exp(j\beta z)$.

In general, $\gamma = \alpha + j\beta$ a complex number, which implies that

$$e^{-\gamma z} = e^{-\alpha z} e^{-j\beta z}$$

so the voltage wave decays as it progresses. (The arguments are same as given in the section on plane waves travelling in lossy media section.)

Distortion-less Transmission Lines

One particular case is of great interest, that is of the distortion-less line where

$$\frac{R}{G} = \frac{L}{C} = Z_0^2 \tag{12.26}$$

In this case

$$\gamma = \sqrt{(j\omega L + R)(j\omega C + G)}$$

$$= \sqrt{(j\omega Z_0^2 C + Z_0^2 G)(j\omega C + G)}$$

$$= Z_0(j\omega C + G)$$

$$= j\omega \sqrt{LC} + \sqrt{RG} \tag{12.27}$$

$$\alpha = \sqrt{RG}$$

$$\beta = \omega \sqrt{LC} \tag{12.28}$$

Note that the real part is *frequency independent* and the imaginary part is proportional to the frequency which means that the velocity of propagation,

$$v = \omega/\beta = 1/\sqrt{LC} \tag{12.29}$$

is constant for all frequencies. Therefore, a pulse sent down the line arrives at the other end undistorted.

Example 12.2 The constants per kilometer of a cable are $R = 42.9\,\Omega$, $L = 0.7\,\text{mH}$, $C = 0.1\,\mu\text{F}$, and $G = 24\,\mu\mho$. Calculate the attenuation factor, characteristic impedance and phase velocity when $\omega = 5000$ rad/s.

(Mumbai University, May 2009)

Solution The per meter values are $R = 42.9 \times 10^{-3}\,\Omega$, $L = 0.7 \times 10^{-6}\,\text{H}$, $C = 0.1 \times 10^{-9}\,\text{F}$ and $G = 24 \times 10^{-9}\,\mho$ which are obtained by dividing by 10^3. The value of the complex propagation constant is

$$\gamma = \sqrt{(R + j\omega L)(G + j\omega C)}$$

$$= 1.0204 \times 10^{-4} + j1.0551 \times 10^{-4}$$

therefore the attenuation factor is

$$\alpha = 1.0204 \times 10^{-4}\ \text{Nepers/m}$$

and the phase constant is

$$\beta = 1.0551 \times 10^{-4} \text{ rad/m}$$

the phase velocity is

$$v_p = \frac{\omega}{\beta} = 4.739 \times 10^7 \text{ m/s}$$

(This formula is remembered as $v_p = 2\pi f /[2\pi/\lambda] = f\lambda$)

The characteristic impedance is

$$Z_0 = \sqrt{\frac{R + j\omega L}{G + j\omega C}} = 220.3 - j193.5 \ \Omega$$

12.3.2 Low-loss Transmission Lines

Generally, for most transmission lines $R \ll \omega L$ and $G \ll \omega C$ then

$$\gamma = \sqrt{(R + j\omega L)(G + j\omega C)}$$

$$\approx \sqrt{(j\omega L)(j\omega C)} \left(1 + \frac{R}{2j\omega L}\right)\left(1 + \frac{G}{2j\omega C}\right)$$

$$\approx \sqrt{(j\omega L)(j\omega C)} \left(1 + \frac{R}{2j\omega L} + \frac{G}{2j\omega C}\right)$$

$$= j\omega \sqrt{LC} + \frac{R}{2}\sqrt{\frac{C}{L}} + \frac{G}{2}\sqrt{\frac{C}{L}}$$

Hence

$$\alpha = \frac{R}{2}\sqrt{\frac{C}{L}} + \frac{G}{2}\sqrt{\frac{C}{L}}$$

and

$$\beta = \omega \sqrt{LC}$$

Also

$$Z_0 = \sqrt{\frac{R + j\omega L}{G + j\omega C}}$$

$$= \sqrt{R + j\omega L} \times (G + j\omega C)^{-1/2}$$

$$\approx \sqrt{\frac{L}{C}} \left(1 + \frac{R}{2j\omega L}\right)\left(1 - \frac{G}{2j\omega C}\right)$$

$$\approx \sqrt{\frac{L}{C}} \left(1 - \frac{jR}{2\omega L} + \frac{jG}{2\omega C}\right)$$

12.3.3 Practical Transmission Lines

How do we get the values of L, C, R, and G for physical lines? We give here a table (Table 12.1) of these values (Jordan & Balmain, 1968).

Example 12.3 Calculate the inner radius of the outer conductor for a 75 Ω coax line whose inner conductor is of radius 1 mm. The dielectric between the inner and outer conductor is polystyrene. Calculate the L and C for this line.

Solution Polystyrene has a dielectric constant of 2.56. So using the formula from Table 12.1, we have

$$Z = (1/2\pi) \sqrt{\mu/\varepsilon} \ \ln(b/a)$$

$$75 = (1/2\pi) \sqrt{\mu_0/\varepsilon_0 \, \varepsilon_r} \ \ln(b/a)$$

Table 12.1 Calculation of L and C for two wires (conductor radius $= a$, spacing between centres $= b$) and the coaxial (radius of inner conductor $= a$, inner radius of outer conductor $= b$) lines

	Two-wire line	Coaxial line
L	$(\mu/\pi)\cosh^{-1}(b/2a)$	$(\mu/2\pi)\ln(b/a)$
C	$(\pi\varepsilon)/\cosh^{-1}(b/2a)$	$(2\pi\varepsilon)/\ln(b/a)$
$Z = \sqrt{L/C}\ (R = G = 0)$	$(1/\pi)\sqrt{\mu/\varepsilon}\ \cosh^{-1}(b/2a)$	$(1/2\pi)\sqrt{\mu/\varepsilon}\ \ln(b/a)$
R	$\dfrac{1}{\pi a}\sqrt{\dfrac{\omega\mu}{2\sigma}}$	$\dfrac{1}{2\pi}\sqrt{\dfrac{\omega\mu}{2\sigma}}\left(\dfrac{1}{a}+\dfrac{1}{b}\right)$
G	$G = \omega C \times$ dissipation factor	$G = \omega C \times$ dissipation factor
$v_p\ (R = G = 0)$	$1/\sqrt{\mu\varepsilon}$	$1/\sqrt{\mu\varepsilon}$
$\beta\ (R = G = 0)$	$\omega\sqrt{\mu\varepsilon}$	$\omega\sqrt{\mu\varepsilon}$
α	$\dfrac{R}{2Z_0}+\dfrac{GZ_0}{2}$	$\dfrac{R}{2Z_0}+\dfrac{GZ_0}{2}$

$$= (60/\sqrt{2.56})\ln(b/a)$$
$$= 37.5\ln(b/a)$$

giving
$$b/a = 7.3891$$
$$b = 7.3891 \text{ mm}$$

The capacitance/m for this line is given by
$$C = (2\pi\varepsilon)/\ln(b/a)$$
$$= 27.81 \text{ pF/m}$$

while the inductance is $\qquad L = 0.156\ \mu\text{H/m}$

Example 12.4 Calculate the distance between the conductors for a 300 Ω two-wire line whose conductor radius is 1 mm.

Solution From Table 12.1
$$Z = (1/\pi)\sqrt{\mu/\varepsilon}\ \cosh^{-1}(b/2a)$$
$$300 = (1/\pi)\sqrt{\mu_0/\varepsilon_0}\ \cosh^{-1}(b/2a)$$
$$= 120\cosh^{-1}(b/2a)$$
$$b/2a = 6.1323$$
$$b = 12.265 \text{ mm}$$

Another transmission line which is used in a large number of applications is the micro-strip line whose cross-section is depicted in Fig. 12.4. The characteristic

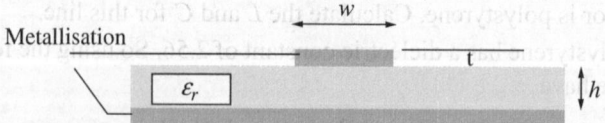

Fig. 12.4 Cross-section of the micro-strip line

impedance of this line is given in Table 12.2 (Sec 12.9.4). The investigation of this line is due to Wheeler (1977).

12.4 | Solutions to the Transmission Line Equation

Since most of the time we will be working with sinusoidal quantities, we will consider the frequency domain equation rather than the time domain equation. The wave equation in the frequency domain, if we recall correctly, is the one-dimensional Helmholtz equation with

$$\beta = \omega \sqrt{LC} = \frac{2\pi}{\lambda_g} \qquad (12.30)$$

We can confirm that β plays the role of the propagation constant in a transmission line by taking a look at the units of β. λ_g is the wavelength of the voltage or current wave in the transmission line. Working with the first equation of the previous Eqn set (12.24), the solution is

$$V(z) = \text{one of} \begin{cases} V_+ e^{-j\beta z}, & \text{for a wave travelling in the } +z \text{ direction, or} \\ V_- e^{+j\beta z}, & \text{for a wave travelling in the } -z \text{ direction} \end{cases}$$

where V_+ and V_- are amplitudes of the forward and backward waves respectively and z is the distance along the line with respect to some origin. An illustration of these waves is shown in Fig. 12.5.

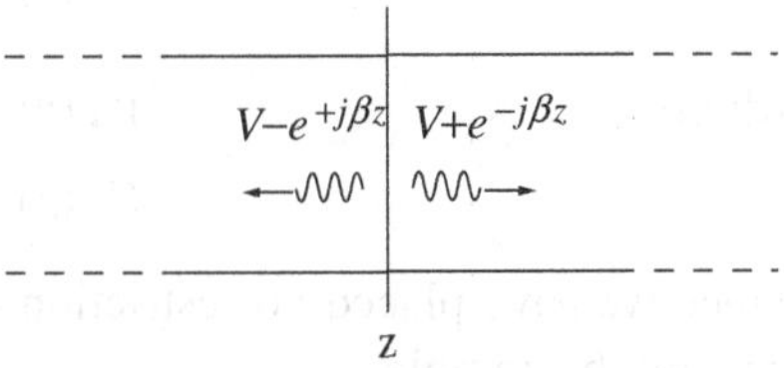

In any section of the line, for a coordinate z along the line, the voltage can be written as a sum of these two solutions, that is, in particular,

Fig. 12.5 Forward and reverse voltage waves

$$V(z) = \underbrace{V_+ e^{-j\beta z}}_{\text{Incident wave}} + \underbrace{V_- e^{+j\beta z}}_{\text{Reflected wave}} \qquad (12.31)$$

For each of these solutions, the corresponding current waves are

$$I(z) = \text{one of} \begin{cases} I_+ e^{-j\beta z} & \text{for a wave travelling in the } +z \text{ direction, or} \\ I_- e^{+j\beta z} & \text{for a wave travelling in the } -z \text{ direction} \end{cases}$$

To get the relation between I_+ and V_+ we use Eqn (12.14)

$$\frac{\partial V(z)}{\partial z} = -j\omega L I(z)$$

$$\frac{\partial}{\partial z} V_+ e^{-j\beta z} = -j\omega L I(z)$$

$$-j\beta V_+ e^{-j\beta z} = -j\omega L I(z)$$

$$\beta V_+ e^{-j\beta z} = \omega L I_+ e^{-j\beta z}$$

$$\omega \sqrt{LC} V_+ = \omega L I_+$$

$$\sqrt{\frac{C}{L}} V_+ = I_+$$

or

$$I_+ = \frac{V_+}{Z_0} \qquad (12.32)$$

where Z_0 is the characteristic impedance of the line. The unit of Z_0

$$Z_0 = \sqrt{\frac{L}{C}} \ (\Omega) \tag{12.33}$$

Note that for a *loss less line*, Z_0 is always *real*. By the same technique, we can show that

$$I_- = -\frac{V_-}{Z_0} \tag{12.34}$$

These two relations are very important, in that the forward and reverse voltage waves are linked to the corresponding current waves, that is, if

$$V(z) = V_+ e^{-j\beta z} + V_- e^{+j\beta z}$$

then

$$I(z) = \frac{1}{Z_0}(V_+ e^{-j\beta z} - V_- e^{+j\beta z}) \tag{12.35}$$

Furthermore, we introduce a new (but important) variable, $\Gamma(z)$, *the reflection coefficient* along the line for any value of z

$$\Gamma(z) = \frac{V_-(z)}{V_+(z)} = \frac{V_-}{V_+}e^{2j\beta z} \tag{12.36}$$

where

$$V_+(z) = V_+ e^{-j\beta z}$$
$$V_-(z) = V_- e^{+j\beta z} \tag{12.37}$$

Since we have placed no restriction on V_+ and V_-, these two coefficients can, in general, be complex.

12.4.1 Power Considerations

Let us take a look at the power in the forward/backward wave. The forward/backward wave carries a total *average* power of (since we are *not* considering rms values of the voltage and current, and Z_0 is assumed to be real)

$$
\begin{aligned}
P_{av} &= \frac{1}{2}\Re\{VI^*\} \\
&= \frac{1}{2}\Re\{(V_\pm e^{\mp j\beta z})(I_\pm e^{\mp j\beta z})^*\} \\
&= \frac{1}{2}\Re\{(V_\pm e^{\mp j\beta z})(I_\pm^* e^{\pm j\beta z})\} \\
&= \frac{1}{2}\Re\left\{V_\pm \times \left(\pm\frac{V_\pm^*}{Z_0}\right)\right\} \\
&= \pm\frac{|V_\pm|^2}{2Z_0} \tag{12.38}
\end{aligned}
$$

A little explanation is needed. The '$\pm$' subscript is the indicator of the forward/backward wave. The power, P_+, in the forward wave is positive, while the power in the reflected wave, P_-, is negative. The ratio of the reflected to the incident power is

$$\frac{P_-}{P_+} = \frac{|V_-|^2/2Z_0}{|V_+|^2/2Z_0} = \left|\frac{V_-}{V_+}\right|^2 = |\Gamma|^2 \tag{12.39}$$

12.4.2 Reflections from Discontinuities

We have obtained the solution to the transmission line equation, but we are not nearer to getting the values of V_+ and V_-. How do we get these, to apply our solution to real problems?

Let us consider the case where the incident electromagnetic wave, characterised by incident voltage and current waves meets a load impedance Z_L, shown in Fig. 12.6. The figure shows that the load is placed at $z = 0$, while the generator is at $z = -L$. The incident parameters are

$$V_+(z) = V_+ e^{-j\beta z}$$

$$I_+(z) = I_+ e^{-j\beta z} \tag{12.40}$$

After meeting the load a reflected electromagnetic wave is set up, characterised by

$$V_-(z) = V_- e^{+j\beta z}$$

$$I_-(z) = I_- e^{+j\beta z} \tag{12.41}$$

The total voltage and currents are

$$V(z) = V_+(z) + V_-(z) \quad \text{for } z < 0$$

$$I(z) = I_+(z) + I_-(z) = \frac{V_+(z) - V_-(z)}{Z_0}, \quad \text{for } z < 0 \tag{12.42}$$

We know that the total voltage and current across the load must satisfy

$$
\begin{aligned}
Z_L &= \frac{V(0)}{I(0)} \\
&= \left. \frac{V_+ e^{-j\beta z} + V_- e^{+j\beta z}}{I_+ e^{-j\beta z} + I_- e^{+j\beta z}} \right|_{z=0} \\
&= \left. \frac{V_+ e^{-j\beta z} + V_- e^{+j\beta z}}{(V_+/Z_0)\, e^{-j\beta z} + (-V_-/Z_0)\, e^{+j\beta z}} \right|_{z=0} \\
&= \left. \frac{Z_0(V_+ e^{-j\beta z} + V_- e^{+j\beta z})}{V_+ e^{-j\beta z} - V_- e^{+j\beta z}} \right|_{z=0} \\
&= \frac{Z_0(V_+ + V_-)}{V_+ - V_-} \\
&= \frac{Z_0 V_+ \{1 + (V_-/V_+)\}}{V_+ \{1 - (V_-/V_+)\}} \\
&= \frac{Z_0(1 + \Gamma_L)}{(1 - \Gamma_L)} \tag{12.43}
\end{aligned}
$$

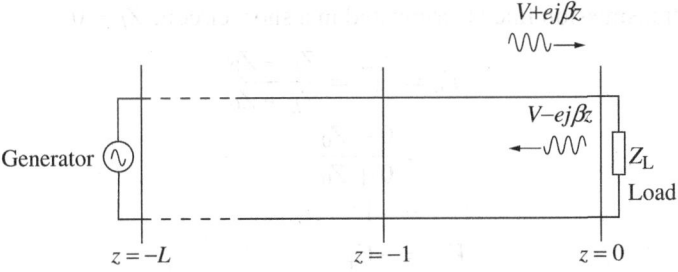

Fig. 12.6 Transmission line with a load impedance

where Γ_L is the reflection coefficient at the load.

$$\Gamma_L = \left.\left(\frac{V_-}{V_+}e^{2j\beta z}\right)\right|_{z=0} = \frac{V_+}{V_-} = \Gamma(0) \tag{12.44}$$

Solving for Γ_L

$$Z_L = \frac{Z_0(1+\Gamma_L)}{(1-\Gamma_L)}$$

$$(1-\Gamma_L)\left(\frac{Z_L}{Z_0}\right) = 1+\Gamma_L$$

$$\left(\frac{Z_L}{Z_0}\right) - \Gamma_L\left(\frac{Z_L}{Z_0}\right) = 1+\Gamma_L$$

$$\left(\frac{Z_L}{Z_0}\right) - 1 = \Gamma_L\left\{\left(\frac{Z_L}{Z_0}\right) + 1\right\}$$

$$\Gamma_L = \frac{\left(\frac{Z_L}{Z_0}\right) - 1}{\left(\frac{Z_L}{Z_0}\right) + 1} \tag{12.45}$$

$$\Gamma_L = \frac{V_-}{V_+} = \frac{Z_L - Z_0}{Z_L + Z_0} \tag{12.46}$$

Example 12.5 Examine the reflection coefficient in the following cases: (a) the termination is an open circuit, (b) the termination is a short circuit, (c) the termination is equal to Z_0 and (d) the termination is a reactive element.

Solution It is interesting to examine this equation in various cases.

When the transmission line is terminated in an open circuit: $Z_L = \infty$

$$\Gamma_L = \frac{V_-}{V_+} = \frac{Z_L - Z_0}{Z_L + Z_0}$$

$$\approx \frac{Z_L}{Z_L} \quad (Z_L \gg Z_0)$$

$$= 1$$

$$V_- = V_+$$

which means that the total incident power is reflected, since

$$\frac{P_-}{P_+} = |\Gamma_L|^2 = 1$$

When the transmission line is terminated in a short circuit: $Z_L = 0$

$$\Gamma_L = \frac{V_-}{V_+} = \frac{Z_L - Z_0}{Z_L + Z_0}$$

$$= \frac{0 - Z_0}{0 + Z_0}$$

$$= -1$$

$$V_- = -V_+$$

again we can see that all the incident power is reflected.

When the transmission line is terminated in the characteristic impedance: $Z_L = Z_0$

$$\Gamma_L = \frac{V_-}{V_+} = \frac{Z_L - Z_0}{Z_L + Z_0}$$

$$= \frac{Z_0 - Z_0}{Z_0 + Z_0}$$

$$= 0$$

$$V_- = 0$$

since there is no reflected wave, all the incident power is consumed. This is the case of maximum power transfer: the condition of a matched load.

We also know that if the load is purely imaginary ($Z_L = jX$, X being real) no power is consumed by the load. In this case too all of the incident power must be reflected.

$$\Gamma_L = \frac{V_-}{V_+} = \frac{Z_L - Z_0}{Z_L + Z_0}$$

$$= \frac{jX - Z_0}{jX + Z_0}$$

$$= 1 e^{-j2 \tan^{-1}(X/Z_0)}$$

$$V_- = \left\{ 1 e^{-j2 \tan^{-1}(X/Z_0)} \right\} \times V_+ \qquad (12.47)$$

The total incident power is reflected here since

$$|\Gamma_L| = 1 \qquad (12.48)$$

which confirms our surmise.

___

Example 12.6 Find the percentage of reflected power when a load of value $Z_L = 40 + j40 \ \Omega$ terminates a 75 Ω ($= Z_0$) transmission line.

Solution

$$\Gamma_L = \frac{Z_L - Z_0}{Z_L + Z_0}$$

$$= -0.16358 + j0.40472$$

Now

$$|\Gamma_L| = \left| \frac{V_-}{V_+} \right| = 0.43653$$

and

$$\left| \frac{V_-}{V_+} \right|^2 = 0.19056$$

So 19% of the power is reflected.

___

The reflection coefficient along the line from an earlier equation (Eqn 12.36) is

$$\Gamma(z) = \frac{V_-}{V_+} e^{2\beta z} = \Gamma_L e^{2\beta z}$$

and when we move towards the generator, $z = -l$

$$\Gamma(-l) = \Gamma_L e^{-j2\beta l} \qquad (12.49)$$

which tells us that the magnitude of $\Gamma(z)$ remains constant, $= |\Gamma_L|$, while the phase changes by $-2\beta l$. Using $\beta = 2\pi / \lambda_g$,

$$\angle \Gamma(-l) = -4\pi l_\lambda \qquad (12.50)$$

where $l_\lambda = l/\lambda$ is the distance in wavelengths towards the generator.

If we study this equation carefully we find that the complex reflection coefficient has a constant magnitude, $|\Gamma_L|$, and it returns to its original value every half a wavelength: that is, for $l_\lambda = 0.5$.

12.4.3 Standing Wave Ratio

Referring to Fig. 12.6, the voltage along the line at a point $z = -l$ is given by

$$V(-l) = V_+e^{-j\beta(-l)} + V_-e^{j\beta(-l)}$$
$$= V_+(\cos\beta l + j\sin\beta l) + V_-(\cos\beta l - j\sin\beta l)$$
$$= (V_+ + V_-)\cos\beta l + j\sin\beta l(V_+ - V_-)$$
$$= V_L\cos\beta l + jZ_0I_L\sin\beta l \tag{12.51}$$
$$= V_L\left\{\cos\beta l + j\left(\frac{Z_0}{Z_L}\right)\sin\beta l\right\} \tag{12.52}$$

This is a mathematical statement of the fact that the total voltage along the line is a sum of the forward and backward voltage waves. Here V_L and I_L are the voltage and current across and through the load, respectively.

$$V_L = V(0) = V_+e^{-j\beta(-l)} + V_-e^{j\beta(-l)}\Big|_{l=0} = V_+ + V_-$$
$$I_L = I(0) = I_+e^{-j\beta(-l)} + I_-e^{j\beta(-l)}\Big|_{l=0} = I_+ + I_- = \frac{V_+ - V_-}{Z_0}$$
$$V_L = I_LZ_L \tag{12.53}$$

A plot of the magnitude of the normalised voltage, $|V(-l)/V_L|$, along the line for $V_L = 1$ V and $Z_L/Z_0 = 2$ is shown in Fig. 12.7. The figure shows the resultant sum of the forward and backward voltages. A study of the figure shows that

1. There is a voltage maxima at the load.
2. The periodicity of the standing wave pattern is $\lambda/2$ instead of the usual λ.
3. $V_{max}/V_{min} = 2$.

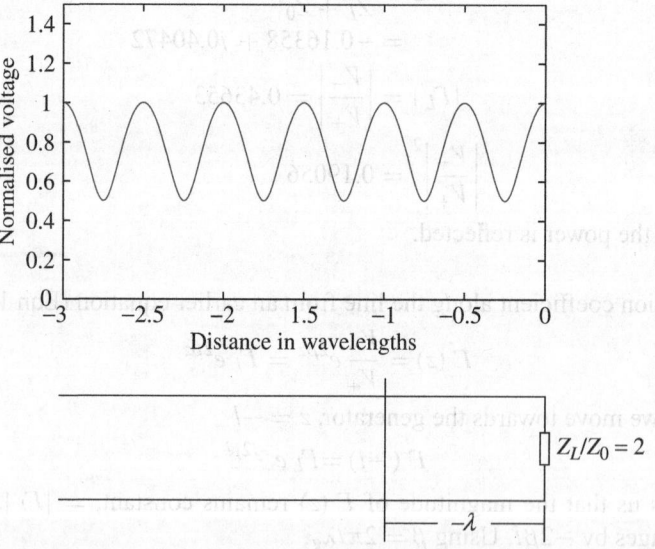

Fig. 12.7 Magnitude of the normalised voltage, $|V(-l)/V_L|$, along a line for $V_L = 1$ V and $Z_L/Z_0 = 2$

Let us investigate this analytically (Z_L is real). From Eqn (12.52),

$$V(-l) = V_L \left\{ \cos \beta l + j \left(\frac{Z_0}{Z_L} \right) \sin \beta l \right\}$$

$$|V(-l)| = V_L \sqrt{\cos^2 \beta l + \left(\frac{Z_0}{Z_L} \right)^2 \sin^2 \beta l} \qquad (12.54)$$

Differentiating this function with respect to l, keeping $z = Z_0/Z_L$

$$\frac{d|V(-l)|}{dl} = \frac{\dfrac{2\beta \cos(\beta l) \sin(\beta l)}{z} - 2\beta \cos(\beta l) \sin(\beta l)}{2\sqrt{\dfrac{\sin^2(\beta l)}{z} + \cos^2(\beta l)}}$$

To find the maximas and minimas, we set the numerator to be zero, which gives

$$\sin(2\beta l) = 0$$
$$2\beta l = n\pi$$
$$l = \frac{n\pi}{2 \times \frac{2\pi}{\lambda}}$$
$$l = \frac{n\lambda}{4} \qquad (12.55)$$

Hence, maximas and minimas occur at $-l = 0, -\lambda/4, -\lambda/2, 3\lambda/4, \ldots$ At these lengths

$$\beta l = \frac{2\pi}{\lambda} \times \frac{n\lambda}{4} = \frac{n\pi}{2}$$

Making a table

βl	0	$-\pi/2$	$-\pi$	$-3\pi/2$	$\ldots$
$|V(-l)|$	1	Z_0/Z_L	1	Z_0/Z_L	$\ldots$

Comparing values in the second row,

if $\begin{cases} Z_L > Z_0, & \text{then there is a maximum at the load } (1 > Z_0/Z_L) \\ Z_L < Z_0, & \text{then there is a minimum at the load } (1 < Z_0/Z_L) \end{cases}$

and maxima and minima alternate thereafter at intervals of $\lambda/4$.

The ratio $V_{\text{max}}/V_{\text{min}}$ is called the voltage standing wave ratio (VSWR)

$$s \text{ (VSWR)} = V_{\text{max}}/V_{\text{min}} \qquad (12.56)$$

We now proceed to obtain an analytic expression for s. The voltage along the line is given by

$$V(z) = V_+ e^{-j\beta z} + V_- e^{j\beta z} \qquad (12.57)$$

Due to some reflection, a standing wave is formed. Let the reflection coefficient be Γ_L. Then

$$V_- = \Gamma_L V_+ \qquad (12.58)$$

and the previous equation may be written as

$$V(z) = V_+ (e^{-j\beta z} + \Gamma_L e^{j\beta z})$$
$$|V(z)| = |V_+| |(e^{-j\beta z} + \Gamma_L e^{j\beta z})| \qquad (12.59)$$

Now the maximum value of the above equation is when the maximum of each individual term adds and the minimum value is when the two terms in the bracket subtract (remember that $|\Gamma_L| \leqslant 1$). That is,

$$|V(z)|_{max} = |V_+||(1 + |\Gamma_L|)|$$
$$|V(z)|_{min} = |V_+||(1 - |\Gamma_L|)| \qquad (12.60)$$

so
$$s\ (VSWR) = \frac{|V(z)|_{max}}{|V(z)|_{min}} = \frac{1 + |\Gamma_L|}{1 - |\Gamma_L|} \qquad (12.61)$$

Applying this formula to the case of Fig. 12.7

$$Z_L/Z_0 = 2$$

$$\Gamma_L = \frac{Z_L/Z_0 - 1}{Z_L/Z_0 + 1}$$

$$= 1/3$$

$$\text{VSWR, } s = \frac{1\frac{1}{3}}{\frac{2}{3}} = 2$$

For the current, we may proceed along similar lines. The current along the line is given by

$$I(-l) = I_+ e^{-j\beta(-l)} + I_- e^{j\beta(-l)}$$
$$= I_+(\cos\beta l + j\sin\beta l) + I_-(\cos\beta l - j\sin\beta l)$$
$$= (I_+ + I_-)\cos\beta l + j\sin\beta l(I_+ - I_-)$$
$$= I_L\cos\beta l + j\frac{V_L}{Z_0}\sin\beta l \qquad (12.62)$$
$$= I_L\left\{\cos\beta l + j\left(\frac{Z_L}{Z_0}\right)\sin\beta l\right\} \qquad (12.63)$$

A plot of the magnitude of the normalised current, $|I(-l)/I_L|$, along the line for $I_L = 0.5$ A and $Z_L/Z_0 = 2$ is shown in Fig. 12.8. These are the voltage and current along the line in terms of the load parameters and load impedance.

Example 12.7 A lossless transmission line is terminated in (a) an open circuit, (b) a short circuit, (c) Z_0, (d) a reactive termination and (e) a resistive termination. Find the VSWR in all these cases.

Solution

(a) For an open circuit,

$$\Gamma_L = 1$$

and
$$s_{open} = \frac{V_{max}}{V_{min}} = \frac{1 + |\Gamma_L|}{1 - |\Gamma_L|} = \infty$$

(b) For a short circuit,

$$\Gamma_L = -1$$

so again
$$s_{short} = \frac{1 + |\Gamma_L|}{1 - |\Gamma_L|} = \infty$$

(c) For $Z_L = Z_0$

$$\Gamma_L = 0$$

which gives
$$s(\text{of } Z_0) = 1$$

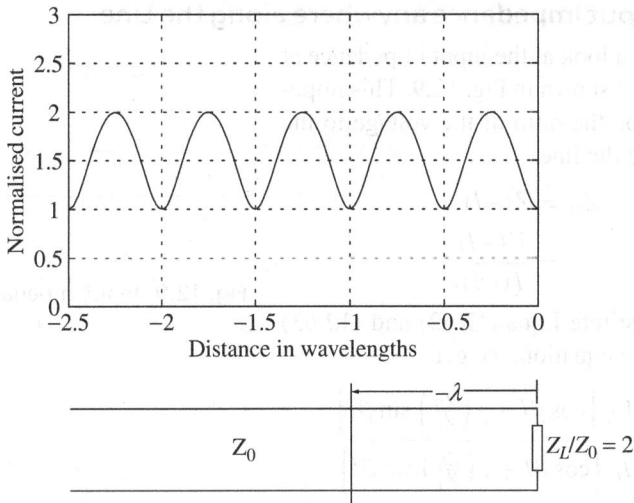

Fig. 12.8 Magnitude of normalised current, $|I(-l)/I_L|$ along the line for $I_L = 0.5$ A and $Z_L/Z_0 = 2$

(d) For a reactive termination

$$|\Gamma_L| = 1$$

which means that $s = \infty$

(e) For a resistive termination

$$\Gamma_L = \frac{R_L - Z_0}{R_L + Z_0}$$

and

$$s = \frac{1 + |\Gamma_L|}{1 - |\Gamma_L|}$$

$$= \frac{1 + |(R_L - Z_0)|/(R_L + Z_0)}{1 - |(R_L - Z_0)|/(R_L + Z_0)}$$

$$= \frac{(R_L + Z_0) + |(R_L - Z_0)|}{(R_L + Z_0) - |(R_L - Z_0)|}$$

Now two cases arise:

Case (i) $R_L > Z_0$ then

$$s = \frac{(R_L + Z_0) + (R_L - Z_0)}{(R_L + Z_0) - (R_L - Z_0)}$$

$$= \frac{R_L}{Z_0} > 1$$

and Case (ii) $R_L < Z_0$ then

$$s = \frac{(R_L + Z_0) + (Z_0 - R_L)}{(R_L + Z_0) - (Z_0 - R_L)}$$

$$= \frac{Z_0}{R_L} > 1$$

The above calculation gives an interesting finding that if two resistors are normalised with respect to the characteristic impedances, and the normalised values are reciprocals of each other then they have the same VSWR!

12.4.4 Input Impedance anywhere along the Line

We now take a look at the input impedance at a point $z = -l$ shown in Fig. 12.9. This impedance must be the ratio of the voltage to the current along the line.

$$Z_{in} = Z(-l)$$
$$= \frac{V(-l)}{I(-l)}$$

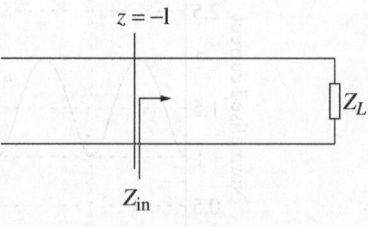

Fig. 12.9 Input impedance, Z_{in}, at $z = -l$

We now substitute Eqns (12.52) and (12.63) into the above equation, we get

$$Z_{in} = \frac{V_L \left\{ \cos \beta l + j \left(\frac{Z_0}{Z_L} \right) \sin \beta l \right\}}{I_L \left\{ \cos \beta l + j \left(\frac{Z_L}{Z_0} \right) \sin \beta l \right\}}$$

$$= \frac{Z_L \left\{ \cos \beta l + j \left(\frac{Z_0}{Z_L} \right) \sin \beta l \right\}}{\left\{ \cos \beta l + j \left(\frac{Z_L}{Z_0} \right) \sin \beta l \right\}} \quad (\text{since } Z_L = V_L/I_L)$$

$$= \frac{\{ Z_L \cos \beta l + j Z_0 \sin \beta l \}}{\frac{1}{Z_0} \{ Z_0 \cos \beta l + j Z_L \sin \beta l \}}$$

$$= Z_0 \left\{ \frac{Z_L + j Z_0 \tan(\beta l)}{Z_0 + j Z_L \tan(\beta l)} \right\} \quad (\text{Dividing the Num. and Den. by } \cos \beta l)$$

$$Z_{in}(-l) = Z_0 \left\{ \frac{Z_L + j Z_0 \tan(\beta l)}{Z_0 + j Z_L \tan(\beta l)} \right\} \qquad (12.64)$$

Example 12.8 Find the input impedance for a lossless line with Z_L equal to (a) an open circuit, (b) a short circuit, and (c) Z_0.

Solution Let us apply Eqn (10.63) to various cases.

(a) Open circuit: $Z_L = \infty$

$$Z_{in} = Z_0 \left\{ \frac{Z_L + j Z_0 \tan(\beta l)}{Z_0 + j Z_L \tan(\beta l)} \right\}$$

$$= Z_0 \left\{ \frac{Z_L}{j Z_L \tan(\beta l)} \right\} \quad (Z_L \gg Z_0)$$

$$= -j Z_0 \cot \beta l$$

(b) Short circuit: $Z_L = 0$

$$Z_{in} = Z_0 \left\{ \frac{Z_L + j Z_0 \tan(\beta l)}{Z_0 + j Z_L \tan(\beta l)} \right\}$$

$$= Z_0 \left\{ \frac{j Z_0 \tan(\beta l)}{Z_0} \right\} \quad (Z_L = 0)$$

$$= j Z_0 \tan \beta l$$

(c) Matched load $Z_L = Z_0$

$$Z_{in} = Z_0 \left\{ \frac{Z_L + j Z_0 \tan(\beta l)}{Z_0 + j Z_L \tan(\beta l)} \right\}$$

$$= Z_0 \left\{ \frac{Z_0 + jZ_0 \tan(\beta l)}{Z_0 + jZ_0 \tan(\beta l)} \right\} \quad (Z_L = Z_0)$$

$$= Z_0$$

Example 12.9 Find the input impedance at $l = \lambda/4$ for a lossless line with Z_L equal to (a) an inductor L, (b) a capacitor C, (c) $R + jX$.

Solution For a lossless line,

$$Z_{in} = Z_0 \left\{ \frac{Z_L + jZ_0 \tan(\beta l)}{Z_0 + jZ_L \tan(\beta l)} \right\}$$

(a) $Z_L = j\omega L$ gives

$$Z_{in} = Z_0 \left\{ \frac{j\omega L + jZ_0 \tan(\beta l)}{Z_0 + j(j\omega L) \tan(\beta l)} \right\} = jZ_0 \left\{ \frac{\omega L + Z_0 \tan(\beta l)}{Z_0 - \omega L \tan(\beta l)} \right\}$$

Putting these values in the above expression for Z_{in}, for $l = \lambda/4$, $\beta l = 2\pi (l/\lambda)$
$= \pi/2$

$$Z_{in}(-\lambda/4) = Z_0 \left\{ \frac{jZ_0}{-\omega L} \right\}$$

$$= -jZ_0 \left\{ \frac{Z_0}{\omega L} \right\}$$

$$= \frac{Z_0^2}{j\omega L}$$

The result shows that an inductor behaves like a capacitor.

(b) Putting $Z_L = 1/(j\omega C)$ in Eqn (12.64)

$$Z_{in} = Z_0 \left\{ \frac{[1/(j\omega C)] + jZ_0 \tan(\beta l)}{Z_0 + j[1/(j\omega C)] \tan(\beta l)} \right\}_{l=\lambda/4}$$

$$= Z_0 \left\{ \frac{jZ_0}{j[1/(j\omega C)]} \right\}$$

$$= Z_0^2 (j\omega C)$$

The result shows that a capacitor behaves like an inductor.

(c) For the third case, if $Z_L = R + jX$

$$Z_{in} = Z_0 \left\{ \frac{Z_L + jZ_0 \tan(\beta l)}{Z_0 + jZ_L \tan(\beta l)} \right\}_{l=\lambda/4}$$

$$= Z_0 \left\{ \frac{jZ_0}{jZ_L} \right\} = Z_0^2 Y_L$$

This shows that the input impedance appears as its reciprocal, Y_L.

 See `zinput.m` in Chapter 12

12.4.5 UHF Circuit Elements: $\lambda/8$, $\lambda/4$, and $\lambda/2$ Lines

At UHF frequencies and above (≥ 1 GHz) circuit dimensions start to be comparable to the wavelength (see the box in Sec. 12.1). If we look at short transmission lines, some very interesting properties emerge. Specifically, we take a look at $\lambda/8$, $\lambda/4$, and $\lambda/2$ lines. First of all let us calculate the value of $\tan\beta l$ for various values of l.

$$\tan\beta l]_{l=\lambda/8} = \tan(\pi/4) = 1$$
$$\tan\beta l]_{l=\lambda/4} = \tan(\pi/2) = \infty$$
$$\tan\beta l]_{l=\lambda/2} = \tan(\pi) = 0 \tag{12.65}$$

Using these values and substituting in Eqn (12.64), we get

$$Z_{in} = Z_0 \left\{ \frac{Z_L + jZ_0 \tan(\beta l)}{Z_0 + jZ_L \tan(\beta l)} \right\}$$

$$Z_{in}(-\lambda/8) = Z_0 \left(\frac{Z_L + jZ_0}{Z_0 + jZ_L} \right) = Z_0[\angle(Z_L + jZ_0) - \angle(Z_0 + jZ_L)] \tag{12.66}$$

$$Z_{in}(-\lambda/4) = Z_0 \left(\frac{Z_0}{Z_L} \right) = Z_0^2/Z_L$$

$$Z_{in}(-\lambda/2) = Z_0 \left(\frac{Z_L}{Z_0} \right) = Z_L \tag{12.67}$$

If we look at Eqn set (12.67) then the most useful equations are the first two equations of Eqn (12.67).

Example 12.10   Create a 1 nH inductor at a frequency of 1 GHz.
Solution   Using a $\lambda/4$ line,

$$Z_{in}(-\lambda/4) = Z_0^2/Z_L$$

Let Z_L be a capacitor,

$$Z_L = 1/(j\omega C)$$

then

$$Z_{in}(-\lambda/4) = j\omega C Z_0^2$$

At 1 GHz, 1 nH inductor has an impedance of

$$Z = j\omega \times 1 \times 10^{-9}$$

Equating

$$j\omega \times 1 \times 10^{-9} = j\omega C Z_0^2$$

$$C = \frac{1 \times 10^{-9}}{Z_0^2}$$

Using a convenient value of Z_0, say 50 Ω,

$$C = 0.4 \text{ pF}$$

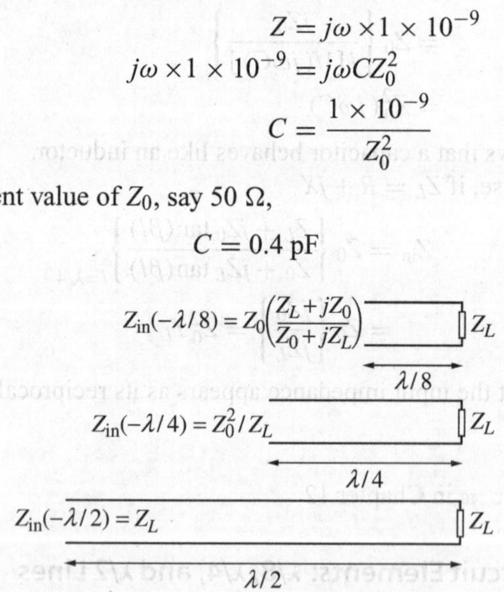

Fig. 12.10 $\lambda/8$, $\lambda/4$, and $\lambda/2$ lines

So if we terminate a $\lambda/4$ line of $Z_0 = 50\,\Omega$ with a capacitance of 0.4 pF, the input to the line is a 1 nH inductor (Fig. 12.10).

Example 12.11 Investigate the properties of $\lambda/8$ lines.

Solution For $\lambda/8$ lines,

$$Z_{in}(-\lambda/8) = Z_0 \left(\frac{Z_L + jZ_0}{Z_0 + jZ_L} \right)$$

Taking the real and imaginary parts of this equation

$$\Re(Z_{in}) = \frac{2 Z_0^2 Z_L}{Z_L^2 + Z_0^2}$$

$$\Im(Z_{in}) = \frac{Z_0 (Z_0^2 - Z_L^2)}{Z_L^2 + Z_0^2} \qquad (12.68)$$

The interesting cases are for $Z_L = 0$ and $Z_L = \infty$. For $Z_L = 0$,

$$\Re(Z_{in}) = 0$$
$$\Im(Z_{in}) = Z_0$$

So Z_{in} behaves like an inductor whose impedance is Z_0,

$$Z_{in} = jZ_0 = j\omega L$$

On the other hand, if we set $Z_L = \infty$ in Eqn set (12.68) then

$$\Re(Z_{in}) = 0$$
$$\Im(Z_{in}) = -Z_0$$

therefore an open-circuited $\lambda/8$ line behaves like a capacitor whose independance is Z_0,

$$Z_{in} = -jZ_0 = \frac{1}{jY_0} = \frac{1}{j\omega C}$$

the value of capacitor is given by

$$\omega C = Y_0$$

12.4.6 Propagation of Plane Waves with Transmission Line Concepts

We have studied plane waves traversing in space and we have studied transmission lines. Is there a link between the two? We study this point using a concrete example.

Example 12.12 Discuss the propagation of plane waves using transmission line concepts.

Solution In general if we examine the propagation of plane waves in any medium, the complex propagation constant γ is given by

$$\gamma = \sqrt{j\omega\mu\,(j\omega\varepsilon + \sigma)} \qquad (12.69)$$

and if we compare this with the complex propagation constant for a transmission line

$$\gamma = \sqrt{(j\omega L + R)\,(j\omega C + G)}$$

then the analogy is clear,

$$L = \mu$$
$$R = 0$$
$$C = \varepsilon$$

and

$$G = \sigma$$

Using this analogy

$$Z = \sqrt{\frac{j\omega\mu}{j\omega\varepsilon + \sigma}}$$

where Z is the intrinsic impedance of the medium for a plane wave. Similarly, for normal incidence on a perfect electric conductor, the reflection coefficient is

$$R = \frac{Z_L - Z}{Z_L + Z}$$

where Z_L is equal to zero (a perfect short circuit) which gives $R = -1$ and therefore sets up a standing wave. Similarly, for normal incidence on a dielctric interface (μ, ε_2), from medium 1, (μ, ε_1)

$$R = \frac{Z_L - Z}{Z_L + Z}$$
$$= \frac{\sqrt{\mu/\varepsilon_2} - \sqrt{\mu/\varepsilon_1}}{\sqrt{\mu/\varepsilon_2} + \sqrt{\mu/\varepsilon_1}}$$
$$= \frac{\sqrt{1/\varepsilon_2} - \sqrt{1/\varepsilon_1}}{\sqrt{1/\varepsilon_2} + \sqrt{1/\varepsilon_1}}$$

This analogy may also be extended to waveguides as well.

12.5 | Transmission Line Charts

It is clear that the number of equations involved in transmission line problems increases the complexity of the computation, hence a graphical procedure has been evolved using transmission line charts. We know that the terminating impedance Z_L is complex, as is the case with Z_{in}, but for any sense to be made of this parameter, it must be normalised with respect to the characteristic impedance, Z_0, of the line. So we define a new parameter

$$\mathbb{Z}_{in}(-l) = \frac{Z_{in}(-l)}{Z_0} \tag{12.70}$$

$$= r + jx \tag{12.71}$$

Using the new parameter

$$\Gamma = \frac{Z_{in} - Z_0}{Z_{in} + Z_0}$$

$$u + jv = \frac{(Z_{in}/Z_0 - 1)}{(Z_{in}/Z_0 + 1)} \quad (\Gamma = u + jv)$$

$$= \frac{\mathbb{Z}_{in} - 1}{\mathbb{Z}_{in} + 1} \tag{12.72}$$

This is a mapping between the complex variables Γ and $\mathbb{Z}_{\text{in}}$. The relations of u and v as functions of r and x are

$$u = \frac{x^2 + (r-1)(r+1)}{x^2 + (r+1)^2} \tag{12.73}$$

$$v = \frac{2x}{x^2 + (r+1)^2} \tag{12.74}$$

Thus for every r and x, we get a u and a (remember $\Gamma = u + jv$). With reference to Fig. 12.11, we know that since generally

$$|\Gamma| = \left| \frac{V_-(z)}{V_+(z)} \right| \leq 1 \tag{12.75}$$

then the region contained within the unit circle (which is $|\Gamma| = 1$) contains all values of r and x. For example, using the above equations,

1. $r = 0$, $x = 0$ maps to $u = -1$, $v = 0$.
2. $r = 1$, $x = 0$ maps to $u = 0$, $v = 0$.
3. $r = \infty$, $x = 0$ maps to $u = 1$, $v = 0$.
4. $r = 0$, $x = 1$ maps to $u = 0$, $v = 1$
5. $r = 0$, $x = -1$ maps to $u = 0$, $v = -1$

The transformation in the other direction is more useful.

$$\mathbb{Z}_{\text{in}} = \frac{1+\Gamma}{1-\Gamma} \tag{12.76}$$

$$r = \frac{(1-u)(u+1) - v^2}{v^2 + (1-u)^2} \tag{12.77}$$

$$x = \frac{2v}{v^2 + (1-u)^2} \tag{12.78}$$

Working with the previous equation,

$$r\{v^2 + (1-u)^2\} = (1-u)(u+1) - v^2$$

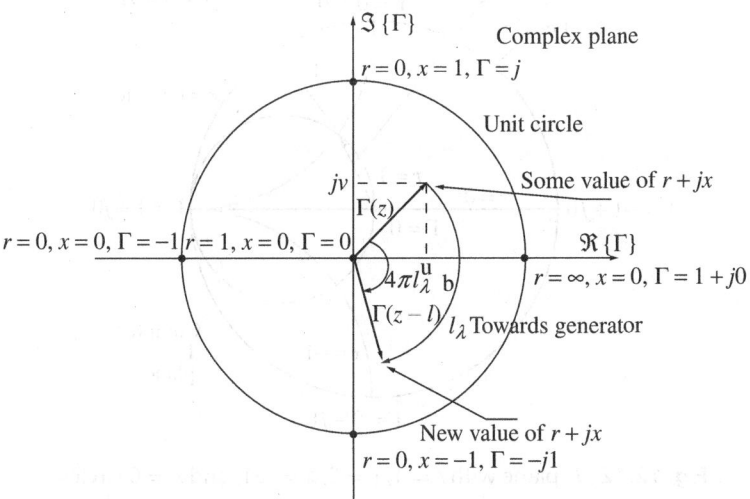

Fig. 12.11 Γ in the complex plane

$$(1+r)v^2 + r + (1+r)u^2 - 2ru = 1$$

$$v^2 + \frac{r}{1+r} + u^2 - \frac{2r}{1+r}u = \frac{1}{1+r}$$

$$v^2 + \left(u - \frac{r}{1+r}\right)^2 - \left(\frac{r}{1+r}\right)^2 = \frac{1}{1+r} - \frac{r}{1+r}$$

$$v^2 + \left(u - \frac{r}{1+r}\right)^2 = \frac{1}{1+r} - \frac{r}{1+r} + \left(\frac{r}{r+1}\right)^2$$

$$v^2 + \left(u - \frac{r}{1+r}\right)^2 = \left(\frac{1}{1+r}\right)^2 \qquad (12.79)$$

which is the equation of a circle with centre $[r/(1+r), 0]$ and radius $1/(1+r)$. Similarly, the other equation gives

$$(u-1)^2 + \left(v - \frac{1}{x}\right)^2 = \left(\frac{1}{x}\right)^2 \qquad (12.80)$$

which is the equation of a circle with centre $(1, 1/x)$ and radius $1/|x|$. On the gamma plane, let us draw the five circles with $r = 1$, $r = 0$, $x = \pm 1$, and $x = 0$ which are shown in Fig. 12.12.

Let us see how these plots were obtained.

1. The $r = 1$ circle by drawing a circle with centre $(1/2, 0)$ and radius $1/2$.
2. The $r = 0$ circle has its centre at $(0, 0)$ and has a radius of 1.
3. The $x = \pm 1$ circles have their centre at $(1, \pm 1)$ and radii of 1.
4. The $x = 0$ 'circle' has its centre at $(1, \infty)$ and radius of ∞, which is a straight line.

In this way we can get every point on the gamma plane in terms of (r, x), which is the Smith chart shown in Fig. 12.13 (Smith, 1939).

 See the folder 'Smith Chart' for coloured Smith charts.

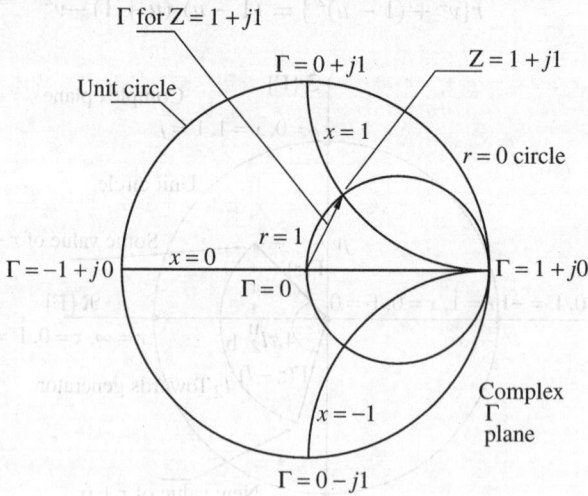

Fig. 12.12 Γ plane with $r = 1, r = 0, x = \pm 1$, and $x = 0$ circles

Example 12.13 For a 50 Ω transmission line, plot the following values of Z_L (a) 0, (b) ∞, (c) 30 Ω, (d) $30 + j30$ Ω, (e) 50 Ω on Smith chart show in Fig. 12.13.

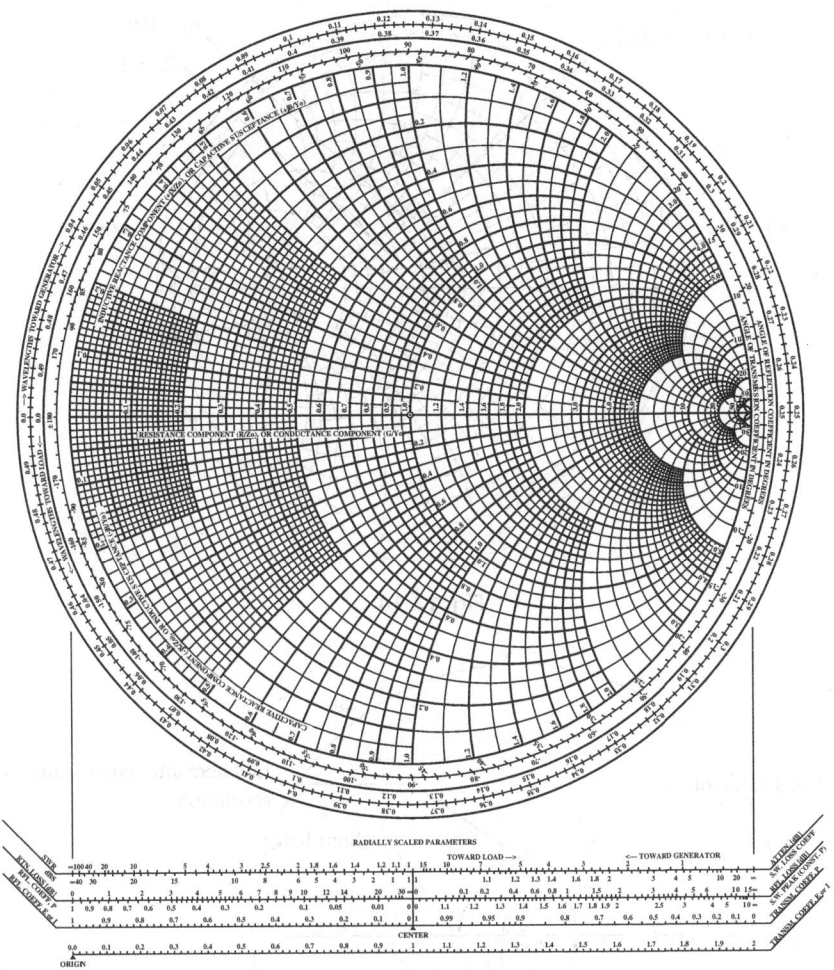

Fig. 12.13 Smith chart

Solution For all these values of Z_L, the normalised values ($\mathbb{Z}_L = Z_L/Z_0$) are (a) 0, (b) ∞, (c) 30/50=0.6, (d) $0.6 + j0.6$, and (e) 50/50=1. These values are plotted in Fig. 12.14.

Let us apply the Smith chart to some simple cases.

Example 12.14 A 20 m 50 Ω line operated at 350 MHz is terminated with a 75 Ω load. Find the input impedance and the complex reflection coefficient at the input end.

Solution At the outset we find the wavelength and the normalised impedances. The wavelength is

$$\lambda = c/f = \frac{3 \times 10^8}{350 \times 10^6} = 0.8571 \text{ m}$$

the load impedance is

$$Z_L = 75 \ \Omega$$

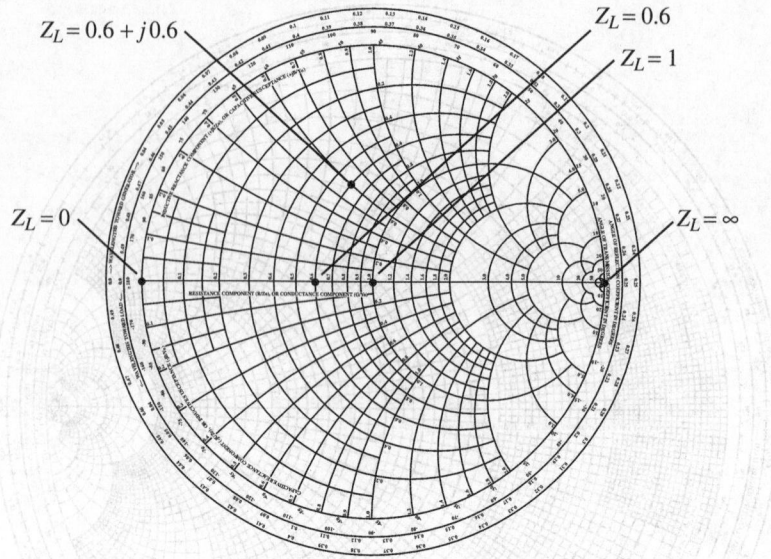

Fig. 12.14

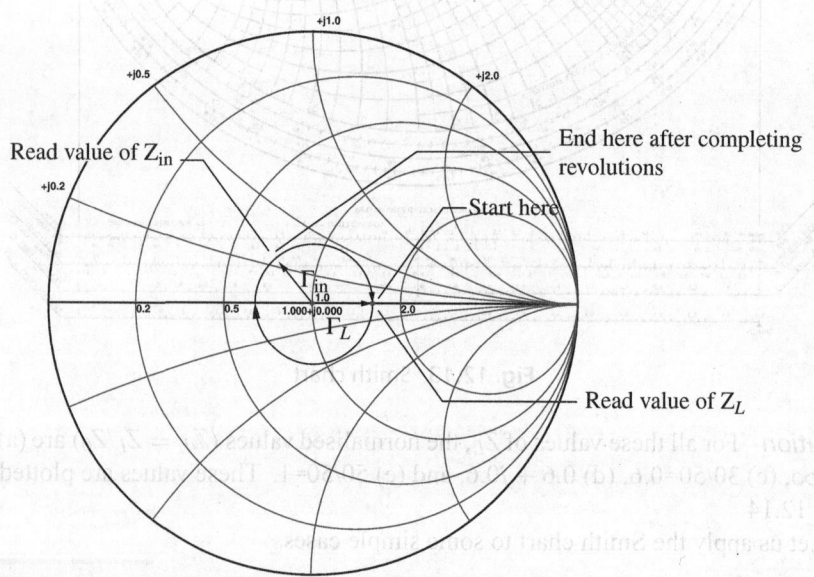

Fig. 12.15 Smith chart

so the normalised impedance is

$$\mathbb{Z}_L = 75/50 = 1.5 \quad (r = 1.5, \; x = 0)$$

the normalised load impedance plotted on the Smith chart gives the position marked 'start' in Fig. 12.15. The origin of the complex gamma plane connected to this point is Γ_L (as shown),

$$\Gamma_L = \frac{Z_L - Z_0}{Z_L + Z_0} = \frac{75 - 50}{75 + 50} = \frac{25}{125}$$

$$\Gamma_L = 0.2$$

We have to move towards the generator by the amount

$$l = 20 \text{ m}$$

to give the new value of the reflection coefficient

$$\Gamma(-l) = \Gamma_L e^{-j2\beta l}$$
$$= \Gamma_L e^{-j(4\pi/\lambda)l}$$
$$= \Gamma_L e^{-j4\pi l_\lambda}$$

The length of the line in wavelengths is

$$l_\lambda = 20/0.85714 = 23.33 \, \lambda$$

and in terms of phase this is

$$\phi = 4\pi l_\lambda = 2\pi \times 46.66 \text{ rad}$$
$$= 2\pi(46 + 0.66) \text{ rad.}$$
$$= 2\pi \times 46 + 2\pi \times 0.66$$
$$= 46(2\pi) + 2 \times 180° \times 0.66$$
$$= 46(2\pi) + 238°$$

which is 46 complete revolutions and followed by $238° = 180° + 58°$ in the clock-wise sense.

Having decided the angle, one moves on the Smith chart in the clockwise sense by ϕ on the circle shown. *One moves on a circle since the magnitude of the reflection coefficient is a constant, as one moves towards the generator.* The final value of $\Gamma(-l)$ is

$$\Gamma_{in} = \Gamma_L e^{-j\phi}$$
$$= -0.11 + j0.17$$

This value of the reflection coefficient falls on the normalised input impedance $\mathbb{Z}_{in}$ which when read out from the Smith chart gives us

$$\mathbb{Z}_{in} = 0.8 + j0.27$$

This is the value of $\mathbb{Z}_{in}$ at the end of the 20 m line. The actual, unnormalised impedance is

$$Z_{in} = 50 \, \mathbb{Z}_{in}$$
$$= 50 \, (0.8 + j0.27)$$
$$= 38 + j13.5 \, \Omega$$

12.5.1 Matching a Load

Example 12.15 With the load given above, and the transmission line, we would like to 'somehow' match the load to the line, so that the power delivered to the load be maximum. How do we go about doing this?

Solution To match the load, we will have to adopt a strategy. Referring to Fig. 12.16 we move along the line towards the generator, by a distance l along the physical line and from a to b on the Smith chart, such that the point b lies on the $r = 1$ circle. Doing this we find that the normalised impedance at b is

$$\mathbb{Z}_{in} = 1 - j0.43$$

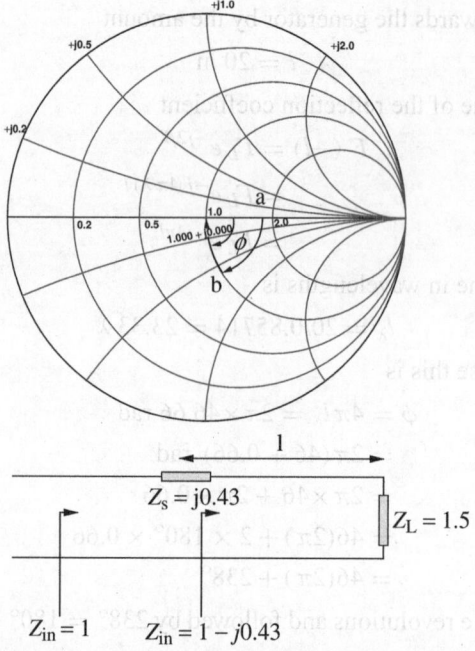

Fig. 12.16 Load matching using a Smith chart

At this point we cancel out the reactive part of $\mathbb{Z}_{in}$ by adding a series reactance

$$\mathbb{Z}_s = j0.43$$

so that beyond that point, $\mathbb{Z}_{in} = 1$. ϕ turns out to be (from the Smith chart)

$$\phi = 39°$$

Corresponding to ϕ, the distance l_λ turns out to be

$$l_\lambda = 0.11\,\lambda$$

If we want more line, l_λ can be increased by half a wavelength.

This is because one revolution of the Smith chart corresponds to exactly half a wavelength

Then

$$l_\lambda = (0.11 + 0.5)\,\lambda = 0.61\,\lambda$$

Three questions arise from the consideration of the above examples.

1. Is it always possible to match any load?
2. How are we going to conveniently obtain a series reactance?
3. Is it possible to use a shunt reactance?

Let us answer these questions one by one.

1. The answer to the first question is yes as by the method outlined above, it is possible to match any load.
2. The next question is answered by considering a transmission line terminated in an open or a short circuit.

    A transmission line terminated in a short circuit has an input impedance given by

$$\mathbb{Z}_{in} = jZ_0 \tan(\beta l)$$

or

$$\mathbb{Z}_{in} = j \tan(\beta l)$$

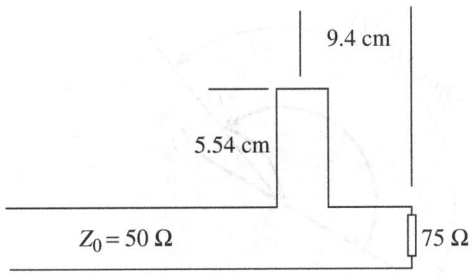

Fig. 12.17

for example to realise

$$\mathbb{Z}_s = j0.43$$

we equate

$$\tan(\beta l) = 0.43$$
$$\beta l = 0.4061$$
$$2\pi l_\lambda = 0.4061$$
$$l_\lambda = 0.064633 \, \lambda$$
$$l = 5.54 \text{ cm}$$

The circuit for the example is shown in Fig. 12.17. In the same manner if we use an open circuited line, the required length would be obtained from

$$-\cot(\beta l) = .43$$
$$-\tan(\beta l) = 2.3256$$
$$\beta l = 1.9769$$
$$l_\lambda = 0.31463 \, \lambda$$
$$l = 26.967 \text{ cm}$$

3. To consider admittances (which for physical reasons is more convenient) we need to consider the Smith chart with admittances as the parameters. To this end, we look at the transformation

$$\mathbb{Z}_{in} = \frac{1+\Gamma}{1-\Gamma}$$

letting $\mathbb{Y}_{in} = 1/\mathbb{Z}_{in}$

$$\mathbb{Y}_{in} = g + jb = \frac{1-\Gamma}{1+\Gamma} \tag{12.81}$$

in this transformation, if we replace Γ by $-\Gamma$ we obtain the original transformation (the previous equation) and the new chart can be read just like the old one with $r \to g$ and $x \to b$.

For example, if we are at the normalised impedance point $\mathbb{Z} = 1 + j2$ then the complex reflection coefficient Γ for this point is the straight line joining the origin (which is $\mathbb{Z} = 1 + j0$) to $\mathbb{Z} = 1 + j2$. Let $\Gamma = |\Gamma| \angle \phi$. To go over to the admittance Smith chart, we plot $-\Gamma$ $(= |\Gamma| \angle(\phi + \pi))$ on the chart and we can read out the normalised admittance value of $\mathbb{Y} = 0.2 - j0.4$ from the chart, treating r as g and x as b. To confirm, we can also do the mathematics $1/(1 + j2) = 0.2 - j0.4$.

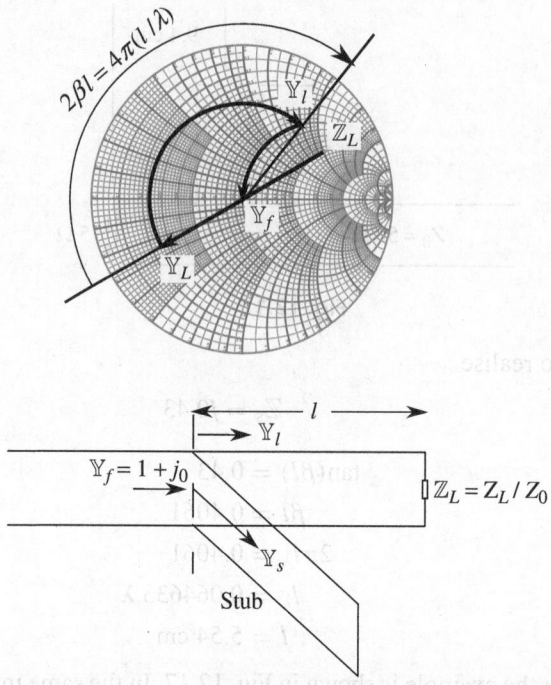

Fig. 12.18 Single stub matching

If however without going from Γ to $-\Gamma$ we want to read the admittance directly, then we need a Smith chart which has both the impedance and admittance on the same chart.

12.5.2 Single Stub Matching

We now consider single stub matching where the load is matched by means of a short circuited stub. If we refer to Fig. 12.18, where we would like to match the load $\mathbb{Z}_L = Z_L/Z_0$ shown in the figure. If the complex reflection coefficient is Γ_L then to go to the admittance chart, we reach the point $\mathbb{Y}_L$ shown at $-\Gamma_L$. From this point we go towards the generator by a distance l such that we reach the $\mathbb{Y} = 1$ circle to reach the admittance $\mathbb{Y}_1 = 1 + jx$ where x is some value. We now add a shunt short-circuited stub of a convenient value so that $\mathbb{Y}_f = \mathbb{Y}_1 + \mathbb{Y}_s = 1 + j0$.

Example 12.16 Match a load of 75 Ω in a 50 Ω line by using a single stub as shown in Fig. 12.19.

Solution The start point is the load of 75 Ω which gives $\mathbb{Z}_L = 1.5$ and $\mathbb{Y}_L = 0.667$. We now go down the line by a distance of $l_\lambda = 0.142\,\lambda$ which brings us to $\mathbb{Y}_l = 1.0 + j0.409$. We now add a short circuited stub of value $\mathbb{Y}_s = -j0.409$ in parallel with the line. The length of the stub is $0.188\,\lambda$. The final end point is $\mathbb{Y}_f = 1.0 + j0$.

12.5.3 Voltage and Currents along a Line

When we observe Eqns (12.31) and (12.35) in Section 12.4, we find that the total voltage and currents may be written, after some manipulation, as,

$$V(-l) = V_+ e^{j\beta l}(1+\Gamma)$$

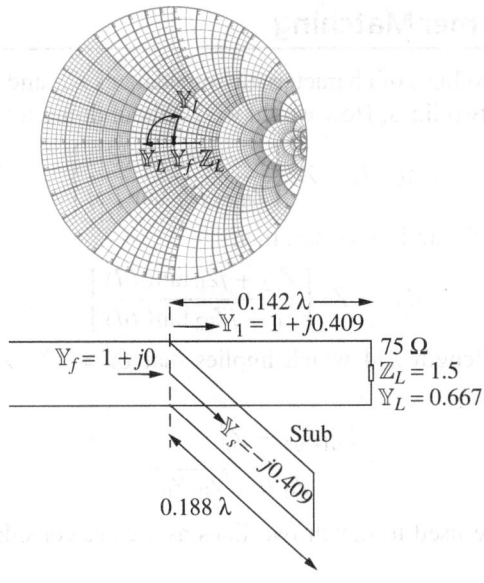

Fig. 12.19 Parallel stub matching

$$I(-l) = \frac{V_+}{Z_0}e^{j\beta l}(1-\Gamma)$$

and therefore

$$|V(-l)| = |V_+||(1+\Gamma)|$$

$$|I(-l)| = \left|\frac{V_+}{Z_0}\right||(1-\Gamma)|$$

the two terms $1+\Gamma$ and $1-\Gamma$ are shown on the Smith chart circle on Fig. 12.20.

Since

$$|V(-l)| \propto |1+\Gamma|$$

and

$$|I(-l)| \propto |1-\Gamma|$$

A voltage maximum occurs when the angle of Γ is zero or an even multiple of π, and a voltage minimum occurs when $\angle \Gamma$ is an odd multiple of π. The statement can be changed appropriately when we talk of the current along the line as we move towards the generator. Note that at a voltage maximum or minimum the input impedance Z_{in} is always real. A similar statement can be made for the current along the line.

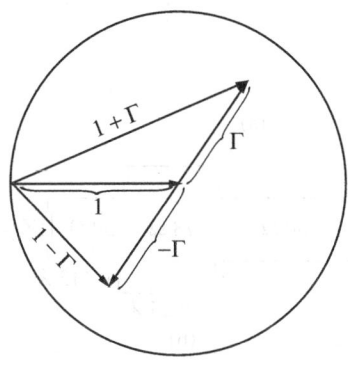

Fig. 12.20 Voltage and current on a Smith chart

12.6 | Transformer Matching

Suppose we have two lines of characteristic impedances Z_{01} and Z_{02} and we would like to match these two lines. How would we go about doing it? Starting from

$$Z_{in}(-l) = Z_0 \left\{ \frac{Z_L + jZ_0 \tan(\beta l)}{Z_0 + jZ_L \tan(\beta l)} \right\}$$

we want $Z_{in}(-l) = Z_{01}$ and $Z_L = Z_{02}$, then

$$Z_{01} = Z_0 \left\{ \frac{Z_{02} + jZ_0 \tan(\beta l)}{Z_0 + jZ_{02} \tan(\beta l)} \right\}$$

Consider a line of length $\lambda/4$ which implies that $\beta l = (2\pi/\lambda)(\lambda/4) = \pi/2$ or $\tan(\pi/2) \to \infty$ then

$$Z_{01}Z_{02} = Z_0^2$$

or

$$Z_0 = \sqrt{Z_{01}Z_{02}} \qquad (12.82)$$

This formula can be used to match two lines as we are considering or even any given load.

Thus to match a load, for example of $Z_L = 40 + j40 \ \Omega$ (which in normalised form is $\mathbb{Z}_L = 0.8 + j0.8$) we move towards the generator a distance of $l_\lambda = 0.113 \lambda$ which brings us to the point $Z_{in1} = 122 + j0$ ($\mathbb{Z}_{in1} = 2.44 + j0$) which lies on the $x = 0$ line on the Smith chart Fig. 12.21(a). Now we take a section of quarter wavelength line of characteristic impedance $Z_0 = \sqrt{50 \times 122} = 78 \ \Omega$ and the load is matched to the 50 Ω line. The results are shown on a Smith chart. The method of transformer matching has been extended to wide band matching by many researchers.

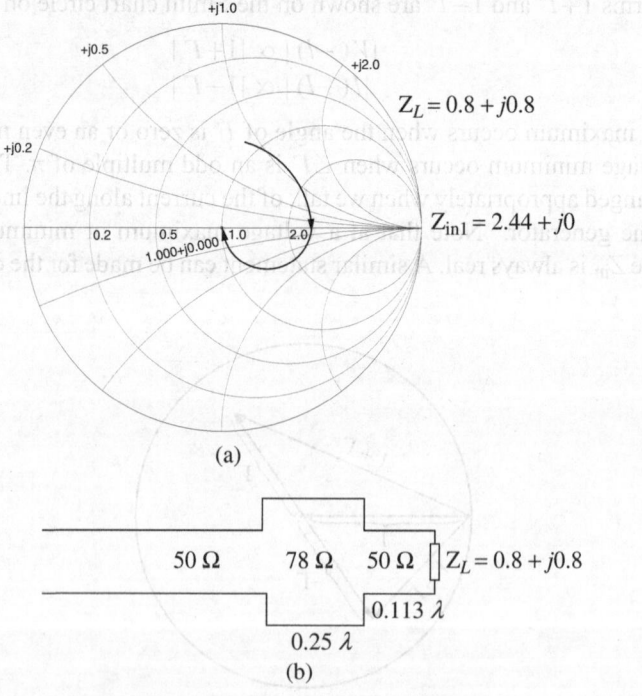

(a)

(b)

Fig. 12.21 Matching with a transformer

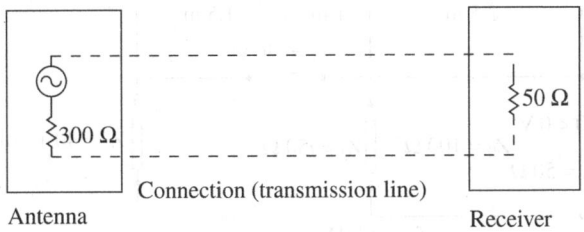

Fig. 12.22

Example 12.17 We would like to look at a practical but simple example of the application of the theory which we have developed here. A folded dipole antenna receives EM energy and must be connected to a television receiver, so that we may be able to see our TV programs. How do we go about it?

Solution The first point which we must decide is which is the generator and which is the load. After a little reflection, we come to the conclusion that the antenna is the generator, since the EM energy must go from the antenna to the TV receiver. Next we must model the TV receiver as the load, and the antenna as the generator.

If we search for the specifications of a folded dipole we find that it has an input impedance of 300 Ω. Most modern televisions have an input jack marked 50 Ω, so the configuration is now as shown in Fig. 12.22.

Since we always start with the load and the transmission line, what should be the characteristic impedance of the line? As a rule of thumb, the transmission line must be matched to the generator. Hence the line should be a two-wire ribbon line with $Z_0 = 300$ Ω. However, the line must be connected to a 50 Ω load, so there must be a match. We now use a transformer. We can use a quarter wavelength transformer with a characteristic impedance of 123 Ω, but the modern solution is to use a $\sqrt{6} : 1$ turns ratio transformer which operates into the MHz region. So the final set-up is shown in Fig. 12.23.

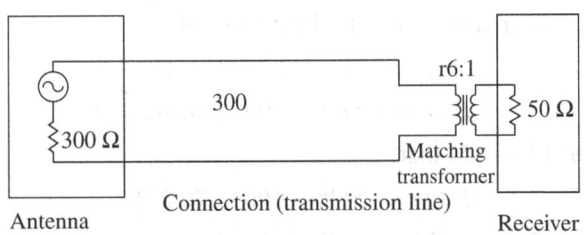

Fig. 12.23

Example 12.18 Analyse the transmission line configuration shown in Fig. 12.24.
Solution From Fig. 12.24

$$\lambda = 3 \times 10^8/3 \times 10^7 = 10 \text{ m}$$

$$\beta = 2\pi/\lambda = 0.628 \text{ rad/m} = 36°/\text{m}$$

At the load

$$\Gamma_L = -0.321 + j.377$$

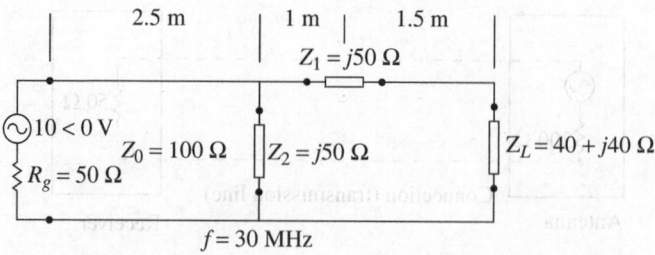

Fig. 12.24

Moving towards the generator by 1.5 m on the Smith chart of Fig. 12.13, we arrive at

$$Z(l = -1.5) = 229 + j114 \ \Omega$$
$$\Gamma \ (l = -1.5) = 0.458 + j0.118$$

Here we add $j50 \ \Omega$ to the impedance in series. The new impedance and reflection coefficient are

$$Z(l = -1.5*) = 229 + j164 \ \Omega$$
$$\Gamma \ (l = -1.5*) = 0.514 + j0.244$$

The star notation $l = -1.5*$ is used to indicate the parameter values on the side of the discontinuity on the generator side. Next we move towards the generator by 1 m. The new parameters are

$$Z(l = -2.5) = 125 - j152 \ \Omega$$
$$Y(l = -2.5) = 0.00322 + j0.00393 \ \mho$$
$$\Gamma \ (l = -2.5) = +0.390 - j0.414$$

At this point we add an admittance of $-j0.2 \ \mho$ in shunt. So the new parameters are

$$Z(l = -2.5*) = 11.97 + j59.77 \ \Omega$$
$$Y(l = -2.5*) = 0.00322 - j0.0161 \ \mho$$
$$\Gamma \ (l = -2.5*) = -0.390 + j0.742$$

Finally we move towards the generator by 2.5 m and

$$Z(l = -5) = 32.21 - j161 \ \Omega$$
$$\Gamma \ (l = -5) = 0.390 - j0.742$$

We know from $\Gamma \ (l = -5)$ that

$$|\Gamma \ (l = -5)|^2 \times 100 = 70.27\%$$

is the percentage of power which is reflected and

$$(1 - |\Gamma \ (l = -5)|^2) \times 100 = 29.73\%$$

is the power absorbed by the system. But, except for the load, all the other elements are reactive, therefore this is also the percentage of the power absorbed by the load. But to compute the incident power, we need to compute $V_+(z) = V_+ \exp(-j\beta z)$ and $V_-(z) = V_- \exp(j\beta z)$ along the line.

The important point to note is to move between transmission line concepts and lumped concepts as and when required.

Let us now start at the generator. The voltage across $Z(l = -5)$ and the current through this impedance is

$$V(l = -5) = \frac{(10\angle 0)\,(32.21 - j161)}{50 + (32.21 - j161)}$$
$$= 8.7422 - j2.4633 = V_+(-5) + V_-(-5)$$

$$I(l = -5) = \frac{(10\angle 0)}{50 + (32.21 - j161)}$$
$$= 0.025156 + j0.049266 = \frac{V_+(-5) - V_-(-5)}{Z_0}$$

$$V_+(-5) - V_-(-5) = 2.5156 + j4.9266$$
$$V_+(-5) = 5.6289 + j1.2317$$
$$V_-(-5) = 3.133 - j3.695$$
$$I_+(-5) = V_+(-5)/Z_0$$
$$I_-(-5) = -V_-(-5)/Z_0$$

Hence the incident power is

$$|V_+(-5)|^2/Z_0 = 0.33 \text{ W}$$

We know that about 30% is absorbed. Therefore, only 0.1 W is delivered to the load.

To continue our calculations, we move towards the load by 2.5 m which corresponds to a length of $\lambda/4$

$$V_+(-2.5*) = V_+(-5)\,e^{-j\beta(\lambda/4)} = 1.2317 - j5.6289$$
$$V_-(-2.5*) = V_-(-5)\,e^{j\beta(\lambda/4)} = 3.695 + j3.133$$

The total voltage is

$$V(-2.5*) = V_+(-2.5*) + V_-(-2.5*)$$

The total current is

$$I(-2.5*) = \{V_+(-2.5*) - V_-(-2.5*)\}/Z_0$$
$$= -0.024633 - j0.087422$$

This current gets divided between two admittances of values $-j0.02\ \mho$ (in shunt) and $0.00322 + j0.00393\ \mho$ which is the input admittance at this point. Using the current divider theorem

$$I(-2.5) = \frac{(-0.024633 - j0.087422) \times (0.00322 + j0.00393)}{-j0.2 + 0.00322 + j0.00393}$$
$$= 0.016119 + j0.023086$$
$$= \{V_+(-2.5) - V_-(-2.5)\}/Z_0$$

While the voltage remains the same beyond the shunt element

$$V(2.5) = V_+(-2.5*) + V_-(-2.5*)$$
$$= V_+(-2.5) + V_-(-2.5)$$

From these equations we compute $V_+(-2.5)$ and $V_-(-2.5)$, and proceed towards the load. In this manner we continue on. The Smith chart is shown in Fig. 12.25.

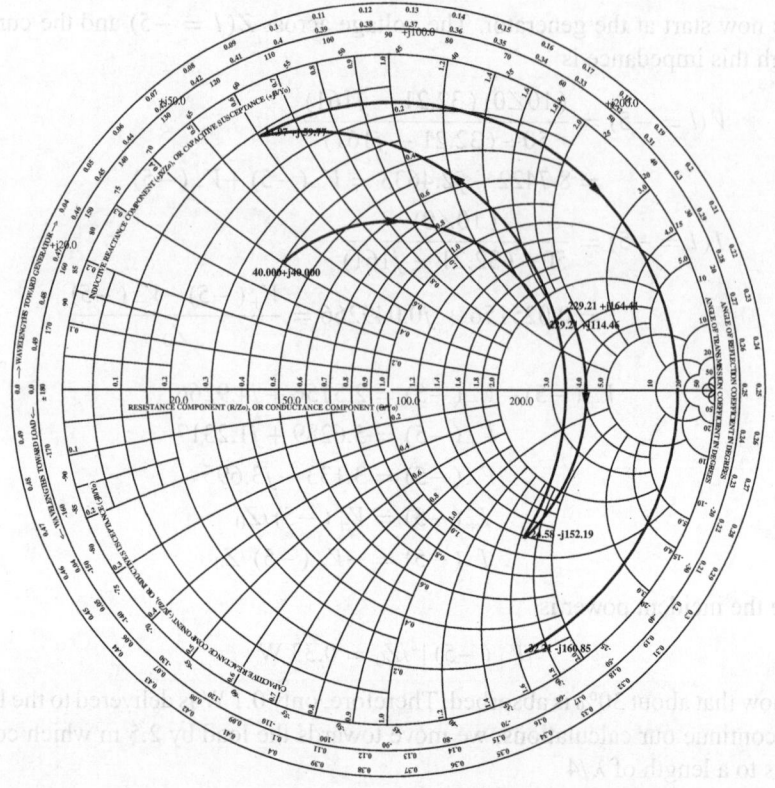

Fig. 12.25 Smith chart

Example 12.19 A 50 Ω transmission line at a frequency whose wavelength is 10 cm has $Z_L = 150 + j50$ Ω. Use the Smith chart to determine (a) the reflection coefficient, (b) VSWR, (c) distance to the first voltage maximum from the load, (d) distance to the first voltage minimum from the load, (e) impedance at V_{max}, (f) impedance at V_{min}.

Solution

(a) First we find the normalised value of the load impedance which is

$$\mathbb{Z}_L = \frac{150 + j50}{50} = 3 + j1$$

when we plot this on the Smith chart (marked '$\mathbb{Z}'_L$'), the distance of this point from the centre of the chart (measured with a ruler) and normalised to the radius of the circle (again measured with a ruler) is about 0.54. This is the magnitude of Γ

$$|\Gamma| = 0.54$$

Similarly, the angle (measured with a protractor) from the *x*-axis is about 13°, as shown in Fig. 12.26. Therefore

$$\Gamma = 0.54\angle 13°$$

(b) From the earlier discussion it is clear that the VSWR is given by

$$s = \frac{V_{max}}{V_{min}} = \frac{1 + |\Gamma|}{1 - |\Gamma|} \approx 3.45$$

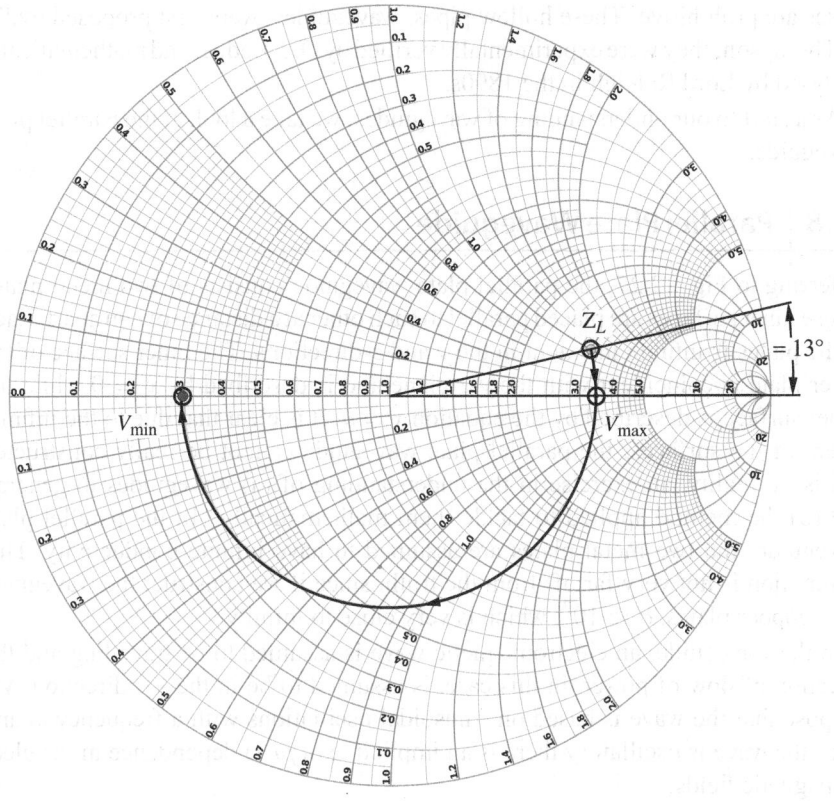

Fig. 12.26

(c) Moving from the load point in the clockwise sense, we move by 13° when we come to the $jx = 0$ line or the x-axis which is point marked 'V'$_{max}$. This is the physical angle along the Smith chart. An angle of 360° corresponds to $\lambda/2$ (= 5 cm) so 13° corresponds to about 0.18 cm. This is the point where the first maximum occurs.

(d) The first minimum occurs at $\lambda/4 + 0.18$ cm = 2.68 cm away from the load and this is the point marked 'V'$_{min}$.

(e) The normalised impedance at the point where the maximum occurs is about

$$\mathbb{Z}_{max} \approx 3.3 \quad \text{or}$$
$$Z_{max} \approx 3.3 \times 50 = 167 \ \Omega$$

(f) The normalised impedance at the point where the minimum occurs is about

$$\mathbb{Z}_{min} \approx 0.3 \quad \text{or}$$
$$Z_{min} \approx 0.3 \times 50 = 15 \ \Omega$$

12.7 | Waveguides

Energy transfer in the gigahertz range must necessarily be done through propagation of waves travelling in hollow pipes called waveguides. The method of energy transfer by using an open structure consisting of wires is inefficient—the radiation losses are prohibitive. These hollow pipes, waveguides, were first proposed by Sir J.J.Thompson; they were experimentally verified by O. J. Lodge, and mathematically analysed by Lord Raleigh in the 1890s.

As a start to our understanding of waveguides, we take a look at the parallel plate waveguide.

12.8 | Parallel Plate Waveguide

Referring to Fig. 12.27, the parallel plate waveguide consists of two metal plates whose inner surfaces are placed parallel to each other at a distance of a m apart. Each of the plates is infinite in extent in the y and z-directions. The inner surface of the lower plate is coincident with the y-z plane and is described by $x = 0$ while the upper surface is described by the equation $x = a$. It is clear that due to the infinite extent of the surfaces, the parallel plate waveguide cannot be really constructed and is, in reality, an exercise in the understanding of (a) waveguides in general, and (b) the method applied to tackle them mathematically. In the parallel plate waveguide, the two metal plates are assumed to have infinite conductivity. This assumption is not very far off from the truth, since if we construct the waveguide from copper plates, $\sigma \simeq 10^7$, which is very close to infinity.

In the waveguide, an electromagnetic wave is assumed to be travelling and the direction of flow of power, in this case, is assumed to be in the $+z$ direction. We suppose that the wave is based on sinusoidal oscillations with a frequency ω and since the wave is oscillatory there is an implicit $\exp(j\omega t)$ dependence in the electromagnetic fields.

Since we are considering a travelling wave we must apply the Helmholtz equation, which applies to waves with sinusoidal oscillations. Equation (10.72), applied to

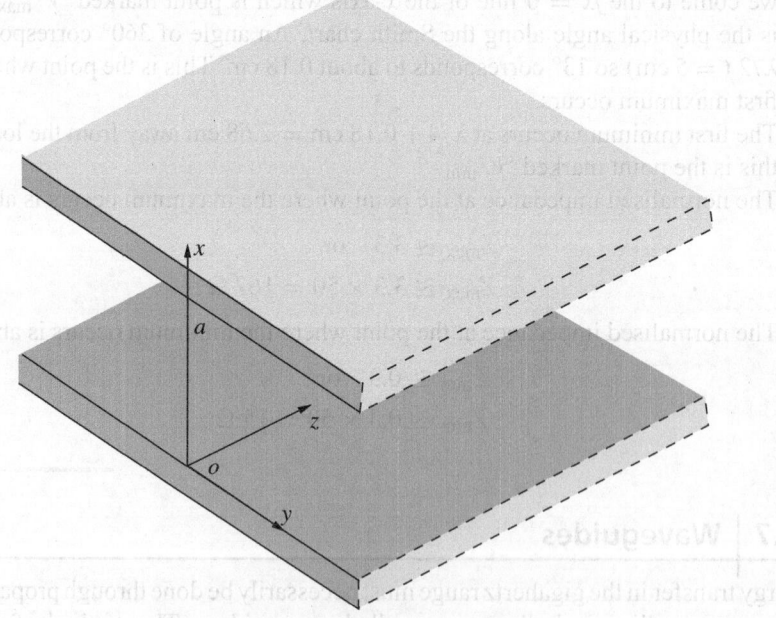

Fig. 12.27 Parallel plate waveguide

the electric $\mathbf{E}$ and magnetic $\mathbf{H}$ fields is

$$\nabla^2 \mathbf{E} = -k^2 \mathbf{E}$$

$$\nabla^2 \mathbf{H} = -k^2 \mathbf{H} \qquad (12.83)$$

where $k(= \omega \sqrt{\mu\varepsilon})$ is the free-space propagation constant and $\mathbf{E}$ and $\mathbf{H}$ are phasors.

The z and time dependence, t, of the $\mathbf{E}$ and $\mathbf{H}$ fields is $e^{j(\omega t - \beta z)}$, which represents a travelling wave moving in the $+z$ direction with a propagation constant $\beta = 2\pi/\lambda_g$. λ_g being the wavelength in the waveguide.

In the y direction, the waveguide is infinite in extent, hence the fields do not depend on the variable y.

Therefore

$$\mathbf{E} = \mathbf{E}_0(x) \, e^{j(\omega t - \beta z)}$$

$$\mathbf{H} = \mathbf{H}_0(x) \, e^{j(\omega t - \beta z)} \qquad (12.84)$$

Working with the x component of the electric field, Helmholtz equation becomes

$$\nabla^2 E_x = -k^2 E_x$$

Since $\partial / \partial y$ gives zero, (no dependence on y) and

$$\frac{\partial}{\partial z} E_x = \frac{\partial}{\partial z} E_{0x}(x) \, e^{j(\omega t - \beta z)} = -j\beta \, E_{0x}(x) \, e^{j(\omega t - \beta z)} = -j\beta \, E_x$$

which reduces to multiplication of E_x by $-j\beta$. The previous equation becomes

$$\nabla^2 E_x = \frac{\partial^2 E_x}{\partial x^2} + \underbrace{\frac{\partial^2 E_x}{\partial y^2}}_{=0} + \underbrace{\frac{\partial^2 E_x}{\partial z^2}}_{=(-j\beta)(-j\beta)E_x} = -k^2 E_x$$

$$\frac{\partial^2 E_x}{\partial x^2} - \beta^2 \, E_x = -k^2 E_x$$

$$\frac{\partial^2 E_x}{\partial x^2} + (k^2 - \beta^2) \, E_x = 0 \qquad (12.85)$$

The solution to this equation is (if $k > \beta$)

$$E_x(x,z) = \left\{ E_{x01} \sin\left(\sqrt{k^2 - \beta^2} \, x \right) + E_{x02} \cos\left(\sqrt{k^2 - \beta^2} \, x \right) \right\} e^{-j\beta z} \qquad (12.86)$$

Note that $\exp(j\omega t)$ is implicit (understood to be present). To simplify the expression we can set

$$k_x = \sqrt{k^2 - \beta^2} \qquad (12.87)$$

The x dependence of the other fields is similar.

Next we apply the boundary conditions of the waveguide. The tangential electric field and the normal magnetic field on a perfect conductor are both zero. Working with E_y which is tangential to the two planes,

$$E_y(x,z) = \{ E_{y01} \sin(k_x x) + E_{y02} \cos(k_x x) \} e^{-j\beta z} \qquad (12.88)$$

must be zero for $x = 0$ and $x = a$ for all values of z. Applying the first condition first,

$$\{ E_{y01} \sin(k_x x) + E_{y02} \cos(k_x x) \} e^{-j\beta z} |_{x=0} = 0$$

which is

$$\{ E_{y01} \sin(k_x \times 0) + E_{y02} \cos(k_x \times 0) \} e^{-j\beta z} = 0$$

$$E_{y02} \cos(k_x \times 0) = 0 \qquad (12.89)$$

$E_{y02} = 0$ since $\cos(\sqrt{k^2-\beta^2}x) = 1$ for $x = 0$. Therefore

$$E_y = E_{y01} \sin(k_x x) e^{-j\beta z} \qquad (12.90)$$

Applying the second boundary condition at $x = a$ gives

$$\sin(k_x a) = 0$$

$$k_x a = m\pi \quad m = 1, 2, \ldots$$

Therefore
$$k_x = \sqrt{k^2 - \beta^2} = \frac{m\pi}{a} \quad m = 1, 2, \ldots \qquad (12.91)$$

or
$$\beta = \sqrt{k^2 - \left(\frac{m\pi}{a}\right)^2} \quad m = 1, 2, \ldots \qquad (12.92)$$

and
$$E_y = E_{y01} \sin\left(\frac{m\pi}{a}x\right) e^{-j\beta z} \quad m = 1, 2 \ldots \qquad (12.93)$$

In the same manner

$$H_x = H_{x01} \sin(k_x x) e^{-j\beta z}$$

$$E_z = E_{z01} \sin(k_x x) e^{-j\beta z} \qquad (12.94)$$

since H_x is normal and E_z is tangential to the two planes and therefore both must be zero on the two planes.

The concept of modes now comes into play. For example when $m = 1$ those fields which have a $\sin(m\pi x/a)$ dependence have a single maximum in the x direction, and they vanish at the end points. While in a similar case when $m = 2$, then the field will have one maximum and one minimum; and so on. (When $m = 0$ for a $\sin(.)$ function means that particular field vanishes and in the case of $\cos(.)$ the field reduces to a constant.)

To proceed further, Maxwell's equations have to be taken into account. Working first with the curl of the electric field

$$\nabla \times \mathbf{E} = -j\omega\mu\,\mathbf{H}$$

which is
$$\frac{\partial E_z}{\partial y} - \frac{\partial E_y}{\partial z} = -j\omega\mu\,H_x$$

$$\frac{\partial E_x}{\partial z} - \frac{\partial E_z}{\partial x} = -j\omega\mu\,H_y$$

$$\frac{\partial E_y}{\partial x} - \frac{\partial E_x}{\partial y} = -j\omega\mu\,H_z \qquad (12.95)$$

Since $\partial/\partial y$ gives zero, and $\partial/\partial z$ reduces to multiplication by $-j\beta$ these equations become

$$j\beta\, E_y = -j\omega\mu\, H_x$$

$$-j\beta\, E_x - \frac{\partial E_z}{\partial x} = -j\omega\mu\, H_y$$

$$\frac{\partial E_y}{\partial x} = -j\omega\mu\, H_z \qquad (12.96)$$

Working next with the curl of $\mathbf{H}$

$$\nabla \times \mathbf{H} = j\omega\varepsilon\,\mathbf{E}$$

which become
$$j\beta\, H_y = j\omega\varepsilon\, E_x$$

$$-j\beta H_x - \frac{\partial H_z}{\partial x} = j\omega\varepsilon \, E_y$$

$$\frac{\partial H_y}{\partial x} = j\omega\varepsilon \, E_z \qquad (12.97)$$

Looking at the internal structure of the fields given above, E_y, H_z, and H_x form one set of fields while E_z, H_y, and E_x form another set. If one set is present the *other set need not be present*. We define the first set as transverse electric ($E_z = 0$) fields (for obvious reasons) while the second set is called the transverse magnetic ($H_z = 0$) field. This general manner of division is useful in analysis of other waveguide structures as well.

1. *TE modes.*

   Let us start with

   $$E_y = A \sin\left(\frac{m\pi}{a}x\right) e^{-j\beta z} \qquad (12.98)$$

   then using Eqn set (12.96)

   $$H_x = -A\frac{\beta}{\omega\mu} \sin\left(\frac{m\pi}{a}x\right) e^{-j\beta z}$$

   $$H_z = -A\frac{m\pi}{j\omega\mu \, a} \cos\left(\frac{m\pi}{a}x\right) e^{-j\beta z} \qquad (12.99)$$

   E_z, E_x, and H_y may be set to 0 and the equations will be consistent.

2. *TM modes*

   Here we start with

   $$H_y = A \cos\left(\frac{m\pi}{a}x\right) e^{-j\beta z} \qquad (12.100)$$

   Then using Eqn set (12.97)

   $$E_x = A\frac{\beta}{\omega\varepsilon} \cos\left(\frac{m\pi}{a}x\right) e^{-j\beta z}$$

   $$E_z = A\frac{m\pi}{j\omega\varepsilon \, a} \sin\left(\frac{m\pi}{a}x\right) e^{-j\beta z} \qquad (12.101)$$

H_z, H_x, and E_y may be set to 0 and the equations will be consistent. Observing the fields of the first TE mode ($m = 1$) we find that E_y is proportional to $\sin(\pi x/a)$ and H_z is proportional to $\cos(\pi x/a)$. These two fields are plotted in Fig. 12.28. E_y is zero at $x = 0$ and $x = a$, as it should be, since the tangential electric field must be zero at the walls of the metallic conductor. On the other hand H_z is zero at $x = a/2$ and has maximum and minimum values at the two walls. Due to the presence of H_z there is a y-directed surface currents on the two walls at $x = 0$ and $x = a$. These currents change direction every half a wavelength.

Going over to the examination of the modes, β is given by the formula

$$\beta = \sqrt{k^2 - \left(\frac{m\pi}{a}\right)^2} \qquad (12.102)$$

What happens when ($k = 2\pi/\lambda = \omega/c$) becomes less than $m\pi/a$? We find that then β becomes imaginary, $\beta = -j\alpha$ and

$$e^{-j\beta z} = e^{-j(-j\alpha)z} = e^{-\alpha z} \qquad (12.103)$$

and the wave decays. Such a mode is called an *evanescent mode* and is said to be *cut off*. What is the cut-off condition? Starting with zero frequency, the value

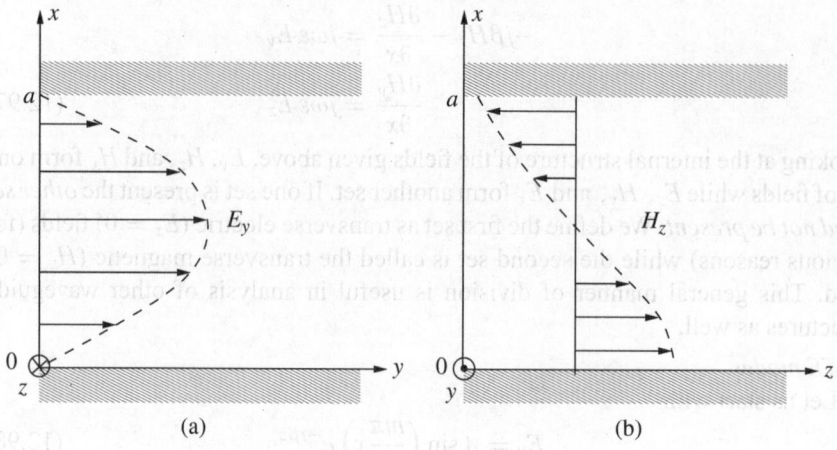

Fig. 12.28 Parallel plate waveguide. (a) E_y (b) H_z

of $k = \omega/c$ is zero, and the wave is cut off. As the frequency increases, there comes one frequency, ω_c when the wave just begins to propagate. That is when β passes from being imaginary to being real. At that frequency $\beta = 0$

$$\beta = \sqrt{k_c^2 - \left(\frac{m\pi}{a}\right)^2} = 0$$

$$k_c = \frac{m\pi}{a} \tag{12.104}$$

$$\frac{2\pi}{\lambda_c} = \frac{m\pi}{a}$$

$$\lambda_c = \frac{2a}{m} \tag{12.105}$$

where λ_c is the free space wavelength corresponding to the cut-off frequency ω_c. When the frequency is lower than the cut-off frequency, that particular mode is evanescent.

3. *TEM mode* (both $E_z = 0$ and $H_z = 0$). Is there a possibility of $m = 0$? If we examine the TE modes we find that there is no travelling wave which satisfies Maxwell's equations. However, when we put $m = 0$ in the TM modes we find that

$$H_y = A e^{-j\beta z} \tag{12.106}$$

and

$$E_x = A \frac{\beta}{\omega\varepsilon} e^{-j\beta z} \tag{12.107}$$

The other fields being zero. Maxwell's equations are indeed satisfied. Furthermore

$$\beta = \sqrt{k^2 - \left(\frac{m\pi}{a}\right)^2}\Bigg|_{m=0} = k \tag{12.108}$$

the propagation constant will coincide with the free space propagation constant. So

$$H_y = A e^{-jkz}$$

$$E_x = A \frac{k}{\omega\varepsilon} e^{-jkz}$$

$$= A \frac{\omega \sqrt{\mu\varepsilon}}{\omega\varepsilon} e^{-jkz}$$

$$= A \sqrt{\frac{\mu}{\varepsilon}} e^{-jkz}$$

$$= A Z e^{-jkz} \qquad (12.109)$$

$Z = Z_0 = 120\pi = 377\ \Omega$ for free space.

Example 12.20 A parallel plate waveguide is formed by two perfect electric conductors and a plate separation of 3 cm. If the frequency of operation is 15 GHz, find the modes that propagate. (a) If a TEM mode propagates find the wave impedance, wavelength in the guide, and the phase velocity. (b) For the other modes, find the wavelength in the guide and the phase velocity.

Solution

(a) For a two-conductor guide, a TEM mode will always propagate. A TEM mode has

$$\beta = k = \omega \sqrt{\mu\varepsilon}$$
$$= 314.2\ \text{rad/m}$$

The wave impedance is

$$Z_0 = \sqrt{\frac{\mu}{\varepsilon}} = 377\ \Omega$$

and the wavelength is

$$\lambda_g = \frac{2\pi}{\beta} = 0.02\ \text{m}$$

(b) *TE modes.* The mode would be TE$_{10}$ mode (here the subscript 1 stands for $m = 1$ and 0 for $n = 0$)

$$\lambda_{cm0} = \frac{2 \times 0.03}{m} = \frac{0.06}{m}\ \text{m}$$

so for $m = 1$ and 2

$$\lambda_{c10} = 0.06\ \text{m}$$
$$\lambda_{c20} = 0.03\ \text{m}$$
$$\lambda_{c30} = 0.02\ \text{m}$$

Since the free space wavelength, $\lambda = 3 \times 10^8 / 15 \times 10^9 = 0.02$ m, and λ must be less than λ_{cm}, only the first two modes propagate and

$$\beta_{10} = \sqrt{k^2 - \left(\frac{\pi}{a}\right)^2}$$
$$= 296.1\ \text{rad/m}$$

$$\beta_{20} = \sqrt{k^2 - \left(\frac{2\pi}{a}\right)^2}$$
$$= 234.15\ \text{rad/m}$$

Since $\lambda_g = 2\pi/\beta$, therefore

$$\lambda_{g10} = 2.12\ \text{cm}$$
$$\lambda_{g20} = 2.68\ \text{cm}$$

The phase velocity is given by the formula $v_p = \omega/\beta$ therefore

$$v_{p10} = 3.182 \times 10^8 \text{ m/s}$$

$$v_{p20} = 4.025 \times 10^8 \text{ m/s}$$

TM modes. The TE and TM modes have identical treatment.

12.9 | TEM Mode Waveguides

As we saw in the previous section that apart from supporting higher order modes, the parallel plate waveguide also supports the TEM mode. TEM modes exist on two conductor lines and may be analysed as transmission lines for frequencies starting from dc to the start of the first higher order mode. These waveguides have equivalent transmission line equivalent circuits. The most important of these are given in the subsequent sections.

12.9.1 Parallel Conductor Line

Here there are two parallel conductors with conductor radius $= a$ and the spacing between centres $= b$. Then the transmission line parameters for the two-wire line are (see Fig. 12.2)

$$L = (\mu/\pi) \cosh^{-1}(b/2a)$$

$$C = (\pi\varepsilon)/\cosh^{-1}(b/2a) \tag{12.110}$$

12.9.2 Coaxial Line

In this waveguide there are two concentric conductors, with the inner conductor having a radius $= a$ and the inner surface of the outer conductor having a radius $= b$ (see Fig. 12.2).

$$L = (\mu/2\pi) \ln(b/a)$$

$$C = (2\pi\varepsilon)/\ln(b/a) \tag{12.111}$$

12.9.3 Parallel Plate Line

Here there are two plates placed a m apart and each of width w. Then the parameters L and C are (approximately) (See Fig. 12.29)

$$L \simeq \mu\, a/w$$

$$C \simeq \varepsilon\, w/a \tag{12.112}$$

12.9.4 Micro-strip Line

The micro-strip line has a characteristic impedance given by the entries as in Table 12.2. (see Fig. 12.4)

Z_m is the characteristic impedance of the micro-strip line
ε_r is the dielectric constant of substrate,
w is the width of the strip and
h is the thickness ('height') of substrate

Practice Problem 12.1
Calculate the L and C values of micro-strip line.

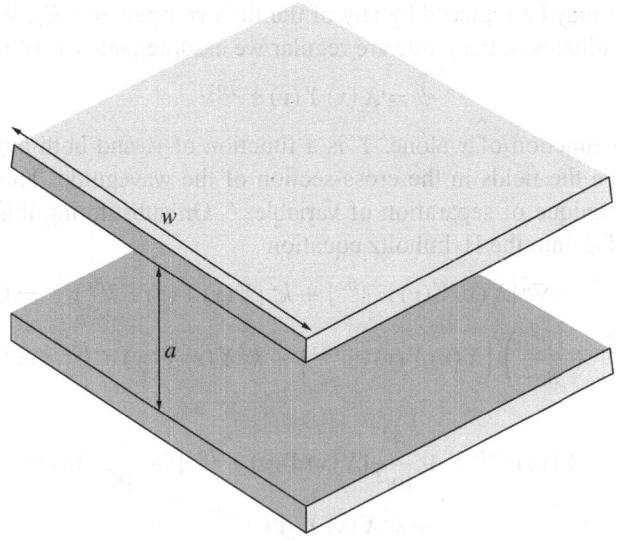

Fig. 12.29 The parallel plate transmission line

Table 12.2 Characteristic impedance of a micro-strip line

$w/h < 1$	$w/h \geq 1$
$Z_m = \frac{60}{\sqrt{\varepsilon_{eff}}}\left[\ln\left(8\frac{h}{w} + 0.25\frac{w}{h}\right)\right]$	$Z_m = \frac{120\pi}{\sqrt{\varepsilon_{eff}}\times\left[\frac{w}{h}+1.393+\frac{2}{3}\ln\left(\frac{w}{h}+1.444\right)\right]}$
$\varepsilon_{eff} = \frac{e_r+1}{2} + \frac{e_r-1}{2}$ $\times\left[\left(1+12\frac{h}{w}\right)^{-1/2} + .04\left(1-\frac{w}{h}\right)^2\right]$	$\varepsilon_{eff} = \frac{e_r+1}{2} + \frac{e_r-1}{2}\left(1+12\frac{h}{w}\right)^{-1/2}$

12.10 | Rectangular Waveguide

The rectangular waveguide is a hollow pipe of rectangular cross-section, capable of conveying travelling waves. A three-dimensional view of the waveguide is shown in Fig. 12.30. The inner surface of the waveguide is made of some high-conductivity material like copper or gold and is rectangular in cross-section. The dimensions of the rect-

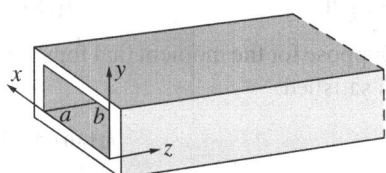

Fig. 12.30 Rectangular waveguide

angle are a in the x-direction and b in the y-direction as shown with $a > b$. The wave is assumed to be travelling in the z direction with a time dependence of the form $\exp(j\omega t)$ and a z dependence of the type $\exp(-j\beta z)$ where β $(= 2\pi/\lambda_g)$ is the propagation constant.

12.10.1 Separation of Variables

To analyse the waveguide mathematically we need to make a slight detour on the analysis of the Helmholtz's equation. We will take a look at the scalar Helmholtz equation

$$\nabla^2 \psi + k^2 \psi = 0 \qquad (12.113)$$

$k = \omega \sqrt{\mu\varepsilon}$. ψ may be replaced by any of the field components, E_x, $E_y \ldots$ or H_z. Since the coordinates of the guide are regular we assume that ψ is of the form

$$\psi = X(x)\, Y(y)\, e^{-j\beta z} \tag{12.114}$$

that is, X is a function of x alone, Y is a function of y, and both are multiplied together to give the fields in the cross-section of the waveguide. This method is called 'the technique of separation of variables.' On substituting this functional dependence of ψ into the Helmholtz equation

$$\nabla^2 \{X(x)\, Y(y)\, e^{-j\beta z}\} + k^2 \{X(x)\, Y(y)\, e^{-j\beta z}\} = 0$$

$$\left(\frac{\partial^2}{\partial x^2} + \frac{\partial^2}{\partial y^2} + \frac{\partial^2}{\partial z^2} \right) \{X(x)\, Y(y)\, e^{-j\beta z}\} + k^2 X(x)\, Y(y)\, e^{-j\beta z} = 0 \tag{12.115}$$

or $\quad \dfrac{\partial^2}{\partial x^2}\{X(x)\, Y(y)\, e^{-j\beta z}\} + \dfrac{\partial^2}{\partial y^2}\{X(x)\, Y(y)\, e^{-j\beta z}\} + \dfrac{\partial^2}{\partial z^2}\{X(x)\, Y(y)\, e^{-j\beta z}\}$

$$+ k^2 X(x)\, Y(y)\, e^{-j\beta z} = 0 \tag{12.116}$$

or $\quad Y(y)\, e^{-j\beta z} \dfrac{\partial^2 X(x)}{\partial x^2} + X(x)\, e^{-j\beta z} \dfrac{\partial^2 Y(y)}{\partial y^2} + (-j\beta)(-j\beta)\, X(x)\, Y(y)\, e^{-j\beta z}$

$$+ k^2 X(x)\, Y(y)\, e^{-j\beta z} = 0 \tag{12.117}$$

Dividing throughout by $X(x)\, Y(y)\, \exp(-j\beta z)$ and simplifying

$$\frac{1}{X(x)} \frac{\partial^2 X(x)}{\partial x^2} + \frac{1}{Y(y)} \frac{\partial^2 Y(y)}{\partial y^2} + (k^2 - \beta^2) = 0 \tag{12.118}$$

Substituting $\qquad\qquad f_1(x) = \dfrac{1}{X(x)} \dfrac{\partial^2 X(x)}{\partial x^2}$

and $\qquad\qquad\qquad f_2(y) = \dfrac{1}{Y(y)} \dfrac{\partial^2 Y(y)}{\partial y^2} \tag{12.119}$

we get $\qquad\qquad f_1(x) + f_2(y) + (k^2 - \beta^2) = 0 \tag{12.120}$

Suppose for the moment that for some value of the pair $(x, y) = (x_0, y_0)$, the equation is satisfied.

$$f_1(x_0) + f_2(y_0) + (k^2 - \beta^2) = 0 \tag{12.121}$$

Now let x increase to $x_0 + \Delta x$, then y *must change* to keep the equation satisfied, *unless* each of the two functions is equal to a constant:

$$f_1(x) = -k_x^2 \text{ (a constant)}$$

$$f_2(y) = -k_y^2 \text{ (another constant)} \tag{12.122}$$

Therefore $\qquad\qquad k^2 = \beta^2 + k_x^2 + k_y^2 \tag{12.123}$

and going back to the earlier equations

$$\frac{1}{X(x)} \frac{\partial^2 X(x)}{\partial x^2} = -k_x^2$$

$$\frac{1}{Y(y)} \frac{\partial^2 Y(y)}{\partial y^2} = -k_y^2 \tag{12.124}$$

The solutions to these equations (as in the case of the parallel plate waveguide) are

$$X(x) = X_{01} \cos(k_x x) + X_{02} \sin(k_x x)$$
$$Y(y) = Y_{01} \cos(k_y y) + Y_{02} \sin(k_y y) \tag{12.125}$$

k_x and k_y will assume definite values on applying the conditions at the boundaries of the waveguide, and this will be clear a little later in the discussion.

We know from earlier chapters [see Eqns (11.72) and (11.80)] that the electric and magnetic fields for travelling waves satisfy the Helmholtz equation. Therefore,

$$\mathbf{E}(x,y,z) = \mathbf{E}(x,y)\,e^{-j\beta z}$$

and

$$\mathbf{H}(x,y,z) = \mathbf{H}(x,y)\,e^{-j\beta z}$$

and the analysis given above applies to these fields.

Let us now consider Maxwell's equations and the electromagnetic fields whose time dependence is $\exp(j\omega t)$ and the z dependence is $\exp(-j\beta z)$ (which is the condition of a wave travelling in the z direction). ω is the frequency of oscillation of the fields in radians/sec, and $\beta\,(= 2\pi/\lambda_g)$ is the propagation constant. The curl of $\mathbf{E}$ is

$$\nabla \times \mathbf{E} = -j\omega\mu\,\mathbf{H}$$

which becomes on replacing $\partial/\partial z$ by $-j\beta$

$$\frac{\partial E_z}{\partial y} + j\beta\, E_y = -j\omega\mu\, H_x$$

$$-j\beta\, E_x - \frac{\partial E_z}{\partial x} = -j\omega\mu\, H_y$$

$$\frac{\partial E_y}{\partial x} - \frac{\partial E_x}{\partial y} = -j\omega\mu\, H_z \tag{12.126}$$

Similarly, the curl of $\mathbf{H}$ is

$$\nabla \times \mathbf{H} = j\omega\varepsilon\,\mathbf{E}$$

$$\frac{\partial H_z}{\partial y} + j\beta\, H_y = j\omega\varepsilon\, E_x$$

$$-j\beta\, H_x - \frac{\partial H_z}{\partial x} = j\omega\varepsilon\, E_y$$

$$\frac{\partial H_y}{\partial x} - \frac{\partial H_x}{\partial y} = j\omega\varepsilon\, E_z \tag{12.127}$$

We manipulate these equations in such a manner that E_x, E_y, H_x, and H_y are written in terms of the derivatives of E_z and H_z (left as an exercise for the reader)

$$E_x = -\frac{j\beta}{(k^2-\beta^2)}\frac{\partial E_z}{\partial x} - \frac{j\omega\mu}{(k^2-\beta^2)}\frac{\partial H_z}{\partial y}$$

$$H_x = -\frac{j\beta}{(k^2-\beta^2)}\frac{\partial H_z}{\partial x} + \frac{j\omega\varepsilon}{(k^2-\beta^2)}\frac{\partial E_z}{\partial y}$$

$$E_y = -\frac{j\beta}{(k^2-\beta^2)}\frac{\partial E_z}{\partial y} + \frac{j\omega\mu}{(k^2-\beta^2)}\frac{\partial H_z}{\partial x}$$

$$H_y = -\frac{j\beta}{(k^2-\beta^2)}\frac{\partial H_z}{\partial y} - \frac{j\omega\varepsilon}{(k^2-\beta^2)}\frac{\partial E_z}{\partial x} \tag{12.128}$$

where $k = \omega\sqrt{\mu\varepsilon}\,(= 2\pi/\lambda_0)$. From these equations it is clear that both E_z and H_z cannot be zero. We work with the transverse magnetic (TM) modes first ($H_z = 0$).

12.10.2 Transverse Magnetic (TM) Modes ($H_z = 0$)

Again we use the separation of variables technique (see Section 12.10.1). Starting from E_z, we get

$$E_z = X_{Ez}(x)\, Y_{Ez}(y)\, e^{-j\beta z}$$

where

$$X_{Ez}(x) = A\cos(k_x x) + B\sin(k_x x)$$

$$Y_{Ez}(y) = C\cos(k_y y) + D\sin(k_y y) \qquad (12.129)$$

where A, B, C, and D are constants. Similarly all the other field components have a similar functional dependence. Applying the boundary condition (tangential E, $E_z = 0$) on guide wall at $y = 0$ for all values of x and z (as in the case of the parallel plate waveguide),

$$C\cos(k_y y) + D\sin(k_y y)|_{y=0} = 0$$

$$C\cos(0) = 0$$

Therefore

$$C = 0$$

Similarly at $y = b$ we set $E_z = 0$

$$D\sin(k_y y)\big|_{y=b} = 0$$

$$D\sin(k_y b) = 0$$

or

$$k_y = n\pi/b \qquad n = 1,2,\ldots \qquad (12.130)$$

Therefore

$$E_z = \{A\cos(k_x x) + B\sin(k_x x)\}D\sin(k_y y)\, e^{-j\beta z} \qquad (12.131)$$

where

$$k_y = n\pi/b$$

We now proceed to apply the boundary condition at $x = 0$ and $x = a$ in a similar manner. Setting $E_z = 0$ at $x = 0$ for all values of y and z

$$\{A\cos(k_x x) + B\sin(k_x x)\}|_{x=0} = 0$$

which implies that

$$A = 0 \qquad (12.132)$$

and for $x = a$ we must have

$$B\sin(k_x x)\,|_{x=a} = 0$$

or

$$k_x = m\pi/a, \quad m = 0,1,2\ldots \qquad (12.133)$$

Hence E_z is

$$E_z = BD\sin(k_x x)\sin(k_y y)\, e^{-j\beta z}$$

$$= E_{z0}\sin\left(\frac{m\pi}{a}x\right)\sin\left(\frac{n\pi}{b}y\right)e^{-j\beta z} \qquad (12.134)$$

Plugging in this value of E_z into Eqn set (12.128)

$$E_x = -E_{z0}\left(\frac{m\pi}{a}\right)\frac{j\beta}{(k^2-\beta^2)}\cos\left(\frac{m\pi}{a}x\right)\sin\left(\frac{n\pi}{b}y\right)e^{-j\beta z}$$

$$H_x = E_{z0}\left(\frac{n\pi}{b}\right)\frac{j\omega\varepsilon}{(k^2-\beta^2)}\sin\left(\frac{m\pi}{a}x\right)\cos\left(\frac{n\pi}{b}y\right)e^{-j\beta z}$$

$$E_y = -E_{z0}\left(\frac{n\pi}{b}\right)\frac{j\beta}{(k^2-\beta^2)}\sin\left(\frac{m\pi}{a}x\right)\cos\left(\frac{n\pi}{b}y\right)e^{-j\beta z}$$

$$H_y = -E_{z0}\left(\frac{m\pi}{a}\right)\frac{j\omega\varepsilon}{(k^2-\beta^2)}\cos\left(\frac{m\pi}{a}x\right)\sin\left(\frac{n\pi}{b}y\right)e^{-j\beta z} \qquad (12.135)$$

where $\quad k^2 = \beta^2 + \left(\dfrac{m\pi}{a}\right)^2 + \left(\dfrac{n\pi}{b}\right)^2 \begin{cases} m = 1, 2, \ldots \\ n = 1, 2, \ldots \end{cases}$ (12.136)

In these formulae neither m nor n is zero since then E_z is zero and all the fields become zero.

> For the rectangular waveguide there is no TM_{01} or TM_{10} mode.

12.10.3 Transverse Electric (TE) Modes ($E_z = 0$)

Here we start with H_z

$$H_z = X_{Hz}(x)\, Y_{Hz}(y)\, e^{-j\beta z} \tag{12.137}$$

where

$$X_{Hz}(x) = A' \cos(k_x x) + B' \sin(k_x x)$$
$$Y_{Hz}(y) = C' \cos(k_y y) + D' \sin(k_y y) \tag{12.138}$$

where A', B', C', and D' are constants. Since the boundary condition of H_z yields no particular results, we obtain the other field components from Eqn set (12.128).

$$E_x = -\frac{j\omega\mu}{(k^2-\beta^2)}\frac{\partial H_z}{\partial y} = -\frac{j\omega\mu}{(k^2-\beta^2)} X_{Hz}(x)\, Y'_{Hz}(y)\, e^{-j\beta z}$$

$$H_x = -\frac{j\beta}{(k^2-\beta^2)}\frac{\partial H_z}{\partial x} = -\frac{j\beta}{(k^2-\beta^2)} X'_{Hz}(x)\, Y_{Hz}(y)\, e^{-j\beta z}$$

$$E_y = \frac{j\omega\mu}{(k^2-\beta^2)}\frac{\partial H_z}{\partial x} = \frac{j\omega\mu}{(k^2-\beta^2)} X'_{Hz}(x)\, Y_{Hz}(y)\, e^{-j\beta z}$$

$$H_y = -\frac{j\beta}{(k^2-\beta^2)}\frac{\partial H_z}{\partial y} = -\frac{j\beta}{(k^2-\beta^2)} X_{Hz}(x)\, Y'_{Hz}(y)\, e^{-j\beta z} \tag{12.139}$$

where

$$X'_{Hz} = \frac{dX_{Hz}}{dx} = k_x\{-A' \sin(k_x x) + B' \cos(k_x x)\}$$

$$Y'_{Hz} = \frac{dY_{Hz}}{dy} = k_y\{-C' \sin(k_y y) + D' \cos(k_y y)\} \tag{12.140}$$

Applying the boundary condition on E_x at $y = 0$ and $y = b$ we find that $D' = 0$ and $k_y = n\pi/b, n = 0, 1, 2 \ldots$. Similarly applying the boundary condition on E_y at $x = 0$ and $x = b$ we find that $B' = 0$ and $k_x = m\pi/b, m = 0, 1, 2 \ldots$. When we apply the boundary conditions on the other magnetic field components (the normal components of the magnetic field are zero on the walls of the waveguide) we do not go any further, since $H_y \propto E_x$ and $H_x \propto E_y$. Therefore

$$H_z = A'C' \cos(k_x x) \cos(k_y y)\, e^{-j\beta z}$$
$$= H_{z0} \cos(k_x x) \cos(k_y y)\, e^{-j\beta z} \tag{12.141}$$

$$k_x = \frac{m\pi}{a} \quad m = 0, 1, 2, \ldots \tag{12.142}$$

$$k_y = \frac{n\pi}{b} \quad n = 0, 1, 2, \ldots \tag{12.143}$$

The complete fields are

$$E_x = H_{z0} \left(\frac{n\pi}{b}\right) \frac{j\omega\mu}{(k^2-\beta^2)} \cos\left(\frac{m\pi}{a}x\right) \sin\left(\frac{n\pi}{b}y\right) e^{-j\beta z}$$

$$H_x = H_{z0} \left(\frac{m\pi}{a}\right) \frac{j\beta}{(k^2-\beta^2)} \sin\left(\frac{m\pi}{a}x\right) \cos\left(\frac{n\pi}{b}y\right) e^{-j\beta z}$$

$$E_y = -H_{z0}\left(\frac{m\pi}{a}\right)\frac{j\omega\mu}{(k^2-\beta^2)}\sin\left(\frac{m\pi}{a}x\right)\cos\left(\frac{n\pi}{b}y\right)e^{-j\beta z}$$

$$H_y = H_{z0}\left(\frac{n\pi}{b}\right)\frac{j\beta}{(k^2-\beta^2)}\cos\left(\frac{m\pi}{a}x\right)\sin\left(\frac{n\pi}{b}y\right)e^{-j\beta z} \qquad (12.144)$$

where
$$k^2 = \beta^2 + \left(\frac{m\pi}{a}\right)^2 + \left(\frac{n\pi}{b}\right)^2 \begin{cases} m = 0, 1, 2, \ldots \\ n = 0, 1, 2, \ldots \\ \text{but both } m, n \neq 0 \end{cases} \qquad (12.145)$$

These fields and equations do appear complex! The propagation constant in the waveguide is

$$\beta = \sqrt{k^2 - \left(\frac{m\pi}{a}\right)^2 - \left(\frac{n\pi}{b}\right)^2} \qquad (12.146)$$

Each mode propagates with a certain cut-off frequency. Since the mode starts propagating when $\beta = 0$, therefore

$$k_{cmn} = \sqrt{\left(\frac{m\pi}{a}\right)^2 + \left(\frac{n\pi}{b}\right)^2} \qquad (12.147)$$

$$\omega_{cmn}\sqrt{\mu\varepsilon} = \sqrt{\left(\frac{m\pi}{a}\right)^2 + \left(\frac{n\pi}{b}\right)^2}$$

$$\frac{\omega_{cmn}}{c} = \sqrt{\left(\frac{m\pi}{a}\right)^2 + \left(\frac{n\pi}{b}\right)^2}$$

$$\omega_{cmn} = c\sqrt{\left(\frac{m\pi}{a}\right)^2 + \left(\frac{n\pi}{b}\right)^2} \qquad (12.148)$$

where ω_{cmn} is the cut-off radian frequency for the m, n^{th} TE or TM mode and c is the velocity of light in the medium filling the waveguide..

Hence
$$f_{cmn} = \frac{c}{2\pi}\sqrt{\left(\frac{m\pi}{a}\right)^2 + \left(\frac{n\pi}{b}\right)^2}$$

Example 12.21 For a rectangular waveguide $a = 2b = 1$ cm, find the dominant mode and its range of frequencies of operation before the next mode starts to operate.

Solution

Step 1. To find out whether a mode is operational, we must study the fields of the TE and TM modes as well as the cut-off frequencies. If we study the TM modes Eqn set (12.135), we find that if $m = 0$ or $n = 0$, then all the fields vanish. Therefore, the TM modes *cannot* be the lowest frequency modes. On the other hand, if we study the TE modes, as given in Eqn set (12.144), then we find that if $m = 0$, $n \neq 0$ then all the fields do not vanish. Similarly, when $m \neq 0$, $n = 0$ the fields do not vanish. So the dominant mode has to be a TE mode.

Step 2. Knowing that the dominant mode is a TE mode then

$$f_{c10} = \frac{c}{2\pi}\left\{\sqrt{\left(\frac{m\pi}{a}\right)^2 + \left(\frac{n\pi}{b}\right)^2}\right\}\Bigg|_{m=1, n=0}$$

$$= \frac{c}{2a}$$

$$= \frac{3 \times 10^8}{0.02} = 15 \text{ GHz}$$

and

$$f_{c01} = \frac{c}{2\pi} \left\{ \sqrt{\left(\frac{m\pi}{a}\right)^2 + \left(\frac{n\pi}{b}\right)^2} \right\} \Bigg|_{m=0,\, n=1}$$

$$= \frac{c}{2b} = \frac{3 \times 10^8}{0.01} = 30 \text{ GHz}$$

Step 3. The dominant mode (that is the first mode with the lowest cut-off frequency) is the TE_{10} mode which starts to propagate at 15 GHz. The next higher order mode, for this waveguide, is the TE_{01} mode. Its cut-off frequency is 30 GHz. Therefore, the range of frequencies when only the dominant modes operates is 15–30 GHz.

In general, for the TE_{10} mode $f_{c10} = c/\lambda_{c10}$. Or $\lambda_{c10} = c/f_{10}$, which gives

$$\lambda_{c01} = 2a$$

Hence the cut-off wavelength is $2a$. The fields of the TE_{10} mode are

$$H_z = H_{z0} \cos\left(\frac{\pi}{a}x\right) e^{-j\beta z} \tag{12.149}$$

$$E_x = 0$$

$$H_x = H_{z0} \left(\frac{\pi}{a}\right) \frac{j\beta}{(k^2 - \beta^2)} \sin\left(\frac{\pi}{a}x\right) e^{-j\beta z}$$

$$E_y = -H_{z0} \left(\frac{\pi}{a}\right) \frac{j\omega\mu}{(k^2 - \beta^2)} \sin\left(\frac{\pi}{a}x\right) e^{-j\beta z}$$

$$H_y = 0 \tag{12.150}$$

Degenerate modes are those TE and TM modes which have the same cut-off frequency but have different values of m and n. For example the TE_{20} ($m = 2$, $n = 0$) mode and the TE_{01} ($m = 0$, $n = 1$) for the $a = 2b$ waveguides are degenerate.

 See `RectWvgdeBeta.m` in Chapter 12

12.11 | Circular Waveguide

The circular waveguide is a hollow pipe of circular cross-section made of some high conductivity material like copper (Fig. 12.31). To analyse such a waveguide we would have to write the Helmholtz equation in cylindrical coordinates. The wave equation in cylindrical coordinates applied to a scalar ψ is

$$\frac{\partial^2 \psi}{\partial \rho^2} + \frac{1}{\rho^2}\frac{\partial^2 \psi}{\partial \phi^2} + \frac{\partial^2 \psi}{\partial z^2} + \frac{1}{\rho}\frac{\partial \psi}{\partial \rho} = -k^2 \psi \tag{12.151}$$

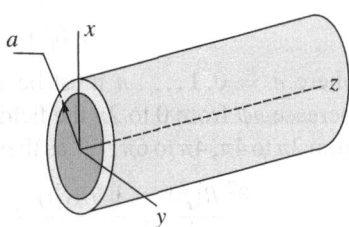

Fig. 12.31 Circular waveguide

where $k = \omega \sqrt{\mu\varepsilon}$. We use the same technique as earlier, namely the method of separation of variables. ψ is written as the multiplication of three functions

$$\psi(\rho, \phi, z) = R(\rho)\, F(\phi)\, e^{-j\beta z} \tag{12.152}$$

The z dependent function has an $\exp\{-j\beta z\}$ dependency since we are considering a travelling wave in the z-direction. Substituting this function in the previous equation

$$\frac{\partial^2 \psi}{\partial \rho^2} + \frac{1}{\rho^2}\frac{\partial^2 \psi}{\partial \phi^2} + \frac{\partial^2 \psi}{\partial z^2} + \frac{1}{\rho}\frac{\partial \psi}{\partial \rho} = -k^2\psi \tag{12.153}$$

or

$$\frac{\partial^2}{\partial \rho^2} R(\rho)\,F(\phi)\,e^{-j\beta z} + \frac{1}{\rho^2}\frac{\partial^2}{\partial \phi^2} R(\rho)\,F(\phi)\,e^{-j\beta z}$$

$$+ \frac{\partial^2}{\partial z^2} R(\rho)\,F(\phi)\,e^{-j\beta z}$$

$$+ \frac{1}{\rho}\frac{\partial}{\partial \rho} R(\rho)\,F(\phi)\,e^{-j\beta z} = -k^2 R(\rho)\,F(\phi)\,e^{-j\beta z} \tag{12.154}$$

which becomes

$$F(\phi)\,e^{-j\beta z}\frac{\partial^2 R(\rho)}{\partial \rho^2} + \frac{R(\rho)\,e^{-j\beta z}}{\rho^2}\frac{\partial^2 F(\phi)}{\partial \phi^2}$$

$$+(-j\beta)(-j\beta)\,R(\rho)\,F(\phi)\,e^{-j\beta z}$$

$$+\frac{F(\phi)\,e^{-j\beta z}}{\rho}\frac{\partial R(\rho)}{\partial \rho} = -k^2 R(\rho)\,F(\phi)\,e^{-j\beta z}$$

Dividing throughout by $R(\rho)\,F(\phi)\,\exp(-j\beta z)$ gives

$$\left\{\frac{1}{R(\rho)}\frac{\partial^2 R(\rho)}{\partial \rho^2} + \frac{1}{\rho R(\rho)}\frac{\partial R(\rho)}{\partial \rho}\right\} + \left\{\frac{1}{\rho^2 F(\phi)}\frac{\partial^2 F(\phi)}{\partial \phi^2}\right\} + (k^2-\beta^2) = 0$$

Multiplying this equation throughout by ρ^2

$$\underbrace{\left\{\frac{\rho^2}{R(\rho)}\frac{\partial^2 R(\rho)}{\partial \rho^2} + \frac{\rho}{R(\rho)}\frac{\partial R(\rho)}{\partial \rho} + \rho^2(k^2-\beta^2)\right\}}_{\text{A function of }\rho} + \underbrace{\left\{\frac{1}{F(\phi)}\frac{\partial^2 F(\phi)}{\partial \phi^2}\right\}}_{\text{A function of }\phi} = 0$$

Hence since the two parts are functions of different variables, therefore

$$\frac{1}{F(\phi)}\frac{\partial^2 F(\phi)}{\partial \phi^2} = -n^2$$

$$\frac{\rho^2}{R(\rho)}\frac{\partial^2 R(\rho)}{\partial \rho^2} + \frac{\rho}{R(\rho)}\frac{\partial R(\rho)}{\partial \rho} + \rho^2(k^2-\beta^2) = n^2 \tag{12.155}$$

The first of these equations integrates to

$$F(\phi) = A\cos(n\phi) + B\sin(n\phi) \tag{12.156}$$

where $n = 0, 1\ldots$. n must be an integer since for any fixed value of ρ, as we increase $n\phi$ from 0 to 2π the fields must repeat themselves: they must be identical from 2π to 4π, 4π to 6π, ... as they were from 0 to 2π. The second equation becomes

$$\frac{\partial^2 R(\rho)}{\partial \rho^2} + \frac{1}{\rho}\frac{\partial R(\rho)}{\partial \rho} + R(\rho)(k^2-\beta^2) = R(\rho)\left(\frac{n}{\rho}\right)^2 \tag{12.157}$$

if we let $\zeta = \sqrt{k^2-\beta^2}$ then

$$\frac{\partial^2 R(\rho)}{\partial \rho^2} + \frac{1}{\rho}\frac{\partial R(\rho)}{\partial \rho} + R(\rho)\,\zeta^2 = R(\rho)\left(\frac{n}{\rho}\right)^2$$

or

$$\frac{\partial^2 R(\rho)}{\zeta^2\,\partial \rho^2} + \frac{1}{\zeta\rho}\frac{\partial R(\rho)}{\zeta\,\partial \rho} + R(\rho) = R(\rho)\left(\frac{n}{\zeta\rho}\right)^2$$

or

$$\frac{\partial^2 R(\zeta\rho)}{\partial(\zeta\rho)^2} + \frac{1}{(\zeta\rho)}\frac{\partial R(\zeta\rho)}{\partial(\zeta\rho)} + R(\zeta\rho) = R(\zeta\rho)\left(\frac{n}{\zeta\rho}\right)^2$$

or

$$\frac{\partial^2 R(\zeta\rho)}{\partial(\zeta\rho)^2} + \frac{1}{(\zeta\rho)}\frac{\partial R(\zeta\rho)}{\partial(\zeta\rho)} + R(\zeta\rho)\left[1 - \left\{\frac{n}{(\zeta\rho)}\right\}^2\right] = 0 \qquad (12.158)$$

This equation is the Bessel equation (Bowman, 1968). Bessel functions, Z_n, are defined by

$$Z_{n+1} + Z_{n-1} = \frac{2n}{x}Z_n$$

$$Z_{n+1} - Z_{n-1} = -2\frac{dZ_n}{dx} \qquad (12.159)$$

and the differential equation which defines Bessel functions is

$$x^2\frac{d^2y}{dx^2} + x\frac{dy}{dx} + (x^2 - n^2)y = 0 \qquad (12.160)$$

This equation, being a second-order equation has two solutions, namely Bessel functions of the first and second kind. Bessel functions of the second kind are infinite at the origin, so they will be of no use to us (this is so because none of the field components are infinite anywhere within the guide). Bessel functions of the first kind are depicted in Fig. 12.32 for the first few orders.

The notation for these functions is $J_n(x)$ (the Bessel function of the first kind of order n). The very first Bessel function is 1 at the origin while it oscillates and decays slowly when the value of x is increased. All the other higher order Bessel functions of the first kind are zero at the origin while showing similar behaviour as we move away from the origin.

Keeping in mind the discussion given above, each field component has the functional form

$$E_\rho, E_\phi \ldots H_z = A\cos(n\phi + \theta)\,J_n(\zeta\rho)\,e^{-j\beta z} \qquad (12.161)$$

where A and θ are constants; $\zeta = \sqrt{k^2 - \beta^2}$. By orienting the x and y axes correctly, θ may be set to be zero. So

$$E_\rho, E_\phi \ldots H_z = A\cos(n\phi)\,J_n(\zeta\rho)\,e^{-j\beta z} \qquad (12.162)$$

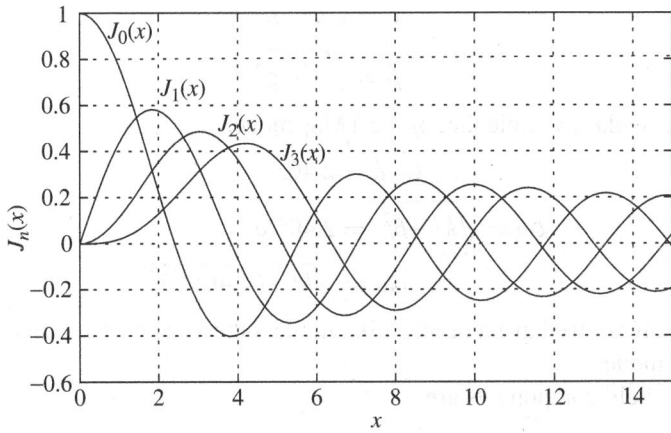

Fig. 12.32 The first few Bessel functions of the first kind

The other field components in terms of E_z and H_z are (see Eqn set (12.128) which are similar)

$$H_\rho = \frac{j\omega\varepsilon}{\zeta^2 \rho} \frac{\partial E_z}{\partial \phi} - \frac{j\beta}{\zeta^2} \frac{\partial H_z}{\partial \rho}$$

$$E_\phi = -\frac{j\beta}{\zeta^2 \rho} \frac{\partial E_z}{\partial \phi} + \frac{j\omega\mu}{\zeta^2} \frac{\partial H_z}{\partial \rho}$$

$$H_\phi = -\frac{j\omega\varepsilon}{\zeta^2} \frac{\partial E_z}{\partial \rho} - \frac{j\beta}{\zeta^2 \rho} \frac{\partial H_z}{\partial \phi}$$

$$E_\rho = -\frac{j\beta}{\zeta^2} \frac{\partial E_z}{\partial \rho} - \frac{j\omega\mu}{\zeta^2 \rho} \frac{\partial H_z}{\partial \phi} \qquad (12.163)$$

12.11.1 Transverse Magnetic Modes (TM) ($H_z = 0$)

In the transverse magnetic modes, we start with E_z.

$$E_z = E_{z0} \cos(n\phi) J_n(\zeta \rho) e^{-j\beta z} \qquad (12.164)$$

from which we derive the other field components. The boundary condition on E_z is

$$E_z|_{\rho=a} = E_{z0} \cos(n\phi) J_n(\zeta a) e^{-j\beta z} = 0$$

which means that

$$J_n(\zeta a) = 0 \qquad (12.165)$$

(see Fig. 12.32 paying attention to the zeros of $J_0(x)$ and $J_1(x)$)

$J_n(\cdot)$	n	First zero	Second zero
$J_0(\zeta a)$	$n = 0$	$\zeta_{01} a = 2.405$	$\zeta_{02} a = 5.52$
$J_1(\zeta a)$	$n = 1$	$\zeta_{11} a = 3.85$	$\zeta_{12} a = 7.02$

These waves are the

$$\text{TM}_{01}(\zeta_{01} a = 2.41), \ \text{TM}_{11}(\zeta_{11} a = 3.85),$$
$$\text{TM}_{02}(\zeta_{02} a = 5.52), \ \text{and} \ \text{TM}_{12}(\zeta_{12} a = 7.02) \qquad (12.166)$$

modes, starting from the lowest to the highest modes. Note that

$$\zeta = \sqrt{k^2 - \beta^2} \qquad (12.167)$$

and

$$\beta = \sqrt{k^2 - \zeta^2} \qquad (12.168)$$

Taking a particular example that of the TM_{01} mode,

$$\zeta_{01} a = 2.405$$

$$\zeta_{01} = \sqrt{k^2 - \beta_{01}^2} = 2.405/a$$

$$\beta_{01} = \sqrt{k^2 - (2.405/a)^2} \qquad (12.169)$$

When $k > 2.405/a$ we have a propagating mode, otherwise we have an evanescent (or cut-off) mode.

The other field components are

$$H_\rho = -E_{z0} \frac{j\omega\varepsilon\, n}{\zeta^2 \rho} \sin(n\phi) J_n(\zeta \rho) e^{-j\beta z}$$

$$E_\phi = E_{z0}\frac{j\beta\, n}{\zeta^2\, \rho}\sin(n\phi)\, J_n(\zeta\rho)\, e^{-j\beta z}$$

$$H_\phi = -E_{z0}\frac{j\omega\varepsilon}{\zeta}\cos(n\phi)\, J_n'(\zeta\rho)\, e^{-j\beta z}$$

$$E_\rho = -E_{z0}\frac{j\beta}{\zeta}\cos(n\phi)\, J_n'(\zeta\rho)\, e^{-j\beta z} \tag{12.170}$$

Here $J_n'(\cdot)$ is the Bessel function differentiated with respect to its argument.

12.11.2 Transverse Electric Modes (TE) ($E_z = 0$)

Here
$$H_z = H_{z0}\cos(n\phi)\, J_n(\zeta\rho)\, e^{-j\beta z} \tag{12.171}$$

and the other field components are

$$H_\rho = -H_{z0}\frac{j\beta}{\zeta}\cos(n\phi)\, J_n'(\zeta\rho)\, e^{-j\beta z}$$

$$E_\phi = H_{z0}\frac{j\omega\mu}{\zeta}\cos(n\phi)\, J_n'(\zeta\rho)\, e^{-j\beta z}$$

$$H_\phi = H_{z0}\frac{j\beta\, n}{\zeta^2\, \rho}\sin(n\phi)\, J_n(\zeta\rho)\, e^{-j\beta z}$$

$$E_\rho = H_{z0}\frac{j\omega\mu\, n}{\zeta^2\, \rho}\sin(n\phi)\, J_n(\zeta\rho)\, e^{-j\beta z} \tag{12.172}$$

Applying the boundary condition on E_ϕ,

$$E_\phi|_{\rho=a} = H_{z0}\frac{j\omega\mu}{\zeta}\cos(n\phi)\, J_n'(\zeta a)\, e^{-j\beta z} = 0$$

Then
$$J_n'(\zeta a) = 0 \tag{12.173}$$

(see Fig. 12.32 paying attention to the maximas and minimas of $J_0(x)$ and $J_1(x)$)

$J_n'(\cdot)$	n	First max or min	Second max or min
$J_0'(\zeta'a)$	$n = 0$	$\zeta_{01}'\, a = 3.83$	$\zeta_{02}'\, a = 7.02$
$J_1'(\zeta'a)$	$n = 1$	$\zeta_{11}'\, a = 1.84$	$\zeta_{12}'\, a = 5.33$

These waves are the

$$\text{TE}_{11}(\zeta_{11}'\, a = 1.84),\ \text{TE}_{01}(\zeta_{01}'\, a = 3.83),$$
$$\text{TM}_{12}(\zeta_{12}'\, a = 5.33)\ \text{and}\ \text{TM}_{02}(\zeta_{02}'\, a = 7.02) \tag{12.174}$$

starting from the lowest to the highest modes.

Comparing the TE and TM modes, we can see that the TE_{11} ($\zeta_{11}'\, a = 1.84$) is the dominant mode of the circular waveguide.

Example 12.22 An air-filled cylindrical waveguide has a diameter of 12 cm. Find the cut-off wavelength (λ_c) of the TE_{01} mode and the modes which propagate at wavelengths longer than $0.85\lambda_c$.

Solution For the circular waveguide, for the TE_{01} mode

$$\zeta_{01}'\, a = 3.83$$

where $a = 6$ cm. Now

$$\zeta_{01}' = \frac{3.83}{a} = 63.83 = \sqrt{k^2 - \beta_{01}^2}$$

at the cut-off frequency, $\beta_{01} = 0$, so $k_c = 63.83$. The cut-off wavelength and frequency are

$$\lambda_c = \frac{2\pi}{k_c} = 9.844 \text{ cm}$$

$$f_c = 3.05 \text{ GHz}$$

If we are to consider $\lambda' = 0.85\lambda_c = 8.37$ cm which gives a value of $k' = 75.068$; then $k'a = 4.5$. So the modes which operate are the TE_{01}, TE_{11}, TM_{01}, and TM_{11} modes, since $4.5 > 3.83$, 1.84, 2.405, and 3.85 but less than the other mode cut-off values (which are 5.33, etc.). From this discussion it is clear that TE_{01}, TE_{11}, TM_{01}, and TM_{11} modes propagate while the others are cut off.

 See `CutOffFreqCircWvguide.m` in Chapter 12

12.12 | Cavity Resonators

A resonator—typically an LC circuit in the lumped device approach—is a kind of filter which allows a frequency to resonate in an electronic circuit. At microwave frequencies, a closed device—*or cavity*—is constructed so that the device resonates typically at a single frequency. If however the cavity is large, a number of frequencies may resonate, giving us a number of resonant frequencies.

Cavity resonators are used in microwave amplifiers and frequency measuring devices so that a single frequency may be amplified (in the first case) or detected (in the second case).

To analyse a cavity resonator of a rectangular type shown in Fig. 12.33 we start with the equations for a waveguide and modify them suitably. We start with the TM modes in a cavity resonator.

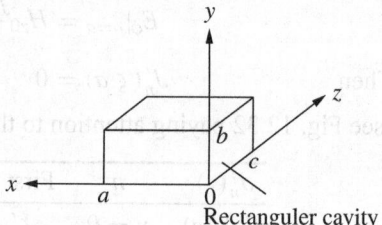

Fig. 12.33 A rectangular cavity resonator

 See `RectCavResonator.m` in Chapter 12

12.12.1 TM Modes in a Cavity ($H_z = 0$)

First of all we have to start with the Helmholtz equation as treated in Section 12.10.1 which shows us how to use the separation of variables technique for rectangular geometries. We therefore start with Eqn (12.134) which is a labour saving device. The equation is

$$E_z = E_{z0} \sin\left(\frac{m\pi}{a}x\right) \sin\left(\frac{n\pi}{b}y\right) e^{-j\beta z}$$

In this equation the wave is a travelling wave, in the $+z$ direction. We now modify this equation so that the z-directed electric field does not vanish on $z = 0, c$, since there is a surface charge where $(D_n = \rho_s)$.

$$E_z = E_{z0} \sin\left(\frac{m\pi}{a}x\right) \sin\left(\frac{n\pi}{b}y\right) \cos\left(\frac{p\pi}{c}z\right)$$

with
$$k_x = \frac{m\pi}{a} \qquad m \neq 0; \ m = 1, 2 \ldots$$

$$k_y = \frac{n\pi}{b} \qquad n \neq 0; \ n = 1, 2, \ldots$$

$$k_z = \frac{p\pi}{c} \qquad p = 0, 1, \ldots$$

and
$$k_r^2 = k_x^2 + k_y^2 + k_z^2 \quad \text{(using } v = \omega_r / k_r)$$

$$\frac{\omega_r^2}{v^2} = \left(\frac{m\pi}{a}\right)^2 + \left(\frac{n\pi}{b}\right)^2 + \left(\frac{p\pi}{c}\right)^2$$

where v is the velocity of light (to avoid using c) and ω_r is the resonant frequency. Therefore the resonant frequencies in Hz are given by

$$f_r = \frac{v}{2}\sqrt{\left(\frac{m}{a}\right)^2 + \left(\frac{n}{b}\right)^2 + \left(\frac{p}{c}\right)^2} \quad m, n = 1, 2, \ldots; \ p = 0, 1, 2, \ldots$$

Example 12.23 Find the resonant frequency for the TM$_{110}$ and TM$_{111}$ modes for $a = 5$ cm, $b = 4$ cm, and $c = 3$ cm.

Solution For this question, $m = 1, n = 1, p = 0, 1$. So

$$f_{r110} = \left(\frac{3 \times 10^8}{2}\right) \times \sqrt{\left(\frac{1}{0.05}\right)^2 + \left(\frac{1}{0.04}\right)^2}$$

$$= 4.802 \text{ GHz}$$

and
$$f_{r110} = \left(\frac{3 \times 10^8}{2}\right) \times \sqrt{\left(\frac{1}{0.05}\right)^2 + \left(\frac{1}{0.04}\right)^2 + \left(\frac{1}{0.03}\right)^2}$$

$$= 6.9327 \text{ GHz}$$

12.12.2 TE Modes in a Cavity ($E_z = 0$)

By analogy we start with the case of H_z.

$$H_z = H_{z0} \cos(k_x x) \cos(k_y y) \sin(k_z z)$$

since $B_z = 0$ at $z = 0, c$. In this equation

$$k_x = \frac{m\pi}{a}, \qquad m = 0, 1, 2 \ldots$$

$$k_y = \frac{n\pi}{b}, \qquad n = 0, 1, 2, \ldots$$

$$k_z = \frac{p\pi}{c}, \qquad p = 1, 2, \ldots$$

and we get the resonant frequencies as obtained earlier

$$f_r = \frac{v}{2}\sqrt{\left(\frac{m}{a}\right)^2 + \left(\frac{n}{b}\right)^2 + \left(\frac{p}{c}\right)^2} \quad m, n = 0, 1, 2, \ldots; \ p = 1, 2, \ldots$$

In this equation both m and n cannot be zero (See the remarks with Eqn set (12.144)).

Example 12.24 Find the resonant frequency for the TE$_{011}$ and TE$_{101}$ modes for $a = 5$ cm, $b = 4$ cm, and $c = 3$ cm.

Solution   For this question, $m = 0, n = 1, p = 1$. So

$$f_{r011} = \left(\frac{3 \times 10^8}{2}\right) \times \sqrt{\left(\frac{1}{0.03}\right)^2 + \left(\frac{1}{0.04}\right)^2}$$

$$= 6.25 \text{ GHz}$$

and $$f_{r101} = \left(\frac{3 \times 10^8}{2}\right) \times \sqrt{\left(\frac{1}{0.05}\right)^2 + \left(\frac{1}{0.03}\right)^2}$$

$$= 5.831 \text{ GHz}$$

POINTS TO REMEMBER

- A common rule of thumb is that the cable or wire should be treated as a transmission line if the length of the line is greater than $\sim(1/10)\,\lambda$.
- The time domain voltage transmission line equation is

$$\frac{\partial^2 V(z,t)}{\partial z^2} = LC \frac{\partial^2 V(z,t)}{\partial t^2}$$

where $v = 1/\sqrt{LC}$. This equation applies to lossless lines. The solution to this equation is $V(z,t) = V(z - vt)$.
- The time domain current transmission line equation is

$$\frac{\partial^2 I(z,t)}{\partial z^2} = LC \frac{\partial^2 I(z,t)}{\partial t^2}$$

This equation applies to lossless lines. The solution to this equation is $I(z,t) = I(z - vt)$.
- The frequency domain transmission line general equations are

$$\frac{d^2 V(z)}{dz^2} = (j\omega L + R)(j\omega C + G)\,V(z)$$

$$\frac{d^2 I(z)}{dz^2} = (j\omega L + R)(j\omega C + G)\,I(z)$$

where $\gamma^2 = (j\omega L + R)(j\omega C + G) = \alpha + j\beta$. α is the attenuation constant and β is the phase constant. The solution to these equations are

$$V(z) = V_{\pm} e^{j\omega t \mp \gamma z}$$

$$I(z) = I_{\pm} e^{j\omega t \mp \gamma z}$$

and where $\exp\{-\gamma z\}$ is the wave travelling in the $+z$ direction, and so on.
- The following equations are important

$$v_p = \frac{\omega}{\beta}$$

$$\beta = \frac{2\pi}{\lambda_g}$$

- For low-loss transmission lines where $R \neq 0$ and $R \ll j\omega L$; $G \neq 0$ and $G \ll j\omega C$

$$\gamma \approx j\omega \sqrt{LC} + \frac{R}{2}\sqrt{\frac{C}{L}} + \frac{G}{2}\sqrt{\frac{C}{L}}$$

$$Z_0 \approx \sqrt{\frac{L}{C}}\left(1 - \frac{jR}{2\omega L} + \frac{jG}{2\omega C}\right)$$

- For distortion-less lines

$$\frac{R}{G} = \frac{L}{C}$$

$$\alpha = \sqrt{RG}$$

$$\beta = \omega\sqrt{LC}$$

- The current and voltage of the forward and reverse waves are related by

$$I_+ = \frac{V_+}{Z_0}$$

$$I_- = -\frac{V_-}{Z_0}$$

$$V(z) = V_+ e^{-j\beta z} - V_- e^{+j\beta z}$$

$$I(z) = \frac{1}{Z_0}(V_+ e^{-j\beta z} - V_- e^{+j\beta z})$$

- The power propagated in the forward (+) and reverse (−) waves is

$$P_+ = \frac{|V_+|^2}{2Z_0}$$

$$P_- = \frac{|V_-|^2}{2Z_0}$$

- The reflection coefficient at the load is

$$\Gamma_L = \frac{V_-}{V_+} = \frac{Z_L - Z_0}{Z_L + Z_0}$$

- The reflection coefficient at $z = -l$ is given by

$$\Gamma(-l) = \Gamma_L e^{-j2\beta l}$$

- The VSWR (s) is given by

$$s = \frac{V_{\max}}{V_{\min}} = \frac{1 + |\Gamma_L|}{1 - |\Gamma_L|}$$

- The input impedance along the line is given by

$$Z_{\text{in}}(-l) = Z_0 \left\{ \frac{Z_L + jZ_0 \tan(\beta l)}{Z_0 + jZ_L \tan(\beta l)} \right\}$$

- The voltage and current along the line are given by the proportional equations

$$V(-l) \propto 1 + \Gamma(-l)$$

$$I(-l) \propto 1 - \Gamma(-l)$$

- The input impedance along the line is given by

$$Z_{\text{in}}(-l) = Z_0 \frac{1 + \Gamma(-l)}{1 - \Gamma(-l)}$$

- $\lambda/4$ and $\lambda/8$ lines with proper termination can be used as circuit elements.
- The Smith chart can be used to solve transmission line problems very conveniently.
- A $\lambda/4$ line of characteristic impedance Z_0 may be used to match two lines of characteristic impedances Z_{01} and Z_{02} by

$$Z_0 = \sqrt{Z_{01}Z_{02}}$$

- In the parallel plate waveguide, the cut-off wavelength is given by

$$\lambda_{cm0} = \frac{2a}{m}$$

where a is the plate separation. This equation applies to both TE as well as TM modes provided the mode exists.

- The propagation constant of the parallel plate waveguide is given by

$$\beta = \sqrt{k^2 - \left(\frac{m\pi}{a}\right)^2}, \quad m = 0, 1, \ldots$$

- In the rectangular waveguide, the cut-off frequency is given by

$$f_{cmn} = \frac{c}{2\pi}\sqrt{\left(\frac{m\pi}{a}\right)^2 + \left(\frac{n\pi}{b}\right)^2} \quad \begin{cases} m = 0, 1 \ldots \\ n = 0, 1 \ldots \\ m \text{ and } n \neq 0 \end{cases}$$

where a and b are the internal dimensions of the waveguide ($a > b$) and c is the velocity of light in vacuum. This equation applies to both TE as well as TM modes provided the mode exists.

- The propagation constant in a rectangular waveguide is given by

$$\beta = \sqrt{k^2 - \left(\frac{m\pi}{a}\right)^2 - \left(\frac{n\pi}{b}\right)^2}$$

- The cut-off condition for a circular waveguide is given by
  - TE modes: $J_n'(k_{cnm}'a) = 0$ where $k_{cnm}'a$ is the mth zero of the derivative of the Bessel function of the first kind of order n. $k_{c01}'a = 3.83$, $k_{c02}'a = 7.02$, and $k_{c11}'a = 1.84$, and so on. a is the radius of the waveguide. The dominant mode is the TE_{11} mode.
  - TM modes: $J_n(k_{cnm}a) = 0$ where $k_{cnm}a$ is the mth zero of the Bessel function of the first kind of order n. $k_{c01}a = 2.41$ and $k_{c02}a = 5.52$ and so on. a is the radius of the waveguide.

- The resonant frequency for TM modes in a cavity are

$$f_r = \frac{v}{2}\sqrt{\left(\frac{m}{a}\right)^2 + \left(\frac{n}{b}\right)^2 + \left(\frac{p}{c}\right)^2} \quad m, n = 1, 2, \ldots; \, p = 0, 1, 2, \ldots$$

and for TE modes are

$$f_r = \frac{v}{2}\sqrt{\left(\frac{m}{a}\right)^2 + \left(\frac{n}{b}\right)^2 + \left(\frac{p}{c}\right)^2} \quad m, n = 0, 1, 2, \ldots; \, p = 1, 2, \ldots$$

where both m and n are not zero simultaneously.

SELF ASSESSMENT

Objective Type Questions

In the following questions one or more choices may be correct. Sometimes choices may be separated with an 'or' and sometimes with an 'and'.

1. For a lossless transmission line the characteristic impedance is
    (a) purely real
    (b) purely imaginary
    (c) in general it will have a real part and an imaginary part
    (d) is a function of frequency

2. For a lossless transmission line
   (a) $L = 0$ (b) $C = 0$ (c) $R = 0$ (d) $G = 0$
3. For a lossless transmission line
   (a) the unit of L is H,   (b) the unit of C may be pf/km
   (c) the SI unit of L is H/m   (d) the unit of R may be kΩ
4. The function $P_T(x - 3 \times 10^8 t)$ (where $P_T(\cdot)$ is a pulse of width T) describes
   (a) a stationary pulse   (b) a travelling pulse
   (c) at $t = 5$s the pulse is at $x = 15 \times 10^8$ m   (d) none of the above
5. The parameter G applicable to a transmission line is
   (a) a shunt lumped element put there for design purposes
   (b) is present due to losses contributed by the metal conductors comprising the line
   (c) is present due to losses contributed by the dielectric filling of the line
   (d) none of the above
6. For a transmission line if $L/C = R/G$ then
   (a) $Z_0 = R/G$
   (b) the line is called a distortionless line
   (c) If a series of pulsed are transmitted they arrive undistorted
   (d) the line is lossless
7. A micro-strip line
   (a) is lossless   (b) is distortionless
   (c) has a phase velocity equal to 3×10^8 m/s   (d) none of the above
8. If there are two conductors parallel to each other separated by a dielectric in a wave-guiding structure
   (a) only a TEM mode propagates in the structure
   (b) TE and TM modes may propagate in the structure
   (c) a TEM mode propagates at all frequencies
   (d) none of the above
9. In a coaxial line if the capacitance per metre is C and inductance per metre is L
   (a) the line may be treated as a transmission line with $Z_0 = \sqrt{L/C}$
   (b) the line may be treated as a transmission line with $\beta = \sqrt{LC}$
   (c) the line may be treated as a transmission line with $v_p = 1/\sqrt{LC}$
   (d) none of the above
10. In the case of the rectangular waveguide
    (a) a mode propagates from $f = 0$ Hz   (b) for TE modes $H_z = 0$
    (c) a TEM mode may propagate   (d) none of the above
11. For a circular waveguide
    (a) modes propagates from $f = 0$ Hz   (b) for TM modes $H_z = 0$
    (c) a TEM mode may propagate   (d) the dominant ode is the TE_{10} mode

Short-Answer Questions

1. Prove that if two load resistors are normalised with respect to the characteristic impedances, and the normalised values are reciprocals of each other then they have the same VSWR.
2. If a resistor is 1 cm long, at about which frequency will it be treated as a distributed element?
3. A 50 Ω transmission line is connected to a 100 Ω load. Find the input impedance for $l = \lambda/2$, $\lambda/4$, $\lambda/8$ and $\lambda/10$. Do the same with $Z_L = 25$ Ω.
4. Explain why a voltage maximum occurs when the angle of Γ is zero or an even multiple of π, and a voltage minimum occurs when $\angle\Gamma$ is an odd multiple of π.

5. Explain why when there is a voltage minimum along the line the current is a maximum.

6. Discuss about how to match an arbitrary load with lumped matching networks as shown in Fig. 12.34 by using different values of B, X, and Z_0.

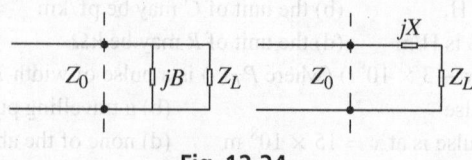

Fig. 12.34

7. A parallel plate waveguide is formed by two perfect electric conductors and a dielectric filling of $\varepsilon_r = 2$, and a plate separation of 3 cm. If the frequency of operation is 8 GHz, find whether the TEM mode propagates. If the TEM mode propagates, find the wave impedance, wavelength in the guide, and the phase velocity.

8. For Numerical Example 7 find (a) which TE modes propagate (if any) (b) which TM modes propagate (if any). In both cases find the wavelength and phase velocity in the guide for the first of each mode.

9. For a micro-strip line with characteristic impedance Z_0 and $v_p = c/\sqrt{\varepsilon_{eff}}$ calculate the L and C of the line ($c = 3 \times 10^8$ m/s).

Review Questions

1. Why are transmission line concepts used as the frequency is increased?
2. In transmission line analysis why do we use the concept of distributed elements?
3. Write down the time domain equation for I applicable to a transmission line.
4. What is the difference between the frequency domain equation and time domain equation?
5. When we want to take into account the losses in the metal conductors of the transmission line, which parameter is affected?
6. Why is it that in a distortion-less line a pulse launched at the input end of a transmission line arrives as undistorted at the output end?
7. Write the advantages and disadvantages of a quarter wave transformer.
8. Derive Eqns (12.94).

Numerical Problems

If the medium is not specified, assume $\varepsilon = \varepsilon_0$ and $\mu = \mu_0$.

1. Determine the transmission line equations for the models of Fig. 12.35.
2. A parallel wire transmission line has per meter values of $L = 2\,\mu H/m$, $C = 6$ pF/m, $G = 0\,\mho/m$, and $R = 3\,m\Omega/m$. Calculate
   (a) the characteristic impedance (b) the propagation constant
   (c) velocity of propagation, and (d) the wavelength for operation at 1 MHz
3. In this problem we do a transmission line analysis and a lumped circuit analysis and compare results. In Fig. 12.36, if $V_s = 10$ V, $R_s = 10\,\Omega$, and $Z_L = 10\,\Omega$ and the source is connected to the load by a 10 cm two-wire transmission line with $L = 2\,\mu H/m$,

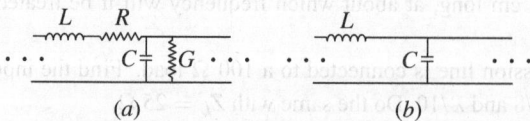

Fig. 12.35 Two transmission line models

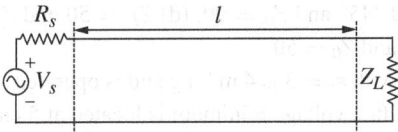

Fig. 12.36 An example to show the importance of transmission line analysis as opposed to lumped circuit analysis

$C = 6$ pF/m. Find (a) the power delivered to the load if we use a 'lumped' approach and (b) the power delivered to the load if we use a transmission line approach. In (b), let the frequencies of operation be (i) 1 kHz (ii) 1 MHz (iii) 1 GHz.

4. A coaxial line has an inner conductor diameter equal to 2 mm and the inside diameter of the outer conductor equal to 6 mm, and the region between two conductors is filled with a dielectric of $\varepsilon_r = 3$. If the conductors and dielectric are both considered to be ideal, calculate the L, R, C, and G and also γ and v_p. Are these parameters frequency-independent?

5. Calculate the constants R, L, C and G for a two-wire line with conductor radius of 1 mm and distance between centres equal to 2 cm. Assume the material of the lines to be copper and the frequency to be 10 MHz.

6. If a coaxial line has to be low loss (a low value of R), what is the thickness of gold plating required at a frequency of 1 GHz? For this line, $a = 2$ mm and $b = 5$ mm. Hint: use the skin depth for gold and Table 12.1.

7. Show that the voltage $v(x, t) = v_0 \cos(\omega t - \phi) \exp\{-j\beta x\}$ satisfies the transmission line equations for a uniform lossless line if $\beta = \omega \sqrt{LC}$.

8. The reflection coefficient of a transmission line of characteristic impedance 50 Ω is $\Gamma = 1/3$ (purely real) at a discontinuity. Guess the nature of the discontinuity (capacitive, inductive or resistive, etc.)

9. The reflection coefficient of a transmission line of characteristic impedance 50 Ω is $\Gamma = 1/3$ at a discontinuity. If the incident voltage is $5 \exp\{-j\beta z\}$ find the reflected voltage, reflected current, and incident current.

10. A coaxial line of 50 Ω is terminated by a 80 Ω resistive load. Find Γ at (a) the load (b) 10 cm from the load if $f = 100$ MHz.

11. A coaxial line of 50 Ω is terminated by a 80 Ω resistive load. Find the first, second, and third maximas of Γ from the load end if $f = 100$ MHz.

12. A coaxial line of 50 Ω is terminated by a 80 Ω resistive load. Find the first, second, and third minimas of Γ from the load end if $f = 100$ MHz.

13. A transmission line of 50 Ω is terminated by an unknown load. If $|\Gamma| = 1/5$, find the possible maximum and minimum load resistance.

14. Find the reflection coefficient if (a) $Z_L = 0$, (b) $Z_L = \infty$, (c) $Z_L = Z_0^*$, (d) $Z_L = jZ_0$, (e) $Z_L = -jZ_0$, (f) $Z_L = Z_0(1 + j)$.

15. Find the VSWR for Problem 14.

16. A 500 Ω transmission line is 200 m long and operates at 500 kHz with $\alpha = 2.4 \times 10^{-3}$ Np/m and $\beta = 0.0212$ rad/m. The load impedance is $Z_L = 424.3 \angle 45°$. Find the length of the line in wavelengths, the reflection coefficient at the load and source, and the input impedance at the source.

17. Use the Smith chart to compute the reflection coefficient and VSWR for the following pairs of impedances in ohms (a) $Z_L = 100 + j150$ and $Z_0 = 50$, (b) $Z_L = 10 + j15$ and

$Z_0 = 50$, (c) $Z_L = 90\angle45°$ and $Z_0 = 90$, (d) $Z_L = 50$ and $Z_0 = 50$, (e) $Z_L = 0$ and $Z_0 = 50$, (f) $Z_L = \infty$ and $Z_0 = 50$.

18. A 100 Ω lossless line with $\varepsilon_r = 3$ is 4 m long and is operated at 500 MHz. The VSWR on the line is 1.5. The first voltage minimum is located at 5 cm from the load. Use the Smith chart to find (a) the load admittance, (b) the reflection coefficient at the load, and (c) the input admittance to the line.

19. Use the Smith chart to find the shortest lengths of an open-circuited 75 Ω line to give the following input impedances (a) $Z_{in} = 0$, (b) $Z_{in} = \infty$, (c) $Z_{in} = j75$ Ω, (d) $Z_{in} = -j50$ Ω, and (e) $Z_{in} = j10$ Ω.

20. Use the Smith chart to find the input impedance of the configuration shown in Fig. 12.37. $\lambda = 30$ cm. Show the impedance/admittance at the important points on the Smith chart.

21. A load impedance of $Z_L = 100 + j50$ Ω is to be matched, and which is connected to a 75 Ω line. A single stub tuner is to be used. Consider the following cases (a) open circuited shunt stub, (b) short circuited shunt stub, (c) open circuited series stub, and (d) short circuited series stub.

22. Use the Smith chart to find (a) VSWR, (b) Γ_L, (c) Z_{in}, (d) the distance from the load to the first voltage maximum in wavelengths, for the circuit shown in Fig. 12.37.

23. Analyse the configuration shown below. Comment on the answer.

24. Derive Eqn set (12.128).

25. For a particular application a rectangular waveguide is used with dimensions $a = 2.3$ cm and $b = 1$ cm, find cut-off wavelength for the first mode. Also find the frequency range of single mode operation and the mode that operates. For this mode operating at 1.5 times the cut-off frequency find (a) the propagation constant, (b) guide wavelength, (c) the phase velocity.

26. A rectangular waveguide operates at 4.2 GHz: its dimensions are $a = 4.75$ and $b = 2.22$ cm. Calculate the guide wavelength and cut-off frequency for the dominant mode.

27. A Ku band rectangular waveguide is to be designed with $a/b = 2$ and is to have a cut-off frequency of 9.5 GHz. Obtain the internal dimensions of the waveguide.

28. Design a rectangular waveguide with an air filling. The waveguide must have a single mode operation over a bandwidth of 1 GHz. Find possible values of a and b.

29. What are the modes which can possibly propagate at 4 GHz in a square waveguide of side 10 cm?

30. Derive Eqn set (12.163) for the circular waveguide.

31. A circular waveguide has a diameter of 5 cm. Find the cut-off frequency, f_c, for the dominant mode.

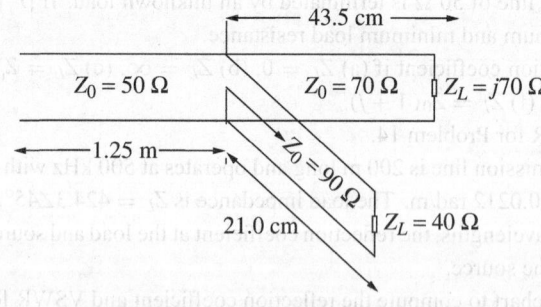

Fig. 12.37

32. A circular waveguide with $a = 1$ cm has a dielectric filling of $\varepsilon_r = 2.3$. Identify the cut-off frequencies of the first four propagating modes. What is the frequency range of the dominant mode?

33. Design a circular waveguide with a filling of $\varepsilon_r = 4$. The waveguide must have a single mode operation over a bandwidth of 2 GHz. Find the radius of the guide.

Answers

Objective Type Questions

1. (a) 2. (c) and (d) 3. (b) and (c) 4. (b) and (c) 5. (c)
6. (a), (b) and (c) 7. (d) 8. (b) and (c) 9. (a) and (c) 10. (d) 11. (b)

Short-Answer Questions

3. For $Z_L = 100\ \Omega$: $Z_{in} = 100\ \Omega$ for $l = \lambda/2$, $Z_{in} = 25\ \Omega$ for $l = \lambda/4$, $Z_{in} = 40 - j30\ \Omega$ for $l = \lambda/8$, $Z_{in} = 49 - j35\ \Omega$ for $l = \lambda/10$. For $Z_L = 25\ \Omega$: $Z_{in} = 25\ \Omega$ for $l = \lambda/2$, $Z_{in} = 100\ \Omega$ for $l = \lambda/4$, $Z_{in} = 40 + j30\ \Omega$ for $l = \lambda/8$, $Z_{in} = 33 + j24\ \Omega$ for $l = \lambda/10$.

7. Yes, a TEM mode propagates. $Z = 266.57\ \Omega$; $\lambda_g = 2.651$ cm; $v_p = 2.12 \times 10^8$.

8. TE_{10}, TM_{10}, TE_{20}, and TM_{20} modes propagate. For TE_{10}, TM_{10} modes $\lambda_g = 2.96$ cm and $v_p = 2.36 \times 10^8$ m/s. For TE_{20}, TM_{20} modes $\lambda_g = 5.67$ cm and $v_p = 4.53 \times 10^8$ m/s.

9. $C = 1/(v_p Z_0)$; $L = Z_0/v_p$.

Numerical Problems

1. See Eqns (12.17) and (12.19).

2. (a) $Z_0 = 577.4 - j0.07\ \Omega$ (b) $\gamma \approx j0.0218$ r/m (c) 2.88×10^8 m/s (d) 288.2 m

3. In the lumped approach power delivered to the load in all three cases is 2.5 W; In the transmission line approach (i) at 1 kHz it is 2.5 W (ii) at 1 MHz the dissipation is ≈ 2.5 W (iii) at 1 GHz it is 0.0597 W.

4. $L = 219.7$ nH/m; $C = 183.27$ pF/m; $\gamma = j6.35 \times 10^{-9}\omega$ r/s; $v_p = 1.576 \times 10^8$ m/s. Only γ is frequency dependent.

5. $L = 1.197\ \mu H/m$; $C = 9.289$ pF/m; $R = 0.2626\ \Omega/m$

6. Skin depth for gold is 2.485 μm. A thickness of 3δ is adequate. $R \approx 1\ \Omega/m$.

8. Resistive

9. $V_-(z) = (5/3)\exp\{+j\beta z\}$; $I_-(z) = (1/30)\exp\{+j\beta z\}$; $I_+(z) = (1/10)\exp\{-j\beta z\}$

10. (a) 0.231 (b) $0.211 - j0.094$.

11. First maximum at load itself; second maximum 1.5 m; third maximum at 3.0 m

12. First minimum at 0.75 m; second minimum 2.25 m; third minimum at 3.75 m

13. $33.3 < R_L < 75$

14. (a) $\Gamma = -1$ (b) $\Gamma = 1$ (c) $\Gamma = (jy)/x$ when $Z_L = x + jy$ and $Z_0 = x - jy$ (d) $\Gamma = j$ (e)

$$\Gamma = \frac{j+1}{j-1}$$

(f)

$$\Gamma = \frac{j}{j+2}$$

15. (a) VSWR$=\infty$ (b) VSWR$= \infty$ (c)

$$VSWR = \frac{|y| + |x|}{|y| - |x|}$$

where $Z_L = x + jy$ (d) VSWR$= \infty$ (e) VSWR$=\infty$

16. Length of the line is $0.6748\lambda_g$; $\Gamma_L = -0.0959 + j0.410$; $\Gamma_s = 0.1488 - j0.0622$; $Z_{in} = 668.5 - j85.4\,\Omega$

17. (a) $\Gamma_L = 0.66 + j0.33$, VSWR= 6.854 (b) $\Gamma_L = -0.568 + j0.392$; VSWR= 5.467 (c) $\Gamma_L = -0.155 + j0.536$; VSWR= 3.52 (d) $\Gamma_L = 0$; VSWR= 1 (e) $\Gamma_L = -1$; VSWR= ∞ (f) $\Gamma_L = 1$; VSWR= ∞.

18. $Y_L = 8.46 + j3.43$ m℧ (b) $\Gamma_L = 0.047 - j0.194$ (c) $Y_{in} = 10.66 + j4.17$ m℧

19. (a) $\lambda/4$ (b) $\lambda/2$ (c) $3\lambda/8$ (d) 0.16λ (e) 0.27λ

20. After 43.5 cm $Z_{in} = j35.67\,\Omega$; the parallel line adds $Z_p = 145.9 + j77.4\,\Omega$; after 1.25 m $Z_{in} = 314.7 + j377.5\,\Omega$

21. (a) move by 0.214λ towards the generator and add o.c. shunt stub of length 0.41λ (b) (a) move by 0.214λ towards the generator and add s.c. shunt stub of length $0.0.16\lambda$ (c) move by 0.16λ towards the generator and add an o.c. series stub of length 0.35λ (d) move by 0.16λ towards the generator and add an s.c. series stub of length 0.09λ

22. (a) VSWR =2.46 (b) $\Gamma_L = 0.246 + j0.342$ (c) $Z_{in} = 24.5 + j20.3\,\Omega$ (d) 0.077λ

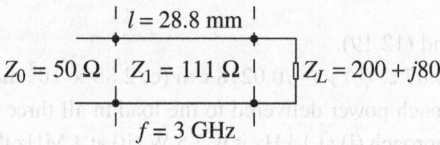

23. When we move towards the generator by 28.8 mm as shown, the input impedance is 50 Ω.

25. $\lambda_c = 4.6$ cm for the TE_{10} mode; frequency range of the fundamental mode is 6.52 GHz to 13.04 GHz. (a) $\beta = 152.64$ r/m (b) $\lambda_g = 4.11$ cm (c) $v_p = 4.026 \times 10^8$ m/s

26. $\lambda_g = 10.83$ cm; $f_c = 3.16$ GHz.

28. Many solutions are possible, one solution is $a = 15$ cm, $b = 6$ cm.

29. TE_{10}, TE_{01}, TE_{11}, TM_{11}, TE_{20}, TE_{02}, TE_{21}, TE_{12}, TM_{21}, and TM_{12} modes.

31. 1.76 GHz

32. TE_{11} has $f_c = 5.79$ GHz; TM_{01} has $f_c = 9.77$ GHz; TE_{01} has $f_c = 12.05$ GHz, and TM_{11} has $f_c = 12.11$ GHz. The fundamental mode operates from 5.79 to 9.77 GHz.

33. Many solutions are possible but one of them is $a = 6.744$ mm.

13 | Radiation from Currents

The illiterate of the 21st century
will not be those who cannot read and write,
but those who cannot learn, unlearn, and relearn.
— Alvin Toffler

CHAPTER OBJECTIVES

To enable the students to understand the following:

- Retarded potentials
- Wave equation due to charges and currents
- Radiation from a current element
- Half wave dipole antenna
- Far field approximation
- Basic antenna concepts
- The Friis transmission formula
- The radar equation

13.1 | Retarded Potentials

The previous chapters considered the cases where

1. The wave is travelling in a region *free of currents and charges*
2. The wave is guided by conductors

If we refer to Section 7.4 we have Poisson's equation where the charges are static (i.e. no variation with time)

$$\nabla^2 V = -\frac{\rho_v}{\varepsilon}$$

and similarly we have Eqn (8.78) given by

$$\nabla^2 \mathbf{A} = -\mu \mathbf{J}$$

The solutions to these equations were already considered and are

$$V(\mathbf{r}) = \frac{1}{4\pi\varepsilon} \iiint\limits_{V'} \frac{\rho_v(\mathbf{r}') \, dV'}{R}$$

$$\mathbf{A}(\mathbf{r}) = \frac{\mu}{4\pi} \iiint\limits_{\mathcal{V}'} \frac{\mathbf{J}(\mathbf{r}')\, dV'}{R} \tag{13.1}$$

where $\mathbf{R} = \mathbf{r} - \mathbf{r}'$ (and therefore $R = |\mathbf{r} - \mathbf{r}'|$); $\mathbf{r}$ is the field point; $\mathbf{r}'$ is the source point, and generally $r \gg r'$ as shown in Fig. 13.1. *The above solutions apply only to the static case.*

We now allow the charge density and current density to vary with time, that is,

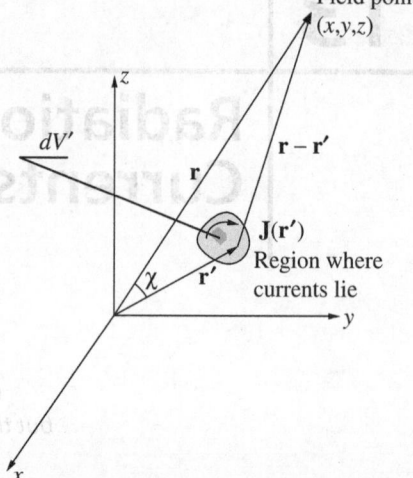

$$\rho_v\,(\mathbf{r}') \to \rho_v\,(\mathbf{r}', t)$$

$$\mathbf{J}(\mathbf{r}') \to \mathbf{J}(\mathbf{r}', t)$$

The question which we may ask is that how can we use the Eqn set (13.1) to predict the radiated far fields. If we look at the fields for the static case

$$\mathbf{E} = -\nabla V$$

and $$\mathbf{H} = \frac{\mathbf{B}}{\mu} = \frac{1}{\mu} \nabla \times \mathbf{A} \tag{13.2}$$

knowing that if a change occurs in the charges and currents at time τ then they will reflect in field changes at a time $\tau + R/c$ later, where c is the velocity of light. This allows for the time required for

Fig. 13.1 Notation applied to the sources and fields

the change which occurs in the sources to be communicated to the field point. Therefore, if we were to observe the fields at the distant point $\mathbf{r}$, at time t, then the values of the sources must be taken into account at an earlier time, namely $t - R/c$. Or, to be specific

$$V(\mathbf{r}, t) = \frac{1}{4\pi\varepsilon} \iiint\limits_{\mathcal{V}'} \frac{\rho_v\,(\mathbf{r}', t - R/c)\, dV'}{R}$$

$$\mathbf{A}(\mathbf{r}, t) = \frac{\mu}{4\pi} \iiint\limits_{\mathcal{V}'} \frac{\mathbf{J}(\mathbf{r}', t - R/c)\, dV'}{R} \tag{13.3}$$

These solutions are called the retarded potentials since they were obtained from charges and currents which were retarded in time. If we obtain the solutions to Eqn set (13.3) using the retarded potentials and obtain the fields through Eqn set (13.2), we find that the electric field is incorrect though the magnetic field is correct. So the procedure we can follow is

1. Obtain the time varying vector potential $\mathbf{A}(\mathbf{r}, t)$.
2. Then obtain $\mathbf{H}(\mathbf{r}, t)$ using the second equation from Eqn set (13.2).
3. Obtain $\mathbf{E}(\mathbf{r}, t)$ from an appropriate Maxwell's equation: $\partial \mathbf{D}/\partial t = \nabla \times \mathbf{H}$

13.2 | Wave Equation due to Charges and Currents

In this case we have to proceed along a slightly different path because the problem becomes more complex. Let us start from Eqn (10.66) which is applicable to sinusoidally varying quantities:

$$\nabla \cdot \mathbf{B} = 0$$

Since the divergence of **B** is *always* zero, **B** must be equal to the curl of some vector, since $\nabla \cdot (\nabla \times \mathbf{A})$ is always zero for any vector **A**.

$$\mathbf{B} = \nabla \times \mathbf{A} \tag{13.4}$$

A is the vector potential. Since $\mathbf{B} = \mu \mathbf{H}$

$$\mathbf{H} = \frac{1}{\mu} \nabla \times \mathbf{A} \tag{13.5}$$

Now from Eqn (10.65)

$$\nabla \times \mathbf{E} = -j\omega\mu \mathbf{H}$$

$$= -j\omega\mu \left(\frac{1}{\mu} \nabla \times \mathbf{A} \right)$$

$$= -j\omega\nabla \times \mathbf{A} \tag{13.6}$$

Therefore
$$\nabla \times \mathbf{E} + j\omega\nabla \times \mathbf{A} = 0$$
$$\nabla \times (\mathbf{E} + j\omega \mathbf{A}) = 0$$
$$\mathbf{E} + j\omega \mathbf{A} = -\nabla\phi$$
$$\mathbf{E} = -j\omega \mathbf{A} - \nabla\phi \tag{13.7}$$

where the term $-\nabla\phi$ is placed on the right for the general case, since the curl of the gradient is always identically zero. That is, $\nabla \times \mathbf{E} = \nabla \times (-j\omega \mathbf{A} - \nabla\phi) = -j\omega\nabla \times \mathbf{A}$. Substituting these equations in Eqn (10.67)

$$\nabla \times \mathbf{H} = j\omega\varepsilon \mathbf{E} + \mathbf{J}$$

$$\nabla \times \left(\frac{1}{\mu} \nabla \times \mathbf{A} \right) = j\omega\varepsilon \left(-j\omega \mathbf{A} - \nabla\phi \right) + \mathbf{J}$$

On simplifying $\qquad \nabla \times (\nabla \times \mathbf{A}) = \omega^2 \mu\varepsilon\mathbf{A} - j\omega\varepsilon\mu\nabla\phi + \mu\mathbf{J} \tag{13.8}$

As usual we can use $\nabla \times (\nabla \times \mathbf{A}) = \nabla(\nabla \cdot \mathbf{A}) - (\nabla \cdot \nabla)\mathbf{A}$ (See Appendix B.2.5, identity 5) to get

$$\nabla(\nabla \cdot \mathbf{A}) - (\nabla \cdot \nabla)\mathbf{A} = \omega^2\mu\varepsilon\mathbf{A} - j\omega\varepsilon\mu\nabla\phi + \mu\mathbf{J} \tag{13.9}$$

So far the curl of **A** has been defined through Eqn (13.4) but not its divergence and *a vector field is not fully defined if both are not defined.* Therefore, we use a particularly appropriate form of the divergence of **A** to simplify the above equation:

$$\nabla \cdot \mathbf{A} = -j\omega\varepsilon\mu\phi \tag{13.10}$$

This equation is called the Lorentz's gauge condition. By cancelling $\nabla(\nabla \cdot \mathbf{A})$ with $-j\omega\varepsilon\nabla\phi$, we get

$$-(\nabla \cdot \nabla)\mathbf{A} = \omega^2\mu\varepsilon\mathbf{A} + \mu\mathbf{J} \tag{13.11}$$

which is the frequency domain wave equation with a *source term* **J**. Once **A** is obtained, we can get **H** from Eqn (13.5). From **H**, applying Maxwell's Eqn (10.67) we can get **E**. More will be said about this when we actually apply the previous equation. Concentrating now on Eqn (13.11), it becomes (remember that $k^2 = \omega^2\mu\varepsilon$)

$$\frac{\partial^2 \mathbf{A}}{\partial x^2} + \frac{\partial^2 \mathbf{A}}{\partial y^2} + \frac{\partial^2 \mathbf{A}}{\partial z^2} + k^2\mathbf{A} = -\mu\mathbf{J} \tag{13.12}$$

which is three equations

$$\frac{\partial^2 A_x}{\partial x^2} + \frac{\partial^2 A_x}{\partial y^2} + \frac{\partial^2 A_x}{\partial z^2} + k^2 A_x = -\mu J_x$$

$$\frac{\partial^2 A_y}{\partial x^2} + \frac{\partial^2 A_y}{\partial y^2} + \frac{\partial^2 A_y}{\partial z^2} + k^2 A_y = -\mu J_y$$

$$\frac{\partial^2 A_z}{\partial x^2} + \frac{\partial^2 A_z}{\partial y^2} + \frac{\partial^2 A_z}{\partial z^2} + k^2 A_z = -\mu J_z \qquad (13.13)$$

We will use these equations in the forthcoming chapters on propagation and radiation since propagating waves as well as radiating waves are governed by them.

Figure 13.1 shows the region of application of the equation.

- $\mathbf{A}(\mathbf{r})$ is the vector potential at the field point $\mathbf{r}$. $\mathbf{r}$ is usually far away from the source. If the source is of the order of a few wavelengths, the field point is usually much more than tens of wavelengths away.
- $\mathbf{J}(\mathbf{r}')$ is the current density, which is the source term, shown shaded in the figure. Usually $\mathbf{J}$ flows in the metal of the conductor comprising the antenna.

We can now hunt for the solution in the following manner. We know that the second equation (of the vector potential) of Eqn set (13.3) is correct. We also know that for sinusoidally varying functions,

$$\mathbf{J}(\mathbf{r}', t) = \mathbf{J}(\mathbf{r}')\, e^{j\omega t} \qquad (13.14)$$

so
$$\mathbf{J}(\mathbf{r}', t - R/c) = \mathbf{J}(\mathbf{r}')\, e^{j\omega(t - R/c)}$$

$$= \mathbf{J}(\mathbf{r}')\, e^{j\omega t} e^{-j(R\omega)/c}$$

$$= \mathbf{J}(\mathbf{r}')\, e^{j\omega t} e^{-j(R\omega)/(\omega/k)} \quad (\text{since } c = \omega/k)$$

$$= \mathbf{J}(\mathbf{r}')\, e^{j\omega t} e^{-jkR}$$

Therefore the solution to Eqn (13.13) for the vector potential, $\mathbf{A}(\mathbf{r})$, is

$$\mathbf{A}(\mathbf{r}) = \frac{\mu}{4\pi} \iiint_{\mathcal{V}'} \frac{\mathbf{J}(\mathbf{r}')\, e^{-jk|\mathbf{r}-\mathbf{r}'|}}{|\mathbf{r} - \mathbf{r}'|} dV' \qquad (13.15)$$

Let us examine this equation in a little more detail. If we convert the integral to a summation (which is the reverse process) then

$$\mathbf{A} \approx \frac{\mu}{4\pi} \sum \frac{\mathbf{J}(\mathbf{r}')\, e^{-jk|\mathbf{r}-\mathbf{r}'|}}{|\mathbf{r} - \mathbf{r}'|} \Delta V'$$

which states that the vector potential $\mathbf{A}$ consists of a summation of outgoing spherical waves of the type

$$\frac{\mathbf{J}(\mathbf{r}')\, e^{-jk|\mathbf{r}-\mathbf{r}'|}}{|\mathbf{r} - \mathbf{r}'|}$$

This factor is a spherical wave with centre $\mathbf{r}'$ (due to the term)

$$e^{-jk|\mathbf{r}-\mathbf{r}'|}$$

the wave is streaming away from the source $\mathbf{J}(\mathbf{r}')$ located at $\mathbf{r}'$. The wave amplitude diminishes far away from the source and is proportional to

$$\frac{1}{|\mathbf{r} - \mathbf{r}'|}$$

The total wave comprising the vector potential is a vector sum of all these tiny wavelets.

Example 13.1 Obtain the equation for the vector potential, similar to Eqn (13.15), for the case of a current filament carrying a current I.

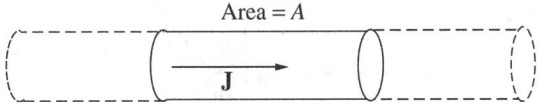

Area = A

J

Fig. 13.2 A current carrying conductor

Solution To do this, consider Fig. 13.2 where a small shaded section of a current carrying conductor is shown. The current density in the conductor is $\mathbf{J}$ assumed to be uniform over the cross-section. Consider the term

$$\mathbf{J}dV$$

Since $dV = dSdl$, where dS is an element of the area of cross-section and dl is a small length along the wire,

$$\mathbf{J}dV = \mathbf{J}dSdl$$

Integrating now over the area of cross-section, A, of the wire

$$\iiint \mathbf{J}dSdl = \int \mathbf{J}Adl = \int JA\hat{\mathbf{J}}dl = \int Id\mathbf{l} \quad (I = JA; \, d\mathbf{l} = \hat{\mathbf{J}}dl)$$

where I is the current in the wire and $d\mathbf{l}$ is a small section of the wire in the direction of the current flow. The integral over the cross-section is

$$\int \mathbf{J}dS = JA$$

since $\mathbf{J}$ is constant over the cross-section. The element $d\mathbf{l}$ may be treated as a very small vector. Equation (13.15) becomes

$$\mathbf{A} = \frac{\mu.}{4\pi} \int_{L'} \frac{I(\mathbf{r}')\,e^{-jk|\mathbf{r}-\mathbf{r}'|}}{|\mathbf{r}-\mathbf{r}'|}d\mathbf{l}' \tag{13.16}$$

13.3 | Radiation from a Current Element

Example 13.2 Apply Eqn (13.16) to a small wire carrying a current. In particular, apply the above equation to an elemental wire of length Δl carrying a current I as shown in Fig. 13.3.

Solution The figure shows the orientation of the wire in the z direction

$$d\mathbf{l}' \approx \Delta l\,\mathbf{a}_z \tag{13.17}$$

Also in the equation for the vector potential $\mathbf{A}$, the integration sign may be removed since the region of integration is small:

$$\mathbf{A} = \frac{\mu}{4\pi} \int_{L'} \frac{I(\mathbf{r}')\,e^{-jk|\mathbf{r}-\mathbf{r}'|}}{|\mathbf{r}-\mathbf{r}'|}d\mathbf{l}' \approx \frac{\mu}{4\pi}\left[\frac{Ie^{-jk|\mathbf{r}-\mathbf{r}'|}}{|\mathbf{r}-\mathbf{r}'|}\right]\mathbf{a}_z(\Delta l)$$

Because the wire is very small lengthwise, the value of

$$|\mathbf{r}-\mathbf{r}'| \approx |\mathbf{r}|$$

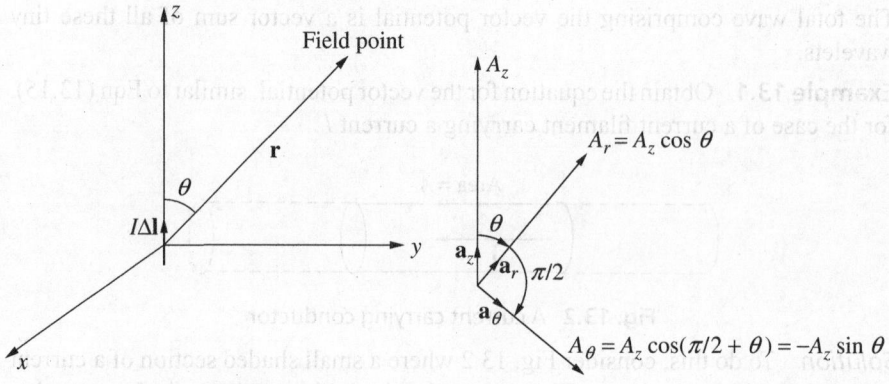

Fig. 13.3 An elemental wire carrying a current

Fig. 13.4 The relationship between A_z, A_θ, and A_r

since $\mathbf{r} \gg \mathbf{r}'(\mathbf{r}' \approx 0)$. Therefore

$$|\mathbf{r} - \mathbf{r}'| \approx |\mathbf{r}| = r \qquad (13.18)$$

The r being considered in the above equation is the r coordinate of spherical coordinates. Therefore

$$\mathbf{A} \approx \frac{\mu}{4\pi} \left[\frac{I e^{-jk|\mathbf{r}-\mathbf{r}'|}}{|\mathbf{r} - \mathbf{r}'|} \right] \mathbf{a}_z (\Delta l)$$

$$\approx \frac{\mu}{4\pi} \left[\frac{I(\Delta l) e^{-jk|\mathbf{r}|}}{|\mathbf{r}|} \right] \mathbf{a}_z = \frac{\mu}{4\pi} \left[\frac{I(\Delta l) e^{-jkr}}{r} \right] \mathbf{a}_z$$

So far we have been considering a mixed coordinate system. In the above equation

$$\mathbf{A} = \mathbf{a}_z A_z \approx \frac{\mu}{4\pi} \left[\frac{I(\Delta l) e^{-jkr}}{r} \right] \mathbf{a}_z$$

Expressing A_z in spherical coordinates we directly obtain (see Fig. 13.4)

$$A_r = A_z \cos \theta \approx \frac{\mu}{4\pi} \left[\frac{I(\Delta l) e^{-jkr}}{r} \right] \cos \theta$$

$$A_\theta = -A_z \sin \theta \approx -\frac{\mu}{4\pi} \left[\frac{I(\Delta l) e^{-jkr}}{r} \right] \sin \theta$$

$$A_\phi = 0 \qquad (13.19)$$

Using the relation $\mathbf{B} = \nabla \times \mathbf{A}$, we can obtain the magnetic flux density $\mathbf{B}$

$$\mathbf{B} = \nabla \times \mathbf{A}$$

$$= \frac{1}{r \sin \theta} \left\{ \frac{\partial (\sin \theta \, A_\phi)}{\partial \theta} - \frac{\partial A_\theta}{\partial \phi} \right\} \mathbf{a}_r + \frac{1}{r} \left\{ \frac{1}{\sin \theta} \frac{\partial A_r}{\partial \phi} - \frac{\partial (r A_\phi)}{\partial r} \right\} \mathbf{a}_\theta$$

$$+ \frac{1}{r} \left\{ \frac{\partial (r A_\theta)}{\partial r} - \frac{\partial A_r}{\partial \theta} \right\} \mathbf{a}_\phi \qquad (13.20)$$

The above equation is the equation of the curl in spherical coordinates. We can proceed to simplify it because we know that $A_\phi = 0$. Therefore

$$\mathbf{B} = \frac{1}{r\sin\theta}\left\{\frac{\partial(\sin\theta\,\cancel{A_\phi})}{\partial\theta} - \frac{\partial A_\theta}{\partial\phi}\right\}\mathbf{a}_r + \frac{1}{r}\left\{\frac{1}{\sin\theta}\frac{\partial A_r}{\partial\phi} - \frac{\partial(r\cancel{A_\phi})}{\partial r}\right\}\mathbf{a}_\theta$$

$$+ \frac{1}{r}\left\{\frac{\partial(rA_\theta)}{\partial r} - \frac{\partial A_r}{\partial\theta}\right\}\mathbf{a}_\phi$$

$$= \frac{1}{r\sin\theta}\left\{-\frac{\partial A_\theta}{\partial\phi}\right\}\mathbf{a}_r + \frac{1}{r}\left\{\frac{1}{\sin\theta}\frac{\partial A_r}{\partial\phi}\right\}\mathbf{a}_\theta + \frac{1}{r}\left\{\frac{\partial(rA_\theta)}{\partial r} - \frac{\partial A_r}{\partial\theta}\right\}\mathbf{a}_\phi$$

Next we know that neither of A_θ nor A_r are functions of ϕ.

Hence $\mathbf{B} = \dfrac{1}{r\sin\theta}\left\{-\cancel{\dfrac{\partial A_\theta}{\partial\phi}}\right\}\mathbf{a}_r + \dfrac{1}{r}\left\{\dfrac{1}{\sin\theta}\cancel{\dfrac{\partial A_r}{\partial\phi}}\right\}\mathbf{a}_\theta + \dfrac{1}{r}\left\{\dfrac{\partial(rA_\theta)}{\partial r} - \dfrac{\partial A_r}{\partial\theta}\right\}\mathbf{a}_\phi$

$$= \frac{1}{r}\left\{\frac{\partial(rA_\theta)}{\partial r} - \frac{\partial A_r}{\partial\theta}\right\}\mathbf{a}_\phi \tag{13.21}$$

Therefore the first term is

$$\frac{1}{r}\left[\frac{\partial(-rA_z\sin\theta)}{\partial r}\right] = \frac{\sin\theta}{r}\left[\frac{\partial(-rA_z)}{\partial r}\right]$$

$$= \frac{\mu\sin\theta}{4\pi r}\left[\frac{\partial(-I(\Delta l)\,e^{-jkr})}{\partial r}\right]$$

$$= \frac{jk\mu\sin\theta}{4\pi r}I(\Delta l)\,e^{-jkr} \tag{13.22}$$

and the second term is

$$-\frac{1}{r}\left[\frac{\partial(A_z(r)\cos\theta)}{\partial\theta}\right] = \frac{A_z(r)\sin\theta}{r}$$

$$= \frac{\mu\sin\theta}{4\pi}\left[\frac{I(\Delta l)\,e^{-jkr}}{r^2}\right]$$

Therefore the magnetic flux density $\mathbf{B}$ at a field point (r,θ,ϕ) due to a current $I\Delta l\mathbf{a}_z$ located at the origin at $r = 0$ is

$$\mathbf{B} = \frac{\mu\sin\theta\,[I(\Delta l)]}{4\pi}\left(\frac{jk}{r} + \frac{1}{r^2}\right)e^{-jkr}\mathbf{a}_\phi \tag{13.23}$$

We know the relation between $\mathbf{B}$ and $\mathbf{H}$: $\mathbf{B} = \mu\mathbf{H}$ so $\mathbf{H}$ is

$$\mathbf{H} = \frac{\sin\theta\,[I(\Delta l)]}{4\pi}\left(\frac{jk}{r} + \frac{1}{r^2}\right)e^{-jkr}\mathbf{a}_\phi \tag{13.24}$$

Using Maxwell's equation

$$\nabla \times \mathbf{H} = j\omega\varepsilon\mathbf{E}$$

$$j\omega\varepsilon\mathbf{E} = \frac{1}{r\sin\theta}\left\{\frac{\partial(\sin\theta\,H_\phi)}{\partial\theta} - \frac{\partial H_\theta}{\partial\phi}\right\}\mathbf{a}_r + \frac{1}{r}\left\{\frac{1}{\sin\theta}\frac{\partial H_r}{\partial\phi} - \frac{\partial(rH_\phi)}{\partial r}\right\}\mathbf{a}_\theta$$

$$+ \frac{1}{r}\left\{\frac{\partial(rH_\theta)}{\partial r} - \frac{\partial H_r}{\partial\theta}\right\}\mathbf{a}_\phi$$

where the right-hand side is the curl of **H** in spherical coordinates. Since $H_\theta = H_r = 0$

$$j\omega\varepsilon\mathbf{E} = \frac{1}{r\sin\theta}\left\{\frac{\partial(\sin\theta\,H_\phi)}{\partial\theta} - \frac{\partial H_\theta}{\partial\phi}\right\}\mathbf{a}_r + \frac{1}{r}\left\{\frac{1}{\sin\theta}\frac{\partial H_r}{\partial\phi} - \frac{\partial(rH_\phi)}{\partial r}\right\}\mathbf{a}_\theta$$

$$+ \frac{1}{r}\left\{\frac{\partial(rH_\theta)}{\partial r} - \frac{\partial H_r}{\partial\theta}\right\}\mathbf{a}_\phi \tag{13.25}$$

$$= \underbrace{\frac{1}{r\sin\theta}\left\{\frac{\partial(\sin\theta\,H_\phi)}{\partial\theta}\right\}\mathbf{a}_r}_{\text{Term 1}} + \underbrace{\frac{1}{r}\left\{-\frac{\partial(rH_\phi)}{\partial r}\right\}\mathbf{a}_\theta}_{\text{Term 2}} \tag{13.26}$$

Considering each term: the first term (Term 1) is

$$\frac{1}{r\sin\theta}\left\{\frac{\partial(\sin\theta\,H_\phi)}{\partial\theta}\right\}\mathbf{a}_r = \frac{1}{r\sin\theta}\frac{\partial}{\partial\theta}\left\{\frac{\sin^2\theta\,[I(\Delta l)]}{4\pi}\left(\frac{jk}{r} + \frac{1}{r^2}\right)e^{-jkr}\right\}\mathbf{a}_r$$

$$= \frac{1}{r\sin\theta}\left\{\frac{2\sin\theta\cos\theta\,[I(\Delta l)]}{4\pi}\left(\frac{jk}{r} + \frac{1}{r^2}\right)e^{-jkr}\right\}\mathbf{a}_r$$

$$= \left\{\frac{\cos\theta\,[I(\Delta l)]}{2\pi}\left(\frac{jk}{r^2} + \frac{1}{r^3}\right)e^{-jkr}\right\}\mathbf{a}_r$$

and the second term (Term 2) is

$$\frac{1}{r}\left\{-\frac{\partial(rH_\phi)}{\partial r}\right\}\mathbf{a}_\theta = \frac{1}{r}\left\{-\frac{\partial}{\partial r}\left[\frac{\sin\theta\,[I(\Delta l)]}{4\pi}\left(jk + \frac{1}{r}\right)e^{-jkr}\right]\right\}\mathbf{a}_\theta$$

$$= -\frac{\sin\theta\,[I(\Delta l)]}{4\pi r}\left\{\frac{\partial}{\partial r}\left[\left(jk + \frac{1}{r}\right)e^{-jkr}\right]\right\}\mathbf{a}_\theta$$

$$= -\frac{\sin\theta\,[I(\Delta l)]}{4\pi r}\left\{k^2 - \frac{jk}{r} - \frac{1}{r^2}\right\}e^{-jkr}\mathbf{a}_\theta$$

$$= -\frac{\sin\theta\,[I(\Delta l)]}{4\pi}\left\{\frac{k^2}{r} - \frac{jk}{r^2} - \frac{1}{r^3}\right\}e^{-jkr}\mathbf{a}_\theta$$

Hence

$$\mathbf{E} = \frac{\nabla\times\mathbf{H}}{j\omega\varepsilon} = \frac{[I(\Delta l)]}{j\omega\varepsilon}\left\{\left[\frac{\cos\theta}{2\pi}\left(\frac{jk}{r^2} + \frac{1}{r^3}\right)e^{-jkr}\right]\mathbf{a}_r\right.$$

$$\left. - \left[\frac{\sin\theta}{4\pi}\left(\frac{k^2}{r} - \frac{jk}{r^2} - \frac{1}{r^3}\right)e^{-jkr}\right]\mathbf{a}_\theta\right\} \tag{13.27}$$

Generally we will be interested in the far fields. Neglecting the components which are proportional to $1/r^2$ and $1/r^3$ (which become small as r becomes large) we find that

$$\mathbf{E} = \frac{[I(\Delta l)]}{j\omega\varepsilon}\left\{\frac{\cos\theta}{2\pi}\underbrace{(jk/r^2 + 1/r^3)}_{\text{neglected}}e^{-jkr}\mathbf{a}_r\right.$$

$$\left. - \frac{\sin\theta}{4\pi}(k^2/r\underbrace{-jk/r^2 - 1/r^3}_{\text{neglected}})\,e^{-jkr}\mathbf{a}_\theta\right\}$$

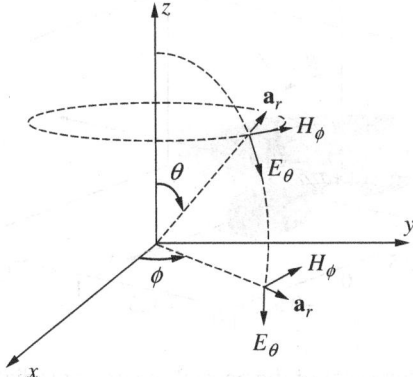

Fig. 13.5 Direction of E_θ and H_ϕ shown for a spherical wave

$$\approx \frac{[I(\Delta l)]}{j\omega\varepsilon}\left[\frac{\sin\theta}{4\pi}\left(\frac{k^2}{r}\right)e^{-jkr}\right]\mathbf{a}_\theta \tag{13.28}$$

$$= j\frac{Z_0 I \sin\theta}{2r}\left(\frac{\Delta l}{\lambda}\right)e^{-jkr}\mathbf{a}_\theta \tag{13.29}$$

and **H** is
$$\mathbf{H} = \frac{\sin\theta\,[I(\Delta l)]}{4\pi}(jk/r + \underbrace{1/r^2}_{\text{neglected}})\,e^{-jkr}\mathbf{a}_\phi$$

$$\approx j\frac{I \sin\theta}{2r}\left(\frac{\Delta l}{\lambda}\right)e^{-jkr}\mathbf{a}_\phi \tag{13.30}$$

So the electromagnetic fields are

$$\mathbf{E} = j\frac{Z_0 I \sin\theta}{2r}\left(\frac{\Delta l}{\lambda}\right)e^{-jkr}\mathbf{a}_\theta$$
$$\mathbf{H} = j\frac{I \sin\theta}{2r}\left(\frac{\Delta l}{\lambda}\right)e^{-jkr}\mathbf{a}_\phi \tag{13.31}$$

The direction of the fields are shown in Fig. 13.5.

 See `EFldHertzianDipole.m` in Chapter 13

Example 13.3 Find the ratio E/H for the far fields of the antenna of Example 13.2.
Solution The ratio is
$$\frac{E_\theta}{H_\phi} = \left\{jZ_0\frac{I \sin\theta}{2r}\left(\frac{\Delta l}{\lambda}\right)e^{-jkr}\right\}\bigg/\left\{j\frac{I \sin\theta}{2r}\left(\frac{\Delta l}{\lambda}\right)e^{-jkr}\right\}$$
$$= Z_0 \qquad \text{(Characteristic impedence for free space is 377 }\Omega\text{.)}$$
which is the same result that we got in the case of a plane wave travelling in free space.

Example 13.4 Find the Poynting vector **P** for the far fields of the antenna of Example 13.2.
Solution The Poynting vector denotes the direction of power flow. Since we are considering phasors, the formula for the Poynting vector is

$$\mathbf{P} = \frac{1}{2}(\mathbf{E} \times \mathbf{H}^*)$$

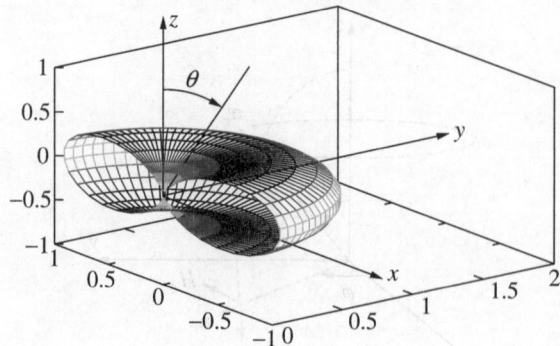

Fig. 13.6 Polar plot of the normalised power radiated by an infinitesimal current element

(Note that since we are *not* considering rms values, the factor 1/2 is present). Substituting the values of **E** and **H**

$$\mathbf{P} = \frac{Z_0}{2}\left[\frac{jI\sin\theta}{2r}\left(\frac{\Delta l}{\lambda}\right)e^{-jkr}\mathbf{a}_\theta\right] \times \left(\frac{jI\sin\theta}{2r}\left(\frac{\Delta l}{\lambda}\right)e^{-jkr}\mathbf{a}_\phi\right)^*$$

where the star indicates complex conjugation. Converting all the j s to $-j$ s in the second bracket

$$\mathbf{P} = \frac{Z_0}{2}\left[\frac{jI\sin\theta}{2r}\left(\frac{\Delta l}{\lambda}\right)e^{-jkr}\mathbf{a}_\theta\right] \times \left(\frac{-jI\sin\theta}{2r}\left(\frac{\Delta l}{\lambda}\right)e^{+jkr}\mathbf{a}_\phi\right)$$

Next we know that $\mathbf{a}_\theta \times \mathbf{a}_\phi = \mathbf{a}_r$ and therefore

$$\mathbf{P} = \frac{1}{2}Z_0\left[\frac{\sin\theta}{2r}\left\{I\left(\frac{\Delta l}{\lambda}\right)\right\}\right]^2\mathbf{a}_r \ (\text{W/m}^2)$$

1. The power streams out in the $\mathbf{a}_r$ direction. It is proportional to the squares of $(I\Delta l)$ and inversely proportional to the squares of $(r\lambda)$.
2. The power streams out directionally: at $\theta = 0$ and π no power is radiated, while at $\theta = \pi/2$ maximum power is radiated. This three-dimensional half-plot of the normalised power density (W/m²) as a function θ is shown in Fig. 13.6.
3. The power radiated per steradian (solid angle) is

$$U(\theta,\phi) = P(r,\theta,\phi)\,r^2 = \frac{1}{2}Z_0\left[\frac{\sin\theta}{2}\left\{I\left(\frac{\Delta l}{\lambda}\right)\right\}\right]^2$$

A solid angle subtended by some area at the centre of a sphere is the area on the surface of the sphere divided by the square of the radius. Thus, an elemental solid angle $d\Omega = dS/r^2 = \sin\theta\,d\theta\,d\phi$.

Example 13.5 Find the total power radiated by the antenna of Example 13.2.
Solution The total power radiated is the flux of the Poynting vector through a sphere of radius r.

$$P_T = \oiint \mathbf{P} \cdot d\mathbf{S}$$

$$= \oiint (P\mathbf{a}_r) \cdot (\mathbf{a}_r r^2 \sin\theta\,d\theta\,d\phi)$$

Since $P = U/r^2$ therefore

$$P_T = \iint U(\theta,\phi) \sin\theta \, d\theta \, d\phi$$

$$= \int_{\theta=0,\phi=0}^{\theta=\pi,\phi=2\pi} \left\{ \frac{1}{2} Z_0 \left[\frac{\sin\theta}{2} \left\{ I \left(\frac{\Delta l}{\lambda} \right) \right\} \right]^2 \right\} \times (\sin\theta \, d\theta \, d\phi)$$

Carrying out the integration over ϕ which is 2π and collecting terms

$$P_T = \frac{1}{2} Z_0 \left\{ I \left(\frac{\Delta l}{2\lambda} \right) \right\}^2 \int_{\theta=0,\phi=0}^{\theta=\pi,\phi=2\pi} \sin^3\theta \, d\theta \, d\phi$$

$$= \pi Z_0 \left\{ I \left(\frac{\Delta l}{2\lambda} \right) \right\}^2 \int_{\theta=0}^{\theta=\pi} \sin^3\theta \, d\theta$$

The last integration is 4/3. Therefore

$$P_T = \pi Z_0 \left\{ I \left(\frac{\Delta l}{2\lambda} \right) \right\}^2 \int_{\theta=0}^{\theta=\pi} \sin^3\theta \, d\theta$$

$$= \frac{\pi}{2} Z_0 \left(\frac{\Delta l}{\lambda} \right)^2 \left(\frac{I}{\sqrt{2}} \right)^2 \frac{4}{3}$$

$$= \frac{2\pi}{3} Z_0 \left(\frac{\Delta l}{\lambda} \right)^2 \left(\frac{I}{\sqrt{2}} \right)^2 \tag{13.32}$$

Example 13.6 Find the Thevenin equivalent of the antenna of Example 13.2.
Solution In Eqn (13.32), $I/\sqrt{2}$ is the rms value of the current, and the rest,

$$\frac{2\pi}{3} Z_0 \left(\frac{\Delta l}{\lambda} \right)^2$$

is a resistance called the *radiation resistance* (R_r). Therefore,

$$P_T = I_{rms}^2 R_r \tag{13.33}$$

$$R_r = \frac{2\pi}{3} Z_0 \left(\frac{\Delta l}{\lambda} \right)^2$$

$$= 80\pi^2 \left(\frac{\Delta l}{\lambda} \right)^2 \tag{13.34}$$

The power dissipated in the radiation resistance of an antenna represents the total power radiated away. If we examine Fig. 13.7, we can see that the source supplies power to the antenna and sees a loss. Therefore, the source and antenna can be modelled as a Thevenin equivalent circuit, which is shown in the figure. Referring to the figure, the total power radiated away is represented by the power dissipated in R_r, the radiation resistance. In actual practice the antenna is modelled as an impedance, $Z = R_r + jX$. See for example (Kraus, 1988, Jordan & Balmain, 1968) for details.

@ See RadResisHertzDipole.m in Chapter 13

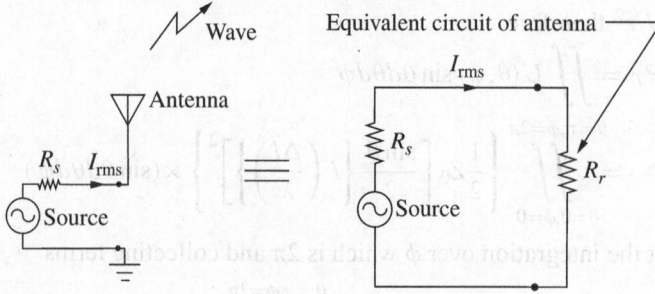

Fig. 13.7 Equivalent circuit of an antenna

13.4 | Half-Wave Dipole Antenna

Example 13.7 Analyse and obtain the fields of a half-wave dipole antenna shown in Fig. 13.8 where $L = \lambda/4$.

Solution The dipole antenna consists of two thin but stiff wires of equal length ($= L$), and which are bent into a shape as shown in Fig. 13.8. The ends of the two wires are embedded in an insulator which insulates them from each other, and are fed by a source of microwave frequency. Generally, the feeder is coaxial line, the centre conductor of which is connected to one of the wires of the dipole, while the outer conductor is connected to the other.

Since the wires are limited in extent, the currents at the ends must be zero. Also the currents are sinusoidally distributed on each of the wires (analogous to the case of transmission lines). The most probable current distribution on the upper wire is therefore

$$I = I_0 \sin(kz + \theta)$$

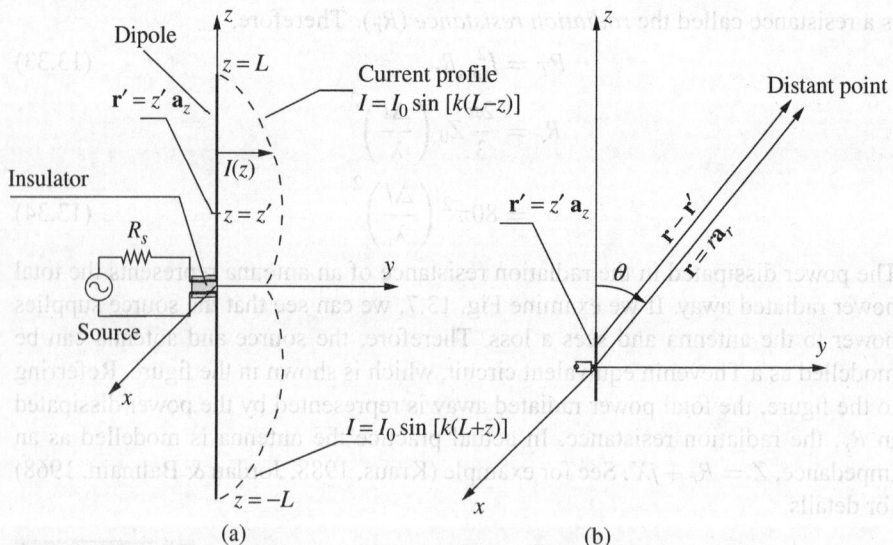

Fig. 13.8 The dipole antenna. (a) close-up details, (b) details with respect to the far field

where k is the free-space propagation constant ($= 2\pi/\lambda$). Now at $z = L$, $I = 0$ so

$$\sin(kL + \theta) = 0, \text{ or}$$

$$kL + \theta = 0, \pi \text{ or}$$

$$\theta = -kL, \text{ or } \pi - kL$$

so
$$\underbrace{I = I_0 \sin(kz - kL)}_{\text{Function 1}} \text{ or } \underbrace{I = I_0 \sin(kz + \pi - kL)}_{\text{Function 2}}$$

Now observing the two functions we know that in the first function (Function 1) as z is decreased from its value at the end, that is, $z < L$, then the argument of the sine function $k(z - L)$ becomes negative and the value of the current, I, is negative. However, this is not true for the second function (Function 2). Therefore, on the upper conductor

$$I = I_0 \sin(kz + \pi - kL)$$
$$= I_0 \sin[k(L - z)], \quad \text{for } 0 \leq z \leq L \qquad (13.35)$$

By the same reasoning for the lower conductor the current, I, is

$$I = I_0 \sin[k(L + z)], \quad \text{for } -L \leq z \leq 0 \qquad (13.36)$$

Once we have obtained the current distribution on the antenna, we can proceed to apply Eqn (13.16)

$$\mathbf{A} = \frac{\mu}{4\pi} \int_{L'} \frac{I(\mathbf{r}') e^{-jk|\mathbf{r}-\mathbf{r}'|}}{|\mathbf{r} - \mathbf{r}'|} d\mathbf{l}'$$

In this equation we use the value for the current we just derived earlier, and $\mathbf{r}$ and $\mathbf{r}'$ are shown in Fig. 13.8(b). $\mathbf{r} = \mathbf{a}_r r$ and $\mathbf{r}' = \mathbf{a}_z z'$. The equation then becomes

$$\mathbf{A} = \frac{\mu}{4\pi} \int_{z'=-L}^{z'=L} \frac{I(z') e^{-jk|\mathbf{a}_r r - \mathbf{a}_z z'|}}{|\mathbf{a}_r r - \mathbf{a}_z z'|} \mathbf{a}_z dz'$$

Two approximations are made, which are both quite reasonable: (Refer Fig. 13.9)

1. The term $|\mathbf{r} - \mathbf{r}'|$ is approximated by $r - z' \cos\theta$. How? We notice from the figure that $\mathbf{r} = \mathbf{a}_r r$ (spherical coordinates), and $\mathbf{r}' = \mathbf{a}_z z'$. But the magnitude $|\mathbf{r} - \mathbf{r}'|$ is the distance AB or approximately the distance CD since $AB \approx CD$ (To corroborate this, let $L = 1$ m, $r = 1000$ m, and $\theta = \pi/2$ then $AB = 1000.0005$ m and $CD = 1000$ m. For the same case for $\theta = 0$ $AB = 999$ m and $CD = 999$ m) therefore

$$|\mathbf{r} - \mathbf{r}'| \approx CD = OD - OC = r - OC = r - z' \cos\theta$$

Using this approximation

$$\mathbf{A} \approx \frac{\mathbf{a}_z \mu}{4\pi} \int_{z'=-L}^{z'=L} \frac{I(z') e^{-jk(r - z' \cos\theta)}}{r - z' \cos\theta} dz'$$

2. In the second part of the approximation, the denominator $r - z' \cos\theta \approx r$ since the integral with or without this approximation gives practically the same result. With the second approximation

$$\mathbf{A} \approx \frac{\mathbf{a}_z \mu}{4\pi} \int_{z'=-L}^{z'=L} \frac{I(z') e^{-jk(r - z' \cos\theta)}}{r} dz' = \frac{\mathbf{a}_z \mu e^{-jkr}}{4\pi r} \int_{z'=-L}^{z'=L} I(z') e^{jkz' \cos\theta} dz'$$

$$(13.37)$$

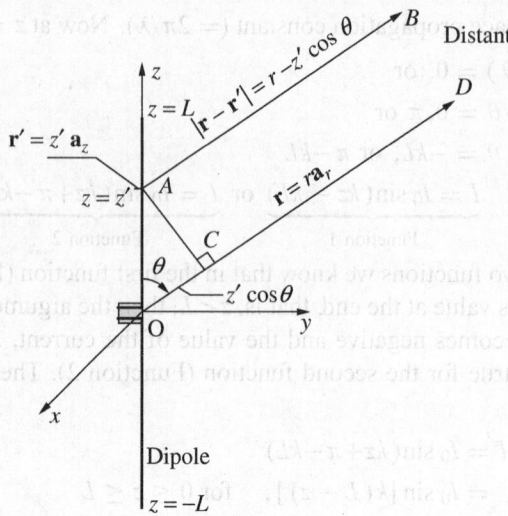

Fig. 13.9 The approximation $|\mathbf{R} - \mathbf{R}'| \approx r - z' \cos \theta$

Now using the value of the current, just calculated

$$
\mathbf{A} = \frac{\mathbf{a}_z \mu \, e^{-jkr}}{4\pi r} \int_{z'=-L}^{z'=0} \underbrace{I_0 \sin \left[k(L+z') \right]}_{\text{Current for } -L \le z' \le 0} e^{jkz' \cos \theta} \, dz'
$$

$$
+ \frac{\mathbf{a}_z \mu \, e^{-jkr}}{4\pi r} \int_{z'=0}^{z'=L} \underbrace{I_0 \sin \left[k(L-z') \right]}_{\text{Current for } 0 \le z' \le L} e^{jkz' \cos \theta} \, dz' \qquad (13.38)
$$

Integrating the first integral (by parts), we get

$$
\frac{\mathbf{a}_z \mu \, e^{-jkr}}{4\pi r} \int_{z'=-L}^{z'=0} I_0 \sin \left[k(L+z') \right] e^{jkz' \cos \theta} \, dz'
$$

$$
= \frac{\mathbf{a}_z \mu \, e^{-jkr} I_0}{4\pi r k} \left\{ \frac{j \cos \theta \, \sin(kL) - \cos(kL)}{\sin^2 \theta} + \frac{e^{-jk \cos \theta L}}{\sin^2 \theta} \right\} \qquad (13.39)
$$

The second integral similarly evaluated is

$$
\frac{\mathbf{a}_z \mu \, e^{-jkr}}{4\pi r} \int_{z'=0}^{z'=L} I_0 \sin[k(L-z')] e^{jkz' \cos \theta} \, dz'
$$

$$
= \frac{\mathbf{a}_z \mu \, e^{-jkr} I_0}{4\pi r k} \left\{ -\frac{j \cos \theta \, \sin(kL) + \cos(kL)}{\sin^2 \theta} + \frac{e^{jk \cos \theta L}}{\sin^2 \theta} \right\} \qquad (13.40)
$$

Adding these two integrals, we get

$$
\mathbf{A} = \frac{\mathbf{a}_z \mu \, e^{-jkr} I_0}{4\pi r k \, \sin^2 \theta} \left\{ e^{-jk \cos \theta L} + e^{jk \cos \theta L} - 2 \cos(kL) \right\}
$$

$$
= \frac{\mathbf{a}_z \mu \, e^{-jkr} I_0}{4\pi r k \, \sin^2 \theta} \left\{ 2 \cos(kL \cos \theta) - 2 \cos(kL) \right\} \qquad (13.41)
$$

Since we are interested in a half-wave dipole, we set $L = \lambda/4$ or $kL = (2\pi/\lambda)$
$(\lambda/4) = \pi/2$

$$A = \frac{a_z \mu\, e^{-jkr} I_0 \cos\left(\frac{\pi}{2}\cos\theta\right)}{2\pi rk \sin^2\theta} \qquad (13.42)$$

We now use the Eqn set (13.19), which gives the value of A_r and A_θ in terms of A_z, to give

$$A_r = \frac{\mu\, e^{-jkr} I_0 \cos\left(\frac{\pi}{2}\cos\theta\right)\cos\theta}{2\pi rk \sin^2\theta}; \quad A_\theta = -\frac{\mu\, e^{-jkr} I_0 \cos\left(\frac{\pi}{2}\cos\theta\right)}{2\pi rk \sin\theta} \qquad (13.43)$$

and then use the definition of the curl, in the spherical coordinate system, (Eqn (13.20))

$$\mu\mathbf{H} = \nabla \times \mathbf{A}$$

$$= \frac{1}{r}\left\{\frac{\partial(rA_\theta)}{\partial r} - \frac{\partial A_r}{\partial\theta}\right\}\mathbf{a}_\phi \qquad (13.44)$$

where we have dropped all the terms which are zero, which are (i) all terms in A_ϕ and (ii) all terms which have $\partial/\partial\phi$. Next we examine each of the two terms carefully. The term $(1/r)(\partial A_r/\partial\theta)$ leads to a $1/r^2$ term, (by inspection) which may be dropped (since the field is expected to decay as $1/r$ as in the case of the infinitesimal current element).

Therefore $\qquad \mu\mathbf{H} = \dfrac{1}{r}\dfrac{\partial(rA_\theta)}{\partial r}\mathbf{a}_\phi = \dfrac{j\mu\, e^{-jkr} I_0 \cos\left(\frac{\pi}{2}\cos\theta\right)}{2\pi r \sin\theta}\mathbf{a}_\phi$

or $\qquad H_\phi = \dfrac{j e^{-jkr} I_0 \cos\left(\frac{\pi}{2}\cos\theta\right)}{2\pi r \sin\theta} \qquad (13.45)$

Next we obtain $\mathbf{E}$ from the magnetic field using Eqn (13.26)

$$j\omega\varepsilon\mathbf{E} = \underbrace{\frac{1}{r\sin\theta}\left\{\frac{\partial(\sin\theta\, H_\phi)}{\partial\theta}\right\}\mathbf{a}_r}_{\text{Term 1}} + \underbrace{\frac{1}{r}\left\{-\frac{\partial(rH_\phi)}{\partial r}\right\}\mathbf{a}_\theta}_{\text{Term 2}}$$

From inspection, Term 1 decays as $1/r^2$, and can be dropped, as before. On the other hand

$$j\omega\varepsilon\mathbf{E} = \frac{1}{r}\left\{-\frac{\partial(rH_\phi)}{\partial r}\right\}\mathbf{a}_\theta = -\frac{k e^{-jkr} I_0 \cos\left(\frac{\pi}{2}\cos\theta\right)}{2\pi r \sin\theta}\mathbf{a}_\theta$$

So $\qquad E_\theta = \dfrac{jZ_0 e^{-jkr} I_0 \cos\left(\frac{\pi}{2}\cos\theta\right)}{2\pi r \sin\theta} \qquad (13.46)$

where $Z_0 = k/\omega\varepsilon = 120\pi = 377\ \Omega$ for free space. Note that

$$\frac{E_\theta}{H_\phi} = Z_0 \qquad (13.47)$$

From the computation of these fields it is clear that (1) the electric and magnetic fields are proportional to each other; (2) far away from the source (which is the dipole) the fields constitute a 'local' plane wave: that is the electromagnetic fields are perpendicular to each other and also perpendicular to the direction of travel. To be specific E_θ is perpendicular to H_ϕ and both are perpendicular to $\mathbf{a}_r$, the direction of travel; and (3) in different directions, the amplitudes of the

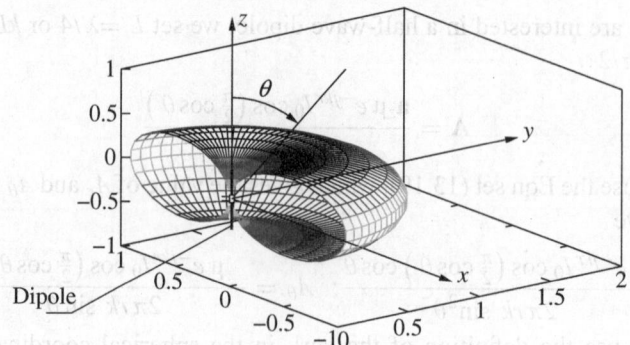

Fig. 13.10 Magnitude of the electric field radiated by a half-wave dipole, shown as a polar plot. Only half the plot is shown

fields are different. That is, in the directions of $\theta = 0$, π there is no radiation at all[1], while in the direction $\theta = \pi/2$ there is maximum radiation. These remarks are illustrated in Fig. 13.10.

 See EF1dH1fWveDipole.m in Chapter 13

Field Probes

To measure the electric field at a distant point from an antenna, field probes are used. Since the introduction of the probe automatically implies that the field is perturbed, they are designed to minimise the perturbation: probes are made to be as small as possible and use an electrically **small dipole** antenna connected to a sensor.

Example 13.8 Compute the Poynting vector of a half-wave dipole antenna shown in Fig. 13.8 where $L = \lambda/4$.

Solution The Poynting vector at a distant point is

$$
\begin{aligned}
\mathbf{P} &= \frac{1}{2}\mathbf{E} \times \mathbf{H}^* \\
&= \frac{1}{2}(E_\theta \mathbf{a}_\theta) \times (H_\phi^* \mathbf{a}_\phi) \\
&= \frac{1}{2}(E_\theta \mathbf{a}_\theta) \times \left(\underbrace{\frac{1}{Z_0} E_\theta^* \mathbf{a}_\phi}_{H_\phi^*} \right) \\
&= \frac{1}{2Z_0}(E_\theta E_\theta^*) \mathbf{a}_r \quad \text{(since } \mathbf{a}_r = \mathbf{a}_\theta \times \mathbf{a}_\phi\text{)}
\end{aligned}
$$ (13.48)

Substituting the value of E_θ

$$
\mathbf{P} = \frac{1}{2Z_0}\left[\frac{Z_0 I_0 \cos\left(\frac{\pi}{2}\cos\theta\right)}{2\pi r \sin\theta} \right]^2 \mathbf{a}_r
$$

[1]Consider the limit of $\cos\left(\frac{\pi}{2}\cos\theta\right)/\sin\theta$ as $\theta \to 0$, π.

$$= \frac{Z_0 I_0^2}{8\pi^2 r^2} \left[\frac{\cos\left(\frac{\pi}{2}\cos\theta\right)}{\sin\theta} \right]^2 \mathbf{a}_r \text{ W/m}^2 \tag{13.49}$$

Thus we find that the power streams radially away from the origin, and it is also directional in nature.

Example 13.9 Obtain the Thevenin equivalent of a half-wave dipole antenna.
Solution To compute the radiation resistance of the antenna, we first compute the the total power radiated away by the dipole. The total power radiated, P_T, is the total outward flux of the Poynting vector integrated over a sphere of radius r where $r \gg \lambda$, which avoids the near fields.

$$P_T = \oiint_{\text{Sphere}} \mathbf{P} \cdot d\mathbf{S} \tag{13.50}$$

Choosing such a sphere, and using Eqn (13.49),

$$P_T = \frac{Z_0 I_0^2}{8\pi^2} \int\int_{\theta=0,\ \phi=0}^{\theta=\pi,\ \phi=2\pi} \left\{ \left[\frac{\cos\left(\frac{\pi}{2}\cos\theta\right)}{r\sin\theta} \right]^2 \mathbf{a}_r \right\} \cdot \underbrace{\mathbf{a}_r r^2 \sin\theta\, d\theta\, d\phi}_{d\mathbf{S}}$$

$$= \frac{Z_0 I_0^2}{8\pi^2} \times 2\pi \times \int_{\theta=0}^{\theta=\pi} \left[\frac{\cos\left(\frac{\pi}{2}\cos\theta\right)}{\sin\theta} \right]^2 \sin\theta\, d\theta \quad \text{(integration over } \phi\text{)}$$

$$= \frac{Z_0 I_0^2}{4\pi} \times \underbrace{\int_{\theta=0}^{\theta=\pi} \frac{\cos^2\left(\frac{\pi}{2}\cos\theta\right)}{\sin\theta}\, d\theta}_{J} \tag{13.51}$$

The integration of J can be done by any standard numerical integration software such as MATLAB using the Simpson's integration

$$J = \int_{\theta=0}^{\theta=\pi} \frac{\cos^2\left(\frac{\pi}{2}\cos\theta\right)}{\sin\theta}\, d\theta = 1.218827 \tag{13.52}$$

So $\quad P_T = \dfrac{Z_0 I_0^2}{4\pi} \times 1.218827$

$$\approx 60 \left(\frac{I_0}{\sqrt{2}} \right)^2 \times 1.218827 \quad (Z_0 \approx 120\pi)$$

$$= 73.129621 I_{\text{rms}}^2 \text{ (W)} \quad (I_{\text{rms}} = I_0/\sqrt{2} \text{ is the rms current)} \tag{13.53}$$

So the radiation resistance, R_r, is

$$R_r = P_T / I_{\text{rms}}^2 \approx 73\ \Omega \tag{13.54}$$

The Thevenin equivalent is shown in Fig. 13.7 where $R_r = 73\ \Omega$.

If $P(r,\theta,\phi)$ is the magnitude of Poynting vector, then the normalised power pattern of the antenna, $P_n(\theta,\phi) = P(r,\theta,\phi)/P_{\max}$ is

$$P_n(\theta,\phi) = \frac{\dfrac{Z_0 I_0^2}{8\pi^2 r^2} \left[\dfrac{\cos\left(\frac{\pi}{2}\cos\theta\right)}{\sin\theta} \right]^2}{\left. \dfrac{Z_0 I_0^2}{8\pi^2 r^2} \left[\dfrac{\cos\left(\frac{\pi}{2}\cos\theta\right)}{\sin\theta} \right]^2 \right|_{\max}} = \frac{\left[\dfrac{\cos\left(\frac{\pi}{2}\cos\theta\right)}{\sin\theta} \right]^2}{\left. \left[\dfrac{\cos\left(\frac{\pi}{2}\cos\theta\right)}{\sin\theta} \right]^2 \right|_{\max}} \tag{13.55}$$

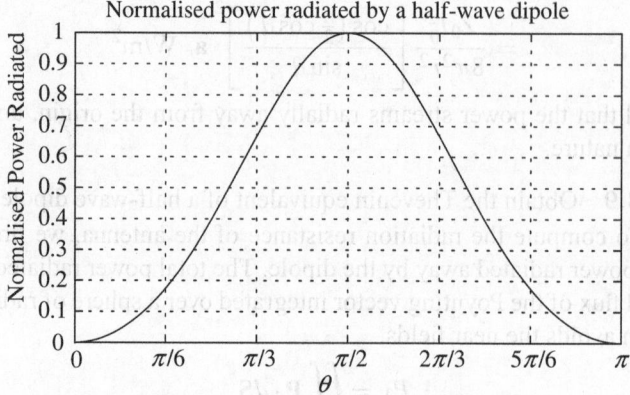

Fig. 13.11 Normalised power pattern of a half-wave dipole as a function of θ

To find the maximum of the denominator, we plot the function for the range $0 \le \theta \le \pi$. The function has zeros at $\theta = 0, \pi$ and has a single maximum at $\theta = \pi/2$, which is equal to 1 (see Fig. 13.10). That is

$$\left[\frac{\cos\left(\frac{\pi}{2}\cos\theta\right)}{\sin\theta}\right]^2\Bigg|_{max} = \left[\frac{\cos\left(\frac{\pi}{2}\cos\theta\right)}{\sin\theta}\right]^2\Bigg|_{\theta=\pi/2} = 1$$

So
$$P_n(\theta,\phi) = \left[\frac{\cos\left(\frac{\pi}{2}\cos\theta\right)}{\sin\theta}\right]^2 \qquad (13.56)$$

The function is plotted in Fig. 13.11.

13.5 | Far Field Approximation

To simplify computations we would like to find the far field approximation of the fields retaining only the $1/r$ terms and dropping all terms which are proportional to $1/r^2$ and higher order terms. Referring to Fig. 13.12, we would like to find the vector potential far away from the source which is shown shaded in the figure. To do this we start from Eqn (13.15)

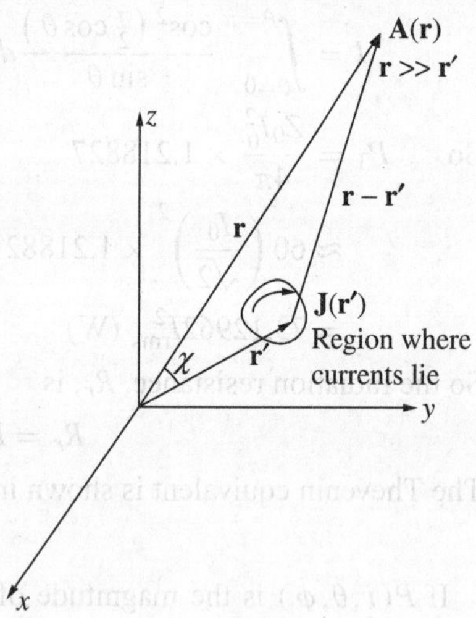

$$\mathbf{A}(\mathbf{r}) = \frac{\mu}{4\pi}\iiint\limits_{V'} \frac{\mathbf{J}(\mathbf{r}')\,e^{-jk|\mathbf{r}-\mathbf{r}'|}}{|\mathbf{r}-\mathbf{r}'|}dV'$$

and use the approximation as derived in Section 8.10

$$|\mathbf{r}-\mathbf{r}'|^{-1} \approx \frac{1}{r} + \frac{r'\cos\chi}{r^2}$$

Fig. 13.12 Far field approximation

$$\approx \frac{1}{r} \quad (\text{dropping the } 1/r^2 \text{ term}) \qquad (13.57)$$

Similarly, $\qquad |\mathbf{r}-\mathbf{r}'| \approx r - \mathbf{r}'\cdot\hat{\mathbf{r}} \qquad (13.58)$

giving us
$$A(r) \approx \frac{\mu e^{-jkr}}{4\pi r} \iiint_{\mathcal{V}'} J(r') e^{jkr' \cdot \hat{r}} dV'$$

or
$$A(r) \approx \frac{\mu e^{-jkr}}{4\pi r} \int_{\mathcal{L}'} I e^{jkr' \cdot \hat{r}} dl' \qquad (13.59)$$

For the far field vector potential,

$$A \approx \frac{\mu e^{-jkr}}{4\pi r} f(\theta, \phi)$$

where
$$f(\theta, \phi) = \iiint_{\mathcal{V}'} J(r') e^{jkr' \cdot \hat{r}} dV'$$

Then the E and H fields are given by

$$H = \frac{jke^{-jkr}}{4\pi r} (a_\theta f_\phi - a_\phi f_\theta)$$

$$E = -\frac{jkZ_0 e^{-jkr}}{4\pi r} (a_\theta f_\theta + a_\phi f_\phi) \qquad (13.60)$$

where $Z_0 = \sqrt{\mu/\varepsilon}$.

13.5.1 Radiated Fields of a Small Current Loop

Example 13.10 Find the far fields of a small current loop of radius $a \ll \lambda$, and carrying a current I.

Solution Using the second of Eqn set (13.59), we have

$$r' = a\cos\phi' \, a_x + a\sin\phi' \, a_y \qquad (13.61)$$

which is the position vector of a point on the loop as shown in Fig. 13.13. The unit vector of a field point is given by

$$\hat{r} = \cos\theta \, a_z + \sin\theta \cos\phi \, a_x + \sin\theta \sin\phi \, a_y$$

and
$$dl' = (a\cos\phi' \, a_y - a\sin\phi' \, a_x) d\phi'$$

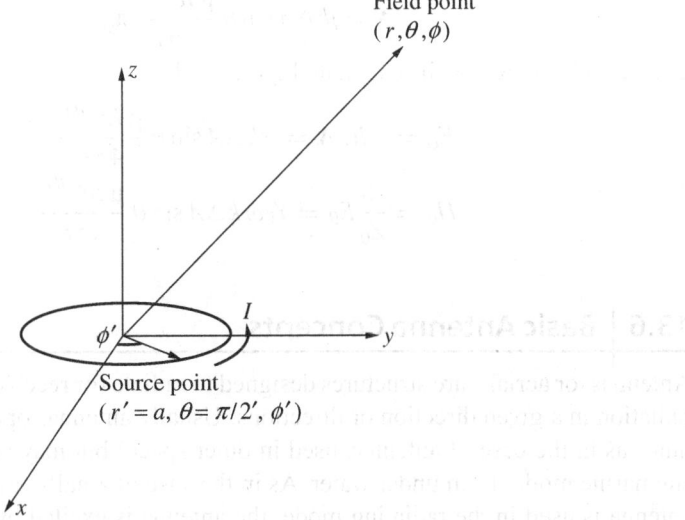

Fig. 13.13 A small radiating current loop

which is a differential linear element. Therefore

$$\mathbf{r}' \cdot \hat{\mathbf{r}} = a \cos\phi' \sin\theta \cos\phi + a \sin\phi' \sin\theta \sin\phi$$
$$= a \sin\theta \cos(\phi' - \phi)$$

So the integral for the vector potential is

$$\mathbf{A}(\mathbf{r}) \approx \frac{\mu e^{-jkr}}{4\pi r} \int_{L'} I e^{jk\mathbf{r}' \cdot \hat{\mathbf{r}}} d\mathbf{l}'$$

$$= \frac{\mu I e^{-jkr}}{4\pi r} \int_0^{2\pi} e^{jka \sin\theta \cos(\phi' - \phi)} \left(a\cos\phi' \, \mathbf{a}_y - a\sin\phi' \, \mathbf{a}_x \right) d\phi'$$

In the integral when $ka = (2\pi a)/\lambda \ll 1$,

$$e^{jka \sin\theta\cos(\phi' - \phi)} \approx 1 + jka \sin\theta \cos(\phi' - \phi)$$

since $\exp\{x\} = 1 + x + \dots$

$$\mathbf{A}(\mathbf{r}) = \frac{\mu I e^{-jkr}}{4\pi r} \int_0^{2\pi} \underbrace{\left[1 + jka \sin\theta \cos(\phi' - \phi)\right]}_{\text{First bracket}} \left(a\cos\phi' \, \mathbf{a}_y - a\sin\phi' \, \mathbf{a}_x\right) d\phi'$$

Evaluating this integral at $\phi = 0$, and multiplying with the first term in the first bracket, which is 1, the integral

$$\int_0^{2\pi} \left(a\cos\phi' \, \mathbf{a}_y - a\sin\phi' \, \mathbf{a}_x\right) d\phi' = 0$$

and for the second term in the first bracket,

$$\mathbf{A}(\mathbf{r}) = \frac{\mu I e^{-jkr}}{4\pi r} \int_0^{2\pi} \left[jka \sin\theta \cos(\phi')\right] \left(a\cos\phi' \, \mathbf{a}_y - a\sin\phi' \, \mathbf{a}_x\right) d\phi'$$

$$= jk \underbrace{\pi a^2}_{\Delta A} \sin\theta \, \frac{\mu I e^{-jkr}}{4\pi r} \mathbf{a}_y$$

where ΔA is the area, and the vector potential $\mathbf{A}$ is

$$\mathbf{A} = jk\Delta A \sin\theta \, \frac{\mu I e^{-jkr}}{4\pi r} \mathbf{a}_\phi$$

The far fields may now be calculated quite easily

$$E_\phi = -j\omega \mathbf{A} = \omega k\Delta A \sin\theta \, \frac{\mu I e^{-jkr}}{4\pi r}$$

$$H_\theta = \frac{1}{Z_0} E_\theta = Y_0 \omega k\Delta A \sin\theta \, \frac{\mu I e^{-jkr}}{4\pi r}$$

13.6 | Basic Antenna Concepts

Antennas, or aerials, are structures designed to radiate (or receive) electromagnetic radiation in a given direction or directions. Usually antennas operate in air or vacuum (as in the case of antennas used in outer space) but may also be operated in submarine mode, from under water. As in the case of a half-wave dipole, when an antenna is used in the radiating mode, the antenna is excited by a source of electromagnetic waves, and the antenna radiates the radiation as efficiently as possible.

In this chapter, we will look at the fundamentals of antennas and those concepts which will be useful to a communication engineer. Typical antennas are shown in Fig. 13.14.

When an antenna radiates, it does so directionally, that is, it radiates more in one direction than another one. The *far field* electromagnetic fields, in the spherical coordinate system and in phasor notation, are invariably of the type (see Fig 13.15)

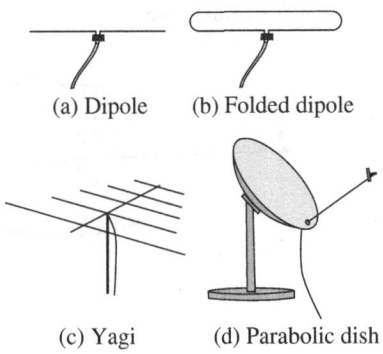

(a) Dipole   (b) Folded dipole

(c) Yagi   (d) Parabolic dish

Fig. 13.14 Examples of antennas

$$E(r,\theta,\phi) = \frac{Kf_n(\theta,\phi)}{r}e^{-jkr}$$

$$H(r,\theta,\phi) = \frac{Kf_n(\theta,\phi)}{Z_0 r}e^{-jkr} \tag{13.62}$$

where
- K is a complex constant;
- $f_n(\theta,\phi)$ is a real function whose maximum value for all values of θ and ϕ is unity. For example, $f_n(\theta,\phi)$ for an infinitesimal current element is $\sin\theta$ and in the case of the half-wave dipole it is $\cos[(\pi/2)\cos\theta]/\sin\theta$;
- $Z_0 = 377\ \Omega = 120\pi$, the intrinsic impedance of free space; and
- k is the free-space propagation constant.

As one can see, the wave decays far away from the source as $1/r$.

The directions of E and H are such that the electric field is perpendicular to the magnetic field, and both are perpendicular to $\mathbf{a}_r$. Furthermore, the Poynting vector, $\mathbf{P} = \mathbf{E} \times \mathbf{H}$ is in the direction of $\mathbf{a}_r$, that is,

$$\mathbf{P} = P(r,\theta,\phi)\,\mathbf{a}_r\ (\text{W/m}^2) \tag{13.63}$$

Moreover, if α is the angle (which is $\pi/2$) from the electric field vector to the magnetic field vector, then

$$P = \frac{1}{2}EH^*\sin\underset{(\pi/2)}{\alpha} = \frac{1}{2}\left[\frac{Kf_n(\theta,\phi)}{r}e^{-jkr}\right]\left[\frac{Kf_n(\theta,\phi)}{Z_0 r}e^{-jkr}\right]^* = \frac{|K|^2}{2Z_0 r^2}[f_n(\theta,\phi)]^2$$

$$\tag{13.64}$$

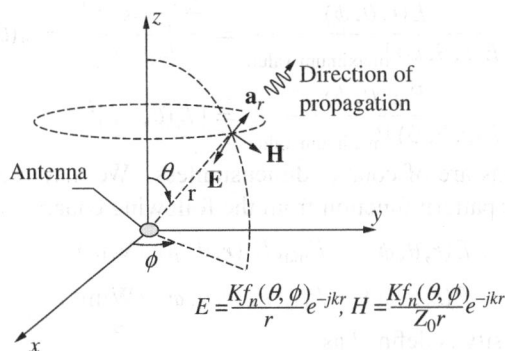

$$E = \frac{Kf_n(\theta,\phi)}{r}e^{-jkr};\ H = \frac{Kf_n(\theta,\phi)}{Z_0 r}e^{-jkr}$$

Fig. 13.15 Far fields of an antenna

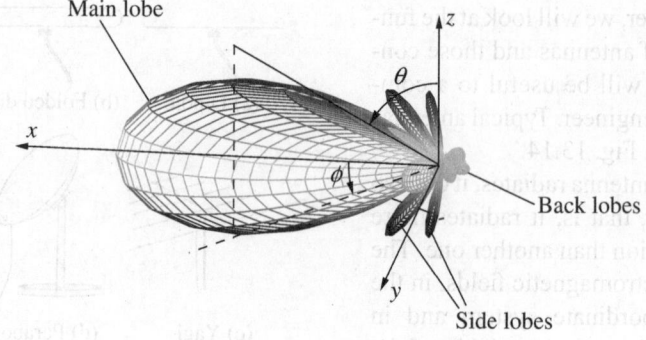

Fig. 13.16 The power pattern of an antenna

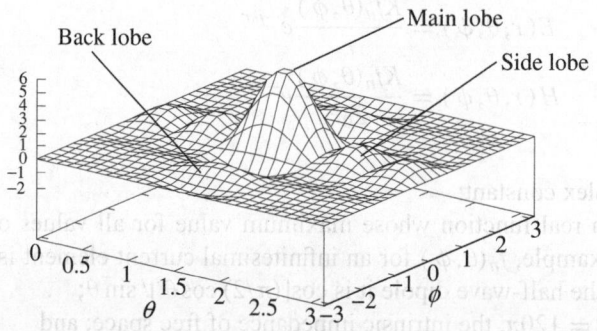

Fig. 13.17 Power pattern of the same antenna shown in 3-D rectangular coordinates

A typical power pattern of an antenna, in a 3-D polar form, is given in Fig. 13.16. The figure shows the *main beam* of an antenna pointing in the direction of the x-axis, and having *side lobes* or *minor lobes* pointing in other directions. Some radiation is also in a direction opposite to the main lobes, and these are called *back lobes*.

Figure 13.17 shows the same pattern in 3-D rectangular coordinates.

Very often the patterns are shown in the *principal planes*. The principal planes are two perpendicular planes which pass through the axis of the main beam. Thus, the principal planes of the pattern given above are the xy ($z = 0$) or xz ($\phi = 0$) planes, and the electric field or power pattern is often given in these planes, and the field or power functions are often given in terms of normalised functions

$$E_n(r,\theta,\phi) = \frac{E(r,\theta,\phi)}{E(r,\theta,\phi)\,|_{\text{maximum value}}} = \frac{\frac{Kf_n(\theta,\phi)}{r}e^{-jkr}}{\frac{K}{r}e^{-jkr}} = f_n(\theta,\phi) \qquad (13.65)$$

$$P_n(r,\theta,\phi) = \frac{P(r,\theta,\phi)}{P(r,\theta,\phi)\,|_{\text{maximum value}}} = [f_n(\theta,\phi)]^2 \qquad (13.66)$$

Both these functions are of course, dimensionless. We may recover the electric field and the power pattern function from the following equations.

$$E(r,\theta,\phi) = E_{\max}E_n(r,\theta,\phi) \text{ (V/m)} \qquad (13.67)$$

$$P(r,\theta,\phi) = P_{\max}P_n(r,\theta,\phi) \text{ (W/m}^2) \qquad (13.68)$$

The radiation intensity is defined as

$$U(\theta,\phi) = r^2 P(\theta,\phi) \quad \text{(Units= W/str)} \qquad (13.69)$$

13.7 | Directivity

The directivity of an antenna is defined by the formula

$$D = \frac{\text{Maximum power density radiated (W/m}^2)}{\text{Average power density radiated (W/m}^2)} \qquad (13.70)$$

The maximum power density radiated will be in some specific direction given by (θ_0, ϕ_0). The average power radiated by an antenna, P_{av}, is

$$P_{av} = \frac{\text{Total power radiated (W)}}{4\pi r^2 (=\text{area of the surface of a sphere})} \quad (\text{W/m}^2) \qquad (13.71)$$

and the total power radiated is an integral of the Poynting vector over the same sphere

$$P_T = \underset{\text{Sphere of radius } r}{\oint\!\!\!\oint} P(r,\theta,\phi) \, \underbrace{r^2 \sin\theta \, d\theta \, d\phi}_{\text{An element of area}} \qquad (13.72)$$

If we examine the term

$$\underbrace{r^2}_{\text{"Area"}} \underbrace{\sin\theta \, d\theta \, d\phi}_{\text{"Angle"} \times \text{"Angle"}} = r^2 d\Omega$$

the area $r^2 \sin\theta d\theta d\phi$ subtends a 'solid angle' $d\Omega = \sin\theta d\theta d\phi$ at the centre of the sphere. Its analogy is the circle. An arc $rd\theta$ subtends an angle $d\theta$ at the centre of a circle; in the same way, the area $r^2 \sin\theta \, d\theta \, d\phi$ subtends a solid angle $\sin\theta \, d\theta \, d\phi$ at the centre of a sphere. These ideas can be better understood with reference to Fig. 13.18.

- Therefore, if Ω is the solid angle subtended at the centre of the sphere, then

$$P_T = \underset{\text{Sphere of radius } r}{\oint\!\!\!\oint} P(r,\theta,\phi) r^2 d\Omega$$

$$= \oint\!\!\!\oint P_{max} P_n(\theta,\phi) r^2 d\Omega$$

and $\qquad P_{av} = \dfrac{\oint\!\!\!\oint P_{max} P_n(\theta,\phi) r^2 d\Omega}{4\pi r^2} = \dfrac{1}{4\pi}\oint\!\!\!\oint P_{max} P_n(\theta,\phi) d\Omega$

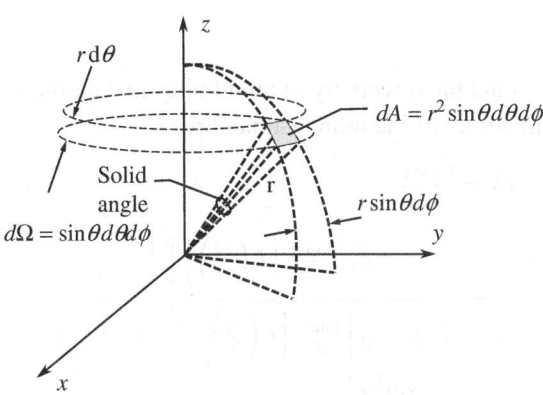

Fig. 13.18 Definition of a solid angle

The directivity D is also equal to

$$D = \frac{P_{\max}}{\frac{1}{4\pi} \iint P_{\max} P_n(\theta, \phi)\, d\Omega} = \frac{4\pi}{\iint P_n(\theta, \phi)\, d\Omega}$$

In this equation the rightmost denominator, is called the *beam area*.

$$\text{Beam area} = \Omega_A = \iint P_n(\theta, \phi)\, d\Omega$$

so

$$D = \frac{4\pi}{\Omega_A} \qquad (13.73)$$

Note that the directivity $D \geq 1$.

Of what use is the directivity? If we examine the definition of directivity, Eqn (13.70), we find that the denominator gives us the power density of an *isotropic* radiator, that is, one where the power spreads equally in all directions.

$$P_{av} = P_{isotropic} = \frac{\text{Total output power (W)}}{4\pi r^2 \,(m^2)} \quad (W/m^2)$$

So if $P_{\max}$ is the power density at a distance r in the direction of the main beam, (θ_0, ϕ_0), then

$$P_{\max} = D P_{isotropic} \qquad (13.74)$$

Example 13.11 An antenna, which radiates 1 MW of power, has a directivity of 23 dB. Find the power density radiated in the direction of the main beam, 10 km from the source.

Solution If the antenna was an isotropic radiator, it would radiate

$$P_{isotropic} = \frac{P_T}{4\pi r^2} = \frac{10^6}{4 \times \pi \times (10^4)^2} = 7.96 \times 10^{-4} \; (W/m^2)$$

The directivity in dB is

$$D_{dB} = 10 \log_{10}(D)$$

So

$$D = 10^{(D_{dB}/10)} = 10^{2.3} = 199.53$$

Using this value of directivity

$$P_{\text{Main beam}} = D P_{isotropic} = 7.96 \times 10^{-4} \times 199.53 = 0.159 \; (W/m^2)$$

Example 13.12 Find the directivity of an infinitesimal dipole of length 0.1λ.

Solution The directivity of the infinitesimal dipole is

$$D = \frac{P_{\max}}{P_{av}}$$

$$= \frac{\frac{1}{2} Z_0 \left[\frac{\sin\theta}{2} \left\{ I\left(\frac{\Delta l}{\lambda}\right) \right\} \right]^2 \Big|_{\max}}{\frac{1}{4\pi} \iint \frac{1}{2} Z_0 \left[\frac{\sin\theta}{2} \left\{ I\left(\frac{\Delta l}{\lambda}\right) \right\} \right]^2 \sin\theta\, d\theta\, d\phi}$$

$$= \frac{\sin^2\theta \big|_{\max}}{\frac{1}{4\pi} \iint \sin^3\theta\, d\theta\, d\phi}$$

$$= \frac{4\pi}{2\pi\,(4/3)} = \frac{3}{2}$$

The directivity is 1.5 regardless of the length.

Example 13.13 Find the beam area and directivity of an isotropic radiator.

Solution An isotropic radiator radiates equally in all directions. So

$$P_{\max} = \frac{P_T}{4\pi r^2}$$

where P_T is the total power radiated. Also the average power radiated is

$$P_{av} = \frac{P_T}{4\pi r^2}$$

Therefore $$D = P_{\max}/P_{av} = 1$$

The beam area may be calculated by considering

$$P_n(\theta, \phi) = 1 \quad \text{(Equal power density in all directions)} \qquad (13.75)$$

Therefore
$$\Omega = \int\limits_{\theta=0,\ \phi=0}^{\theta=\pi,\ \phi=2\pi}\!\!\!\! P_n(\theta, \phi)\, \sin\theta\, d\theta\, d\phi$$

$$= \int\limits_{\theta=0,\ \phi=0}^{\theta=\pi,\ \phi=2\pi}\!\!\!\! \sin\theta\, d\theta\, d\phi$$

$$= 2\pi \int_{\theta=0}^{\theta=\pi} \sin\theta\, d\theta$$

$$= 4\pi$$

 See `DirectivityFrmHPBW.m` in Chapter 13

Practice Problem 13.1

Find the directivity of a half-wave dipole.

[Ans: 1.64]

13.7.1 Directivity from the Beam Pattern

Very often we have an antenna whose pattern we can measure but we may not know the analytical expression for it. So the question we can ask is that can we predict the directivity from the pattern? The answer is yes we can compute the approximate directivity of the antenna using the pattern and by making some approximations.

Let us look at the three-dimensional pattern of Fig. 13.16. If we visualise two cuts in two perpendicular planes: the *xy* and *xz* planes, then in each of the two planes the pattern will look similar to the pattern shown in Fig. 13.19. If we look at the figure carefully we realise that it is a polar pattern with two types of patterns: the field pattern and the power pattern. On both patterns we also observe that the maximum value of the signal is 1.0. Circles are also drawn at the 0.7071 and 0.5 magnitudes. If we draw a line from *O* to *A* and *B*, passing through the half power points then *AOB* is the *half power beam width*, θ_{HP} (HPBW). The two lines *OA* and

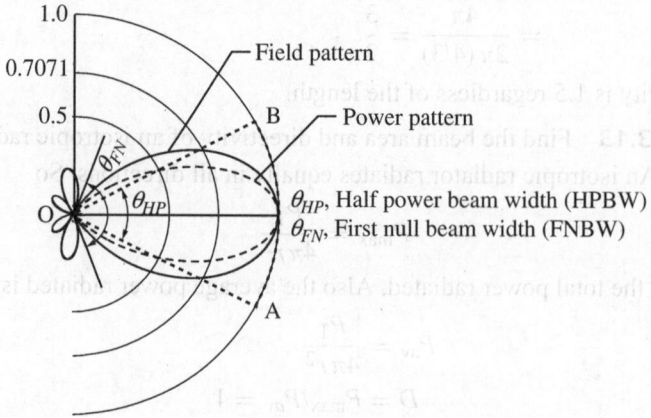

Fig. 13.19 Radiation pattern in one of the two principal planes

OB also cross the field pattern at the 0.7071 magnitude.[2] There are radial lines also drawn at the points in the pattern where there are nulls. The angle between these radial lines is the *first null beam width*, θ_{FN} (FNBW). The half power beam width is approximately equal to half of the FNBW.

$$\theta_{HP} \approx \frac{\theta_{FN}}{2} \qquad (13.76)$$

The arc length AB is given by

$$AB = r\theta_{HP} = \theta_{HP} \qquad (\text{since } r = 1)$$

Having discussed the pattern in one plane, we make a similar cut in the pattern in the perpendicular plane and obtain another half power and first null beam widths, ϕ_{HP} and ϕ_{FN}. Note that the notation of θ and ϕ are not to be confused with the angles of the same name in spherical coordinates. We now use the following approximation: if the pattern in a particular plane is $f(\theta)$ passing through the beam maximum, then

$$\int_{-\pi}^{\pi} f(\theta)\, d\theta \approx \int_{-\theta_{HP}/2}^{\theta_{HP}/2} d\theta = \theta_{HP} \qquad (13.77)$$

since $f(\theta)$ is equal to unity at $\theta = 0$ and then rapidly decays as we move away from $\theta = 0$.

We now examine Fig. 13.20. The figure shows a beam pattern and the half power beam widths in the two perpendicular planes OAB and OCD.

$$\angle AOB = \theta_{HP}, \qquad \angle COD = \phi_{HP}$$

Now AB and CD lie on a sphere with unit radius ($r = 1$). Therefore

$$AB = \theta_{HP}, \qquad CD = \phi_{HP}$$

and area of the rectangle $ABCD$ is

$$AB \times CD = \theta_{HP}\phi_{HP}$$

and the solid angle subtended at the centre O is area $ABCD$ (rectangle) divided by the square of the radius (which is one). Hence the solid angle is

$$\theta_{HP}\phi_{HP} \qquad (13.78)$$

[2]Since the square of the field pattern gives us the power pattern and $0.7071^2 = 0.5$.

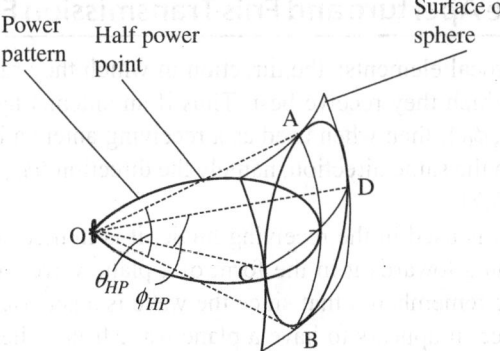

Fig. 13.20 Directivity calculation

If we make a similar argument as we did with Eqn (13.77),

$$\Omega_A = \iint P_n d\Omega \cong \theta_{HP}\phi_{HP} \qquad (13.79)$$

where $P_n(\theta, \phi)$ is the normalised power pattern, Ω is the solid angle, and Ω_A is the beam area. Proceeding on, by Eqn (13.73),

$$D = \frac{4\pi}{\Omega_A} \cong \frac{4\pi}{\theta_{HP}\phi_{HP}} \qquad (13.80)$$

We can refine this approximation further by noticing that if we were to project O along with every point of the half power points of the beam onto the sphere with radius one, we would effectively get the ellipse $ACBD$ with major and minor axes AB and CD (or vice versa) and the area of the ellipse would be

$$\frac{\pi}{4}(AB \times CD) = \frac{\pi}{4}\theta_{HP}\phi_{HP} = \Omega_A$$

and

$$D = \frac{16}{\theta_{HP}\phi_{HP}} \qquad (13.81)$$

13.7.2 Directivity Using Square Degrees

Very often we measure the half power beam widths in degrees rather than radians. How do we use degrees in our formulae? Now one radian is

$$1^r = 57.3°$$

and one steradian is one square radian,

$$1^{str} = (1^r)^2 = 3283°$$

so

$$4\pi^{str} = 41253°$$

Hence from Eqn (13.80)

$$D = \frac{4\pi}{\theta_{HP}\phi_{HP}} \cong \frac{41,000}{\theta°_{HP}\phi°_{HP}} \qquad (13.82)$$

or by the second formula, Eqn (13.81)

$$D = \frac{4\pi}{(\pi/4)\,\theta_{HP}\,\phi_{HP}} = \frac{41,253}{(\pi/4)\,\theta°_{HP}\phi°_{HP}} \cong \frac{52,500}{\theta°_{HP}\phi°_{HP}} \qquad (13.83)$$

13.8 | Effective Aperture and Friis Transmission Formula

Antennas are reciprocal elements: the direction in which they radiate best, is the same direction in which they receive best. Thus if an antenna has its main beam in the direction (θ_0, ϕ_0), then when used as a receiving antenna it registers maximum radiation from the same direction, namely the direction (θ_0, ϕ_0). This is well illustrated in Fig. 13.21.

When an antenna is used in the receiving mode, it is immersed in electromagnetic radiation coming towards it in the form of a plane wave from the direction (θ_0, ϕ_0). It must be remembered that since the wave is a spherical wave, coming from a great distance, it appears to have a plane wave front. The Poynting vector magnitude of the plane wave is P watts/m^2. Hence to trap power (watts), and to have power delivered to its terminals, the antenna must have what is called an *effective aperture, A_e*(m^2). The received power P_r is

$$P_r = A_e P \text{ (W)} \qquad (13.84)$$

It turns out that the effective aperture and directivity are linked by

$$A_e = \lambda^2 \frac{D}{4\pi} \text{ (m}^2) \qquad (13.85)$$

for every antenna. With this formula we can derive the power received by an antenna when it is transmitted by another antenna.

Figure 13.22 illustrates the configuration of two antennas placed a distance r apart directed in such a way that the transmitting antenna points its main beam in the direction of the receiving antenna, and the receiving antenna is oriented to

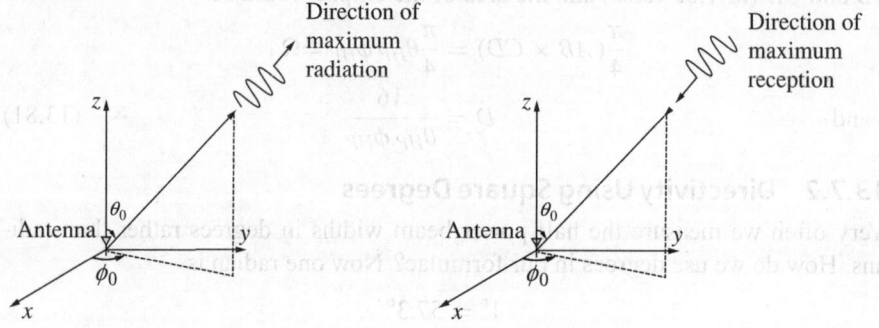

Fig. 13.21 Reciprocity property of antennas

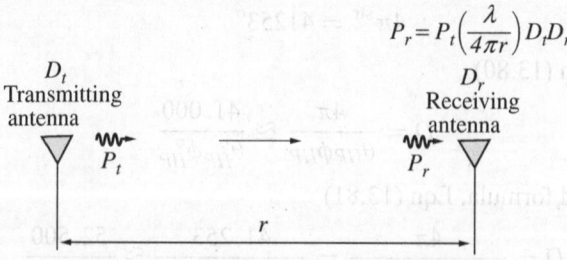

$$P_r = P_t \left(\frac{\lambda}{4\pi r} \right) D_t D_r$$

Fig. 13.22 Friis transmission formula

receive maximum radiation. If the transmitting antenna of directivity D_t broadcasts P_t watts of power, then at the receiving end, the Poynting vector magnitude is

$$\mathbb{P}(r) = \frac{P_t}{4\pi r^2}D_t \ (\text{W/m}^2) \tag{13.86}$$

where $\mathbb{P}$, the power density is used to differentiate it from the received power P_r. Let the receiving antenna have an effective area A_{er}, then the power received, P_r, is

$$P_r = \frac{P_t}{4\pi r^2}D_t A_{er} \ (\text{W}) \tag{13.87}$$

Substituting the value of A_{er} from Eqn (13.85)

$$P_r == \frac{P_t}{4\pi r^2}D_t \underbrace{\left(\frac{D_r \lambda^2}{4\pi}\right)}_{A_{er}} \tag{13.88}$$

which, after simplification becomes

$$P_r = P_t \left(\frac{\lambda}{4\pi r}\right)^2 D_t D_r \ (\text{W}) \tag{13.89}$$

This is the Friis transmission formula.

Example 13.14 Two half-wave dipoles are placed 1 km apart, and the transmitting antenna transmits 1 W of power at 100 MHz. Find the power received by the receiving antenna.

Solution Both dipoles have directivities of 1.64.

$$D_t = D_r = 1.64$$

The wavelength of the wave in air is

$$\frac{c}{f} = \frac{3 \times 10^8}{100 \times 10^6} = 3 \ (\text{m})$$

The effective area of the receiving antenna is

$$A_{er} = \lambda \frac{D_r}{4\pi} = 3 \times \frac{1.64}{4\pi} = 0.39152 \ (\text{m}^2)$$

and the Poynting vector at the receiving antenna is

$$\mathbb{P} = \frac{P_t}{4\pi r^2}D_t = \frac{1}{4\pi \times (10^3)^2} \times 1.64 = 1.3051 \times 10^{-7} \ (\text{W/m}^2)$$

Therefore the received power is

$$P_r = \mathbb{P}A_{er} = 1.3051 \times 10^{-7} \times 0.39152 = 51.11 \ \text{nW}$$

13.9 | Radar Equation

The radar equation is an application of Friis transmission equation. In bistatic radar, pulsed electromagnetic energy is transmitted towards a target as shown in Fig. 13.23 which has a radar cross-section σ. Bistatic radar means that we use different antennas for transmission and reception.

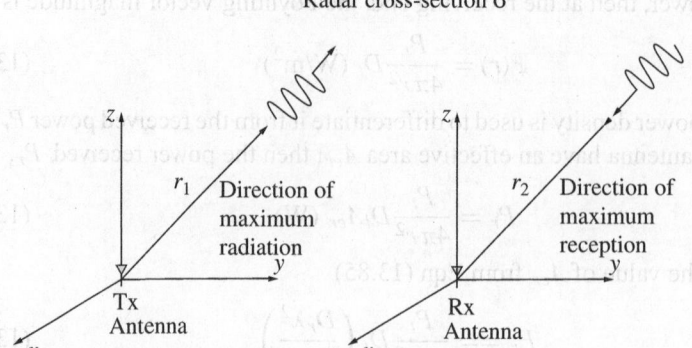

Fig. 13.23 Radar equation

The power density at the target is

$$\mathbb{P}_{\text{target}} = \frac{P_t D_t}{4\pi r_1^2} \tag{13.90}$$

where r_1 is the distance of the transmitting antenna from the target. The scattering of radiation from the target causes a power density at the receiving antenna given by

$$\mathbb{P}_{\text{Rx}} = \mathbb{P}_{\text{target}} \frac{\sigma}{4\pi r_2^2}$$

$$= \left(\frac{P_t D_t}{4\pi r_1^2}\right)\left(\frac{\sigma}{4\pi r_2^2}\right) \tag{13.91}$$

where r_2 is the distance from the target from the receiving antenna. Therefore, the power received at the receiving end is

$$P_r = \left(\frac{P_t D_t}{4\pi r_1^2}\right)\left(\frac{\sigma A_{er}}{4\pi r_2^2}\right) \text{ W} \tag{13.92}$$

In the case of monostatic radar where the transmitting and receiving antennas are same,

$$P_r = \left(\frac{P_t D_t}{4\pi r^2}\right)\left(\frac{\sigma A_{et}}{4\pi r^2}\right) \text{ W} \tag{13.93}$$

Example 13.15 A monostatic radar operating at 10 GHz has its antenna with a directivity of 10 at 5 km from a target which has an RCS of 10 m². What should be the transmitted power so that the received power is 1 μW?

Solution In the first instance, we need to calculate the effective area of the antenna. At 10 GHz, $\lambda = 0.03$ m

$$A_e = \frac{\lambda^2}{4\pi}D$$

$$= 0.7162 \times 10^{-3}$$

The transmitted power is given by

$$P_t = \frac{P_r \times (4\pi \times 5000)^2}{D\sigma A_e}$$

$$= \frac{(1 \times 10^{-6})(4\pi \times 5000)^2}{10 \times 10 \times 0.72 \times 10^{-3}}$$

$$= 55 \text{ kW}$$

POINTS TO REMEMBER

- If we observe the fields at the distant point $\mathbf{r}$, at time t, then the values of the sources must be taken into account at an earlier time, namely $t - R/c$. Or, to be specific

$$\mathbf{A}(\mathbf{r}, t) = \frac{\mu}{4\pi} \iiint_{\mathcal{V}'} \frac{\mathbf{J}(\mathbf{r}', t - R/c)\, dV'}{R}$$

and from this equation

$$\mathbf{H} = \frac{\mathbf{B}}{\mu} = \frac{1}{\mu}\nabla \times \mathbf{A}$$

and then obtain $\mathbf{E}(\mathbf{r}, t)$ from an appropriate Maxwell's equation: $\partial \mathbf{D}/\partial t = \nabla \times \mathbf{H}$

- The wave equation for the vector potential for sinusoidal time dependence is

$$\frac{\partial^2 \mathbf{A}}{\partial x^2} + \frac{\partial^2 \mathbf{A}}{\partial y^2} + \frac{\partial^2 \mathbf{A}}{\partial z^2} + k^2 \mathbf{A} = -\mu \mathbf{J}$$

- The far fields due to a short Hertzian dipole are

$$\mathbf{E} = j\frac{Z_0 I \sin\theta}{2r}\left(\frac{\Delta l}{\lambda}\right) e^{-jkr} \mathbf{a}_\theta$$

$$\mathbf{H} = j\frac{I \sin\theta}{2r}\left(\frac{\Delta l}{\lambda}\right) e^{-jkr} \mathbf{a}_\phi$$

- The ratio of the far fields for a short dipole is

$$\frac{E_\theta}{H_\phi} = Z_0 = 377\ \Omega = 120\pi$$

- The Poynting vector of the short dipole for the far fields is

$$\mathbf{P} = \frac{1}{2} Z_0 \left[\frac{\sin\theta}{2r}\left\{I\left(\frac{\Delta l}{\lambda}\right)\right\}\right]^2 \mathbf{a}_r \quad (\text{W/m}^2)$$

- The power radiated per steradian of a short dipole is

$$U(\theta, \phi) = P(r, \theta, \phi)\, r^2 = \frac{1}{2} Z_0 \left[\frac{\sin\theta}{2}\left\{I\left(\frac{\Delta l}{\lambda}\right)\right\}\right]^2$$

- The radiation resistance of a short dipole is

$$R_r = 80\pi^2 \left(\frac{\Delta l}{\lambda}\right)^2$$

- The far fields of a half-wave dipole are

$$H_\phi = \frac{j e^{-jkr} I_0 \cos\left(\frac{\pi}{2}\cos\theta\right)}{2\pi r \sin\theta}$$

$$E_\theta = \frac{j Z_0 e^{-jkr} I_0 \cos\left(\frac{\pi}{2}\cos\theta\right)}{2\pi r \sin\theta}$$

- The Poynting vector of the half dipole for the far fields is

$$\mathbf{P} = \frac{Z_0 I_0^2}{8\pi^2 r^2}\left[\frac{\cos\left(\frac{\pi}{2}\cos\theta\right)}{\sin\theta}\right]^2 \mathbf{a}_r \quad \text{W/m}^2$$

- The radiation resistance of a half-wave dipole is

$$R_r = 73 \ \Omega$$

- The far field approximation of the vector potential is

$$\mathbf{A(r)} \approx \frac{\mu \, e^{-jkr}}{4\pi r} \iiint_{\mathcal{V}'} \mathbf{J(r')} \, e^{jk\mathbf{r'} \cdot \hat{\mathbf{r}}} dV'$$

or

$$\mathbf{A(r)} \approx \frac{\mu \, e^{-jkr}}{4\pi r} \int_{\mathcal{L}'} I e^{jk\mathbf{r'} \cdot \hat{\mathbf{r}}} d\mathbf{l}'$$

- The *far field* electromagnetic fields, in the spherical coordinate system and in phasor notation, are invariably of the type

$$E(r,\theta,\phi) = \frac{K f_n(\theta,\phi)}{r} e^{-jkr}$$

$$H(r,\theta,\phi) = \frac{K f_n(\theta,\phi)}{Z_0 r} e^{-jkr} \tag{13.94}$$

where K is a complex constant and $Z_0 = 377 \ \Omega = 120\pi$, the intrinsic impedance of free space.

- The directivity of an antenna is defined by the formula

$$D = \frac{\text{Maximum power density radiated } (\text{W/m}^2)}{\text{Average power density radiated } (\text{W/m}^2)}$$

- The average power density radiated is given by

$$P_{av} = \frac{\oiint P_{\max} P_n(\theta,\phi) \, r^2 d\Omega}{4\pi r^2} = \frac{1}{4\pi} \oiint P_{\max} P_n(\theta,\phi) \, d\Omega$$

The directivity D is also equal to

$$D = \frac{P_{\max}}{\frac{1}{4\pi} \oiint P_{\max} P_n(\theta,\phi) \, d\Omega} = \frac{4\pi}{\oiint P_n(\theta,\phi) \, d\Omega} \tag{13.95}$$

- In Eqn (13.95) the rightmost in the denominator, is called the *beam area*, that is,

$$\text{Beam area} = \Omega_A = \oiint P_n(\theta,\phi) \, d\Omega$$

so

$$D = \frac{4\pi}{\Omega_A}$$

- The half power beam widths, θ_{HP} and ϕ_{HP}, are the beam widths between half power points in two perpendicular planes through the beam respectively.
- The beam width between first nulls, θ_{FN} and ϕ_{FN}, are the beam widths between the first nulls in two perpendicular planes through the beam respectively.

$$\theta_{HP} \approx \frac{\theta_{FN}}{2}$$

$$\phi_{HP} \approx \frac{\phi_{FN}}{2}$$

- The directivity is

$$D = \frac{4\pi}{\Omega_A} \approx \frac{4\pi}{\theta_{HP}\,\phi_{HP}}$$

where θ_{HP} and ϕ_{HP} are in radians.

- The directivity is

$$D = \frac{4\pi}{\theta_{HP}\,\phi_{HP}} \approx \frac{41,000}{\theta_{HP}^\circ \, \phi_{HP}^\circ}$$

where θ_{HP}°, and ϕ_{HP}° are in degrees.

- If the Poynting vector magnitude of the plane wave is P watts/m^2 and the antenna has an *effective aperture* $A_e(\text{m}^2)$, then the received power P_r is

$$P_r = A_e P \quad \text{(W)} \tag{13.96}$$

- The effective aperture and directivity are linked by

$$A_e = \lambda^2 \frac{D}{4\pi} \quad (\text{m}^2) \tag{13.97}$$

for every antenna.

- Friis transmission formula is given by

$$P_r = P_t \left(\frac{\lambda}{4\pi r}\right)^2 D_t D_r$$

SELF ASSESSMENT

Objective Type Questions

1. Retarded potentials are so called
   (a) since the potentials are retarded
   (b) the potentials are calculated from the sources at an earlier time
   (c) though they are functions of t they are actually functions of $t - R/c$
   (d) none of the above
2. The retarded potentials $\mathbf{A}(\mathbf{r}, t)$ and $\phi(\mathbf{r}, t)$ can be use to
   (a) calculate the solution to the Helmholtz equation
   (b) correctly calculate the field $\mathbf{H}$
   (c) correctly calculate the field $\mathbf{E}$ in an indirect manner
   (d) none of the above
3. The $\mathbf{r}$ and $\mathbf{r}'$ coordinates
   (a) are coordinates of the field point and source point respectively
   (b) are coordinates of the source point and field point respectively
   (c) can be used to calculate $R = |\mathbf{r} - \mathbf{r}'|$
   (d) none of the above
4. The far field of antennas are characterised by
   (a) $\exp\{-jkr\}/r$ (b) $\exp\{+jkr\}/r$
   (c) $\exp\{-jkr\}/r^2$ (d) $\exp\{+jkr\}/r^2$
5. In the far field, the E/H ratio
   (a) changes from antenna to antenna   (b) is equal to 377 Ω
   (c) is equal to 120π (d) none of the above
6. In the far field there are many antennas with
   (a) an E_r component   (b) an H_r component
   (c) an E_θ component,   (d) an H_θ component
7. The directivity of an isotropic antenna is
   (a) 10   (b) 0.5   (c) 1   (d) none of the above
8. The magnitude of the Poynting vector of the far field is
   (a) $|E|^2/2Z_0$ (b) $|EH^*|/2Z_0$ (c) $|EH^*|/2$ (d) none of the above
9. Directivity
   (a) is equal to the maximum power density radiated, P_m
   (b) is a ratio which is dimensionless
   (c) is equal to the average power density radiated, P_a

(d) P_m/P_a

(e) none of the above

10. The Friis formula
   (a) relates the directivities of two antennas
   (b) relates the transmitted and received powers of two antennas
   (c) takes into account the reflection from the ground
   (d) none of these

Short-Answer Questions

1. Derive the equation of continuity using Maxwell's equations.
2. If $\mathbf{A}(\mathbf{r}, t) = (K/r) T(r - ct) \mathbf{a}_z$ where r is the distance from the source, c is the velocity of light in vacuum, t is time, and $T(\cdot)$ is a triangular wave of width τ as shown below, find the magnetic field $\mathbf{H}$ in spherical coordinates.
3. Find the far fields of a centre fed linear antenna of length L which has a current distribution $\mathbf{I}(z, t) = I_0 \exp\{j\omega t\}\mathbf{a}_z$ where I_0 is a constant.
4. Using the fields calculated in Example 3, compute the power radiated and radiation resistance of the antenna assuming that $kL \ll 1$.
5. If the power delivered by an antenna is 0.5 dBm, and the effective area of an antenna is 0.5 m^2, calculate the incident power density.

Review Questions

1. What are retarded potentials and why are they so called?
2. Name the different types of boundary-value problems which occur in electromagnetics.
3. In the text, the directivity of an antenna is defined as some number 'dBi'. What does this mean?
4. What is radiation resistance? How does it relate to the Thevenin equivalent of an antenna?
5. Observe the relation between the far fields radiated by a short dipole and those radiated by a small current loop. Is there any connection between these fields?

Numerical Problems

1. Obtain the equation

$$\nabla \times \nabla \times \mathbf{E} = k^2 \mathbf{E} - j\omega\mu\mathbf{J}$$

from Maxwell's equations.

2. Using the retarded vector potential equation, Eqn (13.3), find the vector potential far away from the source for $\mathbf{I}(z, t) = I_0 e^{-t/\tau} u(t) \mathbf{a}_z$ with length of the antenna equal to ΔL. Retain terms only proportional to $1/r$. $u(t)$ is the unit step function.
3. Using the vector potential obtained in Example 2, obtain the magnetic field at a great distance from the origin. Retain terms only proportional to $1/r$.
4. For a short dipole, find the distance in wavelengths where the inductive fields are equal to the far fields.
5. A short current element (also called Hertzian dipole) carries a current of 1 A at a frequency of 1 GHz. The length of the antenna is 0.1λ. Find the electric and magnetic field vectors at $\theta = 0$, 45°, and 90°.
6. Find the total power radiated and radiation resistance of the dipole of Example 5.
7. A length of wire of 10 m is erected vertically with a mast, and used as a short dipole at a frequency of 1 MHz. What should be the current in the wire so that the power density

radiated in the $\theta = 90°$ direction is $1\ \mu W/m^2$, 10 km from the source? Is the assumption of a short dipole valid?

8. Find the far fields of a centre fed linear antenna of length L which has a current distribution

$$\mathbf{I}(z,t) = \mathbf{a}_z e^{j\omega t} \times \begin{cases} I_0(1 - 2z/L), & 0 < z < L/2 \\ I_0(1 + 2z/L), & -L/2 < z < 0 \end{cases}$$

where I_0 is a constant. Draw the current distribution.

9. Using the fields calculated in Example 8, compute the power radiated and radiation resistance of the antenna for $kL \ll 1$.

10. Obtain the value of J in Eqn (13.52) by suitable manipulations and using a Taylor series expansion.

11. Find the far field expression of the vector potential for the $\lambda/4$ monopole over a ground plane (Fig 13.24), and show that the fields are identical as that for a $\lambda/2$ dipole.
   Hint: Use the method of images for currents. Move a charge upwards over a ground plane and see what happens to the current so produced.

12. A half-wave dipole is fed by a 10 m transmission line which is connected to a source R_s. Find the required transmission line impedance, and the value of R_s so that maximum power is radiated. The frequency of operation is 100 MHz.

13. A rectangular region in the x-y plane has a current distribution of

$$\mathbf{J}_s(x,y) = \mathbf{a}_x J_{s0} \sin\left(\frac{2\pi x}{a}\right) \sin\left(\frac{2\pi}{b}y\right) e^{j\omega t} \begin{cases} -a/2 < x < a/2 \quad \text{and} \\ -b/2 < y < b/2 \end{cases}$$

Formulate the problem to obtain the vector potential $\mathbf{A}$ far away from the source.

14. For Problem 13, if a and b are both $\ll \lambda$, can you find the far fields? If yes find the far fields.

15. Two identical antennas each of directivity D are separated by a distance of 100λ. If the ratio of the transmitted to received power is 100, find the value of D.

16. Compare the directivities of two antennas with power patterns $P_1 = \sin^2\theta$ and $P_2 = \sin^6\theta$ where the notation of spherical coordinates is being used.

17. If an antenna consists of 100 circular turns of wire of radius 1 in. for radiation at 10 MHz, what is the required current amplitude if the electric field in the direction $\theta = 90°$ and $r = 1$ km is to be $1\ \mu$ V/m?

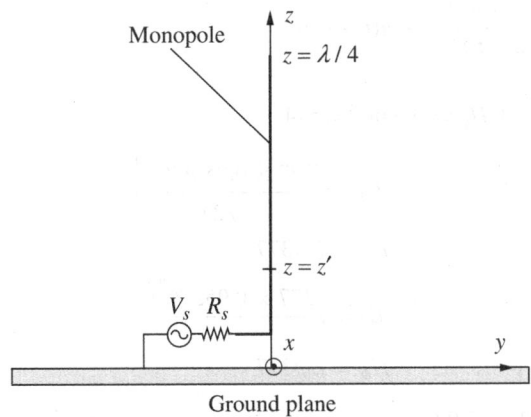

Fig. 13.24

18. If an antenna consists of N circular turns of wire of radius 10 cm for radiation at 10 MHz, and the current supplied is 1 A what is the required value of N if the magnetic field in the direction $\theta = 45°$ and $r = 1$ km, is to be 1 μA/m?

Answers

Objective Type Questions

1. (b) and (c)   2. (a), (b) and (c)   3. (a) and (c)   4. (a)   5. (b) and (c)
6. (c) and (d)   7. (c)   8. (a) and (c)   9. (b) and (d)   10. (b)

Short-Answer Questions

2. $E_\theta = j\omega\,(K/r)\,T(r - ct)\sin\theta$
   $H_\phi = 377E_\theta$

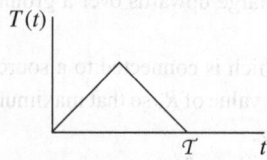

3. See Example 13.8.

4. $P_r = \dfrac{L^2\,I_0{}^2\,Z_0}{3\lambda^2}$

   $R_r = \dfrac{2L^2\,Z_0}{3\lambda^2}$

5. 2.24 mW/m^2

Numerical Problems

2. $A_r = \dfrac{\mu\,I_0 e^{-\frac{t-r/c}{\tau}}\,u(t - r/c)\,\cos(\theta)\,\Delta L}{4\pi r}$,

   $A_\theta = -\dfrac{\mu\,I_0 e^{-\frac{t-r/c}{\tau}}\,u(t - r/c)\,\sin(\theta)\,\Delta L}{4\pi r}$, $A_\phi = 0$

3. $H_\phi = -\dfrac{I_0 e^{-\frac{t-r/c}{\tau}}\,\sin(\theta)\,\Delta L}{4\pi c\,\tau\,r}u(t - r/c)$

4. $\lambda/6$

5. For $\theta = 0$, $E_\theta = 0$, $H_\phi = 0$. For $\theta = \pi/4$,

$$E_\theta = j\dfrac{377 \times 0.05\,e^{-j\frac{20\pi r}{3}}}{\sqrt{2}\,r}$$

$$H_\phi = E_\theta/377$$

For $\theta = \pi/2$, $\qquad E_\theta = j\dfrac{377 \times 0.05 e^{-j\frac{20\pi r}{3}}}{r}$

$$H_\phi = E_\theta/377$$

6. $P_T = 3.95$ W; $R_r = 7.9\ \Omega$.
7. $I = 848.6$ A. Yes the assumption of a short dipole is valid.

8. $H_\phi = j\dfrac{2I_0}{\pi r} e^{-jkr} \dfrac{\sin\theta \sin^2\left(\frac{kL\cos\theta}{4}\right)}{\cos^2\theta}$

$E_\theta = 377 H_\phi$

the distribution is a triangular distribution

9. Power radiated and radiation resistance is

$$P_r = \frac{k^2 I_0^2 L^2 Z_0}{4}$$

$$R_r = \frac{k^2 L^2 Z_0}{2}$$

12. Z_0 of transmission line is 75 Ω; $R_s = 75\ \Omega$.

13. $\mathbf{A} = \dfrac{\mu e^{-jkr}}{4\pi r} \displaystyle\iint\limits_{S'} \mathbf{J}_s(x',y')\, e^{jk(x'\cos\theta\cos\phi + y'\cos\theta\sin\phi)}\, dx'\, dy'$

14. The far fields are zero.

15. $D = 40\pi$

16. $D_1 = 1.5$; $D_2 \approx 2.2$

17. 374.9 μA

18. 20 turns

14 | Introduction to Antennas

CHAPTER OBJECTIVES

To enable the students to understand the following:

- Linear arrays
- Uniform arrays
- Array factor
- Nulls and sidelobes of a uniform array
- Beam pointing angle
- Principle of pattern multiplication
- The far field of a continuous current distribution
- Aperture antennas
- Horn antennas
- Parabolic reflector antenna

14.1 | Introduction

Every communication engineer has, at some point, to deal with using an antenna for the purpose of communication. To broadcast messages, an antenna has to convert a current which is the end point in the communication process to electromagnetic waves, which have to be beamed towards a receiver. On the other hand to receive messages the antenna has to be properly oriented to receive maximum radiation.

The various design criteria which the communication engineer has to take into account are illustrated through Fig. 14.1.

- For the transmitting antenna
  - Has the *shape* of the beam been properly designed? (Point A)
  - Does the antenna have the required *directivity*? The distance to the receiving antenna has to be taken into account so that the received power is sufficient (Point A).
  - Is the *polarisation* correct? For example in some applications a circular polarisation may be required in case there are many receiving antennas with different polarisations (Point A).

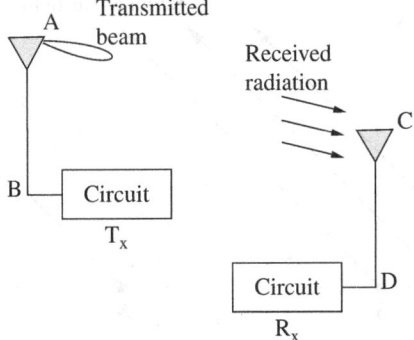

Fig. 14.1 Design criteria of antennas

- Is the *sidelobe level* as it should be? High sidelobes implies that the power is radiated away in other directions (Point A).
- Is the antenna correctly *matched* to the circuit? An incorrect match implies that the transmitting circuit does not deliver adequate power to the antenna (Point B).
- Does the antenna has the required *bandwidth*? If the antenna has to be used for high bandwidth applications then this is an important point (Points A and B).

• For the receiving antenna

- Does the antenna has the required directivity? This point has already been made with respect to the transmitting antenna (Point C).
- Is the polarisation correct? The receiving antenna has to match the polarisation of the transmitting antenna (Point C).
- Is the sidelobe level as it should be? With high sidelobe levels, the receiving antenna will receive radiation from other sources which will behave like noise (Point C).
- Is the antenna correctly matched to the circuit? The received power is precious. With improper match the power received from the antenna will not be delivered to the circuit (Point D).
- Does the antenna has the required bandwidth? If the bandwidth of the received beam is greater than the bandwidth of the receiving antenna then information will be lost (Points C and D).

14.2 | Linear Antenna Array

A linear antenna array is modeled as a series of isotropic radiators linearly placed (on the x-axis for simplicity) with a inter-element spacing of d as shown in Fig. 14.2. Since each radiator is identical, the electric field of the far field produced by each radiator when it is placed at the origin is

$$\mathbf{E}(\mathbf{r}) = \mathbf{E}_{0t} \frac{e^{-jkr}}{r} \qquad (14.1)$$

where the $\exp(j\omega t)$ term has been omitted.

Equation (14.1) says that the far field for each radiator is (a) independent of θ and ϕ coordinates of spherical coordinates, and hence by implication the radiator is isotropic; (b) the wave is propagating in the $\mathbf{a}_r$ direction due to the $\exp(-jkr)$ term,

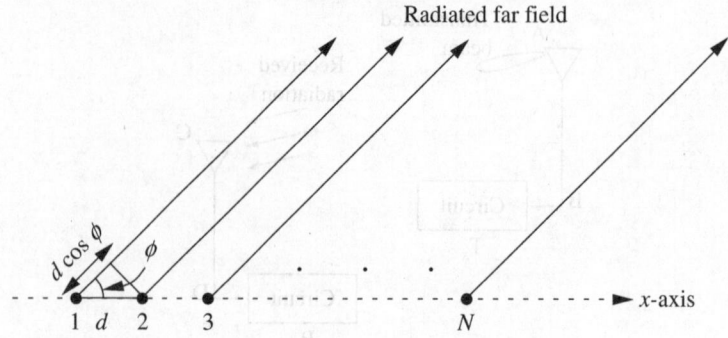

Fig. 14.2 An equi-spaced linear array

and (c) the radiation obeys the inverse square law due to the $1/r$ term. Furthermore, the vector $\mathbf{E}_{0t}$ is a constant vector oriented in some arbitrary direction but transverse to $\mathbf{a}_r$.

After studying the equations in the last chapter, it should be clear that both the electric as well as magnetic fields are proportional to the current feeding the element. So we can re-write Eqn (14.1) as

$$\mathbf{E}(\mathbf{r}) = \hat{\mathbf{t}} I_0 C \frac{e^{-jkr}}{r} \tag{14.2}$$

where $\hat{\mathbf{t}}$ is the unit vector in the direction of $\mathbf{E}_{0t}$, I_0 is the current feeding the antenna and C is some constant. Since each radiator is identical we may normalise the far field without loss of generality and write the far field for each radiator as

$$\mathbf{E}(\mathbf{r}) = \hat{\mathbf{t}} I_0 \frac{e^{-jkr}}{r} \tag{14.3}$$

Now the problem which we have to solve is that when the N radiators are placed in a straight line, as shown, what is the far field pattern? Let us first investigate the problem with only two such radiators.

Let the first radiator be placed at the origin of the x-axis as shown in Fig. 14.2, and let $\mathbf{E}_1$ be its far field electric field. Similarly, let $\mathbf{E}_2$ be the far field electric field of the second radiator when radiating with the position vector $d\mathbf{a}_x$ with respect to the first. Then

$$\mathbf{E}_1(\mathbf{r}) = \hat{\mathbf{t}} I_1 \frac{e^{-jk|\mathbf{r}|}}{|\mathbf{r}|}$$

$$\mathbf{E}_2(\mathbf{r}) = \hat{\mathbf{t}} I_2 \frac{e^{-jk|\mathbf{r}-\mathbf{a}_x d|}}{|\mathbf{r} - d\mathbf{a}_x|}$$

$$\mathbf{E}_T(\mathbf{r}) = \mathbf{E}_1(\mathbf{r}) + \mathbf{E}_2(\mathbf{r}) = \hat{\mathbf{t}} \left(I_1 \frac{e^{-jk|\mathbf{r}|}}{|\mathbf{r}|} + I_2 \frac{e^{-jk|\mathbf{r}-\mathbf{a}_x d|}}{|\mathbf{r} - d\mathbf{a}_x|} \right) \tag{14.4}$$

where $\mathbf{E}_T(\mathbf{r})$ is the total electric field at a distant point. Now

$$|\mathbf{r} - d\mathbf{a}_x| = \sqrt{r^2 - 2rd\cos\phi + d^2}$$

$$= r\sqrt{1 - 2\left(\frac{d}{r}\right)\cos\phi + \left(\frac{d}{r}\right)^2} \tag{14.5}$$

where ϕ is the angle from the x-axis, and is not to be confused with the ϕ of spherical coordinates. Treating d/r as a single variable, for $d \ll r$

$$|\mathbf{r} - d\mathbf{a}_x| \cong r(1 - \frac{d}{r}\cos\phi)$$

and

$$\frac{1}{|\mathbf{r} - d\mathbf{a}_x|} \cong \frac{1}{r(1-(d/r)\cos\phi)} \cong \frac{1}{r}\left(1 + \frac{d}{r}\cos\phi\right)$$

$$\cong \frac{1}{r} \tag{14.6}$$

Using these approximations, we get

$$\mathbf{E}_T(\mathbf{r}) = \hat{\mathbf{t}}\left(I_1 \frac{e^{-jk|\mathbf{r}|}}{|\mathbf{r}|} + I_2 \frac{e^{-jk|\mathbf{r}-\mathbf{a}_x d|}}{|\mathbf{r} - d\mathbf{a}_x|}\right)$$

$$= \hat{\mathbf{t}}\left(I_1 \frac{e^{-jkr}}{r} + I_2 \frac{e^{-jk(r-d\cos\phi)}}{r}\right)$$

$$= \hat{\mathbf{t}}\underbrace{\frac{e^{-jkr}}{r}}_{A}\underbrace{(I_1 + I_2 e^{jkd\cos\phi})}_{B} \tag{14.7}$$

We notice that there are two terms. Term A is common to both elements and it is the field pattern of an isotropic radiator, while term B is what shapes the pattern. For now we will drop using the A term, knowing it is there and its importance will be discussed later. By convention, the B term is called 'E' for the electric field and it is the factor contributed by the geometry of the array (notice d and $\cos\phi$).

Example 14.1 Calculate the error of $|\mathbf{r} - \mathbf{r}'|$ being replaced by $r - r'\cos\theta$ where θ is the angle between $\mathbf{r}$ and $\mathbf{r}'$.

Solution If the angle between $\mathbf{r}$ and $\mathbf{r}'$ is θ, and $|\mathbf{r}'| = |\mathbf{r}|/100$ then

$$|\mathbf{r} - \mathbf{r}'| = \sqrt{r^2 + r'^2 - 2rr'\cos\theta} \quad (\cong r - r'\cos\theta)$$

$$= r\sqrt{1 + x^2 - 2x\cos\theta} \quad x = r'/r = 0.01$$

$$= r\sqrt{1.0001 - .02\cos\theta}$$

comparing the values for $\cos\theta = -1, 0, 1$,

$\cos\theta$	$a = r - r'\cos\theta$	$b = \sqrt{r^2 + r'^2 - 2rr'\cos\theta}$	% of error $100|(b-a)|/a$
-1	$r(1.01)$	$r(1.01)$	0
$+1$	$r(0.99)$	$r(0.99)$	0
0	r	$r(1.0001)$	0.01

Hence we can say with a great deal of confidence that $|\mathbf{r} - \mathbf{r}'| \cong r - 2r'\cos\theta$.

Equation (14.7) was obtained using only two elements. If we consider N elements, then by simple logic

$$E = |I_1 + I_2 e^{jkd\cos\phi} + I_3 e^{j2kd\cos\phi} + \cdots + I_N e^{j(N-1)kd\cos\phi}| \tag{14.8}$$

14.3 | Linear Array with Equal Currents

14.3.1 Array Factor

Let us consider Eqn (14.8) and substitute equal current amplitudes, I_0, but phases which increase in the form of an arithmetical progression: $0, \alpha, 2\alpha, \dots (N-1)\alpha$. Then

$$I_n = I_0 \angle (n-1)\alpha \quad n = 1, 2 \dots N$$

$$= I_0 e^{j(n-1)\alpha} \tag{14.9}$$

Then $E = I_0 |e^{j0\alpha} + e^{j\alpha} e^{jkd\cos\phi} + e^{j2\alpha} e^{j2kd\cos\phi} + \dots + e^{j(N-1)\alpha} e^{j(N-1)kd\cos\phi}|$

$$= I_0 |1 + e^{j(kd\cos\phi+\alpha)} + e^{j2(kd\cos\phi+\alpha)} + \dots + e^{j(N-1)(kd\cos\phi+\alpha)}|$$

$$= I_0 |1 + e^{j\psi} + e^{j2\psi} + \dots + e^{j(N-1)\psi}| \tag{14.10}$$

where $\psi = kd\cos\phi + \alpha$. We realise that the above equation is a geometric progression. For a geometric progression of $a, ar, ar^2 \dots ar^{N-1}$, the sum of N terms is

$$a\frac{1-r^N}{1-r} \tag{14.11}$$

In Eqn (14.10), $a = 1$ and $r = \exp(j\psi)$. So

$$E = \left| \frac{1-e^{jN\psi}}{1-e^{\psi}} \right| \quad \left(\text{Taking } \frac{e^{jN\psi/2}}{e^{j\psi/2}} \text{ out of the bracket} \right)$$

$$= \left| \frac{e^{jN\psi/2}}{e^{j\psi/2}} \times \left(\frac{e^{-jN\psi/2} - e^{jN\psi/2}}{e^{-j\psi/2} - e^{j\psi/2}} \right) \right|$$

$$= \left| \frac{e^{jN\psi/2}}{e^{j\psi/2}} \right| \times \left| \frac{e^{-jN\psi/2} - e^{jN\psi/2}}{e^{-j\psi/2} - e^{j\psi/2}} \right| \quad (\text{since } |e^{jN\psi/2}| = |e^{j\psi/2}| = 1)$$

$$= \left| \frac{\sin(N\psi/2)}{\sin(\psi/2)} \right| \tag{14.12}$$

Example 14.2 Plot the array factor as a function of ψ for $-\pi < \psi < \pi$ for $N = 10$.
Solution The function $(1/N)|\sin(N\psi/2)/\sin(\psi/2)|$ for $N = 10$ is plotted in the top part of Fig. 14.3.

14.3.2 Nulls and Sidelobes

The function $(1/N)|\sin(N\psi/2)/\sin(\psi/2)|$ is shown in the figure connected with Example 14.2, where term $1/N$ multiplies the function to normalise it. It is clear that the function is periodic with a period of 2π [Fig. 14.3(a)] and the numerator and denominator are plotted separately in Fig. 14.3(b).

We can see from the figure that the numerator $|\sin(N\psi/2)|$ for $N = 10$ has 9 maximas, where the function takes the value 1, and 10 minimas, where the function becomes zero in the range $-\pi \le \psi \le \pi$. The maximas of the numerator are at

$$\frac{N\psi}{2} = \pm\frac{\pi}{2}, \pm\frac{3\pi}{2}, \dots, \pm(2m+1)\frac{\pi}{2}, \dots \text{ for } m = 0, 1, \dots$$

$$\psi = \pm\frac{\pi}{N}, \pm 3\frac{\pi}{N}, \dots \pm(2m+1)\frac{\pi}{N}, \dots \tag{14.13}$$

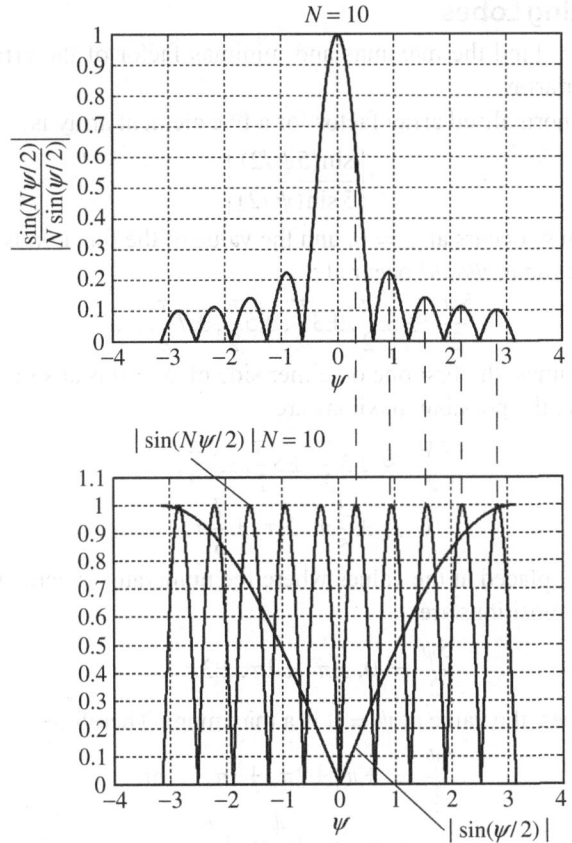

Fig. 14.3 (a) Plot of $|\sin(N\psi/2)/N\sin(\psi/2)|$ for $N = 10$. (b) Numerator and denominator of Eqn (14.12)

which means that the maximas are at *odd multiples* of $\pm\pi/N$. The minimas (zeros) of the numerator occur when

$$\frac{N\psi}{2} = 0, \pm\pi, \pm 2\pi, \dots, \pm m\pi \dots \text{ for } m = 0, 1, \dots$$

$$\psi = 0, \pm 2\frac{\pi}{N}, \pm 4\frac{\pi}{N}, \dots, \pm 2m\frac{\pi'}{N} \dots \tag{14.14}$$

or it is clear the the zeros are at *even multiples* of $\pm\pi/N$.

If we consider the zeros of the field pattern or nulls, then the zeros of the numerator are necessarily zeros of the field pattern, except where the denominator also has a zero. Near $\psi = 0$ the numerator is $\sin(N\psi/2) \approx N\psi/2$ and the denominator is $\sin(\psi/2) \approx \psi/2$. At $\psi = 0$, the field pattern has a magnitude of N, which is a global maximum and constitutes the main beam or principal lobe. The other zeros of the numerator are nulls in the field pattern, which occur at $\psi = (\pm 2m\pi)/N$.

The first maximum of the numerator occurs at $\psi = \pm\pi/N$. But this maximum does not figure in the field pattern, it is absorbed in the main beam. The first sidelobe maxima of the field pattern occur at $\psi \approx \pm 3\pi/N$, as shown by the dashed line in the figure. Similarly, the second sidelobe maxima occur at $\psi \approx \pm 5\pi/N$ and so on. Our interest in the sidelobes is mainly in the first sidelobe adjacent to the principle lobe.

14.3.3 Grating Lobes

Example 14.3 Find the maximas and minimas factor of the array factor of a 5-element linear array.

Solution The normalised array factor for a five element array is

$$\left| \frac{\sin(5\psi/2)}{5\sin(\psi/2)} \right|$$

The first maximum occurs at $\psi = 0$, and the value of the function is 1.

The other maximas *should* occur at

$$\frac{5\psi}{2} \approx \pm\frac{\pi}{2}, \pm3\frac{\pi}{2}, \pm5\frac{\pi}{2}, \pm7\frac{\pi}{2}, \ldots$$

From these maximas, the first one on either side of $\psi = 0$ is absorbed in the main beam. Therefore, the possible maximas are

$$\frac{5\psi}{2} \approx \pm3\frac{\pi}{2}, \pm5\frac{\pi}{2}, \pm7\frac{\pi}{2}, \ldots$$

or

$$\psi \approx \pm\frac{3}{5}\pi, \pm\pi, \pm\frac{7}{5}\pi, \ldots$$

The minimas are placed at the values where the numerator is zero. Therefore, the possible minima positions are

$$\frac{5\psi}{2} = 0, \pm\pi, \pm2\pi, \pm3\pi, \ldots$$

out of these values, the value at $\psi = 0$ is a maximum. Therefore

$$\frac{5\psi}{2} = \pm\pi, \pm2\pi, \pm3\pi, \ldots \text{or}$$

$$\psi = \pm\frac{2}{5}\pi, \pm\frac{4}{5}\pi, \pm\frac{6}{5}\pi, \ldots$$

Figure 14.4 shows the array factor for this example. The figure shows that apart from the main beam at $\psi = 0$ there are other beams of equal intensity at $\psi = -2\pi$ and $\psi = 2\pi$. These are *grating lobes*. In fact grating lobes exist at $\psi = \pm2m\pi$ for $m = 1, 2, \ldots$

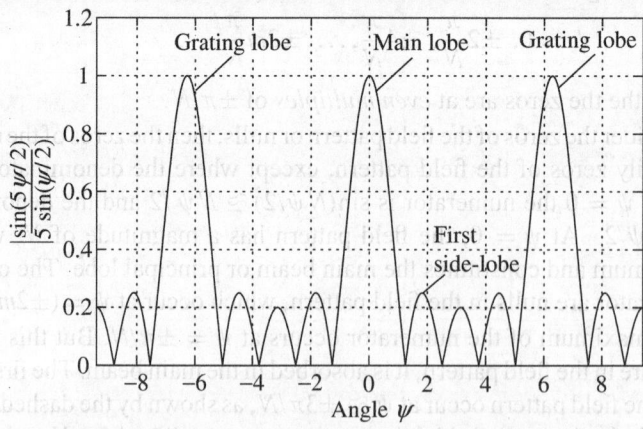

Fig. 14.4 Array factor of a 5-element array plotted in the range $-3\pi < \psi < 3\pi$

14.3.4 Beam Pointing Angle

We are also interested in answering the following important questions:

- What is the direction of the main beam? The main beam is also called the principal lobe.
- What is the angle where the first null appears? The angle at which the first null appears effectively decides the beam width.
- What is the magnitude of the first sidelobe with respect to the main beam?

It should also be clear that *the magnitude* of the first sidelobe can be obtained from this function (Eqn (14.12)) but *all placements* have to be determined from

$$\psi = kd\cos\phi + \alpha \qquad (14.15)$$

The normalised magnitude of the first sidelobe is at

$$E_{sl} \approx \left[\left|\frac{\sin(N\psi/2)}{N\sin(\psi/2)}\right|\right]_{\psi=3\pi/N}$$

$$= \left|\frac{1}{N\sin(3\pi/2N)}\right| \qquad (\text{using } \sin x \approx x, \text{ in the next Eqn})$$

$$\approx \left|\frac{1}{N(3\pi/2N)}\right| \qquad (\text{for large } N)$$

$$= \frac{2}{3\pi}$$

which is 13.46 db down, since $20\log(2/3\pi) = -13.46$.

> For large N, the first sidelobe in a uniform array is about 13.46 db down from the principal lobe.

Example 14.4 Find the first sidelobe level for $N = 4$.

Solution

Step 1. The array factor is

$$E = \left|\frac{\sin(2\psi)}{4\sin(\psi/2)}\right|$$

Step 2. For the first sidelobe, the sidelobe maximum is at $2\psi = 3\pi/2$. So

$$E_{slmax} \approx \left|\frac{1}{4\sin(3\pi/4)}\right| = 0.3536$$

which is -9.031 dB down.

We now look at several aspects of the uniform array. First and foremost what should be the values of the phase progression α to point the beam at a *particular* direction ϕ_0? To do that, we know from the field pattern $E(\psi)$ that the main beam points to $\psi = 0$. So from Eqn (14.15),

$$kd\cos\phi_0 + \alpha = 0$$

or
$$\alpha = -kd\cos\phi_0 \qquad (14.16)$$

Hence, the array factor, Eqn (14.12) becomes

$$E_n(\phi) = \left| \frac{\sin[N(kd\cos\phi + \alpha)/2]}{N\sin[(kd\cos\phi + \alpha)/2]} \right|$$

$$= \left| \frac{\sin[Nkd(\cos\phi - \cos\phi_0)/2]}{N\sin[kd(\cos\phi - \cos\phi_0)/2]} \right| \qquad (14.17)$$

where E_n is the normalised electric field pattern.

Let us find the position of the first null. From Eqn (14.14), the first nulls are at $\psi = \pm 2\pi$. Or

$$kd(\cos\phi_{\mathrm{fn}} - \cos\phi_0) = \pm 2\pi \qquad (14.18)$$

Since ϕ_{fn} is close to ϕ_0, $\phi_{\mathrm{fn}} = \phi_0 + \Delta\phi_{\mathrm{fn}}$. From Taylor's series expansion of $f(x)$ at the point $x = a$

$$f(x) = f(a) + f'(a)(x - a) + f''(a)\frac{(x-a)^2}{2!} + \cdots$$

or $\quad \cos\phi = \cos(\phi_0) - \sin(\phi_0)(\phi - \phi_0) - \cos(\phi_0)(\phi - \phi_0)^2/2 + \cdots \quad (14.19)$

$\qquad \cos\phi_{\mathrm{fn}} = \cos(\phi_0) - \sin(\phi_0)(\phi_{\mathrm{fn}} - \phi_0) - \cos(\phi_0)(\phi_{\mathrm{fn}} - \phi_0)^2/2 + \cdots$

or $\quad \cos\phi_{\mathrm{fn}} = \cos(\phi_0) - \sin(\phi_0)\Delta\phi_{\mathrm{fn}} - \cos(\phi_0)(\Delta\phi_{\mathrm{fn}})^2/2 + \cdots$

Substituting in Eqn (14.18),

$$kd(\cos\phi_{\mathrm{fn}} - \cos\phi_0) = -\frac{2\pi}{N}$$

$$kd\left[-\sin(\phi_0)\Delta\phi_{\mathrm{fn}} - \frac{\cos(\phi_0)(\Delta\phi_{\mathrm{fn}})^2}{2} \right] \approx -\frac{2\pi}{N} \qquad (14.20)$$

The negative sign is taken on the right since when ϕ increases, ψ decreases.

Now for a *broadside array* (where $\phi_0 = \pi/2$), $\sin\phi_0 = 1$, $\cos\phi_0 = 0$

so

$$\Delta\phi_{\mathrm{fn}} = \frac{2\pi}{kdN} = \frac{\lambda}{dN} \qquad (14.21)$$

and the *beam width between first nulls* (BWFN) is

$$2\Delta\phi_{\mathrm{fn}} = \frac{2\lambda}{dN} \qquad (14.22)$$

Note that the width of the beam is inversely proportional to the length of the antenna, ($= dN$), also called the antenna aperture. For an *end fire array* (where $\phi_0 = 0$), $\sin\phi_0 = 0$, $\cos\phi_0 = 1$,

$$\Delta\phi_{\mathrm{fn}} = \sqrt{\frac{4\pi}{kdN}} = \sqrt{\frac{2\lambda}{dN}} \qquad (14.23)$$

and the BWFN is

$$2\Delta\phi_{\mathrm{fn}} = 2\sqrt{\frac{2\lambda}{dN}} \qquad (14.24)$$

Figure 14.5 shows a three-dimensional view of the electric field pattern. Notice that the pattern has circular symmetry since the radiation at an angle ϕ from a linear array is conical. In this pattern, $kd = \pi$, $\alpha = 0$, and $N = 10$.

Next we consider how the main beam points in different directions. If we change the value of α in accordance with Eqn (14.16), we can change the beam pointing

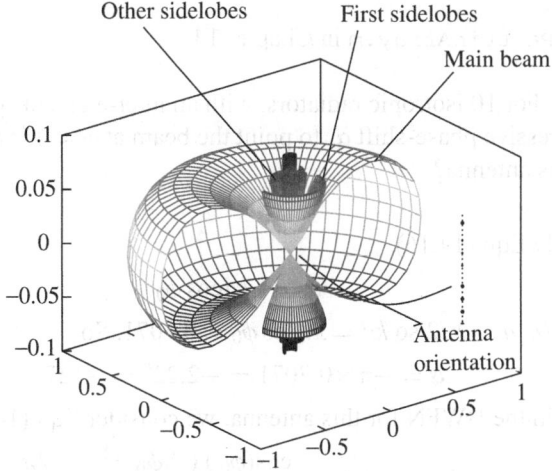

Fig. 14.5 3-D view of a broadside pattern, where the antenna is oriented vertically. Note the difference in the scale of the z-axis which has been broadened to show greater particulars of the main beam and sidelobes

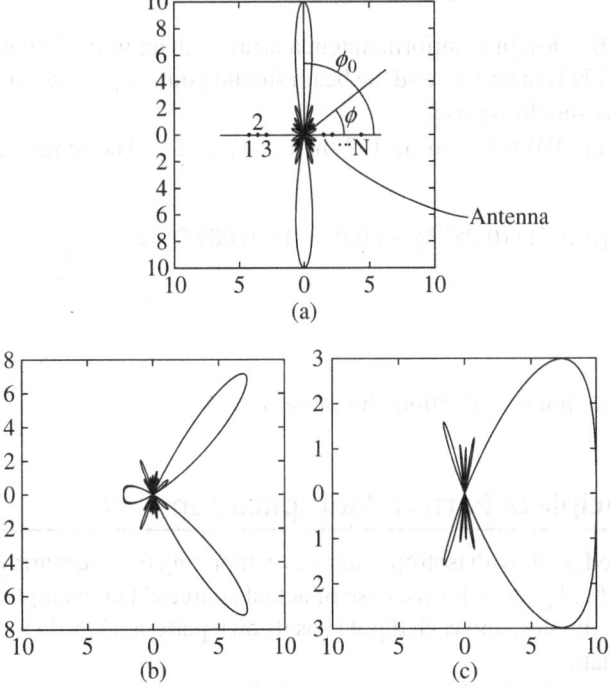

Fig. 14.6 Pointing the main beam in different directions. (a) $\phi_0 = \pi/2$, $\alpha = 0$; (b) $\phi_0 = \pi/4$, $\alpha = -0.7071\,kd$; (c) $\phi_0 = 0$, $\alpha = -kd$. For all these plots, $kd = \pi$

angle. This is shown in Fig. 14.6. In part (a) of the figure, α is set equal to zero ($= -kd\cos\phi_0 = -kd\cos(\pi/2)$) the beam points at the broadside direction. $\alpha = 0$ means that all the currents are in phase. Similarly in (b) and (c), the beam is shown pointing at $\phi = 45°$ and $\phi = 0°$. Notice how the beam broadens when pointing in the end-fire direction ($\phi_0 = 0°$).

See `RadPattLinArray.m` in Chapter 14

Example 14.5 For 10 isotropic radiators, with an inter-element spacing of 0.5λ, what is the progressive phase-shift α, to point the beam at $\phi = 45°$ angle? What is the BWFN of this antenna?

Solution

Step 1. We apply Eqn (14.16),

$$\alpha = -kd\cos\phi_0$$

Step 2. $k = 2\pi/\lambda$, $d = \lambda/2$ so $kd = \pi$. $\cos\phi_0 = 0.7071$. So

$$\alpha = -\pi \times 0.7071 = -2.22^r = -127°$$

Step 3. To obtain the BWFN for this antenna, we consider Eqn (14.20),

$$\sin(\phi_0)\,\Delta\phi_{\text{fn}} + \frac{\cos(\phi_0)\,(\Delta\phi_{\text{fn}})^2}{2} \approx \frac{2\pi}{Nkd}$$

Step 4. The solution to this equation is (for $\phi_0 = \pi/4$, $N = 10$, and $kd = \pi$)

$$\Delta\phi_{\text{fn}} = 0.2512^r = 14.4°$$

Hence the BWFN is 28.8°.

Example 14.6 Design a uniform antenna array with an inter-element spacing of 0.5λ. The BWFN is to be 10°, and the beam should point at $\phi = 45°$ direction. How many elements should we use?

Solution If the BWFN is to be 10° then $\Delta\phi_{\text{fn}} = 5°$. The required equation is Eqn (14.20).

$$(0.7071)(0.0873) + (0.07071)(0.0873)^2/2 = \frac{2\pi}{\pi N}$$

$$N = \frac{2}{0.0620}$$

$$N = 32.25$$

Obviously N cannot be a fraction, therefore $N = 33$.

14.4 | Principle of Pattern Multiplication

We have worked so far with isotropic sources comprising the linear array. How do we obtain the far field pattern for the case of actual sources? For example, the sources may consist of dipoles; and each dipole has its own pattern. How do we incorporate this in our equations?

First of all, we align the linear antenna with the z-axis so that ϕ is replaced by θ of spherical coordinates in our analysis. Now let the electric field of element pattern of the far field be

$$\mathbf{E}(\theta,\phi) = \hat{\mathbf{E}}\frac{e^{-jkr}}{r}f(\theta,\phi) \tag{14.25}$$

then Eqn (14.7) for two radiators will become

$$\mathbf{E}_T(\mathbf{r}) = \hat{\mathbf{E}}\frac{e^{-jkr}}{r}f(\theta,\phi)\,(I_1 + I_2 e^{jkd\cos\theta})$$

where we have replaced ϕ by θ. Similarly, for N radiators

$$\mathbf{E}_T(\mathbf{r}) = \hat{\mathbf{E}}\frac{e^{-jkr}}{r} f(\theta\,,\phi\,)\,(I_1 + I_2 e^{jkd\cos\theta} + I_3 e^{j2kd\cos\theta} \ldots + I_N e^{j(N-1)kd\cos\theta}) \tag{14.26}$$

The array factor now becomes

$$E = \left| \underbrace{f(\theta\,,\phi\,)}_{\text{Element pattern}} \underbrace{\left(I_1 + I_2 e^{jkd\cos\theta} + I_3 e^{j2kd\cos\theta} \ldots + I_N e^{j(N-1)kd\cos\theta}\right)}_{\text{Group pattern}} \right| \tag{14.27}$$

From the principle of pattern multiplication, we can do various things.

Example 14.7 Find the pattern of the far field of a linear antenna where the currents are in the ratio 1:2:3:2:1. The phase of the currents are all equal to 0.

Solution In Fig. 14.7, we define an element (shown as a triangle) consisting of three isotropic radiators each represented by a circular dot. The array factor of the element is

$$E_{\text{element}} = \frac{\sin(3\psi/2)}{\sin(\psi/2)}$$

where $\psi = kd\cos\phi$. Similarly for the three elements (triangles), the array factor is exactly the same as before

$$E_{\text{group}} = \frac{\sin(3\psi/2)}{\sin(\psi/2)}$$

therefore the total array factor is

$$\begin{aligned} E_{\text{total}} &= E_{\text{element}} \times E_{\text{group}} \\ &= \left[\frac{\sin(3\psi/2)}{\sin(\psi/2)}\right]^2 \end{aligned}$$

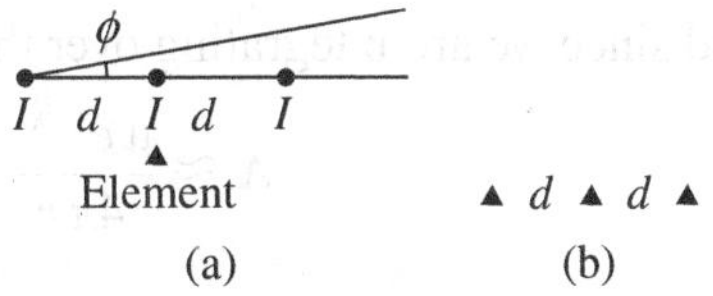

Fig. 14.7 (a) An element consisting of three isotropic radiators each carrying current I, (b) three elements shown in (a) forming an antenna

14.5 Far Field Pattern of a Continuous Current Distribution

We know that as currents oscillate, they radiate electromagnetic waves. The vector potential **A** is given by Eqn (14.16), which applies to Fig. 14.8. The equation is reproduced here for convenience:

$$\mathbf{A} = \frac{\mu}{4\pi}\iiint\limits_{\mathcal{V}'} \frac{\mathbf{J}(\mathbf{r}')\,e^{-jk|\mathbf{r}-\mathbf{r}'|}}{|\mathbf{r}-\mathbf{r}'|}\,dV' \tag{14.28}$$

Using the results of the last section, using Eqn set (14.6)

$$\begin{aligned} |\mathbf{r}-\mathbf{r}'| &\approxeq r - r'\cos\chi \\ |(\mathbf{r}-\mathbf{r}')^{-1}| &\approxeq r \end{aligned} \tag{14.29}$$

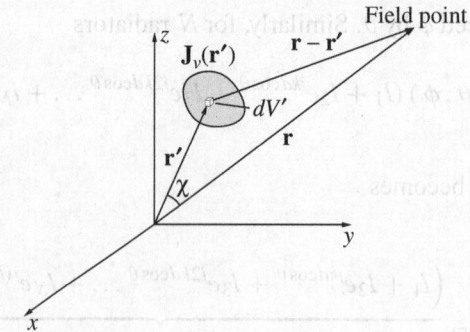

Fig. 14.8 The far field pattern due to current sources

Using these approximations,

$$\mathbf{A} \approx \frac{\mu}{4\pi} \iiint\limits_{V'} \frac{\mathbf{J}(\mathbf{r}')\, e^{-jk(r - r'\cos\chi)}}{r} dV' \qquad (14.30)$$

and since we are integrating over the primed coordinates

$$\mathbf{A} \approx \frac{\mu\, e^{-jkr}}{4\pi r} \iiint\limits_{V'} \mathbf{J}(\mathbf{r}')\, e^{jkr'\cos\chi}\, dV'$$

which gives a very important result in spherical coordinates: since $r'\cos\chi = \mathbf{r}'\cdot\hat{\mathbf{r}}$, and $\hat{\mathbf{r}} = \mathbf{a}_x \sin\theta\cos\phi + \mathbf{a}_y \sin\theta\sin\phi + \mathbf{a}_z\cos\theta$.

$$\mathbf{A} = \frac{\mu\, e^{-jkr}}{4\pi r} \iiint\limits_{V'} \mathbf{J}(x',y',z')\, e^{jk(x'\sin\theta\cos\phi + y'\sin\theta\sin\phi + z'\cos\theta)}\, dx'\,dy'\,dz' \qquad (14.31)$$

Example 14.8 Consider the very simple linear current density

$$\mathbf{J}_l(z') = \begin{cases} c\mathbf{a}_z, & \text{for } -l<z<l;\ c = \text{constant} \\ 0, & \text{elsewhere} \end{cases}$$

find the far field with respect to this current density.
Solution

$$\mathbf{A} = \frac{\mu\, e^{-jkr}}{4\pi r} \iiint\limits_{V'} \mathbf{J}(x',y',z')\, e^{jk(x'\sin\theta\cos\phi + y'\sin\theta\sin\phi + z'\cos\theta)}\, dx'\,dy'\,dz'$$

$$= \frac{\mu\, e^{-jkr}}{4\pi r} \int_{-l}^{l} c\mathbf{a}_z e^{jkz'\cos\theta}\, dz'$$

$$= -\frac{\mu\, e^{-jkr}}{2\pi r} \frac{c\,\sin(kl\cos\theta)}{k\cos\theta}\mathbf{a}_z$$

and using $\mu\mathbf{H} = \nabla \times \mathbf{A}$ first, neglecting all terms proportional to $1/r^2$ and $1/r^3$ then computing $\mathbf{E} = \mathbf{H} \times \mathbf{a}_r$

$$H_\phi = \frac{jk\, e^{-jkr}c\,\sin(\theta)\,\sin(kl\cos(\theta))}{r\cos(\theta)}$$

$$E_\theta = \frac{jk\, e^{-jkr}c\,\sin(\theta)\,\sin(kl\cos(\theta))}{Z_0 r\cos(\theta)}$$

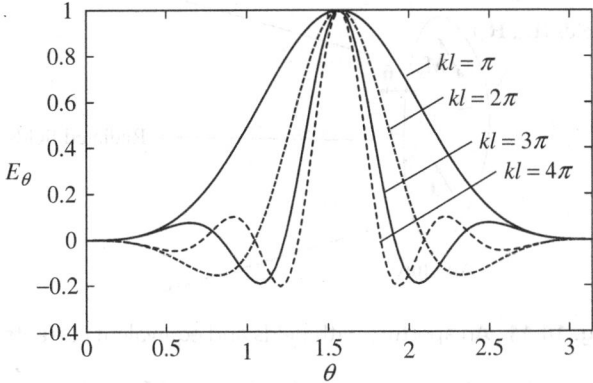

Fig. 14.9 E_θ versus θ for various values of kl

A normalised plot of E_θ vs θ, $0 < \theta < \pi$ is shown in Fig. 14.9. If we scrutinise the figure what strikes straight away is that as the length of current element is increased, the beam becomes narrower, and the directivity of the antenna improves. Notice that the pattern resembles the $\sin x/x$ function in the far field.

14.6 | Aperture Antennas

In his book *Traité de la Lumière* published in 1690, Huygens discussed a new principle whereby if there is a wavefront of light at $t = t_0$ then each point on the wave front acts as a source of a spherical wave which propagates with speed of light. Since each point on the wavefront acts like a source, the new wavefront is an envelope of all these little spheres as shown in Fig. 14.10.

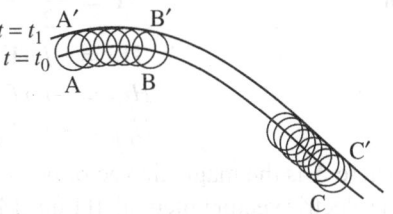

Fig. 14.10 Wave propagation based on Huygens principle

We can see from the figure that at $t = t_0$ the wavefront is ABC. At every point on the wavefront as shown at A, B, and C small spheres are drawn whose radii are $c(t_1 - t_0)$ and the wavefront advances to A', B', and C'. The new wavefront is $A'B'C'$. The surface $A'B'C'$ is slightly different from ABC.

Aperture antennas are based on the fact that if a travelling electromagnetic wavefront is present on some surface then if those fields act like a source of waves then what the far field would like. A good example is that of a waveguide antenna. A waveguide antenna consists of a waveguide which is open and in flush with a metal plane. The open end of the waveguide radiates into space. Essentially the fields which exist at the open end, act like a source which radiate.

The modern theory of diffraction and radiation predicts that if the fields in a plane aperture are $\mathbf{E}_a$ and $\mathbf{H}_a$, then we can define two surface current densities,[1] with reference to Fig. 14.11,

$$\mathbf{J}_s = \hat{\mathbf{n}} \times \mathbf{H}_a$$
$$\mathbf{M}_s = -\hat{\mathbf{n}} \times \mathbf{E}_a \tag{14.32}$$

---

[1] This is given without proof, since the topic is beyond the scope of this book.

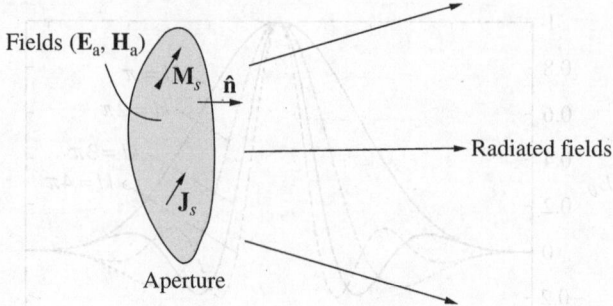

Fig. 14.11 An aperture with fields and equivalent currents

where $\mathbf{J}_s$ is the surface electric current density and $\mathbf{M}_s$ is the surface magnetic current density. We can define the far field vector potentials in terms of either of them by

$$A = \frac{\mu e^{-jkr}}{2\pi r} \iiint \mathbf{J}'_s e^{-jk\mathbf{r}' \cdot \hat{\mathbf{r}}} dV' \tag{14.33}$$

$$\mathbf{E}_f = -j\omega \mathbf{A} \tag{14.34}$$

$$E_{f\theta} = -j\omega A_\theta; \quad H_{f\phi} = E_{f\theta}/Z_0 \tag{14.35}$$

$$E_{f\phi} = -j\omega A_\phi; \quad H_{f\theta} = -E_f/Z_0 \tag{14.36}$$

or

$$F = \frac{\varepsilon e^{-jkr}}{2\pi r} \iiint \mathbf{M}'_s e^{-jk\mathbf{r}' \cdot \hat{\mathbf{r}}} dV' \tag{14.37}$$

$$\mathbf{H}_f = -j\omega \mathbf{F} \tag{14.38}$$

$$H_{f\theta} = -j\omega F_\theta; \quad E_{f\phi} = -Z_0 H_{f\theta} \tag{14.39}$$

$$H_{f\phi} = -j\omega F_\phi; \quad E_{f\theta} = Z_0 H_{f\theta} \tag{14.40}$$

where $\mathbf{A}$ is the magnetic vector potential and $\mathbf{F}$ is the electric vector potential. If Eqn (14.34) is used, then from $\mathbf{E}_f$ we calculate the magnetic field from Eqns (14.35) and (14.36). On the other hand, if we use Eqn set (14.38), then we obtain the fields as shown in those set of equations. This discussion applies to spherical coordinates.

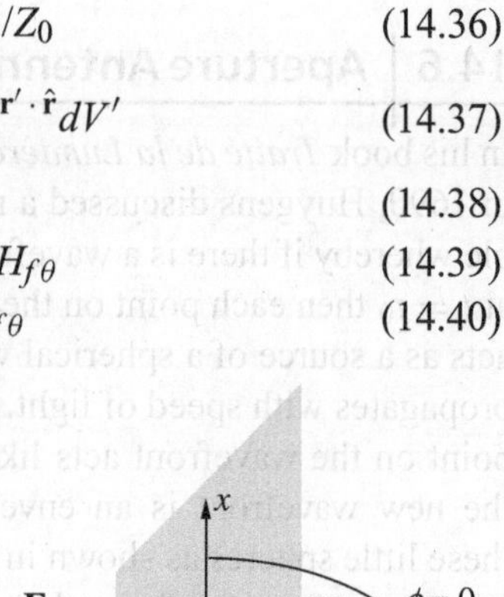

Example 14.9 A plane wave is incident from the region $z < 0$ (see Fig. 14.12) propagating in the $+z$ direction, on an infinite metal sheet, coincident with the x–y plane (the plane is infinite in extent in the x and y directions). The sheet has a rectangular aperture cut into it with dimensions of $-a < x < a$ and $-b < y < b$. Find the radiation pattern in the half space $z > 0$.

Solution

Fig. 14.12 Plane wave incident on a metal sheet with a rectangular aperture

Step 1. The fields of the plane wave are

$$\mathbf{E} = E_0 \mathbf{a}_x e^{-jkz}$$

$$\mathbf{H} = \frac{E_0}{Z_0} \mathbf{a}_y e^{-jkz} \tag{14.41}$$

Step 2. The aperture fields are

$$\mathbf{E}_a = E_0 \mathbf{a}_x \quad (\text{exist only in } -a/2 < x < a/2,$$

$$\mathbf{H}_a = \frac{E_0}{Z_0} \mathbf{a}_y \quad -b/2 < y < b/2 \text{ and } z = 0 \tag{14.42}$$

Step 3. The magnetic surface current in the aperture is

$$\mathbf{M}_s = -\hat{\mathbf{n}} \times \mathbf{E}_a = -\mathbf{a}_z \times \mathbf{a}_x E_0 = -\mathbf{a}_y E_0$$

Step 4. The electric vector potential is

$$\mathbf{F} = \frac{\varepsilon \, e^{-jkr}}{2\pi r} \iiint \mathbf{M}_s' e^{-jk\mathbf{r}' \cdot \hat{\mathbf{r}}} dV'$$

$$= \frac{\varepsilon \, e^{-jkr}}{2\pi r}(-\mathbf{a}_y E_0) \iint_{\text{aperture}} e^{jk(x' \sin\theta \cos\phi + y' \sin\theta \sin\phi)} dx' dy'$$

$$= \frac{\varepsilon \, e^{-jkr}}{2\pi r}(-\mathbf{a}_y E_0) \int_{-a}^{a} e^{jkx' \sin\theta \cos\phi} dx' \int_{-b}^{b} e^{jky' \sin\theta \sin\phi} dy'$$

$$= -\mathbf{a}_y \underbrace{\frac{2\varepsilon E_0 e^{-jkr} ab}{\pi r}}_{C} \times \frac{\sin(ka \sin\theta \cos\phi)}{ka \sin\theta \cos\phi} \times \frac{\sin(kb \sin\theta \sin\phi)}{kb \sin\theta \sin\phi}$$

$$= \mathbf{a}_y F_y \tag{14.43}$$

where C is the term shown in the under-brace.

Step 5. Since $F_y \mathbf{a}_y = F_y(\mathbf{a}_r \sin\theta \sin\phi + \mathbf{a}_\theta \cos\theta \sin\phi + \mathbf{a}_\phi \cos\phi)$

Step 6. Then $F_\theta = \cos\theta \sin\phi \, F_y$

$$F_\phi = \cos\phi \, F_y \tag{14.44}$$

Step 7. So, $\quad H_\theta = -j\omega \cos\theta \sin\phi \, F_y \tag{14.45}$

$$= -j\omega \, C \cos\theta \, \sin\phi \frac{\sin(ka \sin\theta \cos\phi)}{ka \sin\theta \cos\phi} \times \frac{\sin(kb \sin\theta \sin\phi)}{kb \sin\theta \sin\phi}$$

$$= -j\omega \, C \cos\theta \frac{\sin(ka \sin\theta \cos\phi)}{ka \sin\theta \cos\phi} \times \frac{\sin(kb \sin\theta \sin\phi)}{kb \sin\theta} \tag{14.46}$$

$$H_\phi = -j\omega \cos\phi \, F_y \tag{14.47}$$

$$= -j\omega \, C \cos\phi \frac{\sin(ka \sin\theta \cos\phi)}{ka \sin\theta \cos\phi} \times \frac{\sin(kb \sin\theta \sin\phi)}{kb \sin\theta \sin\phi}$$

$$= -j\omega \, C \frac{\sin(ka \sin\theta \cos\phi)}{ka \sin\theta} \times \frac{\sin(kb \sin\theta \sin\phi)}{kb \sin\theta \sin\phi} \tag{14.48}$$

Step 8. Plotting the normalised pattern in the $\phi = 0$ plane, which is the x–z plane ($0 < \theta < \pi/2$)

$$H_{n\theta} = 0; \quad H_{n\phi} = \frac{\sin(ka \sin\theta)}{ka \sin\theta}$$

and in the $\phi = \pi/2$ plane, which is the y–z plane ($0 < \theta < \pi/2$)

$$H_{n\theta} = \cos\theta \frac{\sin(kb \sin\theta)}{kb \sin\theta}; \quad H_{n\phi} = 0$$

Plots are shown in Fig. 14.13.

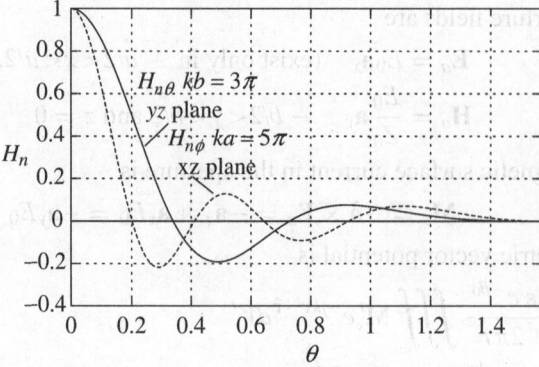

Fig. 14.13 Radiation plots of the normalised magnetic field for Example 14.9

14.7 | Horn Antennas

14.7.1 Introduction

Horn antennas are used in the frequency bands right from UHF (300 MHz, $\lambda = 1$ m) up to the Ku band (18 GHz, $\lambda = 1.7$ cm) where it is used for satellite applications. Horn antennas, when well designed, can have high directivities which can be as high as about 20 dB. As the directivity of the antenna is increased, the length of the horn and the area of the aperture can become very long and large respectively, so convenient designs are limited to directivities of around 10–15 dB. The input impedance of these antennas are fairly frequency insensitive showing bandwidths of about 1–$10 f_0$ with a design frequency of around 3–$4 f_0$. One of the advantages of horn antennas is that they exhibit low loss, so directivity and gain are interchangeable. Typical examples of horn antennas are shown in Fig. 14.14.

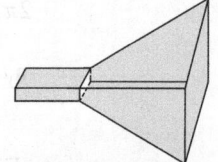

(a) Pyramidal horn

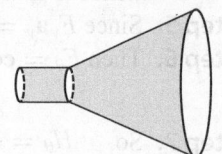

(b) Conical horn

Fig. 14.14 Examples of horn antennas

14.7.2 Far Fields and Patterns for Horn Antennas

If we apply the theory which was presented in Section 14.6 along with Eqn (14.31) the analysis can yield very complicated functions to depict the radiation patterns of horn antennas. To do the analysis, the electric and magnetic fields ($\mathbf{E}_a, \mathbf{H}_a$) are assumed across the mouth of the horn as explained in Section 14.6 and the surface currents ($\mathbf{J}_s, \mathbf{M}_s$) are calculated as per the Eqn set (14.32). The far-field radiation pattern is then computed using the potentials ($\mathbf{A}, \mathbf{F}$), the Eqn sets (14.36) and (14.40). When performing the integrations we find that the far field functions turn out to be extremely complicated and not worth all the trouble.

However, a much simpler approach may be taken which gives fairly accurate results, and that is we assume that the fields in the mouth of the horn are uniform and constant, then it is fairly straightforward to calculate the far fields as explained in Example 14.9.

14.7.3 Optimum Dimensions of Horn Antennas

Horn antennas may be divided into three types

1. H-plane sectoral horn (shown in Fig. 14.15(a))

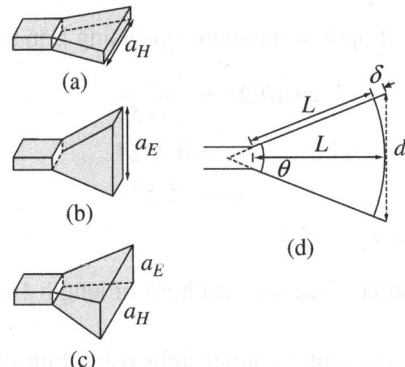

Fig. 14.15 Horn antennas. (a) H-plane sectoral horn, (b) E-plane sectoral horn, (c) pyramidal horn, (d) flare angle diagram

2. E-plane sectoral horn (shown in Fig. 14.15(b))
3. Pyramidal horn (shown in Fig. 14.15(c))

Radiation patterns of horn antenna depend on the flare angle and dimensions a_E and a_H shown in the figure. For a given length L, there is an optimum dimension of a_E and a_H (and therefore θ) shown in Fig. 14.15(d).

Referring to 14.15(d), we can infer

$$\cos\frac{\theta}{2} = \frac{L}{L+\delta}$$

$$\sin\frac{\theta}{2} = \frac{d}{2(L+\delta)} \tag{14.49}$$

from

$$(L+\delta)^2 = \frac{d^2}{4} + L^2$$

$$\cancel{L^2} + 2L\delta + \delta^2 = \frac{d^2}{4} + \cancel{L^2}$$

$$2L\delta \approx \frac{d^2}{4} \quad \text{(neglecting } \delta^2 \text{)}$$

$$\delta = \frac{d^2}{8L} \tag{14.50}$$

In design of the horn we aspire to obtain a uniform aperture distribution, as considered in Example 14.9. The optimum values of δ and L are

$$\delta_o = \frac{L}{\cos(\theta/2)} - L$$

or

$$L = \frac{\delta_o \cos(\theta/2)}{1 - \cos(\theta/2)} \tag{14.51}$$

δ_o lies in the range of 0.1 to 0.4λ. This analysis applies to both E- and H-plane horns, as well as pyramidal horns.

 See DFromA.m in Chapter 14

Example 14.10 For an E-plane sectoral horn of length 5λ, find the flare angle for $\delta_o = 0.25\lambda$.

Solution For the horn of optimum dimensions using Eqn set (14.49)

$$\cos(\theta/2) = \frac{L}{L + \delta_o}$$

$$= 0.9524$$

$$\theta = 35.5°$$

the E-plane flare angle is $\theta_E = 35.5°$.

Example 14.11 For an H-plane sectoral horn of length 10λ, find the flare angle for $\delta_o = 0.15\lambda$.

Solution For the horn of optimum dimensions using Eqn set (14.49)

$$\cos(\theta/2) = \frac{L}{L + \delta_o}$$

$$= 0.9852$$

$$\theta = 19.7°$$

the H-plane flare angle is $\theta_H = 19.7°$.

We know the well known relation

$$D = \frac{4\pi}{\lambda^2} A_e$$

where D is the directivity and A_e is the effective aperture. Writing in terms of the physical aperture

$$A_e = \varepsilon_{ap} A_p \tag{14.52}$$

where A_p is the physical aperture of the horn and ε_{ap} is the aperture efficiency. For a rectangular horn pyramidal horn, $A_p = a_E a_H$ and for a conical horn $A_p = \pi R^2$ where R is the radius of the aperture. Then

$$D = \frac{4\pi}{\lambda^2} \varepsilon_{ap} A_p \tag{14.53}$$

Experimental studies show that ε_{ap} lies between 0.5 and 0.6 as long as the a_E or a_H or R is at least 1λ. Assuming a value of $\varepsilon_{ap} = 0.6$,

$$D \approx \frac{7.5 A_p}{\lambda^2} \tag{14.54}$$

The half power beam widths (in degrees) an E-plane and H-plane horns for a rectangular horn are given by (Kraus, 1988)

$$\mathrm{HPBW(E)} = \frac{56}{a_{E\lambda}}$$

$$\mathrm{HPBW(H)} = \frac{67}{a_{H\lambda}} \tag{14.55}$$

and the beam widths (in degrees) between first nulls are given by

$$\mathrm{BWFN(E)} = \frac{115}{a_{E\lambda}}$$

$$\mathrm{BWFN(H)} = \frac{172}{a_{H\lambda}} \tag{14.56}$$

Example 14.12 A horn has dimensions $a_E = 5\lambda$ and $a_H = 10\lambda$. Find the directivity and half-power beam widths in the two planes assuming that it is an optimum horn.

Solution

Step 1. The physical aperture is given by

$$A_p = a_E a_H$$
$$= 50\lambda^2$$

Step 2. Assuming an $\varepsilon_{ap} = 0.6$

$$A_e = 30\lambda^2$$

Step 3. The directivity is

$$D = \frac{4\pi}{\lambda^2} A_e$$
$$= 120\pi$$
$$= 377$$

Step 4. The HPBW in the E-plane from Eqn set (14.55) is

$$\text{HPBW(E)} = \frac{56}{a_E \lambda}$$
$$= \frac{56}{5}$$
$$= 11.2°$$

similarly we can calculate the HPBW in the H-plane.

Example 14.13 For a pyramidal horn, $L = 12\lambda$, $a_E = 5\lambda$, $a_H = 8\lambda$, find the value of δ in the two planes. Is this an optimum horn?

Solution In each of the two planes by Eqn (14.50)

$$\delta = \frac{d^2}{8L}$$

taking first the case of the E-plane:

$$\delta_E = \frac{a_E^2}{8L}$$
$$= \frac{25\lambda^2}{96\lambda} = 0.26\lambda$$

Next focusing on the H-plane,

$$\delta_H = \frac{a_H^2}{8L}$$
$$= \frac{64\lambda^2}{96\lambda} = 0.67\lambda$$

since the dimensions of δ do *not* lie in the range $[0.1\lambda, 0.4\lambda]$ the horn is not an optimum horn.

Example 14.14 A conical horn antenna of optimum dimensions has an aperture diameter of 3λ. Find the directivity of the horn.

Solution The physical aperture of the horn is

$$A_p = \pi \left(\frac{3\lambda}{2}\right)^2$$

therefore the effective aperture will be

$$A_e = 0.6 A_p$$

$$= 0.6 \times 7.086\lambda^2 = 4.2412\lambda^2$$

the directivity D will be

$$D = \frac{4\pi}{\lambda^2} A_e = 53.29$$

14.8 | Parabolic Reflector

Suppose we have a bulb (a point source) in a flashlight and we want to send out a parallel beam of light using a reflecting surface. How would we do it? Referring to Fig. 14.16(a), suppose the point source were placed at F and the distance to the reflecting surface was L then when a ray would reflect off the surface at O, and reach F after reflection, then the distance travelled would be $2L$: from F to O and back. Let us take any other ray of light, FA. It *should* reach the plane PP'—which passes through the focus—with the same path length, $2L$. By this means *all* rays would reach the plane PP' in the same phase and therefore we would have a plane wave along the plane PP'.

Putting this in mathematical language,

i.e., $$R + R\cos\theta = 2L \quad \text{(solving for } R)$$

$$R = \frac{2L}{1 + \cos\theta} \qquad (14.57)$$

which is the equation of a parabola.

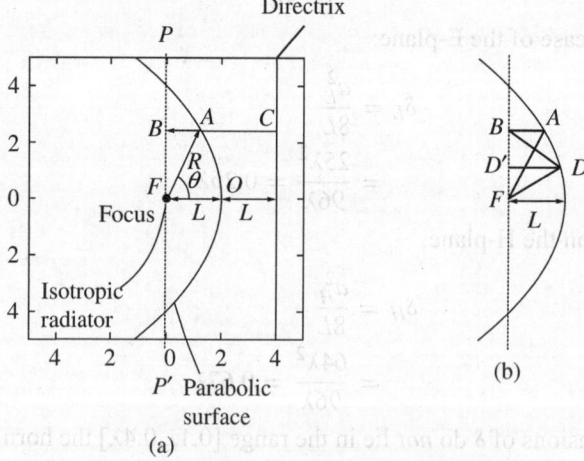

Fig. 14.16 Parabolic reflector

It is well known that the parabola is the locus of a point which is equidistant from the focus and a line. Which means that $FA = AC$ and distance $FAB = CAB$. Hence all rays 'seem' to come from a plane, which is the directrix. So on exiting the parabola we get a parallel beam.

We can apply Fermat's principle[2] and get the same result by referring to Fig. 14.16(b). If a ray were to start from F and reach D and then after reflection reach B^2 then from the property of the parabola

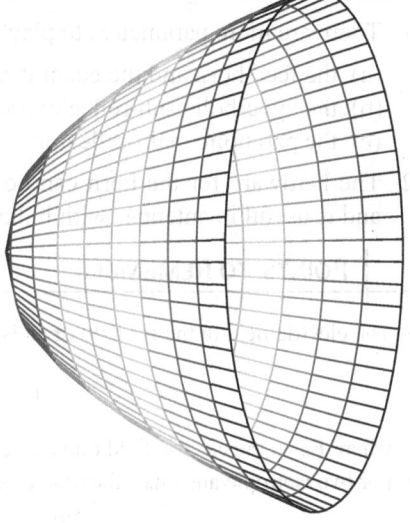

$$FAB = FDD' = 2L$$

and $$FDD' < FDB$$

∴ $$FAB < FDB$$

and since D can be any point on the parabola except A, therefore FAB is the shortest path from F to B.

Today the principle of the parabola is applied to a paraboloid of revolution and the parabolic antenna is most popular as a dish antenna to receive satellite television Direct To Home (DTH) signals. A paraboloid of revolution is shown in Fig. 14.17.

Fig. 14.17 Paraboloid of revolution

 See `ParabolaPlot.m` in Chapter 14

14.9 | Cassagrain Antenna

A Cassagrain antenna is basically a parabolic reflector of a large dimension, fed by a system of a feed horn and a hyperbolic sub-reflector. The arrangement which is used is depicted in Fig. 14.18. Referring to the figure, H is the feed horn which is optimally designed so that it radiates a narrow beam towards a hyperbolic sub-reflector. The horn is mounted in such a manner that the centre of the mouth of the horn is placed at O which is the real focus of the sub-reflector. Rays from the horn (OA and OA') reflect off the subreflector in such a way that when they are extended backward (AF and $A'F$) meet at the virtual focus of the sub-reflector *which is coincident with the focus of the parabolic main-reflector*. This behaviour is in keeping with the

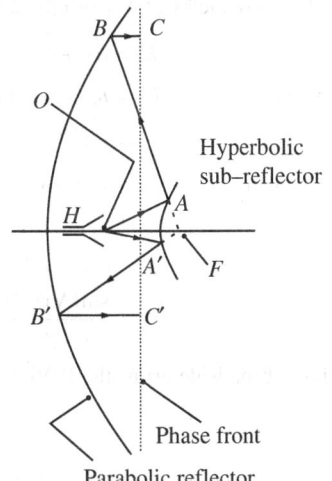

Fig. 14.18 A Cassagrain antenna

property of the hyperbolic surface whereby rays coming from the real focus appear to diverge from the virtual focus of the hyperbola. Now, proceeding on, since the parabolic dish sees the rays emanating from its own focus they are radiated parallel

---

[2]Fermat's principle says that if a ray of light is to go from point A to point B then it will take the shortest path (or least time) if there is no change in medium, otherwise it will take the least time.

to the axis as shown by rays $(F)\,ABC$ and $(F)\,A'B'C'$. The path lengths $OABC$ and $OA'B'C'$ are all identical so that the parabolic dish radiates a pencil beam.

The Cassagrain antenna is sometimes a better alternative to a simple horn-parabolic dish arrangement for two reasons:

1. There are more parameters to play with:
   (a) the feed horn and the beam it radiates
   (b) the hyperbolic sub-reflector and
   (c) the parabolic main-reflector
2. The hardware for the horn can be placed behind the parabolic main-reflector and is useful in antennas with bulky and complex feeding arrangements.

POINTS TO REMEMBER

- The electric field in the far field of an isotropic antenna is given by

$$\mathbf{E}(\mathbf{r}) = \mathbf{E}_{0t}\frac{e^{-jkr}}{r}$$

where $\mathbf{E}_{0t}$ is the electric field transverse to the direction of propagation.

- For two isotropic antennas, the total electric field in the far field is

$$\mathbf{E}_1(\mathbf{r}) = \hat{\mathbf{t}}I_1\frac{e^{-jk|\mathbf{r}|}}{|\mathbf{r}|}$$

$$\mathbf{E}_2(\mathbf{r}) = \hat{\mathbf{t}}I_2\frac{e^{-jk|\mathbf{r}-\mathbf{a}_xd|}}{|\mathbf{r}-d\mathbf{a}_x|}$$

$$\mathbf{E}_T(\mathbf{r}) = \mathbf{E}_1(\mathbf{r}) + \mathbf{E}_2(\mathbf{r}) = \hat{\mathbf{t}}\left(I_1\frac{e^{-jk|\mathbf{r}|}}{|\mathbf{r}|} + I_2\frac{e^{-jk|\mathbf{r}-\mathbf{a}_xd|}}{|\mathbf{r}-d\mathbf{a}_x|}\right)$$

- The array factor of N isotropic antennas with a uniform current distribution is

$$E = |I_1 + I_2e^{jkd\cos\phi} + I_3e^{j2kd\cos\phi} + \cdots + I_Ne^{j(N-1)kd\cos\phi}|$$

where
$$I_n = I_0\angle(n-1)\alpha, \quad n = 1, 2\ldots, N$$
$$= I_0e^{j(n-1)\alpha}$$

- The array factor for a uniform array is

$$E = \left|\frac{1-e^{jN\psi}}{1-e^{\psi}}\right| \quad \left(\text{Taking } \frac{e^{jN\psi/2}}{e^{j\psi/2}} \text{ out of the bracket}\right)$$

$$= \frac{\sin(N\psi/2)}{\sin(\psi/2)}$$

- For a broadside array, the BWFN is

$$2\Delta\phi_{\text{fn}} = \frac{2\lambda}{dN}$$

- For an end fire array, the BWFN is

$$2\Delta\phi_{\text{fn}} = 2\sqrt{\frac{2\lambda}{dN}}$$

- The beam pointing angle is controlled by the two equations:

$$kd\cos\phi_0 + \alpha = 0$$

$$\alpha = -kd\cos\phi_0$$

- From the currents in a region of space, the far field vector potential is given by

$$\mathbf{A} = \frac{\mu e^{-jkr}}{4\pi r} \iiint\limits_{V'} \mathbf{J}(x',y',z') \, e^{jk(x' \sin\theta\cos\phi + y' \sin\theta\sin\phi + z' \cos\theta)} dx'\,dy'\,dz'$$

SELF ASSESSMENT

Objective Type Questions

1. An antenna has a good bandwidth if
   (a) bandwidth has no physical meaning
   (b) the beam pointing angle and directivity is constant with frequency
   (c) the directivity is constant with current changes
   (d) none of the above
2. An antenna
   (a) is a reciprocal device   (b) has a Thevenin equivalent
   (c) may behave like a voltage source   (d) may behave like a current source
3. An antenna is said to be matched if
   (a) the source resistance is absent
   (b) the source resistance is much smaller than the antenna resistance
   (c) the source resistance is the complex conjugate of the antenna resistance
   (d) none of these
4. An antenna may radiate circularly polarised waves:
   (a) True   (b) Always false
5. A uniform linear array radiates in a
   (a) a $\sin(x)/x$ pattern   (b) a $\sin(Nx)/\sin(x)$ pattern
   (c) a $\sin(Nx/2)/\sin(x/2)$ pattern   (d) none of the above
6. For an isotropic radiator, placed at the origin of a coordinate system, the far fields
   (a) are proportional to $\exp\{k|\mathbf{r} - \mathbf{r}'|\}$ (b) are proportional to $\exp\{-jkr\}$
   (c) are inversely proportional to r (d) none of the above
7. For an isotropic radiator, placed at the $d\mathbf{a}_z$ of a coordinate system, the far fields
   (a) are proportional to $\exp\{k|\mathbf{r} - \mathbf{r}'|\}$
   (b) are proportional to $\exp\{-jk(r - d\cos\theta)\}$
   (c) are inversely proportional to $r - d\cos\theta$
   (d) none of the above
8. The phase of the currents, $\angle m\alpha$, $m = 0, 1, \ldots$, in a linear array
   (a) are to be neglected
   (b) determine the magnitude of the main beam
   (c) determine the pointing angle of the main beam
   (d) determine the directivity of the main beam
9. In a uniform array, grating lobes appear when
   (a) $d \gg \lambda$ (b) $d \ll \lambda$ (c) $d = \lambda$ (d) none of the above
10. In a linear antenna with length L, the beamwidth
    (a) increases with L (b) decreases with L
    (c) has no connection with L (d) none of the above

Short-Answer Questions

1. For a linear array of N elements, find the directivity for a broadside array using BWFN.
2. Find the vector potential of the current distribution $\mathbf{J} = \mathbf{a}_z C\delta(\mathbf{r})$, where C is a constant.

3. Find the far field vector potential of the current distribution $\mathbf{J} = \mathbf{a}_z C(z)$ where $C(z) = C_0$, a constant for $-l/2 < z < l/2$ and zero elsewhere.

Review Questions

1. Why is it important to design the shape of a beam?
2. Explain why a horn antenna is an aperture antenna.
3. Show that if a linear, uniform array is aligned along the z direction, then the angle ϕ of the theory which has been outlined is actually the angle θ of spherical coordinates.
4. Why are the E- and H-planes called by these names?
5. Why does a parabolic reflector give us a pencil (narrow) beam?

Numerical Problems

1. For a circularly symmetric beam field pattern which is a function of θ alone (spherical coordinates), find HPBW in the case of

$$E(\theta) = Ae^{-\theta^2}, \quad \text{for } 0 < \theta < \pi, \ 0 < \phi < 2\pi$$

and also find the approximate directivity from the pattern.

 See `PlotForTheta.m` in Chapter 14

2. For a circularly symmetric beam pattern which is a function of θ alone, find the BWFN for

$$f(\theta) = \frac{\sin \theta}{\theta}, \quad \text{for } 0 < \theta < \pi, \ 0 < \phi < 2\pi$$

and compute the *main beam area* given by

$$\Omega = \iiint f(\theta) \, d\Omega$$

approximately.
Hint: Do a numerical integration.

3. What should be the number of elements in a linear array to have a broadside BWFN equal to $10°$ and $d = \lambda/2$?
4. Find the value of the first sidelobe level for $N = 10$ elements for a uniform array.
5. What should be the number of elements in a linear array to have a endfire BWFN equal to $10°$ and $d = \lambda/2$?
6. Obtain the array factor for a linear array of five elements with equal spacing d, the excitation currents are according to $I_n = {}^5C_n$ for $(n = 0, 1, 2, 3, 4, 5)$. Plot the array factor $E(\psi)$.
7. Derive the condition on spacing between the elements of an antenna array such that in the region $(0° < \theta < 180°)$ there is only one main lobe.
8. Find the radiation pattern and plot the array factor, $E(\psi)$, if the excitation for a 4 linear element antenna array are $+1, +1, -1,$ and -1, respectively.
[**Note:** This is a difference pattern]
9. Plot the array factor, $E(\psi)$ for $\psi = kd\cos\phi$, as a function of frequency, $kd = \pi, 1.5\pi, 2\pi$, for a fixed linear antenna array. (Frequency dependence of antenna pattern).
10. Show that the array factor for a linear continuous array is the Fourier transform of the excitation current.
11. Find the far field distribution of the current distribution given in Fig. 14.19 ($\mathbf{I} = \mathbf{a}_z \Delta(z)$) where $\Delta(z)$ is the distribution shown.

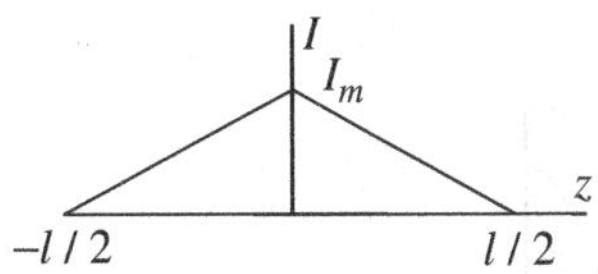

Fig. 14.19

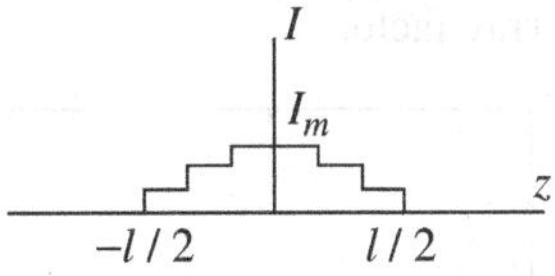

Fig. 14.20

12. Find the far field distribution of the current distribution given in Fig. 14.20 ($I = a_z S(z)$) where $S(z)$ is the distribution shown.
13. What is the approximate maximum power gain of an optimum horn antenna with a rectangular aperture with 7λ in the E-plane and 9λ in the H-plane?
14. (a) Assuming uniform fields in the mouth of the aperture, plot the E-plane pattern of the horn in Problem 13.
    (b) What is the half power beam width and the angle between first nulls of the E-plane pattern?
15. An optimum pyramidal horn operates at 10 GHz with the requirement that the horn has a directivity 20 dBi. Calculate the required aperture area of the horn.
16. An optimum conical horn operates at 10 GHz with the requirement that the horn has a directivity 20 dBi. Calculate the required aperture area of the horn.
17. An optimum pyramidal horn operates at a frequency of f GHz with the requirement that the horn has a directivity D dBi. Calculate the required aperture area of the horn.
18. A conical horn has a length of 10 λ, assuming $\delta = 0.25 \lambda$ what is the required flare angle? Apply your design to a frequency of 30 GHz with a 12dBi gain.
19. If pyramidal horn is to be designed to operate at 10 GHz, with $a_H = 5\lambda$ and directivity $D = 20$ dBi find the required value of a_E. What aperture efficiency will you use?
20. For Problem 19, what should be the length of the horn so that $\delta_{H-\text{plane}} = 0.4\lambda$?
21. For Problem 19 and using the result of Problem 20, what is the value of $\delta_{E-\text{plane}}$?
22. For Problem 19, what are the HPBWs in the E- and H-planes?

Answers

Objective Type Questions

1. (b) and (c) to some extent 2. All 3. (c) 4. (a) 5. (d)
6. (b) and (c) 7. (b) 8. (c) and (d) 9. (a) and (d) 10. (b)

Short-Answer Questions

1. $D \approx 2dN/\lambda$
2. $A = \dfrac{\mu e^{-jkr}}{4\pi r} a_z$
3. See Example 14.8.

Numerical Problems

1. HPBW $= 1.1774^r$
   $$D = \frac{4\pi}{(1.1774)^2}$$
2. BWFN $= 2\pi$; $\Omega = 2\pi \times 1.2188$
3. 23 elements
4. -13.14 dB down from the main lobe
5. 526 elements

6. Plot of array factor

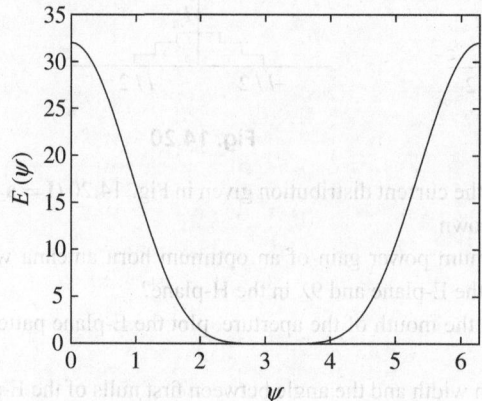

7. $d = \lambda / (1 + |\cos\theta_0|)$
8. Plot of array factor

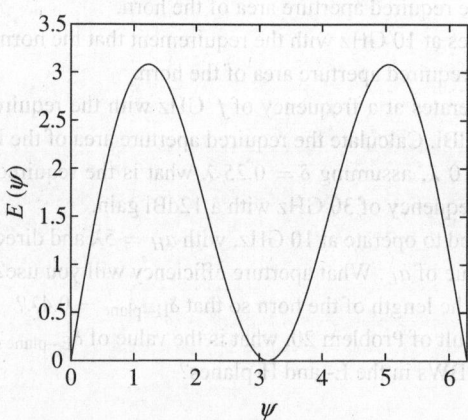

11. See Chapter 13, Numerical Problem 8
12. See Example 14.8 and use the principle of superposition for the currents
13. $D \approx 472$
14. (a) See Example 14.9.
    (b) HPBW: $\approx 8°$; BWFN: $16.4°$.
15. $A_p \approx 120$ cm^2
16. $A_p \approx 120$ cm^2
17. $A_p(\text{cm}^2) \approx \left(\dfrac{30}{f}\right)^2 \dfrac{10^{D/10}}{7.5}$
18. $\theta = 25.36°$; $r = 2.25\lambda$; at 30 GHz, $L = 10$ cm, $r = 2.25$ cm.
19. $a_E = 2.67\lambda$; $\varepsilon_{\text{app}} \approx 0.6$.
20. $L = 7.81\lambda$
21. $\delta_{\text{E-plane}} = 0.114\lambda$.
22. HPBW(E) $\approx 20.9°$; HPBW(H) $\approx 13.4°$

15

Radio Wave Propagation

Education is a progressive discovery of our ignorance.
— Will Durant

CHAPTER OBJECTIVES

To enable the students to understand the following:

- Ground wave propogation
- Wave tilt of the surface wave
- Tropospheric propagation and the bending of rays due to variation of the refractive index with height.
- The physical and radio horizon
- Duct propagation
- Ionospheric propagation

15.1 | Introduction

In 1860s, while predicting wave motion, Maxwell (Maxwell's equations) proposed that the electromagnetic waves may be made to propagate over great distances. This prediction was verified experimentally by Heinrich Hertz in the 1880s using frequencies in the UHF range (for explanation of UHF, see Fig. 11.3), distances which Hertz used were of the order of a few feet.

G. Marconi (Guglielmo Marconi (1874–1937), Italian inventor, shared the 1909 Nobel Prize in Physics with Karl Ferdinand Braun in recognition of their contributions to the development of wireless telegraphy), in another set of experiments he succeeded in his first trans-Atlantic communication using the Morse code in 1901. He used waves of about $\lambda = 300$ m in length, or the MF band as is now designated. At that time there was scepticism about this achievement as there was no theoretical understanding of the phenomenon of long distance communication. Today we understand this achievement in terms of ionospheric wave propagation. Just as in the case of Marconi, the theory of radio propagation sometimes offers no proper explanation of experiments, and even today there are many phenomena which are not well understood or predicted by theory.

Power transmitted through electromagnetic waves by a transmitting antenna on the surface of the earth is quite common. For example, a radio or television station may transmit radio signals to homes nearby. Each radio or television station will have its own frequency band tailored to the requirements of the bandwidth of the baseband signal. Another application would be the transmission of telephonic signals over large distances where copper wire may have not been laid, as in the case of transmission over hilly regions and so on. Some other applications are

- long distance point to point communications
- radar
- radio and television broadcasting
- navigational aids

Propagation characteristics which include parameters and methods have the potential of influencing communication systems design. They therefore merit attention and are considered by the communication systems and propagation specialists.

Once transmission takes place from an antenna, transmitted power may reach the receiving antenna in many ways. Typically these may be divided into (refer Fig. 15.1):

1. **Ground wave propagation**

   (a) **Line of sight** When the receiver and transmitter are within line-of-sight (LOS) of each other, the signal is assumed to travel by 'straight line paths' between the Tx and Rx, which is similar to the concept of waves travelling by rays. The direct path, is the one shown in Fig. 15.2 and is indicated by p_d. This method is the one used by ground to satellite communication links since the two are in a direct path. The wave which travels by LOS include what is known as sky waves.

   (b) **Ground reflection** Along with LOS communication, signals travel from the transmitter to the receiver by reflection from the ground and may modify the

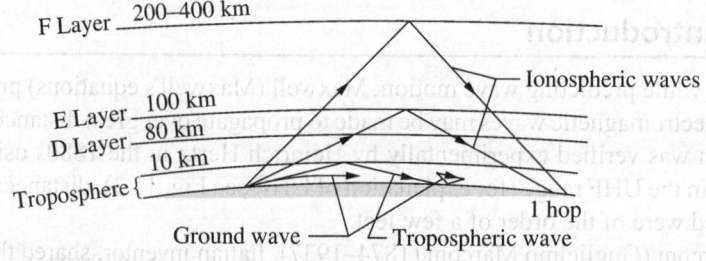

Fig. 15.1 Radio wave propagation paths over the earth

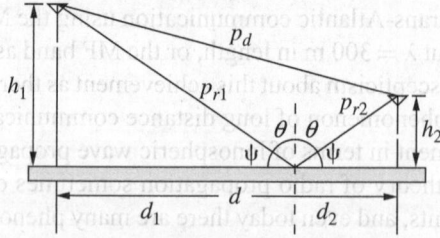

Fig. 15.2 Two antennas of heights h_1 and h_2 in communication

signal which is received from the direct path. For example if the difference in path length plus the change of phase in reflection is half a wavelength, the signals cancel completely. Therefore, the direct and reflected signals (p_d and p_r) must be considered in calculating link performance.

(c) When (i) the frequency is low (of the order of 10–1000 kHz) and (ii) when antennas operate near the ground, the direct (through p_d) and reflected wave (through p_r) cancel almost completely (the reader will understand this point when we consider Example 15.1). In such a case, communication at low frequencies takes place by what is known as a *surface wave*. The signal is guided by the ground itself. Since λ is large at low frequencies (such as at MF) transmitting antennas are big structures erected on the ground or they may be slightly raised, the surface wave is the dominant mechanism. The surface wave is the chief method of communication for AM transmission.

2. **Tropospheric Communication**

(a) **Tropospheric refraction** As we go vertically up from the ground, we find that atmosphere becomes more rarefied than that at sea level. The atmosphere is more dense at lower altitudes than at higher ones, the pressure falls exponentially and the refractive index of air also falls. This effect causes a bending of radio waves when waves travel horizontally in the troposphere. This is known as tropospheric refraction.

(b) **Tropospheric scatter** It takes place when sufficiently high power signals in the proper frequency range are beamed upward at the proper angle at this region of the Earth's atmosphere. The signal is incident on the troposphere and a small part of it is reflected or scattered back to the Tx. The region of the troposphere must be within the LOS of both Tx and Rx (See Fig. 15.1).

3. **Ionospheric wave propagation** In it the waves are reflected back from the ionosphere which lies from 100–400 km above the earth's surface. This effect occurs in the MHz range (typically 10–30 MHz or short wave). Reflection takes place in various regions (called D, E, and F layers as shown in Fig. 15.1) of the ionosphere and at varying heights.

Each of these modes of propagation have their own properties and frequency band requirements.

15.2 | Ground Wave Propagation

Ground wave propagation (or propagation close to the ground) consists of three modes of communication:

1. The direct path between the two antennas.
2. A reflection from the surface of the earth going from the transmitting antenna to the receiving antenna.
3. Surface wave communication between the two.

The electric field at the receiving antenna consists of the sum of three waves (terms):

$$\mathbf{E}_R = \mathbf{E}_D + \mathbf{E}_R + \mathbf{E}_S \qquad (15.1)$$

where $\mathbf{E}_R$ is the electric field at the receiving antenna; $\mathbf{E}_D$ is the electric field of the direct wave; $\mathbf{E}_R$ is the electric field of the reflected wave and $\mathbf{E}_S$ is the electric field of the surface wave. At low frequencies the electric fields due to the direct and reflected waves cancel out and communication is due to the surface wave alone.

Example 15.1 Consider two antennas (both omnidirectional, for the sake of simplicity) separated by a distance of 10 km, both (h) 10 m above the ground, operating at 100 kHz. Find the resultant electric field due to the direct and reflected waves. Assume a reflection coefficient R of -1.

Solution

Step 1. The direct path length p_d is 10 km $= 10^4$ m (see Fig. 15.2).

Step 2. The reflected path length, $p_r = p_{r1} + p_{r2}$ is given by

$$2 \times \sqrt{p_d^2/4 + h^2} = 10000.02 \text{ m}$$

Step 3. The phase difference $\Delta\phi$ is $(2\pi/\lambda) \times (p_d - p_r)$

$$\Delta\phi = -0.02 \times \left(\frac{2\pi}{\lambda}\right) \text{ rad}$$

Step 4. Since $\lambda = 3 \times 10^8/10^5 = 3000$ m, $\Delta\phi = 4.2 \times 10^{-5}$ rad. The phasors at the receiving antennas add as

$$\mathbf{E}_D + \mathbf{E}_R \approx \frac{E_0}{p_d^2}(e^{j\theta} + Re^{j\theta+\Delta\phi})$$

$$= \frac{E_0}{p_d^2}(e^{j\theta} - e^{j\theta+\Delta\phi}) \quad (R = -1)$$

$$= \frac{E_0}{p_d^2}e^{j(\theta+\Delta\phi/2)}(e^{-j\Delta\phi/2} - e^{j\Delta\phi/2})$$

$$= -\frac{E_0}{p_d^2}2je^{j(\theta+\Delta\phi/2)}\sin\Delta\phi/2$$

$$\approx 0$$

We will now deal with the direct path communication, combined with the ground reflection aspect of communication.

15.3 | Ground Reflection

Smooth ground behaves like a imperfectly conducting dielectric with a dielectric constant ε_r and a conductivity σ. Many factors determine the amount of reflection from the ground. These are (a) the roughness of the ground and (b) ε_r, σ, and ω (the frequency of operation) all taken together. (c) The polarisation of the wave: whether the $\mathbf{E}$ field is perpendicular or parallel to the plane of incidence.

To consider the first factor, (a), given above, we simply consider the roughness of the ground. The roughness factor R (also called the Rayleigh roughness criterion) is given by

$$R = 4\pi \underbrace{\sin\psi}_{A} \underbrace{\left(\frac{\sigma}{\lambda}\right)}_{B} \tag{15.2}$$

where R is the 'roughness' parameter, θ is the angle of incidence measured from the grazing angle ψ (shown in Fig. 15.2), λ is the wavelength and σ (not to be confused with the conductivity) is the standard deviation of the surface irregularities.

For $R < 0.1$, we may assume the surface to be 'smooth', while for $R > 10$ the surface may be considered rough. When the surface is smooth we can consider the wave to be reflected in accordance with the laws of EM-waves. While if the surface is rough, we may treat the reflected wave to be absent or negligible.

In Eqn (15.2), the A factor may be small if the distance between the transmitting and receiving antenna is large as compared with the height of the antenna. For example, if each antenna is at a height of 10 m and the distance between the two is 1 km, then $\sin\psi = 0.01$.

The B factor will be dependent on the soil condition: if the earth is 'smooth' and strewn with pebbles a few centimetres in size, then the ground may be considered smooth for frequencies from the kHz (for 3 kHz, $\lambda = 100$ km) range up to a few GHz (for 3 GHz, $\lambda = 10$ cm).

We now take into consideration the electrical parameters of the earth, σ and ε to formulate Maxwell's equations. If we refer to Eqns (11.41) and (11.43), we realise that the ground may be modelled as a lossy dielectric as considered in Section 11.3.

Writing out Maxwell's equations with a consideration of σ incorporated into the equations is

$$\nabla \times \mathbf{H} = j\omega\varepsilon\,\mathbf{E} + \mathbf{J} = j\omega\varepsilon\,\mathbf{E} + \sigma\,\mathbf{E}$$

$$= j\omega\left(\varepsilon - \frac{j\sigma}{\omega}\right)\mathbf{E}$$

$$= j\omega\varepsilon\left(1 - \frac{j\sigma}{\omega\varepsilon}\right)\mathbf{E}$$

$$= j\omega\varepsilon_C\mathbf{E} \qquad (15.3)$$

where $\varepsilon_C = (\varepsilon - j\sigma/\omega)$ is the complex permittivity, the other Maxwell's equations remaining the same as before.

For the case of reflection from the ground when the wave is obliquely incident upon it, we refer to Fig. 15.3. The wave is incident in two configurations (a) with the electric field perpendicular to the plane of incidence, and (b) with the electric field parallel to the plane of incidence. Both types of waves are shown in the figure. When the wave is incident on the dielectric interface, the angle it makes with the normal is θ_i and the angle it makes with the ground is ψ. In all cases, the direction of the vector $\mathbf{E} \times \mathbf{H}$ must be in the direction of wave propagation. After the wave front meets the ground, a part of it is reflected with $\theta_r = \theta_i$ and a part is transmitted into the earth. If $\mathbf{E}_i$ is the incident electric field and $\mathbf{E}_r$ is the reflected electric field then

$$\mathbf{E}_r = R\mathbf{E}_i \qquad (15.4)$$

where R is the reflection coefficient, which, in general, is complex.

15.3.1 Perpendicular Polarisation

Considering the ground as a dielectric which has a complex dielectric constant, the reflection coefficient for the perpendicular polarisation is given by

$$R_\perp = \frac{\cos\theta_i - \sqrt{(\varepsilon_2/\varepsilon_1) - \sin^2\theta_i}}{\cos\theta_i + \sqrt{(\varepsilon_2/\varepsilon_1) - \sin^2\theta_i}} = \frac{\sin\psi - \sqrt{(\varepsilon_2/\varepsilon_1) - \cos^2\psi}}{\sin\psi + \sqrt{(\varepsilon_2/\varepsilon_1) - \cos^2\psi}} \qquad (15.5)$$

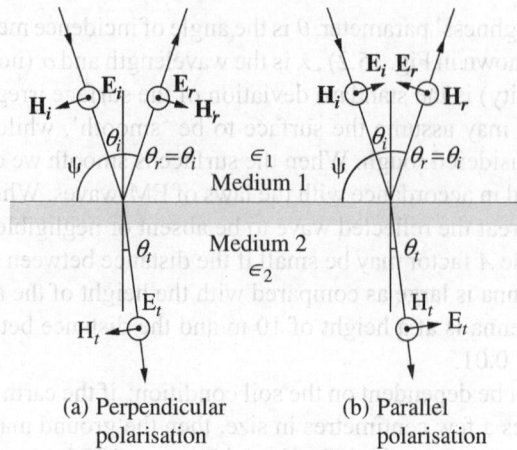

(a) Perpendicular
polarisation

(b) Parallel
polarisation

Fig. 15.3 A plane wave obliquely incident on a dielectric interface

when medium 1 is air and medium 2 is the ground, then

$$\varepsilon_2 = \varepsilon_C = \varepsilon_0\,\varepsilon_r - j(\sigma/\omega)$$

$$\varepsilon_1 = \varepsilon_0 \tag{15.6}$$

where ε_0, ε_r, and σ are the permittivity of vacuum, the relative dielectric constant, and conductivity of the ground respectively. Then

$$\varepsilon_2/\varepsilon_1 = \varepsilon_r - j(\sigma/\varepsilon_0\,\omega)$$

if we write $\varepsilon_0 \approx (1/36\pi) \times 10^{-9}$ and specifying $f = 2\pi\omega$ then $\omega\varepsilon_0 \approx (1/18) f_{MHz} \times 10^{-3}$ where f_{MHz} is the frequency in MHz. Further following the notation used in Jordan & Balmain (1968) we use $x = \sigma/\varepsilon_0\,\omega = 18\sigma \times 10^3/f_{MHz}$. Using this notation

$$R_\perp = \frac{\sin\psi - \sqrt{(\varepsilon_r - jx) - \cos^2\psi}}{\sin\psi + \sqrt{(\varepsilon_r - jx) - \cos^2\psi}}$$

In the case of the ground ε_r vary from 4–50 (see Table 15.1) but a typical value is that of $\varepsilon_r = 15$. Conductivities of the ground vary from 1×10^{-3} S/m to about 30×10^{-3} S/m but a typical value would be about $\sigma = 12 \times 10^{-3}$ S/m.

In Fig. 15.4, we plot the reflection coefficient as a function of the grazing angle of incidence, ψ. Referring to figure, we can observe for low values of ψ (which is the case for long distance communication), the values of the reflection coefficient is close to $1\angle 180°$ for all frequencies. Further graphs are shown in Fig. 15.5. In these graphs the value of $R_\perp$ are shown as a function of frequency but with ψ as a parameter. These graphs show that the reflection coefficient for small values of ψ, never departs much from the value of $1\angle 180°$. For example, the range of $|R_\perp|$ for $0 < \psi < 5°$ is [0.95, 1], and the range of $\angle R_\perp$ for $0 < \psi < 5°$ is [179°, 180°]. For small values of ψ

$$R_\perp \approx -1 - \frac{2j\,\psi}{\sqrt{jx - \varepsilon_r + 1}} \tag{15.7}$$

Example 15.2 Find the magnitude and phase of the reflection coefficient when a plane wave approaches the surface of the earth when the angle of grazing incidence ψ is 10° and the electric field is perpendicular to the plane of incidence. The frequency is 4 MHz. Let $\varepsilon_r = 15$; $\sigma = 12 \times 10^{-3}$ for the ground.

Table 15.1 Typical relative dielectric constants of various geological materials comprising the surface of the earth Davis & Annan (1980)

Material	ε_r
Air	1
Dry sand	3–5
Dry silt	3–30
Asphalt	3–5
Clay	5–40
Concrete	6
Saturated silt	10–40
Dry sandy coastal land	10
Average organic-rich surface soil	12
Marsh or forested land	12
Organic rich agricultural land	15
Saturated sand	20–30
Fresh water	80
Sea water	81–88

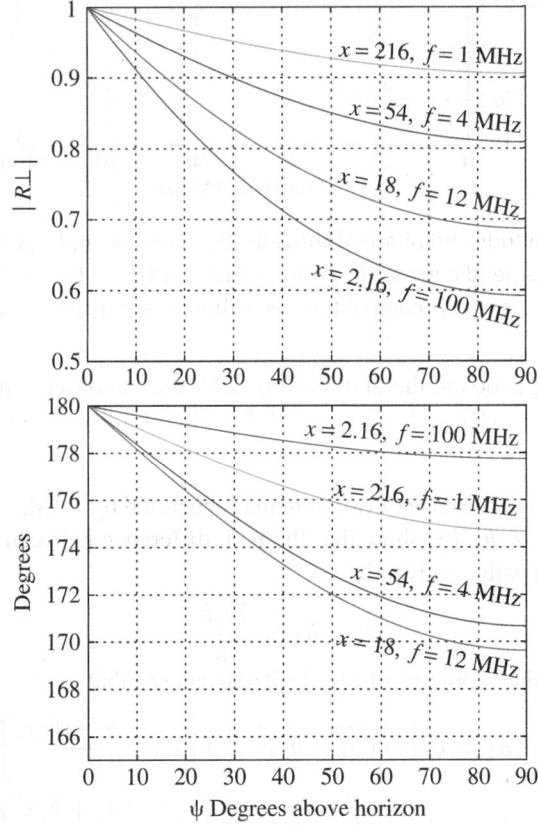

Fig. 15.4 Magnitude and phase of the reflection coefficient, $R_\perp$, of a plane wave whose E-field is perpendicular to the plane of incidence. $\varepsilon_r = 15$; $\sigma = 12 \times 10^{-3}$; $x = 18 \times 10^3 \sigma / f_{MHz}$. ψ is the approach of the wave above the horizon

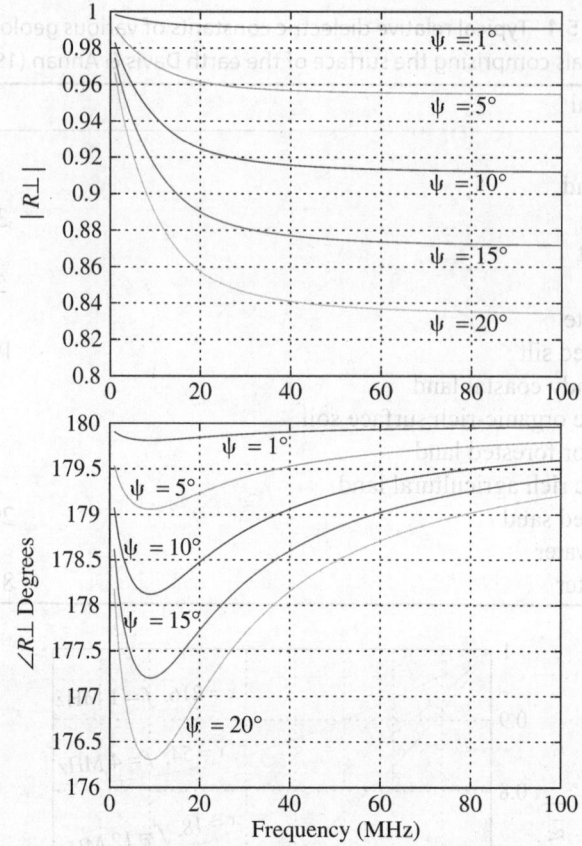

Fig. 15.5 Magnitude and phase of the reflection coefficient, $R_\perp$, of a plane wave whose E-field is perpendicular to the plane of incidence. $\varepsilon_r = 15$; $\sigma = 12 \times 10^{-3}$. ψ is the approach of the wave above the horizon

Solution Taking a look at the graph of Fig. 15.4 we can observe that $f = 4$ MHz curves that $|R_\perp| \cong 0.97$ and $\angle R_\perp \cong 178.5°$. Therefore, we have almost perfect reflection.

Example 15.3 For the case of two antennas separated by a distance d, at heights h_1 and h_2 with $d \gg h_1, h_2$, show that the path difference between the direct (R_1) and reflected (R_2) paths is given by

$$R_2 - R_1 \approx \frac{2h_1 h_2}{d}$$

Solution From the geometry of Fig. 15.2, we can see that

$$R_1 = \sqrt{d^2 + (h_1 - h_2)^2} \approx d\left[1 + \frac{(h_1 - h_2)^2}{2d^2}\right]$$

$$R_2 = \sqrt{d^2 + (h_1 + h_2)^2} \approx d\left[1 + \frac{(h_1 + h_2)^2}{2d^2}\right]$$

$$R_2 - R_1 \approx \frac{(h_1 + h_2)^2}{2d} - \frac{(h_1 - h_2)^2}{2d} = \frac{2h_1 h_2}{d}$$

where we have used the expansion

$$\sqrt{1+x} \approx 1 + \frac{x}{2}$$

when $x \ll 1$.

15.3.2 Parallel Polarisation

As earlier, if the ground is considered to be a dielectric which has a complex dielectric constant, the reflection coefficient for the parallel polarisation is given by

$$
\begin{aligned}
R_{\|} &= \frac{(\varepsilon_2/\varepsilon_1)\cos\theta_i - \sqrt{(\varepsilon_2/\varepsilon_1) - \sin^2\theta_i}}{(\varepsilon_2/\varepsilon_1)\cos\theta_i + \sqrt{(\varepsilon_2/\varepsilon_1) - \sin^2\theta_i}} \\
&= \frac{(\varepsilon_2/\varepsilon_1)\sin\psi - \sqrt{(\varepsilon_2/\varepsilon_1) - \cos^2\psi}}{(\varepsilon_2/\varepsilon_1)\sin\psi + \sqrt{(\varepsilon_2/\varepsilon_1) - \cos^2\psi}} \\
&= \frac{(\varepsilon_r - jx)\sin\psi - \sqrt{(\varepsilon_r - jx) - \cos^2\psi}}{(\varepsilon_r - jx)\sin\psi + \sqrt{(\varepsilon_r - jx) - \cos^2\psi}}
\end{aligned}
\qquad (15.8)
$$

where $x = \sigma/\varepsilon_0\,\omega$, and where ε_0, ε_r, and σ are the permittivity of vacuum, the relative dielectric constant, and conductivity of the ground respectively.

In Fig. 15.6, we again plot the reflection coefficient ($R_{\|}$) as a function of the grazing angle of incidence, ψ. Referring to Fig. 15.6 we can observe for low values of

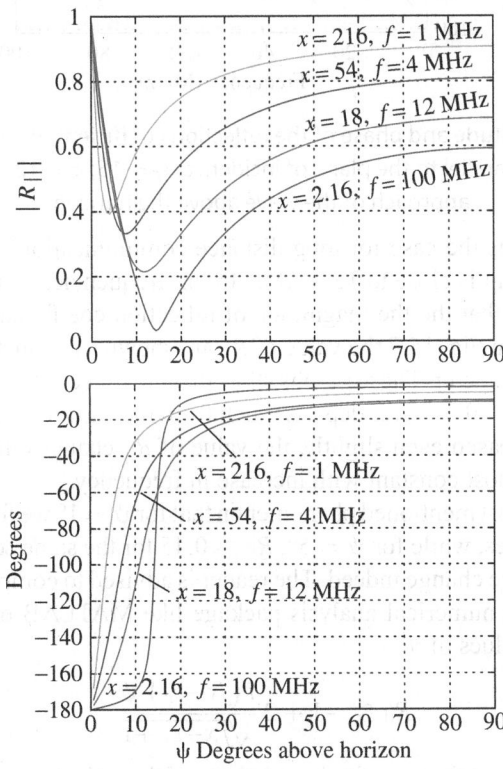

Fig. 15.6 Magnitude and phase of the reflection coefficient, $R_{\|}$, of a plane wave whose E-field is parallel to the plane of incidence. $\varepsilon_r = 15$; $\sigma = 12 \times 10^{-3}$; $x = 18 \times 10^3 \sigma/f_{MHz}$. ψ is the angle of grazing incidence of the wave above the horizon

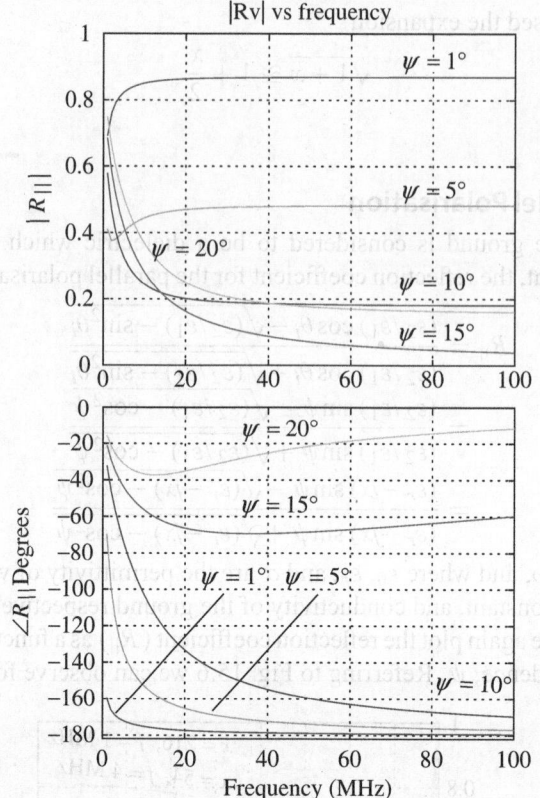

Fig. 15.7 Magnitude and phase of the reflection coefficient, $R_{||}$, of a plane wave whose E-field is parallel to the plane of incidence. $\varepsilon_r = 15$; $\sigma = 12 \times 10^{-3}$. ψ is the approach of the wave above the horizon

$\psi \ (\approx 0°)$ (which is the case for long distance communication) the values of the reflection coefficient is close to $1\angle -180°$ for all frequencies. As the frequency is increased, we find that the the magnitude of reflection coefficient falls rapidly. In the graphs shown in Fig. 15.7 the value of $R_{||}$ are shown as a function of frequency but with ψ as a parameter. These graphs show that the reflection coefficient for very small values of $\psi \ (\approx 0°)$, never departs much from the value of $1\angle 180°$. But as the value of ψ is increased even slightly, the value of $R_{||}$ changes drastically, though then it remains almost constant with increase in frequency.

We can see this last mentioned phenomenon that for $\psi = 1°$ we find that $R_{||} \approx 0.85$ for large frequencies, while for $\psi = 5°$, $R_{||} \approx 0.45$ for the same set of frequencies, which is a very large change indeed. The reader is advised to compute the reflection coefficient using a numerical analysis package like MATLAB on a case by case basis. For small values of ψ,

$$R_{||} \approx -1 + 2\frac{j\left(jx - \varepsilon_r\right)\psi}{\sqrt{jx - \varepsilon_r + 1}} \tag{15.9}$$

Example 15.4 Find the magnitude and phase of the reflection coefficient when a plane wave approaches the surface of the earth when the angle of grazing incidence ψ is 1° and when the electric field is parallel to the plane of incidence. The frequency is 40 MHz. Let $\varepsilon_r = 15$; $\sigma = 12 \times 10^{-3}$ S/m for the ground.

Solution From Fig. 15.7 we can see that $|R_{||}| \approx 0.86$ and $\angle R_{||} \approx -180°$.

Practice Problem 15.1
Show that a dipole held vertically above the ground produces a parallelly polarised wave while a dipole held horizontally above the ground produces a perpendicularly polarised wave.

15.4 | Surface Wave

Apart from the two modes of propagation (direct and reflected) we also have a third
mode of propagation, namely the surface wave which was first proposed by various
researchers Norton (1936, 1937), Sommerfeld (1909). The surface wave is guided
by the ground which acts like a dielectric medium being diffracted over the surface
of the earth, which acts as a boundary. These diffraction effects are greater as the
radiation frequency goes towards the lower frequencies (the kHz range and slightly
higher). The surface wave is such that most of the energy of the wave is away from
the ground.

15.4.1 Surface Wave for the Vertical Dipole

For a vertical dipole the surface wave radiation field when $R \gg \lambda$ is given by[1]

$$\mathbf{E}_{su} = j30kIdl(1 - R_{||}) F \left(\frac{e^{-jkR}}{R} \right)$$

$$\times \left\{ \mathbf{a}_k(1 - u^2) + \mathbf{a}_r \cos\psi' \left(1 + \frac{\sin^2\psi'}{2} \right) u\sqrt{1 - u^2 \cos^2\psi'} \right\} \quad (15.10)$$

where $\mathbf{E}_{su}$ is the electric field of the surface wave at the receiving point as per
Fig. 15.8; $k = 2\pi/\lambda$,

$$R_{||} = \frac{(\varepsilon_r - jx)\sin\psi' - \sqrt{(\varepsilon_r - jx) - \cos^2\psi'}}{(\varepsilon_r - jx)\sin\psi' + \sqrt{(\varepsilon_r - jx) - \cos^2\psi'}} \quad (15.11)$$

which has already been introduced;

$$u^2 = \frac{1}{(\varepsilon_r - jx)} \quad (15.12)$$

$$x = \frac{18 \times 10^3 \sigma}{f_{MHz}} \quad (15.13)$$

$$F = 1 - j\sqrt{\pi\omega}e^{-\omega}\text{erfc}(j\sqrt{\omega}) \quad (15.14)$$

$$\omega = -jkR\frac{u^2(1 - u^2\cos^2\psi')}{2}\left[1 + \frac{\sin\psi'}{u\sqrt{1 - u^2\cos^2\psi'}}\right]^2 \quad (15.15)$$

and

$$\text{erfc}(j\omega) = \frac{2}{\sqrt{\omega}} \int_{j\omega}^{\infty} e^{-v^2} dv \quad (15.16)$$

---

[1]The physical reality of space and surface waves in the radiation field of radio antennas, K A
Norton FCC Washington DC.

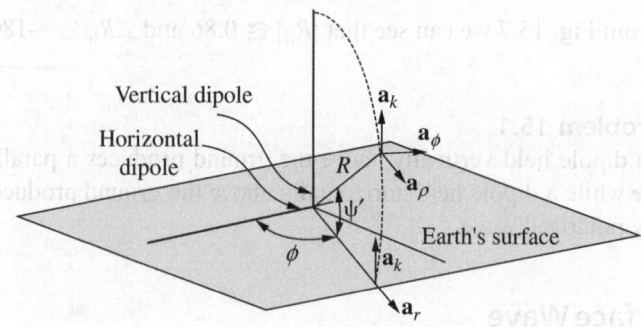

Fig. 15.8 Coordinate system for the surface wave

we can see from the formulae that the factor F is of great complexity and decays rapidly with increasing value of kR.

We also take a look at the surface wave as $\psi' \approx 0$ which is the case for long-distance communication. Then

$$\omega_{\psi'=0} = p_1 = -jkR\frac{u^2(1-u^2)}{2} = -\frac{jkR}{2}\left\{\frac{1}{\varepsilon_r-jx}\left(1-\frac{1}{\varepsilon_r-jx}\right)\right\} \quad (15.17)$$

$$= -j\left(\frac{R}{\lambda}\right)\underbrace{\left\{\frac{\pi}{\varepsilon_r-jx}\left(1-\frac{1}{\varepsilon_r-jx}\right)\right\}}_{\text{The factor } a} \quad (15.18)$$

$$= pe^{jb} \quad (15.19)$$

where p is known as the *numerical distance* and b is the *phase constant* and both are real and greater than zero. The factor

$$a = \frac{\pi}{\varepsilon_r-jx}\left(1-\frac{1}{\varepsilon_r-jx}\right) \quad (15.20)$$

$$p = \frac{R}{\lambda}|a| \quad (15.21)$$

$$b \approx \tan^{-1}\left(\frac{\varepsilon_r+1}{x}\right) \quad (15.22)$$

is the complex part. A plot of the magnitude of a and the approximate value of the phase b are shown in Fig. 15.9. A perusal of these graphs shows that the factor a has a magnitude which increases from 0 to a maximum of $(\pi/\varepsilon_r)(1-1/\varepsilon_r)$ as frequency is increased. The phase b starts at about $0°$ for $f \approx 0$ to $90°$ as $f \to \infty$.

If we consider the function $F_{\psi'=0}$, then

$$A = |F_{\psi'=0}| = |1 - j\sqrt{\pi p_1}e^{-p_1}\text{erfc}(j\sqrt{p_1})| \quad (15.23)$$

where A is called the *ground-wave attenuation factor*. Since in the calculation of the power carried by the surface only a good approximation is the requirement,

$$A \approx A_1 - \sin(b)\sqrt{\frac{p}{2}}e^{-(5/8)p} \quad (15.24)$$

where

$$A_1 = \frac{2+0.3p}{2+p+0.6p^2} \quad (15.25)$$

The approximate function A has been plotted in Fig. 15.10. The function is almost equal to 1 (no attenuation) for the range $0 < p < 0.1$. A lies in the range $0.1 < A < 1$

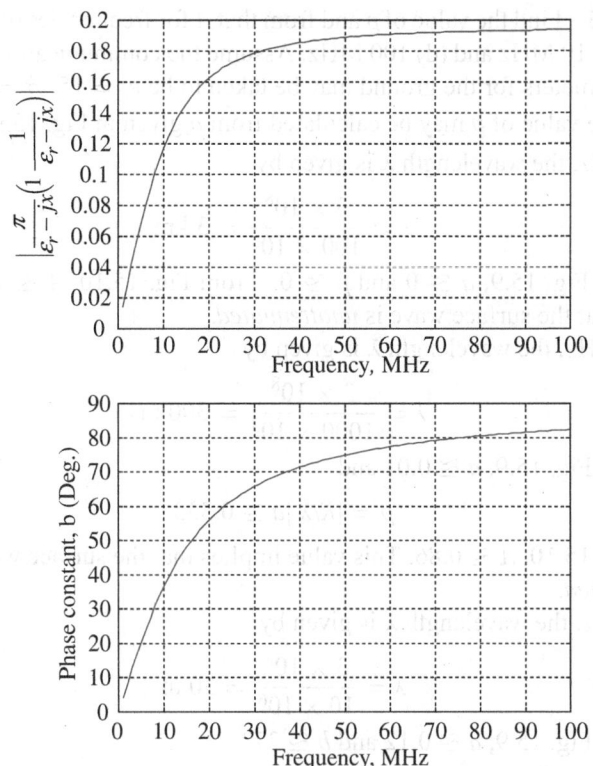

Fig. 15.9 (Upper graph) Magnitude of the factor a. $p = (R/\lambda)\, a$; (Lower graph) The phase constant b. For these graphs, $\varepsilon_r = 15$, $\sigma = 0.012$ S/m

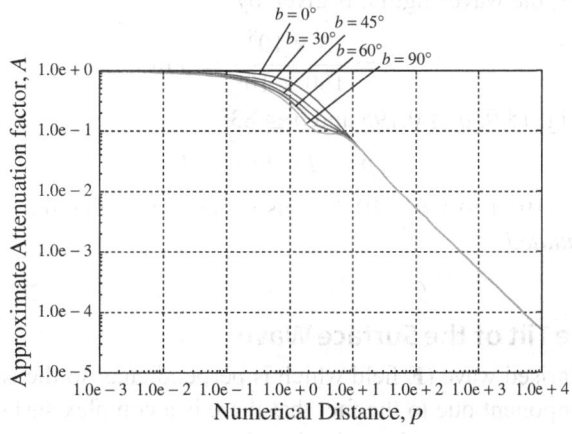

Fig. 15.10 Graph of the approximate value of the ground wave attenuation factor A versus the numerical distance p for various values of b

for $0.1 < p < 10$ for various values of b; and for greater values of p, A decreases linearly on the logarithmic scale. In the last range

$$A \approx \frac{1}{2p - 3.7} \tag{15.26}$$

Example 15.5 Find the value of p and from that A for frequencies of (a) 100 kHz, (b) 1 MHz, (c) 10 MHz and (d) 100 MHz. Assume the communication distance, R, is 10 km. Parameters for the ground may be taken to be $\varepsilon_r = 15$, $\sigma = 0.012$ S/m.

Solution The value of p may be calculated from a given in Fig. 15.9.

(a) At 100 kHz, the wavelength λ is given by

$$\lambda = \frac{3 \times 10^8}{100 \times 10^3} = 3 \text{ km}$$

then from Fig. 15.9, $a \approx 0$ and $p \approx 0$. From Fig. 15.10, $A \approx 1$. This value implies that the surface wave is *unattenuated*.

(b) At 1000 kHz, the wavelength λ is given by

$$\lambda = \frac{3 \times 10^8}{1000 \times 10^3} = 300 \text{ m}$$

then from Fig. 15.9, $a \approx 0.01$ and

$$p = |R/\lambda|a \approx 0.333$$

From Fig. 15.10, $A \approx 0.86$. This value implies that the surface wave is almost *unattenuated*.

(c) At 10 MHz, the wavelength λ is given by

$$\lambda = \frac{3 \times 10^8}{10 \times 10^6} = 30 \text{ m}$$

then from Fig. 15.9, $a \approx 0.12$ and $b \approx 37°$

$$p = |R/\lambda|a \approx 40$$

From Fig. 15.10, $A \approx 0.14$. This value implies that the surface wave is *attenuated*.

(d) At 100 MHz, the wavelength λ is given by

$$\lambda = \frac{3 \times 10^8}{100 \times 10^6} = 3 \text{ m}$$

then from Fig. 15.9, $a \approx 0.195$ and $b \approx 83°$

$$p = |R/\lambda|a \approx 650$$

From Fig. 15.10, $A \approx 7.8 \times 10^{-4}$. This value implies that the surface wave is *highly attenuated*.

15.4.2 Wave Tilt of the Surface Wave

A vertically polarised wave (**E** field which is perpendicular to the earth) develops a horizontal component due to the fact that there is a complex surface impedance of the earth. The surface impedance is given by

$$Z_s \approx \sqrt{\frac{\omega \mu_0}{\sqrt{\sigma^2 + \omega^2 \left(\varepsilon_r \varepsilon_0\right)^2}}} \angle \left\{ \frac{1}{2} \tan^{-1}\left(\frac{\sigma}{\omega \varepsilon_r \varepsilon_0}\right) \right\}$$

$$= \sqrt{\frac{\mu_0}{\varepsilon_0 \sqrt{\left(\frac{\sigma}{\omega \varepsilon_0}\right)^2 + \varepsilon_r^2}}} \angle \left\{ \frac{1}{2} \tan^{-1}\left(\frac{x}{\varepsilon_r}\right) \right\}$$

$$Z_s = Z_0 \sqrt{\frac{1}{\sqrt{x^2+\varepsilon_r^2}}} \angle \left\{ \frac{1}{2} \tan^{-1}\left(\frac{x}{\varepsilon_r}\right) \right\} \qquad (15.27)$$

the manner by which this happens is explained through Fig. 15.11. E_v is the electric field which is vertical and the associated magnetic field is

$$\mathbf{H}_v = \frac{E_v}{Z_0} = \frac{E_v}{377} \qquad (15.28)$$

Also a surface current (which is parallel to the ground) is formed due to the presence of this magnetic field

$$\mathbf{J}_s = \hat{\mathbf{n}} \times \mathbf{H}_v$$
$$J_s = H_v \qquad (15.29)$$

The surface current gives rise to a horizontal component of the electric field, $\mathbf{E}_h$ which is directed towards the direction of propagation

$$\mathbf{J}_s = \frac{\mathbf{E}_h}{Z_s} \qquad (15.30)$$

or taking magnitudes all throughout,

$$H_v = \frac{E_h}{Z_s} = \frac{E_v}{Z_0} \qquad (15.31)$$

$$\frac{E_h}{E_v} = \frac{Z_s}{Z_0} = \frac{1}{\sqrt[4]{x^2+\varepsilon_r^2}} \angle \left\{ \frac{1}{2} \tan^{-1}\left(\frac{x}{\varepsilon_r}\right) \right\} \qquad (15.32)$$

The results are shown in Fig. 15.12 where the exact polarisation ellipses are shown at three values of frequencies: 0.5, 10, and 100 MHz. For the three cases, if $E_v = \sin t$ then E_h is the function shown in the figure, namely, $E_h = 0.05 \sin(t + 44°)$, $0.2 \sin(t + 28°)$, and $0.26 \sin(t + 4°)$.

We notice from these functions that (a) the magnitude of E_h increases with increasing frequency and (b) the phase decreases with increasing frequency.

For a surface wave with a vertical electric field, a rough idea of the distance that is covered is approximately given by

$$R(\text{km}) \approx \frac{200}{\sqrt{f_{\text{MHz}}}}$$

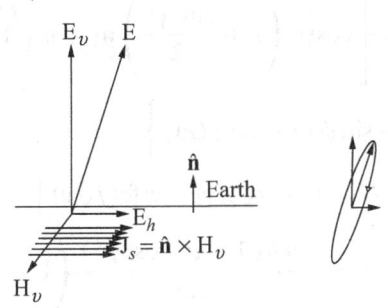

Fig. 15.11 Wave tilt for a surface wave

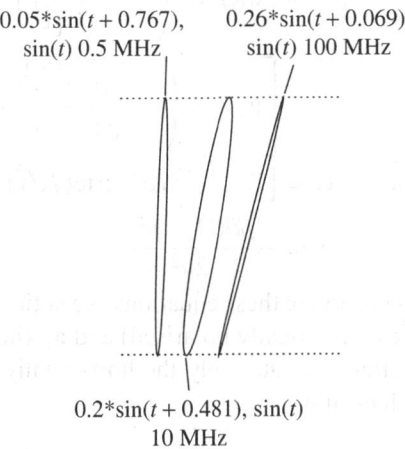

0.05*sin(t + 0.767), sin(t) 0.5 MHz

0.26*sin(t + 0.069), sin(t) 100 MHz

0.2*sin(t + 0.481), sin(t) 10 MHz

Fig. 15.12 Wave tilt for a surface wave—calculation for three frequencies: 0.5, 10, and 100 MHz. $\varepsilon_r = 15$, $\sigma = 0.012$

15.4.3 Surface Wave for a Horizontal Dipole

The surface wave for a horizontal dipole on the surface of the earth is given by (refer Fig. 15.8),

$$\mathbf{E}_{su} = j30kIdlF\left(\frac{e^{-jkR}}{R}\right)\left\{\left[(1-R_{||})\,Fu\cos\phi\,\sqrt{1-u^2\cos^2\psi'}\right]\right.$$

$$\times\left[\cos\psi'\left(1+\frac{\sin^2\psi'}{2}\right)\mathbf{a}_k + u\,\sqrt{1-u^2\cos^2\psi'}\left(\frac{1-\sin^4\psi' - \frac{(1-R_\perp)G}{(1-R_{||})u^2F}}{1-u^2\cos^2\psi'}\right)\mathbf{a}_\rho\right]$$

$$\left. + \sin\phi\,(1-R_\perp)\,G\mathbf{a}_\phi\right\} \tag{15.33}$$

where $G = \left[1 - j\sqrt{\pi v}\,e^{-v}\mathrm{erfc}(j\sqrt{v})\right]$ (15.34)

$$v = -\frac{jkR(1-u^2\cos^2\psi')}{2u^2}\left(1+\frac{u\sin\psi'}{\sqrt{1-u^2\cos^2\psi'}}\right)^2 \tag{15.35}$$

from these expressions, it is clear that for $\psi' = 0$ the electric field becomes

$$\mathbf{E}_{su} = j30kIdlF\left(\frac{e^{-jkR}}{R}\right)\left\{\left[(1-R_{||})\,Fu\cos\phi\,\sqrt{1-u^2}\right]\times\right.$$

$$\left[\mathbf{a}_k + u\left(\frac{1-\frac{(1-R_\perp)G}{(1-R_{||})u^2F}}{\sqrt{1-u^2}}\right)\mathbf{a}_\rho\right] + \sin\phi\,(1-R_\perp)\,G\mathbf{a}_\phi\right\} \tag{15.36}$$

and $\quad G = \left[1 - j\sqrt{\pi v}\,e^{-v}\mathrm{erfc}(j\sqrt{v})\right]$ (15.37)

$$v = -\frac{jkR(1-u^2)}{2u^2} \tag{15.38}$$

by observing these equations, we notice that in the direction $\phi = 0$ both components the $\mathbf{a}_k$ (vertically polarised) and $\mathbf{a}_\rho$ (horizontally polarised) are present. In the direction $\phi = 90°$ only the horizontally polarised component is present. For large values of p

$$G \cong u^4 F \tag{15.39}$$

and the wave attenuates much more rapidly.

To compute the attenuation at a distance R from the source, the same factor A of Eqn (15.23) is used with

$$p = \frac{\pi R}{\lambda}\frac{x}{\cos b'} \tag{15.40}$$

$$b = 180° - b' \tag{15.41}$$

$$b' = \tan^{-1}\frac{\varepsilon_r - 1}{x} \tag{15.42}$$

and $x = 18\times10^3\sigma/f_{\mathrm{MHz}}$. Computations and experimental results show that the surface wave contributes very slightly to communication using horizontally placed dipoles and instead elevated antennas with considerations of the direct and reflected waves are more suitable.

15.5 | Approximations for Ground Wave Propagation

Consider two vertical dipole antennas elevated above the surface of the earth at heights of h_1 and h_2 (Fig. 15.2); the *frequency which is used is in the high and very high frequency range* (HF and VHF). Then the vertical component of the electric field at the receiving antenna is

$$E_z = j30kIl_{eff}\left\{\cos^2\psi\left[\underbrace{\frac{e^{-jkR_1}}{R_1} + R_{||}\frac{e^{-jkR_2}}{R_2}}_{I} + \underbrace{\frac{(1 - R_{||})\,F\frac{e^{-jkR_2}}{R_2}}{II}}_{}\right]\right\}$$ (15.43)

where the first term is the space wave, and the second one is the surface wave. Here $R_1 = p_d$ and $R_2 = p_{r1} + p_{r2}$ of the figure. This expression is accurate for distances from the antenna larger than a few wavelengths. Here l_{eff} is the effective length of the dipole and I is the current in the dipole.

However since we need to make an approximate computation, we make the following approximations:

$$\cos\psi \approx 1 \tag{15.44}$$
$$R_1 = R_2 \approx d \quad (\text{for the denominator only}) \tag{15.45}$$

for large numerical distances

$$F \approx -1/2\omega \tag{15.46}$$

where

$$ud \gg h_1 + h_2 \tag{15.47}$$
$$|\omega| > 20 \tag{15.48}$$

With these approximations, the received wave for vertical dipoles is

$$E_z = \frac{j30kIl_{eff}}{d}\left\{e^{-jkR_1} + R_{||}e^{-jkR_2} - (1 - R_{||})\frac{e^{-jkR_2}}{2\omega}\right\} \tag{15.49}$$

for small values of ψ

$$R_{||} \approx -1 + 2\frac{j\left(jx - \varepsilon_r\right)\psi}{\sqrt{jx - \varepsilon_r + 1}}$$

also

$$R_1 = \sqrt{d^2 + (h_1 - h_2)^2} \approx d\left[1 + \frac{(h_1 - h_2)^2}{2d^2}\right]$$

$$R_2 = \sqrt{d^2 + (h_1 + h_2)^2} \approx d\left[1 + \frac{(h_1 + h_2)^2}{2d^2}\right]$$

$$R_2 - R_1 \approx \frac{(h_1 + h_2)^2}{2d} - \frac{(h_1 - h_2)^2}{2d} = \frac{2h_1 h_2}{d}$$

so Eqn (15.49) becomes

$$E_z = \frac{j30kIl_{eff}\,e^{-jkR_1}}{d}\left\{1 + R_{||}e^{-jk(R_2 - R_1)} - (1 - R_{||})\frac{e^{-jk(R_2 - R_1)}}{2\omega}\right\}$$

$$= \frac{j30kIl_{eff}\,e^{-jkR_1}}{d}\left\{1 + \underbrace{R_{||}}_{\approx -1}\,e^{-jk(2h_1 h_2/d)} - \underbrace{(1 - R_{||})\frac{e^{-jk(2h_1 h_2/d)}}{2\omega}}_{\text{neglected}}\right\}$$

We now use the following approximations:

$$R_{||} \approx -1$$

$$e^{-jk(2h_1 h_2/d)} = \cos(2kh_1 h_2/d) - j\sin(2kh_1 h_2/d)$$

$$\approx 1 - j\frac{2kh_1h_2}{d} \qquad (15.50)$$

which gives

$$|E_z| \approx \frac{60k^2 Il_{\text{eff}} h_1 h_2}{d^2} \qquad (15.51)$$

Practically the same result may be applied to horizontal dipoles.

15.6 | Tropospheric Propagation

15.6.1 Spherical Earth Considerations

Though we have considered ground wave propagation based on formulae of a surface and space wave, these ideas are limited by the fact that we did not take into account the curvature of the earth. The distances up to which the surface wave formulae developed earlier give fairly correct results is given by

$$d = 50/f_{\text{MHz}}^{1/3} \quad \text{miles} \qquad (15.52)$$

and the results of Fig. 15.10 may be applied only up to these distances.

The reason for an error in the results are due to the following. (a) These results start diverging from reality in that the surface wave reaches the receiver due to diffraction and refraction from the lower atmosphere rather than from the consideration

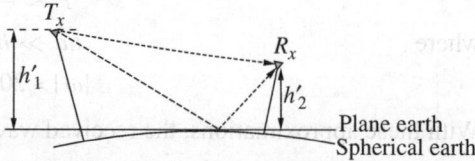

Fig. 15.13 Spherical and plane earth

that the earth is a flat plane. (b) The reflected part of the space wave is reflected from a curved surface rather than a plane one which causes the energy to be diverged as in the case of a convex mirror and which introduces inaccuracies in the final result; and finally, (c) For a spherical earth, heights h_1 and h_2 of the transmitting and receiving antennas are greater than the corresponding heights for a 'plane earth' (h_1' and h_2' as shown in Fig. 15.13).

Example 15.6 Find the distance in kilometers for a surface wave of 10 MHz may be used in communication.

Solution From Eqn (15.52)

$$d = \frac{1.6 \times 50}{\sqrt[3]{10}}$$

$$= 37.13 \text{ km}$$

15.6.2 Tropospheric Waves

The troposphere is that portion of the earth's atmosphere which is closest to the ground, extending up to about 10 miles (16 km). It is the region where clouds are formed and aircraft travel. The rate of change of temperature is about $-6.5°$/km till a minimum of $-55°$ to $-45°$C; the pressure too decreases as one goes up. Since there is a change in the constituents of the troposphere, wave propagation can be of several types: diffraction, refraction, abnormal reflection, and refraction.

As we go up, there is a variation in the physical parameters in the troposphere like temperature, pressure, and percentage of water vapour content. The dielectric constant (and therefore the refractive index, $n = \sqrt{\varepsilon_r}$) is a function of height. We know that the dielectric constant of outer space is one. From outer space as we come towards the surface of the earth and enter the troposphere, the dielectric constant increases very slightly. If we define a function, N, *the refractivity* to be

$$N = (n - 1) \, 10^6 \tag{15.53}$$

where n is the refractive index, then (see Jordan & Balmain, 1968) N is found to be

$$N = \frac{77.6}{T} p + 4810 \frac{e}{T} \tag{15.54}$$

Another somewhat more accurate formula is

$$N = \frac{77.6}{T} p + 3.73 \times 10^5 \frac{e}{T^2} \tag{15.55}$$

where p and e are the atmospheric pressure and the partial pressure of water vapour in millibars respectively, and T is the absolute temperature (degrees Kelvin). The nominal value of the pressure at the surface of the earth is about 1 bar or 1000 millibars and the partial pressure of water vapour is around 6–40 millibars (more details are available in Hall & Hewitt 1996). The atmospheric pressure p falls exponentially with height, falling to $1/e$ of the surface value at a height of about 8000 m. Or

$$p \cong 1000 e^{-(h/8000)} \tag{15.56}$$

And the temperature near the surface of the earth is given by

$$T \cong -6.5 \times 10^{-3} h + T_0$$

where T_0 is the temperature at the surface of the earth and h is in meters. The refractive index of the atmosphere has a value of approximately 1.0003 at the earth's surface.

Example 15.7   Find the refractive index of the atmosphere at a height of 1000 m at 27°C. Neglect the water vapour term.

Solution   At a height of 1000 m,

$$p = 1000 e^{-(1/8)}$$
$$= 882.5$$

then

$$N = \frac{77.6}{300} \times 882.5$$
$$= 228.3$$

so the refractive index is

$$n = 1 + 228.3 \times 10^{-6}$$
$$= 1.0002283$$

It has been observed that a radio wave which is launched horizontally above the surface of the earth in the troposphere, follows a curved path as shown by AB' in Fig. 15.14 rather than the straight path AB. The path AB' has a radius of curvature $\rho > R$, the radius of the earth. Why does this happen? The reason is that as we move along the straight line from A to B the wave enters a region where the medium is rarer, so the wave bends *away* from the normal as shown in the inset.

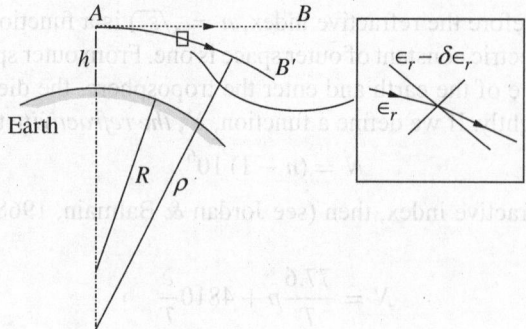

Fig. 15.14 Bending of rays in the troposphere

To calculate the radius of curvature ρ, we adopt the following method. Referring to Fig. 15.15, ρ which is the distance OA is the radius of curvature of the path of the radiated beam; $O'B$ is the radius of the earth. In a medium with dielectric constant ε_r, the velocity v is given by

$$v = 1/\sqrt{\mu_0 \varepsilon_0 \varepsilon_r} = c/n \qquad (15.57)$$

where c is the velocity of light in vacuum. If the refractive index at A is n and at C is $n + dn$ then ray path from these two points is

$$cdt/n = \rho\, d\theta \qquad (15.58)$$

$$cdt/(n + dn) = (\rho + dh)\, d\theta \qquad (15.59)$$

$$(1/n - dn/n^2)\, (c\, dt) = (\rho + dh)\, d\theta \qquad (15.60)$$

since $1/(n + dn) \approx 1/n - dn/n^2$. Subtracting the first equation from the second

$$-(dn/n^2)\, (c\, dt) = dh\, d\theta \qquad (15.61)$$

since $c\, dt = \rho\, d\theta$

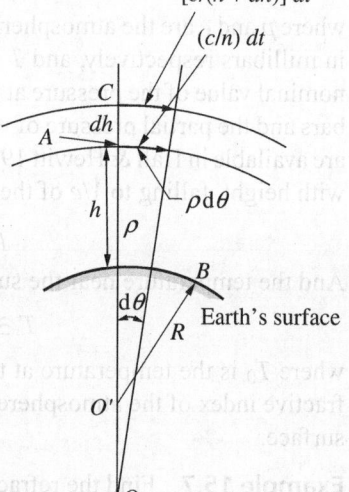

Fig. 15.15 Calculation of the curvature of rays in the troposphere

$$-(dn/n^2)\, (\rho\, d\theta) = dh\, d\theta \qquad (15.62)$$

or

$$\rho = -n^2 \frac{dh}{dn} \qquad (15.63)$$

and since $n^2 \approx 1$

$$\rho = -\frac{dh}{dn} = -\frac{1}{dn/dh} \qquad (15.64)$$

Example 15.8 Calculate the 'nominal' radius for the bending of rays in the troposphere at the surface of the earth.

Solution We know that

$$N = (n - 1)\, 10^6 = \frac{77.6}{T} p + 4810 \frac{e}{T}$$

omitting the water vapour term (the second term on the right)

$$n \approx \frac{77.6}{T} p \times 10^{-6} + 1 \qquad (15.65)$$

also
$$p \approx 1000e^{-(h/8000)} \qquad (15.66)$$

hence
$$n \approx \frac{77.6}{T} \times 10^{-3} \times e^{-(h/8000)} + 1 \qquad (15.67)$$

so for $h = 0$ and $T = 300°$ K

$$\left.\frac{dn}{dh}\right]_{h=0,\,T=300} = -\frac{77.6}{300} \times 10^{-6} \times \frac{1}{8}$$

$$= -3.233 \times 10^{-8}$$

so
$$\rho = -\frac{1}{dn/dh} = 3.092 \times 10^{7}\ \text{m}$$

since R, the radius of the earth is 6378.1 km

$$\rho \approx 4.85R \qquad (15.68)$$

From this calculation, the radius of curvature of the path ρ is a function of the rate of change of the dielectric constant with height, which is continuously varying. But for calculations a nominal radius of $\rho = 4R$ is used.

We realise from this that the problem of bending of waves in calculations become more complex. We would like to deal with straight line paths and not curved paths which are predicted by our theory. The ideal solution would be to make curved paths into straight paths for simplicity. A little thought would tell us is that we would need a mapping where the curved ray paths are made straight.

Let us visualise a situation where we draw Fig. 15.16(a) on a plane piece of plasticine and then deform the plasticine by bending it upward. By doing this the earth would become flatter and the curved ray paths would become straight lines as in Fig. 15.16(b). To obtain the transformation we proceed as follows. (Referring to Fig. 15.16) we require that when the wave travels from A to B the ray remains at the same height in both cases, h and $h + dh$, (a) and (b), while travelling a distance D. The diagram to the left, (a) is the actual situation while the one on the right (b) is where the radius of the earth has been changed to kR. Now from the (b) diagram,

$$OA = kR + h$$
$$OB = kR + h + dh$$
$$kR + h = (kR + h + dh)\cos\theta_e \qquad (15.69)$$

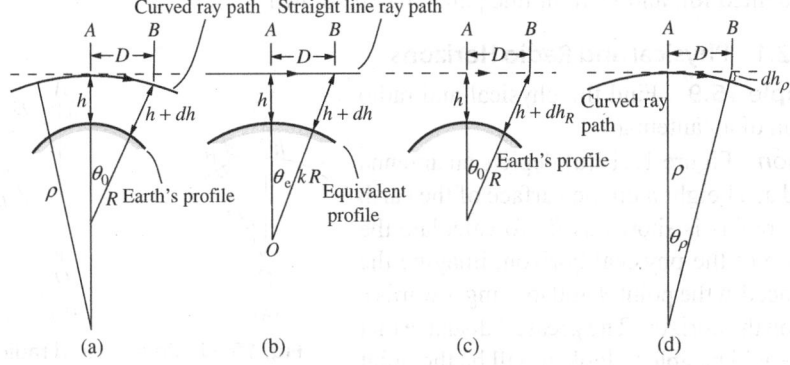

Fig. 15.16 Conversion of curved paths to straight paths using an effective radius for the earth

from these equations

$$dh = (kR + h)\left(\frac{1}{\cos\theta_e} - 1\right) \tag{15.70}$$

for small values of θ_e, $\cos\theta_e \approx 1 - \theta_e^2/2$ and so $1/\cos\theta_e \approx 1 + \theta_e^2/2$. Using these results,

$$dh = (kR + h)\,\theta_e^2/2 \approx kR\theta_e^2/2 \tag{15.71}$$

since we expect that $kR \gg h$.

$$\theta_e \approx \sin\theta_e = \frac{D}{kR} \tag{15.72}$$

hence

$$dh = \frac{D^2}{2kR} \tag{15.73}$$

From Fig. 15.16 (a), we do exactly the same calculation as for (b) but for two cases, (i) where the radius is earth's radius shown in (c) which gives

$$dh_R = \frac{D^2}{2R} \tag{15.74}$$

and the second case (ii) where we use the radius as ρ. ((d) part of the figure). In this case

$$dh_\rho = \frac{D^2}{2\rho} \tag{15.75}$$

then

$$dh = dh_R - dh_\rho$$

$$\frac{D^2}{2kR} = \frac{D^2}{2R} - \frac{D^2}{2\rho}$$

which gives

$$\frac{1}{kR} = \frac{1}{R} - \frac{1}{\rho} \tag{15.76}$$

and therefore

$$k = \frac{1}{1 - R/\rho} \tag{15.77}$$

using $\rho = 4R$

$$k = 4/3 \tag{15.78}$$

By using the effective radius (kR) instead of the actual radius (R) in making computations involving the radius of the earth and the atmosphere close to the ground (and effectively also the troposphere) the bending of the waves in the atmosphere is accounted for, and straight line paths may be drawn.

15.6.2.1 Physical and Radio Horizons

Example 15.9 Find the physical and radio horizon of an antenna.

Solution Figure 15.17(a) depicts an antenna placed at a height h on the surface of the earth whose radius is shown as R. To calculate the distance of the physical horizon, imagine the eye placed at the point A and looking towards a point on the horizon. The greatest distance that the eye will be able to look at will be the point B. It is obvious from elementary geometry that AB will be a tangent to the earth's surface and

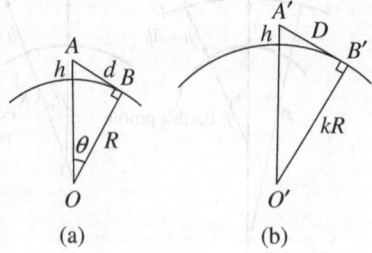

Fig. 15.17 Physical and radio horizons

OAB is a right-angled triangle. Therefore

$$OA^2 = AB^2 + OB^2$$

or

$$(R + h)^2 = R^2 + d^2$$

$$R^2 + 2Rh + h^2 = R^2 + d^2 \quad \text{(since } R \gg h)$$

$$2Rh = d^2$$

Now from the figure it is clear that the distance along the surface, $R\theta \approx d$, therefore the physical distance along the surface of the earth is

$$d \approx \sqrt{2Rh}$$

$$= 3.57\sqrt{h} \text{ km}$$

We now apply the same arguments to Fig. 15.17(b) where the effective radius of the earth is considered. Now the new radius is kR, and for the same height h, the distance covered by the waves is

$$D = \sqrt{2kRh} = \sqrt{\frac{4}{3} \times 2Rh}$$

which is

$$D = 1.155\sqrt{2Rh} = 4.13\sqrt{h} \text{ km}$$

here the height h is in meters.

15.6.2.2 Duct Propagation

The value of N on the surface of the earth, neglecting the water vapour term is about 220–280. Assuming that water vapour is present (e.g., near the sea) or for any other reason, N becomes unduly large and changes suddenly with height, the refractive index decreases much more rapidly than some value per meter and then waves travelling parallel or those at a small angle to the earth's surface are bent downward. The waves which bend downward, strike the surface of the earth and from there they are reflected back. The condition for reflection is considered in Example 15.10.

In this way, this refraction and reflection continues a number of times, and radio waves are trapped in a '*duct*'. Sometimes the reflection takes place above the surface of the Earth, in which case we have an elevated duct.

Example 15.10 Find the value of dn/dh such that rays bend towards the earth's surface.

Solution From Eqn (15.64) we know that

$$\rho = -\frac{1}{dn/dh}$$

and for rays to bend towards the surface of the earth,

$$\rho < R$$

where R is the earth's radius. Therefore

$$-\frac{1}{dn/dh} < R = 6378 \times 10^3$$

$$-\frac{dn}{dh} > \frac{1}{6378 \times 10^3} = 1.57 \times 10^{-7}$$

or
$$\frac{dn}{dh} < -1.57 \times 10^{-7}$$

15.7 | Ionospheric Propagation

In 1901, Marconi managed to broadcast the letter 'S' by Morse code across the Atlantic which was many thousands of miles away. It is obvious that all calculations based on tropospheric propagation fail to account for this successful result. Today the mechanism of wave propagation using the ionosphere accounts for this fact.

Regarding the history related to the ionosphere, in 1902 Oliver Heaviside proposed the existence of the ionosphere as a part of earth's atmosphere; later in 1912 HF ionospheric wave propagation characteristics were discovered. In a much later development, Vitally Ginzberg formulated the mechanism of radio wave propagation in the ionosphere which was seen to be a plasma.

15.7.1 Ionosphere

The earth is enveloped by an atmosphere consisting of about 78% nitrogen, 21% oxygen and about 1% consisting of carbon dioxide, argon and other gases by volume. Radiation from the Sun reaches the earth on a perpetual basis and is filtered through the earth's atmosphere. Sunlight consists of electromagnetic radiation which covers almost the full spectrum and is very similar to that emitted by a black body elevated to a temperature of about 5800°K. Radiation from the Sun consists of radio waves, infrared, the visible region, UV and even X-rays.

The atmosphere of the earth can be roughly divided into the troposphere (0–12 km), the stratosphere (12–45 km) and the ionosphere (50–1000 km). The pressure of air as we go up vertically falls exponentially as given in Eqn (15.56).

As the Sun's rays enter the ionosphere, where the atmosphere is so thin, the high energy radiation component (UV and above) dislodge electrons from the molecules of air creating a plasma or an electronic gas. A plasma is a mixture of positive ions and free electrons which are attracted to each other but due to thermal motion they do not always stay together. Generally these dislodged free electrons recombine with the positive ions, but a further influx of radiation creates more positive-negative ion pairs. This process is a statistical one where the atmosphere comes to be in an equilibrium where there is a sustained density of free electrons.

The number of electrons at any point in the ionosphere depends on a number of factors which are

1. The amount of radiation received which further depends on
   (a) A diurnal variation—whether it is day or night.
   (b) A seasonal variation—whether it is summer or winter.

2. Geographical zone—whether the point being considered is vertically above the equator or the pole, and

3. Disturbances such as Solar flares, which flood the earth with high energy charged particles.

The density of the the free electrons (and therefore positive ions) is of importance since it has an effect on electromagnetic wave propagation in the ionosphere.

The vertical composition of the ionosphere is shown in Fig. 15.18. Electron density starts at about 100–200 electrons/cc at a height of about 90 km above the earth's

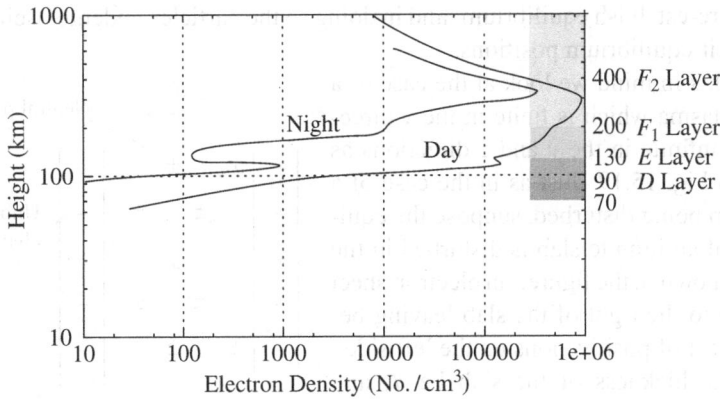

Fig. 15.18 Vertical composition of the ionosphere

Table 15.2 Daytime properties of the ionospheric layers. In all the layers there is a great day-night variation of the electron density. Note that on the ground, the pressure is about 1000 mB. (The pressure is measure of the number of molecules/cc)

Layer	Height (km)	Pressure (mB)	Electron density (No/cc)
D	70–90	0.16–0.045	80–200
E	90–130	0.045–8.8×10^{-5}	$\simeq 10^4$
F_1	130–200	$8.8 \times 10^{-5} - 1.4 \times 10^{-8}$	$\simeq 3 \times 10^4$
F_2	200–1000	$1.4 \times 10^{-8} - 5.2 \times 10^{-52}$	$\simeq 5 \times 10^4$

surface and rises to a maximum of daytime electron density of 10^6 electrons/cc at 300 km. As one goes further up, the density declines to 10^4 electrons/cc at 1000 km above the earth's surface. These figures stand for daytime electron densities.

The ionosphere is divided into layers: D, E, F_1, and F_2 layers. The D layer is accepted as being from 70–90 km, the E layer is from 90–130 km, the F_1 layer is from 130–200 km, and the the F_2 layer is from 130–1000 km. All these are shown in the form of a table (Table 15.2).

The density of electrons gradually increases with height through the D, E, and F_1 layers, undergoes a peak around the F_2 layer (where it reaches a maximum of around 10^6 electrons/cc) and then falls with height. The explanation is that at great heights the ionising capability of the radiation is intense but the number of normal (non-ionised) molecules are few. On the other hand at low heights the number of high energy photons have been considerably reduced due to passage through the upper layers, and therefore density of electrons/cc is low. On the surface of the earth where the pressure is highest there is almost no free electron density. At night, the D layer virtually disappears but the E and F layers are still present.

15.7.2 Plasma Oscillations

The word *plasma* denotes a gas of charged particles where the number of electrons and positive ions are equal but not bound together and free to move like the molecules in a gas. Another way of talking about the charge neutrality is

$$\rho_v(\mathbf{r}) \approx 0 \qquad (15.79)$$

If the distribution of the particles in the plasma is altered so that Eqn (15.79) is invalidated, then the charges move back quickly under the influence of restoring forces to re-establish equilibrium; and in doing so the particles undergo oscillations about their equilibrium positions.

With this in mind we look at the case of a slab of plasma which is finite in the x direction and infinite in the y and z directions as shown in Fig. 15.19. Just as in the case of a pendulum being disturbed, suppose the equilibrium of an infinite slab is disturbed in the manner shown in the figure: an electron sheet is moved to the right of the slab leaving behind a sheet of positive ions on the left side.

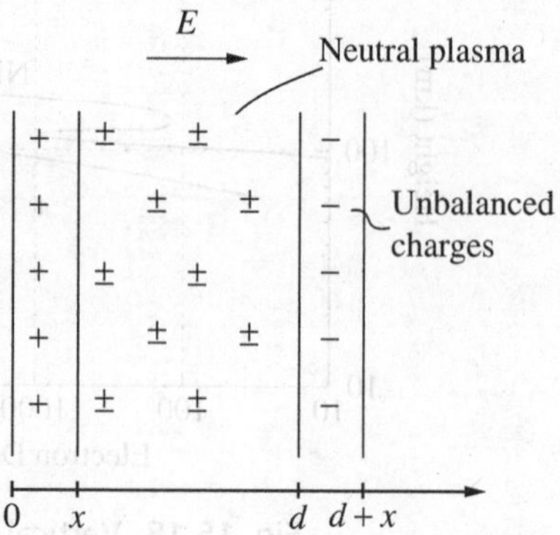

Let the thickness of the slab be d, and the tiny displacement of the charges be x as shown, then surface charge density is

Fig. 15.19 Plasma oscillations of an infinite slab

$$|\rho_s| = N|e|x \quad \text{C/m}^2 \quad (15.80)$$

where N is the charged particle density, and e is the electronic charge. The time varying, but constant field $\mathbf{E}$ within the slab is given by

$$\mathbf{E} = \frac{N|e|x}{\varepsilon_0}\mathbf{a}_x \quad (15.81)$$

for a given value of x. Note that this is the same field as that inside a capacitor. The permittivity ε_0 is used since the plasma is assumed to be similar to air. The surface charges on each side feel the force due to the electric field, but the positive ions being much heavier remain almost stationary. An electron on the other hand feels a force

$$\mathbf{F}_e = -\frac{Ne^2x}{\varepsilon_0}\mathbf{a}_x \quad (15.82)$$

and an acceleration $\ddot{x} = F_e/m_e$. The above equation states that the restoring force on the electron is proportional to the displacement. Writing the equation of motion,

$$m_e\ddot{x} = -\frac{Ne^2}{\varepsilon_0}x$$

or

$$\ddot{x} = -\frac{Ne^2}{m_e\varepsilon_0}x = -\omega_p^2 x$$

where

$$\omega_p = \sqrt{\frac{Ne^2}{m_e\varepsilon_0}} \quad (15.83)$$

is called the plasma frequency. The solution to this equation is

$$x = C_1 \sin\omega_p t + C_2 \cos\omega_p t \quad (15.84)$$

This equation implies that an electron in a plasma undergoes oscillations with frequency ω_p when disturbed slightly.

Example 15.11 Find the plasma frequency for the number of ionized particles, $N = 3 \times 10^4$/cc.

Solution The number of ionised particles per m³ is given by

$$N = 3 \times 10^{10}/\text{m}^3$$

Therefore
$$\omega_p = \sqrt{\frac{Ne^2}{m_e \varepsilon_0}}$$
$$= 9.77 \times 10^6 \text{ rad/sec}$$

15.7.3 Wave Propagation in a Plasma

Let us consider a plasma through which an electromagnetic wave is propagating. Since the wave has an $\mathbf{E}$ and $\mathbf{H}$ vector, both of which are sinusoidal in nature, the electrons in the plasma are disturbed about their equilibrium positions, sinusoidally. With this in mind, the charge densities at a point are

$$n(t) = N_0 + \text{Re}\{n_0 e^{j\omega t}\}$$
$$\rho(t) = en(t) \tag{15.85}$$

where N_0 is the statistically constant value of the electron density and n_0 is maximum change in this constant value. Therefore

$$N_0 + n_0 > n(t) > N_0 - n_0 \tag{15.86}$$

Similarly, the velocity of an electron is strictly speaking

$$\mathbf{v}(t) = \mathbf{V}_0 + \text{Re}\{\mathbf{v}_0 e^{j\omega t}\} \tag{15.87}$$

but we assume for sake of simplicity that the *drift* component V_0 (statistically constant value) is 0,[2] or

$$\mathbf{v}(t) = \text{Re}\{\mathbf{v}_0 e^{j\omega t}\} \tag{15.88}$$

The current density is therefore

$$\mathbf{J}(t) \approx N_0 e \text{Re}\{\mathbf{v}_0 e^{j\omega t}\} \tag{15.89}$$

The equation of motion and Maxwell's equations now may be written for sinusoidal oscillations for phasors as

$$e\mathbf{E} = j\omega\, m_e \mathbf{v} \tag{15.90}$$
$$\nabla \times \mathbf{H} = j\omega\varepsilon_0\, \mathbf{E} + Ne\mathbf{v}$$
$$\nabla \times \mathbf{E} = -j\omega\, \mu_0\, \mathbf{H}$$
$$\nabla \cdot \mathbf{E} = \frac{ne}{\varepsilon_0}$$
$$\nabla \cdot \mathbf{H} = 0 \tag{15.91}$$

where we have dropped the subscript '0' on the charge density N. Substituting Eqn (15.90) into the first of the equations of Eqn set (15.91),

$$\nabla \times \mathbf{H} = j\omega\varepsilon_0\, \mathbf{E} + Ne\mathbf{v}$$
$$= j\omega\varepsilon_0\, \mathbf{E} + Ne\left(\frac{e\mathbf{E}}{j\omega\, m_e}\right)$$
$$= j\omega\left(\varepsilon_0 - \frac{Ne^2}{\omega^2\, m_e}\right)\mathbf{E}$$

---

[2]Since a constant electric field is absent.

$$= j\omega \left(\varepsilon_0 - \varepsilon_0 \frac{\omega_p^2}{\omega^2}\right) \mathbf{E} \tag{15.92}$$

$$= j\omega\varepsilon \, \mathbf{E} \tag{15.93}$$

or that a plasma has a dielectric permittivity

$$\varepsilon = \varepsilon_0 \left(1 - \frac{\omega_p^2}{\omega^2}\right) \tag{15.94}$$

In this expression we have not taken into account collisional effects. Equation (15.94) suggests that $\varepsilon \to -\infty$ as $\omega \to 0$, but this is not the case. If v is the collisional frequency of the electrons with the positive ions then the more accurate formula for a plasma is

$$\varepsilon = \varepsilon_C = \varepsilon_0 \left(\varepsilon_r' - j\varepsilon_r''\right) \tag{15.95}$$

where

$$\varepsilon_r' = \left(1 - \frac{\omega_p^2}{v^2 + \omega^2}\right) \tag{15.96}$$

$$\varepsilon_r'' = \frac{\sigma}{\omega\varepsilon_0} \tag{15.97}$$

$$\sigma = \frac{\omega_p^2 \, v}{v^2 + \omega^2} \tag{15.98}$$

An approximate formula for the collisional frequency is given by

$$v = v_0 \, e^{-h/H} \tag{15.99}$$

where v_0 is a constant. The measured collisional frequency at various heights is given in the following table (Manheimer & Stix, 1977)

Height, h (km)	200	130	95	800
Collisional frequency, v (s^{-1})	10^3	10^4	10^5	50

We can see from this table that v plays an important role only at the lower heights and that too for frequencies in the kHz range. Once we cross 250 km, v may be neglected when compared to ω.

15.7.4 Low Frequency Propagation

For an electromagnetic wave

$$\mathbf{E} = \mathbf{E}_0 e^{-jkz}$$

with

$$k = \omega \sqrt{\mu_0 \, \varepsilon_C} \tag{15.100}$$

since ε_C is complex, therefore k is a complex number. If k has an imaginary part, then the wave decays while it progresses into the ionosphere as in the case of wave travelling in a metal. Also just as in the case of an air metal boundary, reflection takes place from a layer where there is a change in the properties of a medium, and the reflection coefficients are given by Eqns (15.5) and (15.8). This happens for low frequencies ω, especially when

$$v^2 + \omega^2 < \omega_p^2 \tag{15.101}$$

and $\varepsilon_r' < 0$ and k becomes purely imaginary. Conditions for this type of behaviour are right at heights at the upper end of the D layer and the start of the E layer.

15.7.5 High Frequency Propagation

Due to changes in ionisation densities, the refractive index of the ionosphere continually changes, and becomes smaller and smaller at least up to the F layer. This can be observed from the ionisation density profile shown in Fig. 15.18 and application of Eqns (15.94) and (15.96). As the height increases, N increases and the factor ω_p (which is proportional to $\sqrt{N}$) increases. It is because of this that the wave bends away from the normal. This situation is shown in Fig. 15.20. Let the ray move through the troposphere from the point marked S, and enter the ionosphere at an angle ϕ_i with respect to the normal at the start of the ionosphere, AA'. Then we find that the ray bends gradually as shown. At any point F the angle of incidence is ϕ_f. This can be proved as follows.

The diagram on the right shows a ray entering a change of medium from the point A with an angle of incidence ϕ. The medium has a refractive index n. The ray emerges slightly deviated with angle of $\phi + d\phi$ and the medium has a refractive index $n + dn$. Therefore from Snell's law

$$n\sin\phi = (n + dn)\sin(\phi + d\phi)$$
$$= n\sin(\phi + d\phi) + dn\,\sin(\phi + d\phi)$$
$$\approx n(\sin\phi + d\phi\,\cos\phi) + dn\,(\sin\phi + d\phi\,\cos\phi)$$

where $\cos(d\phi) \approx 1$ and $\sin(d\phi) \approx d\phi$; cancelling terms and neglecting double differential terms like $dn\,d\phi$,

$$-n\cos\phi\,d\phi = dn\,\sin\phi$$
$$\frac{\cos\phi\,d\phi}{\sin\phi} = -\frac{dn}{n}$$
$$\frac{d(\sin\phi)}{\sin\phi} = -\frac{dn}{n}$$

integrating this equation, with the initial and final angles of incidences: ϕ_i, ϕ_f and initial and final refractive indices n_i and n_f

$$\ln\frac{\sin\phi_f}{\sin\phi_i} = -\ln\frac{n_f}{n_i} = \ln\frac{n_i}{n_f}$$

therefore $\qquad\qquad n_f\sin\phi_f = n_i\sin\phi_i \qquad\qquad (15.102)$

This equation is a very powerful equation which says that starting from some point in the ionosphere with refractive index n_i and angle of incidence ϕ_i, if the final

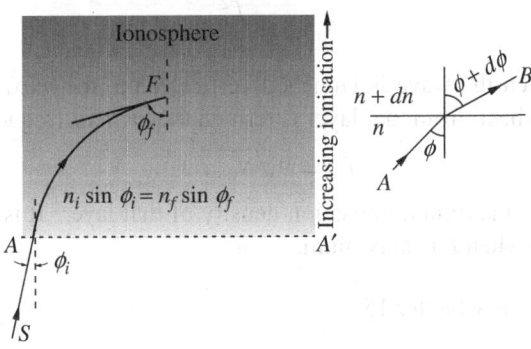

Fig. 15.20 A high frequency wave travelling in the ionosphere

refractive index is n_f, the final angle ϕ_f—with respect to the normal—the wave still obeys Snell's law.

Let us apply the result which we have arrived to the ionosphere. Since at any point in the ionosphere,

$$n = \sqrt{\varepsilon_r} = \sqrt{1 - \frac{\omega_p^2}{v^2 + \omega^2}} \qquad (15.103)$$

Above 200–250 km, $v \ll \omega$ (ω is in the order of 10^7 and v is of the order of 10^3)

$$n \approx \sqrt{1 - \frac{\omega_p^2}{\omega^2}} \qquad (15.104)$$

$$= \sqrt{1 - \frac{81N}{f^2}} \qquad (15.105)$$

where N is expressed in electrons/cc and f in kHz[3].

Since a wave is launched from the surface of the earth with an angle of incidence ϕ_i, the refractive index is 1. Therefore

$$n_f \sin\phi_f = \sin\phi_i \qquad (15.106)$$

now as the wave propagates through the ionosphere, the angle of refraction increases since the ionisation density increases. Therefore there is a point above the earth where the ray is parallel to the earth's surface and the angle of refraction becomes 90° or $\sin\phi_f = 1$. At that point

$$n_f]_{\text{turn back}} = \sin\phi_i$$

or

$$\sqrt{1 - \frac{81N_{tb}}{f}} = \sin\phi_i$$

or

$$N_{tb} = \frac{f^2 \cos^2\phi_i}{81} \qquad (15.107)$$

where N_{tb} can be termed as that electron density where the wave 'turns back.' Note that anywhere in the ionosphere where Eqn (15.107) is satisfied, the wave turns back and returns to the earth's surface.

Suppose we launch the wave vertically upwards, the N_{tb} will take its maximum value, $N_{\max} = (N_{tb})_{max}$ that is

$$N_{\max} = \frac{f^2}{81} \qquad (15.108)$$

For any layer, when the wave is launched vertically up, the *maximum* frequency which is reflected back from that layer is termed the *critical* frequency and is

$$f_{cr} = 9\sqrt{N_{\max}} \qquad (15.109)$$

where $N_{\max}$ is the maximum ionisation density of that layer. It is obvious that N will be maximum when f is maximum.

 See Fcr.m in Chapter 15

---

[3]If N is in electrons/m³ then f is in Hz.

We look at the problem of reflection from the ionosphere in another way. From Eqn (15.107),

$$f = 9\sqrt{N_{tb}}\sec\phi_i$$

In this equation, if we allow allow N_{tb} to take on the value of N_{max} then the factor $9\sqrt{N_{max}}$ may be replaced by f_{cr} the critical frequency which means that f takes on a value higher than the critical frequency. This maximum frequency is

$$f_{max} \overset{\Delta}{=} f_{MUF} = f_{cr}\sec\phi_i$$

which is greater than the critical frequency and is called the *maximum usable frequency*. However note that ϕ_i has the meaning of an angle which the ray makes with the normal at the point of reflection in the ionosphere as shown in Fig. 15.21.

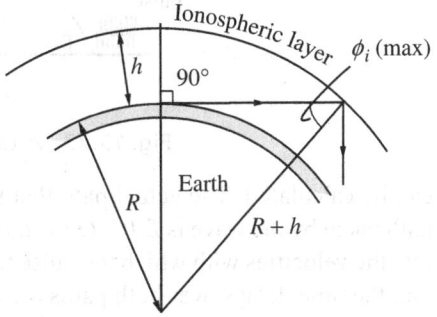

It turns out that this frequency is still not the maximum usable frequency. If we launch a wave horizontally it meets the ionospheric layer at height h as shown in Fig. 15.21 and suffers reflection. Then

Fig. 15.21 Diagram for $(f_{MUF})_{max}$

$$\sin[\phi_i(max)] = \frac{R}{R+h} \qquad (15.110)$$

and

$$f_{MUF,\,max} = f_{cr}\sec[\phi_i(max)] \qquad (15.111)$$

thus if the reflection occurs at 200 km, then

$$\phi_i(max) = 1.323 \text{ radians } (75.8°)$$

and

$$\sec\phi_i(max) = 4.08$$

which makes

$$f_{MUF,\,max} = 4.08 f_{cr}$$

 See NTurnBack.m and fmufmax.m in Chapter 15

Example 15.12 A wave is launched at an angle of 45° with a carrier frequency of 10 MHz. Find the value of the electron density in electrons/cc at which the wave is reflected back.

Solution Using Eqn (15.107) we calculate the value of N_{tb} as 6.17×10^5 electrons per cc.

Example 15.13 Find the critical frequency of the E layer where $N = 2 \times 10^{12}$ electrons/m³.

Solution From Eqn (15.109), we have

$$f_{cr} = 9\sqrt{N_{max}} = 1.27 \times 10^7 \text{Hz}$$

15.7.6 Virtual Height

To experimentally determine the heights of the various layers and their ionisation densities, a radio pulse of a known frequency is launched towards the ionosphere as shown in Fig. 15.22. Knowing the angle, ϕ_i, with respect to the vertical at which the pulse is launched and the time it takes from Tx to Rx, the velocity of the wave

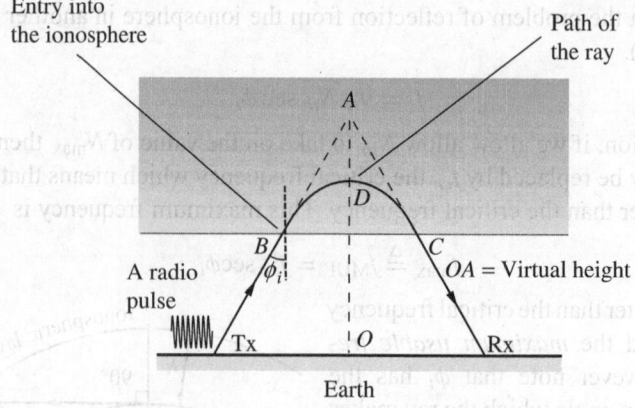

Fig. 15.22 Virtual height of a layer

can be calculated. The actual path that the pulse takes is BDC while the apparent path taken by the wave is BAC. OA is the virtual height of the layer. It can be shown that the velocities with which the pulse travels over the path BDC and BAC are such that the time delays over both paths are equal.

Example 15.14 A radio pulse of frequency 5 MHz is launched vertically upwards and received after 1 ms. Calculate the approximate values of the electron density and height of the layer.

Solution Since the wave is received after 1 ms, the distance travelled is

$$d = 3 \times 10^8 \times 10^{-3}$$
$$= 3 \times 10^5$$
$$= 300 \text{ km}$$

hence the virtual height is 150 km. Assuming that the virtual and actual heights are almost the same, we can say that 150 km is the height of the layer.

Now using Eqn (15.108), where the $\phi_i = 0$, the value of N is

$$N = \frac{f^2}{81}$$
$$= 3.1 \times 10^5 \text{ electrons/cc}$$

POINTS TO REMEMBER

• The Rayleigh roughness coefficient is

$$R = 4\pi \sin\psi \left(\frac{\sigma}{\lambda}\right)$$

where ψ is the grazing angle of incidence and σ is the standard deviation of the surface irregularities.

• The reflection coefficient for perpendicular polarisation is

$$R_\perp = \frac{\sin\psi - \sqrt{(\varepsilon_2/\varepsilon_1) - \cos^2\psi}}{\sin\psi + \sqrt{(\varepsilon_2/\varepsilon_1) - \cos^2\psi}}$$

where ψ is the grazing angle of incidence, and

$$\varepsilon_2/\varepsilon_1 = \varepsilon_r - j(\sigma/\varepsilon_0\, \omega)$$

- The reflection coefficient for parallel polarisation is

$$R_{\|} = \frac{(\varepsilon_2/\varepsilon_1)\sin\psi - \sqrt{(\varepsilon_2/\varepsilon_1) - \cos^2\psi}}{(\varepsilon_2/\varepsilon_1)\sin\psi + \sqrt{(\varepsilon_2/\varepsilon_1) - \cos^2\psi}}$$

- The ground wave attenuation factor A is

$$A \approx A_1 - \sin(b)\sqrt{\frac{p}{2}}e^{-(5/8)p}$$

where

$$A_1 = \frac{2+0.3p}{2+p+0.6p^2}$$

and

$$a = \frac{\pi}{\varepsilon_r - jx}\left(1 - \frac{1}{\varepsilon_r - jx}\right)$$

$$p = \frac{R}{\lambda}|a|$$

$$b \approx \tan^{-1}\left(\frac{\varepsilon_r + 1}{x}\right)$$

- The surface impedance of the earth is

$$Z_s \approx Z_0\frac{1}{\sqrt[4]{\left(\dfrac{\sigma}{\omega\varepsilon_0}\right)^2 + \varepsilon_r^2}}\angle\left\{\frac{1}{2}\tan^{-1}\left(\frac{\sigma}{\omega\varepsilon_r\,\varepsilon_0}\right)\right\}$$

- For two vertical dipoles operating over the ground in the VHF range

$$|E_z| \approx \frac{60k^2 I l_{\text{eff}} h_1 h_2}{d^2}$$

where $h_{1,2}$ are the heights of the two dipoles above the ground, and l_{eff} are the effective lengths of the dipoles.
- The radius ρ of the bending of rays in the troposphere is given by

$$\rho = -\frac{1}{dn/dh}$$

where n is the refractive index as a function of height h.
- In tropospheric propagation, the effective radius of the earth is $4R/3$ where R is the radius of the earth.
- The plasma frequency is given by

$$\omega_p = \sqrt{\frac{Ne^2}{m_e\varepsilon_0}}$$

where N is the electron density, e is the electronic charge, m_e is the electron mass and ε_0 is the permittivity of vacuum.
- The refractivity N for the earth's atmosphere is given by

$$N = (n-1)\,10^6$$

where n is the refractive index.
- The complex permittivity for the ionosphere is given by $\varepsilon_C = \varepsilon_0\,(\varepsilon'_r - j\varepsilon''_r)$, where

$$\varepsilon'_r = \left(1 - \frac{\omega_p^2}{\nu^2 + \omega^2}\right)$$

$$\varepsilon''_r = \frac{\sigma}{\omega\varepsilon_0}$$

$$\sigma = \frac{\omega_p^2\,\nu}{\nu^2 + \omega^2}$$

and ν is the collisional frequency of the electrons.

- Snell's law for the ionosphere is given by

$$n_f \sin\phi_f = n_i \sin\phi_i$$

where $n_{i,f}$ are the refractive indices of the starting and ending points of the ray and $\phi_{i,f}$ are the angles of incidence of the initial and final point.
- The refractive index of the ionosphere is given by

$$n = \sqrt{\varepsilon_r} = \sqrt{1 - \frac{\omega_p^2}{v^2 + \omega^2}}$$

- The electron density where the ray is parallel to the earth's surface, and 'turns back' is given by

$$N_{tb} = \frac{f^2 \cos^2\phi_i}{81}$$

where f is in Hz and N_{tb} is in electrons/m^3 or f is in kHz and N_{tb} is in electrons/cc. ϕ_i is the angle of incidence.
- The critical frequency is given by

$$f_{cr} = 9\sqrt{N_{max}}$$

- The maximum usable frequency is given by

$$f_{MUF} = f_{cr} \sec\phi_i$$

where ϕ_i is given by

$$\sin[\phi_i] = \frac{R}{R+h}$$

where R is the radius of the earth and h is the height at which reflection takes place.
- The absolute maximum usable frequency is obtained when the wave is launched parallel to the earth's surface, and is given by

$$f_{MUF, max} = f_{cr} \sec[\phi_i(\max)] \approx 3.6 f_{cr}$$

SELF ASSESSMENT

Objective Type Questions

1. Ground wave propagation has the following factors:
   (a) a direct path between Tx and Rx   (b) a reflection from the sky
   (c) a surface wave   (d) a reflected path between Tx and Rx
2. In ground wave propagation when we consider the reflection from the earth
   (a) reflections at low frequencies are almost perfect reflections
   (b) only some of the energy is reflected, the rest penetrates into the earth
   (c) at the Brewster angle a sharp dip in reflected energy is seen
   (d) the ground acts like a lossy medium
3. When we consider the surface wave
   (a) attenuation is high for low frequencies
   (b) attenuation is high for high frequencies
   (c) attenuation is greater for fields radiated by a vertical dipole than a horizontal dipole
   (d) attenuation is greater for fields radiated by a horizontal dipole than a vertical dipole
4. When we consider the surface wave
   (a) the wave tilt is greater at low frequencies than at high frequencies

(b) the wave tilt is greater at high frequencies than at low frequencies
(c) there is no tilt
5. In tropospheric propagation
   (a) the wave curves due to gravity
   (b) the wave curves due to pressure change
   (c) the wave curves due to temperature change
   (d) the wave curves due to a change in dielectric constant
6. (a) Waves may reach over the physical horizon in tropospheric propagation
   (b) Waves may reach over the physical horizon in surface wave propagation
   (c) Waves may reach over the physical horizon in ionospheric propagation
   (d) none of these
7. In duct propagation, when waves are reflected in the troposphere, the radius of bending
   of waves ρ, must
   (a) be greater than the radius of the earth (b) be equal to the radius of the earth
   (c) be less than the radius of the earth (d) none of these
8. The ionosphere is caused in part due to
   (a) low energy cosmic rays (b) high energy cosmic rays (c) none of these
9. When rays from the earth are beamed towards the ionosphere
   (a) they bend towards the normal
   (b) they bend away from the normal
   (c) none of these
10. Complex permittivity ε_C of the ground is real, if
    (a) the ground is smooth (b) the ground is rough
    (c) the ground has a conductivity $\sigma = 0$ (d) none of the above

Short-Answer Questions

1. Find the radius of curvature of rays in the troposphere where the height is 5 km and the partial pressure of water vapour is 40 millibars. The temperature is $0°C$.
2. Find the radius of curvature of rays in the troposphere if at a certain height $dn/dh = -1.8 \times 10^{-7}$.
3. Find the physical horizon and radio horizon of an antenna of height 20 m.
4. Find the critical frequency for $N = 1.6 \times 10^{12}$ electrons/m^3.
5. Find the critical frequency of the F-layer where $N = 2.5 \times 10^{12}$ electrons/m^3

Review Questions

1. Discuss the Rayleigh roughness criterion.
2. What is the concept of complex permittivity?
3. Explain why in the case of perpendicular as well as parallel polarisations, for low angles of incidence $(0 - 10°)$, the reflection from the earth is almost like perfect reflection.
4. In which frequency band is surface wave transmission most successful?
5. Explain the phenomenon of duct propagation.
6. Explain how the ionosphere is formed.
7. In the ionosphere, why is the refracive index less than 1?

Numerical Problems

Where the ground parameters are not given, assume $\varepsilon_r = 15$ and $\sigma = 12 \times 10^{-3}$ S/m Where dipole antennas are mentioned, assume half wave dipoles with $l_{\text{eff}} = \lambda/2$.

1. For $\sigma = 0.2\lambda$, for what angles of grazing incidence, ψ, is the ground considered smooth.
   Hint: use the Raleigh roughness criterion.

2. Find the complex permittivity of the ground if $\varepsilon_r = 80$, $\sigma = 10^{-2}$. In which frequency range will the ground behave like a good conductor?

3. Find the magnitude and phase of the reflection coefficient when a plane wave approaches the surface of the earth when the angle of grazing incidence, ψ, is 5° and when the electric field is perpendicular to the plane of incidence. The frequency is 4 MHz.

4. Find the magnitude and phase of the reflection coefficient when a plane wave approaches the surface of the earth when the angle of grazing incidence ψ is 5° and when the electric field is perpendicular to the plane of incidence. The frequency is 40 MHz.

5. Find the magnitude and phase of the reflection coefficient when a dipole is held vertical to the ground and the far field is like a plane wave which approaches the surface of the earth at an angle of grazing incidence of $\psi = 1°$. The frequency is 50 MHz.

6. Find the magnitude and phase of the reflection coefficient when a dipole is held parallel to the ground and the far field is like a plane wave which approaches the surface of the earth at an angle of grazing incidence of $\psi = 1°$. The frequency is 50 MHz.

7. For a vertical dipole, find the value of p and b and from these values, the value of A for frequencies of (a) 50 kHz, (b) 500 kHz, (c) 5 mHz, and (d) 500 MHz. Assume the communication distance R is 50 km. Parameters for the ground may be taken to be $\varepsilon_r = 15$, $\sigma = 0.012$ S/m.

8. Find Z_s of the ground for $\sigma = 10^{-2}$, $\varepsilon_r = 10$, and $f = 500$ kHz.

9. Find the E_v to E_h ratio for $\sigma = 10^{-2}$, $\varepsilon_r = 10$, and $f = 500$ kHz.

10. Find the radius of curvature of rays in the troposphere where the height is 7 km and and the partial pressure of water vapour is 20 millibars. The temperature is 5°C.

11. Find the maximum distance at which two antennas of height 20 m may be placed so that they can communicate with each other through the direct wave.

12. Two half wave dipole antennas (which determines the polarisation) are kept vertical to the surface of the ground and are separated by a distance of 10 km. The current of the broadcasting antenna is 100 A. If both antennas are at 20 m above the ground, find the electric field at the receiving antenna from the sum of the direct and reflected waves. The frequency of operation is 100 MHz.
Hint: Use Eqn (15.51).

13. Find the plasma frequencies for $N = 10^2$, 10^3, 10^4, and 10^5 electrons/cc.

14. A wave of 10 MHz is launched at an angle $\phi_i = 45°$ to the vertical. What is the value of the refractive index when the wave is directed parallel to the earth's surface when it is in the ionosphere?

15. For Problem 14, find the value of N_{tb} when the wave turns back and find the layer (daytime) at which this happens.

16. For $N = 1.8 \times 10^{12}$ electrons/m^3, find the critical frequency.

17. For Problem 16, find the MUF(max) for the E layer at 100 km. Assume $R = 6400$ km.

Answers

Objective Type Questions

1. (a), (c) and (d)   2. All four are correct   3. (b) and (d)   4. (b)
5. (b), (c) and (d)   6. (a), (b) and (c)   7. (c)   8. (b)   9. (b)   10. (c)

Short-Answer Questions

1. $\rho \approx 5.26 \times 10^4$ km
2. 5555.6 km

3. Physical horizon = 15.96 km; radio horizon = 18.5 km
4. 11.38 MHz
5. 14.23 MHz

Numerical Problems

1. $\psi < 2.28°$
2. $f < 22.5$ MHz
3. $R_\perp = 0.982\angle 3.127^r$
4. $R_\perp = 0.957\angle 3.133^r$
5. $R_{\parallel} = 0.8681\angle - 3.122^r$
6. $R_\perp = 0.991\angle 3.140^r$
7. (a) $p \approx 0$; $b \approx 0$; $A \approx 1$ (b) $p \approx 0.083$; $b \approx 0$ $A \approx 1$ (c) $p \approx 50$; $b \approx 18°$ $A \approx 0.011$
   (d) $p \approx 1.7 \times 10^4$; $b \approx 90°$ $A \approx 3 \times 10^{-5}$
8. $19.88\angle 0.7715^r$
9. $0.05273\angle 0.7715^r$
10. $\rho = 6.875 \times 10^4$ km
11. 37 km
12. $E_z \approx 0.158$ V/m
13. 89.8 kHz, 284 kHz, 897 kHz and 2.84 MHz
14. $n_f = 0.7071$
15. 6.17×10^5 electrons/cc; F layer
16. $f_{cr} = 12.1$ MHz
17. $\phi_i \, (max) \approx 80°$; MUF(max) ≈ 68 MHz

PART V

Appendices

PART V

Appendices

APPENDIX A
Coordinate Systems

APPENDIX B
Mathematical Reference

APPENDIX C
Some Key Equations

APPENDIX D
Sample Question Papers

A

Coordinate Systems

A.1 | Rectangular to Cylindrical and Cylindrical to Rectangular Coordinate Systems

The rectangular coordinate system is represented by (x, y, z) while cylindrical coordinates are (ρ, ϕ, z)

$$\hat{a}_i \cdot \hat{a}_j = \delta_{ij} \qquad \hat{a}_i, \hat{a}_j = \mathbf{a}_x \text{ or } \mathbf{a}_y \text{ or } \mathbf{a}_z \tag{A.1}$$

$$\hat{a}_i \cdot \hat{a}_j = \delta_{ij} \qquad \hat{a}_i, \hat{a}_j = \mathbf{a}_\rho \text{ or } \mathbf{a}_\phi \text{ or } \mathbf{a}_z \tag{A.2}$$

where $\delta_{ij} = 1$ for $i = j$ otherwise it is zero.

$\times$	$\mathbf{a}_x$	$\mathbf{a}_y$	$\mathbf{a}_z$
$\mathbf{a}_x$	0	$\mathbf{a}_z$	$-\mathbf{a}_y$
$\mathbf{a}_y$	$-\mathbf{a}_z$	0	$\mathbf{a}_x$
$\mathbf{a}_z$	$\mathbf{a}_y$	$-\mathbf{a}_x$	0

$\times$	$\mathbf{a}_\rho$	$\mathbf{a}_\phi$	$\mathbf{a}_z$
$\mathbf{a}_\rho$	0	$\mathbf{a}_z$	$-\mathbf{a}_\phi$
$\mathbf{a}_\phi$	$\mathbf{a}_z$	0	$\mathbf{a}_\rho$
$\mathbf{a}_z$	$\mathbf{a}_\phi$	$-\mathbf{a}_\rho$	0

$$\begin{aligned} x &= \rho \cos \phi \\ y &= \rho \sin \phi \\ z &= z \end{aligned} \tag{A.3}$$

$$\begin{aligned} \rho &= \sqrt{x^2 + y^2} \\ \phi &= \arctan(y/x) \\ z &= z \end{aligned} \tag{A.4}$$

$$\begin{pmatrix} \mathbf{a}_\rho \\ \mathbf{a}_\phi \\ \mathbf{a}_z \end{pmatrix} = \begin{pmatrix} \cos(\phi) & \sin(\phi) & 0 \\ -\sin(\phi) & \cos(\phi) & 0 \\ 0 & 0 & 1 \end{pmatrix} \begin{pmatrix} \mathbf{a}_x \\ \mathbf{a}_y \\ \mathbf{a}_z \end{pmatrix} \tag{A.5}$$

$$\begin{pmatrix} \mathbf{a}_x \\ \mathbf{a}_y \\ \mathbf{a}_z \end{pmatrix} = \begin{pmatrix} \cos(\phi) & -\sin(\phi) & 0 \\ \sin(\phi) & \cos(\phi) & 0 \\ 0 & 0 & 1 \end{pmatrix} \begin{pmatrix} \mathbf{a}_\rho \\ \mathbf{a}_\phi \\ \mathbf{a}_z \end{pmatrix} \tag{A.6}$$

A.2 | Rectangular to Spherical and Spherical to Rectangular Coordinate Systems

The rectangular coordinate system is represented by (x, y, z) while spherical coordinates are (r, θ, ϕ)

$$\hat{a}_i \cdot \hat{a}_j = \delta_{ij} \qquad \hat{a}_i, \hat{a}_j = \mathbf{a}_r \text{ or } \mathbf{a}_\theta \text{ or } \mathbf{a}_\phi \tag{A.7}$$

where $\delta_{ij} = 1$ for $i = j$ otherwise it is zero.

$\times$	$\mathbf{a}_r$	$\mathbf{a}_\theta$	$\mathbf{a}_\phi$
$\mathbf{a}_r$	0	$\mathbf{a}_\phi$	$-\mathbf{a}_\theta$
$\mathbf{a}_\theta$	$-\mathbf{a}_\phi$	0	$\mathbf{a}_r$
$\mathbf{a}_\phi$	$\mathbf{a}_\theta$	$-\mathbf{a}_r$	0

$$r \equiv |\mathbf{R}| = \sqrt{x^2 + y^2 + z^2}$$

$$\theta = \arccos(z/r) = \arccos\left(\frac{z}{\sqrt{x^2 + y^2 + z^2}}\right) \tag{A.8}$$

$$\phi = \arccos(x/\rho) = \arccos\left(\frac{x}{\sqrt{x^2 + y^2}}\right)$$

$$x = r\sin\theta\cos\phi$$
$$y = r\sin\theta\sin\phi \tag{A.9}$$
$$z = r\cos\theta$$

$$\begin{pmatrix} \mathbf{a}_r \\ \mathbf{a}_\theta \\ \mathbf{a}_\phi \end{pmatrix} = \begin{pmatrix} \sin\theta\cos\phi & \sin\theta\sin\phi & \cos\theta \\ -\cos\theta\cos\phi & \cos\theta\sin\phi & -\sin\theta \\ -\sin\phi & \cos\phi & 0 \end{pmatrix} \begin{pmatrix} \mathbf{a}_x \\ \mathbf{a}_y \\ \mathbf{a}_z \end{pmatrix} \tag{A.10}$$

$$\begin{pmatrix} \mathbf{a}_x \\ \mathbf{a}_y \\ \mathbf{a}_z \end{pmatrix} = \begin{pmatrix} \sin\theta\cos\phi & \cos\theta\cos\phi & -\sin\phi \\ \sin\theta\sin\phi & \cos\theta\sin\phi & \cos\phi \\ \cos\theta & -\sin\theta & 0 \end{pmatrix} \begin{pmatrix} \mathbf{a}_r \\ \mathbf{a}_\theta \\ \mathbf{a}_\phi \end{pmatrix} \tag{A.11}$$

A.3 | Spherical and Cylindrical Coordinates

The cylindrical coordinate system is represented by (ρ, ϕ, z) while spherical coordinates are (r, θ, ϕ). The ϕ coordinate in both systems are identical

$$\rho = r\sin\theta$$
$$\phi = \phi$$
$$z = r\cos\theta \tag{A.12}$$

$$r = \sqrt{\rho^2 + z^2}$$

$$\theta = \arctan(\rho/z) \tag{A.13}$$

$$\phi = \phi$$

$$\mathbf{a}_r = \mathbf{a}_z \cos\theta + \mathbf{a}_\rho \sin\theta$$

$$\mathbf{a}_\theta = -\mathbf{a}_z \sin\theta + \mathbf{a}_\rho \cos\theta$$

$$\mathbf{a}_\phi = \mathbf{a}_\phi \tag{A.14}$$

$$\mathbf{a}_\rho = \mathbf{a}_\theta \cos\theta + \mathbf{a}_r \sin\theta$$

$$\mathbf{a}_\phi = \mathbf{a}_\phi$$

$$\mathbf{a}_z = -\mathbf{a}_\theta \sin\theta + \mathbf{a}_r \cos\theta \tag{A.15}$$

A.4 | Grad, Div, Curl, and Laplacian in Different Coordinate Systems

A.4.1 Cartesian Coordinates

$$\nabla\Phi = \frac{\partial\Phi}{\partial x}\mathbf{a}_x + \frac{\partial\Phi}{\partial y}\mathbf{a}_y + \frac{\partial\Phi}{\partial z}\mathbf{a}_z \tag{A.16}$$

$$\nabla \times \mathbf{A} = \left(\frac{\partial A_z}{\partial y} - \frac{\partial A_y}{\partial z}\right)\mathbf{a}_x + \left(\frac{\partial A_x}{\partial z} - \frac{\partial A_z}{\partial x}\right)\mathbf{a}_y + \left(\frac{\partial A_y}{\partial x} - \frac{\partial A_x}{\partial y}\right)\mathbf{a}_z \tag{A.17}$$

$$\nabla \times \mathbf{A} = \begin{vmatrix} \mathbf{a}_x & \mathbf{a}_y & \mathbf{a}_z \\ \partial/\partial x & \partial/\partial y & \partial/\partial z \\ A_x & A_y & A_z \end{vmatrix} \tag{A.18}$$

$$\nabla \cdot \mathbf{A} = \frac{\partial A_x}{\partial x} + \frac{\partial A_y}{\partial y} + \frac{\partial A_z}{\partial z} \tag{A.19}$$

$$\nabla^2\Phi = \frac{\partial^2\Phi}{\partial x^2} + \frac{\partial^2\Phi}{\partial y^2} + \frac{\partial^2\Phi}{\partial z^2} \tag{A.20}$$

$$\nabla^2\mathbf{A} = \nabla^2 A_x \mathbf{a}_x + \nabla^2 A_y \mathbf{a}_y + \nabla^2 A_z \mathbf{a}_z \tag{A.21}$$

A.4.2 Cylindrical Coordinates

$$\nabla\Phi = \frac{\partial\Phi}{\partial\rho}\mathbf{a}_\rho + \frac{1}{\rho}\frac{\partial\Phi}{\partial\phi}\mathbf{a}_\phi + \frac{\partial\Phi}{\partial z}\mathbf{a}_z \tag{A.22}$$

$$\nabla \times \mathbf{A} = \left(\frac{1}{\rho}\frac{\partial A_z}{\partial\phi} - \frac{\partial A_\phi}{\partial z}\right)\mathbf{a}_\rho + \left(\frac{\partial A_\rho}{\partial z} - \frac{\partial A_z}{\partial\rho}\right)\mathbf{a}_\phi + \frac{1}{\rho}\left\{\frac{\partial(\rho A_\phi)}{\partial\rho} - \frac{\partial A_\rho}{\partial\phi}\right\}\mathbf{a}_z \tag{A.23}$$

$$\nabla \cdot \mathbf{A} = \frac{1}{\rho}\frac{\partial(\rho A_\rho)}{\partial\rho} + \frac{1}{\rho}\frac{\partial A_\phi}{\partial\phi} + \frac{\partial A_z}{\partial z} \tag{A.24}$$

$$\nabla^2\Phi = \frac{1}{\rho}\frac{\partial}{\partial\rho}\left(\rho\frac{\partial\Phi}{\partial\rho}\right) + \frac{1}{\rho^2}\frac{\partial^2\Phi}{\partial\phi^2} + \frac{\partial^2\Phi}{\partial z^2} \tag{A.25}$$

A.4.3 Spherical Coordinates

$$\nabla\Phi = \frac{\partial\Phi}{\partial r}\mathbf{a}_r + \frac{1}{r}\frac{\partial\Phi}{\partial\theta}\mathbf{a}_\theta + \frac{1}{r\sin\theta}\frac{\partial\Phi}{\partial\phi}\mathbf{a}_\phi \tag{A.26}$$

$$\nabla \times \mathbf{A} = \frac{1}{r \sin\theta}\left\{\frac{\partial(\sin\theta\, A_\phi)}{\partial\theta} - \frac{\partial A_\theta}{\partial\phi}\right\}\mathbf{a}_r + \frac{1}{r}\left\{\frac{1}{\sin\theta}\frac{\partial A_r}{\partial\phi} - \frac{\partial(rA_\phi)}{\partial r}\right\}\mathbf{a}_\theta$$

$$+ \frac{1}{r}\left\{\frac{\partial(rA_\theta)}{\partial r} - \frac{\partial A_r}{\partial\theta}\right\}\mathbf{a}_\phi \tag{A.27}$$

$$\nabla \cdot \mathbf{A} = \frac{1}{r^2}\frac{\partial(r^2 A_r)}{\partial r} + \frac{1}{r\sin\theta}\frac{\partial(\sin\theta\, A_\theta)}{\partial\theta} + \frac{1}{r\sin\theta}\frac{\partial A_\phi}{\partial\phi} \tag{A.28}$$

$$\nabla^2\Phi = \frac{1}{r^2}\frac{\partial}{\partial r}\left(r^2\frac{\partial\Phi}{\partial r}\right) + \frac{1}{r^2\sin\theta}\frac{\partial}{\partial\theta}\left(\sin\theta\frac{\partial\Phi}{\partial\theta}\right) + \frac{1}{r^2\sin^2\theta}\frac{\partial^2\Phi}{\partial\phi^2} \tag{A.29}$$

B

Mathematical Reference

B.1 | General

B.1.1 Important Constants
$e = 2.7183$
$\pi = 3.1461$

B.1.2 Taylor's Series Expansion

1. About $x = 0, f(x) = f(0) + f'(0)x + f''(0)\dfrac{x^2}{2}\cdots$

2. About $x = a, f(x) = f(a) + f'(a)(x-a) + f''(a)\dfrac{(x-a)^2}{2}\cdots$
3. $1/\sqrt{1+\alpha x} \approx 1 - \alpha x + \alpha^2 x^2 - \alpha^3 x^3 + \ldots$

B.1.3 $|\mathbf{r} - \mathbf{r}'|$ in Various Coordinate Systems

1. Rectangular, $\quad |\mathbf{r} - \mathbf{r}'| = \sqrt{(x-x')^2 + (y-y')^2 + (z-z')^2}$
2. Spherical,

$$|\mathbf{r} - \mathbf{r}'| = \sqrt{r'^2 + r^2 - 2r\,r'\cos(\phi - \phi')\sin(\theta)\sin(\theta') - 2\,r\,r'\cos(\theta)\cos(\theta')}$$

B.2 | Vector Identities

In the following equations $\mathbf{A}$, $\mathbf{B}$, and $\mathbf{C}$ are vector fields, while a, b, and c are scalar fields. When a vector field is a constant then $\nabla \cdot (\cdots)$ and $\nabla \times (\cdots)$ yield zero results. When a scalar field is a constant then $\nabla(\cdots)$ gives a zero result.

B.2.1 General
1. $\mathbf{A} \cdot (\mathbf{B} \times \mathbf{C}) = \mathbf{B} \cdot (\mathbf{C} \times \mathbf{A}) = \mathbf{C} \cdot (\mathbf{A} \times \mathbf{B})$
2. $\mathbf{A} \times (\mathbf{B} \times \mathbf{C}) = (\mathbf{A} \cdot \mathbf{C})\mathbf{B} - (\mathbf{A} \cdot \mathbf{B})\mathbf{C}$

B.2.2 Gradient

1. $\nabla(a + b) = \nabla a + \nabla b$
2. $\nabla(ab) = b\nabla a + a\nabla b$
3. $\nabla(a/b) = \dfrac{1}{b}\nabla a - \dfrac{a}{b^2}\nabla b = \dfrac{b\nabla a - a\nabla b}{b^2}$
4. $\nabla(a^n) = na^{(n-1)}\nabla a$
5. $\nabla(\mathbf{A} \cdot \mathbf{B}) = (\mathbf{A} \cdot \nabla)\mathbf{B} + (\mathbf{B} \cdot \nabla)\mathbf{A} + \mathbf{A} \times (\nabla \times \mathbf{B}) + \mathbf{B} \times (\nabla \times \mathbf{A})$

B.2.3 Curl

1. $\nabla \times (\mathbf{A} + \mathbf{B}) = \nabla \times \mathbf{A} + \nabla \times \mathbf{B}$
2. $\nabla \times (a\mathbf{A}) = \nabla a \times \mathbf{A} + a\nabla \times \mathbf{A}$

B.2.4 Divergence

1. $\nabla \cdot (\mathbf{A} + \mathbf{B}) = \nabla \cdot \mathbf{A} + \nabla \cdot \mathbf{B}$
2. $\nabla \cdot (a\mathbf{A}) = (\nabla a) \cdot \mathbf{A} + a\nabla \cdot \mathbf{A}$
3. $\nabla \cdot (\mathbf{A} \times \mathbf{B}) = \mathbf{B} \cdot \nabla \times \mathbf{A} - \mathbf{A} \cdot \nabla \times \mathbf{B}$
4. $\nabla \times (\mathbf{A} \times \mathbf{B}) = \mathbf{A}\nabla \cdot \mathbf{B} - \mathbf{B}\nabla \cdot \mathbf{A} - (\mathbf{A} \cdot \nabla)\mathbf{B} + (\mathbf{B} \cdot \nabla)\mathbf{A}$

B.2.5 Double

1. $\nabla \cdot (\nabla a \times \nabla b) = 0$
2. $\nabla \cdot (\nabla a) = (\nabla \cdot \nabla) a = \nabla^2 a$
3. $\nabla \cdot (\nabla \times \mathbf{A}) = 0$
4. $\nabla \times (\nabla a) = 0$
5. $\nabla \times (\nabla \times \mathbf{A}) = \nabla(\nabla \cdot \mathbf{A}) - (\nabla \cdot \nabla)\mathbf{A}$
6. $\nabla \cdot (a\nabla b - b\nabla a) = a\nabla^2 b - b\nabla^2 a$
7. $\nabla^2(a + b) = \nabla^2 a + \nabla^2 b$
8. $\nabla^2(ab) = a\nabla^2 b + 2\nabla a \cdot \nabla b + b\nabla^2 a$

B.3 | Complex Variables

B.3.1 General

If $z_1 = a + jb = r_1 e^{j\theta_1} = r_1(\cos\theta_1 + j\sin\theta_1)$ and
$z_2 = c + jd = r_2 e^{j\theta_2} = r_2(\cos\theta_2 + j\sin\theta_2)$
where $j = \sqrt{-1}$, $r_1 = \sqrt{a^2 + b^2}$, $r_2 = \sqrt{c^2 + d^2}$, $\theta_1 = \tan^{-1}(b/a)$ and
$\theta_2 = \tan^{-1}(d/c)$ then

1. $z_1 + z_2 = (a + jb) + (c + jd) = (a + c) + j(b + d)$
2. $z_1 - z_2 = (a + jb) - (c + jd) = (a - c) + j(b - d)$
3. $z_1 \cdot z_2 = (a + jb) \times (c + jd) = ac + jbc + jad + bdj^2 = (ac - bd) + j(bc + ad)$
   $= r_1 r_2 e^{j(\theta_1 + \theta_2)}$
4. $z_1/z_2 = \dfrac{(a + jb)}{(c + jd)} = \left(\dfrac{ac + bd}{c^2 + d^2}\right) + j\left(\dfrac{bc - ad}{c^2 + d^2}\right) = (r_1/r_2)\, e^{j(\theta_1 - \theta_2)}$

B.3.2 Inequalities

If z and w are two complex numbers, then the absolute value has three important properties:

1. $|z| \geq 0$, where $|z| = 0$ if and only if $z = 0$
2. $|z + w| \leq |z| + |w|$ (triangle inequality)
3. $|z \cdot w| = |z| \cdot |w|$

B.3.3 Complex Conjugates

The complex conjugate of the complex number $z = z^* = (x + jy)^* = x - jy$. z^* has the properties

1. $(z + w)^* = z^* + w^*$
2. $(z \cdot w)^* = z^* \cdot w^*$
3. $(z/w)^* = z^*/w^*$
4. $(z^*)^* = z$
5. $\Re(z) = \dfrac{1}{2}(z + z^*)$
6. $\Im(z) = \dfrac{1}{2}(z - z^*)$
7. $|z| = |z^*|$
8. $|z|^2 = z \cdot z^*$

B.3.4 Euler's Identity

$\cos\theta + j\sin\theta = e^{i\theta}$

B.4 | Trigonometry

B.4.1 Basic Formulae

1. $\sin^2\theta + \cos^2\theta = 1$
2. $\tan^2\theta + 1 = \sec^2\theta$
3. $1 + \cot^2\theta = \csc^2\theta$
4. $\csc\theta = 1/\sin\theta; \sec\theta = 1/\cos\theta; \tan\theta = 1/\cot\theta$
5. $\tan\theta = \sin\theta/\cos\theta$
6. $\sin(\pi/2 - \theta) = \cos\theta; \cos(\pi/2 - \theta) = \sin\theta; \tan(\pi/2 - \theta) = \cot\theta$
7. $\sin(-\theta) = -\sin\theta; \cos(-\theta) = \cos\theta; \tan(-\theta) = -\tan\theta$

B.4.2 Sum and Difference Formulae

1. $\cos(a \pm b) = \cos a \cos b \mp \sin a \sin b$
2. $\sin(a \pm b) = \sin a \cos b \pm \cos a \sin b$
3. $\tan(a \pm b) = (\tan a \pm \tan b)/(1 \mp \tan a \tan b)$

B.4.3 Double Angle Formulae

1. $\sin(2a) = 2\sin a \cos a = 2\tan a/(1 + \tan^2 a)$
2. $\cos(2a) = \cos^2 a - \sin^2 a = 1 - 2\sin^2 a = 2\cos^2 a - 1 = (1 - \tan^2 a)/(1 + \tan^2 a)$
3. $\tan(2a) = (2\tan a)/(1 - \tan^2 a)$

B.4.4 Half Angle Formulae

1. $\sin(a) = \sqrt{\dfrac{1 - \cos(2a)}{2}}; \quad \sin^2(a) = \left[\dfrac{1 - \cos(2a)}{2}\right]$
2. $\cos(a) = \sqrt{\dfrac{1 + \cos(2a)}{2}}; \quad \cos^2(a) = \left[\dfrac{1 + \cos(2a)}{2}\right]$
3. $\tan(a) = \sqrt{\dfrac{1 - \cos(2a)}{1 + \cos(2a)}}; \quad \tan^2(a) = \left[\dfrac{1 - \cos(2a)}{1 + \cos(2a)}\right]$

B.4.5 Product to Sum Formulae

1. $\sin(a)\sin(b) = \dfrac{1}{2}[\cos(a - b) - \cos(a + b)]$

2. $\cos(a)\cos(b) = \dfrac{1}{2}[\cos(a-b) + \cos(a+b)]$

3. $\sin(a)\cos(b) = \dfrac{1}{2}[\sin(a-b) + \sin(a+b)]$

B.4.6 Sum and Difference to Product

1. $\sin(a) + \sin(b) = 2\sin\left(\dfrac{a+b}{2}\right)\cos\left(\dfrac{a-b}{2}\right)$

2. $\sin(a) - \sin(b) = 2\cos\left(\dfrac{a+b}{2}\right)\sin\left(\dfrac{a-b}{2}\right)$

3. $\cos(a) + \cos(b) = 2\cos\left(\dfrac{a+b}{2}\right)\cos\left(\dfrac{a-b}{2}\right)$

4. $\cos(a) - \cos(b) = 2\sin\left(\dfrac{a+b}{2}\right)\sin\left(\dfrac{a-b}{2}\right)$

5. $A\sin(a) + B\cos(a) = C\sin(a+\phi)$ where $C = \sqrt{A^2 + B^2}$ and $\phi = \tan^{-1}\left(\dfrac{B}{A}\right)$

B.4.7 Triangle Formulae

If in a triangle of sides a, b, and c, the angles opposite these sides are A, B, and C; then

1. $a/\sin A = b/\sin B = c/\sin C$
2. $a^2 = b^2 + c^2 - 2bc\,\cos A$
3. $b^2 = a^2 + c^2 - 2ac\,\cos B$
4. $c^2 = b^2 + a^2 - 2ba\,\cos C$

B.4.8 Powers of the Trigonometric Functions

1. $\sin^3\theta = \dfrac{1}{4}(3\sin\theta - \sin 3\theta)$

2. $\cos^3\theta = \dfrac{1}{4}(3\cos\theta + \cos 3\theta)$

3. $\sin^4\theta = \dfrac{1}{8}(\cos(4\theta) - 4\cos(2\theta) + 3)$

4. $\cos^4\theta = \dfrac{1}{8}(\cos(4\theta) + 4\cos(2\theta) + 3)$

B.5 | Differentiation

c is a constant, f and g are functions

B.5.1 Rules

1. $(cf)' = cf'$
2. $(f + g)' = f' + g'$
3. $(fg)' = f'g + fg'$
4. $\left(\dfrac{1}{f}\right)' = \dfrac{-f'}{f^2}$
5. $\left(\dfrac{f}{g}\right)' = \dfrac{f'g - fg'}{g^2}$
6. $[f(g)]' = f'g'$
7. Derivative of inverse function: $(f^{-1})' = \dfrac{1}{f'(f^{-1})}\left[\text{example } (\ln x)' = \dfrac{1}{e^{\ln x}} = \dfrac{1}{x}\right]$

8. Generalised power rule $(f^g)' = f^g \left(g' \ln f + \dfrac{g}{f} f' \right)$

B.5.2 Differentiation of Functions

1. $c' = 0$
2. $x' = 1$
3. $(cx)' = c$
4. $(x^c)' = cx^{c-1}$
5. $\left(\dfrac{1}{x} \right)' = (x^{-1})' = -x^{-2} = -\dfrac{1}{x^2}$
6. $\left(\dfrac{1}{x^c} \right)' = (x^{-c})' = -cx^{-c-1} = -\dfrac{c}{x^{c+1}}$
7. $(c^x)' = c^x \ln c, \quad c > 0$
8. $(e^x)' = e^x$
9. $(\log_c x)' = \dfrac{1}{x \ln c}, \quad c > 0, c \neq 1$
10. $(\ln x)' = \dfrac{1}{x}, \quad x > 0$
11. $(x^x)' = x^x (1 + \ln x)$
12. $(\sin x)' = \cos x$
13. $(\arcsin x)' = \dfrac{1}{\sqrt{1 - x^2}}$
14. $(\cos x)' = -\sin x$
15. $(\arccos x)' = \dfrac{-1}{\sqrt{1 - x^2}}$
16. $(\tan x)' = \sec^2 x = \dfrac{1}{\cos^2 x}$
17. $(\arctan x)' = \dfrac{1}{1 + x^2}$
18. $(\sec x)' = \sec x \tan x$
19. $(\operatorname{arcsec} x)' = \dfrac{1}{|x| \sqrt{x^2 - 1}}$
20. $(\csc x)' = -\csc x \cot x$
21. $(\operatorname{arccsc} x)' = \dfrac{-1}{|x| \sqrt{x^2 - 1}}$
22. $(\cot x)' = -\csc^2 x = \dfrac{-1}{\sin^2 x}$
23. $(\operatorname{arccot} x)' = \dfrac{-1}{1 + x^2}$
24. $(\sinh x)' = \cosh x = \dfrac{e^x + e^{-x}}{2}$
25. $(\operatorname{arcsinh} x)' = \dfrac{1}{\sqrt{x^2 + 1}}$
26. $(\cosh x)' = \sinh x = \dfrac{e^x - e^{-x}}{2}$
27. $(\operatorname{arccosh} x)' = \dfrac{1}{\sqrt{x^2 - 1}}$
28. $(\tanh x)' = \operatorname{sech}^2 x$

29. $(\operatorname{arctanh} x)' = \dfrac{1}{1 - x^2}$

30. $(\operatorname{sech} x)' = -\tanh x$

31. $(\operatorname{sech} x)' = -\tanh x \operatorname{sech} x$

32. $(\operatorname{arcsech} x)' = \dfrac{-1}{x\sqrt{1 - x^2}}$

33. $(\operatorname{csch} x)' = -\coth x \operatorname{csch} x$

34. $(\operatorname{arccsch} x)' = \dfrac{-1}{x\sqrt{1 + x^2}}$

35. $(\coth x)' = -\operatorname{csch}^2 x$

36. $(\operatorname{arccoth} x)' = \dfrac{1}{1 - x^2}$

B.6 | Integration

B.6.1 Common Substitutions

1. $\displaystyle\int f(ax + b)\, dx = \dfrac{1}{a}\int f(u)\, du$ where $u = ax + b$

2. $\displaystyle\int f\left(\sqrt[n]{ax + b}\right) dx = \dfrac{n}{a}\int u^{n-1} f(u)\, du$ where $u = \sqrt[n]{ax + b}$

3. $\displaystyle\int f\left(\sqrt{a^2 - x^2}\right) dx = a\int f(a\cos u)\cos u\, du$ where $x = a\sin u$

4. $\displaystyle\int f\left(\sqrt{x^2 + a^2}\right) dx = a\int f(a\sec u)\sec^2 u\, du$ where $x = a\tan u$

5. $\displaystyle\int f\left(\sqrt{x^2 - a^2}\right) dx = a\int f(a\tan u)\sec u \tan u\, du$ where $x = a\sec u$

B.6.2 Indefinite Integrals

a, b, c are constants, u, v, w are functions of t

1. $\displaystyle\int a\, dt = at$

2. $\displaystyle\int a f(t)\, dt = a\int f(t)\, dt$

3. $\displaystyle\int (u \pm v \pm w \pm \cdots)\, dt = \int u\, dt \pm \int v\, dt \pm \int w\, dt \pm \cdots$

4. $\displaystyle\int u\, dv = uv - \int v\, du$

5. $\displaystyle\int f(at)\, dt = \dfrac{1}{a}\int f(u)\, du$

6. $\displaystyle\int F\{f(t)\}\, dt = \int F(u)\dfrac{dxt}{du}\, du = \int \dfrac{F(u)}{f'(t)}\, du$

7. $\displaystyle\int t^n\, dt = \begin{cases} \dfrac{t^{n+1}}{n + 1} & n \neq -1 \\ \ln t & n = -1 \end{cases}$

8. $\displaystyle\int e^{-at}\, dt = -\dfrac{e^{-at}}{a}$

9. $\displaystyle\int a^{-bt}\, dt = -\dfrac{a^{-bt}}{b \log a}$

10. $\displaystyle\int \sin(at+b)\ dt = -\frac{1}{a}\cos(at+b)$

11. $\displaystyle\int \cos(at+b)\ dt = \frac{1}{a}\sin(at+b)$

12. $\displaystyle\int \frac{f'(t)}{f(t)}dt = \ln[f(t)]$

13. $\displaystyle\int \tan(at+b) = \frac{\ln[\sec(at+b)]}{a} = -\frac{\ln[\cos(at+b)]}{a}$

14. $\displaystyle\int \cot(at+b) = \frac{\ln[\sin(at+b)]}{a}$

15. $\displaystyle\int \sec(at+b) = \frac{\ln[\tan(at+b)+\sec(at+b)]}{a}$

16. $\displaystyle\int t^n e^{-at}\,dt \quad n = \text{positive integer; use } \int u\,dv = uv - \int v\,du$

17. $\displaystyle\int \frac{1}{t^2+a^2}\ dt = \frac{1}{a}\arctan\left(\frac{t}{a}\right)$

18. $\displaystyle\int \frac{t}{t^2 \pm a^2}\ dt = \frac{\log(t^2 \pm a^2)}{2}$

19. $\displaystyle\int \frac{t^2}{t^2+a^2}dt = t - a\arctan\left(\frac{t}{a}\right)$

20. $\displaystyle\int \frac{1}{t^2-a^2}\ dt = \frac{\log(t-a)}{2a} - \frac{\log(t+a)}{2a}$

21. $\displaystyle\int \frac{t^2}{t^2-a^2}dt - \frac{a\log(t+a)}{2} + \frac{a\log(t-a)}{2} + t$

22. $\displaystyle\int \frac{1}{\sqrt{t^2+a^2}}dt = \sinh^{-1}\left(\frac{x}{|a|}\right)$

23. $\displaystyle\int \frac{1}{\sqrt{t^2-a^2}}dt = \log(2\sqrt{t^2-a^2}+2t)$

24. $\displaystyle\int \frac{t}{\sqrt{t^2 \pm a^2}}dt = \sqrt{t^2 \pm a^2}$

25. $\displaystyle\int \frac{1}{\sqrt{a^2-t^2}}dt = \arcsin\left(\frac{t}{a}\right)$

26. $\displaystyle\int \frac{1}{\sqrt{t^2+a^2}}dt = \sinh^{-1}\left(\frac{t}{|a|}\right)$

27. $\displaystyle\int \frac{1}{(a^2-t^2)^{\frac{3}{2}}}dt = \frac{t}{a^2\sqrt{a^2-t^2}}$

28. $\displaystyle\int \frac{1}{(t^2-a^2)^{\frac{3}{2}}}dt = -\frac{t}{a^2\sqrt{t^2-a^2}}$

29. $\displaystyle\int \frac{1}{(t^2+a^2)^{\frac{3}{2}}}dt = \frac{t}{a^2\sqrt{t^2+a^2}}$

30. $\displaystyle\int \frac{t}{(t^2+a^2)^{\frac{3}{2}}}dt = -\frac{1}{\sqrt{t^2+a^2}}$

31. $\displaystyle\int \frac{t^2}{(t^2+a^2)^{\frac{3}{2}}}dt = \sinh^{-1}\left(\frac{t}{|a|}\right) - \frac{t}{\sqrt{t^2+a^2}}$

32. $\displaystyle\int \frac{t^3}{\left(t^2 + a^2\right)^{\frac{3}{2}}} dt = \frac{2\,a^2}{\sqrt{a^2 + x^2}} + \frac{t^2}{\sqrt{a^2 + t^2}}$

33. $\displaystyle\int \frac{1}{\sqrt{t^2 + bt + c}}\, dt = \begin{cases} \log\left(t + \dfrac{b}{2}\right), & \text{for } 4c - b^2 = 0 \\[2mm] \sinh^{-1}\left(\dfrac{2t + b}{\sqrt{4c - b^2}}\right), & \text{for } 4c - b^2 > 0 \\[2mm] \log\left(2\sqrt{t^2 + bt + c} + 2t + b\right), & \text{for } 4c - b^2 < 0 \end{cases}$

34. $\displaystyle\int \frac{1}{\sqrt{-t^2 + bt + c}}\, dt = \begin{cases} -\sin^{-1}\left(\dfrac{b - 2t}{\sqrt{4c + b^2}}\right), & \text{for } b^2 + 4c > 0 \\[2mm] -j\,\sinh^{-1}\left(\dfrac{2t - b}{\sqrt{-4c - b^2}}\right), & \text{for } b^2 + 4c < 0 \\[2mm] j\,\ln\left(x - \dfrac{b}{2}\right), & \text{for } b^2 + 4c = 0 \end{cases}$

35. $\displaystyle\int \frac{1}{(t^2 + bt + c)^{\frac{3}{2}}}\, dt$

$= \begin{cases} \dfrac{4\,t}{(4c - b^2)\,\sqrt{t^2 + bt + c}} + \dfrac{2\,b}{(4c - b^2)\,\sqrt{t^2 + bt + c}}, & \text{for } 4c - b^2 \neq 0 \\[4mm] -\dfrac{1}{2\left(t + \dfrac{b}{2}\right)^2}, & \text{for } 4c - b^2 = 0 \end{cases}$

36. $\displaystyle\int \frac{1}{(-t^2 + bt + c)^{\frac{3}{2}}}\, dt$

$= \begin{cases} \dfrac{2\,b}{(-4c - b^2)\,\sqrt{-t^2 + bt + c}} - \dfrac{4\,t}{(-4c - b^2)\,\sqrt{-t^2 + bt + c}}, & \text{for } 4c + b^2 \neq 0 \\[4mm] -\dfrac{j}{2\left(t - \frac{b}{2}\right)^2}, & \text{for } 4c + b^2 = 0 \end{cases}$

37. $\displaystyle\int \frac{1}{t^2 + bt + c}\, dt$

$= \begin{cases} \left(\sqrt{4c - b^2}\right)^{-1} 2 \arctan\left(\dfrac{2t + b}{\sqrt{4c - b^2}}\right), & \text{for } 4c - b^2 > 0 \\[4mm] \left(\sqrt{b^2 - 4c}\right)^{-1} \log\left(\dfrac{2t - \sqrt{b^2 - 4c} + b}{2t + \sqrt{b^2 - 4c} + b}\right), & \text{for } 4c - b^2 < 0 \end{cases}$

38. $\displaystyle\int e^{\beta x} \sin(\alpha\, t + \phi)\, dt = \frac{e^{\beta x}\,[\beta\,\sin(\alpha\, t + \phi) - \alpha\,\cos(\alpha\, t + \phi)]}{\beta^2 + \alpha^2}$

39. $\displaystyle\int e^{\beta x} \cos(\alpha\, t + \phi)\, dt = \frac{e^{\beta x}\,[\alpha\,\sin(\alpha\, t + \varphi) + \beta\,\cos(\alpha\, t + \varphi)]}{\beta^2 + \alpha^2}$

C

Some Key Equations

In electromagnetics, a big stumbling block is the memorisation of key equations. This appendix tries to remedy this situation somewhat.

Some equations just *have* to be memorised. Among them are the all important Maxwell's equations:

$$\nabla \cdot \mathbf{D} = \rho_v$$

$$\nabla \times \mathbf{E} = -\partial \mathbf{B}/\partial t$$

$$\nabla \cdot \mathbf{B} = 0$$

$$\nabla \times \mathbf{H} = \mathbf{J} + \partial \mathbf{D}/\partial t$$

where $\partial \mathbf{B}/\partial t$ and $\partial \mathbf{D}/\partial t$ are the time-dependant terms.

The first and third equations are Gauss's laws (or divergence) for electric and magnetic fields. The second equation is Faraday's law and the last equation is Ampere's circuital law. From these laws, the second equation leads to the scalar potential and third equation defines the vector potential. That is for $\partial/\partial t = 0$,

$$\mathbf{E} = -\nabla V$$

$$\mathbf{B} = \nabla \times \mathbf{A}$$

from $\nabla \times \mathbf{E} = 0$ and $\nabla \cdot \mathbf{B} = 0$, respectively.

Next starting from the electric field and potential due to a point charge, we arrive at a number of key equations. The electric field due to a *point* charge Q placed at $\mathbf{r}'$ is

$$\mathbf{E}(\mathbf{r}) = \left(\frac{Q}{4\pi\varepsilon_0}\right) \frac{(\mathbf{r} - \mathbf{r}')}{|\mathbf{r} - \mathbf{r}'|^3}$$

from which we get the general electric field due to an infinitesimal charge

$$d\mathbf{E}(\mathbf{r}) = \left(\frac{dQ}{4\pi\varepsilon_0}\right) \frac{(\mathbf{r} - \mathbf{r}')}{|\mathbf{r} - \mathbf{r}'|^3}$$

and from this equation,

$$\int d\mathbf{E}(\mathbf{r}) = \int \left[\left(\frac{dQ}{4\pi\varepsilon_0}\right) \frac{(\mathbf{r} - \mathbf{r}')}{|\mathbf{r} - \mathbf{r}'|^3}\right] \tag{C.1}$$

Similarly, the potential due to a point charge placed at $\mathbf{r}'$ is

$$V(\mathbf{r}) = \left(\frac{Q}{4\pi\varepsilon_0}\right) \frac{1}{|\mathbf{r} - \mathbf{r}'|}$$

from which we get the general potential field due to an infinitesimal charge

$$dV(\mathbf{r}) = \left(\frac{dQ}{4\pi\varepsilon_0}\right) \frac{1}{|\mathbf{r} - \mathbf{r}'|}$$

and from this equation

$$\int dV(\mathbf{r}) = \int \left[\left(\frac{dQ}{4\pi\varepsilon_0}\right) \frac{1}{|\mathbf{r} - \mathbf{r}'|}\right] \tag{C.2}$$

From Eqn (C.1), we can get the Biot-Savart law

$$\int d\mathbf{E}(\mathbf{r}) = \int \left[\left(\frac{dQ}{4\pi\varepsilon_0}\right) \frac{(\mathbf{r} - \mathbf{r}')}{|\mathbf{r} - \mathbf{r}'|^3}\right] \Rightarrow \int d\mathbf{H}(\mathbf{r}) = \int \left[\left(\frac{\mathbf{J}dV'}{4\pi}\right) \times \frac{(\mathbf{r} - \mathbf{r}')}{|\mathbf{r} - \mathbf{r}'|^3}\right]$$

where dQ $(= \rho_v \, dV')$ has been substituted by $\mathbf{J}dV'$ and the *regular product* by the *cross-product*. (Note that ε_0 has been replaced by '1'). In the same manner starting from Eqn (C.2)

$$\int dV(\mathbf{r}) = \int \left[\left(\frac{dQ}{4\pi\varepsilon_0}\right) \frac{1}{|\mathbf{r} - \mathbf{r}'|}\right] \Rightarrow \int d\mathbf{A}(\mathbf{r}) = \int \left[\left(\frac{\mu_0 \, \mathbf{J}dV'}{4\pi}\right) \frac{1}{|\mathbf{r} - \mathbf{r}'|}\right]$$

D

Sample Question Papers

Question Paper 1

1. Attempt any six questions.

   (a) State and prove the divergence theorem.

   (b) Points P and Q are located at $(0, 2, 4)$ and $(-3, 1, 5)$. Calculate

      (i) The position vector of P

      (ii) The vector from P to Q

      (iii) The distance between P and Q

      (iv) A vector parallel to PQ with magnitude 5.

   (c) Define Biot-Savart's law using the concept of magnetic vector potential.

   (d) Explain how the magnetic dipoles act as a distributed source for the magnetic field.

   (e) State and prove Poynting's theorem.

   (f) Briefly explain the plane wave propogation.

   (g) Write a short note on TM and TE modes of propogation.

   (h) State the significance and features of Smith chart. (Calicut University, 2004)

2. Attempt any four.

   (a) State and prove Gauss's theorem. Explain why is it called the divergence theorem.

   (b) Explain the concept of displacement current. How is this current different from conduction current?

   (c) Write Maxwell's equation in free space for the time varying fields both in differential and integral form. Why are these equations not completely symmetrical?

   (d) Prove that in a travelling plane electromagnetic wave there is a definite ratio between the amplitudes of **E** and **H**. Find this ratio.

   (e) Explain briefly the oblique incidence of wave on a perfect electric conductor.
   (Punjab Technical University)

3. Attempt any two.

   (a) Attempt both questions.

      (i) Discuss analogies between electric and magnetic fields.

570 | Appendix D: Sample Question Papers

(ii) Develop an expression for the potential difference at any point between spherical shells in terms of the applied potential employing Laplace's equation.

(b) Attempt both questions.

(i) Differentiate between phase velocity and group velocity. Calculate the velocity of electromagnetic wave in a medium whose dielectric constant is 2.56.

(ii) A plane electromagnetic wave travelling in free space has an amplitude of E_0 equal to $50\,\mu V/m$ and the electric field vector at any point varies sinusoidally with time. What are the peak and average value of Poynting vector?

(c) Write short notes on

(i) Uniqueness theorem.

(ii) Reflection at surface of conducting medium.

(Punjab Technical University)

4. The semi-infinite regions with $z < 0$ and $z > 1$ are free space. For $0 < z < 1$, $\varepsilon_r = 4$, $\mu_r = 1$, and $\sigma = 0$, a uniform plane wave travels towards the $z = 0$ interface in the $\mathbf{a}_z$ direction with $\omega = 4 \times 10^8$ rev/s.

(a) Find the VSWR in all three regions.

(b) Find the location of the maximum of the $|\mathbf{E}|$ for $z < 0$ but nearest to $z = 0$

(Jaypee University of Information Technology, 2009)

Question Paper 2

1. Answer all questions.

(a) Mention the importance of a unit vector.

(b) State Faraday's law of elctromagnetic induction.

(c) Write Laplace's equation in cylindrical coordinates.

(d) State boundary conditions.

(e) What are the conditions for the field to be irrotational?

(f) Define Poynting vector.

(g) State the Stokes's theorem. What do you infer from it?

(h) What is meant by homogenous and isotropic mediums?

(i) Define propogation constant.

(j) State the uniqueness theorem. (Punjab Technical University)

2. Answer any two.

(a) (i) Justify that the net electric field within a conductor is always zero.

(ii) Derive the equation of continuity for time varying fields.

(b) Define uniform plane wave propogation. Discuss its properties. A uniform plane electromagnetic wave propogating in air is given by

$$E = j \cos[\omega t - (2\pi/\lambda)y]$$

Derive by using Maxwell's equations, the expression for the vector magnetic field.

(c) Write short notes on

(i) Magnetic vector potential.

(ii) Helmholtz equation.   (Punjab Technical University)

3. Solve any one question of the following:

  (a) Given point $P(-2, 6, 3)$ and vector $\mathbf{A} = y\mathbf{a}_x + (x+z)\,\mathbf{a}_y$ in rectangular coordinates, express P and $\mathbf{A}$ in cylindrical and spherical coordinates. Evaluate $\mathbf{A}$ at P in the Cartesian, cylindrical, and spherical systems.

  (b) Derive the boundary conditions for the following interfaces:

    (i) Dielectric (ε_{r_1}) and dielectric (ε_{r_2}).

    (ii) Conductor and dielectric.   (Calicut University, 2004)

4. Attempt any four.

  (a) Differentiate between scalar quantity and a scalar field and vector quantity and and vector field.

  (b) What is the physical definition of a curl of a vector field?

  (c) Discuss the Cartesian coordinate system.

  (d) Given

    (i) $\mathbf{A} = 2\mathbf{a}_x - 3\mathbf{a}_y + \mathbf{a}_z$

    (ii) $\mathbf{B} = 2\mathbf{a}_x - \mathbf{a}_y + 3\mathbf{a}_z$

    (iii) $\mathbf{C} = 4\mathbf{a}_x - 2\mathbf{a}_y - 2\mathbf{a}_z$

    Prove that $\mathbf{C}$ is perpendiculer to both $\mathbf{B}$ and $\mathbf{A}$.

  (e) Given

    $\mathbf{A} = \mathbf{a}_x(2x + 3y) - \mathbf{a}_y(2y + 3z) + \mathbf{a}_z(3x - y)$

    Determine the unit vector parallel to $\mathbf{A}$ at point $P(1, -1, 2)$

  (f) Find out the gradient of a scalar $\phi = x^2 + y^2 + 2xz$

                         (UP Technical University, 2006–2007)

Question Paper 3

1. Attempt all questions.

  (a) Differentiate between scalar fields and vector fields. Give examples.

  (b) Define divergence theorem.

  (c) State Biot-Savart's law.

  (d) What do you mean by equipotential surfaces?

  (e) State the condition at a boundary between dielectric and conducting surfaces.

  (f) Give the expression for energy stored in a static electric field.

  (g) What is Ampere's law?

  (h) What is displacement current? Does it exist in free space?

  (i) Define magnetic field intensity and give its relation with magnetic flux density.

  (j) What do you understand by homogenous and isotropic medium?

                         (Punjab Technical University)

2. Attempt any four.

  (a) State and prove Gauss's law.

  (b) Write Maxwell's equation in time varying fields and give their interpretation.

  (c) Find an expression for the magnetic flux density $\mathbf{B}$ at a distance h above the center of rectangular loop of wire b meter on one side and a meter on the other side. The loop carries a current of 1 A.

  (d) State and prove Poynting theorem.

   (e) Differentiate between linear, elliptical, and circular polarisations.

(Punjab Technical University)

3. Do any one question of the following.

  (a) (i) Write a short note on boundary conditions for magnetic fields.

      (ii) Determine the boundary conditions for the normal components of the magnetic flux density at the interface between the two regions.

  (b) (i) Explain in detail about Faraday's laws of electromagnetic induction.

      (ii) Derive an expression for the induced emf of an open coil with N turns.

(Calicut University, 2004)

4. Attempt any four.

  (a) What is the boundary condition for electrostatic potential at an interface between two different dielectric media?

  (b) State and prove Laplace's equation for a simple medium in vector notation.

  (c) Find a mathematical expression for electrostatic energy in terms of field quantities.

  (d) Four capacitators $C_1 = 1\,\mu F$, $C_2 = 2\,\mu F$, $C_3 = 3\,\mu F$, $C_4 = 4\,\mu F$ are connected as in Fig. D.1. A DC voltage of 100 V is applied to the external terminal $a - b$. Determine

    (i) Total equivalent capacitance between ab.

    (ii) Charge on each capacitator.

    (iii) The potential difference accross each capacitator.

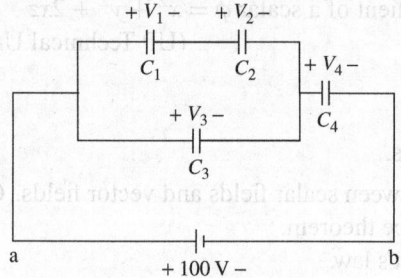

Fig. D.1 Circuit for Problem 4(d)

  (e) State and explain Coulomb's law.

  (f) Is Gauss's law useful in finding the electric field vector of a finite line charge? Explain. (UP Technical University, 2006–2007)

Question Paper 4

1. Attempt all questions.

  (a) (i) State and explain the Lorentz force equation.

      (ii) State and explain Ampere's circuital law.

  (b) The potential difference at any point in a space containing dielectric material of relative permittivity 2.1 is given by $V = 5x^2y + 3yz^2 + 6xz$ V where x, y, z are in meters. Find the volume charge density at any point.

  (c) Derive an expression of magnetic energy in terms of field quantities.

(UP Technical University, 2006–2007)

2. Attempt all questions.
   (a) Write the differential form of Maxwell's equations. Are all four Maxwell's equation independent? Explain.
   (b) State and explain the Poynting theorem.
   (c) Derive the formula

$$P = \frac{1 + |\Gamma|}{1 - |\Gamma|}$$

   where P is the SWR.   (UP Technical University, 2006–2007)

3. Attempt both questions.
   (a) The electric field intensity in a dielectric medium (perfect) is given as $\mathbf{E} = E_0 \cos(\omega t - kz)\, \mathbf{a}_x$ V/m, where E_0 is its peak value and k is a constant quantity. Determine
      (i) The magnetic field intensity of the region.
      (ii) The direction of power flow.
      (iii) The average power density.
   (b) Derive the boundary conditions for time varying fields.
                                          (Calicut University, 2004)

4. Attempt both questions.
   (a) (i) What are the possible reasons for transients along transmission lines?
       (ii) Express the propogation constant and the characteristic impedence of a transmission line in terms of its parameters.
   (b) Explain in detail about the phenemenon called the skin effect.
                                          (Calicut University, 2004)

References

1. Bowman, F. (1968). *Introduction to Bessel Functions*, Dover Publications.

2. Davis, J. L. & P. Annan (1980). Electromagnetic determination of soil water content: Measurement in coaxial transmission lines, *Water Resources Research* 16(3), 574–582.

3. Feynman , R. P., R. B. Leighton, & M. Sands (2001). *The Feynman Lectures on Physics— Volume* 2, Narosa Publishing House.

4. Hall, M. P. M. & M. T. Hewitt (1996). *Propagation of Radio Waves*, IEE.

5. Hayt Jr., W. H. & J. A. Buck (2001). *Engineering Electromagnetics*, 6th Edn, Tata-Mcgraw Hill.

6. Jackson, J. D. (1999). *Classical Electrodynamics*, John Wiley and Sons.

7. Jordan, E. C. & K. G. Balmain (1968). *Electromagnetic Waves and Radiating Systems*, 2nd edn, Prentice Hall of India.

8. Kraus, J. D. (1988). *Antennas*, 2nd edn, McGraw Hill International Editions.

9. Kreyszig, E. (2003). *Advanced Engineering Mathematics*, 8th edn, John Wiley and Sons (Asia).

10. Manheimer, N., L. Sugiyama, & T. Stix (eds) (1977). *Plasma Science and the Environment*, AIP Press.

11. Narayan, S. (2001). *Theory of Functions of a Complex Variable*, 7th Edn, S. Chand and Company.

12. Norton, K. A. (1936). The propagation of waves over the surface of the Earth and in the upper atmosphere, *Proc. IRE* **24**, 1367.

13. Norton, K. A. (1937). The propagation of waves over the surface of the Earth and in the upper atmosphere, *Proceedings of the IRE* **25**, 1203.

14. Sadiku, M. (2006). *Elements of Electromagnetic*, Oxford University Press.

15. Slalskaya, I. P., N. N. Lebedev, & Y. S. Uflyand (1979). *Worked Problems in Applied Mathematics*, Dover Publications.

16. Smith, P. H. (1939). A transmission line calculator, *Electronics* 12(1), 29–31.

17. Sommerfeld, A. (1909). The propagation of waves in wireless telegraphy, *Annelen der Physik* **28**, 665.

18. Spiegel, M. R. (1974). *Vector Analysis and an Introduction to Tensor Analysis*, Schaum Outline Series, McGraw Hill Book Co., Singapore.

19. Sze, S. M. (1969). *Physics of Semiconductor Devices*, John Wiley.

20. Thomas, G. B., & R. L. Finney (1996). *Calculus and Analytical Geometry*, 9th edn, Pearson Education, Asia.

21. Tikhonov, A. N. & A. A. Samarskii (1963). *Equations of Mathematical Physics*, Dover Publications.

22. Tyagi, M. S. (2004). *Introduction to Semiconductor Materials and Devices*, John Wiley and Sons.

23. Wheeler, H. A. (1977). Transmission-line properties of a strip on a dielectric sheet on a plane, *IEEE Transactions MTT* **25**, 631–647.

Index

About the Author

Sunil Bhooshan is graduated from IIT Delhi and obtained his MS and Ph D degrees from the University of Illinois, Urbana-Champaign.

Dr Bhooshan has served both in the industry and the academia. He has been a member of the technical staff at Hewlett Packard (now Agilent Technologies) at Santa Rosa. Besides having taught at IIT Kanpur, he has also served as an editor of *IT* (an EFY magazine). He has published more than fifty research papers in reputed journals and presented in international conferences. He is also a member of IEEE (Institute of Electrical and Electronics Engineers) and WSEAS (World Scientific and Engineering Academy and Society).

Related Titles

MATLAB AND SIMULINK FOR ENGINEERS | 9780198072447

Agam Kumar Tyagi *College of Engineering Studies, University of Petroleum & Energy Studies, Dehradun.*

MATLAB and Simulink for Engineers is designed to serve as a self-study material for students of electrical, computer, and mechanical engineering who require to use MATLAB and Simulink for varied courses. Based on version 2010a of MATLAB, the book begins with an introduction to MATLAB programming describing the MATLAB toolbar and Simulink toolboxes. It goes on to discuss various MATLAB operators, functions, and graphics in detail. Applications of Simulink and MATLAB in electrical engineering, electrical machines and power system projects, simulation of rectifiers, inverters, choppers, and cycloconverters are presented in detail. The Online Resource Centre of the book contains user interactive programs, modeling and simulation projects, and programming and simulation exercises.

DIGITAL ELECTRONICS | 9780198061830

G K Kharate *Dean, Pune University and Principal, Matoshri College of Engineering and Research Centre, Nashik*

Digital Electronics is specially designed as a textbook for undergraduate students of electronics, communication, computer science, and electrical and instrumentation engineering for an introductory course on digital electronics or digital system design. Beginning with the fundamental concepts such a logic families, number systems and codes, Boolean algebra and logic gates, and combinational circuits, the book moves on to cover the applied aspects such as sequential logic, asynchronous sequential circuits, algorithmic state machines, programmable logic devices, converters, and semiconductor memories.

SIGNALS AND SYSTEMS | 9780198066798

Tarun Kumar Rawat *Netaji Subhas Institute of Technology (NSIT), Delhi*

Signals and Systems is a comprehensive textbook designed for undergraduate students of engineering for a course on signals and systems. The book provides a simultaneous coverage of continuous-time and discrete-time signals and systems. Beginning with classifying signals, the book sequentially covers important topics such as convolution and correlation of signals, continuous-time Fourier series, discrete-time Fourier series, continuous-time Fourier transforms, sampling, Hilbert transform, Laplace transform and z-transform. A chapter on MATLAB programs presenting the applicability to the software to problems of signals and systems has also been included

Other Related Titles

9780198061854 V Chandra Sekar: *Analog Communication*
9780198075493 R N Mutagi: *Digital Communication*
9780198067665 Debaprasad Das: *VLSI Design*
9780195686661 Harish and Sachidananda: *Antennas and Wave Propagation*
9780198070788 S Sridhar: *Digital Image Processing*
9780198066477 Senthil Kumar, Saravanan, Jeevananthan: *Microprocessors and Microcontrollers*
9780198060666 Upena Dalal: Wireless Communication
9780195670929 V R Moorthi: *Power Electronics*